Elementary Linear Algebra

Elementary Linear Algebra

SECOND EDITION

STANLEY I. GROSSMAN

Wadsworth Publishing Company
Belmont, California
A Division of Wadsworth, Inc.

Mathematics Editor: Richard Jones
Production: Del Mar Associates
Designer: Al Burkhardt
Cover Designer: Louis Neiheisel
Technical Illustrators: Kim Fraley and Pam Posey

Printed in the United States of America

4 5 6 7 8 9 10—88 87 86 85

ISBN 0-534-02738-5

Library of Congress Cataloging in Publication Data

Grossman, Stanley I.
Elementary linear algebra.

Includes index.
1. Algebra, Linear. I. Title.
QA184.G76 1984 512′.5 83-6623
ISBN 0-534-02738-5

To Kerstin, Erik, and Aaron

Contents

CHAPTER 4 Vectors in $\mathbb{R}^2$ and $\mathbb{R}^3$

CHAPTER 5 Vector Spaces

CHAPTER 6 Linear Transformations

CHAPTER 7 Eigenvalues, Eigenvectors, and Canonical Forms

Preface

Even as recently as twenty years ago, the study of linear algebra was largely confined to mathematics and physics majors and those who needed a knowledge of matrix theory to work in technical areas such as multivariate statistics. Linear algebra is now studied by students in a wide variety of disciplines due to the invention of high-speed computers and a general increase in the application of mathematics in traditionally nontechnical areas.

In writing this text I have had two goals in mind. I have tried to make a large number of linear algebra topics accessible to a wide variety of students who need only a good knowledge of high school algebra. Because many students will have had a year of calculus, I have also included several examples involving topics from calculus. These are indicated by the symbol □. One optional section (Section 7.7) requires calculus, but *calculus is not a prerequisite.*

My second goal was to convince students of the importance of linear algebra in their fields of study. Thus, especially in the early chapters, examples are drawn from a variety of disciplines. These examples are necessarily short, but I hope they convey the use to which the mathematics can be put.

The approach I have used in this text is gradual. Chapter 1 starts with the familiar notion of a straight line. In Sections 1.3 and 1.4, the techniques of Gauss–Jordan and Gaussian Elimination are developed to solve systems of equations. These techniques are seen as generalizations of the case of two equations in two unknowns (two straight lines in the plane). A major feature of the text is the frequent appearance of the Summing Up Theorem, which ties together seemingly disparate topics in the study of matrices and linear transformations. The theorem is first encountered in Section 1.2.

In Chapters 2 and 3, standard properties of vectors, matrices, and determinants are discussed. In each of Sections 2.4 through 2.8, an important idea in matrix theory is introduced, while in Chapter 3, a rather complete discussion of determinants is given.

Chapter 4 discusses vectors in the plane and in space. Many of the topics in this chapter are covered in a calculus sequence. Since much of linear algebra is concerned with a discussion of abstract vector spaces, students need a storehouse of concrete examples, most easily provided by the study of vectors in the plane and in space. The more difficult and more abstract material in Chapters 5 and 6 is illustrated with examples from the very concrete Chapter 4.

Chapter 5 contains an introduction to general vector spaces and is, necessarily, more abstract than the earlier chapters. I have tried, however, to present the material as a natural extension of properties of vectors in the plane—which is really how the subject evolved. Chapter 6 continues the discussion begun in Chapter 5 with an introduction to linear transformations from one vector space to another. The chapter begins with two examples showing how such transformations can arise in a natural way.

Chapter 7, the longest chapter in the book, describes the theory of eigenvalues and eigenvectors. These are introduced in Section 7.1, and a detailed biological application is given in Section 7.2. Sections 7.3, 7.4, and 7.5 all involve the diagonalization of a matrix, while Section 7.6 illustrates, in a few cases, how a matrix can be reduced to its Jordan-canonical form. Section 7.7 discusses matrix differential equations and is the only chapter of the book that requires a knowledge of freshman calculus. This section provides an illustration of the usefulness of reducing a matrix to its Jordan-canonical form (which is usually a diagonal matrix). In Section 7.8, I have introduced two of my favorite results from matrix theory: the Cayley-Hamilton Theorem and the Gershgorin Circle Theorem. The latter result, rarely discussed in an elementary linear algebra text, provides an easy way to estimate the eigenvalues of any matrix.

In Chapter 7, I had to make a difficult decision: whether or not to discuss complex eigenvalues and eigenvectors. I decided to include them because it seemed to be the only honest thing to do. Some of the "nicest" matrices have complex eigenvalues. To define an eigenvalue as a real number only may at first make things seem simpler, but it is certainly wrong. Moreover, in many applications involving eigenvalues (including some of the material in Section 7.7), the most interesting models result in periodic phenomena—and these involve complex eigenvalues. Complex numbers are not avoided in this book. For students who have not encountered them before, the few properties they need are fully discussed in Appendix 2.

In this era, when students have access to computers and/or hand-held calculators, it is important in a linear algebra text at any level to describe how computations are done in "real life." In the earlier chapters, many computations are done using a calculator. Chapter 8 contains an introduction to several numerical techniques used for solving systems of equations and computing eigenvalues and eigenvectors. Section 8.1 discusses some of the problems that can arise when solving problems on a computer. The techniques in Sections 8.1, 8.2, and 8.3 can be covered anytime after Chapter 2. The material in Section 8.4 (on computing eigenvalues and eigenvectors) uses material presented in Section 7.1.

This book has two appendices—one on mathematical induction and one on complex numbers. Some of the proofs in the book use mathematical induction, and for students who have not used this important technique, Appendix 1 provides a brief introduction.

A word on chapter interdependence: The book is written sequentially. Each chapter depends on the ones that precede it, with two exceptions. Chapter 7 can be covered without most of the material in Chapter 6. Chapter 8 (except Section 8.4) can be covered after Chapter 2. Moreover, sections marked "If Time Permits" can be omitted without losing continuity.

To me, the most important part of any elementary mathematics textbook is its use of examples and problems. I learned much of my linear algebra by solving problems, and I believe that this holds for the majority of students. Thus, in most sections, I have included as many examples as is reasonable, followed by large numbers of both drill type problems and problems of a more theoretical nature. These are supplemented by review exercises at the end of each chapter. The more difficult problems are marked with * and a few exceptionally difficult with **.

Changes in the Second Edition

During the four years that the first edition of *Linear Algebra* was in print, many readers wrote in their criticisms and suggestions. Further, several individuals read portions of the manuscript for the second edition and suggested a number ways to make the text more accessible to students. As a result, I have made hundreds of small changes that should make this book easier to read.

I have also made several larger changes. Section 2.7, "The Inverse of a Square Matrix," was basically rewritten. It is my hope that now students will not only know how to find a matrix inverse but will also understand why the technique works. Sections 5.6 ("Change of Basis") and 6.4 ("The Matrix Representation of a Linear Transformation") were rewritten to make this material more comprehensible. In addition, I have added a number of simple applications to the first two chapters of the book to answer my students' frequently asked question, "Why am I studying linear algebra?"

Finally, the book has undergone a major change in format. Color is used to highlight steps in computations and proofs. New terms are introduced in boldface print rather than italics. Page references to earlier material will make previous examples, definitions, and theorems easier to find.

ACKNOWLEDGMENTS

I am very grateful to many people who helped during the time this book was written. I received much help in the preparation of problems and solution sets from Robert Hollister, a graduate student at the University of Montana, and from Robert Osterheld, a graduate student at Purdue. I am also grateful to Academic Press, Inc., for permission to use material from my book *Calculus* in Chapter 4. The biography of Karl Friedrich Gauss on page 9 is taken from Howard Eves, *An Introduction to the History of Mathematics, Fifth Edition,* copyright © 1983 by CBS College Publishing. I am grateful to CBS College Publishing for permission to use this material.

I owe a considerable debt to Richard Jones, mathematics editor at Wadsworth, whose guidance and insight helped me through the most difficult parts of the writing process. I also want to thank the following reviewers who have made invaluable contributions to the reliability and teachability of this text.

First Edition Reviewers

Donald F. Bailey
Cornell College

Richard W. Ball
Auburn University

Sandra A. Bollinger
Longwood College

James Bradley
Roberts Wesleyan College

Joel V. Brawley
Clemson University

Paul Bugl
University of Hartford

Gary Chartrand
Western Michigan University

Richard J. Easton
Indiana State University

Gary L. Eerkes
Gonzaga University

Garret J. Etgen
University of Houston

William R. Fuller
Purdue University

John D. Fulton
Clemson University

Lillian Gough
University of Wisconsin

Barbara Ann Greim
University of North Carolina
at Wilmington

Charles H. Haggard
Transylvania University

David Kinsey
Indiana State University

Thomas Lupton
University of North Carolina
at Wilmington

J. J. Malone
Worcester Polytechnic Institute

Edwin L. Marsden
Norwich University

Martin E. Nass
Iowa Central Community College

John Petro
Western Michigan University

William B. Rundberg
College of San Mateo

William Rundell
Texas A & M University

David Schedler
Virginia Commonwealth University

Jack C. Slay
Mississippi State University

Douglas D. Smith
Central Michigan University

T. W. Tucker
Dartmouth College

Marcellus E. Waddill
Wake Forrest University

James Wall
Auburn University

Cary Webb
Chicago State University

James J. Woeppel
Indiana University Southeast

Dennis G. Zill
Loyola Marymount University

Second Edition Reviewers

Joseph M. Cavanaugh
East Stroudsburg State College

Graham A. Chambers
University of Alberta

P. A. Binding
University of Calgary

Louis Florence
University of Toronto

John Sawka
University of Santa Clara

Keith Stumpff
Central Missouri State University

Eugene Johnson
University of Iowa

Ronald Eldringhoff
St. Louis Community College

1 Systems of Linear Equations

1.1 Introduction

This is a book about linear algebra. If you look up the word "linear" in a dictionary, you will find something like the following: lin-e-ar (lin′ ē ər), adj. 1. of, consisting of, or using lines.† In mathematics, the word "linear" means a good deal more than that. Nevertheless, much of the theory of elementary linear algebra is in fact a generalization of properties of straight lines. As a review, here are some fundamental facts about straight lines:

i. The **slope** m of a line passing through the points (x_1, y_1) and (x_2, y_2) is given by (if $x_2 \neq x_1$)

$$m = \frac{y_2 - y_1}{x_2 - x_1} = \frac{\Delta y}{\Delta x}$$

ii. If $x_2 - x_1 = 0$ and $y_2 \neq y_1$, then the line is vertical and the slope is said to be **undefined**.‡

iii. Any line (except one with infinite slope) can be described by writing its equation in the slope-intercept form $y = mx + b$, where m is the slope of the line and b is the y-intercept of the line (the value of y at the point where the line crosses the y-axis).

iv. Two lines are parallel if and only if they have the same slope.

v. If the equation of a line is written in the form $ax + by = c$ $(b \neq 0)$, then, as is easily computed, $m = -a/b$.

vi. If m_1 is the slope of line L_1, m_2 is the slope of line L_2, $m_1 \neq 0$, and L_1 and L_2 are perpendicular, then $m_2 = -1/m_1$.

vii. Lines parallel to the x-axis have a slope of zero.

viii. Lines parallel to the y-axis have an undefined slope.

In the next section we shall illustrate the relationship between solving systems of equations and finding points of intersection of pairs of straight lines.

---

† Taken from the pocket edition of *The Random House Dictionary*.

‡ In some textbooks a vertical line is said to have "an infinite slope."

1.2 Two Linear Equations in Two Unknowns

Consider the following system of two linear equations in the two unknowns x_1 and x_2:

$$\begin{aligned} a_{11}x_1 + a_{12}x_2 &= b_1 \\ a_{21}x_1 + a_{22}x_2 &= b_2 \end{aligned} \tag{1}$$

where a_{11}, a_{12}, a_{21}, a_{22}, b_1, and b_2 are given numbers. Each of these equations is the equation of a straight line (in the x_1x_2-plane instead of the xy-plane). The slope of the first line is $-a_{11}/a_{12}$; the slope of the second line is $-a_{21}/a_{22}$ (if $a_{12} \neq 0$ and $a_{22} \neq 0$). A **solution** to system (1) is a pair of numbers, denoted (x_1, x_2), that satisfies (1). The questions that naturally arise are whether (1) has any solutions and if so, how many? We shall answer these questions after looking at some examples. In these examples we will make use of two important facts from elementary algebra:

Fact A If $a = b$ and $c = d$, then $a + c = b + d$.

Fact B If $a = b$ and c is any real number, then $ca = cb$.

Fact A states that if we add two equations together, we obtain a third, valid equation. Fact B states that if we multiply both sides of an equation by a constant, we obtain a second, valid equation.

EXAMPLE 1 Consider the system

$$\begin{aligned} x_1 - x_2 &= 7 \\ x_1 + x_2 &= 5 \end{aligned} \tag{2}$$

Adding the two equations together gives us, by Fact A, the following equation: $2x_1 = 12$ (or $x_1 = 6$). Then, from the second equation, $x_2 = 5 - x_1 = 5 - 6 = -1$. Thus the pair $(6, -1)$ satisfies system (2) and the way we found the solution shows that it is the only pair of numbers to do so. That is, system (2) has a **unique solution**.

EXAMPLE 2 Consider the system

$$\begin{aligned} x_1 - x_2 &= 7 \\ 2x_1 - 2x_2 &= 14 \end{aligned} \tag{3}$$

It is apparent that these two equations are equivalent. To see this multiply the first by 2. (This is permitted by Fact B.) Then $x_1 - x_2 = 7$ or $x_2 = x_1 - 7$. Thus the pair $(x_1, x_1 - 7)$ is a solution to system (3) for any real number x_1. That is, system (3) has an **infinite number of solutions**. For example, the following pairs are solutions: $(7, 0)$, $(0, -7)$, $(8, 1)$, $(1, -6)$, $(3, -4)$, and $(-2, -9)$.

EXAMPLE 3 Consider the system

$$\begin{aligned} x_1 - x_2 &= 7 \\ 2x_1 - 2x_2 &= 13 \end{aligned} \tag{4}$$

Multiplying the first equation by 2 (which, again, is permitted by Fact B) gives us $2x_1 - 2x_2 = 14$. This contradicts the second equation. Thus system (4) has **no solution**.

It is easy to explain, geometrically, what is going on in the preceding examples. First we repeat that the equations in system (1) are both equations of straight lines. A solution to (1) is a point (x_1, x_2) that lies on both lines. If the two lines are not parallel, then they intersect at a single point. If they are parallel, then either they never intersect (no points in common) or they are the same line (infinite number of points in common). In Example 1 the lines have slopes of 1 and -1, respectively. Thus they are not parallel. They have the single point $(6, -1)$ in common. In Example 2 the lines are parallel (slope of 1) and coincident. In Example 3 the lines are parallel and distinct. These relationships are all illustrated in Figure 1.1.

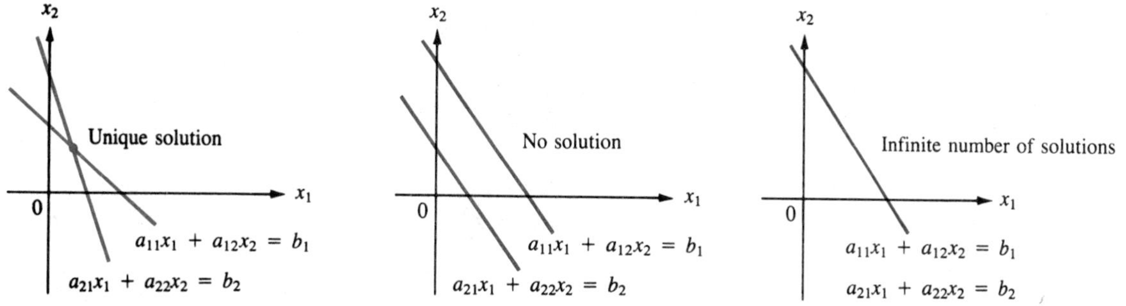

(a) Lines not parallel; one point of intersection

(b) Lines parallel; no points of intersection

(c) Lines coincide; infinite number of points of intersection

Figure 1.1

Let us now solve system (1) formally. We have

$$\begin{aligned} a_{11}x_1 + a_{12}x_2 &= b_1 \\ a_{21}x_1 + a_{22}x_2 &= b_2 \end{aligned}$$

Multiplying the first equation by a_{22} and the second by a_{12} yields

$$\begin{aligned} a_{11}a_{22}x_1 + a_{12}a_{22}x_2 &= a_{22}b_1 \\ a_{12}a_{21}x_1 + a_{12}a_{22}x_2 &= a_{12}b_2 \end{aligned} \tag{5}$$

Before continuing we note that system (1) and system (5) are **equivalent**. By that we mean that any solution to system (1) is a solution to system (5) and vice versa. This follows immediately from Fact B. Next we subtract the second equation from the first to obtain

$$(a_{11}a_{22}-a_{12}a_{21})x_1=a_{22}b_1-a_{12}b_2 \tag{6}$$

At this point we must pause. If $a_{11}a_{22}-a_{12}a_{21}\neq 0$, then we can divide by it to obtain

$$x_1=\frac{a_{22}b_1-a_{12}b_2}{a_{11}a_{22}-a_{12}a_{21}}$$

Then we can "plug" this value of x_1 into system (1) to solve for x_2 and we have found the unique solution to the system. We define the **determinant** of system (1) by

$$\text{Determinant of system (1)}=a_{11}a_{22}-a_{12}a_{21} \tag{7}$$

and we have shown the following:

> If the determinant of system (1) $\neq 0$, then the system has a unique solution. (8)

How does this statement relate to what we discussed earlier? In system (1) we see that the slope of the first line is $-a_{11}/a_{12}$ and the slope of the second is $-a_{21}/a_{22}$.† In Problems 31, 32, and 33 you are asked to show that the determinant of system (1) is zero if and only if the lines are parallel (have the same slope). So, if the determinant is *not* zero, the lines are not parallel and the system has a unique solution.

We now put the facts discussed above together in a theorem. It is a theorem that will be generalized in later sections of this chapter and in subsequent chapters. We shall keep track of our progress by referring to the theorem as our "Summing Up Theorem." When all its parts have been proved, we shall see a remarkable relationship among several important concepts in linear algebra.

THEOREM 1 **SUMMING UP THEOREM—VIEW 1** The system

$$a_{11}x_1 + a_{12}x_2 = b_1$$
$$a_{21}x_1 + a_{22}x_2 = b_2$$

of two equations in the two unknowns x_1 and x_2 has no solution, a unique solution, or an infinite number of solutions. It has:

i. A unique solution if and only if its determinant is not zero.
ii. No solution or an infinite number of solutions if and only if its determinant is zero.

---

† Here we are assuming that neither a_{12} nor a_{22} equals zero. If either a_{12} or a_{22} is zero, then system (1) can be easily solved. If $a_{12}=0$, for example, then $x_1=b_1/a_{11}$. Thus we shall not worry about this possibility.

In the next section we will discuss systems of m equations in n unknowns and will see that always there is either no solution, one solution, or an infinite number of solutions. In Chapter 3 we define and calculate determinants for systems of n equations in n unknowns and shall find that our Summing Up Theorem—Theorem 1—is true in this general setting.

PROBLEMS 1.2

In Problems 1–12, find all solutions (if any) to the given systems. In each case calculate the determinant.

1. $x_1 - 3x_2 = 4$, $-4x_1 + 2x_2 = 6$

2. $2x_1 - x_2 = -3$, $5x_1 + 7x_2 = 4$

3. $2x_1 - 8x_2 = 5$, $-3x_1 + 12x_2 = 8$

4. $2x_1 - 8x_2 = 6$, $-3x_1 + 12x_2 = -9$

5. $6x_1 + x_2 = 3$, $-4x_1 - x_2 = 8$

6. $3x_1 + x_2 = 0$, $2x_1 - 3x_2 = 0$

7. $4x_1 - 6x_2 = 0$, $-2x_1 + 3x_2 = 0$

8. $5x_1 + 2x_2 = 3$, $2x_1 + 5x_2 = 3$

9. $2x_1 + 3x_2 = 4$, $3x_1 + 4x_2 = 5$

10. $ax_1 + bx_2 = c$, $ax_1 - bx_2 = c$

11. $ax_1 + bx_2 = c$, $bx_1 + ax_2 = c$

12. $ax_1 - bx_2 = c$, $bx_1 + ax_2 = d$

13. Find conditions on a and b such that the system in Problem 10 has a unique solution.

14. Find conditions on a, b, and c such that the system in Problem 11 has an infinite number of solutions.

15. Find conditions on a, b, c, and d such that the system in Problem 12 has no solutions.

In Problems 16–21, find the point of intersection (if there is one) of the two lines.

16. $x - y = 7$; $2x + 3y = 1$

17. $y - 2x = 4$; $4x - 2y = 6$

18. $4x - 6y = 7$; $6x - 9y = 12$

19. $4x - 6y = 10$; $6x - 9y = 15$

20. $3x + y = 4$; $y - 5x = 2$

21. $3x + 4y = 5$; $6x - 7y = 8$

Let L be a line and let $L_\perp$ denote the line perpendicular to L that passes through a given point P. The **distance** from L to P is defined to be the distance† between P and the point of intersection of L and $L_\perp$. In Problems 22–27, find the distance between the given line and point.

22. $x - y = 6$; $(0, 0)$

23. $2x + 3y = -1$; $(0, 0)$

24. $3x + y = 7$; $(1, 2)$

25. $5x - 6y = 3$; $(2, \frac{16}{5})$

26. $2y - 5x = -2$; $(5, -3)$

27. $6y + 3x = 3$; $(8, -1)$

28. Find the distance between the line $2x - y = 6$ and the point of intersection of the lines $2x - 3y = 1$ and $3x + 6y = 12$.

***29.** Prove that the distance between the point (x_1, y_1) and the line $ax + by = c$ is given by

$$d = \frac{|ax_1 + by_1 - c|}{\sqrt{a^2 + b^2}}$$

30. A zoo keeps birds (two-legged) and beasts (four-legged). If the zoo contains 60 heads and 200 feet, how many birds and how many beasts live there?

31. Suppose that the determinant of system (1) is zero. Show that the lines given in (1) are parallel.

---

† Recall that if (x_1, y_1) and (x_2, y_2) are two points in the xy-plane, then the distance d between them is given by $d = \sqrt{(x_1 - x_2)^2 + (y_1 - y_2)^2}$.

32. If there is a unique solution to system (1), show that its determinant is nonzero.
33. If the determinant of system (1) is nonzero, show that the system has a unique solution.
34. The Sunrise Porcelain Company manufactures ceramic cups and saucers. For each cup or saucer a worker measures a fixed amount of material and puts it into a forming machine, from which it is automatically glazed and dried. On the average, a worker needs 3 minutes to get the process started for a cup and 2 minutes for a saucer. The material for a cup costs 25¢ and the material for a saucer costs 20¢. If \$44 is allocated daily for production of cups and saucers, how many of each can be manufactured in an 8-hour work day if a worker is working every minute and exactly \$44 is spent on materials?
35. Answer the question of Problem 34 if the materials for a cup and saucer cost 15¢ and 10¢, respectively, and \$24 is spent in an 8-hour day.
36. Answer the question of Problem 35 if \$25 is spent in an 8-hour day.
37. An ice-cream shop sells only ice-cream sodas and milk shakes. It puts 1 ounce of syrup and 4 ounces of ice cream in an ice-cream soda, and 1 ounce of syrup and 3 ounces of ice cream in a milk shake. If the store used 4 gallons of ice cream and 5 quarts of syrup in a day, how many ice-cream sodas and milk shakes did it sell? [Hint: 1 quart = 32 ounces; 1 gallon = 128 ounces]

1.3 *m* Equations in *n* Unknowns: Gauss-Jordan and Gaussian Elimination

In this section we describe a method for finding all solutions (if any) to a system of m linear equations in n unknowns. In doing so we shall see that, like the 2×2 case, such a system has no solutions, one solution, or an infinite number of solutions. Before launching into the general method, let us look at some simple examples.

EXAMPLE 1 Solve the system

$$\begin{aligned} 2x_1 + 4x_2 + 6x_3 &= 18 \\ 4x_1 + 5x_2 + 6x_3 &= 24 \\ 3x_1 + x_2 - 2x_3 &= 4 \end{aligned} \tag{1}$$

Solution Here we seek three numbers x_1, x_2, and x_3 such that the three equations in (1) are satisfied. Our method of solution will be to simplify the equations as we did in the last section so that solutions can be readily identified. We begin by dividing the first equation by 2. This gives us

$$\begin{aligned} x_1 + 2x_2 + 3x_3 &= 9 \\ 4x_1 + 5x_2 + 6x_3 &= 24 \\ 3x_1 + x_2 - 2x_3 &= 4 \end{aligned} \tag{2}$$

As we saw in the last section (Fact A), adding two equations together leads to a third, valid equation. This equation may replace either of the two equations used to obtain it in the system. We begin simplifying system (2) by multiplying both sides of the first equation in (2) by −4 and adding this new equation to the second equation. This gives us

$$\begin{array}{r} -4x_1 - 8x_2 - 12x_3 = -36 \\ 4x_1 + 5x_2 + 6x_3 = 24 \\ \hline - 3x_2 - 6x_3 = -12 \end{array}$$

The equation $-3x_2 - 6x_3 = -12$ is our new second equation and the system is now

$$\begin{aligned} x_1 + 2x_2 + 3x_3 &= 9 \\ - 3x_2 - 6x_3 &= -12 \\ 3x_1 + x_2 - 2x_3 &= 4 \end{aligned}$$

We then multiply the first equation by −3 and add it to the third equation:

$$\begin{aligned} x_1 + 2x_2 + 3x_3 &= 9 \\ - 3x_2 - 6x_3 &= -12 \\ - 5x_2 - 11x_3 &= -23 \end{aligned} \qquad (3)$$

Note that in system (3) the variable x_1 has been eliminated from the second and third equations. Next we divide the second equation by -3:

$$\begin{aligned} x_1 + 2x_2 + 3x_3 &= 9 \\ x_2 + 2x_3 &= 4 \\ - 5x_2 - 11x_3 &= -23 \end{aligned}$$

We multiply the second equation by −2 and add it to the first and then multiply the second equation by 5 and add it to the third:

$$\begin{aligned} x_1 \qquad - x_3 &= 1 \\ x_2 + 2x_3 &= 4 \\ - x_3 &= -3 \end{aligned}$$

We multiply the third equation by -1:

$$\begin{aligned} x_1 \qquad - x_3 &= 1 \\ x_2 + 2x_3 &= 4 \\ x_3 &= 3 \end{aligned}$$

Finally, we add the third equation to the first and then multiply the third equation by -2 and add it to the second to obtain the following system (which is equivalent to system (1)):

$$\begin{aligned} x_1 \qquad\qquad &= 4 \\ x_2 \qquad &= -2 \\ x_3 &= 3 \end{aligned}$$

This is the unique solution to the system. We write it in the form $(4, -2, 3)$. The method we used here is called **Gauss-Jordan elimination.**†

Before going on to another example, let us summarize what we have done in this example:

i. We divided to make the coefficient of x_1 in the first equation equal to 1.
ii. We "eliminated" the x_1 terms in the second and third equations. That is, we made the coefficients of these terms equal to zero by multiplying the first equation by appropriate numbers and then adding it to the second and third equations, respectively.
iii. We divided to make the coefficient of the x_2 term in the second equation equal to 1 and then proceeded to use the second equation to eliminate the x_2 terms in the first and third equations.
iv. We divided to make the coefficient of the x_3 term in the third equation equal to 1 and then proceeded to use the third equation to eliminate the x_3 terms in the first and second equations.

We emphasize that, at every step, we obtained systems that were equivalent. That is, each system had the same set of solutions as the one that preceded it. This follows from Facts A and B on page 2.

Before solving other systems of equations, we introduce notation that makes it easier to write down each step in our procedure. A **matrix** is a rectangular array of numbers. For example, the coefficients of the variables x_1, x_2, x_3 in system (1) can be written as the entries of a matrix A, called the **coefficient matrix** of the system:

$$A = \begin{pmatrix} 2 & 4 & 6 \\ 4 & 5 & 6 \\ 3 & 1 & -2 \end{pmatrix} \tag{4}$$

The study of matrices will take a large part of the remaining chapters of this text. We introduce them here for convenience of notation.

Using matrix notation, system (1) can be written as the **augmented matrix**

$$\left(\begin{array}{ccc|c} 2 & 4 & 6 & 18 \\ 4 & 5 & 6 & 24 \\ 3 & 1 & -2 & 4 \end{array}\right) \tag{5}$$

---

† Named after the great German mathematician Karl Friedrich Gauss (1777–1855) and the French mathematician Camille Jordan (1838–1922).

Carl Friedrich Gauss
(Library of Congress)

A man of awesome mathematical stature and talent, Carl Friedrich Gauss straddled the eighteenth and nineteenth centuries like a mathematical Colossus of Rhodes. He is universally regarded as the greatest mathematician of the nineteenth century and, along with Archimedes and Isaac Newton, as one of the three greatest mathematicians of all time.

Carl was born in Brunswick, Germany, in 1777. His father was a hardworking laborer with stubborn and unappreciative views of formal education. His mother, however, though uneducated herself, encouraged the boy in his studies and maintained a lifelong pride in her son's achievements.

Carl was one of those remarkable infant prodigies who appear from time to time. They tell of him the incredible story that at the age of three he detected an arithmetical error in his father's bookkeeping. And there is the often-told story that when ten years old and in the public schools, his teacher, to keep the class occupied, set the pupils to adding the numbers from 1 to 100. Almost immediately Carl placed his slate, writing side down, on the annoyed teacher's desk. When all the slates were finally turned in, the amazed teacher found that Carl alone had the correct answer, 5050, but with no accompanying calculation. Carl had mentally summed the arithmetic progression $1 + 2 + 3 + \cdots + 98 + 99 + 100$ by noting that $100 + 1 = 101$, $99 + 2 = 101$, $98 + 3 = 101$, and so on for 50 such pairs, whence the answer is 50×101, or 5050. Later in life Gauss used to claim jocularly that he could figure before he could talk.

Gauss's precocity came to the attention of the Duke of Brunswick, who, as a kindly and understanding patron, saw the boy enter the college in Brunswick at the age of 15, and then Göttingen University at the age of 18. Vacillating between becoming a philologist or a mathematician (though he had already devised the method of least squares a decade before Legendre independently published it), his mind was dramatically made up in favor of mathematics on March 30, 1796, when he was still a month short of his nineteenth birthday. The event was his surprising contribution to the theory of the Euclidean construction of regular polygons and, in particular, the discovery that a regular polygon of 17 sides can be so constructed.

In his doctorial dissertation, at the University of Helmstädt and written at the age of twenty, Gauss gave the first wholly satisfactory proof of the *fundamental theorem of algebra* (that a polynomial equation with complex coefficients and of degree n has at least one complex root). Unsuccessful attempts to prove this theorem had been made by Newton, Euler, d'Alembert, and Lagrange.

Gauss contributed noteworthily to astronomy, geodesy, and electricity. In 1801 he calculated, by a new procedure and from meager data, the orbit of the then recently discovered planetoid Ceres, and in the following year that of the planetoid Pallas. In 1807 he became professor of mathematics and director of the observatory at Göttingen, a post that he held until his death. In 1821 he carried out a triangulation of Hannover, measured a meridional arc, and invented the heliotrope (or heliograph). In 1831 he commenced collaboration with his colleague Wilhelm Weber (1804–1891) in basic research in electricity and magnetism, and in 1833 the two scientists devised the electromagnetic telegraph.

Famous is Gauss's assertion that "mathematics is the queen of the sciences, and the theory of numbers is the queen of mathematics." Gauss has been described as "the mathematical giant who from his lofty heights embraces in one view the stars and the abysses." In his scientific writing, Gauss was a perfectionist. Claiming that a cathedral is not a cathedral until the last piece of scaffolding is removed, he strove to make each of his works complete, concise, polished, and convincing, with every trace of the analysis by which he reached his results removed. He accordingly adopted as his seal a tree bearing only a few fruits and carrying the motto: *Pauca sed matura* (*Few, but ripe*). Gauss chose for his second motto the following lines from *King Lear*:

> Thou, nature, art my goddess; to thy laws
> My services are bound . . . .

Gauss thus believed that mathematics, for inspiration, must touch the real world. As Wordsworth put it, "Wisdom oft is nearer when we stoop than when we soar."

Gauss died in his home at the Göttingen Observatory in 1855.

For example, the first row in the augmented matrix (5) is read $2x_1+4x_2+6x_3=18$. Note that each row of the augmented matrix corresponds to one of the equations in the system.

We now introduce some terminology. We have seen that multiplying (or dividing) the sides of an equation by a nonzero number gives us a new, valid equation. Moreover, adding a multiple of one equation to another equation in a system gives us another valid equation. Finally, if we interchange two equations in a system of equations, we obtain an equivalent system. These three operations, when applied to the rows of the augmented matrix representation of a system of equations, are called **elementary row operations**.

To sum up, the three elementary row operations applied to the augmented matrix representation of a system of equations are:

ELEMENTARY ROW OPERATIONS

i. Multiply (or divide) one row by a nonzero number.
ii. Add a multiple of one row to another row.
iii. Interchange two rows.

The process of applying elementary row operations to simplify an augmented matrix is called **row reduction**.

Notation **i.** $M_i(c)$ stands for "multiply the ith row of a matrix by the number c."
ii. $A_{i,j}(c)$ stands for "multiply the ith row by c and add it to the jth row."
iii. $P_{i,j}$ stands for "interchange (permute) rows i and j."
iv. $A \to B$ indicates that the augmented matrices A and B are equivalent; that is, the systems they represent have the same solution.

In Example 1 we saw that by using the elementary row operations (i) and (ii) several times we could obtain a system in which the solutions to the system were given explicitly. We now repeat the steps in Example 1, using the notation just introduced:

$$\left(\begin{array}{ccc|c} 2 & 4 & 6 & 18 \\ 4 & 5 & 6 & 24 \\ 3 & 1 & -2 & 4 \end{array}\right) \xrightarrow{M_1(\frac{1}{2})} \left(\begin{array}{ccc|c} 1 & 2 & 3 & 9 \\ 4 & 5 & 6 & 24 \\ 3 & 1 & -2 & 4 \end{array}\right) \xrightarrow[A_{1,3}(-3)]{A_{1,2}(-4)} \left(\begin{array}{ccc|c} 1 & 2 & 3 & 9 \\ 0 & -3 & -6 & -12 \\ 0 & -5 & -11 & -23 \end{array}\right)$$

$$\xrightarrow{M_2(-\frac{1}{3})} \left(\begin{array}{ccc|c} 1 & 2 & 3 & 9 \\ 0 & 1 & 2 & 4 \\ 0 & -5 & -11 & -23 \end{array}\right) \xrightarrow[A_{2,3}(5)]{A_{2,1}(-2)} \left(\begin{array}{ccc|c} 1 & 0 & -1 & 1 \\ 0 & 1 & 2 & 4 \\ 0 & 0 & -1 & -3 \end{array}\right)$$

$$\xrightarrow{M_3(-1)} \left(\begin{array}{ccc|c} 1 & 0 & -1 & 1 \\ 0 & 1 & 2 & 4 \\ 0 & 0 & 1 & 3 \end{array}\right) \xrightarrow[A_{3,2}(-2)]{A_{3,1}(1)} \left(\begin{array}{ccc|c} 1 & 0 & 0 & 4 \\ 0 & 1 & 0 & -2 \\ 0 & 0 & 1 & 3 \end{array}\right)$$

Again we can easily "see" the solution $x_1=4$, $x_2=-2$, $x_3=3$.

EXAMPLE 2 Solve the system

$$\begin{aligned} 2x_1 + 4x_2 + 6x_3 &= 18 \\ 4x_1 + 5x_2 + 6x_3 &= 24 \\ 2x_1 + 7x_2 + 12x_3 &= 30 \end{aligned}$$

Solution We proceed as in Example 1, first writing the system as an augmented matrix:

$$\left(\begin{array}{ccc|c} 2 & 4 & 6 & 18 \\ 4 & 5 & 6 & 24 \\ 2 & 7 & 12 & 30 \end{array}\right)$$

We then obtain, successively,

$$\xrightarrow{M_1(\frac{1}{2})} \left(\begin{array}{ccc|c} 1 & 2 & 3 & 9 \\ 4 & 5 & 6 & 24 \\ 2 & 7 & 12 & 30 \end{array}\right) \xrightarrow[A_{1,3}(-2)]{A_{1,2}(-4)} \left(\begin{array}{ccc|c} 1 & 2 & 3 & 9 \\ 0 & -3 & -6 & -12 \\ 0 & 3 & 6 & 12 \end{array}\right)$$

$$\xrightarrow{M_2(-\frac{1}{3})} \left(\begin{array}{ccc|c} 1 & 2 & 3 & 9 \\ 0 & 1 & 2 & 4 \\ 0 & 3 & 6 & 12 \end{array}\right) \xrightarrow[A_{2,3}(-3)]{A_{2,1}(-2)} \left(\begin{array}{ccc|c} 1 & 0 & -1 & 1 \\ 0 & 1 & 2 & 4 \\ 0 & 0 & 0 & 0 \end{array}\right)$$

This is equivalent to the system of equations

$$\begin{aligned} x_1 \qquad - x_3 &= 1 \\ x_2 + 2x_3 &= 4 \end{aligned}$$

This is as far as we can go. There are now only two equations in the three unknowns x_1, x_2, x_3 and there are an infinite number of solutions. To see this, let x_3 be chosen. Then $x_2 = 4 - 2x_3$ and $x_1 = 1 + x_3$. This will be a solution for any number x_3. We write these solutions in the form $(1 + x_3, 4 - 2x_3, x_3)$. For example, if $x_3 = 0$ we obtain the solution $(1, 4, 0)$. For $x_3 = 10$ we obtain the solution $(11, -16, 10)$.

EXAMPLE 3 Solve the system

$$\begin{aligned} 2x_1 + 4x_2 + 6x_3 &= 18 \\ 4x_1 + 5x_2 + 6x_3 &= 24 \\ 2x_1 + 7x_2 + 12x_3 &= 40 \end{aligned} \tag{6}$$

Solution We use the augmented-matrix form and proceed exactly as in Example 2 to obtain, successively, the following systems. (Note how, in each step, we use either elementary row operation (*i*) or (*ii*).)

$$\left(\begin{array}{ccc|c} 2 & 4 & 6 & 18 \\ 4 & 5 & 6 & 24 \\ 2 & 7 & 12 & 40 \end{array}\right) \xrightarrow{M_1(\frac{1}{2})} \left(\begin{array}{ccc|c} 1 & 2 & 3 & 9 \\ 4 & 5 & 6 & 24 \\ 2 & 7 & 12 & 40 \end{array}\right)$$

$$\xrightarrow[A_{1,3}(-2)]{A_{1,2}(-4)} \left(\begin{array}{ccc|c} 1 & 2 & 3 & 9 \\ 0 & -3 & -6 & -12 \\ 0 & 3 & 6 & 22 \end{array}\right) \xrightarrow{M_2(-\frac{1}{3})} \left(\begin{array}{ccc|c} 1 & 2 & 3 & 9 \\ 0 & 1 & 2 & 4 \\ 0 & 3 & 6 & 22 \end{array}\right)$$

$$\xrightarrow[A_{2,3}(-3)]{A_{2,1}(-2)} \left(\begin{array}{ccc|c} 1 & 0 & -1 & 1 \\ 0 & 1 & 2 & 4 \\ 0 & 0 & 0 & 10 \end{array}\right) \xrightarrow{M_3(\frac{1}{10})} \left(\begin{array}{ccc|c} 1 & 0 & -1 & 1 \\ 0 & 1 & 2 & 4 \\ 0 & 0 & 0 & 1 \end{array}\right)$$

The last equation now reads $0x_1+0x_2+0x_3=1$, which is impossible since $0 \neq 1$. Thus system (6) has *no* solution.

Let us take another look at these three examples. In Example 1 we began with the matrix

$$A_1 = \begin{pmatrix} 2 & 4 & 6 \\ 4 & 5 & 6 \\ 3 & 1 & -2 \end{pmatrix}.$$

In the process of row reduction A_1 was "reduced" to the matrix

$$R_1 = \begin{pmatrix} 1 & 0 & 0 \\ 0 & 1 & 0 \\ 0 & 0 & 1 \end{pmatrix}.$$

In Example 2 we started with

$$A_2 = \begin{pmatrix} 2 & 4 & 6 \\ 4 & 5 & 6 \\ 2 & 7 & 12 \end{pmatrix}$$

and ended up with

$$R_2 = \begin{pmatrix} 1 & 0 & -1 \\ 0 & 1 & 2 \\ 0 & 0 & 0 \end{pmatrix}.$$

In Example 3 we began with

$$A_3 = \begin{pmatrix} 2 & 4 & 6 \\ 4 & 5 & 6 \\ 2 & 7 & 12 \end{pmatrix}$$

and again ended up with

$$R_3 = \begin{pmatrix} 1 & 0 & -1 \\ 0 & 1 & 2 \\ 0 & 0 & 0 \end{pmatrix}.$$

The matrices R_1, R_2, and R_3 are called the *reduced row echelon forms* of the matrices A_1, A_2, and A_3, respectively. In general, we have the following definition.

DEFINITION 1 **REDUCED ROW ECHELON FORM** A matrix is in **reduced row echelon form** if the following four conditions hold:

i. All rows (if any) consisting entirely of zeros appear at the bottom of the matrix.
ii. The first nonzero number (starting from the left) in any row not consisting entirely of zeros is 1.
iii. If two successive rows do not consist entirely of zeros, then the first 1 in the lower row occurs farther to the right than the first 1 in the higher row.
iv. Any column containing the first 1 in a row has zeros everywhere else.

EXAMPLE 4 The following matrices are in reduced row echelon form:

i. $\begin{pmatrix} 1 & 0 & 0 \\ 0 & 1 & 0 \\ 0 & 0 & 1 \end{pmatrix}$ **ii.** $\begin{pmatrix} 1 & 0 & 0 & 0 \\ 0 & 1 & 0 & 0 \\ 0 & 0 & 0 & 1 \end{pmatrix}$ **iii.** $\begin{pmatrix} 1 & 0 & 0 & 5 \\ 0 & 0 & 1 & 2 \end{pmatrix}$

iv. $\begin{pmatrix} 1 & 0 \\ 0 & 1 \end{pmatrix}$ **v.** $\begin{pmatrix} 1 & 0 & 2 & 5 \\ 0 & 1 & 3 & 6 \\ 0 & 0 & 0 & 0 \end{pmatrix}$

DEFINITION 2 **ROW ECHELON FORM** A matrix is in **row echelon form** if conditions (*i*), (*ii*), and (*iii*) hold in Definition 1.

EXAMPLE 5 The following matrices are in row echelon form:

i. $\begin{pmatrix} 1 & 2 & 3 \\ 0 & 1 & 5 \\ 0 & 0 & 1 \end{pmatrix}$ **ii.** $\begin{pmatrix} 1 & -1 & 6 & 4 \\ 0 & 1 & 2 & -8 \\ 0 & 0 & 0 & 1 \end{pmatrix}$

iii. $\begin{pmatrix} 1 & 0 & 2 & 5 \\ 0 & 0 & 1 & 2 \end{pmatrix}$ **iv.** $\begin{pmatrix} 1 & 2 \\ 0 & 1 \end{pmatrix}$ **v.** $\begin{pmatrix} 1 & 3 & 2 & 5 \\ 0 & 1 & 3 & 6 \\ 0 & 0 & 0 & 0 \end{pmatrix}$

Remark 1. The difference between these two forms should be clear from the examples. In row echelon form, all the numbers below the first 1 in a row are zero. In reduced row echelon form, all the numbers above and below the first 1 in a row are zero. Thus reduced row echelon form is more exclusive. That is, every matrix in reduced row echelon form is in row echelon form, but not conversely.

Remark 2. We can always reduce a matrix to reduced row echelon form or row echelon form by performing elementary row operations. We saw this reduction to reduced row echelon form in Examples 1, 2, and 3.

As we saw in Examples 1, 2, and 3, there is a strong connection between the reduced row echelon form of a matrix and the existence of a unique solution to the system. In Example 1, the reduced row echelon form of the **coefficient matrix** (that is, the first three columns of the augmented matrix) had a 1 in each row and there was a unique solution. In Examples 2 and 3, the reduced row echelon form of the coefficient matrix had a row of zeros and the system had either no solution or an infinite number of solutions. This turns out always to be true in any system with the same number of equations as unknowns. But before turning to the general case, let us discuss the usefulness of the row echelon form of a matrix. It is possible to solve the system in Example 1 by reducing the coefficient matrix to its row echelon form.

EXAMPLE 6 Solve the system of Example 1 by reducing the coefficient matrix to row echelon form.

Solution We begin as before:

$$\left(\begin{array}{ccc|c} 2 & 4 & 6 & 18 \\ 4 & 5 & 6 & 24 \\ 3 & 1 & -2 & 4 \end{array}\right) \xrightarrow{M_1(\frac{1}{2})} \left(\begin{array}{ccc|c} 1 & 2 & 3 & 9 \\ 4 & 5 & 6 & 24 \\ 3 & 1 & -2 & 4 \end{array}\right) \xrightarrow[A_{1,3}(-3)]{A_{1,2}(-4)}$$

$$\left(\begin{array}{ccc|c} 1 & 2 & 3 & 9 \\ 0 & -3 & -6 & -12 \\ 0 & -5 & -11 & -23 \end{array}\right) \xrightarrow{M_2(-\frac{1}{3})} \left(\begin{array}{ccc|c} 1 & 2 & 3 & 9 \\ 0 & 1 & 2 & 4 \\ 0 & -5 & -11 & -23 \end{array}\right)$$

So far, this process is identical to our earlier one. Now, however, we only make zero the number (-5) below the first 1 in the second row:

$$\xrightarrow{A_{2,3}(5)} \left(\begin{array}{ccc|c} 1 & 2 & 3 & 9 \\ 0 & 1 & 2 & 4 \\ 0 & 0 & -1 & -3 \end{array}\right) \xrightarrow{M_3(-1)} \left(\begin{array}{ccc|c} 1 & 2 & 3 & 9 \\ 0 & 1 & 2 & 4 \\ 0 & 0 & 1 & 3 \end{array}\right)$$

The augmented matrix of the system (and the coefficient matrix) are now in row echelon form and we immediately see that $x_3 = 3$. We then use **back**

substitution to solve for x_2 and then x_1. The second equation reads $x_2 + 2x_3 = 4$. Thus $x_2 + 2(3) = 4$ and $x_2 = -2$. Similarly, from the first equation we obtain $x_1 + 2(-2) + 3(3) = 9$ or $x_1 = 4$. Thus we again obtain the solution $(4, -2, 3)$. The method of solution just employed is called **Gaussian elimination.**

We therefore have two methods for solving our sample systems of equations:

i. GAUSS-JORDAN ELIMINATION:
Row-reduce the coefficient matrix to reduced row echelon form.

ii. GAUSSIAN ELIMINATION:
Row-reduce the coefficient matrix to row echelon form, solve for the last unknown, and then use back substitution to solve for the other unknowns.

Which method is more useful? It depends. In solving systems of equations on a computer, Gaussian elimination is the preferred method because it involves fewer elementary row operations. We discuss the numerical solution of systems of equations in Sections 8.2 and 8.3. On the other hand, there are times when it is essential to obtain the reduced row echelon form of a matrix (one of these is discussed in Section 2.7). In these cases Gauss-Jordan elimination is the preferred method.

We now turn to the solution of a general system of m equations in n unknowns. Because of our need to do so in Section 2.7, we shall be solving most of the systems by Gauss-Jordan elimination. Keep in mind, however, that Gaussian elimination is sometimes the preferred approach.

The general $m \times n$ system of m linear equations in n unknowns is given by

$$\begin{aligned}
a_{11}x_1 + a_{12}x_2 + a_{13}x_3 + \cdots + a_{1n}x_n &= b_1 \\
a_{21}x_1 + a_{22}x_2 + a_{23}x_3 + \cdots + a_{2n}x_n &= b_2 \\
a_{31}x_1 + a_{32}x_2 + a_{33}x_3 + \cdots + a_{3n}x_n &= b_3 \\
\vdots \qquad \vdots \qquad \vdots \qquad & \qquad \vdots \\
a_{m1}x_1 + a_{m2}x_2 + a_{m3}x_3 + \cdots + a_{mn}x_n &= b_m
\end{aligned} \tag{7}$$

In system (7) all the a's and b's are given real numbers. The problem is to find all sets of n numbers, denoted $(x_1, x_2, x_3, \ldots, x_n)$, that satisfy each of the m equations in (7). The number a_{ij} is the coefficient of the variable x_j in the ith equation.

We solve system (7) by writing the system as an augmented matrix and row-reducing the matrix to its reduced row echelon form. We start by dividing the first row by a_{11} (elementary row operation (i)). If $a_{11} = 0$ then we rearrange† the equations so that, with rearrangement, the new $a_{11} \neq 0$. We then use the first equation to eliminate the x_1 term in each of the other equations (using elementary row operation (*ii*)). Then the new second equation is divided by the new a_{22} term and the new, new second equation is used to eliminate the x_2 terms in all the other equations. The process is continued until one of three situations occurs:

i. The last nonzero‡ equation reads $x_n = c$ for some constant c. Then there is either a unique solution or an infinite number of solutions to the system.

ii. The last nonzero equation reads $a'_{ij}x_j + a'_{i,j+1}x_{j+1} + \cdots + a'_{i,j+k}x_n = c$ for some constant c where at least two of the a's are nonzero. That is, the last equation is a linear equation in two or more of the variables. Then there are an infinite number of solutions.

iii. The last equation reads $0 = c$, where $c \neq 0$. Then there is no solution. In this case the system is called **inconsistent.** In cases (*i*) and (*ii*) the system is called **consistent.**

EXAMPLE 7 Solve the system

$$x_1 + 3x_2 - 5x_3 + x_4 = 4$$

$$2x_1 + 5x_2 - 2x_3 + 4x_4 = 6$$

Solution We write this system as an augmented matrix and row reduce:

$$\left(\begin{array}{cccc|c} 1 & 3 & -5 & 1 & 4 \\ 2 & 5 & -2 & 4 & 6 \end{array}\right) \xrightarrow{A_{1,2}(-2)} \left(\begin{array}{cccc|c} 1 & 3 & -5 & 1 & 4 \\ 0 & -1 & 8 & 2 & -2 \end{array}\right) \xrightarrow{M_2(-1)}$$

$$\left(\begin{array}{cccc|c} 1 & 3 & -5 & 1 & 4 \\ 0 & 1 & -8 & -2 & 2 \end{array}\right) \xrightarrow{A_{2,1}(-3)} \left(\begin{array}{cccc|c} 1 & 0 & 19 & 7 & -2 \\ 0 & 1 & -8 & -2 & 2 \end{array}\right)$$

This is as far as we can go. The coefficient matrix is in reduced row echelon form—case (*ii*) above. There are evidently an infinite number of solutions. The variables x_3 and x_4 can be chosen arbitrarily. Then $x_2 = 2 + 8x_3 + 2x_4$ and $x_1 = -2 - 19x_3 - 7x_4$. All solutions are, therefore, represented by $(-2 - 19x_3 - 7x_4, 2 + 8x_3 + 2x_4, x_3, x_4)$. For example, if $x_3 = 1$ and $x_4 = 2$, we obtain the solution $(-35, 14, 1, 2)$.

As you will see if you do a lot of system solving, the computations can become very messy. It is a good rule of thumb to use a calculator whenever the fractions become unpleasant. It should be noted, however, that if computations are carried out on a computer or calculator, "round-off" errors can be introduced. This problem is discussed in Section 8.1.

† To rearrange a system of equations we simply write the same equations in a different order. For example, the first equation can become the fourth equation, the third equation can become the second equation, and so on. This is a sequence of elementary row operations (*iii*).

‡ The "zero equation" is the equation $0 = 0$.

We close this section with three examples illustrating how a system of linear equations can arise in a practical situation.

EXAMPLE 8 A model that is often used in economics is the **Leontief input-output model.**† Suppose an economic system has n industries. There are two kinds of demands on each industry. First there is the *external* demand from outside the system. If the system is a country, for example, then the external demand could be from another country. Second there is the demand placed on one industry by another industry in the same system. In the United States, for example, there is a demand on the output of the steel industry by the automobile industry.

Let e_i represent the external demand placed on the ith industry. Let a_{ij} represent the internal demand placed on the ith industry by the jth industry. More precisely, a_{ij} represents the number of units of the output of industry i needed to produce 1 unit of the output of industry j. Let x_i represent the output of industry i. Now we assume that the output of each industry is equal to its demand (that is, there is no overproduction). The total demand is equal to the sum of the internal and external demands. To calculate the internal demand on industry 2, for example, we note that $a_{21}x_1$ is the demand on industry 2 made by industry 1. Thus the total internal demand on industry 2 is $a_{21}x_1 + a_{22}x_2 + \cdots + a_{2n}x_n$.

We are led to the following system of equations obtained by equating the total demand with the output of each industry:

$$\begin{array}{ccccccccccc}
a_{11}x_1 & + & a_{12}x_2 & + \cdots + & a_{1n}x_n & + & e_1 & = & x_1 \\
a_{21}x_1 & + & a_{22}x_2 & + \cdots + & a_{2n}x_n & + & e_2 & = & x_2 \\
\vdots & & \vdots & & \vdots & & \vdots & & \vdots \\
a_{n1}x_1 & + & a_{n2}x_2 & + \cdots + & a_{nn}x_n & + & e_n & = & x_n
\end{array} \tag{8}$$

Or, rewriting (8) so it looks like system (7), we get

$$\begin{array}{ccccccc}
(1-a_{11})x_1 & - & a_{12}x_2 & - \cdots - & a_{1n}x_n & = & e_1 \\
-a_{21}x_1 & + & (1-a_{22})x_2 & - \cdots - & a_{2n}x_n & = & e_2 \\
\vdots & & \vdots & & \vdots & & \vdots \\
-a_{n1}x_1 & - & a_{n2}x_2 & - \cdots + & (1-a_{nn})x_n & = & e_n
\end{array} \tag{9}$$

System (9) of n equations in n unknowns is very important in economic analysis.

---

†Named after American economist Wassily W. Leontief. This model was used in his pioneering paper "Qualitative Input and Output Relations in the Economic System of the United States" in *Review of Economic Statistics* 18(1936):105–125. An updated version of this model appears in Leontief's book *Input-Output Analysis* (New York: Oxford University Press, 1966). Leontief won the Nobel Prize in economics in 1973 for his development of input-output analysis.

EXAMPLE 9 In an economic system with three industries, suppose that the external demands are, respectively, 10, 25, and 20. Suppose that $a_{11}=0.2$, $a_{12}=0.5$, $a_{13}=0.15$, $a_{21}=0.4$, $a_{22}=0.1$, $a_{23}=0.3$, $a_{31}=0.25$, $a_{32}=0.5$, and $a_{33}=0.15$. Find the output in each industry such that supply exactly equals demand.

Solution Here $n=3$, $1-a_{11}=0.8$, $1-a_{22}=0.9$, and $1-a_{33}=0.85$. Then system (9) is

$$\begin{aligned} 0.8x_1-0.5x_2-0.15x_3&=10\\ -0.4x_1+0.9x_2-\ 0.3x_3&=25\\ -0.25x_1-0.5x_2+0.85x_3&=20 \end{aligned}$$

Solving this system by using a calculator, we obtain successively (using five-decimal-place accuracy and Gauss-Jordan elimination)

$$\left(\begin{array}{ccc|c} 0.8 & -0.5 & -0.15 & 10\\ -0.4 & 0.9 & -0.3 & 25\\ -0.25 & -0.5 & 0.85 & 20 \end{array}\right) \xrightarrow{M_1(\frac{1}{0.8})} \left(\begin{array}{ccc|c} 1 & -0.625 & -0.1875 & 12.5\\ -0.4 & 0.9 & -0.3 & 25\\ -0.25 & -0.5 & 0.85 & 20 \end{array}\right)$$

$$\xrightarrow[A_{1,3}(0.25)]{A_{1,2}(0.4)} \left(\begin{array}{ccc|c} 1 & -0.625 & -0.1875 & 12.5\\ 0 & 0.65 & -0.375 & 30\\ 0 & -0.65625 & 0.80313 & 23.125 \end{array}\right)$$

$$\xrightarrow{M_2(\frac{1}{0.65})} \left(\begin{array}{ccc|c} 1 & -0.625 & -0.1875 & 12.5\\ 0 & 1 & -0.57692 & 46.15385\\ 0 & -0.65625 & 0.80313 & 23.125 \end{array}\right)$$

$$\xrightarrow[A_{2,3}(0.65625)]{A_{2,1}(0.625)} \left(\begin{array}{ccc|c} 1 & 0 & -0.54808 & 41.34616\\ 0 & 1 & -0.57692 & 46.15385\\ 0 & 0 & 0.42453 & 53.41346 \end{array}\right)$$

$$\xrightarrow{M_3(1/0.42453)} \left(\begin{array}{ccc|c} 1 & 0 & -0.54808 & 41.34616\\ 0 & 1 & -0.57692 & 46.15385\\ 0 & 0 & 1 & 125.81787 \end{array}\right)$$

$$\xrightarrow[A_{3,2}(0.57692)]{A_{3,1}(0.54808)} \left(\begin{array}{ccc|c} 1 & 0 & 0 & 110.30442\\ 0 & 1 & 0 & 118.74070\\ 0 & 0 & 1 & 125.81787 \end{array}\right)$$

We conclude that the outputs needed for supply to equal demand are, approximately, $x_1=110$, $x_2=119$, and $x_3=126$.

EXAMPLE 10 A State Fish and Game Department supplies three types of food to a lake that supports three species of fish. Each fish of Species 1 consumes, each week, an average of 1 unit of Food 1, 1 unit of Food 2, and 2 units of Food 3. Each fish of Species 2 consumes, each week, an average of 3 units of Food 1, 4 units of Food 2, and 5 units of Food 3. For a fish of Species 3, the average weekly consumption is 2 units of Food 1, 1 unit of Food 2, and 5 units of Food 3. Each week 25,000 units of Food 1, 20,000 units of Food 2, and 55,000 units of

Food 3 are supplied to the lake. If we assume that all food is eaten, how many fish of each species can coexist in the lake?

Solution We let x_1, x_2, and x_3 denote the numbers of fish of the three species being supported by the lake environment. Using the information in the problem, we see that x_1 fish of Species 1 consume x_1 units of Food 1, x_2 fish of Species 2 consume $3x_2$ units of Food 1, and x_3 fish of Species 3 consume $2x_3$ units of Food 1. Thus $x_1 + 3x_2 + 2x_3 = 25{,}000$ = total weekly supply of Food 1. Obtaining a similar equation for each of the other two foods, we are led to the following system:

$$\begin{aligned} x_1 + 3x_2 + 2x_3 &= 25{,}000 \\ x_1 + 4x_2 + x_3 &= 20{,}000 \\ 2x_1 + 5x_2 + 5x_3 &= 55{,}000 \end{aligned}$$

Upon solving, we obtain

$$\left(\begin{array}{ccc|c} 1 & 3 & 2 & 25{,}000 \\ 1 & 4 & 1 & 20{,}000 \\ 2 & 5 & 5 & 55{,}000 \end{array}\right)$$

$$\xrightarrow[A_{1,3}(-2)]{A_{1,2}(-1)} \left(\begin{array}{ccc|c} 1 & 3 & 2 & 25{,}000 \\ 0 & 1 & -1 & -5{,}000 \\ 0 & -1 & 1 & 5{,}000 \end{array}\right) \xrightarrow[A_{2,3}(1)]{A_{2,1}(-3)} \left(\begin{array}{ccc|c} 1 & 0 & 5 & 40{,}000 \\ 0 & 1 & -1 & -5{,}000 \\ 0 & 0 & 0 & 0 \end{array}\right)$$

Thus, if x_3 is chosen arbitrarily, we have an infinite number of solutions given by $(40{,}000 - 5x_3,\ x_3 - 5{,}000,\ x_3)$. Of course, we must have $x_1 \ge 0$, $x_2 \ge 0$ and $x_3 \ge 0$. Since $x_2 = x_3 - 5{,}000 \ge 0$, we have $x_3 \ge 5{,}000$. This means that $0 \le x_1 \le 40{,}000 - 5(5{,}000) = 15{,}000$. Finally, since $40{,}000 - 5x_3 \ge 0$, we see that $x_3 \le 8{,}000$. This means that the populations that can be supported by the lake with all food consumed are

$$\begin{aligned} x_1 &= 40{,}000 - 5x_3 \\ x_2 &= x_3 - 5{,}000 \\ 5{,}000 &\le x_3 \le 8{,}000 \end{aligned}$$

For example, if $x_3 = 6{,}000$, then $x_1 = 10{,}000$ and $x_2 = 1{,}000$.

PROBLEMS 1.3

In Problems 1–20, use Gauss-Jordan and Gaussian elimination to find all solutions, if any, to the given systems.

1. $\begin{aligned} x_1 - 2x_2 + 3x_3 &= 11 \\ 4x_1 + x_2 - x_3 &= 4 \\ 2x_1 - x_2 + 3x_3 &= 10 \end{aligned}$

2. $\begin{aligned} -2x_1 + x_2 + 6x_3 &= 18 \\ 5x_1 + 8x_3 &= -16 \\ 3x_1 + 2x_2 - 10x_3 &= -3 \end{aligned}$

3. $\begin{aligned} 3x_1 + 6x_2 - 6x_3 &= 9 \\ 2x_1 - 5x_2 + 4x_3 &= 6 \\ -x_1 + 16x_2 - 14x_3 &= -3 \end{aligned}$

4. $\begin{aligned} 3x_1 + 6x_2 - 6x_3 &= 9 \\ 2x_1 - 5x_2 + 4x_3 &= 6 \\ 5x_1 + 28x_2 - 26x_3 &= -8 \end{aligned}$

5. $\begin{aligned} x_1 + x_2 - x_3 &= 7 \\ 4x_1 - x_2 + 5x_3 &= 4 \\ 2x_1 + 2x_2 - 3x_3 &= 0 \end{aligned}$

6. $\begin{aligned} x_1 + x_2 - x_3 &= 7 \\ 4x_1 - x_2 + 5x_3 &= 4 \\ 6x_1 + x_2 + 3x_3 &= 18 \end{aligned}$

7. $$\begin{aligned} x_1+x_2-x_3&=7\\ 4x_1-x_2+5x_3&=4\\ 6x_1+x_2+3x_3&=20 \end{aligned}$$

8. $$\begin{aligned} x_1-2x_2+3x_3&=0\\ 4x_1+x_2-x_3&=0\\ 2x_1-x_2+3x_3&=0 \end{aligned}$$

9. $$\begin{aligned} x_1+x_2-x_3&=0\\ 4x_1-x_2+5x_3&=0\\ 6x_1+x_2+3x_3&=0 \end{aligned}$$

10. $$\begin{aligned} 2x_2+5x_3&=6\\ x_1\qquad -2x_3&=4\\ 2x_1+4x_2\qquad&=-2 \end{aligned}$$

11. $$\begin{aligned} x_1+2x_2-x_3&=4\\ 3x_1+4x_2-2x_3&=7 \end{aligned}$$

12. $$\begin{aligned} x_1+2x_2-4x_3&=4\\ -2x_1-4x_2+8x_3&=-8 \end{aligned}$$

13. $$\begin{aligned} x_1+2x_2-4x_3&=4\\ -2x_1-4x_2+8x_3&=-9 \end{aligned}$$

14. $$\begin{aligned} x_1+2x_2-x_3+x_4&=7\\ 3x_1+6x_2-3x_3+3x_4&=21 \end{aligned}$$

15. $$\begin{aligned} 2x_1+6x_2-4x_3+2x_4&=4\\ x_1\qquad -x_3+x_4&=5\\ -3x_1+2x_2-2x_3\qquad&=-2 \end{aligned}$$

16. $$\begin{aligned} x_1-2x_2+x_3+x_4&=2\\ 3x_1\qquad +2x_3-2x_4&=-8\\ 4x_2-x_3-x_4&=1\\ -x_1+6x_2-2x_3\qquad&=7 \end{aligned}$$

17. $$\begin{aligned} x_1-2x_2+x_3+x_4&=2\\ 3x_1\qquad +2x_3-2x_4&=-8\\ 4x_2-x_3-x_4&=1\\ 5x_1\qquad +3x_3-x_4&=-3 \end{aligned}$$

18. $$\begin{aligned} x_1-2x_2+x_3+x_4&=2\\ 3x_1\qquad +2x_3-2x_4&=-8\\ 4x_2-x_3-x_4&=1\\ 5x_1\qquad +3x_3-x_4&=0 \end{aligned}$$

19. $$\begin{aligned} x_1+x_2&=4\\ 2x_1-3x_2&=7\\ 3x_1+2x_2&=8 \end{aligned}$$

20. $$\begin{aligned} x_1+x_2&=4\\ 2x_1-3x_2&=7\\ 3x_1-2x_2&=11 \end{aligned}$$

In Problems 21–29 determine whether the given matrix is in row echelon form (but not reduced row echelon form), reduced row echelon form, or neither.

21. $\begin{pmatrix} 1 & 1 & 0\\ 0 & 1 & 1\\ 0 & 0 & 1 \end{pmatrix}$ **22.** $\begin{pmatrix} 2 & 0 & 0\\ 0 & 1 & 0\\ 0 & 0 & -1 \end{pmatrix}$ **23.** $\begin{pmatrix} 1 & 0 & 1 & 0\\ 0 & 1 & 1 & 0\\ 0 & 0 & 0 & 0 \end{pmatrix}$

24. $\begin{pmatrix} 1 & 0 & 0 & 0\\ 0 & 0 & 1 & 0\\ 0 & 0 & 0 & 1 \end{pmatrix}$ **25.** $\begin{pmatrix} 0 & 1 & 0 & 0\\ 1 & 0 & 0 & 0\\ 0 & 0 & 0 & 0 \end{pmatrix}$ **26.** $\begin{pmatrix} 1 & 0 & 1 & 2\\ 0 & 1 & 3 & 4 \end{pmatrix}$

27. $\begin{pmatrix} 1 & 0\\ 0 & 1\\ 0 & 0 \end{pmatrix}$ **28.** $\begin{pmatrix} 1 & 0 & 0\\ 0 & 0 & 0\\ 0 & 0 & 1 \end{pmatrix}$ **29.** $\begin{pmatrix} 1 & 0 & 0 & 4\\ 0 & 1 & 0 & 5\\ 0 & 1 & 1 & 6 \end{pmatrix}$

In Problems 30–35 use the elementary row operations to reduce the given matrices to row echelon form and reduced row echelon form.

30. $\begin{pmatrix} 1 & 1\\ 2 & 3 \end{pmatrix}$ **31.** $\begin{pmatrix} -1 & 6\\ 4 & 2 \end{pmatrix}$ **32.** $\begin{pmatrix} 1 & -1 & 1\\ 2 & 4 & 3\\ 5 & 6 & -2 \end{pmatrix}$

33. $\begin{pmatrix} 2 & -4 & 8\\ 3 & 5 & 8\\ -6 & 0 & 4 \end{pmatrix}$ **34.** $\begin{pmatrix} 2 & -4 & -2\\ 3 & 1 & 6 \end{pmatrix}$ **35.** $\begin{pmatrix} 2 & -7\\ 3 & 5\\ 4 & -3 \end{pmatrix}$

36. In the Leontief input-output model of Example 8, suppose that there are three industries. Suppose further that $e_1=10$, $e_2=15$, $e_3=30$, $a_{11}=\frac{1}{3}$, $a_{12}=\frac{1}{2}$, $a_{13}=\frac{1}{6}$, $a_{21}=\frac{1}{4}$, $a_{22}=\frac{1}{4}$, $a_{23}=\frac{1}{8}$, $a_{31}=\frac{1}{12}$, $a_{32}=\frac{1}{3}$, and $a_{33}=\frac{1}{6}$. Find the output of each industry such that supply exactly equals demand.

37. In Example 10, assume that there are 15,000 units of the first food, 10,000 units of the second, and 35,000 units of the third supplied to the lake each week. Assuming that all three foods are consumed, what populations of the three species can coexist in the lake? Is there a unique solution?

38. A traveler just returned from Europe spent \$30 a day for housing in England, \$20 a day in France and \$20 a day in Spain. For food the traveler spent \$20 a day in England, \$30 a day in France, and \$20 a day in Spain. The traveler spent \$10 a day in each country for incidental expenses. The traveler's records of the trip indicate a total of \$340 spent for housing, \$320 for food, and \$140 for incidental expenses while traveling in these countries. Calculate the number of days the traveler spent in each of the countries or show that the records must be incorrect, because the amounts spent are incompatible with each other.

39. An investor remarks to a stockbroker that all her stock holdings are in three companies, Eastern Airlines, Hilton Hotels, and McDonald's, and that 2 days ago the value of her stocks went down \$350 but yesterday the value increased by \$600. The broker recalls that 2 days ago the price of Eastern Airlines stock dropped by \$1 a share, Hilton Hotels dropped \$1.50, but the price of McDonald's stock rose by \$0.50. The broker also remembers that yesterday the price of Eastern Airlines stock rose \$1.50, there was a further drop of \$0.50 a share in Hilton Hotels stock, and McDonald's stock rose \$1. Show that the broker does not have enough information to calculate the number of shares the investor owns of each company's stock, but that when the investor says that she owns 200 shares of McDonald's stock, the broker can calculate the number of shares of Eastern Airlines and Hilton Hotels.

40. An intelligence agent knows that 60 aircraft, consisting of fighter planes and bombers, are stationed at a certain secret airfield. The agent wishes to determine how many of the 60 are fighter planes and how many are bombers. There is a type of rocket carried by both sorts of planes; the fighter carries six of these rockets, the bomber only two. The agent learns that 250 rockets are required to arm every plane at this airfield. Furthermore, the agent overhears a remark that there are twice as many fighter planes as bombers at the base (that is, the number of fighter planes minus twice the number of bombers equals zero). Calculate the number of fighter planes and bombers at the airfield or show that the agent's information must be incorrect, because it is inconsistent.

41. Consider the system

$$\begin{aligned} 2x_1 - x_2 + 3x_3 &= a \\ 3x_1 + x_2 - 5x_3 &= b \\ -5x_1 - 5x_2 + 21x_3 &= c \end{aligned}$$

Show that the system is inconsistent if $c \neq 2a - 3b$.

42. Consider the system

$$\begin{aligned} 2x_1 + 3x_2 - x_3 &= a \\ x_1 - x_2 + 3x_3 &= b \\ 3x_1 + 7x_2 - 5x_3 &= c \end{aligned}$$

Find conditions on a, b, and c such that the system is consistent.

***43.** Consider the general system of three linear equations in three unknowns:

$$\begin{aligned} a_{11}x_1 + a_{12}x_2 + a_{13}x_3 &= b_1 \\ a_{21}x_1 + a_{22}x_2 + a_{23}x_3 &= b_2 \\ a_{31}x_1 + a_{32}x_2 + a_{33}x_3 &= b_3 \end{aligned}$$

Find conditions on the coefficients a_{ij} such that the system has a unique solution.

44. Solve the following system using a hand calculator and carrying 5 decimal places of accuracy:

$$\begin{aligned} 2x_2 - x_3 - 4x_4 &= 2 \\ x_1 - x_2 + 5x_3 + 2x_4 &= -4 \\ 3x_1 + 3x_2 - 7x_3 - x_4 &= 4 \\ -x_1 - 2x_2 + 3x_3 \qquad &= -7 \end{aligned}$$

45. Do the same for the system

$$\begin{aligned} 3.8x_1 + 1.6x_2 + 0.9x_3 &= 3.72 \\ -0.7x_1 + 5.4x_2 + 1.6x_3 &= 3.16 \\ 1.5x_1 + 1.1x_2 - 3.2x_3 &= 43.78 \end{aligned}$$

1.4 Homogeneous Systems of Equations

The general $m \times n$ system of linear equations (system (1.3.7), page 15) is called **homogeneous** if all the constants $b_1, b_2, \ldots, b_m$ are zero. That is, the general homogeneous system is given by

$$\begin{aligned} a_{11}x_1 + a_{12}x_2 + \cdots + a_{1n}x_n &= 0 \\ a_{21}x_1 + a_{22}x_2 + \cdots + a_{2n}x_n &= 0 \\ \vdots \qquad \vdots \qquad\quad & \vdots \\ a_{m1}x_1 + a_{m2}x_2 + \cdots + a_{mn}x_n &= 0 \end{aligned} \tag{1}$$

Homogeneous systems arise in a variety of ways. We shall see one of these in the next chapter (in Section 2.6). In this section we shall solve some homogeneous systems—again by the method of Gauss-Jordan elimination.

For the general linear system there are three possibilities: no solution, one solution, or an infinite number of solutions. For the general homogeneous system the situation is simpler. Since $x_1 = x_2 = \cdots = x_n = 0$ is always a solution (called the **trivial solution** or **zero solution**), there are only two possibilities: either the zero solution is the only solution or there are an infinite number of solutions in addition to the zero solution. (Solutions other than the zero solution are called **nontrivial solutions**.

EXAMPLE 1 Solve the homogeneous system

$$\begin{aligned} 2x_1 + 4x_2 + 6x_3 &= 0 \\ 4x_1 + 5x_2 + 6x_3 &= 0 \\ 3x_1 + x_2 - 2x_3 &= 0 \end{aligned}$$

Solution This is the homogeneous version of the system in Example 1.3.1, page 6.

Reducing successively, we obtain (after dividing the first equation by 2)

$$\left(\begin{array}{ccc|c}1 & 2 & 3 & 0\\ 4 & 5 & 6 & 0\\ 3 & 1 & -2 & 0\end{array}\right) \xrightarrow[A_{1,3}(-3)]{A_{1,2}(-4)} \left(\begin{array}{ccc|c}1 & 2 & 3 & 0\\ 0 & -3 & -6 & 0\\ 0 & -5 & -11 & 0\end{array}\right) \xrightarrow{M_2(-\frac{1}{3})} \left(\begin{array}{ccc|c}1 & 2 & 3 & 0\\ 0 & 1 & 2 & 0\\ 0 & -5 & -11 & 0\end{array}\right)$$

$$\xrightarrow[A_{2,3}(5)]{A_{2,1}(-2)} \left(\begin{array}{ccc|c}1 & 0 & -1 & 0\\ 0 & 1 & 2 & 0\\ 0 & 0 & -1 & 0\end{array}\right) \xrightarrow{M_3(-1)} \left(\begin{array}{ccc|c}1 & 0 & -1 & 0\\ 0 & 1 & 2 & 0\\ 0 & 0 & 1 & 0\end{array}\right) \xrightarrow[A_{3,2}(-2)]{A_{3,1}(1)} \left(\begin{array}{ccc|c}1 & 0 & 0 & 0\\ 0 & 1 & 0 & 0\\ 0 & 0 & 1 & 0\end{array}\right)$$

Thus the system has the unique solution (0, 0, 0). That is, the system has only the trivial solution.

EXAMPLE 2 Solve the homogeneous system

$$\begin{aligned} x_1 + 2x_2 - x_3 &= 0\\ 3x_1 - 3x_2 + 2x_3 &= 0\\ -x_1 - 11x_2 + 6x_3 &= 0 \end{aligned}$$

Solution Using Gauss-Jordan elimination we obtain, successively,

$$\left(\begin{array}{ccc|c}1 & 2 & -1 & 0\\ 3 & -3 & 2 & 0\\ -1 & -11 & 6 & 0\end{array}\right) \xrightarrow[A_{1,3}(1)]{A_{1,2}(-3)} \left(\begin{array}{ccc|c}1 & 2 & -1 & 0\\ 0 & -9 & 5 & 0\\ 0 & -9 & 5 & 0\end{array}\right)$$

$$\xrightarrow{M_2(-\frac{1}{9})} \left(\begin{array}{ccc|c}1 & 2 & -1 & 0\\ 0 & 1 & -\frac{5}{9} & 0\\ 0 & -9 & 5 & 0\end{array}\right) \xrightarrow[A_{2,3}(9)]{A_{2,1}(-2)} \left(\begin{array}{ccc|c}1 & 0 & \frac{1}{9} & 0\\ 0 & 1 & -\frac{5}{9} & 0\\ 0 & 0 & 0 & 0\end{array}\right)$$

The augmented matrix is now in reduced row echelon form and, evidently, there are an infinite number of solutions given by $(-\frac{1}{9}x_3, \frac{5}{9}x_3, x_3)$. If $x_3 = 0$, for example, we obtain the trivial solution. If $x_3 = 1$ we obtain the solution $(-\frac{1}{9}, \frac{5}{9}, 1)$.

EXAMPLE 3 Solve the system

$$\begin{aligned} x_1 + x_2 - x_3 &= 0\\ 4x_1 - 2x_2 + 7x_3 &= 0 \end{aligned} \tag{2}$$

Solution Row-reducing, we obtain

$$\left(\begin{array}{ccc|c}1 & 1 & -1 & 0\\ 4 & -2 & 7 & 0\end{array}\right) \xrightarrow{A_{1,2}(-4)} \left(\begin{array}{ccc|c}1 & 1 & -1 & 0\\ 0 & -6 & 11 & 0\end{array}\right)$$

$$\xrightarrow{M_2(-\frac{1}{6})} \left(\begin{array}{ccc|c}1 & 1 & -1 & 0\\ 0 & 1 & -\frac{11}{6} & 0\end{array}\right) \xrightarrow{A_{2,1}(-1)} \left(\begin{array}{ccc|c}1 & 0 & \frac{5}{6} & 0\\ 0 & 1 & -\frac{11}{6} & 0\end{array}\right)$$

Thus there are an infinite number of solutions given by $(-\frac{5}{6}x_3, \frac{11}{6}x_3, x_3)$. This is not surprising since system (2) contains three unknowns and only two equations.

In fact, if there are more unknowns than equations, the homogeneous system (1) will always have an infinite number of solutions. To see this, note that if there were only the trivial solution, then row reduction would lead us to the system

$$\begin{aligned} x_1 \qquad\qquad &= 0 \\ x_2 \qquad &= 0 \\ &\vdots \\ x_n &= 0 \end{aligned}$$

and, possibly, additional equations of the form $0=0$. But this system has at least as many equations as unknowns. Since row reduction does not change either the number of equations or the number of unknowns, we have a contradiction of our assumption that there were more unknowns than equations. Thus we have Theorem 1.

THEOREM 1 The homogeneous system (1) has an infinite number of solutions if $n > m$.

PROBLEMS 1.4 In Problems 1–13 find all solutions to the homogeneous systems.

1. $\begin{aligned} 2x_1 - x_2 &= 0 \\ 3x_1 + 4x_2 &= 0 \end{aligned}$

2. $\begin{aligned} x_1 - 5x_2 &= 0 \\ -x_1 + 5x_2 &= 0 \end{aligned}$

3. $\begin{aligned} x_1 + x_2 - x_3 &= 0 \\ 2x_1 - 4x_2 + 3x_3 &= 0 \\ 3x_1 + 7x_2 - x_3 &= 0 \end{aligned}$

4. $\begin{aligned} x_1 + x_2 - x_3 &= 0 \\ 2x_1 - 4x_2 + 3x_3 &= 0 \\ -x_1 - 7x_2 + 6x_3 &= 0 \end{aligned}$

5. $\begin{aligned} x_1 + x_2 - x_3 &= 0 \\ 2x_1 - 4x_2 + 3x_3 &= 0 \\ -5x_1 + 13x_2 - 10x_3 &= 0 \end{aligned}$

6. $\begin{aligned} 2x_1 + 3x_2 - x_3 &= 0 \\ 6x_1 - 5x_2 + 7x_3 &= 0 \end{aligned}$

7. $\begin{aligned} 4x_1 - x_2 &= 0 \\ 7x_1 + 3x_2 &= 0 \\ -8x_1 + 6x_2 &= 0 \end{aligned}$

8. $\begin{aligned} x_1 - x_2 + 7x_3 - x_4 &= 0 \\ 2x_1 + 3x_2 - 8x_3 + x_4 &= 0 \end{aligned}$

9. $\begin{aligned} x_1 - 2x_2 + x_3 + x_4 &= 0 \\ 3x_1 \qquad + 2x_3 - 2x_4 &= 0 \\ 4x_2 - x_3 - x_4 &= 0 \\ 5x_1 \qquad + 3x_3 - x_4 &= 0 \end{aligned}$

10. $\begin{aligned} -2x_1 \qquad\qquad + 7x_4 &= 0 \\ x_1 + 2x_2 - x_3 + 4x_4 &= 0 \\ 3x_1 \qquad - x_3 + 5x_4 &= 0 \\ 4x_1 + 2x_2 + 3x_3 \qquad &= 0 \end{aligned}$

11. $\begin{aligned} 2x_1 - x_2 &= 0 \\ 3x_1 + 5x_2 &= 0 \\ 7x_1 - 3x_2 &= 0 \\ -2x_1 + 3x_2 &= 0 \end{aligned}$

12. $\begin{aligned} x_1 - 3x_2 &= 0 \\ -2x_1 + 6x_2 &= 0 \\ 4x_1 - 12x_2 &= 0 \end{aligned}$

13. $\begin{aligned} x_1 + x_2 - x_3 &= 0 \\ 4x_1 - x_2 + 5x_3 &= 0 \\ -2x_1 + x_2 - 2x_3 &= 0 \\ 3x_1 + 2x_2 - 6x_3 &= 0 \end{aligned}$

14. Show that the homogeneous system

$$\begin{aligned} a_{11}x_1 + a_{12}x_2 &= 0 \\ a_{21}x_1 + a_{22}x_2 &= 0 \end{aligned}$$

has an infinite number of solutions if and only if the determinant of the system is zero.

15. Consider the system

$$\begin{aligned} 2x_1 - 3x_2 + 5x_3 &= 0 \\ -x_1 + 7x_2 - x_3 &= 0 \\ 4x_1 - 11x_2 + kx_3 &= 0 \end{aligned}$$

For what value of k will the system have nontrivial solutions?

***16.** Consider the 3×3 homogeneous system

$$\begin{aligned} a_{11}x_1 + a_{12}x_2 + a_{13}x_3 &= 0 \\ a_{21}x_1 + a_{22}x_2 + a_{23}x_3 &= 0 \\ a_{31}x_1 + a_{32}x_2 + a_{33}x_3 &= 0 \end{aligned}$$

Find conditions on the coefficients a_{ij} such that the zero solution is the only solution.

Review Exercises for Chapter 1

In Exercises 1–14, find all solutions (if any) to the given systems.

1. $$\begin{aligned} 3x_1 + 6x_2 &= 9 \\ -2x_1 + 3x_2 &= 4 \end{aligned}$$

2. $$\begin{aligned} 3x_1 + 6x_2 &= 9 \\ 2x_1 + 4x_2 &= 6 \end{aligned}$$

3. $$\begin{aligned} 3x_1 - 6x_2 &= 9 \\ -2x_1 + 4x_2 &= 6 \end{aligned}$$

4. $$\begin{aligned} x_1 + x_2 + x_3 &= 2 \\ 2x_1 - x_2 + 2x_3 &= 4 \\ -3x_1 + 2x_2 + 3x_3 &= 8 \end{aligned}$$

5. $$\begin{aligned} x_1 + x_2 + x_3 &= 0 \\ 2x_1 - x_2 + 2x_3 &= 0 \\ -3x_1 + 2x_2 + 3x_3 &= 0 \end{aligned}$$

6. $$\begin{aligned} x_1 + x_2 + x_3 &= 2 \\ 2x_1 - x_2 + 2x_3 &= 4 \\ -x_1 + 4x_2 + x_3 &= 2 \end{aligned}$$

7. $$\begin{aligned} x_1 + x_2 + x_3 &= 2 \\ 2x_1 - x_2 + 2x_3 &= 4 \\ -x_1 + 4x_2 + x_3 &= 3 \end{aligned}$$

8. $$\begin{aligned} x_1 + x_2 + x_3 &= 0 \\ 2x_1 - x_2 + 2x_3 &= 0 \\ -x_1 + 4x_2 + x_3 &= 0 \end{aligned}$$

9. $$\begin{aligned} 2x_1 + x_2 - 3x_3 &= 0 \\ 4x_1 - x_2 + x_3 &= 0 \end{aligned}$$

10. $$\begin{aligned} x_1 + x_2 &= 0 \\ 2x_1 + x_2 &= 0 \\ 3x_1 + x_2 &= 0 \end{aligned}$$

11. $$\begin{aligned} x_1 + x_2 &= 1 \\ 2x_1 + x_2 &= 3 \\ 3x_1 + x_2 &= 4 \end{aligned}$$

12. $$\begin{aligned} x_1 + x_2 + x_3 + x_4 &= 4 \\ 2x_1 - 3x_2 - x_3 + 4x_4 &= 7 \\ -2x_1 + 4x_2 + x_3 - 2x_4 &= 1 \\ 5x_1 - x_2 + 2x_3 + x_4 &= -1 \end{aligned}$$

13. $$\begin{aligned} x_1 + x_2 + x_3 + x_4 &= 0 \\ 2x_1 - 3x_2 - x_3 + 4x_4 &= 0 \\ -2x_1 + 4x_2 + x_3 - 2x_4 &= 0 \\ 5x_1 - x_2 + 2x_3 + x_4 &= 0 \end{aligned}$$

14. $$\begin{aligned} x_1 + x_2 + x_3 + x_4 &= 0 \\ 2x_1 - 3x_2 - x_3 + 4x_4 &= 0 \\ -2x_1 + 4x_2 + x_3 - 2x_4 &= 0 \end{aligned}$$

15. Find the distance from the point $(3, -2)$ to the line $x - 2y = 6$.

In Exercises 16–20 determine whether the given matrix is in row echelon form (but not reduced row echelon form), reduced row echelon form, or neither.

16. $\begin{pmatrix} 1 & 0 & 0 & 0 \\ 0 & 1 & 0 & 2 \\ 0 & 0 & 1 & 3 \end{pmatrix}$ **17.** $\begin{pmatrix} 1 & 8 & 1 & 0 \\ 0 & 1 & 5 & -7 \\ 0 & 0 & 1 & 4 \end{pmatrix}$ **18.** $\begin{pmatrix} 1 & 0 \\ 0 & 3 \\ 0 & 0 \end{pmatrix}$

19. $\begin{pmatrix} 1 & 0 & 2 & 0 \\ 0 & 1 & 3 & 0 \end{pmatrix}$ **20.** $\begin{pmatrix} 1 & 1 & 1 & 1 \\ 0 & 1 & 1 & 1 \end{pmatrix}$

In Exercises 21 and 22, reduce the matrix to row echelon form and reduced row echelon form.

21. $\begin{pmatrix} 2 & 8 & -2 \\ 1 & 0 & -6 \end{pmatrix}$ **22.** $\begin{pmatrix} 1 & -1 & 2 & 4 \\ -1 & 2 & 0 & 3 \\ 2 & 3 & -1 & 1 \end{pmatrix}$

2 Vectors and Matrices

2.1 Vectors

William Rowan Hamilton (Granger Collection)

The study of vectors began essentially with the work of the great Irish mathematician Sir William Rowan Hamilton (1805–1865). Hamilton was a genius who, by the age of twelve, had mastered not only the languages of continental Europe but also Greek, Latin, Sanskrit, Hebrew, Chinese, Persian, Arabic, Malay, Hindi, Bengali, and several others as well. In his twenties, Hamilton turned to science, and his treatises on mechanics and optics provide the basis for much of modern physics.

In his thirties this remarkable man began his research in mathematics. His desire to find a way to represent certain objects in the plane and in space led to the discovery of what he called "quaternions." This notion led to the development of what we now call *vectors*. Throughout Hamilton's life, and for the remainder of the nineteenth century, there was considerable debate over the usefulness of quaternions and vectors. At the end of the century, the great British physicist Lord Kelvin wrote that quaternions, "although beautifully ingenious, have been an unmixed evil to those who have touched them in any way [and] vectors. . . have never been of the slightest use to any creature."

But Kelvin was wrong. Today nearly all branches of classical and modern physics are represented by means of the language of vectors. Vectors are also used with increasing frequency in the social and biological sciences.†

On page 2 we described the solution to a system of two equations in two unknowns to be a pair of numbers written (x_1, x_2). In Example 1.3.1 on page 8 we wrote the solution of the system of three equations in three unknowns as the triple of numbers $(2, -1, 4)$. Both (x_1, x_2) and $(2, -1, 4)$ are **vectors**.

***n*-COMPONENT ROW VECTOR**

We define an ***n*-component row vector** to be an **ordered** set of n numbers written as

$$(x_1, x_2, \ldots, x_n) \tag{1}$$

† For interesting discussions of the development of modern vector analysis, consult the book by M. J. Crowe, *A History of Vector Analysis* (Notre Dame: University of Notre Dame Press, 1967) or Morris Kline's excellent book *Mathematical Thought from Ancient to Modern Times* (New York: Oxford University Press, 1972), chap. 32.

n-COMPONENT COLUMN VECTOR

An ***n*-component column vector** is an **ordered** set of n numbers written as

$$\begin{pmatrix} x_1 \\ x_2 \\ \vdots \\ x_n \end{pmatrix} \tag{2}$$

In (1) or (2), x_1 is called the **first component** of the vector, x_2 is the **second component**, and so on. In general, x_k is called the ***k*th component** of the vector.

For simplicity, we shall often refer to an n-component row vector as a **row vector** or an ***n*-vector**. Similarly, we shall use the term **column vector** (or n-vector) to denote an n-component column vector. Any vector whose entries are all zero is called a **zero vector**.

EXAMPLE 1

The following are vectors:

i. $(3, 6)$ is a row vector (or a 2-vector).

ii. $\begin{pmatrix} 2 \\ -1 \\ 5 \end{pmatrix}$ is a column vector (or a 3-vector).

iii. $(2, -1, 0, 4)$ is a row vector (or a 4-vector).

iv. $\begin{pmatrix} 0 \\ 0 \\ 0 \\ 0 \\ 0 \end{pmatrix}$ is a column vector and a zero vector.

Warning. The word "ordered" in the definition of a vector is essential. Two vectors with the same components written in different orders are *not* the same. Thus, for example, the row vectors $(1, 2)$ and $(2, 1)$ are not equal.

For the remainder of this text we shall denote vectors with boldface lowercase letters like **u**, **v**, **a**, **b**, **c**, and so on. A zero vector is denoted **0**.

Vectors arise in a great number of ways. Suppose that the buyer for a manufacturing plant must order different quantities of steel, aluminum, oil, and paper. He can keep track of the quantities to be ordered with a single vector. The vector $\begin{pmatrix} 10 \\ 30 \\ 15 \\ 60 \end{pmatrix}$ indicates that he would order 10 units of steel, 30 units of aluminum, and so on.

Remark. We see here why the order in which the components of a vector are written is important. It is clear that the vectors $\begin{pmatrix} 30 \\ 15 \\ 60 \\ 10 \end{pmatrix}$ and $\begin{pmatrix} 10 \\ 30 \\ 15 \\ 60 \end{pmatrix}$ mean very different things to the buyer.

It is time to describe some properties of vectors. Since it would be repetitive to do so first for row vectors and then for column vectors, we shall give all definitions in terms of column vectors. Similar definitions hold for row vectors.

The components of all the vectors in this text are either real or complex numbers.† We use the symbol $\mathbb{R}^n$ to denote the set of all n-vectors $\begin{pmatrix} a_1 \\ a_2 \\ \vdots \\ a_n \end{pmatrix}$, where each a_i is a real number. Similarly, we use the symbol $\mathbb{C}^n$ to denote the set of all n-vectors $\begin{pmatrix} c_1 \\ c_2 \\ \vdots \\ c_n \end{pmatrix}$, where each c_i is a complex number. In Chapter 4 we shall discuss the sets $\mathbb{R}^2$ (vectors in the plane) and $\mathbb{R}^3$ (vectors in space). In Chapter 5 we shall examine arbitrary sets of vectors.

DEFINITION 1 **EQUALITY OF VECTORS** Two column (or row) vectors **a** and **b** are **equal** if and only if‡ they have the same number of components and their corresponding components are equal. In symbols, the vectors $\mathbf{a} = \begin{pmatrix} a_1 \\ a_2 \\ \vdots \\ a_n \end{pmatrix}$ and $\mathbf{b} = \begin{pmatrix} b_1 \\ b_2 \\ \vdots \\ b_n \end{pmatrix}$ are equal if and only if $a_1 = b_1, a_2 = b_2, \ldots, a_n = b_n$.

DEFINITION 2 **ADDITION OF VECTORS** Let $\mathbf{a} = \begin{pmatrix} a_1 \\ a_2 \\ \vdots \\ a_n \end{pmatrix}$ and $\mathbf{b} = \begin{pmatrix} b_1 \\ b_2 \\ \vdots \\ b_n \end{pmatrix}$ be n-vectors. Then the sum of **a** and **b** is defined by

$$\mathbf{a}+\mathbf{b} = \begin{pmatrix} a_1+b_1 \\ a_2+b_2 \\ \vdots \\ a_n+b_n \end{pmatrix} \tag{3}$$

† A complex number is a number of the form $a+ib$, where a and b are real numbers and $i=\sqrt{-1}$. A description of complex numbers is given in Appendix 2. We shall not encounter complex vectors again until Chapter 5; they will be especially useful in Chapter 7. Therefore, unless otherwise stated, we assume, for the time being, that all vectors have real components.

‡ The term "if and only if" applies to two statements. "Statement A if and only if statement B" means that statements A and B are equivalent. That is, if statement A is true then statement B is true and if statement B is true then statement A is true. Put another way, it means that you cannot have one without the other.

EXAMPLE 2

$$\begin{pmatrix}1\\2\\4\end{pmatrix}+\begin{pmatrix}-6\\7\\5\end{pmatrix}=\begin{pmatrix}-5\\9\\9\end{pmatrix}$$

EXAMPLE 3

$$\begin{pmatrix}2\\-1\end{pmatrix}+\begin{pmatrix}-2\\1\end{pmatrix}=\begin{pmatrix}0\\0\end{pmatrix}$$

Warning. It is essential that **a** and **b** have the same number of components. For example, the sum $\begin{pmatrix}2\\3\end{pmatrix}+\begin{pmatrix}1\\2\\3\end{pmatrix}$ is not defined since 2-vectors and 3-vectors are different kinds of objects and cannot be added together. Moreover, it is not possible to add a row and a column vector together. For example, the sum $\begin{pmatrix}1\\2\end{pmatrix}+(3, 5)$ is *not* defined.

When dealing with vectors, we shall refer to numbers as **scalars** (which may be real or complex depending on whether the vectors in question are real or complex).†

DEFINITION 3 **SCALAR MULTIPLICATION OF VECTORS** Let $\mathbf{a}=\begin{pmatrix}a_1\\a_2\\\vdots\\a_n\end{pmatrix}$ be a vector and α a scalar. Then the product $\alpha\mathbf{a}$ is given by

$$\alpha\mathbf{a}=\begin{pmatrix}\alpha a_1\\\alpha a_2\\\vdots\\\alpha a_n\end{pmatrix} \tag{4}$$

That is, to multiply a vector by a scalar, we simply multiply each component of the vector by the scalar.

† *Historical Note:* The term "scalar" originated with Hamilton. His definition of the quaternion included what he called a "real part" and an "imaginary part." In his paper "On Quaternions, or on a New System of Imaginaries in Algebra," in *Philosophical Magazine*, 3rd series, 25(1844): 26–27 he wrote: "The algebraically *real* part may receive... all values contained on the one *scale* of progression of numbers from negative to positive infinity; we shall call it therefore the *scalar part*, or simply the *scalar* of the quaternion...." In the same paper, Hamilton went on to define the imaginary part of his quaternion as the *vector* part. Although this was not the first usage of the word "vector," it was the first time it was used in the context of the definitions in this section. It is fair to say that the paper from which the preceding quotation was taken marks the beginning of modern vector analysis.

EXAMPLE 4

$$3\begin{pmatrix}2\\-1\\4\end{pmatrix}=\begin{pmatrix}6\\-3\\12\end{pmatrix}$$

Note. Putting Definition 1 and Definition 2 together, we can define the difference of two vectors by

$$\mathbf{a}-\mathbf{b}=\mathbf{a}+(-1)\mathbf{b} \tag{5}$$

This means that if $\mathbf{a}=\begin{pmatrix}a_1\\a_2\\\vdots\\a_n\end{pmatrix}$ and $\mathbf{b}=\begin{pmatrix}b_1\\b_2\\\vdots\\b_n\end{pmatrix}$, then $\mathbf{a}-\mathbf{b}=\begin{pmatrix}a_1-b_1\\a_2-b_2\\\vdots\\a_n-b_n\end{pmatrix}$.

EXAMPLE 5

Let $\mathbf{a}=\begin{pmatrix}4\\6\\1\\3\end{pmatrix}$ and $\mathbf{b}=\begin{pmatrix}-2\\4\\-3\\0\end{pmatrix}$. Calculate $2\mathbf{a}-3\mathbf{b}$.

Solution $$2\mathbf{a}-3\mathbf{b}=2\begin{pmatrix}4\\6\\1\\3\end{pmatrix}+(-3)\begin{pmatrix}-2\\4\\-3\\0\end{pmatrix}=\begin{pmatrix}8\\12\\2\\6\end{pmatrix}+\begin{pmatrix}6\\-12\\9\\0\end{pmatrix}=\begin{pmatrix}14\\0\\11\\6\end{pmatrix}$$

Once we know how to add vectors and multiply them by scalars, we can prove a number of facts relating these operations. Several of these facts are given in Theorem 1 below. We prove parts (*ii*) and (*iii*) and leave the remaining parts as exercises (see Problems 21–23).

THEOREM 1

Let **a**, **b**, and **c** be n-vectors and let α and β be scalars. Then:

i. $\mathbf{a}+\mathbf{0}=\mathbf{a}$
ii. $0\mathbf{a}=\mathbf{0}$ (Note that the zero on the left is the number zero, whereas the zero on the right is a zero vector.)
iii. $\mathbf{a}+\mathbf{b}=\mathbf{b}+\mathbf{a}$ **(commutative law)**
iv. $(\mathbf{a}+\mathbf{b})+\mathbf{c}=\mathbf{a}+(\mathbf{b}+\mathbf{c})$ **(associative law)**
v. $\alpha(\mathbf{a}+\mathbf{b})=\alpha\mathbf{a}+\alpha\mathbf{b}$ **(distributive law for scalar multiplication)**
vi. $(\alpha+\beta)\mathbf{a}=\alpha\mathbf{a}+\beta\mathbf{a}$
vii. $(\alpha\beta)\mathbf{a}=\alpha(\beta\mathbf{a})$

Proof of (ii) and (iii)

ii. If $\mathbf{a}=\begin{pmatrix}a_1\\a_2\\\vdots\\a_n\end{pmatrix}$, then $0\mathbf{a}=0\begin{pmatrix}a_1\\a_2\\\vdots\\a_n\end{pmatrix}=\begin{pmatrix}0\cdot a_1\\0\cdot a_2\\\vdots\\0\cdot a_n\end{pmatrix}=\begin{pmatrix}0\\0\\\vdots\\0\end{pmatrix}=\mathbf{0}$.

iii. Let $\mathbf{b}=\begin{pmatrix} b_1 \\ b_2 \\ \vdots \\ b_n \end{pmatrix}$. Then $\mathbf{a}+\mathbf{b}=\begin{pmatrix} a_1+b_1 \\ a_2+b_2 \\ \vdots \\ a_n+b_n \end{pmatrix}=\begin{pmatrix} b_1+a_1 \\ b_2+a_2 \\ \vdots \\ b_n+a_n \end{pmatrix}=\mathbf{b}+\mathbf{a}$.

Here we used the fact that for any two numbers x and y, $x+y=y+x$ and $0\cdot x=0$. ■

Note. In (*ii*), the zero on the left is the scalar zero (that is, the real number 0) and the zero on the right is the zero vector. These two things are different.

EXAMPLE 6 To illustrate the associative law, we note that

$$\left[\begin{pmatrix} 3 \\ 1 \\ 2 \end{pmatrix}+\begin{pmatrix} -2 \\ 4 \\ -1 \end{pmatrix}\right]+\begin{pmatrix} 6 \\ -3 \\ 5 \end{pmatrix}=\begin{pmatrix} 1 \\ 5 \\ 1 \end{pmatrix}+\begin{pmatrix} 6 \\ -3 \\ 5 \end{pmatrix}=\begin{pmatrix} 7 \\ 2 \\ 6 \end{pmatrix}$$

while

$$\begin{pmatrix} 3 \\ 1 \\ 2 \end{pmatrix}+\left[\begin{pmatrix} -2 \\ 4 \\ -1 \end{pmatrix}+\begin{pmatrix} 6 \\ -3 \\ 5 \end{pmatrix}\right]=\begin{pmatrix} 3 \\ 1 \\ 2 \end{pmatrix}+\begin{pmatrix} 4 \\ 1 \\ 4 \end{pmatrix}=\begin{pmatrix} 7 \\ 2 \\ 6 \end{pmatrix}$$

Example 6 illustrates the importance of the associative law of vector addition, since if we wish to add together three or more vectors, we can only do so by adding them together two at a time. The associative law tells us that we can do this in two different ways and still come up with the same answer. If this were not the case, the sum of three or more vectors would be more difficult to define since we would have to specify whether we wanted $(\mathbf{a}+\mathbf{b})+\mathbf{c}$ or $\mathbf{a}+(\mathbf{b}+\mathbf{c})$ to be the definition of the sum $\mathbf{a}+\mathbf{b}+\mathbf{c}$.

PROBLEMS 2.1 In Problems 1–10 perform the indicated computation with $\mathbf{a}=\begin{pmatrix} -3 \\ 1 \\ 4 \end{pmatrix}$, $\mathbf{b}=\begin{pmatrix} 5 \\ -4 \\ 7 \end{pmatrix}$, and $\mathbf{c}=\begin{pmatrix} 2 \\ 0 \\ -2 \end{pmatrix}$.

1. $\mathbf{a}+\mathbf{b}$
2. $3\mathbf{b}$
3. $-2\mathbf{c}$
4. $\mathbf{b}+3\mathbf{c}$
5. $2\mathbf{a}-5\mathbf{b}$
6. $-3\mathbf{b}+2\mathbf{c}$
7. $0\mathbf{c}$
8. $\mathbf{a}+\mathbf{b}+\mathbf{c}$
9. $3\mathbf{a}-2\mathbf{b}+4\mathbf{c}$
10. $3\mathbf{b}-7\mathbf{c}+2\mathbf{a}$

In Problems 11–20 perform the indicated computation with $\mathbf{a}=(3,-1,4,2)$, $\mathbf{b}=(6,0,-1,4)$, and $\mathbf{c}=(-2,3,1,5)$. Of course, it is first necessary to extend the definitions in this section to row vectors.

11. $\mathbf{a}+\mathbf{c}$
12. $\mathbf{b}-\mathbf{a}$
13. $4\mathbf{c}$
14. $-2\mathbf{b}$
15. $2\mathbf{a}-\mathbf{c}$
16. $4\mathbf{b}-7\mathbf{a}$

17. $\mathbf{a}+\mathbf{b}+\mathbf{c}$ **18.** $\mathbf{c}-\mathbf{b}+2\mathbf{a}$ **19.** $3\mathbf{a}-2\mathbf{b}+4\mathbf{c}$

20. $\alpha\mathbf{a}+\beta\mathbf{b}+\gamma\mathbf{c}$

21. Let $\mathbf{a}=\begin{pmatrix}a_1\\a_2\\\vdots\\a_n\end{pmatrix}$ and let $\mathbf{0}$ denote the n-component zero column vector. Use Definitions 2 and 3 to show that $\mathbf{a}+\mathbf{0}=\mathbf{a}$ and $0\mathbf{a}=\mathbf{0}$.

22. Let $\mathbf{a}=\begin{pmatrix}a_1\\a_2\\\vdots\\a_n\end{pmatrix}$, $\mathbf{b}=\begin{pmatrix}b_1\\b_2\\\vdots\\b_n\end{pmatrix}$, and $\mathbf{c}=\begin{pmatrix}c_1\\c_2\\\vdots\\c_n\end{pmatrix}$. Compute $(\mathbf{a}+\mathbf{b})+\mathbf{c}$ and $\mathbf{a}+(\mathbf{b}+\mathbf{c})$ and show that they are equal.

23. Let $\mathbf{a}$ and $\mathbf{b}$ be as in Problem 22 and let α and β be scalars. Compute $\alpha(\mathbf{a}+\mathbf{b})$ and $\alpha\mathbf{a}+\alpha\mathbf{b}$ and show that they are equal. Similarly, compute $(\alpha+\beta)\mathbf{a}$ and $\alpha\mathbf{a}+\beta\mathbf{a}$ and show that they are equal. Finally, show that $(\alpha\beta)\mathbf{a}=\alpha(\beta\mathbf{a})$.

24. Find numbers α, β, and γ such that $(2, -1, 4)+(\alpha, \beta, \gamma)=\mathbf{0}$.

25. In the manufacture of a certain product, four raw materials are needed. The vector $\mathbf{d}=\begin{pmatrix}d_1\\d_2\\d_3\\d_4\end{pmatrix}$ represents a given factory's demand for each of the four raw materials to produce 1 unit of its product. If $\mathbf{d}_1$ is the demand vector for factory 1 and $\mathbf{d}_2$ is the demand vector for factory 2, what is represented by the vectors $\mathbf{d}_1+\mathbf{d}_2$ and $2\mathbf{d}_1$?

26. Let $\mathbf{a}=\begin{pmatrix}1\\3\\2\end{pmatrix}$, $\mathbf{b}=\begin{pmatrix}-2\\4\\1\end{pmatrix}$, and $\mathbf{c}=\begin{pmatrix}0\\1\\4\end{pmatrix}$. Find a vector $\mathbf{v}$ such that $2\mathbf{a}-\mathbf{b}+3\mathbf{v}=4\mathbf{c}$.

27. With $\mathbf{a}$, $\mathbf{b}$, and $\mathbf{c}$ as in Problem 26, find a vector $\mathbf{w}$ such that $\mathbf{a}-\mathbf{b}+\mathbf{c}-\mathbf{w}=\mathbf{0}$.

2.2 The Scalar Product of Two Vectors

In Section 2.1 we saw how two vectors could be added and multiplied by scalars. There are several ways that two vectors can be multiplied together. Two of these ways are discussed in this text. In this section we define a product of two vectors the result of which is a scalar. In Section 4.4 we show how the product of two vectors can yield a vector.

DEFINITION 1 **SCALAR PRODUCT** Let $\mathbf{a}=\begin{pmatrix}a_1\\a_2\\\vdots\\a_n\end{pmatrix}$ and $\mathbf{b}=\begin{pmatrix}b_1\\b_1\\\vdots\\b_n\end{pmatrix}$ be two n-vectors. Then the **scalar product** of $\mathbf{a}$ and $\mathbf{b}$, denoted $\mathbf{a}\cdot\mathbf{b}$, is given by

$$\mathbf{a}\cdot\mathbf{b}=a_1b_1+a_2b_2+\cdots+a_nb_n \tag{1}$$

Because of the notation in (1), the scalar product of two vectors is often called the **dot product** of the vectors. Note that the scalar product of two n-vectors is a scalar (that is, a number).

Warning. When taking the scalar product of $\mathbf{a}$ and $\mathbf{b}$, it is necessary that $\mathbf{a}$ and $\mathbf{b}$ have the same number of components.

We shall often be taking the scalar product of a row vector and column vector. In this case we have

$$(a_1, a_2, \ldots, a_n) \cdot \begin{pmatrix} b_1 \\ b_2 \\ \vdots \\ b_n \end{pmatrix} = a_1b_1 + a_2b_2 + \cdots + a_nb_n \tag{2}$$

EXAMPLE 1 Let $\mathbf{a} = \begin{pmatrix} 1 \\ -2 \\ 3 \end{pmatrix}$ and $\mathbf{b} = \begin{pmatrix} 3 \\ -2 \\ 4 \end{pmatrix}$. Calculate $\mathbf{a} \cdot \mathbf{b}$.

Solution $\mathbf{a} \cdot \mathbf{b} = (1)(3) + (-2)(-2) + (3)(4) = 3 + 4 + 12 = 19$

EXAMPLE 2 Let $\mathbf{a} = (2, -3, 4, -6)$ and $\mathbf{b} = \begin{pmatrix} 1 \\ 2 \\ 0 \\ 3 \end{pmatrix}$. Compute $\mathbf{a} \cdot \mathbf{b}$.

Solution Here $\mathbf{a} \cdot \mathbf{b} = (2)(1) + (-3)(2) + (4)(0) + (-6)(3) = 2 - 6 + 0 - 18 = -22$.

EXAMPLE 3 Suppose that a manufacturer produces four items. The demand for the items is given by the demand vector $\mathbf{d} = (30, 20, 40, 10)$. The price per unit that he receives for the items is given by the price vector $\mathbf{p} = (\$20, \$15, \$18, \$40)$. If he meets his demand, how much money will he receive?

Solution His demand for the first item is 30 and he receives \$20 for each of the first item sold. He therefore receives $(30)(20) = \$600$ from the sale of the first item. By continuing this reasoning we see that the total cash received is given by $\mathbf{d} \cdot \mathbf{p}$. Thus income received $= \mathbf{d} \cdot \mathbf{p} = (30)(20) + (20)(15) + (40)(18) + (10)(40) = 600 + 300 + 720 + 400 = \$2020$.

The next result follows directly from the definition of the scalar product (see Problem 22).

THEOREM 1 Let $\mathbf{a}$, $\mathbf{b}$, and $\mathbf{c}$ be n-vectors and let α and β be scalars. Then:

i. $\mathbf{a}\cdot\mathbf{0}=0$
ii. $\mathbf{a}\cdot\mathbf{b}=\mathbf{b}\cdot\mathbf{a}$ **(commutative law for scalar product)**
iii. $\mathbf{a}\cdot(\mathbf{b}+\mathbf{c})=\mathbf{a}\cdot\mathbf{b}+\mathbf{a}\cdot\mathbf{c}$ **(distributive law for scalar product)**
iv. $(\alpha\mathbf{a})\cdot\mathbf{b}=\alpha(\mathbf{a}\cdot\mathbf{b})$

Note that there is *no* associative law for the scalar product. The expression $(\mathbf{a}\cdot\mathbf{b})\cdot\mathbf{c}=\mathbf{a}\cdot(\mathbf{b}\cdot\mathbf{c})$ does not make sense because neither side of the equation is defined. For the left side, this follows from the fact that $\mathbf{a}\cdot\mathbf{b}$ is a scalar and the scalar product of the scalar $\mathbf{a}\cdot\mathbf{b}$ and the vector $\mathbf{c}$ is not defined.

EXAMPLE 4 Let $\mathbf{a}=\begin{pmatrix}1\\2\\4\end{pmatrix}$, $\mathbf{b}=\begin{pmatrix}-3\\1\\7\end{pmatrix}$, and $\mathbf{c}=\begin{pmatrix}2\\5\\-1\end{pmatrix}$. Then $\mathbf{b}+\mathbf{c}=\begin{pmatrix}-1\\6\\6\end{pmatrix}$. We verify that the distributive law for scalar products (part (*iii*) of Theorem 1) holds for these vectors. We have

$$\mathbf{a}\cdot\mathbf{b}=-3+2+28=27 \quad\text{and}\quad \mathbf{a}\cdot\mathbf{c}=2+10-4=8$$

Next we compute

$$\mathbf{a}\cdot(\mathbf{b}+\mathbf{c})=-1+12+24=35$$

and

$$\mathbf{a}\cdot\mathbf{b}+\mathbf{a}\cdot\mathbf{c}=27+8=35$$

Finally, we observe that

$$\mathbf{a}\cdot(\mathbf{b}+\mathbf{c})=\mathbf{a}\cdot\mathbf{b}+\mathbf{a}\cdot\mathbf{c}$$

PROBLEMS 2.2 In Problems 1–7 calculate the scalar product of the two vectors.

1. $\begin{pmatrix}2\\3\\-5\end{pmatrix};\ \begin{pmatrix}3\\0\\4\end{pmatrix}$

2. $(1,2,-1,0);\ (3,-7,4,-2)$

3. $\begin{pmatrix}5\\7\end{pmatrix};\ \begin{pmatrix}3\\-2\end{pmatrix}$

4. $(8,3,1);\ (7,-4,3)$

5. $(a,b);\ (c,d)$

6. $\begin{pmatrix}x\\y\\z\end{pmatrix};\ \begin{pmatrix}y\\z\\x\end{pmatrix}$

7. $(-1,-3,4,5);\ (-1,-3,4,5)$

8. Let $\mathbf{a}$ be an n-vector. Show that $\mathbf{a}\cdot\mathbf{a}\geq 0$.

9. Find conditions on a vector $\mathbf{a}$ such that $\mathbf{a}\cdot\mathbf{a}=0$.

In Problems 10–14 perform the indicated computation with $\mathbf{a}=\begin{pmatrix}1\\-2\\4\end{pmatrix}$, $\mathbf{b}=\begin{pmatrix}0\\-3\\-7\end{pmatrix}$, and $\mathbf{c}=\begin{pmatrix}4\\-1\\5\end{pmatrix}$.

10. $(2\mathbf{a})\cdot(3\mathbf{b})$ **11.** $\mathbf{a}\cdot(\mathbf{b}+\mathbf{c})$ **12.** $\mathbf{c}\cdot(\mathbf{a}-\mathbf{b})$
13. $(2\mathbf{b})\cdot(3\mathbf{c}-5\mathbf{a})$ **14.** $(\mathbf{a}-\mathbf{c})\cdot(3\mathbf{b}-4\mathbf{a})$

ORTHOGONAL VECTORS

Two vectors **a** and **b** are said to be **orthogonal** if $\mathbf{a}\cdot\mathbf{b}=0$. In Problems 15–19 determine which pairs of vectors are orthogonal.†

15. $\begin{pmatrix}2\\-3\end{pmatrix}$; $\begin{pmatrix}3\\2\end{pmatrix}$ **16.** $\begin{pmatrix}2\\-3\end{pmatrix}$; $\begin{pmatrix}-3\\2\end{pmatrix}$ **17.** $\begin{pmatrix}1\\4\\-7\end{pmatrix}$; $\begin{pmatrix}2\\3\\2\end{pmatrix}$

18. $(1, 0, 1, 0)$; $(0, 1, 0, 1)$ **19.** $\begin{pmatrix}a\\0\\b\\0\\c\end{pmatrix}$; $\begin{pmatrix}0\\d\\0\\e\\0\end{pmatrix}$

20. Determine a number α such that $(1, -2, 3, 5)$ is orthogonal to $(-4, \alpha, 6, -1)$.

21. Determine all numbers α and β such that the vectors $\begin{pmatrix}1\\-\alpha\\2\\3\end{pmatrix}$ and $\begin{pmatrix}4\\5\\-2\beta\\7\end{pmatrix}$ are orthogonal.

22. Using the definition of the scalar product, prove Theorem 1.

23. A manufacturer of custom-designed jewelry has orders for two rings, three pairs of earrings, five pins, and one necklace. The manufacturer estimates that it takes 1 hour of labor to make a ring, $1\frac{1}{2}$ hours to make a pair of earrings, $\frac{1}{2}$ hour for each pin, and 2 hours to make a necklace.
(a) Express the manufacturer's orders as a row vector.
(b) Express the hourly requirements for the various types of jewelry as a column vector.
(c) Use the scalar product to calculate the total number of hours it will require to complete all the orders.

24. A tourist returned from a European trip with the following foreign currency: 1000 Austrian schillings, 20 British pounds, 100 French francs, 5000 Italian lire, and 50 German marks. In American money, a schilling was worth \$0.055, the pound \$1.80, the franc \$0.20, the lira \$0.001, and the mark \$0.40.
(a) Express the quantity of each currency by means of a row vector.
(b) Express the value of each currency in American money by means of a column vector.
(c) Use the scalar product to compute how much the tourist's foreign currency was worth in American money.

---

† We shall be dealing extensively with orthogonal vectors in Chapters 4 and 5.

2.3 Matrices

MATRIX

An $m \times n$ **matrix**† A is a rectangular array of mn numbers arranged in m rows and n columns:‡

$$A = \begin{pmatrix} a_{11} & a_{12} & \cdots & a_{1j} & \cdots & a_{1n} \\ a_{21} & a_{22} & \cdots & a_{2j} & \cdots & a_{2n} \\ \vdots & \vdots & & \vdots & & \vdots \\ a_{i1} & a_{i2} & \cdots & a_{ij} & \cdots & a_{in} \\ \vdots & \vdots & & \vdots & & \vdots \\ a_{m1} & a_{m2} & \cdots & a_{mj} & \cdots & a_{mn} \end{pmatrix} \quad (1)$$

The ijth component of A, denoted a_{ij}, is the number appearing in the ith row and jth column of A. We will sometimes write the matrix A as $A = (a_{ij})$. Usually, matrices will be denoted by capital letters.

If A is an $m \times n$ matrix with $m = n$, then A is called a **square matrix**. An $m \times n$ matrix with all components equal to zero is called the $m \times n$ **zero matrix**.

An $m \times n$ matrix is said to have the **size** $m \times n$. Two matrices $A = (a_{ij})$ and $B = (b_{ij})$ are **equal** if (i) they have the same size and (ii) corresponding components are equal.

EXAMPLE 1

Five matrices of different sizes are given below:

i. $A = \begin{pmatrix} 1 & 3 \\ 4 & 2 \end{pmatrix}$, 2×2 (square) **ii.** $A = \begin{pmatrix} -1 & 3 \\ 4 & 0 \\ 1 & -2 \end{pmatrix}$, 3×2

iii. $\begin{pmatrix} -1 & 4 & 1 \\ 3 & 0 & 2 \end{pmatrix}$, 2×3 **iv.** $\begin{pmatrix} 1 & 6 & -2 \\ 3 & 1 & 4 \\ 2 & -6 & 5 \end{pmatrix}$, 3×3 (square)

v. $\begin{pmatrix} 0 & 0 & 0 & 0 \\ 0 & 0 & 0 & 0 \end{pmatrix}$, 2×4 zero matrix

---

Each vector is a special kind of matrix. Thus, for example, the n-component row vector $(a_1, a_2, \ldots, a_n)$ is a $1 \times n$ matrix whereas the n-component column vector $\begin{pmatrix} a_1 \\ a_2 \\ \vdots \\ a_n \end{pmatrix}$ is an $n \times 1$ matrix.

† *Historical note:* The term "matrix" was first used in 1850 by the British mathematician James Joseph Sylvester (1814–1897) to distinguish matrices from determinants (which we shall discuss in Chapter 3). In fact, the term "matrix" was intended to mean "mother of determinants."

‡ As with vectors, we shall always assume, unless stated otherwise, that the numbers in a matrix are real.

Matrices, like vectors, arise in a great number of practical situations. For example, we saw in Section 2.1 how the vector $\begin{pmatrix} 10 \\ 30 \\ 15 \\ 60 \end{pmatrix}$ could represent order quantities for four different products used by one manufacturer. Suppose that there were five different plants. Then the 4×5 matrix

$$Q = \begin{pmatrix} 10 & 20 & 15 & 16 & 25 \\ 30 & 10 & 20 & 25 & 22 \\ 15 & 22 & 18 & 20 & 13 \\ 60 & 40 & 50 & 35 & 45 \end{pmatrix}$$

could represent the orders for the four products in each of the five plants. We can see, for example, that plant 4 orders 25 units of the second product while plant 2 orders 40 units of the fourth product.

Matrices, like vectors, can be added and multiplied by scalars.†

DEFINITION 1 **ADDITION OF MATRICES** Let $A = (a_{ij})$ and $B = (b_{ij})$ be two $m \times n$ matrices. Then the sum of A and B is the $m \times n$ matrix $A + B$ given by

$$A + B = (a_{ij} + b_{ij}) = \begin{pmatrix} a_{11} + b_{11} & a_{12} + b_{12} & \cdots & a_{1n} + b_{1n} \\ a_{21} + b_{21} & a_{22} + b_{22} & \cdots & a_{2n} + b_{2n} \\ \vdots & \vdots & & \vdots \\ a_{m1} + b_{m1} & a_{m2} + b_{m2} & \cdots & a_{mn} + b_{mn} \end{pmatrix}$$

That is, $A + B$ is the $m \times n$ matrix obtained by adding the corresponding components of A and B.

Arthur Cayley
(Library of Congress)

† The algebra of matrices, that is, the rules by which matrices can be added and multiplied, was developed by the English mathematician Arthur Cayley (1821–1895) in 1857. Matrices arose with Cayley in connection with linear transformations of the type

$$x' = ax + by,$$
$$y' = cx + dy,$$

where a, b, c, d are real numbers, and which may be thought of as mapping the point (x, y) into the point (x', y'). Clearly, the above transformation is completely determined by the four coefficients a, b, c, d, and so the transformation can be symbolized by the square array

$$\begin{pmatrix} a & b \\ c & d \end{pmatrix},$$

which we have called a (*square*) *matrix*. We shall discuss linear transformations in Chapter 6.

Warning. The sum of two matrices is defined only when both matrices have the same size. Thus, for example, it is not possible to add together the matrices $\begin{pmatrix} 1 & 2 & 3 \\ 4 & 5 & 6 \end{pmatrix}$ and $\begin{pmatrix} -1 & 0 \\ 2 & -5 \\ 4 & 7 \end{pmatrix}$.

EXAMPLE 2

$$\begin{pmatrix} 2 & 4 & -6 & 7 \\ 1 & 3 & 2 & 1 \\ -4 & 3 & -5 & 5 \end{pmatrix} + \begin{pmatrix} 0 & 1 & 6 & -2 \\ 2 & 3 & 4 & 3 \\ -2 & 1 & 4 & 4 \end{pmatrix} = \begin{pmatrix} 2 & 5 & 0 & 5 \\ 3 & 6 & 6 & 4 \\ -6 & 4 & -1 & 9 \end{pmatrix}$$

DEFINITION 2 **MULTIPLICATION OF A MATRIX BY A SCALAR** If $A = (a_{ij})$ is an $m \times n$ matrix and if α is a scalar, then the $m \times n$ matrix αA is given by

$$\alpha A = (\alpha a_{ij}) = \begin{pmatrix} \alpha a_{11} & \alpha a_{12} & \cdots & \alpha a_{1n} \\ \alpha a_{21} & \alpha a_{22} & \cdots & \alpha a_{2n} \\ \vdots & \vdots & & \vdots \\ \alpha a_{m1} & \alpha a_{m2} & \cdots & \alpha a_{mn} \end{pmatrix} \tag{3}$$

In other words, $\alpha A = (\alpha a_{ij})$ is the matrix obtained by multiplying each component of A by α.

EXAMPLE 3 Let $A = \begin{pmatrix} 1 & -3 & 4 & 2 \\ 3 & 1 & 4 & 6 \\ -2 & 3 & 5 & 7 \end{pmatrix}$. Then $2A = \begin{pmatrix} 2 & -6 & 8 & 4 \\ 6 & 2 & 8 & 12 \\ -4 & 6 & 10 & 14 \end{pmatrix}$,

$-3A = \begin{pmatrix} -3 & 9 & -12 & -6 \\ -9 & -3 & -12 & -18 \\ 6 & -9 & -15 & -21 \end{pmatrix}$, and $0A = \begin{pmatrix} 0 & 0 & 0 & 0 \\ 0 & 0 & 0 & 0 \\ 0 & 0 & 0 & 0 \end{pmatrix}$.

EXAMPLE 4 Let $A = \begin{pmatrix} 1 & 2 & 4 \\ -7 & 3 & -2 \end{pmatrix}$ and $B = \begin{pmatrix} 4 & 0 & 5 \\ 1 & -3 & 6 \end{pmatrix}$. Calculate $-2A + 3B$.

Solution $-2A + 3B = (-2)\begin{pmatrix} 1 & 2 & 4 \\ -7 & 3 & -2 \end{pmatrix} + (3)\begin{pmatrix} 4 & 0 & 5 \\ 1 & -3 & 6 \end{pmatrix} =$

$$\begin{pmatrix} -2 & -4 & -8 \\ 14 & -6 & 4 \end{pmatrix} + \begin{pmatrix} 12 & 0 & 15 \\ 3 & -9 & 18 \end{pmatrix} = \begin{pmatrix} 10 & -4 & 7 \\ 17 & -15 & 22 \end{pmatrix}$$

The next theorem is similar to Theorem 2.1.1 on page 31. Its proof is left as an exercise (see Problems 21–24).

THEOREM 1 Let A, B, and C be $m \times n$ matrices and let α be a scalar. Then:

i. $A + 0 = A$
ii. $0A = 0$ (Note that the zero on the left is the number zero and the zero on the right is a zero matrix.)
iii. $A + B = B + A$ **(commutative law for matrix addition)**
iv. $(A + B) + C = A + (B + C)$ **(associative law for matrix addition)**
v. $\alpha(A + B) = \alpha A + \alpha B$ **(distributive law for scalar multiplication)**
vi. $1A = A$

Note. The zero in part (*i*) of the theorem is the $m \times n$ zero matrix. In part (*ii*) the zero on the left is a scalar while the zero on the right is the $m \times n$ zero matrix.

EXAMPLE 5 To illustrate the associative law we note that

$$\left[\begin{pmatrix} 1 & 4 & -2 \\ 3 & -1 & 0 \end{pmatrix} + \begin{pmatrix} 2 & -2 & 3 \\ 1 & -1 & 5 \end{pmatrix}\right] + \begin{pmatrix} 3 & -1 & 2 \\ 0 & 1 & 4 \end{pmatrix} = \begin{pmatrix} 3 & 2 & 1 \\ 4 & -2 & 5 \end{pmatrix}$$

$$+ \begin{pmatrix} 3 & -1 & 2 \\ 0 & 1 & 4 \end{pmatrix} = \begin{pmatrix} 6 & 1 & 3 \\ 4 & -1 & 9 \end{pmatrix}$$

Similarly,

$$\begin{pmatrix} 1 & 4 & -2 \\ 3 & -1 & 0 \end{pmatrix} + \left[\begin{pmatrix} 2 & -2 & 3 \\ 1 & -1 & 5 \end{pmatrix} + \begin{pmatrix} 3 & -1 & 2 \\ 0 & 1 & 4 \end{pmatrix}\right] = \begin{pmatrix} 1 & 4 & -2 \\ 3 & -1 & 0 \end{pmatrix}$$

$$+ \begin{pmatrix} 5 & -3 & 5 \\ 1 & 0 & 9 \end{pmatrix} = \begin{pmatrix} 6 & 1 & 3 \\ 4 & -1 & 9 \end{pmatrix}$$

As with vectors, the associative law for matrix addition enables us to define the sum of three or more matrices.

PROBLEMS 2.3 In Problems 1–12 perform the indicated computation with $A = \begin{pmatrix} 1 & 3 \\ 2 & 5 \\ -1 & 2 \end{pmatrix}$, $B = \begin{pmatrix} -2 & 0 \\ 1 & 4 \\ -7 & 5 \end{pmatrix}$, and $C = \begin{pmatrix} -1 & 1 \\ 4 & 6 \\ -7 & 3 \end{pmatrix}$.

1. $3A$
2. $A+B$
3. $A-C$
4. $2C-5A$
5. $0B$ (0 is the scalar zero)
6. $-7A+3B$
7. $A+B+C$
8. $C-A-B$
9. $2A-3B+4C$
10. $7C-B+2A$

11. Find a matrix D such that $2A+B-D$ is the 3×2 zero matrix.
12. Find a matrix E such that $A+2B-3C+E$ is the 3×2 zero matrix.

In Problems 13–20 perform the indicated computation with $A=\begin{pmatrix}1 & -1 & 2\\ 3 & 4 & 5\\ 0 & 1 & -1\end{pmatrix}$, $B=\begin{pmatrix}0 & 2 & 1\\ 3 & 0 & 5\\ 7 & -6 & 0\end{pmatrix}$, and $C=\begin{pmatrix}0 & 0 & 2\\ 3 & 1 & 0\\ 0 & -2 & 4\end{pmatrix}$.

13. $A-2B$ **14.** $3A-C$ **15.** $A+B+C$
16. $2A-B+2C$ **17.** $C-A-B$ **18.** $4C-2B+3A$
19. Find a matrix D such that $A+B+C+D$ is the 3×3 zero matrix.
20. Find a matrix E such that $3C-2B+8A-4E$ is the 3×3 zero matrix.
21. Let $A=(a_{ij})$ be an $m\times n$ matrix and let $\bar{0}$ denote the $m\times n$ zero matrix. Use Definitions 1 and 2 to show that $0A=\bar{0}$ and $\bar{0}+A=A$. Similarly, show that $1A=A$.
22. Let $A=(a_{ij})$ and $B=(b_{ij})$ be $m\times n$ matrices. Compute $A+B$ and $B+A$ and show that they are equal.
23. If α is a scalar and A and B are as in Problem 22, compute $\alpha(A+B)$ and $\alpha A+\alpha B$ and show that they are equal.
24. If $A=(a_{ij})$, $B=(b_{ij})$, and $C=(c_{ij})$ are $m\times n$ matrices, compute $(A+B)+C$ and $A+(B+C)$ and show that they are equal.
25. Consider the "graph" joining the four points in the figure. Construct a 4×4 matrix having the property that $a_{ij}=0$ if point i is not connected (joined by a line) to point j and $a_{ij}=1$ if point i is connected to point j.

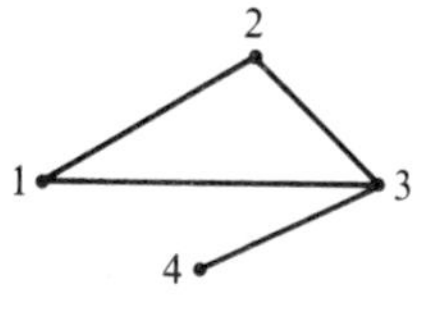

26. Do the same (this time constructing a 5×5 matrix) for the accompanying graph.

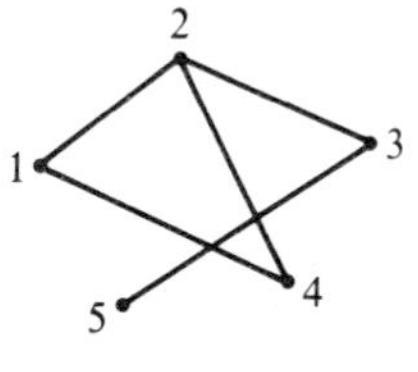

2.4 Matrix Products

In this section we see how two matrices can be multiplied together. Quite obviously, we could define the product of two $m\times n$ matrices $A=(a_{ij})$ and $B=(b_{ij})$ to be the $m\times n$ matrix whose ijth component is $a_{ij}b_{ij}$. However, for just about all the important applications involving matrices, another kind of product is needed. It comes as the generalization of the scalar product.

DEFINITION 1 **PRODUCT OF TWO MATRICES** Let $A = (a_{ij})$ be an $m \times n$ matrix whose ith row is denoted $\mathbf{a}_i$. Let $B = (b_{ij})$ be an $n \times p$ matrix whose jth column is denoted $\mathbf{b}_j$. Then the product of A and B is an $m \times p$ matrix $C = (c_{ij})$, where

$$c_{ij} = \mathbf{a}_i \cdot \mathbf{b}_j \tag{1}$$

That is, the ijth element of AB is the scalar product of the ith row of A ($\mathbf{a}_i$) and the jth column of B ($\mathbf{b}_j$). If we write this out, we obtain

$$c_{ij} = a_{i1}b_{1j} + a_{i2}b_{2j} + \cdots + a_{in}b_{nj} \tag{2}$$

Warning. Two matrices can be multiplied together only if the number of columns of the first is equal to the number of rows of the second. Otherwise the vectors $\mathbf{a}_i$ and $\mathbf{b}_j$ will have different numbers of components and the scalar product in Equation (1) will not be defined.

EXAMPLE 1 If $A = \begin{pmatrix} 1 & 3 \\ -2 & 4 \end{pmatrix}$ and $B = \begin{pmatrix} 3 & -2 \\ 5 & 6 \end{pmatrix}$, calculate AB and BA.

Solution Let $C = (c_{ij}) = AB$. Then $c_{11} = \mathbf{a}_1 \cdot \mathbf{b}_1 = (1 \quad 3) \cdot \begin{pmatrix} 3 \\ 5 \end{pmatrix} = 3 + 15 = 18$; $c_{12} = \mathbf{a}_1 \cdot \mathbf{b}_2 = (1 \quad 3) \cdot \begin{pmatrix} -2 \\ 6 \end{pmatrix} = -2 + 18 = 16$; $c_{21} = (-2 \quad 4) \cdot \begin{pmatrix} 3 \\ 5 \end{pmatrix} = -6 + 20 = 14$; and $c_{22} = (-2 \quad 4) \cdot \begin{pmatrix} -2 \\ 6 \end{pmatrix} = 4 + 24 = 28$. Thus $C = AB = \begin{pmatrix} 18 & 16 \\ 14 & 28 \end{pmatrix}$. Similarly, leaving out the intermediate steps, we see that

$$C' = BA = \begin{pmatrix} 3 & -2 \\ 5 & 6 \end{pmatrix}\begin{pmatrix} 1 & 3 \\ -2 & 4 \end{pmatrix} = \begin{pmatrix} 3+4 & 9-8 \\ 5-12 & 15+24 \end{pmatrix} = \begin{pmatrix} 7 & 1 \\ -7 & 39 \end{pmatrix}$$

Remark. Example 1 illustrates an important fact: *Matrix products do not, in general, commute.* That is, $AB \neq BA$ in general. It sometimes happens that $AB = BA$, but this will be the exception, not the rule. In fact, as the next example illustrates, it may occur that AB is defined while BA is not. Thus we must be careful of *order* when multiplying two matrices together.

EXAMPLE 2 Let $A = \begin{pmatrix} 2 & 0 & -3 \\ 4 & 1 & 5 \end{pmatrix}$ and $B = \begin{pmatrix} 7 & -1 & 4 & 7 \\ 2 & 5 & 0 & -4 \\ -3 & 1 & 2 & 3 \end{pmatrix}$. Calculate AB.

Solution We first note that A is a 2×3 matrix and B is a 3×4 matrix. Hence the number of columns of A equals the number of rows of B. The product AB is therefore defined and is a 2×4 matrix. Let $AB = C = (c_{ij})$. Then

$$c_{11} = (2 \quad 0 \quad -3)\cdot\begin{pmatrix}7\\2\\-3\end{pmatrix} = 23 \qquad c_{12} = (2 \quad 0 \quad -3)\cdot\begin{pmatrix}-1\\5\\1\end{pmatrix} = -5$$

$$c_{13} = (2 \quad 0 \quad -3)\cdot\begin{pmatrix}4\\0\\2\end{pmatrix} = 2 \qquad c_{14} = (2 \quad 0 \quad -3)\cdot\begin{pmatrix}7\\-4\\3\end{pmatrix} = 5$$

$$c_{21} = (4 \quad 1 \quad 5)\cdot\begin{pmatrix}7\\2\\-3\end{pmatrix} = 15 \qquad c_{22} = (4 \quad 1 \quad 5)\cdot\begin{pmatrix}-1\\5\\1\end{pmatrix} = 6$$

$$c_{23} = (4 \quad 1 \quad 5)\cdot\begin{pmatrix}4\\0\\2\end{pmatrix} = 26 \qquad c_{24} = (4 \quad 1 \quad 5)\cdot\begin{pmatrix}7\\-4\\3\end{pmatrix} = 39$$

Hence $AB = \begin{pmatrix}23 & -5 & 2 & 5\\ 15 & 6 & 26 & 39\end{pmatrix}$. This completes the problem. Note that the product BA is *not* defined since the number of columns of B (four) is not equal to the number of rows of A (two).

EXAMPLE 3 **Direct and Indirect Contact with a Contagious Disease.** In this example we show how matrix multiplication can be used to model the spread of a contagious disease. Suppose that four individuals have contracted such a disease. This group has contacts with six people in a second group. We can represent these contacts, called *direct contacts*, by a 4×6 matrix. An example of such a matrix is given below:

DIRECT CONTACT MATRIX: First and second groups

$$A = \begin{pmatrix}0 & 1 & 0 & 0 & 1 & 0\\ 1 & 0 & 0 & 1 & 0 & 1\\ 0 & 0 & 0 & 1 & 1 & 0\\ 1 & 0 & 0 & 0 & 0 & 1\end{pmatrix}$$

Here we set $a_{ij} = 1$ if the ith person in the first group has made contact with the jth person in the second group. For example, the 1 in the 2,4 position means that the second person in the first (infected) group has been in contact with the fourth person in the second group. Now suppose that a third group of five people has had a variety of direct contact with individuals of the second group. We can also represent this by a matrix.

DIRECT CONTACT MATRIX: Second and third groups

$$B = \begin{pmatrix} 0 & 0 & 1 & 0 & 1 \\ 0 & 0 & 0 & 1 & 0 \\ 0 & 1 & 0 & 0 & 0 \\ 1 & 0 & 0 & 0 & 1 \\ 0 & 0 & 0 & 1 & 0 \\ 0 & 0 & 1 & 0 & 0 \end{pmatrix}$$

Note that $b_{64} = 0$, which means that the sixth person in the second group has had no contact with the fourth person in the third group.

The *indirect* or *second-order* contacts between the individuals in the first and third groups is represented by the 4×5 matrix $C = AB$. To see this, observe that a person in group 3 can be infected from someone in group 2 who, in turn, has been infected by someone in group 1. For example, since $a_{24} = 1$ and $b_{45} = 1$, we see that, indirectly, the fifth person in group 3 has contact (through the fourth person in group 2) with the second person in group 1. The total number of indirect contacts between the second person in group 1 and the fifth person in group 3 is given by

$$\begin{aligned} c_{25} &= a_{21}b_{15} + a_{22}b_{25} + a_{23}b_{35} + a_{24}b_{45} + a_{25}b_{55} + a_{26}b_{26} \\ &= 1 \cdot 1 + 0 \cdot 0 + 0 \cdot 0 + 1 \cdot 1 + 0 \cdot 0 + 1 \cdot 0 = 2 \end{aligned}$$

We now compute.

INDIRECT CONTACT MATRIX: First and third groups

$$C = AB = \begin{pmatrix} 0 & 0 & 0 & 2 & 0 \\ 1 & 0 & 2 & 0 & 2 \\ 1 & 0 & 0 & 1 & 1 \\ 0 & 0 & 2 & 0 & 1 \end{pmatrix}$$

We observe that only the second person in Group 3 has no indirect contacts with the disease. The fifth person in this group has $2 + 1 + 1 = 4$ indirect contacts.

We have seen that for matrix multiplication the commutative law does not hold. The next theorem shows that the associative law does hold.

THEOREM 1

ASSOCIATIVE LAW FOR MATRIX MULTIPLICATION Let $A = (a_{ij})$ be an $n \times m$ matrix, $B = (b_{ij})$ an $m \times p$ matrix, and $C = (c_{ij})$ a $p \times q$ matrix. Then the **associative law**

$$A(BC) = (AB)C \tag{3}$$

holds and ABC, defined by either side of (3), is an $n \times q$ matrix.

The proof of this theorem is not difficult, but it is somewhat tedious. It is best given using the summation notation. (If this is unfamiliar take a look at Problems 36–56.) For that reason let us defer it until the end of the section.

EXAMPLE 4 Verify the associative law for $A=\begin{pmatrix}1 & -3\\ 0 & 2\end{pmatrix}$, $B=\begin{pmatrix}2 & -1 & 4\\ 3 & 1 & 5\end{pmatrix}$, and $C=\begin{pmatrix}0 & -2 & 1\\ 4 & 3 & 2\\ -5 & 0 & 6\end{pmatrix}$.

Solution We first note that A is 2×2, B is 2×3, and C is 3×3. Hence all products used in the statement of the associative law are defined and the resulting product will be a 2×3 matrix. We then calculate

$$AB=\begin{pmatrix}1 & -3\\ 0 & 2\end{pmatrix}\begin{pmatrix}2 & -1 & 4\\ 3 & 1 & 5\end{pmatrix}=\begin{pmatrix}-7 & -4 & -11\\ 6 & 2 & 10\end{pmatrix}$$

$$(AB)C=\begin{pmatrix}-7 & -4 & -11\\ 6 & 2 & 10\end{pmatrix}\begin{pmatrix}0 & -2 & 1\\ 4 & 3 & 2\\ -5 & 0 & 6\end{pmatrix}=\begin{pmatrix}39 & 2 & -81\\ -42 & -6 & 70\end{pmatrix}$$

Similarly,

$$BC=\begin{pmatrix}2 & -1 & 4\\ 3 & 1 & 5\end{pmatrix}\begin{pmatrix}0 & -2 & 1\\ 4 & 3 & 2\\ -5 & 0 & 6\end{pmatrix}=\begin{pmatrix}-24 & -7 & 24\\ -21 & -3 & 35\end{pmatrix}$$

$$A(BC)=\begin{pmatrix}1 & -3\\ 0 & 2\end{pmatrix}\begin{pmatrix}-24 & -7 & 24\\ -21 & -3 & 35\end{pmatrix}=\begin{pmatrix}39 & 2 & -81\\ -42 & -6 & 70\end{pmatrix}$$

Thus $(AB)C=A(BC)$.

From now on we shall write the product of three matrices simply as ABC. We can do this because $(AB)C=A(BC)$; thus we get the same answer no matter how the multiplication is carried out (provided that we do not commute any of the matrices).

The associative law can be extended to longer products. For example, suppose that AB, BC, and CD are defined. Then

$$ABCD = A(B(CD)) = ((AB)C)D = A(BC)D = (AB)(CD) \tag{4}$$

There are two distributive laws for matrix multiplication.

THEOREM 2 **DISTRIBUTIVE LAWS FOR MATRIX MULTIPLICATION** If all the following sums and products are defined, then

$$A(B+C) = AB + AC \tag{5}$$

and

$$(A+B)C = AC + BC \tag{6}$$

Proofs of Theorems 1 and 2 **Associative Laws.** Since A is $n \times m$ and B is $m \times p$, AB is $n \times p$. Thus $(AB)C = (n \times p) \times (p \times q)$ is an $n \times q$ matrix. Similarly BC is $m \times q$ and $A(BC)$ is $n \times q$ so that $(AB)C$ and $A(BC)$ are both of the same size. We must show that the ijth component of $(AB)C$ equals the ijth component of $A(BC)$. Define $D = (d_{ij}) = AB$. Then $d_{ij} = \sum_{k=1}^{m} a_{ik}b_{kj}$. The ijth component of $(AB)C = DC$ is $\sum_{l=1}^{p} d_{il}c_{lj} = \sum_{l=1}^{p}\left(\sum_{k=1}^{m} a_{ik}b_{kl}\right)c_{lj} = \sum_{k=1}^{m}\sum_{l=1}^{p} a_{ik}b_{kl}c_{lj}$. Next we define $E = (e_{ij}) = BC$. Then $e_{ij} = \sum_{l=1}^{p} b_{il}c_{lj}$ and the ijth component of $A(BC) = AE$ is $\sum_{k=1}^{m} a_{ik}e_{kj} = \sum_{k=1}^{m}\sum_{l=1}^{p} a_{ik}b_{kl}c_{lj}$. Thus the ijth component of $(AB)C$ is equal to the ijth component of $A(BC)$. This proves the associative law. ■

Distributive Laws. We prove the first distributive law (Equation 5). The proof of the second one (Equation 6) is virtually identical and is therefore omitted. Let A be $n \times m$ and let B and C be $m \times p$. Then the kjth component of $B+C$ is $b_{kj}+c_{kj}$ and the ijth component of $A(B+C)$ is $\sum_{k=1}^{m} a_{ik}(b_{kj}+c_{kj}) = \sum_{k=1}^{m} a_{ik}b_{kj} + \sum_{k=1}^{m} a_{ik}c_{kj} =$ ijth component of AB plus the ijth component of AC and this proves Equation (5). ■

PROBLEMS 2.4

In Problems 1–15 perform the indicated computation.

1. $\begin{pmatrix} 2 & 3 \\ -1 & 2 \end{pmatrix}\begin{pmatrix} 4 & 1 \\ 0 & 6 \end{pmatrix}$ **2.** $\begin{pmatrix} 3 & -2 \\ 1 & 4 \end{pmatrix}\begin{pmatrix} -5 & 6 \\ 1 & 3 \end{pmatrix}$ **3.** $\begin{pmatrix} 1 & -1 \\ 1 & 1 \end{pmatrix}\begin{pmatrix} -1 & 0 \\ 2 & 3 \end{pmatrix}$

4. $\begin{pmatrix} -5 & 6 \\ 1 & 3 \end{pmatrix}\begin{pmatrix} 3 & -2 \\ 1 & 4 \end{pmatrix}$ **5.** $\begin{pmatrix} -4 & 5 & 1 \\ 0 & 4 & 2 \end{pmatrix}\begin{pmatrix} 3 & -1 & 1 \\ 5 & 6 & 4 \\ 0 & 1 & 2 \end{pmatrix}$ **6.** $\begin{pmatrix} 7 & 1 & 4 \\ 2 & -3 & 5 \end{pmatrix}\begin{pmatrix} 1 & 6 \\ 0 & 4 \\ -2 & 3 \end{pmatrix}$

7. $\begin{pmatrix} 1 & 6 \\ 0 & 4 \\ -2 & 3 \end{pmatrix}\begin{pmatrix} 7 & 1 & 4 \\ 2 & -3 & 5 \end{pmatrix}$ **8.** $\begin{pmatrix} 1 & 4 & -2 \\ 3 & 0 & 4 \end{pmatrix}\begin{pmatrix} 0 & 1 \\ 2 & 3 \end{pmatrix}$ **9.** $\begin{pmatrix} 1 & 4 & 6 \\ -2 & 3 & 5 \\ 1 & 0 & 4 \end{pmatrix}\begin{pmatrix} 2 & -3 & 5 \\ 1 & 0 & 6 \\ 2 & 3 & 1 \end{pmatrix}$

10. $\begin{pmatrix} 2 & -3 & 5 \\ 1 & 0 & 6 \\ 2 & 3 & 1 \end{pmatrix}\begin{pmatrix} 1 & 4 & 6 \\ -2 & 3 & 5 \\ 1 & 0 & 4 \end{pmatrix}$ **11.** $\begin{pmatrix} 1 & 4 & 0 & 2 \end{pmatrix}\begin{pmatrix} 3 & -6 \\ 2 & 4 \\ 1 & 0 \\ -2 & 3 \end{pmatrix}$

12. $\begin{pmatrix} 3 & 2 & 1 & -2 \\ -6 & 4 & 0 & 3 \end{pmatrix}\begin{pmatrix} 1 \\ 4 \\ 0 \\ 2 \end{pmatrix}$ **13.** $\begin{pmatrix} 3 & -2 & 1 \\ 4 & 0 & 6 \\ 5 & 1 & 9 \end{pmatrix}\begin{pmatrix} 1 & 0 & 0 \\ 0 & 1 & 0 \\ 0 & 0 & 1 \end{pmatrix}$

14. $\begin{pmatrix} 1 & 0 & 0 \\ 0 & 1 & 0 \\ 0 & 0 & 1 \end{pmatrix}\begin{pmatrix} 3 & -2 & 1 \\ 4 & 0 & 6 \\ 5 & 1 & 9 \end{pmatrix}$ **15.** $\begin{pmatrix} a & b & c \\ d & e & f \\ g & h & j \end{pmatrix}\begin{pmatrix} 1 & 0 & 0 \\ 0 & 1 & 0 \\ 0 & 0 & 1 \end{pmatrix}$, where $a, b, c, d, e, f, g, h, j$ are real numbers.

16. Find a matrix $A = \begin{pmatrix} a & b \\ c & d \end{pmatrix}$ such that $A\begin{pmatrix} 2 & 3 \\ 1 & 2 \end{pmatrix} = \begin{pmatrix} 1 & 0 \\ 0 & 1 \end{pmatrix}$.

***17.** Let a_{11}, a_{12}, a_{21}, and a_{22} be given real numbers such that $a_{11}a_{22} - a_{12}a_{21} \neq 0$. Find numbers b_{11}, b_{12}, b_{21}, and b_{22} such that $\begin{pmatrix} a_{11} & a_{12} \\ a_{21} & a_{22} \end{pmatrix}\begin{pmatrix} b_{11} & b_{12} \\ b_{21} & b_{22} \end{pmatrix} = \begin{pmatrix} 1 & 0 \\ 0 & 1 \end{pmatrix}$.

18. Verify the associative law for multiplication for the matrices $A = \begin{pmatrix} 2 & -1 & 4 \\ 1 & 0 & 6 \end{pmatrix}$, $B = \begin{pmatrix} 1 & 0 & 1 \\ 2 & -1 & 2 \\ 3 & -2 & 0 \end{pmatrix}$, and $C = \begin{pmatrix} 1 & 6 \\ -2 & 4 \\ 0 & 5 \end{pmatrix}$.

19. As in Example 3, suppose that a group of people have contracted a contagious disease. These persons have contacts with a second group who in turn have contacts with a third group. Let $A = \begin{pmatrix} 1 & 0 & 1 & 0 \\ 0 & 1 & 1 & 0 \\ 1 & 0 & 0 & 1 \end{pmatrix}$ represent the contacts between the contagious group and the members of group 2, and let

$$B = \begin{pmatrix} 1 & 0 & 1 & 0 & 0 \\ 0 & 0 & 0 & 1 & 0 \\ 1 & 1 & 0 & 0 & 0 \\ 0 & 0 & 1 & 0 & 1 \end{pmatrix}$$

represent the contacts between groups 2 and 3. (a) How many people are in each group? (b) Find the matrix of indirect contacts between groups 1 and 3.

20. Answer the questions of Problem 19 for $A = \begin{pmatrix} 1 & 0 & 1 & 1 & 0 \\ 0 & 1 & 0 & 1 & 1 \end{pmatrix}$ and

$$B = \begin{pmatrix} 1 & 0 & 0 & 0 & 0 & 0 & 1 \\ 0 & 1 & 0 & 1 & 0 & 0 & 0 \\ 1 & 1 & 0 & 0 & 1 & 1 & 1 \\ 0 & 0 & 0 & 1 & 1 & 0 & 1 \\ 0 & 1 & 0 & 0 & 0 & 0 & 0 \end{pmatrix}$$

21. A company pays its executives a salary and gives them shares of its stock as an annual bonus. Last year, the president of the company received \$80,000 and 50 shares of stock, each of the three vice-presidents were paid \$45,000 and 20 shares of stock, and the treasurer was paid \$40,000 and 10 shares of stock.

(a) Express the payments to the executives in money and stock by means of a 2×3 matrix.

(b) Express the number of executives of each rank by means of a column vector.

(c) Use matrix multiplication to calculate the total amount of money and the total number of shares of stock the company paid these executives last year.

22. Sales, unit gross profits, and unit taxes for sales of a large corporation are given in the table below.

	Product					
	Sales of Item					
Month	I	II	III	Item	Unit Profit (in hundreds of dollars)	Unit Taxes (in hundreds of dollars)
January	4	2	20	I	3.5	1.5
February	6	1	9	II	2.75	2
March	5	3	12	III	1.5	0.6
April	8	2.5	20			

Find a matrix that shows total profits and taxes in each of the 4 months.

23. Let A be a square matrix. Then A^2 is defined simply as AA. Calculate $\begin{pmatrix} 2 & -1 \\ 4 & 6 \end{pmatrix}^2$.

24. Calculate A^2, where $A = \begin{pmatrix} 1 & -2 & 4 \\ 2 & 0 & 3 \\ 1 & 1 & 5 \end{pmatrix}$.

25. Calculate A^3, where $A = \begin{pmatrix} -1 & 2 \\ 3 & 4 \end{pmatrix}$.

26. Calculate A^2, A^3, A^4, and A^5, where

$$A = \begin{pmatrix} 0 & 1 & 0 & 0 \\ 0 & 0 & 1 & 0 \\ 0 & 0 & 0 & 1 \\ 0 & 0 & 0 & 0 \end{pmatrix}$$

27. Calculate A^2, A^3, A^4, and A^5, where

$$A = \begin{pmatrix} 0 & 1 & 0 & 0 & 0 \\ 0 & 0 & 1 & 0 & 0 \\ 0 & 0 & 0 & 1 & 0 \\ 0 & 0 & 0 & 0 & 1 \\ 0 & 0 & 0 & 0 & 0 \end{pmatrix}$$

28. An $n \times n$ matrix A has the property that AB is the zero matrix for any $n \times n$ matrix B. Prove that A is the zero matrix.

29. A **probability matrix** is a square matrix having two properties: (*i*) every component is nonnegative (≥ 0) and (*ii*) the sum of the elements in each row is 1. The following are probability matrices:

$$P = \begin{pmatrix} \frac{1}{3} & \frac{1}{3} & \frac{1}{3} \\ \frac{1}{4} & \frac{1}{2} & \frac{1}{4} \\ 0 & 0 & 1 \end{pmatrix} \quad \text{and} \quad Q = \begin{pmatrix} \frac{1}{6} & \frac{1}{6} & \frac{2}{3} \\ 0 & 1 & 0 \\ \frac{1}{5} & \frac{1}{5} & \frac{3}{5} \end{pmatrix}$$

Show that PQ is a probability matrix.

* **30.** Let P be a probability matrix. Show that P^2 is a probability matrix.

**31. Let P and Q be probability matrices of the same size. Prove that PQ is a probability matrix.

32. Prove formula (4) by using the associative law (Equation 3).

*33. A round robin tennis tournament can be organized in the following way. Each of the n players plays all the others, and the results are recorded in an $n\times n$ matrix R as follows:

$$R_{ij}=\begin{cases}1 & \text{if the } i\text{th player beats the } j\text{th player}\\ 0 & \text{if the } i\text{th player loses to the } j\text{th player}\\ 0 & \text{if } i=j\end{cases}$$

The ith player is then assigned the score

$$S_i=\sum_{j=1}^{n} R_{ij}+\frac{1}{2}\sum_{j=1}^{n}(R^2)_{ij}\dagger$$

a. In a tournament between four players

$$R=\begin{pmatrix}0 & 1 & 0 & 0\\ 0 & 0 & 1 & 1\\ 1 & 0 & 0 & 0\\ 1 & 0 & 1 & 0\end{pmatrix}.$$

Rank the players according to their scores.

b. Interpret the meaning of the score.

34. Let O be the $m\times n$ zero matrix and let A be an $n\times p$ matrix. Show that $OA=O_1$, where O_1 is the $m\times p$ zero matrix.

35. Verify the distributive law (Equation 5) for the matrices

$$A=\begin{pmatrix}1 & 2 & 4\\ 3 & -1 & 0\end{pmatrix}\quad B=\begin{pmatrix}2 & 7\\ -1 & 4\\ 0 & 0\end{pmatrix}\quad C=\begin{pmatrix}-1 & 2\\ 3 & 7\\ 4 & 1\end{pmatrix}$$

Since we shall be using the $\sum$ notation in several parts of this book, you should become familiar with it. The following problems are designed to help you with the symbol. In Problems 36–43 evaluate the given sums.

36. $\sum_{k=1}^{4} 2k$ 37. $\sum_{i=1}^{3} i^3$ 38. $\sum_{k=0}^{6} 1$ 39. $\sum_{k=1}^{8} 3^k$

40. $\sum_{i=2}^{5}\frac{1}{1+i}$ 41. $\sum_{j=5}^{7}\frac{2j+3}{j-2}$ 42. $\sum_{i=1}^{3}\sum_{j=1}^{4} ij$ 43. $\sum_{k=1}^{3}\sum_{j=2}^{4} k^2j^3$

In Problems 44–56 write each sum using the $\sum$ notation.

44. $1+2+4+8+16$ 45. $1-3+9-27+81-243$

46. $\frac{2}{3}+\frac{3}{4}+\frac{4}{5}+\frac{5}{6}+\frac{6}{7}+\frac{7}{8}+\cdots+\frac{n}{n+1}$

47. $1+2^{1/2}+3^{1/3}+4^{1/4}+5^{1/5}+\cdots+n^{1/n}$

48. $1+x^3+x^6+x^9+x^{12}+x^{15}+x^{18}+x^{21}$

49. $-1+\frac{1}{a}-\frac{1}{a^2}+\frac{1}{a^3}-\frac{1}{a^4}+\frac{1}{a^5}-\frac{1}{a^6}+\frac{1}{a^7}-\frac{1}{a^8}+\frac{1}{a^9}$

---

†$(R^2)_{ij}$ is the ijth component of the matrix R^2.

50. $1 \cdot 3+3 \cdot 5+5 \cdot 7+7 \cdot 9+9 \cdot 11+11 \cdot 13+13 \cdot 15+15 \cdot 17$
51. $2^2 \cdot 4+3^2 \cdot 6+4^2 \cdot 8+5^2 \cdot 10+6^2 \cdot 12+7^2 \cdot 14$
52. $a_{11}+a_{12}+a_{13}+a_{21}+a_{22}+a_{23}$
53. $a_{11}+a_{12}+a_{21}+a_{22}+a_{31}+a_{32}$
54. $a_{21}+a_{22}+a_{23}+a_{24}+a_{31}+a_{32}+a_{33}+a_{34}+a_{41}+a_{42}+a_{43}+a_{44}$
55. $a_{31}b_{12}+a_{32}b_{22}+a_{33}b_{32}+a_{34}b_{42}+a_{35}b_{52}$
56. $a_{21}b_{11}c_{15}+a_{21}b_{12}c_{25}+a_{21}b_{13}c_{35}+a_{21}b_{14}c_{45}$
$+a_{22}b_{21}c_{15}+a_{22}b_{22}c_{25}+a_{22}b_{23}c_{35}+a_{22}b_{24}c_{45}$
$+a_{23}b_{31}c_{15}+a_{23}b_{32}c_{25}+a_{23}b_{33}c_{35}+a_{23}b_{34}c_{45}$

2.5 Matrices and Linear Systems of Equations

In Section 1.3, page 15, we discussed the following systems of m equations in n unknowns:

$$\begin{aligned} a_{11}x_1+a_{12}x_2+\cdots+a_{1n}x_n &= b_1 \\ a_{21}x_1+a_{22}x_2+\cdots+a_{2n}x_n &= b_2 \\ \vdots \qquad \vdots \qquad\quad \vdots \quad &\quad \vdots \\ a_{m1}x_1+a_{m2}x_2+\cdots+a_{mn}x_n &= b_m \end{aligned} \tag{1}$$

We define the matrix

$$A=\begin{pmatrix} a_{11} & a_{12} & \cdots & a_{1n} \\ a_{21} & a_{22} & \cdots & a_{2n} \\ \vdots & \vdots & & \vdots \\ a_{m1} & a_{m2} & \cdots & a_{mn} \end{pmatrix},$$

the vector $\mathbf{x}=\begin{pmatrix} x_1 \\ x_2 \\ \vdots \\ x_n \end{pmatrix}$, and the vector $\mathbf{b}=\begin{pmatrix} b_1 \\ b_2 \\ \vdots \\ b_m \end{pmatrix}$. Since A is an $m \times n$ matrix and $\mathbf{x}$ is an $n \times 1$ matrix, the matrix product $A\mathbf{x}$ is defined as an $m \times 1$ matrix. It is not difficult to see that system (1) can be written as

$$A\mathbf{x}=\mathbf{b} \tag{2}$$

EXAMPLE 1 Consider the system

$$\begin{aligned} 2x_1 + 4x_2 + 6x_3 &= 18 \\ 4x_1 + 5x_2 + 6x_3 &= 24 \\ 3x_1 + x_2 - 2x_3 &= 4 \end{aligned} \tag{3}$$

(See Example 1.3.1 on page 6.) This can be written in the form $A\mathbf{x} = \mathbf{b}$ with $A =$

$$\begin{pmatrix} 2 & 4 & 6 \\ 4 & 5 & 6 \\ 3 & 1 & -2 \end{pmatrix}, \mathbf{x} = \begin{pmatrix} x_1 \\ x_2 \\ x_3 \end{pmatrix}, \text{ and } \mathbf{b} = \begin{pmatrix} 18 \\ 24 \\ 4 \end{pmatrix}.$$

It is obviously easier to write out system (1) in the form $A\mathbf{x} = \mathbf{b}$. There are many other advantages, too. In Section 2.7 we shall see how a square system can be solved almost at once if we know a matrix called the *inverse* of A. Even without that, as we saw in Chapter 1, computations are much easier to write down by using an augmented matrix. Let us repeat the computations of Example 1.3.1 starting with the augmented matrix:

$$\left(\begin{array}{ccc|c} 2 & 4 & 6 & 18 \\ 4 & 5 & 6 & 24 \\ 3 & 1 & -2 & 4 \end{array}\right) \xrightarrow{M_1(\frac{1}{2})} \left(\begin{array}{ccc|c} 1 & 2 & 3 & 9 \\ 4 & 5 & 6 & 24 \\ 3 & 1 & -2 & 4 \end{array}\right) \xrightarrow[A_{1,3}(-3)]{A_{1,2}(-4)} \left(\begin{array}{ccc|c} 1 & 2 & 3 & 9 \\ 0 & -3 & -6 & -12 \\ 0 & -5 & -11 & -23 \end{array}\right)$$

$$\xrightarrow{M_2(-\frac{1}{3})} \left(\begin{array}{ccc|c} 1 & 2 & 3 & 9 \\ 0 & 1 & 2 & 4 \\ 0 & -5 & -11 & -23 \end{array}\right) \xrightarrow[A_{2,3}(5)]{A_{2,1}(-2)} \left(\begin{array}{ccc|c} 1 & 0 & -1 & 1 \\ 0 & 1 & 2 & 4 \\ 0 & 0 & -1 & -3 \end{array}\right)$$

$$\xrightarrow{M_3(-1)} \left(\begin{array}{ccc|c} 1 & 0 & -1 & 1 \\ 0 & 1 & 2 & 4 \\ 0 & 0 & 1 & 3 \end{array}\right) \xrightarrow[A_{3,2}(-2)]{A_{3,1}(1)} \left(\begin{array}{ccc|c} 1 & 0 & 0 & 4 \\ 0 & 1 & 0 & -2 \\ 0 & 0 & 1 & 3 \end{array}\right)$$

The last augmented matrix tells us that $x_1 = 4$, $x_2 = -2$, and $x_3 = 3$, as we already knew.

In this last example it is important to note that the last system of equations can be written as

$$I\mathbf{x} = \mathbf{s} \tag{4}$$

where $I = \begin{pmatrix} 1 & 0 & 0 \\ 0 & 1 & 0 \\ 0 & 0 & 1 \end{pmatrix}$ and $\mathbf{s}$ is the solution vector $\begin{pmatrix} 4 \\ -2 \\ 3 \end{pmatrix}$. We shall be making use of this fact in Section 2.7.

PROBLEMS 2.5

In Problems 1–6 write the given system in the form $A\mathbf{x} = \mathbf{b}$.

1. $2x_1 - x_2 = 3$
$4x_1 + 5x_2 = 7$

2. $x_1 - x_2 + 3x_3 = 11$
$4x_1 + x_2 - x_3 = -4$
$2x_1 - x_2 + 3x_3 = 10$

3. $\begin{aligned} 3x_1+6x_2-7x_3&=0\\ 2x_1-\ x_2+3x_3&=1 \end{aligned}$

4. $\begin{aligned} 4x_1-x_2+\ x_3-\ x_4&=-7\\ 3x_1+x_2-5x_3+6x_4&=\ \ 8\\ 2x_1-x_2+\ x_3\qquad&=\ \ 9 \end{aligned}$

5. $\begin{aligned} x_2-x_3&=\ \ 7\\ x_1\qquad+x_3&=\ \ 2\\ 3x_1+2x_2\qquad&=-5 \end{aligned}$

6. $\begin{aligned} 2x_1+3x_2-\ x_3&=0\\ -4x_1+2x_2+\ x_3&=0\\ 7x_1+3x_2-9x_3&=0 \end{aligned}$

In Problems 7–15 write out the system of equations represented by the given augmented matrix.

7. $\left(\begin{array}{ccc|c} 1 & 1 & -1 & 7\\ 4 & -1 & 5 & 4\\ 6 & 1 & 3 & 20 \end{array}\right)$

8. $\left(\begin{array}{cc|c} 0 & 1 & 2\\ 1 & 0 & 3 \end{array}\right)$

9. $\left(\begin{array}{ccc|c} 2 & 0 & 1 & 2\\ -3 & 4 & 0 & 3\\ 0 & 5 & 6 & 5 \end{array}\right)$

10. $\left(\begin{array}{ccc|c} 2 & 3 & 1 & 2\\ 0 & 4 & 1 & 3\\ 0 & 0 & 0 & 0 \end{array}\right)$

11. $\left(\begin{array}{cccc|c} 1 & 0 & 0 & 0 & 2\\ 0 & 1 & 0 & 0 & 3\\ 0 & 0 & 1 & 0 & -5\\ 0 & 0 & 0 & 1 & 6 \end{array}\right)$

12. $\left(\begin{array}{ccc|c} 2 & 3 & 1 & 0\\ 4 & -1 & 5 & 0\\ 3 & 6 & -7 & 0 \end{array}\right)$

13. $\left(\begin{array}{ccc|c} 6 & 2 & 1 & 2\\ -2 & 3 & 1 & 4\\ 0 & 0 & 0 & 2 \end{array}\right)$

14. $\left(\begin{array}{ccc|c} 3 & 1 & 5 & 6\\ 2 & 3 & 2 & 4 \end{array}\right)$

15. $\left(\begin{array}{cc|c} 7 & 2 & 1\\ 3 & 1 & 2\\ 6 & 9 & 3 \end{array}\right)$

16. Solve the system represented by the augmented matrix of Problem 9.

17. Solve the system represented by $\left(\begin{array}{ccc|c} 1 & 2 & -4 & 4\\ -2 & -4 & 8 & -8 \end{array}\right)$.

18. Solve the system represented by $\left(\begin{array}{ccc|c} 1 & 2 & -4 & 4\\ -2 & -4 & 8 & -9 \end{array}\right)$.

19. Solve the homogeneous system represented by $\left(\begin{array}{ccc|c} 1 & -2 & 3 & 0\\ 4 & 1 & -1 & 0\\ 2 & -1 & 3 & 0 \end{array}\right)$.

20. Solve the homogeneous system represented by $\left(\begin{array}{ccc|c} 1 & 1 & -1 & 0\\ 4 & -1 & 5 & 0\\ 6 & 1 & 3 & 0 \end{array}\right)$.

21. Solve the system represented by the augmented matrix

$$\left(\begin{array}{cccc|c} 1 & 3 & -2 & 1 & 3\\ 2 & -6 & 4 & -1 & 2\\ 4 & 12 & -8 & 2 & 4\\ -3 & 0 & 6 & -2 & -8 \end{array}\right)$$

22. Solve the homogeneous system represented by the augmented matrix

$$\left(\begin{array}{ccccc|c} 1 & 2 & -3 & 5 & 4 & 0\\ -2 & 4 & 7 & -3 & 5 & 0\\ -4 & 0 & 13 & -13 & -3 & 0 \end{array}\right)$$

23. Find a matrix A and vectors $\mathbf{x}$ and $\mathbf{b}$ such that the system represented by the following augmented matrix can be written in the form $A\mathbf{x}=\mathbf{b}$ and solve the system.

$$\left(\begin{array}{ccc|c} 2 & 0 & 0 & 3\\ 0 & 4 & 0 & 5\\ 0 & 0 & -5 & 2 \end{array}\right)$$

2.6 Linear Independence and Homogeneous Systems

In the study of linear algebra, one of the central ideas is that of the linear dependence or independence of vectors. In this section we shall define what we mean by linear independence and show how it is related to the theory of homogeneous systems of equations. We return to linear independence in Chapters 3 and 4 and, in Chapter 5, we shall see how this concept is central to the theory of vector spaces.

Is there a special relationship between the vectors $\mathbf{v}_1 = \begin{pmatrix} 1 \\ 2 \end{pmatrix}$ and $\mathbf{v}_2 = \begin{pmatrix} 2 \\ 4 \end{pmatrix}$? Of course, we see that $\mathbf{v}_2 = 2\mathbf{v}_1$ or, writing this equation in another way,

$$2\mathbf{v}_1 - \mathbf{v}_2 = \mathbf{0} \tag{1}$$

What is special about the vectors $\mathbf{v}_1 = \begin{pmatrix} 1 \\ 2 \\ 3 \end{pmatrix}$, $\mathbf{v}_2 = \begin{pmatrix} -4 \\ 1 \\ 5 \end{pmatrix}$, and $\mathbf{v}_3 = \begin{pmatrix} -5 \\ 8 \\ 19 \end{pmatrix}$? This question is more difficult to answer at first glance. It is easy to verify, however, that $\mathbf{v}_3 = 3\mathbf{v}_1 + 2\mathbf{v}_2$, or, rewriting,

$$3\mathbf{v}_1 + 2\mathbf{v}_2 - \mathbf{v}_3 = \mathbf{0} \tag{2}$$

It appears that the two vectors in Equation (1) and the three vectors in (2) are more closely related than an arbitrary pair of 2-vectors or an arbitrary triple of 3-vectors. In each case we say that the vectors are *linearly dependent.*† In general, we have the following important definition.

DEFINITION 1 **LINEARLY DEPENDENT VECTORS** The set of vectors $\mathbf{v}_1, \mathbf{v}_2, \ldots, \mathbf{v}_n$ is **linearly dependent** if there exist scalars $c_1, c_2, \ldots, c_n$ *not all zero* such that

$$c_1\mathbf{v}_1 + c_2\mathbf{v}_2 + \cdots + c_n\mathbf{v}_n = \mathbf{0} \tag{3}$$

With this definition we see that the vectors in Equation (1) $[c_1 = 2, c_2 = -1]$ and Equation (2) $[c_1 = 3, c_2 = 2, c_3 = -1]$ are linearly dependent.

DEFINITION 2 **LINEARLY INDEPENDENT VECTORS** The set of vectors $\mathbf{v}_1, \mathbf{v}_2, \ldots, \mathbf{v}_n$ is **linearly independent** if it is not linearly dependent.

Putting this another way, $\mathbf{v}_1, \mathbf{v}_2, \ldots, \mathbf{v}_n$ are linearly independent if the equation $c_1\mathbf{v}_1 + c_2\mathbf{v}_2 + \cdots + c_n\mathbf{v}_n = \mathbf{0}$ holds only for $c_1 = c_2 = \cdots = c_n = 0$.

How do we determine whether a set of vectors is linearly dependent or independent? The case for two vectors is easy.

† In Chapter 4 (Sections 4.1 and 4.3) we shall see that two linearly dependent vectors in the xy-plane are collinear and that three linearly dependent vectors in space are coplanar.

THEOREM 1 Two vectors are linearly dependent if and only if one is a scalar multiple of the other.

Proof First suppose that $\mathbf{v}_2 = c\mathbf{v}_1$ for some scalar $c \neq 0$. Then $c\mathbf{v}_1 - \mathbf{v}_2 = \mathbf{0}$ and $\mathbf{v}_1$ and $\mathbf{v}_2$ are linearly dependent. On the other hand, suppose that $\mathbf{v}_1$ and $\mathbf{v}_2$ are dependent. Then there are constants c_1 and c_2, not both zero, such that $c_1\mathbf{v}_1 + c_2\mathbf{v}_2 = \mathbf{0}$. If $c_1 \neq 0$, then, dividing by c_1, we obtain $\mathbf{v}_1 + (c_2/c_1)\mathbf{v}_2 = \mathbf{0}$ or

$$\mathbf{v}_1 = \left(-\frac{c_2}{c_1}\right)\mathbf{v}_2$$

That is, $\mathbf{v}_1$ is a scalar multiple of $\mathbf{v}_2$. If $c_1 = 0$ then $c_2 \neq 0$ and hence $\mathbf{v}_2 = \mathbf{0} = 0\mathbf{v}_1$. ■

EXAMPLE 1 The vectors $\mathbf{v}_1 = \begin{pmatrix} 2 \\ -1 \\ 0 \\ 3 \end{pmatrix}$ and $\mathbf{v}_2 = \begin{pmatrix} -6 \\ 3 \\ 0 \\ -9 \end{pmatrix}$ are linearly dependent since $\mathbf{v}_2 = -3\mathbf{v}_1$.

EXAMPLE 2 The vectors $\begin{pmatrix} 1 \\ 2 \\ 4 \end{pmatrix}$ and $\begin{pmatrix} 2 \\ 5 \\ -3 \end{pmatrix}$ are linearly independent; if they were not, we would have $\begin{pmatrix} 2 \\ 5 \\ -3 \end{pmatrix} = c\begin{pmatrix} 1 \\ 2 \\ 4 \end{pmatrix} = \begin{pmatrix} c \\ 2c \\ 4c \end{pmatrix}$. Then $2 = c$, $5 = 2c$, and $-3 = 4c$, which is clearly impossible for any number c.

There are several techniques for determining whether a set of vectors is linearly independent. Let us examine one of these techniques here. Another is given in Chapter 3 (see Example 3.4.6 on page 111).

EXAMPLE 3 Determine whether the vectors $\begin{pmatrix} 1 \\ -2 \\ 3 \end{pmatrix}$, $\begin{pmatrix} 2 \\ -2 \\ 0 \end{pmatrix}$, and $\begin{pmatrix} 0 \\ 1 \\ 7 \end{pmatrix}$ are linearly dependent or independent.

Solution Suppose that $c_1\begin{pmatrix} 1 \\ -2 \\ 3 \end{pmatrix} + c_2\begin{pmatrix} 2 \\ -2 \\ 0 \end{pmatrix} + c_3\begin{pmatrix} 0 \\ 1 \\ 7 \end{pmatrix} = \mathbf{0} = \begin{pmatrix} 0 \\ 0 \\ 0 \end{pmatrix}$. Then, multiplying through and adding, we have $\begin{pmatrix} c_1 + 2c_2 \\ -2c_1 - 2c_2 + c_3 \\ 3c_1 \qquad + 7c_3 \end{pmatrix} = \begin{pmatrix} 0 \\ 0 \\ 0 \end{pmatrix}$. This yields a homogeneous system of three equations in the three unknowns c_1, c_2, and c_3:

$$\begin{aligned} c_1+2c_2 \qquad &=0 \\ -2c_1-2c_2+\ c_3&=0 \\ 3c_1 \qquad +7c_3&=0 \end{aligned} \tag{4}$$

Thus the vectors will be linearly dependent if and only if system (4) has nontrivial solutions. We write system (4) using an augmented matrix and then row-reduce:

$$\left(\begin{array}{rrr|r} 1 & 2 & 0 & 0 \\ -2 & -2 & 1 & 0 \\ 3 & 0 & 7 & 0 \end{array}\right) \xrightarrow[A_{1,3}(-3)]{A_{1,2}(2)} \left(\begin{array}{rrr|r} 1 & 2 & 0 & 0 \\ 0 & 2 & 1 & 0 \\ 0 & -6 & 7 & 0 \end{array}\right)$$

$$\xrightarrow{M_2(\frac{1}{2})} \left(\begin{array}{rrr|r} 1 & 2 & 0 & 0 \\ 0 & 1 & \frac{1}{2} & 0 \\ 0 & -6 & 7 & 0 \end{array}\right) \xrightarrow[A_{2,3}(6)]{A_{2,1}(-2)} \left(\begin{array}{rrr|r} 1 & 0 & -1 & 0 \\ 0 & 1 & \frac{1}{2} & 0 \\ 0 & 0 & 10 & 0 \end{array}\right)$$

$$\xrightarrow{M_3(\frac{1}{10})} \left(\begin{array}{rrr|r} 1 & 0 & -1 & 0 \\ 0 & 1 & \frac{1}{2} & 0 \\ 0 & 0 & 1 & 0 \end{array}\right) \xrightarrow[A_{3,2}(-\frac{1}{2})]{A_{3,1}(1)} \left(\begin{array}{rrr|r} 1 & 0 & 0 & 0 \\ 0 & 1 & 0 & 0 \\ 0 & 0 & 1 & 0 \end{array}\right)$$

The last system of equations reads $c_1=0$, $c_2=0$, $c_3=0$. Hence (4) has no nontrivial solutions and the given vectors are linearly independent.

EXAMPLE 4 Determine whether the vectors $\begin{pmatrix} 1 \\ -3 \\ 0 \end{pmatrix}$, $\begin{pmatrix} 3 \\ 0 \\ 4 \end{pmatrix}$, and $\begin{pmatrix} 11 \\ -6 \\ 12 \end{pmatrix}$ are linearly dependent or independent.

Solution The equation $c_1\begin{pmatrix} 1 \\ -3 \\ 0 \end{pmatrix}+c_2\begin{pmatrix} 3 \\ 0 \\ 4 \end{pmatrix}+c_3\begin{pmatrix} 11 \\ -6 \\ 12 \end{pmatrix}=\begin{pmatrix} 0 \\ 0 \\ 0 \end{pmatrix}$ leads to the homogeneous system

$$\begin{aligned} c_1+3c_2+11c_3&=0 \\ -3c_1 \qquad -\ 6c_3&=0 \\ 4c_2+12c_3&=0 \end{aligned} \tag{5}$$

Writing system (5) in augmented matrix form and row reducing we obtain, successively,

$$\left(\begin{array}{rrr|r} 1 & 3 & 11 & 0 \\ -3 & 0 & -6 & 0 \\ 0 & 4 & 12 & 0 \end{array}\right) \xrightarrow{A_{1,2}(3)} \left(\begin{array}{rrr|r} 1 & 3 & 11 & 0 \\ 0 & 9 & 27 & 0 \\ 0 & 4 & 12 & 0 \end{array}\right)$$

$$\xrightarrow{M_2(\frac{1}{9})} \left(\begin{array}{rrr|r} 1 & 3 & 11 & 0 \\ 0 & 1 & 3 & 0 \\ 0 & 4 & 12 & 0 \end{array}\right) \xrightarrow[A_{2,3}(-4)]{A_{2,1}(-3)} \left(\begin{array}{rrr|r} 1 & 0 & 2 & 0 \\ 0 & 1 & 3 & 0 \\ 0 & 0 & 0 & 0 \end{array}\right)$$

We can stop here since the theory of Section 1.4 shows us that system (5) has an infinite number of solutions. For example, the last augmented matrix reads

$$\begin{aligned} c_1 \qquad +2c_3 &= 0 \\ c_2+3c_3 &= 0 \end{aligned}$$

If we choose $c_3=1$, we have $c_2=-3$ and $c_1=-2$ so that, as is easily verified, $-2\begin{pmatrix}1\\-3\\0\end{pmatrix}-3\begin{pmatrix}3\\0\\4\end{pmatrix}+\begin{pmatrix}11\\-6\\12\end{pmatrix}=\begin{pmatrix}0\\0\\0\end{pmatrix}$ and the vectors are linearly dependent.

The theory of homogeneous systems can tell us something about the linear dependence or independence of vectors.

THEOREM 2 A set of n m-vectors is always linearly dependent if $n>m$.

Proof Let $\mathbf{v}_1, \mathbf{v}_2, \ldots, \mathbf{v}_n$ be n m-vectors and let us try to find constants $c_1, c_2, \ldots, c_n$ not all zero such that

$$c_1\mathbf{v}_1+c_2\mathbf{v}_2+\cdots+c_n\mathbf{v}_n=\mathbf{0} \tag{6}$$

Let $\mathbf{v}_1=\begin{pmatrix}a_{11}\\a_{21}\\\vdots\\a_{m1}\end{pmatrix}$, $\mathbf{v}_2=\begin{pmatrix}a_{12}\\a_{22}\\\vdots\\a_{m2}\end{pmatrix}, \ldots, \mathbf{v}_n=\begin{pmatrix}a_{1n}\\a_{2n}\\\vdots\\a_{mn}\end{pmatrix}$. Then Equation (6) becomes

$$\begin{aligned} a_{11}c_1+a_{12}c_2+\cdots+a_{1n}c_n&=0\\ a_{21}c_1+a_{22}c_2+\cdots+a_{2n}c_n&=0\\ \vdots\qquad\qquad\qquad&\;\;\vdots\\ a_{m1}c_1+a_{m2}c_2+\cdots+a_{mn}c_n&=0 \end{aligned} \tag{7}$$

But system (7) is system (1.4.1) on page 22 and, according to Theorem 1.4.1, this system has an infinite number of solutions if $n>m$. Thus there are scalars $c_1, c_2, \ldots, c_n$ not all zero that satisfy (7) and the vectors $\mathbf{v}_1, \mathbf{v}_2, \ldots, \mathbf{v}_n$ are therefore linearly dependent. ■

EXAMPLE 5 The vectors $\begin{pmatrix}2\\-3\\4\end{pmatrix}$, $\begin{pmatrix}4\\7\\-6\end{pmatrix}$, $\begin{pmatrix}18\\-11\\4\end{pmatrix}$, and $\begin{pmatrix}2\\-7\\3\end{pmatrix}$ are linearly dependent since they comprise a set of four 3-vectors.

There is a very important (and obvious) corollary to Theorem 2.

COROLLARY A set of linearly independent n-vectors contains at most n vectors.

Note. We can rephrase the corollary as follows: If we have n linearly independent n-vectors, then we cannot add any more vectors without making the set linearly dependent. This fact will be very important in Chapter 5.

From system (7) we can make another important observation whose proof is left as an exercise (see Problem 17).

THEOREM 3 Let

$$A = \begin{pmatrix} a_{11} & a_{12} & \cdots & a_{1n} \\ a_{21} & a_{22} & \cdots & a_{2n} \\ \vdots & \vdots & & \vdots \\ a_{m1} & a_{m2} & \cdots & a_{mn} \end{pmatrix}$$

Then the columns of A, considered as vectors, are linearly dependent if and only if system (7), which can be written $A\mathbf{c}=\mathbf{0}$, has an infinite number of solutions. Here $\mathbf{c} = \begin{pmatrix} c_1 \\ c_2 \\ \vdots \\ c_n \end{pmatrix}$.

EXAMPLE 6 Consider the homogeneous system

$$\begin{aligned} x_1 + 2x_2 - x_3 + 2x_4 &= 0 \\ 3x_1 + 7x_2 + x_3 + 4x_4 &= 0 \end{aligned} \tag{8}$$

We solve this by row reduction:

$$\left(\begin{array}{cccc|c} 1 & 2 & -1 & 2 & 0 \\ 3 & 7 & 1 & 4 & 0 \end{array}\right) \xrightarrow{A_{1,2}(-3)} \left(\begin{array}{cccc|c} 1 & 2 & -1 & 2 & 0 \\ 0 & 1 & 4 & -2 & 0 \end{array}\right)$$

$$\xrightarrow{A_{2,1}(-2)} \left(\begin{array}{cccc|c} 1 & 0 & -9 & 6 & 0 \\ 0 & 1 & 4 & -2 & 0 \end{array}\right)$$

The last system is

$$\begin{aligned} x_1 \qquad - 9x_3 + 6x_4 &= 0 \\ x_2 + 4x_3 - 2x_4 &= 0 \end{aligned}$$

We see that this system has an infinite number of solutions, which we write as a column vector:

$$\begin{pmatrix} x_1 \\ x_2 \\ x_3 \\ x_4 \end{pmatrix} = \begin{pmatrix} 9x_3 - 6x_4 \\ -4x_3 + 2x_4 \\ x_3 \\ x_4 \end{pmatrix} = x_3 \begin{pmatrix} 9 \\ -4 \\ 1 \\ 0 \end{pmatrix} + x_4 \begin{pmatrix} -6 \\ 2 \\ 0 \\ 1 \end{pmatrix} \tag{9}$$

Note that $\begin{pmatrix} 9 \\ -4 \\ 1 \\ 0 \end{pmatrix}$ and $\begin{pmatrix} -6 \\ 2 \\ 0 \\ 1 \end{pmatrix}$ are linearly independent solutions to (8) because neither one is a multiple of the other. (You should verify that they are solutions.) Since x_3 and x_4 are arbitrary real numbers, we see, from (9), that we can express all solutions to the system (8) in terms of two linearly independent solution vectors.

PROBLEMS 2.6

In Problems 1–12 determine whether the given set of vectors is linearly dependent or independent.

1. $\begin{pmatrix}1\\2\end{pmatrix}$; $\begin{pmatrix}-1\\-3\end{pmatrix}$

2. $\begin{pmatrix}2\\-1\\4\end{pmatrix}$; $\begin{pmatrix}4\\-2\\7\end{pmatrix}$

3. $\begin{pmatrix}2\\-1\\4\end{pmatrix}$; $\begin{pmatrix}4\\-2\\8\end{pmatrix}$

4. $\begin{pmatrix}-2\\3\end{pmatrix}$; $\begin{pmatrix}4\\7\end{pmatrix}$

5. $\begin{pmatrix}-3\\2\end{pmatrix}$; $\begin{pmatrix}1\\10\end{pmatrix}$; $\begin{pmatrix}4\\-5\end{pmatrix}$

6. $\begin{pmatrix}1\\0\\1\end{pmatrix}$; $\begin{pmatrix}0\\1\\1\end{pmatrix}$; $\begin{pmatrix}1\\1\\0\end{pmatrix}$

7. $\begin{pmatrix}1\\0\\0\end{pmatrix}$; $\begin{pmatrix}0\\1\\0\end{pmatrix}$; $\begin{pmatrix}0\\0\\1\end{pmatrix}$

8. $\begin{pmatrix}-3\\4\\2\end{pmatrix}$; $\begin{pmatrix}7\\-1\\3\end{pmatrix}$; $\begin{pmatrix}1\\2\\8\end{pmatrix}$

9. $\begin{pmatrix}-3\\4\\2\end{pmatrix}$; $\begin{pmatrix}7\\-1\\3\end{pmatrix}$; $\begin{pmatrix}1\\1\\8\end{pmatrix}$

10. $\begin{pmatrix}1\\-2\\1\\1\end{pmatrix}$; $\begin{pmatrix}3\\0\\2\\-2\end{pmatrix}$; $\begin{pmatrix}0\\4\\-1\\-1\end{pmatrix}$; $\begin{pmatrix}5\\0\\3\\-1\end{pmatrix}$

11. $\begin{pmatrix}1\\-2\\1\\1\end{pmatrix}$; $\begin{pmatrix}3\\0\\2\\-2\end{pmatrix}$; $\begin{pmatrix}0\\4\\-1\\1\end{pmatrix}$; $\begin{pmatrix}5\\0\\3\\-1\end{pmatrix}$

12. $\begin{pmatrix}1\\-1\\2\end{pmatrix}$; $\begin{pmatrix}4\\0\\0\end{pmatrix}$; $\begin{pmatrix}-2\\3\\5\end{pmatrix}$; $\begin{pmatrix}7\\1\\2\end{pmatrix}$

13. Determine a condition on the numbers a, b, c, and d such that the vectors $\begin{pmatrix}a\\b\end{pmatrix}$ and $\begin{pmatrix}c\\d\end{pmatrix}$ are linearly dependent.

***14.** Find a condition on the numbers a_{ij} such that the vectors $\begin{pmatrix}a_{11}\\a_{21}\\a_{31}\end{pmatrix}$, $\begin{pmatrix}a_{12}\\a_{22}\\a_{32}\end{pmatrix}$, and $\begin{pmatrix}a_{13}\\a_{23}\\a_{33}\end{pmatrix}$ are linearly dependent.

15. For what value(s) of α will the vectors $\begin{pmatrix}1\\2\\3\end{pmatrix}$, $\begin{pmatrix}2\\-1\\4\end{pmatrix}$, $\begin{pmatrix}3\\\alpha\\4\end{pmatrix}$ be linearly dependent?

16. For what value(s) of α are the vectors $\begin{pmatrix}2\\-3\\1\end{pmatrix}$, $\begin{pmatrix}-4\\6\\-2\end{pmatrix}$, $\begin{pmatrix}\alpha\\1\\2\end{pmatrix}$ linearly dependent? [*Hint:* Look carefully.]

17. Prove Theorem 3. [*Hint:* Look closely at system (7).]

18. Prove that if the vectors $\mathbf{v}_1, \mathbf{v}_2, \ldots, \mathbf{v}_n$ are linearly dependent m-vectors and if $\mathbf{v}_{n+1}$ is any other m-vector, then the set $\mathbf{v}_1, \mathbf{v}_2, \ldots, \mathbf{v}_n, \mathbf{v}_{n+1}$ is linearly dependent.

19. Show that if $\mathbf{v}_1, \mathbf{v}_2, \ldots, \mathbf{v}_n$ $(n \geq 2)$ are linearly independent, then so too are $\mathbf{v}_1, \mathbf{v}_2, \ldots, \mathbf{v}_k$, where $k < n$.

20. Show that if the nonzero vectors $\mathbf{v}_1$ and $\mathbf{v}_2$ are orthogonal (see Problem 2.2.15 page 36), then the set $\{\mathbf{v}_1, \mathbf{v}_2\}$ is linearly independent.

***21.** Suppose that $\mathbf{v}_1$ is orthogonal to $\mathbf{v}_2$ and $\mathbf{v}_3$ and that $\mathbf{v}_2$ is orthogonal to $\mathbf{v}_3$. If $\mathbf{v}_1$, $\mathbf{v}_2$, and $\mathbf{v}_3$ are nonzero, show that the set $\{\mathbf{v}_1, \mathbf{v}_2, \mathbf{v}_3\}$ is linearly independent.

22. Let A be a square matrix whose columns are the vectors $\mathbf{v}_1, \mathbf{v}_2, \ldots, \mathbf{v}_n$. Show that $\mathbf{v}_1, \mathbf{v}_2, \ldots, \mathbf{v}_n$ are linearly independent if and only if the row echelon form of A does not contain a row of zeros.

In Problems 23–26 write the solutions to the given homogeneous systems in terms of one or more linearly independent vectors.

23. $x_1 + x_2 + x_3 = 0$

24. $x_1 - x_2 + 7x_3 - x_4 = 0$
$2x_1 + 3x_2 - 8x_3 + x_4 = 0$

25. $x_1 + 2x_2 - x_3 = 0$
$2x_1 + 5x_2 + 4x_3 = 0$

26. $x_1 + x_2 + x_3 - x_4 - x_5 = 0$
$-2x_1 + 3x_2 + x_3 + 4x_4 - 6x_5 = 0$

2.7 The Inverse of a Square Matrix

In this section we define two kinds of matrices that are central to matrix theory. We begin with a simple example. Let $A = \begin{pmatrix} 2 & 5 \\ 1 & 3 \end{pmatrix}$ and $B = \begin{pmatrix} 3 & -5 \\ -1 & 2 \end{pmatrix}$. Then an easy computation shows that $AB = BA = I_2$, where $I_2 = \begin{pmatrix} 1 & 0 \\ 0 & 1 \end{pmatrix}$. The matrix I_2 is called the 2×2 *identity matrix*. The matrix B is called the *inverse* of A and is written A^{-1}.

DEFINITION 1 **IDENTITY MATRIX** The $n \times n$ **identity matrix** is the $n \times n$ matrix with 1's down the **main diagonal**† and 0's everywhere else. That is,

$$I_n = (b_{ij}) \quad \text{where} \quad b_{ij} = \begin{cases} 1 & \text{if } i = j \\ 0 & \text{if } i \neq j \end{cases} \tag{1}$$

†The main diagonal of $A = (a_{ij})$ consists of the components a_{11}, a_{22}, a_{33}, and so on. Unless otherwise stated, we shall refer to the main diagonal simply as the **diagonal**.

EXAMPLE 1

$$I_3 = \begin{pmatrix} 1 & 0 & 0 \\ 0 & 1 & 0 \\ 0 & 0 & 1 \end{pmatrix} \quad \text{and} \quad I_5 = \begin{pmatrix} 1 & 0 & 0 & 0 & 0 \\ 0 & 1 & 0 & 0 & 0 \\ 0 & 0 & 1 & 0 & 0 \\ 0 & 0 & 0 & 1 & 0 \\ 0 & 0 & 0 & 0 & 1 \end{pmatrix}$$

THEOREM 1 Let A be a square $n \times n$ matrix. Then

$$AI_n = I_nA = A$$

That is, I_n commutes with every $n \times n$ matrix and leaves it unchanged after multiplication on the left or right.

Note. I_n functions for $n \times n$ matrices the way the number 1 functions for real numbers (since $1 \cdot a = a \cdot 1 = a$ for every real number a).

Proof Let c_{ij} be the ijth element of AI_n. Then

$$c_{ij} = a_{i1}b_{1j} + a_{i2}b_{2j} + \cdots + a_{ij}b_{jj} + \cdots + a_{in}b_{nj}.$$

But, from (1), this sum is equal to a_{ij}. Thus $AI_n = A$. In a similar fashion we can show that $I_nA = A$, and this proves the theorem. ■

Notation From now on we shall write the identity matrix simply as I, since if A is $n \times n$, the products IA and AI are defined only if I is also $n \times n$.

DEFINITION 2 **THE INVERSE OF A MATRIX** Let A and B be $n \times n$ matrices. Suppose that

$$AB = BA = I$$

Then B is called the **inverse** of A and is written as A^{-1}. We then have

$$AA^{-1} = A^{-1}A = I$$

If A has an inverse, then A is said to be **invertible**.

Remark 1. From this definition it immediately follows that $(A^{-1})^{-1} = A$ if A is invertible.

Remark 2. This definition does *not* state that every square matrix has an inverse. In fact there are many square matrices that have no inverse. (See, for instance, Example 2 below.)

In Definition 2 we defined *the* inverse of a matrix. This statement suggests that inverses are unique. This is indeed the case, as the following theorem shows.

THEOREM 2 If a square matrix A is invertible, then its inverse is unique.

Proof Suppose B and C are two inverses for A. We can show that $B = C$. By definition, we have $AB = BA = I$ and $AC = CA = I$. Then $B(AC) = BI = B$ and $(BA)C = IC = C$. But $B(AC) = (BA)C$ by the associative law of matrix multiplication. Hence $B = C$ and the theorem is proved. ■

Another important fact about inverses is given below.

THEOREM 3 Let A and B be invertible $n \times n$ matrices. Then AB is invertible and

$$(AB)^{-1} = B^{-1}A^{-1}$$

Proof To prove this result, we refer to Definition 2. That is, $B^{-1}A^{-1} = (AB)^{-1}$ if and only if $B^{-1}A^{-1}(AB) = (AB)(B^{-1}A^{-1}) = I$. But this follows since

Equation (4) on page 45

$$(B^{-1}A^{-1})(AB) = B^{-1}(A^{-1}A)B = B^{-1}IB = B^{-1}B = I$$

and

$$(AB)(B^{-1}A^{-1}) = A(BB^{-1})A^{-1} = AIA^{-1} = AA^{-1} = I. \quad ■$$

Consider the system of n equations in n unknowns

$$A\mathbf{x} = \mathbf{b},$$

and suppose that A is invertible. Then

$$A^{-1}A\mathbf{x} = A^{-1}\mathbf{b} \qquad \text{we multiplied on the left by } A^{-1}$$

$$I\mathbf{x} = A^{-1}\mathbf{b} \qquad A^{-1}A = I$$

$$\mathbf{x} = A^{-1}\mathbf{b} \qquad I\mathbf{x} = \mathbf{x}$$

That is,

If A is invertible, the system $A\mathbf{x} = \mathbf{b}$ has the unique solution $\mathbf{x} = A^{-1}\mathbf{b}$. (2)

This is one of the reasons we study matrix inverses.

There are two basic questions that come to mind once we have defined the inverse of a matrix.

Question 1. What matrices do have inverses?

Question 2. If a matrix has an inverse, how can we compute it?

We answer both questions in this section. Rather than starting by giving you what seems to be a set of arbitrary rules, we look first at what happens in the 2×2 case.

EXAMPLE 2 Let $A = \begin{pmatrix} 2 & -3 \\ -4 & 5 \end{pmatrix}$. Compute A^{-1} if it exists.

Solution Suppose that A^{-1} exists. We write $A^{-1} = \begin{pmatrix} x & y \\ z & w \end{pmatrix}$ and use the fact that $AA^{-1} = I$. Then

$$AA^{-1} = \begin{pmatrix} 2 & -3 \\ -4 & 5 \end{pmatrix}\begin{pmatrix} x & y \\ z & w \end{pmatrix} = \begin{pmatrix} 2x - 3z & 2y - 3w \\ -4x + 5z & -4y + 5w \end{pmatrix} = \begin{pmatrix} 1 & 0 \\ 0 & 1 \end{pmatrix}$$

The last two matrices can be equal only if each of their corresponding components are equal. This means that

$$2x \qquad - 3z \qquad = 1 \tag{3}$$

$$2y \qquad - 3w = 0 \tag{4}$$

$$-4x \qquad + 5z \qquad = 0 \tag{5}$$

$$-4y \qquad + 5w = 1 \tag{6}$$

This is a system of four equations in four unknowns. Note that there are two equations involving x and z only (equations (3) and (5)) and two equations involving y and w only (equations (4) and (6)). We write these two systems in augmented matrix form:

$$\left(\begin{array}{rr|r} 2 & -3 & 1 \\ -4 & 5 & 0 \end{array}\right) \tag{7}$$

$$\left(\begin{array}{rr|r} 2 & -3 & 0 \\ -4 & 5 & 1 \end{array}\right). \tag{8}$$

Now, we know from Section 1.3 that if system (7) (in the variables x and z) has a unique solution, then Gauss-Jordan elimination of (7) will result in

$$\left(\begin{array}{cc|c} 1 & 0 & x \\ 0 & 1 & z \end{array}\right)$$

where (x, z) is the unique pair of numbers that satisfies $2x - 3z = 1$ and $-4x + 5z = 0$. Similarly, row reduction of (8) will result in

$$\left(\begin{array}{cc|c} 1 & 0 & y \\ 0 & 1 & w \end{array}\right)$$

where (y, w) is the unique pair of numbers that satisfies $2y - 3w = 0$ and $-4y + 5w = 1$.

Since the coefficient matrices in (7) and (8) are the same, we can perform the row reductions on the two augmented matrices simultaneously, by considering the new augmented matrix

$$\left(\begin{array}{rr|rr} 2 & -3 & 1 & 0 \\ -4 & 5 & 0 & 1 \end{array}\right). \tag{9}$$

If A^{-1} is invertible, then the system defined by (3), (4), (5), and (6) has a unique solution and, by what we said above, Gauss-Jordan elimination will result in

$$\left(\begin{array}{rr|rr} 1 & 0 & x & y \\ 0 & 1 & z & w \end{array}\right).$$

We now carry out the computation, noting that the matrix on the left in (9) is A and the matrix on the right in (9) is I:

$$\left(\begin{array}{rr|rr} 2 & -3 & 1 & 0 \\ -4 & 5 & 0 & 1 \end{array}\right) \xrightarrow{M_1(\frac{1}{2})} \left(\begin{array}{rr|rr} 1 & -\frac{3}{2} & \frac{1}{2} & 0 \\ -4 & 5 & 0 & 1 \end{array}\right)$$

$$\xrightarrow{A_{1,2}(4)} \left(\begin{array}{rr|rr} 1 & -\frac{3}{2} & \frac{1}{2} & 0 \\ 0 & -1 & 2 & 1 \end{array}\right)$$

$$\xrightarrow{M_2(-1)} \left(\begin{array}{rr|rr} 1 & -\frac{3}{2} & \frac{1}{2} & 0 \\ 0 & 1 & -2 & -1 \end{array}\right)$$

$$\xrightarrow{A_{2,1}(\frac{3}{2})} \left(\begin{array}{rr|rr} 1 & 0 & -\frac{5}{2} & -\frac{3}{2} \\ 0 & 1 & -2 & -1 \end{array}\right).$$

Thus $x = -\frac{5}{2}$, $y = -\frac{3}{2}$, $z = -2$, $w = -1$, and $A^{-1} = \begin{pmatrix} -\frac{5}{2} & -\frac{3}{2} \\ -2 & -1 \end{pmatrix}$. We still must check our answer. We have

$$AA^{-1} = \begin{pmatrix} 2 & -3 \\ -4 & 5 \end{pmatrix}\begin{pmatrix} -\frac{5}{2} & -\frac{3}{2} \\ -2 & -1 \end{pmatrix} = \begin{pmatrix} 1 & 0 \\ 0 & 1 \end{pmatrix}$$

and

$$A^{-1}A = \begin{pmatrix} -\frac{5}{2} & -\frac{3}{2} \\ -2 & -1 \end{pmatrix}\begin{pmatrix} 2 & -3 \\ -4 & 5 \end{pmatrix} = \begin{pmatrix} 1 & 0 \\ 0 & 1 \end{pmatrix}.$$

Thus A is invertible and $A^{-1} = \begin{pmatrix} -\frac{5}{2} & -\frac{3}{2} \\ -2 & -1 \end{pmatrix}$.

EXAMPLE 3 Let $A = \begin{pmatrix} 1 & 2 \\ -2 & -4 \end{pmatrix}$. Calculate A^{-1} if it exists.

Solution If $A^{-1} = \begin{pmatrix} x & y \\ z & w \end{pmatrix}$ exists, then

$$AA^{-1} = \begin{pmatrix} 1 & 2 \\ -2 & -4 \end{pmatrix}\begin{pmatrix} x & y \\ z & w \end{pmatrix} = \begin{pmatrix} x + 2z & y + 2w \\ -2x - 4z & -2y - 4w \end{pmatrix} = \begin{pmatrix} 1 & 0 \\ 0 & 1 \end{pmatrix}.$$

This leads to the system

$$\begin{aligned} x \qquad\quad + 2z \qquad\quad &= 1 \\ y \qquad\quad + 2w &= 0 \\ -2x \qquad\quad - 4z \qquad\quad &= 0 \\ -2y \qquad\quad - 4w &= 1. \end{aligned} \tag{10}$$

Using the same reasoning as in Example 2, we can write this system in the augmented matrix form $(A|I)$ and row-reduce.

$$\left(\begin{array}{cc|cc} 1 & 2 & 1 & 0 \\ -2 & -4 & 0 & 1 \end{array}\right) \xrightarrow{A_{1,2}(2)} \left(\begin{array}{cc|cc} 1 & 2 & 1 & 0 \\ 0 & 0 & 2 & 1 \end{array}\right)$$

This is as far as we can go. The last line reads $0 = 2$ or $0 = 1$, depending on which of the two systems of equations (in x and z or in y and w) is being solved. Thus system (10) is inconsistent and A is not invertible.

The last two examples illustrate a procedure that always works when you are trying to find the inverse of a matrix.

PROCEDURE FOR COMPUTING THE INVERSE OF A SQUARE MATRIX A

Step 1. Write the augmented matrix $(A|I)$.

Step 2. Use row reduction to reduce the matrix A to its reduced row echelon form.

Step 3. Decide if A is invertible.

(a) If A can be reduced to the identity matrix I, then A^{-1} will be the matrix to the right of the vertical bar.

(b) If the row reduction of A leads to a row of zeros to the left of the vertical bar, then A is not invertible.

Remark We can rephrase (a) and (b) as follows.

A square matrix A is invertible if and only if its reduced row echelon form is the identity matrix.

Let $A = \begin{pmatrix} a_{11} & a_{12} \\ a_{21} & a_{22} \end{pmatrix}$. Then, as in Equation (1.2.7, page 4), we define

$$\text{Determinant of } A = a_{11}a_{22} - a_{12}a_{21} \tag{11}$$

We abbreviate the determinant of A by $\det A$.

THEOREM 4 Let A be a 2×2 matrix. Then:

i. A is invertible if and only if $\det A \neq 0$.
ii. If $\det A \neq 0$, then

$$A^{-1} = \frac{1}{\det A}\begin{pmatrix} a_{22} & -a_{12} \\ -a_{21} & a_{11} \end{pmatrix}^{\dagger} \tag{12}$$

Proof First suppose that $\det A \neq 0$ and let $B = (1/\det A)\begin{pmatrix} a_{22} & -a_{12} \\ -a_{21} & a_{11} \end{pmatrix}$. Then

$$BA = \frac{1}{\det A}\begin{pmatrix} a_{22} & -a_{12} \\ -a_{21} & a_{11} \end{pmatrix}\begin{pmatrix} a_{11} & a_{12} \\ a_{21} & a_{22} \end{pmatrix}$$

$$= \frac{1}{a_{11}a_{22} - a_{12}a_{21}}\begin{pmatrix} a_{22}a_{11} - a_{12}a_{21} & 0 \\ 0 & -a_{21}a_{12} + a_{11}a_{22} \end{pmatrix} = \begin{pmatrix} 1 & 0 \\ 0 & 1 \end{pmatrix} = I$$

Similarly $AB = I$, which shows that A is invertible and that $B = A^{-1}$. We still must show that if A is invertible, then $\det A \neq 0$. To do so, we consider the system

$$\begin{aligned} a_{11}x_1 + a_{12}x_2 &= b_1 \\ a_{21}x_1 + a_{22}x_2 &= b_2 \end{aligned} \tag{13}$$

We do this because we know from our Summing Up Theorem (Theorem 1.2.1, page 4) that if this system has a unique solution, then its determinant is nonzero. The system can be written in the form

$$A\mathbf{x} = \mathbf{b} \tag{14}$$

with $\mathbf{x} = \begin{pmatrix} x_1 \\ x_2 \end{pmatrix}$ and $\mathbf{b} = \begin{pmatrix} b_1 \\ b_2 \end{pmatrix}$. Then, since A is invertible, we see from (2) that the system (14) has a unique solution given by

$$\mathbf{x} = A^{-1}\mathbf{b}$$

But by Theorem 1.2.1, the fact that system (13) has a unique solution implies that $\det A \neq 0$. This completes the proof. ■

† This formula can be obtained directly by applying our procedure for computing an inverse (see Problem 46).

EXAMPLE 4 Let $A = \begin{pmatrix} 2 & -4 \\ 1 & 3 \end{pmatrix}$. Calculate A^{-1} if it exists.

Solution We find that det $A = (2)(3) - (-4)(1) = 10$; hence A^{-1} exists. From Equation (12), we get

$$A^{-1} = \frac{1}{10}\begin{pmatrix} 3 & 4 \\ -1 & 2 \end{pmatrix} = \begin{pmatrix} \frac{3}{10} & \frac{4}{10} \\ -\frac{1}{10} & \frac{2}{10} \end{pmatrix}.$$

Check.

$$A^{-1}A = \frac{1}{10}\begin{pmatrix} 3 & 4 \\ -1 & 2 \end{pmatrix}\begin{pmatrix} 2 & -4 \\ 1 & 3 \end{pmatrix} = \frac{1}{10}\begin{pmatrix} 10 & 0 \\ 0 & 10 \end{pmatrix} = \begin{pmatrix} 1 & 0 \\ 0 & 1 \end{pmatrix}$$

and

$$AA^{-1} = \begin{pmatrix} 2 & -4 \\ 1 & 3 \end{pmatrix}\begin{pmatrix} \frac{3}{10} & \frac{4}{10} \\ -\frac{1}{10} & \frac{2}{10} \end{pmatrix} = \begin{pmatrix} 1 & 0 \\ 0 & 1 \end{pmatrix}.$$

EXAMPLE 5 Let $A = \begin{pmatrix} 1 & 2 \\ -2 & -4 \end{pmatrix}$. Calculate A^{-1} if it exists.

Solution We find that det $A = (1)(-4) - (2)(-2) = -4 + 4 = 0$, so that A^{-1} does not exist, as we saw in Example 3.

The procedure described above works for $n \times n$ matrices where $n > 2$. We illustrate this with a number of examples.

EXAMPLE 6 Let $A = \begin{pmatrix} 2 & 4 & 6 \\ 4 & 5 & 6 \\ 3 & 1 & -2 \end{pmatrix}$ (see Example 2.5.1 on page 50). Calculate A^{-1} if it exists.

Solution We first put I next to A in an augmented matrix form

$$\left(\begin{array}{ccc|ccc} 2 & 4 & 6 & 1 & 0 & 0 \\ 4 & 5 & 6 & 0 & 1 & 0 \\ 3 & 1 & -2 & 0 & 0 & 1 \end{array}\right)$$

and then carry out the row reduction.

$$\xrightarrow{M_1(\frac{1}{2})}\left(\begin{array}{ccc|ccc}1 & 2 & 3 & \frac{1}{2} & 0 & 0\\ 4 & 5 & 6 & 0 & 1 & 0\\ 3 & 1 & -2 & 0 & 0 & 1\end{array}\right)\xrightarrow[A_{1,3}(-3)]{A_{1,2}(-4)}\left(\begin{array}{ccc|ccc}1 & 2 & 3 & \frac{1}{2} & 0 & 0\\ 0 & -3 & -6 & -2 & 1 & 0\\ 0 & -5 & -11 & -\frac{3}{2} & 0 & 1\end{array}\right)$$

$$\xrightarrow{M_2(-\frac{1}{3})}\left(\begin{array}{ccc|ccc}1 & 2 & 3 & \frac{1}{2} & 0 & 0\\ 0 & 1 & 2 & \frac{2}{3} & -\frac{1}{3} & 0\\ 0 & -5 & -11 & -\frac{3}{2} & 0 & 1\end{array}\right)\xrightarrow[A_{2,3}(5)]{A_{2,1}(-2)}\left(\begin{array}{ccc|ccc}1 & 0 & -1 & -\frac{5}{6} & \frac{2}{3} & 0\\ 0 & 1 & 2 & \frac{2}{3} & -\frac{1}{3} & 0\\ 0 & 0 & -1 & \frac{11}{6} & -\frac{5}{3} & 1\end{array}\right)$$

$$\xrightarrow{M_3(-1)}\left(\begin{array}{ccc|ccc}1 & 0 & -1 & -\frac{5}{6} & \frac{2}{3} & 0\\ 0 & 1 & 2 & \frac{2}{3} & -\frac{1}{3} & 0\\ 0 & 0 & 1 & -\frac{11}{6} & \frac{5}{3} & -1\end{array}\right)\xrightarrow[A_{3,2}(-2)]{A_{3,1}(1)}\left(\begin{array}{ccc|ccc}1 & 0 & 0 & -\frac{8}{3} & \frac{7}{3} & -1\\ 0 & 1 & 0 & \frac{13}{3} & -\frac{11}{3} & 2\\ 0 & 0 & 1 & -\frac{11}{6} & \frac{5}{3} & -1\end{array}\right)$$

Since A has now been reduced to I, we have

$$A^{-1}=\begin{pmatrix}-\frac{8}{3} & \frac{7}{3} & -1\\ \frac{13}{3} & -\frac{11}{3} & 2\\ -\frac{11}{6} & \frac{5}{3} & -1\end{pmatrix}=\frac{1}{6}\begin{pmatrix}-16 & 14 & -6\\ 26 & -22 & 12\\ -11 & 10 & -6\end{pmatrix}$$

We factor out $\frac{1}{6}$ to make computations easier.

Check. $$A^{-1}A=\frac{1}{6}\begin{pmatrix}-16 & 14 & -6\\ 26 & -22 & 12\\ -11 & 10 & -6\end{pmatrix}\begin{pmatrix}2 & 4 & 6\\ 4 & 5 & 6\\ 3 & 1 & -2\end{pmatrix}=\frac{1}{6}\begin{pmatrix}6 & 0 & 0\\ 0 & 6 & 0\\ 0 & 0 & 6\end{pmatrix}=I.$$

We can also verify that $AA^{-1}=I$.

Warning. It is easy to make numerical errors in computing A^{-1}. Therefore it is essential to check the computations by verifying that $A^{-1}A=I$.

EXAMPLE 7 Let $A=\begin{pmatrix}2 & 4 & 3\\ 0 & 1 & -1\\ 3 & 5 & 7\end{pmatrix}$. Calculate A^{-1} if it exists.

Solution Proceeding as in Example 6 we obtain, successively, the following augmented matrices:

$$\left(\begin{array}{ccc|ccc}2 & 4 & 3 & 1 & 0 & 0\\ 0 & 1 & -1 & 0 & 1 & 0\\ 3 & 5 & 7 & 0 & 0 & 1\end{array}\right)\xrightarrow{M_1(\frac{1}{2})}\left(\begin{array}{ccc|ccc}1 & 2 & \frac{3}{2} & \frac{1}{2} & 0 & 0\\ 0 & 1 & -1 & 0 & 1 & 0\\ 3 & 5 & 7 & 0 & 0 & 1\end{array}\right)$$

$$\xrightarrow{A_{1,3}(-3)}\left(\begin{array}{ccc|ccc}1 & 2 & \frac{3}{2} & \frac{1}{2} & 0 & 0\\ 0 & 1 & -1 & 0 & 1 & 0\\ 0 & -1 & \frac{5}{2} & -\frac{3}{2} & 0 & 1\end{array}\right)\xrightarrow[A_{2,3}(1)]{A_{2,1}(-2)}\left(\begin{array}{ccc|ccc}1 & 0 & \frac{7}{2} & \frac{1}{2} & -2 & 0\\ 0 & 1 & -1 & 0 & 1 & 0\\ 0 & 0 & \frac{3}{2} & -\frac{3}{2} & 1 & 1\end{array}\right)$$

$$\xrightarrow{M_3(\frac{2}{3})}\left(\begin{array}{ccc|ccc}1 & 0 & \frac{7}{2} & \frac{1}{2} & -2 & 0\\ 0 & 1 & -1 & 0 & 1 & 0\\ 0 & 0 & 1 & -1 & \frac{2}{3} & \frac{2}{3}\end{array}\right)\xrightarrow[A_{3,2}(1)]{A_{3,1}(-\frac{7}{2})}\left(\begin{array}{ccc|ccc}1 & 0 & 0 & 4 & -\frac{13}{3} & -\frac{7}{3}\\ 0 & 1 & 0 & -1 & \frac{5}{3} & \frac{2}{3}\\ 0 & 0 & 1 & -1 & \frac{2}{3} & \frac{2}{3}\end{array}\right)$$

Thus

$$A^{-1} = \begin{pmatrix} 4 & -\frac{13}{3} & -\frac{7}{3} \\ -1 & \frac{5}{3} & \frac{2}{3} \\ -1 & \frac{2}{3} & \frac{2}{3} \end{pmatrix}$$

Check. $A^{-1}A = \begin{pmatrix} 4 & -\frac{13}{3} & -\frac{7}{3} \\ -1 & \frac{5}{3} & \frac{2}{3} \\ -1 & \frac{2}{3} & \frac{2}{3} \end{pmatrix}\begin{pmatrix} 2 & 4 & 3 \\ 0 & 1 & -1 \\ 3 & 5 & 7 \end{pmatrix} = \begin{pmatrix} 1 & 0 & 0 \\ 0 & 1 & 0 \\ 0 & 0 & 1 \end{pmatrix}$

EXAMPLE 8 Let $A = \begin{pmatrix} 1 & -3 & 4 \\ 2 & -5 & 7 \\ 0 & -1 & 1 \end{pmatrix}$. Calculate A^{-1} if it exists.

Solution Proceeding as before we obtain, successively,

$$\left(\begin{array}{ccc|ccc} 1 & -3 & 4 & 1 & 0 & 0 \\ 2 & -5 & 7 & 0 & 1 & 0 \\ 0 & -1 & 1 & 0 & 0 & 1 \end{array}\right) \xrightarrow{A_{1,2}(-2)} \left(\begin{array}{ccc|ccc} 1 & -3 & 4 & 1 & 0 & 0 \\ 0 & 1 & -1 & -2 & 1 & 0 \\ 0 & -1 & 1 & 0 & 0 & 1 \end{array}\right)$$

$$\xrightarrow[A_{2,3}(1)]{A_{2,1}(3)} \left(\begin{array}{ccc|ccc} 1 & 0 & 1 & -5 & 3 & 0 \\ 0 & 1 & -1 & -2 & 1 & 0 \\ 0 & 0 & 0 & -2 & 1 & 1 \end{array}\right)$$

This is as far as we can go. The matrix A *cannot* be reduced to the identity matrix and we can conclude that A is *not* invertible.

There is another way to see the result of the last example. Let **b** be any 3-vector and consider the system $A\mathbf{x} = \mathbf{b}$. If we tried to solve this by Gaussian elimination, we would end up with an equation that reads $0 = c \neq 0$ as in Example 3, or $0 = 0$. This is case (*ii*) or (*iii*) of Section 1.3 (see page 16). That is, the system either has no solution or it has an infinite number of solutions. The one possibility ruled out is the case in which the system has a unique solution. But if A^{-1} existed, then there would be a unique solution given by $\mathbf{x} = A^{-1}\mathbf{b}$. We are left to conclude that:

> If in the row reduction of A we end up with a row of zeros, then A is *not* invertible.

DEFINITION 3 **ROW EQUIVALENT MATRICES** Suppose that by elementary row operations we can transform the matrix A into the matrix B. Then A and B are said to be **row equivalent**.

The reasoning used above can be used to prove the following theorem (see Problem 47).

THEOREM 5 Let A be an $n \times n$ matrix.

i. A is invertible if and only if A is row equivalent to the identity matrix I_n; that is, the reduced row echelon form of A is I_n.
ii. A is invertible if and only if the system $A\mathbf{x}=\mathbf{b}$ has a unique solution for every n-vector $\mathbf{b}$.
iii. If A is invertible, then this unique solution is given by $\mathbf{x}=A^{-1}\mathbf{b}$.

EXAMPLE 9 Solve the system

$$\begin{aligned} 2x_1+4x_2+3x_3 &= 6 \\ x_2-x_3 &= -4 \\ 3x_1+5x_2+7x_3 &= 7 \end{aligned}$$

Solution This system can be written as $A\mathbf{x}=\mathbf{b}$, where $A=\begin{pmatrix} 2 & 4 & 3 \\ 0 & 1 & -1 \\ 3 & 5 & 7 \end{pmatrix}$ and $\mathbf{b}=\begin{pmatrix} 6 \\ -4 \\ 7 \end{pmatrix}$. In Example 7 we found that A^{-1} exists and

$$A^{-1}=\begin{pmatrix} 4 & -\frac{13}{3} & -\frac{7}{3} \\ -1 & \frac{5}{3} & \frac{2}{3} \\ -1 & \frac{2}{3} & \frac{2}{3} \end{pmatrix}$$

Thus the unique solution is given by

$$\mathbf{x}=\begin{pmatrix} x_1 \\ x_2 \\ x_3 \end{pmatrix}=A^{-1}\mathbf{b}=\begin{pmatrix} 4 & -\frac{13}{3} & -\frac{7}{3} \\ -1 & \frac{5}{3} & \frac{2}{3} \\ -1 & \frac{2}{3} & \frac{2}{3} \end{pmatrix}\begin{pmatrix} 6 \\ -4 \\ 7 \end{pmatrix}=\begin{pmatrix} 25 \\ -8 \\ -4 \end{pmatrix}$$

EXAMPLE 10 In the Leontief input-output model described in Example 1.3.8 on page 17, we obtained the system

$$\begin{aligned} a_{11}x_1 + a_{12}x_2 + \cdots + a_{1n}x_n + e_1 &= x_1 \\ a_{21}x_1 + a_{22}x_2 + \cdots + a_{2n}x_n + e_2 &= x_2 \\ \vdots \qquad \vdots \qquad\quad \vdots \qquad \vdots \quad &\quad \vdots \\ a_{n1}x_1 + a_{n2}x_2 + \cdots + a_{nn}x_n + e_n &= x_n \end{aligned} \tag{15}$$

which can be written as

$$A\mathbf{x} + \mathbf{e} = \mathbf{x} = I\mathbf{x}$$

or

$$(I - A)\mathbf{x} = \mathbf{e} \tag{16}$$

The matrix A of internal demands is called the **technology matrix**, and the matrix $I - A$ is called the **Leontief matrix**. If the Leontief matrix is invertible, then systems (15) and (16) have unique solutions.

Leontief used his model to analyze the 1958 American economy.† He divided the economy into 81 sectors and grouped them into six families of related sectors. For simplicity, we treat each family of sectors as a single sector so we can treat the American economy as an economy with six industries. These industries are listed in Table 1.

Table 1

Sector	Examples
Final nonmetal (FN)	Furniture, processed food
Final metal (FM)	Household appliances, motor vehicles
Basic metal (BM)	Machine-shop products, mining
Basic nonmetal (BN)	Agriculture, printing
Energy (E)	Petroleum, coal
Services (S)	Amusements, real estate

The input-output table, Table 2, gives internal demands in 1958 based on Leontief's figures. The units in the table are millions of dollars. Thus, for example, the number 0.173 in the 6,5 position means that in order to produce \$1 million worth of energy, it is necessary to provide \$0.173 million = \$173,000 worth of services. Similarly, the 0.037 in the 4,2 position means that in order to produce \$1 million worth of final metal, it is necessary to expend \$0.037 million = \$37,000 on basic nonmetal products.

Table 2

Internal Demands in 1958 U.S. Economy

	FN	FM	BM	BN	E	S
FN	0.170	0.004	0	0.029	0	0.008
FM	0.003	0.295	0.018	0.002	0.004	0.016
BM	0.025	0.173	0.460	0.007	0.011	0.007
BN	0.348	0.037	0.021	0.403	0.011	0.048
E	0.007	0.001	0.039	0.025	0.358	0.025
S	0.120	0.074	0.104	0.123	0.173	0.234

Finally, Leontief estimated the following external demands on the 1958 American economy (in millions of dollars).

Table 3

External Demands on 1958 U.S. Economy (Millions of Dollars)

FN	\$99,640
FM	\$75,548
BM	\$14,444
BN	\$33,501
E	\$23,527
S	\$263,985

† *Scientific American* (April 1965), pp. 26–27.

In order to run the American economy in 1958 and meet all external demands, how many units in each of the six sectors had to be produced?

Solution The technology matrix is given by

$$A = \begin{pmatrix} 0.170 & 0.004 & 0 & 0.029 & 0 & 0.008 \\ 0.003 & 0.295 & 0.018 & 0.002 & 0.004 & 0.016 \\ 0.025 & 0.173 & 0.460 & 0.007 & 0.011 & 0.007 \\ 0.348 & 0.037 & 0.021 & 0.403 & 0.011 & 0.048 \\ 0.007 & 0.001 & 0.039 & 0.025 & 0.358 & 0.025 \\ 0.120 & 0.074 & 0.104 & 0.123 & 0.173 & 0.234 \end{pmatrix}$$

and

$$\mathbf{e} = \begin{pmatrix} 99{,}640 \\ 75{,}548 \\ 14{,}444 \\ 33{,}501 \\ 23{,}527 \\ 263{,}985 \end{pmatrix}$$

To obtain the Leontief matrix, we subtract to obtain

$$I - A = \begin{pmatrix} 1 & 0 & 0 & 0 & 0 & 0 \\ 0 & 1 & 0 & 0 & 0 & 0 \\ 0 & 0 & 1 & 0 & 0 & 0 \\ 0 & 0 & 0 & 1 & 0 & 0 \\ 0 & 0 & 0 & 0 & 1 & 0 \\ 0 & 0 & 0 & 0 & 0 & 1 \end{pmatrix}$$

$$- \begin{pmatrix} 0.170 & 0.004 & 0 & 0.029 & 0 & 0.008 \\ 0.003 & 0.295 & 0.018 & 0.002 & 0.004 & 0.016 \\ 0.025 & 0.173 & 0.460 & 0.007 & 0.011 & 0.007 \\ 0.348 & 0.037 & 0.021 & 0.403 & 0.011 & 0.048 \\ 0.007 & 0.001 & 0.039 & 0.025 & 0.358 & 0.025 \\ 0.120 & 0.074 & 0.104 & 0.123 & 0.173 & 0.234 \end{pmatrix}$$

$$= \begin{pmatrix} 0.830 & -0.004 & 0 & -0.029 & 0 & -0.008 \\ -0.003 & 0.705 & -0.018 & -0.002 & -0.004 & -0.016 \\ -0.025 & -0.173 & 0.540 & -0.007 & -0.011 & -0.007 \\ -0.348 & -0.037 & -0.021 & 0.597 & -0.011 & -0.048 \\ -0.007 & -0.001 & -0.039 & -0.025 & 0.642 & -0.025 \\ -0.120 & -0.074 & -0.104 & -0.123 & -0.173 & 0.766 \end{pmatrix}$$

The computation of the inverse of a 6 × 6 matrix is a tedious affair. Carrying three decimal places on a calculator, we obtain the matrix below. Intermediate steps are omitted.

$$(I - A)^{-1} = \begin{pmatrix} 1.234 & 0.014 & 0.006 & 0.064 & 0.007 & 0.018 \\ 0.017 & 1.436 & 0.057 & 0.012 & 0.020 & 0.032 \\ 0.071 & 0.465 & 1.877 & 0.019 & 0.045 & 0.031 \\ 0.751 & 0.134 & 0.100 & 1.740 & 0.066 & 0.124 \\ 0.060 & 0.045 & 0.130 & 0.082 & 1.578 & 0.059 \\ 0.339 & 0.236 & 0.307 & 0.312 & 0.376 & 1.349 \end{pmatrix}$$

Therefore the "ideal" output vector is given by

$$\mathbf{x} = (I - A)^{-1}\mathbf{e} = \begin{pmatrix} 1.234 & 0.014 & 0.006 & 0.064 & 0.007 & 0.018 \\ 0.017 & 1.436 & 0.057 & 0.012 & 0.020 & 0.032 \\ 0.071 & 0.465 & 1.877 & 0.019 & 0.045 & 0.031 \\ 0.751 & 0.134 & 0.100 & 1.740 & 0.066 & 0.124 \\ 0.060 & 0.045 & 0.130 & 0.082 & 1.578 & 0.059 \\ 0.339 & 0.236 & 0.307 & 0.312 & 0.376 & 1.349 \end{pmatrix} \begin{pmatrix} 99{,}640 \\ 75{,}548 \\ 14{,}444 \\ 33{,}501 \\ 23{,}527 \\ 263{,}985 \end{pmatrix}$$

$$= \begin{pmatrix} 131{,}161 \\ 120{,}324 \\ 79{,}194 \\ 178{,}936 \\ 66{,}703 \\ 426{,}542 \end{pmatrix}$$

This means that it would require 131,161 units ($131,161 million worth) of final nonmetal products, 120,324 units of final metal products, 79,194 units of basic metal products, 178,936 units of basic nonmetal products, 66,703 units of energy and 426,542 service units to run the U.S. economy and meet the external demands in 1958.

In Section 1.2 we encountered the first form of our Summing Up Theorem (Theorem 1.2.1, page 4). We are now ready to improve upon it. The next theorem states that several statements involving inverse, uniqueness of solutions, row equivalence, linear independence, and determinants are equivalent. This means that if one of the statements is true, all the statements are true. At this point we can prove the equivalence of parts (*i*), (*ii*), (*iii*), and (*iv*). We shall finish the proof after we have developed some basic theory about determinants (see Theorem 3.4.4 on page 110).

THEOREM 6 **SUMMING UP THEOREM—VIEW 2** Let A be an $n \times n$ matrix. Then each of the following six statements implies the other five (that is, if one is true, all are true):

i. A is invertible.

ii. The only solution to the homogeneous system $A\mathbf{x}=\mathbf{0}$ is the trivial solution ($\mathbf{x}=\mathbf{0}$).

iii. The system $A\mathbf{x}=\mathbf{b}$ has a unique solution for every n-vector $\mathbf{b}$.
iv. A is row equivalent to the $n\times n$ identity matrix I_n.
v. The rows (and columns) of A are linearly independent.
vi. $\det A \neq 0$. (So far, $\det A$ is only defined if A is a 2×2 matrix.)

Proof We have already seen that statements (*i*) and (*iii*) are equivalent (Theorem 5 (part *ii*)) and that (*i*) and (*iv*) are equivalent (Theorem 5 (part *i*)). We shall see that (*ii*) and (*iv*) are equivalent. Suppose that (*ii*) holds. That is, suppose that $A\mathbf{x}=\mathbf{0}$ has only the trivial solution $\mathbf{x}=\mathbf{0}$. If we write out this system we obtain

$$\begin{aligned} a_{11}x_1+a_{12}x_2+\cdots+a_{1n}x_n&=0\\ a_{21}x_1+a_{22}x_2+\cdots+a_{2n}x_n&=0\\ \vdots\qquad\quad \vdots\qquad\qquad\quad \vdots\quad&\\ a_{n1}x_1+a_{n2}x_2+\cdots+a_{nn}x_n&=0 \end{aligned} \tag{17}$$

If A were not equivalent to I_n, then row reduction of the augmented matrix associated with (17) would leave us with a row of zeros. But if, say, the last row is zero, then the last equation reads $0=0$. Then, the homogeneous system reduces to one with $n-1$ equations in n unknowns which, by Theorem 1.4.1 on page 24 has an infinite number of solutions. But we assumed that $\mathbf{x}=\mathbf{0}$ was the only solution to system (17). This contradiction shows that A is row equivalent to I_n. Conversely, suppose that (*iv*) holds; that is, suppose that A is row equivalent to I_n. Then by Theorem 5 (part *i*), A is invertible and by Theorem 5 (part *iii*) the unique solution to $A\mathbf{x}=\mathbf{0}$ is $\mathbf{x}=A^{-1}\mathbf{0}=\mathbf{0}$. Thus (*ii*) and (*iv*) are equivalent. In Theorem 1.2.1 we showed that (*i*) and (*vi*) are equivalent in the 2×2 case. It is not difficult to prove the equivalence of (*v*) and (*vi*) in the 2×2 case (see Problem 38). We shall prove the equivalence of (*ii*), (*v*), and (*vi*) in Section 3.4. ■

Remark. We could add another statement to the theorem. Suppose the system $A\mathbf{x}=\mathbf{b}$ has a unique solution. Let R be a matrix in row echelon form that is row equivalent to A. Then R cannot have a row of zeros because if it did, it could not be reduced to the identity matrix.† Thus the row echelon form of A must look like this:

$$\begin{pmatrix} 1 & r_{12} & r_{13} & \cdots & r_{1n}\\ 0 & 1 & r_{23} & \cdots & r_{2n}\\ 0 & 0 & 1 & \cdots & r_{3n}\\ \vdots & \vdots & \vdots & & \vdots\\ 0 & 0 & 0 & \cdots & 1 \end{pmatrix} \tag{18}$$

That is, R is a matrix with 1's down the diagonal and 0's below it. We thus have Theorem 7.

---

† Note that if the ith row of R contains only zeros, then the homogeneous system $R\mathbf{x}=\mathbf{0}$ contains more unknowns than equations (since the ith equation is the zero equation) and the system has an infinite number of solutions. But then $A\mathbf{x}=\mathbf{0}$ has an infinite number of solutions, which is a contradiction of our assumption.

THEOREM 7 If any of the statements in Theorem 6 holds, then the row echelon form of A has the form of matrix (18).

We have seen that in order to verify that $B = A^{-1}$, we have to check that $AB = BA = I$. It turns out that only half this work has to be done.

THEOREM 8 Let A and B be $n \times n$ matrices. Then A is invertible and $B = A^{-1}$ if (*i*) $BA = I$ or (*ii*) $AB = I$.

Remark. This theorem simplifies the work in checking that one matrix is the inverse of another.

Proof **i.** We assume that $BA = I$. Consider the homogeneous system $A\mathbf{x} = \mathbf{0}$. Multiplying both sides of this equation on the left by B, we obtain

$$BA\mathbf{x} = B\,\mathbf{0} \tag{19}$$

But $BA = I$ and $B\mathbf{0} = \mathbf{0}$, so (19) becomes $I\mathbf{x} = \mathbf{0}$ or $\mathbf{x} = \mathbf{0}$. This shows that $\mathbf{x}=\mathbf{0}$ is the only solution to $A\mathbf{x} = \mathbf{0}$ and, by Theorem 6 (parts *i* and *ii*), this means that A is invertible. We still have to show that $B = A^{-1}$. Let $A^{-1} = C$. Then $AC = I$. Thus $BAC = B(AC) = BI = B$ and $BAC = (BA)C = IC = C$. Hence $B = C$ and part (*i*) is proved.

ii. Let $AB = I$. Then, from part (*i*), $A = B^{-1}$. From Definition 2 this means that $AB = BA = I$, which proves that A is invertible and that $B = A^{-1}$. This completes the proof. ■

PROBLEMS 2.7 In Problems 1–15 determine whether the given matrix is invertible. If it is, calculate the inverse.

1. $\begin{pmatrix} 2 & 1 \\ 3 & 2 \end{pmatrix}$ **2.** $\begin{pmatrix} -1 & 6 \\ 2 & -12 \end{pmatrix}$ **3.** $\begin{pmatrix} 0 & 1 \\ 1 & 0 \end{pmatrix}$

4. $\begin{pmatrix} 1 & 1 \\ 3 & 3 \end{pmatrix}$ **5.** $\begin{pmatrix} a & a \\ b & b \end{pmatrix}$ **6.** $\begin{pmatrix} 1 & 1 & 1 \\ 0 & 2 & 3 \\ 5 & 5 & 1 \end{pmatrix}$

7. $\begin{pmatrix} 3 & 2 & 1 \\ 0 & 2 & 2 \\ 0 & 0 & -1 \end{pmatrix}$ **8.** $\begin{pmatrix} 1 & 1 & 1 \\ 0 & 1 & 1 \\ 0 & 0 & 1 \end{pmatrix}$ **9.** $\begin{pmatrix} 1 & 6 & 2 \\ -2 & 3 & 5 \\ 7 & 12 & -4 \end{pmatrix}$

10. $\begin{pmatrix} 3 & 1 & 0 \\ 1 & -1 & 2 \\ 1 & 1 & 1 \end{pmatrix}$ **11.** $\begin{pmatrix} 2 & -1 & 4 \\ -1 & 0 & 5 \\ 19 & -7 & 3 \end{pmatrix}$ **12.** $\begin{pmatrix} 1 & 2 & 3 \\ 1 & 1 & 2 \\ 0 & 1 & 2 \end{pmatrix}$

13. $\begin{pmatrix} 1 & 1 & 1 & 1 \\ 1 & 2 & -1 & 2 \\ 1 & -1 & 2 & 1 \\ 1 & 3 & 3 & 2 \end{pmatrix}$ **14.** $\begin{pmatrix} 1 & 0 & 2 & 3 \\ -1 & 1 & 0 & 4 \\ 2 & 1 & -1 & 3 \\ -1 & 0 & 5 & 7 \end{pmatrix}$ **15.** $\begin{pmatrix} 1 & -3 & 0 & -2 \\ 3 & -12 & -2 & -6 \\ -2 & 10 & 2 & 5 \\ -1 & 6 & 1 & 3 \end{pmatrix}$

16. Show that if A, B, and C are invertible matrices, then ABC is invertible and $(ABC)^{-1} = C^{-1}B^{-1}A^{-1}$.

17. If $A_1, A_2, \ldots, A_m$ are invertible $n \times n$ matrices, show that $A_1 A_2 \cdots A_m$ is invertible and calculate its inverse.

18. Show that the matrix $\begin{pmatrix} 3 & 4 \\ -2 & -3 \end{pmatrix}$ is equal to its own inverse.

19. Show that the matrix $\begin{pmatrix} a_{11} & a_{12} \\ a_{21} & a_{22} \end{pmatrix}$ is equal to its own inverse if $A = \pm I$ or if $a_{11} = -a_{22}$ and $a_{21}a_{12} = 1 - a_{11}^2$.

20. Find the output vector $\mathbf{x}$ in the Leontief input–output model if $n = 3$, $\mathbf{e} = \begin{pmatrix} 30 \\ 20 \\ 40 \end{pmatrix}$, and $A = \begin{pmatrix} \frac{1}{5} & \frac{1}{5} & 0 \\ \frac{2}{5} & \frac{2}{5} & \frac{3}{5} \\ \frac{1}{5} & \frac{1}{10} & \frac{2}{5} \end{pmatrix}$.

***21.** Suppose that A is $n \times m$ and B is $m \times n$ so that AB is $n \times n$. Show that AB is not invertible if $n > m$. [*Hint:* Show that there is a nonzero vector $\mathbf{x}$ such that $AB\mathbf{x} = \mathbf{0}$ and then apply Theorem 6.]

***22.** Use the methods of this section to find the inverses of the following matrices with complex entries:

a. $\begin{pmatrix} i & 2 \\ 1 & -i \end{pmatrix}$ **b.** $\begin{pmatrix} 1-i & 0 \\ 0 & 1+i \end{pmatrix}$ **c.** $\begin{pmatrix} 1 & i & 0 \\ -i & 0 & 1 \\ 0 & 1+i & 1-i \end{pmatrix}$

23. Show that for every real number θ the matrix $\begin{pmatrix} \sin\theta & \cos\theta & 0 \\ \cos\theta & -\sin\theta & 0 \\ 0 & 0 & 1 \end{pmatrix}$ is invertible and find its inverse.

24. Calculate the inverse of $A = \begin{pmatrix} 2 & 0 & 0 \\ 0 & 3 & 0 \\ 0 & 0 & 4 \end{pmatrix}$.

25. A square matrix $A = (a_{ij})$ is called **diagonal** if all its elements off the main diagonal are zero. That is, $a_{ij} = 0$ if $i \neq j$. (The matrix of Problem 24 is diagonal.) Show that a diagonal matrix is invertible if and only if each of its diagonal components is nonzero.

26. Let

$$A = \begin{pmatrix} a_{11} & 0 & \cdots & 0 \\ 0 & a_{22} & \cdots & 0 \\ \vdots & & \ddots & \\ 0 & 0 & \cdots & a_{nn} \end{pmatrix}$$

be a diagonal matrix such that each of its diagonal components is nonzero. Calculate A^{-1}.

27. Calculate the inverse of $A = \begin{pmatrix} 2 & 1 & -1 \\ 0 & 3 & 4 \\ 0 & 0 & 5 \end{pmatrix}$.

28. Show that the matrix $A = \begin{pmatrix} 1 & 0 & 0 \\ -2 & 0 & 0 \\ 4 & 6 & 1 \end{pmatrix}$ is not invertible.

***29.** A square matrix is called **upper (lower) triangular** if all its elements below (above) the main diagonal are zero. (The matrix of Problem 27 is upper triangular and the matrix of Problem 28 is lower triangular.) Show that an upper or lower triangular matrix is invertible if and only if each of its diagonal elements is nonzero.

***30.** Show that the inverse of an invertible upper triangular matrix is upper triangular. [*Hint:* First prove the result for a 3×3 matrix.]

In Problems 31 and 32 a matrix is given. In each case show that the matrix is not invertible by finding a nonzero vector **x** such that $A\mathbf{x} = \mathbf{0}$.

31. $\begin{pmatrix} 2 & -1 \\ -4 & 2 \end{pmatrix}$

32. $\begin{pmatrix} 1 & -1 & 3 \\ 0 & 4 & -2 \\ 2 & -6 & 8 \end{pmatrix}$

33. A factory for the construction of quality furniture has two divisions: a machine shop where the parts of the furniture are fabricated, and an assembly and finishing division where the parts are put together into the finished product. Suppose there are 12 employees in the machine shop and 20 in the assembly and finishing division and that each employee works an 8-hour day. Suppose further that the factory produces only two products: chairs and tables. A chair requires $\frac{384}{17}$ hours of machine shop time and $\frac{480}{17}$ hours of assembly and finishing time. A table requires $\frac{240}{17}$ hours of machine shop time and $\frac{640}{17}$ hours of assembly and finishing time. Assuming that there is an unlimited demand for these products and that the manufacturer wishes to keep all employees busy, how many chairs and how many tables can this factory produce each day?
34. A witch's magic cupboard contains 10 oz of ground four-leaf clovers and 14 oz of powdered mandrake root. The cupboard will replenish itself automatically provided she uses up exactly all her supplies. A batch of love potion requires $3\frac{1}{13}$ oz of ground four-leaf clovers and $2\frac{2}{13}$ oz of powdered mandrake root. One recipe of a well-known (to witches) cure for the common cold requires $5\frac{5}{13}$ oz of four-leaf colvers and $10\frac{10}{13}$ oz of mandrake root. How much of the love potion and the cold remedy should the witch make in order to use up the supply in the cupboard exactly?
35. A farmer feeds his cattle a mixture of two types of feed. One standard unit of type A feed supplies a steer with 10% of its minimum daily requirement of protein and 15% of its requirement of carbohydrates. Type B feed contains 12% of the requirement of protein and 8% of the requirement of carbohydrates in a standard unit. If the farmer wishes to feed his cattle exactly 100% of their minimum daily requirement of protein and carbohydrates, how many units of each type of feed should he give a steer each day?
36. A much simplified version of an input-output table for the 1958 Israeli economy divides that economy into three sectors—agriculture, manufacturing, and energy—with the following result.†

	Agriculture	Manufacturing	Energy
Agriculture	0.293	0	0
Manufacturing	0.014	0.207	0.017
Energy	0.044	0.010	0.216

(a) How many units of agricultural production are required to produce one unit of agricultural output?

(b) How many units of agricultural production are required to produce 200,000 units of agricultural output?

† Wassily Leontief, *Input-Output Economics* (New York: Oxford University Press, 1966), 54–57.

(c) How many units of agricultural product go into the production of 50,000 units of energy?

(d) How many units of energy go into the production of 50,000 units of agricultural products?

37. Continuing Problem 36, exports (in thousands of Israeli pounds) in 1958 were

Agriculture	13,213
Manufacturing	17,597
Energy	1,786

(a) Compute the technology and Leontief matrices.

(b) Determine the number of Israeli pounds worth of agricultural products, manufactured goods, and energy required to run this model of the Israeli economy and export the stated value of products.

38. Let $A = \begin{pmatrix} a & b \\ c & d \end{pmatrix}$. Show that $\det A \neq 0$ if and only if the rows of A are linearly independent. [*Hint:* See Theorem 2.6.1, page 54]

In Problems 39–45 compute the row echelon form of the given matrix and use it to determine directly whether the given matrix is invertible.

39. The matrix of Problem 1.
40. The matrix of Problem 4.
41. The matrix of Problem 7.
42. The matrix of Problem 9.
43. The matrix of Problem 11.
44. The matrix of Problem 13.
45. The matrix of Problem 14.

46. Let $A = \begin{pmatrix} a_{11} & a_{12} \\ a_{21} & a_{22} \end{pmatrix}$ and assume that $a_{11}a_{22} - a_{12}a_{21} \neq 0$. Derive formula (12) by row reducing the augmented matrix $\left(\begin{array}{cc|cc} a_{11} & a_{12} & 1 & 0 \\ a_{21} & a_{22} & 0 & 1 \end{array}\right)$.

47. Prove parts (*i*) and (*ii*) of Theorem 5.

2.8 The Transpose of a Matrix

Corresponding to every matrix is another matrix which, as we shall see in Chapter 3, has properties very similar to those of the original matrix.

DEFINITION 1 **TRANSPOSE** Let $A = (a_{ij})$ be an $m \times n$ matrix. Then the **transpose** *of* A, written A^t, is the $n \times m$ matrix obtained by interchanging the rows and columns of A. Succinctly, we may write $A^t = (a_{ji})$. In other words,

$$\text{if } A = \begin{pmatrix} a_{11} & a_{12} & \cdots & a_{1n} \\ a_{21} & a_{22} & \cdots & a_{2n} \\ \vdots & \vdots & & \vdots \\ a_{m1} & a_{m2} & \cdots & a_{mn} \end{pmatrix}, \quad \text{then } A^t = \begin{pmatrix} a_{11} & a_{21} & \cdots & a_{m1} \\ a_{12} & a_{22} & \cdots & a_{m2} \\ \vdots & \vdots & & \vdots \\ a_{1n} & a_{2n} & \cdots & a_{mn} \end{pmatrix} \tag{1}$$

Simply put, the ith row of A is the ith column of A^t and the jth column of A is the jth row of A^t.

EXAMPLE 1 Find the transposes of the matrices

$$A = \begin{pmatrix} 2 & 3 \\ 1 & 4 \end{pmatrix} \quad B = \begin{pmatrix} 2 & 3 & 1 \\ -1 & 4 & 6 \end{pmatrix} \quad C = \begin{pmatrix} 1 & 2 & -6 \\ 2 & -3 & 4 \\ 0 & 1 & 2 \\ 2 & -1 & 5 \end{pmatrix}$$

Solution Interchanging the rows and columns of each matrix, we obtain

$$A^t = \begin{pmatrix} 2 & 1 \\ 3 & 4 \end{pmatrix} \quad B^t = \begin{pmatrix} 2 & -1 \\ 3 & 4 \\ 1 & 6 \end{pmatrix} \quad C^t = \begin{pmatrix} 1 & 2 & 0 & 2 \\ 2 & -3 & 1 & -1 \\ -6 & 4 & 2 & 5 \end{pmatrix}$$

Note, for example, that 4 is the component in row 2 and column 3 of C while 4 is the component in row 3 and column 2 of C^t. That is, the 23 element of C is the 32 element of C^t.

THEOREM 1 Suppose $A = (a_{ij})$ is an $n \times m$ matrix and $B = (b_{ij})$ is an $m \times p$ matrix. Then:

i. $(A^t)^t = A$. (2)
ii. $(AB)^t = B^tA^t$ (3)
iii. If A and B are $n \times m$, then $(A+B)^t = A^t + B^t$. (4)

Proof **i.** This follows directly from the definition of the transpose.
ii. First we note that AB is an $n \times p$ matrix, so $(AB)^t$ is $p \times n$. Also, B^t is $p \times m$ and A^t is $m \times n$, so B^tA^t is $p \times n$. Thus both matrices in Equation (3) have the same size. Now the ijth element of AB is $\sum_{k=1}^{m} a_{ik}b_{kj}$ and this is the jith element of $(AB)^t$. Let $C = B^t$ and $D = A^t$. Then the ijth element c_{ij} of C is b_{ji} and the ijth element d_{ji} of D is a_{ji}. Thus the jith element of CD = the jith element of $B^tA^t = \sum_{k=1}^{m} c_{jk}d_{ki} = \sum_{k=1}^{m} b_{kj}a_{ik} = \sum_{k=1}^{m} a_{ik}b_{kj}$ = the jith element of $(AB)^t$. This completes the proof of part (*ii*).
iii. This part is left as an exercise (see Problem 11). ■

The transpose plays an important role in matrix theory. We shall see in succeeding chapters that A and A^t have many properties in common. Since columns of A^t are rows of A, we shall be able to use facts about the transpose to conclude that just about anything which is true about the rows of a matrix is true about its columns. We conclude this section with an important definition.

DEFINITION 2 **SYMMETRIC MATRIX** The $n \times n$ (square) matrix A is called **symmetric** if $A^t = A$.

EXAMPLE 2 The following four matrices are symmetric:

$$I \qquad A = \begin{pmatrix} 1 & 2 \\ 2 & 3 \end{pmatrix} \qquad B = \begin{pmatrix} 1 & -4 & 2 \\ -4 & 7 & 5 \\ 2 & 5 & 0 \end{pmatrix} \qquad C = \begin{pmatrix} -1 & 2 & 4 & 6 \\ 2 & 7 & 3 & 5 \\ 4 & 3 & 8 & 0 \\ 6 & 5 & 0 & -4 \end{pmatrix}$$

We shall see the importance of symmetric matrices in Chapters 6 and 7.

PROBLEMS 2.8 In Problems 1–10 find the transpose of the given matrix.

1. $\begin{pmatrix} -1 & 4 \\ 6 & 5 \end{pmatrix}$ **2.** $\begin{pmatrix} 3 & 0 \\ 1 & 2 \end{pmatrix}$ **3.** $\begin{pmatrix} 2 & 3 \\ -1 & 2 \\ 1 & 4 \end{pmatrix}$ **4.** $\begin{pmatrix} 2 & -1 & 0 \\ 1 & 5 & 6 \end{pmatrix}$

5. $\begin{pmatrix} 1 & 2 & 3 \\ -1 & 0 & 4 \\ 1 & 5 & 5 \end{pmatrix}$ **6.** $\begin{pmatrix} 1 & 2 & 3 \\ 2 & 4 & -5 \\ 3 & -5 & 7 \end{pmatrix}$ **7.** $\begin{pmatrix} 1 & 0 & 1 & 0 \\ 0 & 1 & 0 & 1 \end{pmatrix}$ **8.** $\begin{pmatrix} 2 & -1 \\ 2 & 4 \\ 1 & 6 \\ 1 & 5 \end{pmatrix}$

9. $\begin{pmatrix} a & b & c \\ d & e & f \\ g & h & j \end{pmatrix}$ **10.** $\begin{pmatrix} 0 & 0 & 0 \\ 0 & 0 & 0 \end{pmatrix}$

11. Let A and B be $n \times m$ matrices. Show, using Definition 1, that $(A+B)^t = A^t + B^t$.

12. Find numbers α and β such that $\begin{pmatrix} 2 & \alpha & 3 \\ 5 & -6 & 2 \\ \beta & 2 & 4 \end{pmatrix}$ is symmetric.

13. If A and B are symmetric $n \times n$ matrices, prove that $A+B$ is symmetric.

14. If A and B are symmetric $n \times n$ matrices, show that $(AB)^t = BA$.

15. For any matrix A, show that the product matrix AA^t is defined and is a symmetric matrix.

16. Show that every diagonal matrix (see Problem 2.7.25, page 75) is symmetric.

17. Show that the transpose of every upper triangular matrix (see Problem 2.7.29) is lower triangular.

18. A square matrix is called **skew-symmetric** if $A^t = -A$ (that is, $a_{ij} = -a_{ji}$). Which of the following matrices are skew-symmetric?

a. $\begin{pmatrix} 1 & -6 \\ 6 & 0 \end{pmatrix}$ **b.** $\begin{pmatrix} 0 & -6 \\ 6 & 0 \end{pmatrix}$ **c.** $\begin{pmatrix} 2 & -2 & -2 \\ 2 & 2 & -2 \\ 2 & 2 & 2 \end{pmatrix}$ **d.** $\begin{pmatrix} 0 & 1 & -1 \\ -1 & 0 & 2 \\ 1 & -2 & 0 \end{pmatrix}$

19. Let A and B be $n \times n$ skew-symmetric matrices. Show that $A+B$ is skew-symmetric.

20. If A is skew-symmetric, show that every component on the main diagonal of A is zero.

21. If A and B are skew-symmetric $n \times n$ matrices, show that $(AB)^t = BA$, so that AB is symmetric if and only if A and B commute.

***22.** Let $A = \begin{pmatrix} a_{11} & a_{12} \\ a_{21} & a_{22} \end{pmatrix}$ be a matrix with nonnegative entries having the properties that (*i*) $a_{11}^2 + a_{21}^2 = 1$ and $a_{12}^2 + a_{22}^2 = 1$ and (*ii*) $\begin{pmatrix} a_{11} \\ a_{21} \end{pmatrix} \cdot \begin{pmatrix} a_{12} \\ a_{22} \end{pmatrix} = 0$. Show that A is invertible and that $A^{-1} = A^t$.

Review Exercises for Chapter 2

In Exercises 1–7 compute using $\mathbf{a} = \begin{pmatrix} -1 \\ 4 \\ 6 \end{pmatrix}$, $\mathbf{b} = \begin{pmatrix} 5 \\ 2 \\ -1 \end{pmatrix}$, and $\mathbf{c} = \begin{pmatrix} -1 \\ 3 \\ 6 \end{pmatrix}$.

1. $\mathbf{a}+\mathbf{b}$ **2.** $2\mathbf{a}-3\mathbf{c}$ **3.** $\mathbf{a}+\mathbf{b}-2\mathbf{c}$ **4.** $-4\mathbf{a}+3\mathbf{b}+5\mathbf{c}$

5. $\mathbf{a}\cdot\mathbf{b}$ **6.** $\mathbf{b}\cdot\mathbf{c}$ **7.** $(\mathbf{a}-\mathbf{b})\cdot(2\mathbf{c}-3\mathbf{b})$

8. Find a number α such that $\begin{pmatrix} 2 \\ -1 \\ 4 \\ 6 \end{pmatrix}$ and $\begin{pmatrix} 1 \\ 5 \\ \alpha \\ 4 \end{pmatrix}$ are orthogonal.

In Exercises 9–16 perform the indicated computations.

9. $3\begin{pmatrix} -2 & 1 \\ 0 & 4 \\ 2 & 3 \end{pmatrix}$

10. $\begin{pmatrix} 1 & 0 & 3 \\ 2 & -1 & 6 \end{pmatrix} + \begin{pmatrix} 2 & 0 & 4 \\ -2 & 5 & 8 \end{pmatrix}$

11. $5\begin{pmatrix} 2 & 1 & 3 \\ -1 & 2 & 4 \\ -6 & 1 & 5 \end{pmatrix} - 3\begin{pmatrix} -2 & 1 & 4 \\ 5 & 0 & 7 \\ 2 & -1 & 3 \end{pmatrix}$

12. $\begin{pmatrix} 2 & 3 \\ -1 & 4 \end{pmatrix}\begin{pmatrix} 5 & -1 \\ 2 & 7 \end{pmatrix}$

13. $\begin{pmatrix} 2 & 3 & 1 & 5 \\ 0 & 6 & 2 & 4 \end{pmatrix}\begin{pmatrix} 5 & 7 & 1 \\ 2 & 0 & 3 \\ 1 & 0 & 0 \\ 0 & 5 & 6 \end{pmatrix}$

14. $\begin{pmatrix} 2 & 3 & 5 \\ -1 & 6 & 4 \\ 1 & 0 & 6 \end{pmatrix}\begin{pmatrix} 0 & -1 & 2 \\ 3 & 1 & 2 \\ -7 & 3 & 5 \end{pmatrix}$

15. $\begin{pmatrix} 1 & 0 & 3 & -1 & 5 \\ 2 & 1 & 6 & 2 & 5 \end{pmatrix}\begin{pmatrix} 7 & 1 \\ 2 & 3 \\ -1 & 0 \\ 5 & 6 \\ 2 & 3 \end{pmatrix}$

16. $\begin{pmatrix} 1 & -1 & 2 \\ 3 & 5 & 6 \\ 2 & 4 & -1 \end{pmatrix}\begin{pmatrix} 2 \\ 1 \\ 3 \end{pmatrix}$

17. Verify the associative law of matrix multiplication for the matrices $A=\begin{pmatrix}2&3&1\\0&4&6\end{pmatrix}$, $B=\begin{pmatrix}1&0&2\\0&3&3\\5&1&-1\end{pmatrix}$, and $C=\begin{pmatrix}5&6\\-1&2\\0&1\end{pmatrix}$.

In Exercises 18–22 determine whether the given set of vectors is linearly dependent or independent.

18. $\begin{pmatrix}2\\3\end{pmatrix}$; $\begin{pmatrix}4\\-6\end{pmatrix}$

19. $\begin{pmatrix}2\\3\end{pmatrix}$; $\begin{pmatrix}4\\6\end{pmatrix}$

20. $\begin{pmatrix}1\\-1\\2\end{pmatrix}$; $\begin{pmatrix}3\\0\\1\end{pmatrix}$; $\begin{pmatrix}0\\0\\0\end{pmatrix}$

21. $\begin{pmatrix}1\\-4\\2\end{pmatrix}$; $\begin{pmatrix}0\\2\\-1\end{pmatrix}$; $\begin{pmatrix}2\\-10\\5\end{pmatrix}$

22. $\begin{pmatrix}1\\0\\0\\0\end{pmatrix}$; $\begin{pmatrix}0\\1\\0\\0\end{pmatrix}$; $\begin{pmatrix}0\\0\\1\\0\end{pmatrix}$; $\begin{pmatrix}0\\0\\0\\1\end{pmatrix}$

In Exercises 23–27 calculate the row echelon form and the inverse of the given matrix (if the inverse exists).

23. $\begin{pmatrix}2&3\\-1&4\end{pmatrix}$

24. $\begin{pmatrix}-1&2\\2&-4\end{pmatrix}$

25. $\begin{pmatrix}1&2&0\\2&1&-1\\3&1&1\end{pmatrix}$

26. $\begin{pmatrix}-1&2&0\\4&1&-3\\2&5&-3\end{pmatrix}$

27. $\begin{pmatrix}2&0&4\\-1&3&1\\0&1&2\end{pmatrix}$

In Exercises 28–30 first write the system in the form $A\mathbf{x}=\mathbf{b}$, then calculate A^{-1}, and, finally, use matrix multiplication to obtain the solution vector.

28. $\begin{aligned}x_1-3x_2&=4\\2x_1+5x_2&=7\end{aligned}$

29. $\begin{aligned}x_1+2x_2\phantom{{}-x_3}&=3\\2x_1+x_2-x_3&=-1\\3x_1+x_2+x_3&=7\end{aligned}$

30. $\begin{aligned}2x_1\phantom{{}+3x_2}+4x_3&=7\\-x_1+3x_2+x_3&=-4\\x_2+2x_3&=5\end{aligned}$

In Exercises 31–36 calculate the transpose of the given matrix and determine whether the matrix is symmetric or skew-symmetric.†

31. $\begin{pmatrix}2&3&1\\-1&0&2\end{pmatrix}$

32. $\begin{pmatrix}4&6\\6&4\end{pmatrix}$

33. $\begin{pmatrix}2&3&1\\3&-6&-5\\1&-5&9\end{pmatrix}$

34. $\begin{pmatrix}0&5&6\\-5&0&4\\-6&-4&0\end{pmatrix}$

35. $\begin{pmatrix}1&-1&4&6\\-1&2&5&7\\4&5&3&-8\\6&7&-8&9\end{pmatrix}$

36. $\begin{pmatrix}0&1&-1&1\\-1&0&1&-2\\1&1&0&1\\1&-2&-1&0\end{pmatrix}$

† From Problem 2.8.18 on page 80 we have: A is skew-symmetric if $A^t=-A$.

3 Determinants

3.1 Definitions

Let $A = \begin{pmatrix} a_{11} & a_{12} \\ a_{21} & a_{22} \end{pmatrix}$ be a 2 × 2 matrix. In Section 2.7 on page 65 we defined the determinant of A by

$$\det A = a_{11}a_{22} - a_{12}a_{21} \tag{1}$$

We shall often denote det A by

$$|A| = \begin{vmatrix} a_{11} & a_{12} \\ a_{21} & a_{22} \end{vmatrix} \tag{2}$$

We showed that A is invertible if and only if $\det A \neq 0$. As we shall see, this important theorem is valid for $n \times n$ matrices.

In this chapter we shall develop some of the basic properties of determinants and see how they can be used to calculate inverses and solve systems of n linear equations in n unknowns.

We shall define the determinant of an $n \times n$ matrix *inductively*. In other words, we use our knowledge of a 2×2 determinant to define a 3×3 determinant, use this to define a 4×4 determinant, and so on. We start by defining a 3×3 determinant.†

DEFINITION 1 **3 × 3 DETERMINANT** Let $A = \begin{pmatrix} a_{11} & a_{12} & a_{13} \\ a_{21} & a_{22} & a_{23} \\ a_{31} & a_{32} & a_{33} \end{pmatrix}$. Then

$$\det A = |A| = a_{11}\begin{vmatrix} a_{22} & a_{23} \\ a_{32} & a_{33} \end{vmatrix} - a_{12}\begin{vmatrix} a_{21} & a_{23} \\ a_{31} & a_{33} \end{vmatrix} + a_{13}\begin{vmatrix} a_{21} & a_{22} \\ a_{31} & a_{32} \end{vmatrix} \tag{3}$$

Note the minus sign before the second term on the right side of (3).

† There are several ways to define a determinant and this is one of them. It is important to realize that "det" is a function which assigns a *number* to a *square* matrix.

EXAMPLE 1 Let $A = \begin{pmatrix} 3 & 5 & 2 \\ 4 & 2 & 3 \\ -1 & 2 & 4 \end{pmatrix}$. Calculate $|A|$.

Solution

$$|A| = \begin{vmatrix} 3 & 5 & 2 \\ 4 & 2 & 3 \\ -1 & 2 & 4 \end{vmatrix} = 3\begin{vmatrix} 2 & 3 \\ 2 & 4 \end{vmatrix} - 5\begin{vmatrix} 4 & 3 \\ -1 & 4 \end{vmatrix} + 2\begin{vmatrix} 4 & 2 \\ -1 & 2 \end{vmatrix}$$
$$= 3 \cdot 2 - 5 \cdot 19 + 2 \cdot 10 = -69$$

EXAMPLE 2 Calculate $\begin{vmatrix} 2 & -3 & 5 \\ 1 & 0 & 4 \\ 3 & -3 & 9 \end{vmatrix}$.

Solution

$$\begin{vmatrix} 2 & -3 & 5 \\ 1 & 0 & 4 \\ 3 & -3 & 9 \end{vmatrix} = 2\begin{vmatrix} 0 & 4 \\ -3 & 9 \end{vmatrix} - (-3)\begin{vmatrix} 1 & 4 \\ 3 & 9 \end{vmatrix} + 5\begin{vmatrix} 1 & 0 \\ 3 & -3 \end{vmatrix}$$
$$= 2 \cdot 12 + 3(-3) + 5(-3) = 0$$

There is a simpler method for calculating 3×3 determinants. From Equation (3) we have

$$\begin{vmatrix} a_{11} & a_{12} & a_{13} \\ a_{21} & a_{22} & a_{23} \\ a_{31} & a_{32} & a_{33} \end{vmatrix} = \begin{aligned} &a_{11}(a_{22}a_{33} - a_{23}a_{32}) - a_{12}(a_{21}a_{33} - a_{23}a_{31}) \\ &+ a_{13}(a_{21}a_{32} - a_{22}a_{31}) \end{aligned}$$

or

$$|A| = a_{11}a_{22}a_{33} + a_{12}a_{23}a_{31} + a_{13}a_{21}a_{32} - a_{13}a_{22}a_{31} - a_{12}a_{21}a_{33} - a_{11}a_{32}a_{23} \tag{4}$$

We write A and adjoin to it its first two columns

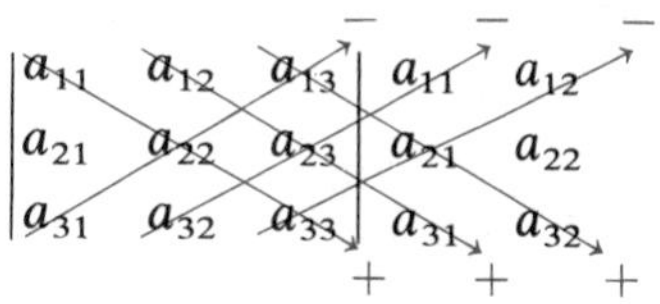

We then calculate the six products, put minus signs before the products with arrows pointing upward, and add. This gives the sum in Equation (4).

EXAMPLE 3 Calculate $\begin{vmatrix} 3 & 5 & 2 \\ 4 & 2 & 3 \\ -1 & 2 & 4 \end{vmatrix}$ by using this new method.

Solution Writing $\begin{vmatrix} 3 & 5 & 2 \\ 4 & 2 & 3 \\ -1 & 2 & 4 \end{vmatrix}\begin{matrix} 3 & 5 \\ 4 & 2 \\ -1 & 2 \end{matrix}$ and multiplying as indicated, we obtain

$$|A| = (3)(2)(4) + (5)(3)(-1) + (2)(4)(2) - (-1)(2)(2) - 2(3)(3) - (4)(4)(5)$$
$$= 24 - 15 + 16 + 4 - 18 - 80 = -69.$$

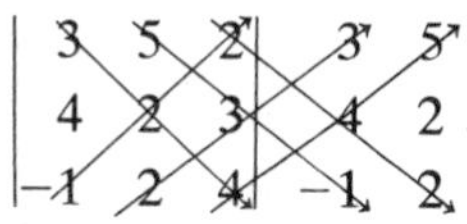

Warning. The method given above will *not* work for $n \times n$ determinants if $n \neq 3$. If you try something analogous for 4×4 or higher-order determinants, you will get the wrong answer.

Before defining $n \times n$ determinants, we first note that in Equation (3), $\begin{pmatrix} a_{22} & a_{23} \\ a_{32} & a_{33} \end{pmatrix}$ is the matrix obtained by deleting the first row and first column of A; $\begin{pmatrix} a_{21} & a_{23} \\ a_{31} & a_{33} \end{pmatrix}$ is the matrix obtained by deleting the first row and second column of A; and $\begin{pmatrix} a_{21} & a_{22} \\ a_{31} & a_{32} \end{pmatrix}$ is the matrix obtained by deleting the first row and third column of A. If we denote these three matrices by M_{11}, M_{12}, and M_{13}, respectively, and if $A_{11} = \det M_{11}$, $A_{12} = -\det M_{12}$, and $A_{13} = \det M_{13}$, then Equation (3) can be written

$$\det A = |A| = a_{11}A_{11} + a_{12}A_{12} + a_{13}A_{13} \tag{5}$$

DEFINITION 2 **MINOR** Let A be an $n \times n$ matrix and let M_{ij} be the $(n-1) \times (n-1)$ matrix obtained from A by deleting the ith row and jth column of A. M_{ij} is called the **ijth minor** of A.

EXAMPLE 4 Let $A = \begin{pmatrix} 2 & -1 & 4 \\ 0 & 1 & 5 \\ 6 & 3 & -4 \end{pmatrix}$. Find M_{13} and M_{32}.

Solution Deleting the first row and third column of A, we obtain $M_{13} = \begin{pmatrix} 0 & 1 \\ 6 & 3 \end{pmatrix}$. Similarly, by eliminating the third row and second column we obtain $M_{32} = \begin{pmatrix} 2 & 4 \\ 0 & 5 \end{pmatrix}$.

EXAMPLE 5 Let $A = \begin{pmatrix} 1 & -3 & 5 & 6 \\ 2 & 4 & 0 & 3 \\ 1 & 5 & 9 & -2 \\ 4 & 0 & 2 & 7 \end{pmatrix}$ Find M_{32} and M_{24}.

Solution Deleting the third row and second column of A, we find that $M_{32} = \begin{pmatrix} 1 & 5 & 6 \\ 2 & 0 & 3 \\ 4 & 2 & 7 \end{pmatrix}$; similarly, $M_{24} = \begin{pmatrix} 1 & -3 & 5 \\ 1 & 5 & 9 \\ 4 & 0 & 2 \end{pmatrix}$.

DEFINITION 3 **COFACTOR** Let A be an $n \times n$ matrix. The **ijth cofactor** of A, denoted A_{ij}, is given by

$$A_{ij} = (-1)^{i+j}|M_{ij}| \tag{6}$$

That is, the ijth cofactor of A is obtained by taking the determinant of the ijth minor and multiplying it by $(-1)^{i+j}$. Note that

$$(-1)^{i+j} = \begin{cases} 1 & \text{if } i+j \text{ is even} \\ -1 & \text{if } i+j \text{ is odd} \end{cases}$$

Remark. Definition 3 makes sense because we are going to define an $n \times n$ determinant with the assumption that we already know what an $(n-1)\times(n-1)$ determinant is.

EXAMPLE 6 In Example 5 we have

$$A_{32} = (-1)^{3+2}|M_{32}| = -\begin{vmatrix} 1 & 5 & 6 \\ 2 & 0 & 3 \\ 4 & 2 & 7 \end{vmatrix} = -8$$

$$A_{24} = (-1)^{2+4}\begin{vmatrix} 1 & -3 & 5 \\ 1 & 5 & 9 \\ 4 & 0 & 2 \end{vmatrix} = -192$$

We now consider the general $n \times n$ matrix. Here

$$A = \begin{pmatrix} a_{11} & a_{12} & \cdots & a_{1n} \\ a_{21} & a_{22} & \cdots & a_{2n} \\ \vdots & \vdots & & \vdots \\ a_{n1} & a_{n2} & \cdots & a_{nn} \end{pmatrix} \tag{7}$$

DEFINITION 4 **$n \times n$ DETERMINANT** Let A be an $n \times n$ matrix. Then the determinant of A, written det A or $|A|$, is given by

$$\begin{aligned} \det A = |A| &= a_{11}A_{11} + a_{12}A_{12} + a_{13}A_{13} + \cdots + a_{1n}A_{1n} \\ &= \sum_{k=1}^{n} a_{1k}A_{1k} \end{aligned} \tag{8}$$

The expression on the right side of (8) is called an **expansion of cofactors.**

In Equation (8) we defined the determinant by expanding by cofactors using components of A in the first row. We shall see in the next section (Theorem 3.2.1) that we get the same answer if we expand by cofactors in any row or column.

EXAMPLE 7 Calculate det A, where

$$A=\begin{pmatrix}1 & 3 & 5 & 2\\ 0 & -1 & 3 & 4\\ 2 & 1 & 9 & 6\\ 3 & 2 & 4 & 8\end{pmatrix}$$

Solution

$$\begin{vmatrix}1 & 3 & 5 & 2\\ 0 & -1 & 3 & 4\\ 2 & 1 & 9 & 6\\ 3 & 2 & 4 & 8\end{vmatrix}=a_{11}A_{11}+a_{12}A_{12}+a_{13}A_{13}+a_{14}A_{14}$$

$$=1\begin{vmatrix}-1 & 3 & 4\\ 1 & 9 & 6\\ 2 & 4 & 8\end{vmatrix}-3\begin{vmatrix}0 & 3 & 4\\ 2 & 9 & 6\\ 3 & 4 & 8\end{vmatrix}+5\begin{vmatrix}0 & -1 & 4\\ 2 & 1 & 6\\ 3 & 2 & 8\end{vmatrix}-2\begin{vmatrix}0 & -1 & 3\\ 2 & 1 & 9\\ 3 & 2 & 4\end{vmatrix}$$

$$=1(-92)-3(-70)+5(2)-2(-16)=160$$

It is clear that calculating the determinant of an $n\times n$ matrix can be tedious. To calculate a 4×4 determinant, we must calculate four 3×3 determinants. To calculate a 5×5 determinant, we must calculate five 4×4 determinants—which is the same as calculating twenty 3×3 determinants. Fortunately, techniques exist for greatly simplifying these computations. Some of these methods are discussed in the next section. There are, however, some matrices whose determinants can easily be calculated.

DEFINITION 5 A square matrix is called **upper triangular** if all its components below the diagonal are zero. It is **lower triangular** if all its components above the diagonal are zero. A matrix is called **diagonal** if all its elements not on the diagonal are zero; that is, $A=(a_{ij})$ is upper triangular if $a_{ij}=0$ for $i>j$, lower triangular if $a_{ij}=0$ for $i<j$, and diagonal if $a_{ij}=0$ for $i\neq j$. Note that a diagonal matrix is both upper and lower triangular.

EXAMPLE 8 The matrices $A=\begin{pmatrix}2 & 1 & 7\\ 0 & 2 & -5\\ 0 & 0 & 1\end{pmatrix}$ and $B=\begin{pmatrix}-2 & 3 & 0 & 1\\ 0 & 0 & 2 & 4\\ 0 & 0 & 1 & 3\\ 0 & 0 & 0 & -2\end{pmatrix}$ are upper triangular; $C=\begin{pmatrix}5 & 0 & 0\\ 2 & 3 & 0\\ -1 & 2 & 4\end{pmatrix}$ and $D=\begin{pmatrix}0 & 0\\ 1 & 0\end{pmatrix}$ are lower triangular; I and $E=\begin{pmatrix}2 & 0 & 0\\ 0 & -7 & 0\\ 0 & 0 & -4\end{pmatrix}$ are diagonal.

EXAMPLE 9 Let

$$A=\begin{pmatrix} a_{11} & 0 & 0 & 0 \\ a_{21} & a_{22} & 0 & 0 \\ a_{31} & a_{32} & a_{33} & 0 \\ a_{41} & a_{42} & a_{43} & a_{44} \end{pmatrix}$$

be lower triangular. Compute det A.

Solution

$$\det A = a_{11}A_{11}+0A_{12}+0A_{13}+0A_{14}=a_{11}A_{11}$$

$$= a_{11}\begin{vmatrix} a_{22} & 0 & 0 \\ a_{32} & a_{33} & 0 \\ a_{42} & a_{43} & a_{44} \end{vmatrix}$$

$$= a_{11}a_{22}\begin{vmatrix} a_{33} & 0 \\ a_{43} & a_{44} \end{vmatrix}$$

$$= a_{11}a_{22}a_{33}a_{44}$$

Example 9 can easily be generalized to prove the following.

THEOREM 1 Let $A=(a_{ij})$ be an upper† or lower triangular $n\times n$ matrix. Then

$$\det A = a_{11}a_{22}a_{33}\cdots a_{nn} \tag{9}$$

That is: The determinant of a triangular matrix equals the product of its diagonal components.

EXAMPLE 10 The determinants of the six matrices in Example 8 are $|A|=2\cdot 2\cdot 1=4$; $|B|=(-2)(0)(1)(-2)=0$; $|C|=5\cdot 3\cdot 4=60$; $|D|=0$; $|I|=1$; $|E|=(2)(-7)(-4)=56$.

PROBLEMS 3.1 In Problems 1–10 calculate the determinant.

1. $\begin{vmatrix} 1 & 0 & 3 \\ 0 & 1 & 4 \\ 2 & 1 & 0 \end{vmatrix}$ **2.** $\begin{vmatrix} -1 & 1 & 0 \\ 2 & 1 & 4 \\ 1 & 5 & 6 \end{vmatrix}$ **3.** $\begin{vmatrix} 3 & -1 & 4 \\ 6 & 3 & 5 \\ 2 & -1 & 6 \end{vmatrix}$

4. $\begin{vmatrix} -1 & 0 & 6 \\ 0 & 2 & 4 \\ 1 & 2 & -3 \end{vmatrix}$ **5.** $\begin{vmatrix} -2 & 3 & 1 \\ 4 & 6 & 5 \\ 0 & 2 & 1 \end{vmatrix}$ **6.** $\begin{vmatrix} 5 & -2 & 1 \\ 6 & 0 & 3 \\ -2 & 1 & 4 \end{vmatrix}$

† The proof for the upper triangular case is more difficult at this stage, but it will be just the same once we know that det A can be evaluated by expanding in any column (Theorem 3.2.1).

7. $\begin{vmatrix} 2 & 0 & 3 & 1 \\ 0 & 1 & 4 & 2 \\ 0 & 0 & 1 & 5 \\ 1 & 2 & 3 & 0 \end{vmatrix}$ **8.** $\begin{vmatrix} -3 & 0 & 0 & 0 \\ -4 & 7 & 0 & 0 \\ 5 & 8 & -1 & 0 \\ 2 & 3 & 0 & 6 \end{vmatrix}$ **9.** $\begin{vmatrix} -2 & 0 & 0 & 7 \\ 1 & 2 & -1 & 4 \\ 3 & 0 & -1 & 5 \\ 4 & 2 & 3 & 0 \end{vmatrix}$

10. $\begin{vmatrix} 2 & 3 & -1 & 4 & 5 \\ 0 & 1 & 7 & 8 & 2 \\ 0 & 0 & 4 & -1 & 5 \\ 0 & 0 & 0 & -2 & 8 \\ 0 & 0 & 0 & 0 & 6 \end{vmatrix}$

11. Show that if A and B are diagonal $n \times n$ matrices, then $\det AB = \det A \det B$.

***12.** Show that if A and B are lower triangular matrices, then $\det AB = \det A \det B$.

13. Show that, in general, it is not true that $\det(A+B) = \det A + \det B$.

14. Show that if A is triangular, then $\det A \neq 0$ if and only if all the diagonal components of A are nonzero.

15. Prove Theorem 1 for a lower triangular matrix.

***16.** We say that the vectors $\begin{pmatrix} 1 \\ 0 \end{pmatrix}$ and $\begin{pmatrix} 0 \\ 1 \end{pmatrix}$ *generate the area* 1 in the plane since if we construct a square with three of its vertices at (0, 0), (1, 0), and (0, 1), we see that the area is 1. (See Figure 3.1*a*.) More generally, if $\begin{pmatrix} x_1 \\ y_1 \end{pmatrix}$ and $\begin{pmatrix} x_2 \\ y_2 \end{pmatrix}$ are two linearly independent 2-vectors, then they generate an area defined to be the area of the parallelogram with three of its four vertices at $(0, 0)$, (x_1, y_1), and (x_2, y_2). (See Figure 3.1*b*.)

Figure 3.1

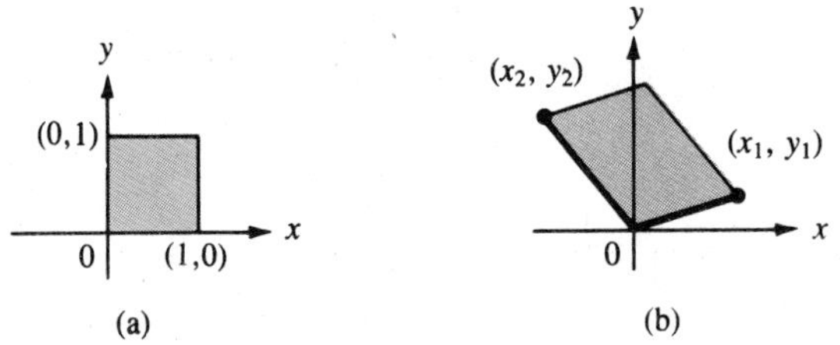

Let A be a 2×2 matrix. If k denotes the area generated by $\begin{pmatrix} x_1 \\ y_1 \end{pmatrix}$ and $\begin{pmatrix} x_2 \\ y_2 \end{pmatrix}$, where $\begin{pmatrix} x_1 \\ y_1 \end{pmatrix} = A\begin{pmatrix} 1 \\ 0 \end{pmatrix}$ and $\begin{pmatrix} x_2 \\ y_2 \end{pmatrix} = A\begin{pmatrix} 0 \\ 1 \end{pmatrix}$, show that $k = |\det A|$.

****17.** Let $\mathbf{u}_1$ and $\mathbf{u}_2$ be two 2-vectors and let $\mathbf{v}_1 = A\mathbf{u}_1$ and $\mathbf{v}_2 = A\mathbf{u}_2$. Show that

$$(\text{area generated by } \mathbf{v}_1 \text{ and } \mathbf{v}_2) = (\text{area generated by } \mathbf{u}_1 \text{ and } \mathbf{u}_2)\,|\det A|$$

This provides a geometric interpretation of the determinant.

3.2 Properties of Determinants

Determinants have many properties that can make computations easier. We begin to describe these properties by stating a theorem from which everything else follows. The proof of this theorem is difficult and is deferred to the next section.

THEOREM 1 **Basic Theorem.** Let

$$A = \begin{pmatrix} a_{11} & a_{12} & \cdots & a_{1n} \\ a_{21} & a_{22} & \cdots & a_{2n} \\ \vdots & \vdots & & \vdots \\ a_{n1} & a_{n2} & \cdots & a_{nn} \end{pmatrix}$$

be an $n \times n$ matrix. Then

$$\det A = a_{i1}A_{i1} + a_{i2}A_{i2} + \cdots + a_{in}A_{in} = \sum_{k=1}^{n} a_{ik}A_{ik} \tag{1}$$

for $i = 1, 2, \ldots, n$. That is, we can calculate det A by expanding by cofactors in *any* row of A. Furthermore:

$$\det A = a_{1j}A_{1j} + a_{2j}A_{2j} + \cdots + a_{nj}A_{nj} = \sum_{k=1}^{n} a_{kj}A_{kj} \tag{2}$$

Since the jth column of A is $\begin{pmatrix} a_{1j} \\ a_{2j} \\ \vdots \\ a_{nj} \end{pmatrix}$, Equation (2) indicates that we can calculate det A by expanding by cofactors in any column of A.

EXAMPLE 1 For $A = \begin{pmatrix} 3 & 5 & 2 \\ 4 & 2 & 3 \\ -1 & 2 & 4 \end{pmatrix}$, we saw in Example 3.1.1 that det $A = -69$. Expanding in the second row we obtain

$$\begin{aligned} \det A &= 4A_{21} + 2A_{22} + 3A_{23} \\ &= 4(-1)^{2+1}\begin{vmatrix} 5 & 2 \\ 2 & 4 \end{vmatrix} + 2(-1)^{2+2}\begin{vmatrix} 3 & 2 \\ -1 & 4 \end{vmatrix} + 3(-1)^{2+3}\begin{vmatrix} 3 & 5 \\ -1 & 2 \end{vmatrix} \\ &= -4(16) + 2(14) - 3(11) = -69 \end{aligned}$$

Similarly, if we expand in the third column, say, we obtain

$$\begin{aligned} \det A &= 2A_{13} + 3A_{23} + 4A_{33} \\ &= 2(-1)^{1+3}\begin{vmatrix} 4 & 2 \\ -1 & 2 \end{vmatrix} + 3(-1)^{2+3}\begin{vmatrix} 3 & 5 \\ -1 & 2 \end{vmatrix} + 4(-1)^{3+3}\begin{vmatrix} 3 & 5 \\ 4 & 2 \end{vmatrix} \\ &= 2(10) - 3(11) + 4(-14) = -69 \end{aligned}$$

You should verify that we get the same answer if we expand in the third row or the first or second column.

We now list and prove some additional properties of determinants. In each case we assume that A is an $n \times n$ matrix.† We shall see that these properties can be used to reduce greatly the work involved in evaluating a determinant.

Property 1 If any row or column of A is the zero vector, then $\det A = 0$.

Proof Suppose the ith row of A contains all zeros. That is, $a_{ij} = 0$ for $j = 1, 2, \ldots, n$. Then $\det A = a_{i1}A_{i1} + a_{i2}A_{i2} + \cdots + a_{in}A_{in} = 0 + 0 + \cdots + 0 = 0$. The same proof works if the jth column is the zero vector. ■

EXAMPLE 2 It is easy to verify that

$$\begin{vmatrix} 2 & 3 & 5 \\ 0 & 0 & 0 \\ 1 & -2 & 4 \end{vmatrix} = 0 \quad \text{and} \quad \begin{vmatrix} -1 & 3 & 0 & 1 \\ 4 & 2 & 0 & 5 \\ -1 & 6 & 0 & 4 \\ 2 & 1 & 0 & 1 \end{vmatrix} = 0$$

Property 2 If the ith row or the jth column of A is multiplied by the constant c, then $\det A$ is multiplied by c. That is, if we call this new matrix B, then

$$|B| = \begin{vmatrix} a_{11} & a_{12} & \cdots & a_{1n} \\ a_{21} & a_{22} & \cdots & a_{2n} \\ \vdots & \vdots & & \vdots \\ ca_{i1} & ca_{i2} & \cdots & ca_{in} \\ \vdots & \vdots & & \vdots \\ a_{n1} & a_{n2} & \cdots & a_{nn} \end{vmatrix} = c \begin{vmatrix} a_{11} & a_{12} & \cdots & a_{1n} \\ a_{21} & a_{22} & \cdots & a_{2n} \\ \vdots & \vdots & & \vdots \\ a_{i1} & a_{i2} & \cdots & a_{in} \\ \vdots & \vdots & & \vdots \\ a_{n1} & a_{n2} & \cdots & a_{nn} \end{vmatrix} = c|A| \qquad (3)$$

Proof To prove (3) we expand in the ith row of A to obtain

$$\begin{aligned} \det B &= ca_{i1}A_{i1} + ca_{i2}A_{i2} + \cdots ca_{in}A_{in} \\ &= c(a_{i1}A_{i1} + a_{i2}A_{i2} + \cdots + a_{in}A_{in}) = c \det A \end{aligned}$$

A similar proof works for columns. ■

† The proofs of these properties are given in terms of the rows of a matrix. Using Theorem 1 the same properties can be proved for columns.

EXAMPLE 3 Let $A = \begin{pmatrix} 1 & -1 & 2 \\ 3 & 1 & 4 \\ 0 & -2 & 5 \end{pmatrix}$. Then $\det A = 16$. If we multiply the second row by 4, we have $B = \begin{pmatrix} 1 & -1 & 2 \\ 12 & 4 & 16 \\ 0 & -2 & 5 \end{pmatrix}$ and $\det B = 64 = 4 \det A$. If the third column is multiplied by -3, we obtain $C = \begin{pmatrix} 1 & -1 & -6 \\ 3 & 1 & -12 \\ 0 & -2 & -15 \end{pmatrix}$ and $\det C = -48 = -3 \det A$.

Remark. Using Property 2 we can prove (see Problem 28) the following interesting fact: For any scalar α and $n \times n$ matrix A, $\det \alpha A = \alpha^n \det A$.

Property 3 Let

$$A = \begin{pmatrix} a_{11} & a_{12} & \cdots & a_{1j} & \cdots & a_{1n} \\ a_{21} & a_{22} & \cdots & a_{2j} & \cdots & a_{2n} \\ \vdots & \vdots & & \vdots & & \vdots \\ a_{n1} & a_{n2} & \cdots & a_{nj} & \cdots & a_{nn} \end{pmatrix}, \quad B = \begin{pmatrix} a_{11} & a_{12} & \cdots & \alpha_{1j} & \cdots & a_{1n} \\ a_{21} & a_{22} & \cdots & \alpha_{2j} & \cdots & a_{2n} \\ \vdots & \vdots & & \vdots & & \vdots \\ a_{n1} & a_{n2} & \cdots & \alpha_{nj} & \cdots & a_{nn} \end{pmatrix},$$

and

$$C = \begin{pmatrix} a_{11} & a_{12} & \cdots & a_{1j}+\alpha_{1j} & \cdots & a_{1n} \\ a_{21} & a_{22} & \cdots & a_{2j}+\alpha_{2j} & \cdots & a_{2n} \\ \vdots & \vdots & & \vdots & & \vdots \\ a_{n1} & a_{n2} & \cdots & a_{nj}+\alpha_{nj} & \cdots & a_{nn} \end{pmatrix}$$

Then

$$\boxed{\det C = \det A + \det B} \tag{4}$$

In other words, suppose that A, B, and C are identical except for the jth column and that the jth column of C is the sum of the jth columns of A and B. Then $\det C = \det A + \det B$. The same statement is true for rows.

Proof We expand $\det C$ in the jth column to obtain

$$\begin{aligned} \det C &= (a_{1j}+\alpha_{1j})A_{1j} + (a_{2j}+\alpha_{2j})A_{2j} + \cdots + (a_{nj}+\alpha_{nj})A_{nj} \\ &= (a_{1j}A_{1j} + a_{2j}A_{2j} + \cdots + a_{nj}A_{nj}) \\ &\quad + (\alpha_{1j}A_{1j} + \alpha_{2j}A_{2j} + \cdots + \alpha_{nj}A_{nj}) = \det A + \det B \quad \blacksquare \end{aligned}$$

EXAMPLE 4 Let $A = \begin{pmatrix} 1 & -1 & 2 \\ 3 & 1 & 4 \\ 0 & -2 & 5 \end{pmatrix}$, $B = \begin{pmatrix} 1 & -6 & 2 \\ 3 & 2 & 4 \\ 0 & 4 & 5 \end{pmatrix}$, and $C = \begin{pmatrix} 1 & -1-6 & 2 \\ 3 & 1+2 & 4 \\ 0 & -2+4 & 5 \end{pmatrix} =$

$\begin{pmatrix} 1 & -7 & 2 \\ 3 & 3 & 4 \\ 0 & 2 & 5 \end{pmatrix}$. Then $\det A = 16$, $\det B = 108$, and $\det C = 124 = \det A + \det B$.

Property 4 Interchanging any two rows (or columns) of A has the effect of multiplying $\det A$ by -1.

Proof We prove the statement for rows and assume first that two adjacent rows are interchanged. That is, we assume that the ith and $(i+1)$st rows are interchanged. Let

$$A = \begin{pmatrix} a_{11} & a_{12} & \cdots & a_{1n} \\ a_{21} & a_{22} & \cdots & a_{2n} \\ \vdots & \vdots & & \vdots \\ a_{i1} & a_{i2} & \cdots & a_{in} \\ a_{i+1,1} & a_{i+1,2} & \cdots & a_{i+1,n} \\ \vdots & \vdots & & \vdots \\ a_{n1} & a_{n2} & & a_{nn} \end{pmatrix} \quad \text{and} \quad B = \begin{pmatrix} a_{11} & a_{12} & \cdots & a_{1n} \\ a_{21} & a_{22} & \cdots & a_{2n} \\ \vdots & \vdots & & \vdots \\ a_{i+1,1} & a_{i+1,2} & \cdots & a_{i+1,n} \\ a_{i1} & a_{i2} & \cdots & a_{in} \\ \vdots & \vdots & & \vdots \\ a_{n1} & a_{n2} & \cdots & a_{nn} \end{pmatrix}$$

Then, expanding $\det A$ in its ith row and $\det B$ in its $(i+1)$st row, we obtain

$$\begin{aligned} \det A &= a_{i1}A_{i1} + a_{i2}A_{i2} + \cdots + a_{in}A_{in} \\ \det B &= a_{i1}B_{i+1,1} + a_{i2}B_{i+1,2} + \cdots + a_{in}B_{i+1,n} \end{aligned} \tag{5}$$

Here $A_{ij} = (-1)^{i+j}|M_{ij}|$, where M_{ij} is obtained by crossing off the ith row and jth column of A. Notice now that if we cross off the $(i+1)$st row and jth column of B, we obtain the same M_{ij}. Thus

$$B_{i+1,j} = (-1)^{i+1+j}|M_{ij}| = -(-1)^{i+j}|M_{ij}| = -A_{ij}$$

so that, from Equations (5), $\det B = -\det A$.

Now suppose that $i<j$ and that the ith and jth rows are to be interchanged. We can do this by interchanging adjacent rows several times. It will take $j-i$ interchanges to move row j into the ith row. Then row i will be in the $(i+1)$st row and it will take an additional $j-i-1$ interchanges to

move row i into the jth row. To illustrate, we interchange rows 2 and 6:†

$$\begin{matrix} 1 & & 1 & & 1 & & 1 & & 1 & & 1 & & 1 & & 1 \\ 2 & & 2 & & 2 & & 2 & & 6 & & 6 & & 6 & & 6 \\ 3 & \to & 3 & \to & 3 & \to & 6 & \to & 2 & \to & 3 & \to & 3 & \to & 3 \\ 4 & & 4 & & 6 & & 3 & & 3 & & 2 & & 4 & & 4 \\ 5 & & 6 & & 4 & & 4 & & 4 & & 4 & & 2 & & 5 \\ 6 & & 5 & & 5 & & 5 & & 5 & & 5 & & 5 & & 2 \\ 7 & & 7 & & 7 & & 7 & & 7 & & 7 & & 7 & & 7 \end{matrix}$$

$6-2=4$ interchanges to move the 6 into the 2 position; $6-2-1=3$ interchanges to get the 2 into the 6 position

Finally, the total number of interchanges of adjacent rows is $(j-i)+(j-i-1)=2j-2i-1$, which is odd. Thus det A is multiplied by -1 an odd number of times, which is what we needed to show. ■

EXAMPLE 5 Let $A=\begin{pmatrix} 1 & -1 & 2 \\ 3 & 1 & 4 \\ 0 & -2 & 5 \end{pmatrix}$. By interchanging the first and third rows we obtain $B=\begin{pmatrix} 0 & -2 & 5 \\ 3 & 1 & 4 \\ 1 & -1 & 2 \end{pmatrix}$. By interchanging the first and second columns of A we obtain $C=\begin{pmatrix} -1 & 1 & 2 \\ 1 & 3 & 4 \\ -2 & 0 & 5 \end{pmatrix}$. Then, by direct calculation, we find that det $A=16$ and det $B=\det C=-16$.

Property 5 If A has two equal rows or columns, then det $A=0$.

Proof Suppose the ith and jth rows of A are equal. By interchanging these rows we get a matrix B having the property that det $B=-\det A$ (from Property 4). But since row $i=$ row j, interchanging them gives us the same matrix. Thus $A=B$ and det $A=\det B=-\det A$. Thus 2 det $A=0$, which can happen only if det $A=0$. ■

EXAMPLE 6 By direct calculation we can verify that for $A=\begin{pmatrix} 1 & -1 & 2 \\ 5 & 7 & 3 \\ 1 & -1 & 2 \end{pmatrix}$ [two equal rows] and $B=\begin{pmatrix} 5 & 2 & 2 \\ 3 & -1 & -1 \\ -2 & 4 & 4 \end{pmatrix}$ [two equal columns], det $A=\det B=0$.

---

† Note that all the numbers here refer to rows.

Property 6 If one row (column) of A is a constant multiple of another row (column), then det $A=0$.

Proof Let $(a_{j1}, a_{j2}, \ldots, a_{jn}) = c(a_{i1}, a_{i2}, \ldots, a_{in})$. Then, from Property 2,

$$\det A = c\begin{vmatrix} a_{11} & a_{12} & \cdots & a_{1n} \\ a_{21} & a_{22} & \cdots & a_{2n} \\ \vdots & \vdots & & \vdots \\ a_{i1} & a_{i2} & \cdots & a_{in} \\ \vdots & \vdots & & \vdots \\ a_{i1} & a_{i2} & \cdots & a_{in} \\ \vdots & \vdots & & \vdots \\ a_{n1} & a_{n2} & \cdots & a_{nn} \end{vmatrix} = 0 \quad \text{(from Property 5)}$$

(jth row $\rightarrow$ indicates the second row $a_{i1}, a_{i2}, \ldots, a_{in}$) ■

EXAMPLE 7 $\begin{vmatrix} 2 & -3 & 5 \\ 1 & 7 & 2 \\ -4 & 6 & -10 \end{vmatrix} = 0$ since the third row is -2 times the first row.

EXAMPLE 8 $\begin{vmatrix} 2 & 4 & 1 & 12 \\ -1 & 1 & 0 & 3 \\ 0 & -1 & 9 & -3 \\ 7 & 3 & 6 & 9 \end{vmatrix} = 0$ since the fourth column is three times the second column.

Property 7 If a multiple of one row (column) of A is added to another row (column) of A, then the determinant is unchanged.

Proof Let B be the matrix obtained by adding c times the ith row of A to the jth row of A. Then

$$\det B = \begin{vmatrix} a_{11} & a_{12} & \cdots & a_{1n} \\ a_{21} & a_{22} & \cdots & a_{2n} \\ \vdots & \vdots & & \vdots \\ a_{i1} & a_{i2} & \cdots & a_{in} \\ \vdots & \vdots & & \vdots \\ a_{j1}+ca_{i1} & a_{j2}+ca_{i2} & \cdots & a_{jn}+ca_{in} \\ \vdots & \vdots & & \vdots \\ a_{n1} & a_{n2} & \cdots & a_{nn} \end{vmatrix}$$

$$\overset{\text{(from Property 3)}}{=}\begin{vmatrix} a_{11} & a_{12} & \cdots & a_{1n} \\ a_{21} & a_{22} & \cdots & a_{2n} \\ \vdots & \vdots & & \vdots \\ a_{i1} & a_{i2} & \cdots & a_{in} \\ \vdots & \vdots & & \vdots \\ a_{j1} & a_{j2} & \cdots & a_{jn} \\ \vdots & \vdots & & \vdots \\ a_{n1} & a_{n2} & \cdots & a_{nn} \end{vmatrix} + \begin{vmatrix} a_{11} & a_{12} & \cdots & a_{1n} \\ a_{21} & a_{22} & \cdots & a_{2n} \\ \vdots & \vdots & & \vdots \\ a_{i1} & a_{i2} & \cdots & a_{in} \\ \vdots & \vdots & & \vdots \\ ca_{i1} & ca_{i2} & \cdots & ca_{in} \\ \vdots & \vdots & & \vdots \\ a_{n1} & a_{n2} & \cdots & a_{nn} \end{vmatrix}$$

$= \det A + 0 = \det A$ (the zero comes from Property 6) ■

EXAMPLE 9 Let $A = \begin{pmatrix} 1 & -1 & 2 \\ 3 & 1 & 4 \\ 0 & -2 & 5 \end{pmatrix}$. Then $\det A = 16$. If we multiply the third row by 4 and add it to the second row, we obtain a new matrix B given by

$$B = \begin{pmatrix} 1 & -1 & 2 \\ 3+4(0) & 1+4(-2) & 4+5(4) \\ 0 & -2 & 5 \end{pmatrix} = \begin{pmatrix} 1 & -1 & 2 \\ 3 & -7 & 24 \\ 0 & -2 & 5 \end{pmatrix}$$

and $\det B = 16 = \det A$.

The properties discussed above make it much easier to evaluate high-order determinants. We simply "row-reduce" the determinant, using Property 7, until the determinant is in an easily evaluated form. The most common goal will be to use Property 7 repeatedly until either (*i*) the new determinant has a row (column) of zeros or one row (column) a multiple of another row (column)—in which case the determinant is zero—or (*ii*) the new matrix is triangular so that its determinant is the product of its diagonal elements.

EXAMPLE 10 Calculate

$$|A| = \begin{vmatrix} 1 & 3 & 5 & 2 \\ 0 & -1 & 3 & 4 \\ 2 & 1 & 9 & 6 \\ 3 & 2 & 4 & 8 \end{vmatrix}$$

Solution (See Example 3.1.7, page 86.)

There is already a zero in the first column, so it is simplest to reduce other elements in the first column to zero. We then continue to reduce, aiming for a triangular matrix:

Multiply the first row by −2 and add it to the third row and multiply the first row by −3 and add it to the fourth row.

$$|A| = \begin{vmatrix} 1 & 3 & 5 & 2 \\ 0 & -1 & 3 & 4 \\ 0 & -5 & -1 & 2 \\ 0 & -7 & -11 & 2 \end{vmatrix}$$

Multiply the second row by −5 and −7 and add it to the third and fourth rows, respectively.

$$= \begin{vmatrix} 1 & 3 & 5 & 2 \\ 0 & -1 & 3 & 4 \\ 0 & 0 & -16 & -18 \\ 0 & 0 & -32 & -26 \end{vmatrix}$$

Factor out −16 from the third row (using Property 2).

$$= -16 \begin{vmatrix} 1 & 3 & 5 & 2 \\ 0 & -1 & 3 & 4 \\ 0 & 0 & 1 & \frac{9}{8} \\ 0 & 0 & -32 & -26 \end{vmatrix}$$

Multiply the third row by 32 and add it to the fourth row.

$$= -16 \begin{vmatrix} 1 & 3 & 5 & 2 \\ 0 & -1 & 3 & 4 \\ 0 & 0 & 1 & \frac{9}{8} \\ 0 & 0 & 0 & 10 \end{vmatrix}$$

Now we have an upper triangular matrix and $|A| = -16(1)(-1)(1)(10) = (-16)(-10) = 160$.

EXAMPLE 11 Calculate

$$|A| = \begin{vmatrix} -2 & 1 & 0 & 4 \\ 3 & -1 & 5 & 2 \\ -2 & 7 & 3 & 1 \\ 3 & -7 & 2 & 5 \end{vmatrix}$$

Solution There are a number of ways to proceed here and it is not apparent which way will get us the answer most quickly. However, since there is already one zero in the first row, we begin our reduction in that row.

Multiply the second column by 2 and −4 and add it to the first and fourth columns, respectively.

$$|A| = \begin{vmatrix} 0 & 1 & 0 & 0 \\ 1 & -1 & 5 & 6 \\ 12 & 7 & 3 & -27 \\ -11 & -7 & 2 & 33 \end{vmatrix}$$

Interchange the first two columns.

$$= - \begin{vmatrix} 1 & 0 & 0 & 0 \\ -1 & 1 & 5 & 6 \\ 7 & 12 & 3 & -27 \\ -7 & -11 & 2 & 33 \end{vmatrix}$$

Multiply the second column by -5 and -6 and add it to the third and fourth columns, respectively.

$$= -\begin{vmatrix} 1 & 0 & 0 & 0 \\ -1 & 1 & 0 & 0 \\ 7 & 12 & -57 & -99 \\ -7 & -11 & 57 & 99 \end{vmatrix}$$

Since the fourth column is now a multiple of the third column (column $4 = \frac{99}{57} \times$ column 3), we see that $|A| = 0$.

EXAMPLE 12 Calculate

$$|A| = \begin{vmatrix} 1 & -2 & 3 & -5 & 7 \\ 2 & 0 & -1 & -5 & 6 \\ 4 & 7 & 3 & -9 & 4 \\ 3 & 1 & -2 & -2 & 3 \\ -5 & -1 & 3 & 7 & -9 \end{vmatrix}$$

Solution Adding first row 2 and then row 4 to row 5, we obtain

$$|A| = \begin{vmatrix} 1 & -2 & 3 & 5 & 7 \\ 2 & 0 & -1 & -5 & 6 \\ 4 & 7 & 3 & -9 & 4 \\ 3 & 1 & -2 & -2 & 3 \\ 0 & 0 & 0 & 0 & 0 \end{vmatrix} = 0 \qquad \text{(from Property 1)}$$

This example illustrates the fact that a little looking before beginning the computations can simplify matters considerably.

There are three additional facts about determinants that will be very useful to us.

THEOREM 2 Let A be an $n \times n$ matrix. Then

$$a_{i1}A_{j1} + a_{i2}A_{j2} + \cdots + a_{in}A_{jn} = 0 \qquad \text{if } i \neq j \tag{6}$$

Note. From Theorem 1 the sum in Equation (6) equals det A if $i = j$.

Proof Let

$$B = \begin{pmatrix} a_{11} & a_{12} & \cdots & a_{1n} \\ a_{21} & a_{22} & \cdots & a_{2n} \\ \vdots & \vdots & & \vdots \\ a_{i1} & a_{i2} & \cdots & a_{in} \\ \vdots & \vdots & & \vdots \\ a_{i1} & a_{i2} & \cdots & a_{in} \\ \vdots & \vdots & & \vdots \\ a_{n1} & a_{n2} & \cdots & a_{nn} \end{pmatrix} \begin{matrix} \\ \\ \\ \\ \\ \leftarrow j\text{th row} \\ \\ \\ \end{matrix}$$

Then, since two rows of B are equal, $\det B = 0$. But $B = A$ except in the jth row. Thus if we calculate $\det B$ by expanding in the jth row of B, we obtain the sum in (6) and the theorem is proved. Note that when we expand in the jth row, the jth row is deleted in computing the cofactors of B. Thus $B_{jk} = A_{jk}$ for $k = 1, 2, \ldots, n$. ■

THEOREM 3 Let A be an $n \times n$ matrix. Then

$$\det A = \det A^t \tag{7}$$

Proof This proof uses mathematical induction. If you are unfamiliar with this important method of proof, refer to Appendix 1. We first prove the theorem in the case $n = 2$. If

$$|A| = \begin{vmatrix} a_{11} & a_{12} \\ a_{21} & a_{22} \end{vmatrix} = a_{11}a_{22} - a_{12}a_{21}$$

then

$$|A^t| = \begin{vmatrix} a_{11} & a_{21} \\ a_{12} & a_{22} \end{vmatrix} = a_{11}a_{22} - a_{21}a_{12} = |A|$$

so the theorem is true for $n = 2$. Next we assume the theorem to be true for $(n-1)\times(n-1)$ matrices and prove it for $n \times n$ matrices. This will prove the theorem. Let $B = A^t$. Then

$$|A| = \begin{vmatrix} a_{11} & a_{12} & \cdots & a_{1n} \\ a_{21} & a_{22} & \cdots & a_{2n} \\ \vdots & \vdots & & \vdots \\ a_{n1} & a_{n2} & \cdots & a_{nn} \end{vmatrix} \quad \text{and} \quad |A^t| = |B| = \begin{vmatrix} a_{11} & a_{21} & \cdots & a_{n1} \\ a_{12} & a_{22} & \cdots & a_{n2} \\ \vdots & \vdots & & \vdots \\ a_{1n} & a_{2n} & \cdots & a_{nn} \end{vmatrix}$$

We expand $|A|$ in the first row and expand $|B|$ in the first column. This gives us

$$|A| = a_{11}A_{11} + a_{12}A_{12} + \cdots + a_{1n}A_{1n}$$
$$|B| = a_{11}B_{11} + a_{12}B_{21} + \cdots + a_{1n}B_{n1}$$

We need to show that $A_{1k} = B_{k1}$ for $k = 1, 2, \ldots, n$. But $A_{1k} = (-1)^{1+k}|M_{1k}|$ and $B_{k1} = (-1)^{k+1}|N_{k1}|$, where M_{1k} is the $1k$th minor of A and N_{k1} is the $k1$st minor of B. Then

$$|M_{1k}| = \begin{vmatrix} a_{21} & a_{22} & \cdots & a_{2,k-1} & a_{2,k+1} & \cdots & a_{2n} \\ a_{31} & a_{32} & \cdots & a_{3,k-1} & a_{3,k+1} & \cdots & a_{3n} \\ \vdots & \vdots & & \vdots & \vdots & & \vdots \\ a_{n1} & a_{n2} & \cdots & a_{n,k-1} & a_{n,k+1} & \cdots & a_{nn} \end{vmatrix}$$

and

$$|N_{k1}| = \begin{vmatrix} a_{21} & a_{31} & \cdots & a_{n1} \\ a_{22} & a_{32} & \cdots & a_{n2} \\ \vdots & \vdots & & \vdots \\ a_{2,k-1} & a_{3,k-1} & \cdots & a_{n,k-1} \\ a_{2,k+1} & a_{3,k+1} & \cdots & a_{n,k+1} \\ \vdots & \vdots & & \vdots \\ a_{2n} & a_{3n} & \cdots & a_{nn} \end{vmatrix}$$

Clearly $M_{1k} = N_{k1}^t$, and since both are $(n-1)\times(n-1)$ matrices, the induction hypothesis tells us that $|M_{1k}| = |N_{k1}|$. Thus $A_{1k} = B_{k1}$ and the proof is complete. ■

EXAMPLE 13 Let $A = \begin{pmatrix} 1 & -1 & 2 \\ 3 & 1 & 4 \\ 0 & -2 & 5 \end{pmatrix}$. Then $A^t = \begin{pmatrix} 1 & 3 & 0 \\ -1 & 1 & -2 \\ 2 & 4 & 5 \end{pmatrix}$ and it is easy to verify that $|A| = |A^t| = 16$.

THEOREM 4 Let A and B be $n \times n$ matrices. Then

$$\det AB = \det A \det B \tag{8}$$

That is: *The determinant of the product is the product of the determinants.*

Remark. The proof of this theorem is not conceptually difficult, but, as you might imagine from having worked with matrix products, it is extremely cumbersome. For that reason we shall simply prove the theorem in the case that A and B are 2×2 matrices. Let $A = \begin{pmatrix} a_{11} & a_{12} \\ a_{21} & a_{22} \end{pmatrix}$ and $B = \begin{pmatrix} b_{11} & b_{12} \\ b_{21} & b_{22} \end{pmatrix}$.

Then

$$\begin{aligned} \det A \det B &= (a_{11}a_{22} - a_{12}a_{21})(b_{11}b_{22} - b_{12}b_{21}) \\ &= a_{11}a_{22}b_{11}b_{22} - a_{11}a_{22}b_{12}b_{21} - a_{12}a_{21}b_{11}b_{22} + a_{12}a_{21}b_{12}b_{21} \end{aligned}$$

and $AB = \begin{pmatrix} a_{11}b_{11} + a_{12}b_{21} & a_{11}b_{12} + a_{12}b_{22} \\ a_{21}b_{11} + a_{22}b_{21} & a_{21}b_{12} + a_{22}b_{22} \end{pmatrix}$. Hence

$$\begin{aligned} \det AB &= (a_{11}b_{11} + a_{12}b_{21})(a_{21}b_{12} + a_{22}b_{22}) - (a_{11}b_{12} + a_{12}b_{22})(a_{21}b_{11} + a_{22}b_{21}) \\ &= a_{11}b_{11}a_{21}b_{12} + a_{11}b_{11}a_{22}b_{22} + a_{12}b_{21}a_{21}b_{12} + a_{12}b_{21}a_{22}b_{22} \\ &\quad - a_{11}b_{12}a_{21}b_{11} - a_{11}b_{12}a_{22}b_{21} - a_{12}b_{22}a_{21}b_{11} - a_{12}b_{22}a_{22}b_{21} \\ &= a_{11}b_{11}a_{22}b_{22} + a_{12}b_{21}a_{21}b_{12} - a_{11}b_{12}a_{22}b_{21} - a_{12}b_{22}a_{21}b_{11} \\ &= \det A \det B \quad \blacksquare \end{aligned}$$

EXAMPLE 14 Verify Equation (8) for $A = \begin{pmatrix} 1 & -1 & 2 \\ 3 & 1 & 4 \\ 0 & -2 & 5 \end{pmatrix}$ and $B = \begin{pmatrix} 1 & -2 & 3 \\ 0 & -1 & 4 \\ 2 & 0 & -2 \end{pmatrix}$.

Solution $\det A = 16$ and $\det B = -8$. We calculate

$$AB = \begin{pmatrix} 1 & -1 & 2 \\ 3 & 1 & 4 \\ 0 & -2 & 5 \end{pmatrix}\begin{pmatrix} 1 & -2 & 3 \\ 0 & -1 & 4 \\ 2 & 0 & -2 \end{pmatrix} = \begin{pmatrix} 5 & -1 & -5 \\ 11 & -7 & 5 \\ 10 & 2 & -18 \end{pmatrix}$$

and $\det AB = -128 = (16)(-8) = \det A \det B$.

PROBLEMS 3.2 In Problems 1–20 evaluate the determinant by using the methods of this section.

1. $\begin{vmatrix} 3 & -5 \\ 2 & 6 \end{vmatrix}$ **2.** $\begin{vmatrix} 4 & 1 \\ 0 & -3 \end{vmatrix}$ **3.** $\begin{vmatrix} -1 & 0 & 2 \\ 3 & 1 & 4 \\ 2 & 0 & -6 \end{vmatrix}$

4. $\begin{vmatrix} 2 & 1 & -1 \\ 3 & -2 & 0 \\ 5 & 1 & 6 \end{vmatrix}$ **5.** $\begin{vmatrix} -3 & 2 & 4 \\ 1 & -1 & 2 \\ -1 & 4 & 0 \end{vmatrix}$ **6.** $\begin{vmatrix} 0 & -2 & 3 \\ 1 & 2 & -3 \\ 4 & 0 & 5 \end{vmatrix}$

7. $\begin{vmatrix} -2 & 3 & 6 \\ 4 & 1 & 8 \\ -2 & 0 & 0 \end{vmatrix}$ **8.** $\begin{vmatrix} 2 & -1 & 3 \\ 4 & 0 & 6 \\ 5 & -2 & 3 \end{vmatrix}$ **9.** $\begin{vmatrix} 1 & -1 & 2 & 4 \\ 0 & -3 & 5 & 6 \\ 1 & 4 & 0 & 3 \\ 0 & 5 & -6 & 7 \end{vmatrix}$

10. $\begin{vmatrix} 2 & -3 & 1 & 4 \\ 0 & -2 & 0 & 0 \\ 3 & 7 & -1 & 2 \\ 4 & 1 & -3 & 8 \end{vmatrix}$ **11.** $\begin{vmatrix} 1 & 1 & -1 & 0 \\ -3 & 4 & 6 & 0 \\ 2 & 5 & -1 & 3 \\ 4 & 0 & 3 & 0 \end{vmatrix}$

12. $\begin{vmatrix} 3 & -1 & 2 & 1 \\ 4 & 3 & 1 & -2 \\ -1 & 0 & 2 & 3 \\ 6 & 2 & 5 & 2 \end{vmatrix}$

13. $\begin{vmatrix} 2 & 0 & 0 & 0 \\ 0 & 0 & 3 & 0 \\ 0 & -1 & 0 & 0 \\ 0 & 0 & 0 & 4 \end{vmatrix}$

14. $\begin{vmatrix} 0 & a & 0 & 0 \\ b & 0 & 0 & 0 \\ 0 & 0 & 0 & c \\ 0 & 0 & d & 0 \end{vmatrix}$

15. $\begin{vmatrix} 1 & 2 & 0 & 0 \\ 3 & -2 & 0 & 0 \\ 0 & 0 & 1 & -5 \\ 0 & 0 & 7 & 2 \end{vmatrix}$

16. $\begin{vmatrix} a & b & 0 & 0 \\ c & d & 0 & 0 \\ 0 & 0 & a & -b \\ 0 & 0 & c & d \end{vmatrix}$

17. $\begin{vmatrix} 2 & -1 & 0 & 4 & 1 \\ 3 & 1 & -1 & 2 & 0 \\ 3 & 2 & -2 & 5 & 1 \\ 0 & 0 & 4 & -1 & 6 \\ 3 & 2 & 1 & -1 & 1 \end{vmatrix}$

18. $\begin{vmatrix} 1 & -1 & 2 & 0 & 0 \\ 3 & 1 & 4 & 0 & 0 \\ 2 & -1 & 5 & 0 & 0 \\ 0 & 0 & 0 & 2 & 3 \\ 0 & 0 & 0 & -1 & 4 \end{vmatrix}$

19. $\begin{vmatrix} a & 0 & 0 & 0 & 0 \\ 0 & 0 & b & 0 & 0 \\ 0 & 0 & 0 & 0 & c \\ 0 & 0 & 0 & d & 0 \\ 0 & e & 0 & 0 & 0 \end{vmatrix}$

20. $\begin{vmatrix} 2 & 5 & -6 & 8 & 0 \\ 0 & 1 & -7 & 6 & 0 \\ 0 & 0 & 0 & 4 & 0 \\ 0 & 2 & 1 & 5 & 1 \\ 4 & -1 & 5 & 3 & 0 \end{vmatrix}$

In Problems 21–27 compute the determinant assuming that

$$\begin{vmatrix} a_{11} & a_{12} & a_{13} \\ a_{21} & a_{22} & a_{23} \\ a_{31} & a_{32} & a_{33} \end{vmatrix} = 8$$

21. $\begin{vmatrix} a_{31} & a_{32} & a_{33} \\ a_{21} & a_{22} & a_{23} \\ a_{11} & a_{12} & a_{13} \end{vmatrix}$

22. $\begin{vmatrix} a_{31} & a_{32} & a_{33} \\ a_{11} & a_{12} & a_{13} \\ a_{21} & a_{22} & a_{23} \end{vmatrix}$

23. $\begin{vmatrix} a_{11} & a_{12} & a_{13} \\ 2a_{21} & 2a_{22} & 2a_{23} \\ a_{31} & a_{32} & a_{33} \end{vmatrix}$

24. $\begin{vmatrix} -3a_{11} & -3a_{12} & -3a_{13} \\ 2a_{21} & 2a_{22} & 2a_{23} \\ 5a_{31} & 5a_{32} & 5a_{33} \end{vmatrix}$

25. $\begin{vmatrix} a_{11} & 2a_{13} & a_{12} \\ a_{21} & 2a_{23} & a_{22} \\ a_{31} & 2a_{33} & a_{32} \end{vmatrix}$

26. $\begin{vmatrix} a_{11}-a_{12} & a_{12} & a_{13} \\ a_{21}-a_{22} & a_{22} & a_{23} \\ a_{31}-a_{32} & a_{32} & a_{33} \end{vmatrix}$

27. $\begin{vmatrix} 2a_{11}-3a_{21} & 2a_{12}-3a_{22} & 2a_{13}-3a_{23} \\ a_{31} & a_{32} & a_{33} \\ a_{21} & a_{22} & a_{23} \end{vmatrix}$

28. Using Property 2, show that if α is a number and A is an $n \times n$ matrix, then $\det \alpha A = \alpha^n \det A$.

*29. Show that

$$\begin{vmatrix} 1+x_1 & x_2 & x_3 & \cdots & x_n \\ x_1 & 1+x_2 & x_3 & \cdots & x_n \\ x_1 & x_2 & 1+x_3 & \cdots & x_n \\ \vdots & \vdots & \vdots & & \vdots \\ x_1 & x_2 & x_3 & \cdots & 1+x_n \end{vmatrix} = 1+x_1+x_2+\cdots+x_n$$

*30. A matrix is **skew-symmetric** if $A^t = -A$. If A is an $n \times n$ skew-symmetric matrix, show that $\det A = (-1)^n \det A$.

31. Using the result of Problem 30, show that if A is a skew-symmetric $n \times n$ matrix and n is odd, then $\det A = 0$.

32. A matrix A is called **orthogonal** if A is invertible and $A^{-1} = A^t$. Show that if A is orthogonal, then $\det A = \pm 1$.

**33. Let Δ denote the triangle in the plane with vertices at (x_1, y_1), (x_2, y_2), and (x_3, y_3). Show that the area of the triangle is given by

$$\text{Area of } \Delta = \pm\tfrac{1}{2} \begin{vmatrix} 1 & x_1 & y_1 \\ 1 & x_2 & y_2 \\ 1 & x_3 & y_3 \end{vmatrix}$$

Under what circumstances will this determinant equal zero?

**34. Three lines, no two of which are parallel, determine a triangle in the plane. Suppose that the lines are given by

$$a_{11}x + a_{12}y + a_{13} = 0$$

$$a_{21}x + a_{22}y + a_{23} = 0$$

$$a_{31}x + a_{32}y + a_{33} = 0$$

Show that the area determined by the lines is

$$\frac{\pm 1}{2A_{13}A_{23}A_{33}} \begin{vmatrix} A_{11} & A_{12} & A_{13} \\ A_{21} & A_{22} & A_{23} \\ A_{31} & A_{32} & A_{33} \end{vmatrix}$$

35. The 3×3 Vandermonde† determinant is given by

$$D_3 = \begin{vmatrix} 1 & 1 & 1 \\ a_1 & a_2 & a_3 \\ a_1^2 & a_2^2 & a_3^2 \end{vmatrix}$$

Show that $D_3 = (a_2 - a_1)(a_3 - a_1)(a_3 - a_2)$.

36. $D_4 = \begin{vmatrix} 1 & 1 & 1 & 1 \\ a_1 & a_2 & a_3 & a_4 \\ a_1^2 & a_2^2 & a_3^2 & a_4^2 \\ a_1^3 & a_2^3 & a_3^3 & a_4^3 \end{vmatrix}$ is the 4×4 Vandermonde determinant.†

Show that $D_4 = (a_2 - a_1)(a_3 - a_1)(a_4 - a_1)(a_3 - a_2)(a_4 - a_2)(a_4 - a_3)$.

37. **a. Define the $n \times n$ Vandermonde determinant D_n.

† A. T. Vandermonde (1735–1796) was a French mathematician.

b. Show that $D_n = \prod_{\substack{i=1 \\ j>i}}^{n} (a_j - a_i)$, where Π stands for the word "product." Note that the product in Problem 36 can be written $\prod_{\substack{i=1 \\ j>i}}^{4} (a_j - a_i)$.

3.3 If Time Permits: Proof of the Basic Theorem

THEOREM 1 **BASIC THEOREM** Let $A = (a_{ij})$ be an $n \times n$ matrix. Then

$$\begin{aligned}\det A &= a_{11}A_{11} + a_{12}A_{12} + \cdots + a_{1n}A_{1n} \\ &= a_{i1}A_{i1} + a_{i2}A_{i2} + \cdots + a_{in}A_{in} \qquad (1) \\ &= a_{1j}A_{1j} + a_{2j}A_{2j} + \cdots + a_{nj}A_{nj} \qquad (2)\end{aligned}$$

for $i = 1, 2, \ldots, n$ and $j = 1, 2, \ldots, n$.

Note. The first equality is Definition 3.1.3 of the determinant by cofactor expansion in the first row; the second equality says that the expansion by cofactors in any other row yields the determinant; the third equality says that expansion by cofactors in any column gives the determinant.

Proof We prove equality (1) by mathematical induction. For the 2×2 matrix $A = \begin{pmatrix} a_{11} & a_{12} \\ a_{21} & a_{22} \end{pmatrix}$, we first expand the first row by cofactors: $\det A = a_{11}A_{11} + a_{12}A_{12} = a_{11}(a_{22}) + a_{12}(-a_{21}) = a_{11}a_{22} - a_{12}a_{21}$. Similarly, expanding in the second row, we obtain $a_{21}A_{21} + a_{22}A_{22} = a_{21}(-a_{12}) + a_{22}(a_{11}) = a_{11}a_{22} - a_{12}a_{21}$. Thus we get the same result by expanding in any row of a 2×2 matrix and this proves equality (1) in the 2×2 case.

We now assume that equality (1) holds for all $(n-1) \times (n-1)$ matrices. We must show that it holds for $n \times n$ matrices. Our procedure will be to expand by cofactors in the first and ith rows and show that the expansions are identical. If we expand in the first row, then a typical term in the cofactor expansion is

$$a_{1k}A_{1k} = (-1)^{1+k}a_{1k}|M_{1k}| \qquad (3)$$

Note that this is the only place in the expansion of $|A|$ that the term a_{1k} occurs since another typical term is $a_{1m}A_{1m} = (-1)^{1+m}|M_{1m}|$, $k \neq m$, and M_{1m} is obtained by deleting the first row and mth column of A (and a_{1k} is in the first row of A). Since M_{1k} is an $(n-1) \times (n-1)$ matrix, we can, by the induction hypothesis, calculate $|M_{1k}|$ by expanding in the ith row of A (which is the $(i-1)$st row of M_{1k}). A typical term in this expansion is

$$a_{il} \text{ (cofactor of } a_{il} \text{ in } M_{1k}) \qquad (k \neq l) \qquad (4)$$

For the reasons outlined above, this is the only term in the expansion of $|M_{1k}|$ in the ith row of A that contains the term a_{il}. Substituting (4) into (3), we find that

$$(-1)^{1+k}a_{1k}a_{il} \text{ (cofactor of } a_{il} \text{ in } M_{1k}) \qquad (k \neq l) \qquad (5)$$

is the only occurrence of the term $a_{1k}a_{il}$ in the cofactor expansion of det A in the first row.

Now if we expand by cofactors in the ith row of A (where $i \neq 1$), a typical term is

$$(-1)^{i+l}a_{il}\,|M_{il}| \tag{6}$$

and a typical term in the expansion of $|M_{il}|$ in the first row of M_{il} is

$$a_{1k}\ (\text{cofactor of } a_{1k} \text{ in } M_{il}) \qquad (k \neq l) \tag{7}$$

and, inserting (7) in (6), we find that the only occurrence of the term $a_{il}a_{1k}$ in the expansion of det A along its ith row is

$$(-1)^{i+l}a_{1k}a_{il}\,(\text{cofactor of } a_{1k} \text{ in } M_{il}) \qquad (k \neq l) \tag{8}$$

If we can show that the expressions in (5) and (8) are the same, then (1) will be proved, for the term in (5) is the only occurrence of $a_{1k}a_{il}$ in the first row expansion, the term in (8) is the only occurrence of $a_{1k}a_{il}$ in the ith row expansion, and k, i, and l are arbitrary. This will show that the sums of the terms in the first and ith row expansions are the same.

Now let $M_{1i,kl}$ denote the $(n-2)\times(n-2)$ matrix obtained by deleting the first and ith rows and kth and lth columns of A. (This is called a **second-order minor** of A.) We first suppose that $k < l$. Then

$$M_{1k} = \begin{pmatrix} a_{21} & \cdots & a_{2,k-1} & a_{2,k+1} & \cdots & a_{2l} & \cdots & a_{2n} \\ \vdots & & \vdots & \vdots & & \vdots & & \vdots \\ a_{i1} & \cdots & a_{i,k-1} & a_{i,k+1} & \cdots & a_{il} & \cdots & a_{in} \\ \vdots & & \vdots & \vdots & & \vdots & & \vdots \\ a_{n1} & \cdots & a_{n,k-1} & a_{n,k+1} & \cdots & a_{nl} & \cdots & a_{nn} \end{pmatrix} \tag{9}$$

$$M_{il} = \begin{pmatrix} a_{11} & \cdots & a_{1k} & \cdots & a_{1,l-1} & a_{1,l+1} & \cdots & a_{1n} \\ \vdots & & \vdots & & \vdots & \vdots & & \vdots \\ a_{i-1,1} & \cdots & a_{i-1,k} & \cdots & a_{i-1,l-1} & a_{i-1,l+1} & \cdots & a_{i-1,n} \\ a_{i+1,1} & \cdots & a_{i+1,k} & \cdots & a_{i+1,l-1} & a_{i+1,l+1} & \cdots & a_{i+1,n} \\ \vdots & & \vdots & & \vdots & \vdots & & \vdots \\ a_{n1} & \cdots & a_{nk} & \cdots & a_{n,l-1} & a_{n,l+1} & \cdots & a_{nn} \end{pmatrix} \tag{10}$$

From (9) and (10), we see that

$$\text{Cofactor of } a_{il} \text{ in } M_{1k} = (-1)^{(i-1)+(l-1)}|M_{1i,kl}| \tag{11}$$

$$\text{Cofactor of } a_{1k} \text{ in } M_{il} = (-1)^{1+k}\,|M_{1i,kl}| \tag{12}$$

Thus (5) becomes

$$(-1)^{1+k}a_{1k}a_{il}(-1)^{(i-1)+(l-1)}|M_{1i,kl}| = (-1)^{i+k+l-1}a_{1k}a_{il}\,|M_{1i,kl}| \tag{13}$$

and (8) becomes

$$(-1)^{i+l}a_{1k}a_{il}(-1)^{1+k}|M_{1i,kl}| = (-1)^{i+k+l+1}a_{1k}a_{il}|M_{1i,kl}| \tag{14}$$

But $(-1)^{i+k+l-1}=(-1)^{i+k+l+1}$, so the right sides of Equations (13) and (14) are equal. Hence expressions (5) and (8) are equal and (1) is proved in the case $k<l$. If $k>l$, then, by similar reasoning, we find that

$$\text{Cofactor of } a_{il} \text{ in } M_{1k} = (-1)^{(i-1)+l}|M_{1i,kl}|$$

$$\text{Cofactor of } a_{1k} \text{ in } M_{il} = (-1)^{1+(k-1)}|M_{1i,kl}|$$

so that (5) becomes

$$(-1)^{1+k}a_{1k}a_{il}(-1)^{(i-1)+l}|M_{1i,kl}| = (-1)^{i+k+l}a_{1k}a_{il}|M_{1i,kl}|$$

and (8) becomes

$$(-1)^{i+l}a_{1k}a_{il}(-1)^{1+k}|M_{1i,kl}| = (-1)^{i+k+l+1}a_{1k}a_{il}|M_{1i,kl}|$$

This completes the proof of Equation (1).

To prove Equation (2) we go through a similar process. If we expand in the kth and lth columns, we find that the only occurrences of the term $a_{1k}a_{il}$ will be given by (5) and (8). (See Problems 1 and 2.) This shows that the expansion by cofactors in any two columns is the same and that each is equal to the expansion along any row. This completes the proof. ■

PROBLEMS 3.3

1. Show that if A is expanded along its kth column, then the only occurrence of the term $a_{1k}a_{il}$ is given by Equation (5).
2. Show that if A is expanded along its lth column, then the only occurrence of the term $a_{1k}a_{il}$ is given by Equation (8).
3. Show that if A is expanded along its kth column, then the only occurrence of the term $a_{ik}a_{jl}$ is $(-1)^{i+k}a_{ik}a_{jl}$ (cofactor of a_{jl} in M_{ik}) for $l \neq k$.
4. Let $A = \begin{pmatrix} 1 & 5 & 7 \\ 2 & -1 & 3 \\ 4 & 5 & -2 \end{pmatrix}$. Compute det A by expanding in each of the rows and columns.
5. Do the same for the matrix $A = \begin{pmatrix} 1 & -1 & 4 \\ 0 & 1 & 5 \\ -3 & 7 & 2 \end{pmatrix}$.

3.4 Determinants and Inverses

In this section we shall see how matrix inverses can be calculated by using determinants. Moreover, we shall complete the task, begun in Chapter 1, of proving the important Summing Up Theorem 2.7.6 (see page 72) showing the equivalence of various properties of matrices. We begin with a simple result.

THEOREM 1 If A is invertible, then $\det A \neq 0$ and

$$\det A^{-1} = \frac{1}{\det A} \tag{1}$$

Proof From Theorems 3.2.4, page 99, and 3.1.1, page 87, we have

$$1 = \det I = \det AA^{-1} = \det A \det A^{-1} \tag{2}$$

If $\det A$ were equal to zero, then Equation (2) would read $1=0$. Thus $\det A \neq 0$ and $\det A^{-1} = 1/\det A$. ■

Before using determinants to calculate inverses, we need to define the *adjoint* of a matrix $A = (a_{ij})$. Let $B = (A_{ij})$ be the matrix of cofactors of A. (Remember that a cofactor, defined on page 85, is a number.) Then

$$B = \begin{pmatrix} A_{11} & A_{12} & \cdots & A_{1n} \\ A_{21} & A_{22} & \cdots & A_{2n} \\ \vdots & \vdots & & \vdots \\ A_{n1} & A_{n2} & \cdots & A_{nn} \end{pmatrix} \tag{3}$$

DEFINITION 1 **THE ADJOINT** Let A be an $n \times n$ matrix and let B, given by (3), denote the matrix of its cofactors. Then the **adjoint** ***of*** **A, written adj A, is the transpose** of the $n \times n$ matrix B; that is,

$$\text{adj } A = B^t = \begin{pmatrix} A_{11} & A_{21} & \cdots & A_{n1} \\ A_{12} & A_{22} & \cdots & A_{n2} \\ \vdots & \vdots & & \vdots \\ A_{1n} & A_{2n} & \cdots & A_{nn} \end{pmatrix} \tag{4}$$

EXAMPLE 1 Let $A = \begin{pmatrix} 2 & 4 & 3 \\ 0 & 1 & -1 \\ 3 & 5 & 7 \end{pmatrix}$. Compute adj A.

Solution We have $A_{11} = \begin{vmatrix} 1 & -1 \\ 5 & 7 \end{vmatrix} = 12$, $A_{12} = -\begin{vmatrix} 0 & -1 \\ 3 & 7 \end{vmatrix} = -3$, $A_{13} = -3$, $A_{21} = -13$, $A_{22} = 5$, $A_{23} = 2$, $A_{31} = -7$, $A_{32} = 2$, and $A_{33} = 2$. Thus $B = \begin{pmatrix} 12 & -3 & -3 \\ -13 & 5 & 2 \\ -7 & 2 & 2 \end{pmatrix}$ and $\text{adj } A = B^t = \begin{pmatrix} 12 & -13 & -7 \\ -3 & 5 & 2 \\ -3 & 2 & 2 \end{pmatrix}$.

EXAMPLE 2 Let

$$A=\begin{pmatrix}1 & -3 & 0 & -2\\ 3 & -12 & -2 & -6\\ -2 & 10 & 2 & 5\\ -1 & 6 & 1 & 3\end{pmatrix}$$

Calculate adj A.

Solution This is more tedious since we have to compute sixteen 3×3 determinants. For example, we have $A_{12}=-\begin{vmatrix}3 & -2 & -6\\ -2 & 2 & 5\\ -1 & 1 & 3\end{vmatrix}=-1$, $A_{24}=\begin{vmatrix}1 & -3 & 0\\ -2 & 10 & 2\\ -1 & 6 & 1\end{vmatrix}=-2$, and $A_{43}=-\begin{vmatrix}1 & -3 & -2\\ 3 & -12 & -6\\ -2 & 10 & 5\end{vmatrix}=3$. Completing these calculations, we find that

$$B=\begin{pmatrix}0 & -1 & 0 & 2\\ -1 & 1 & -1 & -2\\ 0 & 2 & -3 & -3\\ -2 & -2 & 3 & 2\end{pmatrix}$$

$$\text{adj } A=B^t=\begin{pmatrix}0 & -1 & 0 & -2\\ -1 & 1 & 2 & -2\\ 0 & -1 & -3 & 3\\ 2 & -2 & -3 & 2\end{pmatrix}$$

EXAMPLE 3 Let $A=\begin{pmatrix}a_{11} & a_{12}\\ a_{21} & a_{22}\end{pmatrix}$. Then adj $A=\begin{pmatrix}A_{11} & A_{21}\\ A_{12} & A_{22}\end{pmatrix}=\begin{pmatrix}a_{22} & -a_{12}\\ -a_{21} & a_{11}\end{pmatrix}$.

Warning. In taking the adjoint of a matrix, do not forget to transpose the matrix of cofactors.

THEOREM 2 Let A be an $n\times n$ matrix. Then

$$(A)(\text{adj } A)=\begin{pmatrix}\det A & 0 & 0 & \cdots & 0\\ 0 & \det A & 0 & \cdots & 0\\ 0 & 0 & \det A & \cdots & 0\\ \vdots & \vdots & \vdots & & \vdots\\ 0 & 0 & 0 & \cdots & \det A\end{pmatrix}=(\det A)I \qquad (5)$$

Proof Let $C = (c_{ij}) = (A)(\text{adj } A)$. Then

$$C = \begin{pmatrix} a_{11} & a_{12} & \cdots & a_{1n} \\ a_{21} & a_{22} & \cdots & a_{2n} \\ \vdots & \vdots & & \vdots \\ a_{n1} & a_{n2} & \cdots & a_{nn} \end{pmatrix} \begin{pmatrix} A_{11} & A_{21} & \cdots & A_{n1} \\ A_{12} & A_{22} & \cdots & A_{n2} \\ \vdots & \vdots & & \vdots \\ A_{1n} & A_{2n} & \cdots & A_{nn} \end{pmatrix} \tag{6}$$

We have

$$c_{ij} = (i\text{th row of } A) \cdot (j\text{th column of adj } A)$$

$$= (a_{i1} \quad a_{i2} \cdots a_{in}) \cdot \begin{pmatrix} A_{j1} \\ A_{j2} \\ \vdots \\ A_{jn} \end{pmatrix}$$

Thus

$$c_{ij} = a_{i1}A_{j1} + a_{i2}A_{j2} + \cdots + a_{in}A_{jn} \tag{7}$$

Now if $i = j$, the sum in (7) equals $a_{i1}A_{i1} + a_{i2}A_{i2} + \cdots + a_{in}A_{in}$, which is the expansion of det A in the ith row of A. On the other hand, if $i \neq j$, then from Theorem 3.2.2 on page 97, the sum in (7) equals zero. Thus

$$c_{ij} = \begin{cases} \det A & \text{if } i = j \\ 0 & \text{if } i \neq j \end{cases}$$

This proves the theorem. ■

We can now state the main result.

THEOREM 3 Let A be an $n \times n$ matrix. Then A is invertible if and only if $\det A \neq 0$. If $\det A \neq 0$, then

$$A^{-1} = \frac{1}{\det A} \text{adj } A \tag{8}$$

Note that Theorem 2.7.4 on page 65 for 2 × 2 matrices is a special case of this theorem.

Proof If A is invertible, then $\det A \neq 0$ by Theorem 1. If $\det A \neq 0$, then,

$$(A)\left(\frac{1}{\det A} \text{adj } A\right) = \frac{1}{\det A}[A(\text{adj } A)] \overset{\text{Theorem 2}}{=} \frac{1}{\det A}(\det A)I = I$$

But, by Theorem 2.7.8 on page 74, if $AB = I$, then $B = A^{-1}$. Thus

$$(1/\det A) \text{ adj } A = A^{-1}. \quad ■$$

EXAMPLE 4 Let $A = \begin{pmatrix} 2 & 4 & 3 \\ 0 & 1 & -1 \\ 3 & 5 & 7 \end{pmatrix}$. Determine whether A is invertible and calculate A^{-1} if it is.

Solution Since $\det A = 3 \neq 0$, we see that A is invertible. From Example 1,

$$\operatorname{adj} A = \begin{pmatrix} 12 & -13 & -7 \\ -3 & 5 & 2 \\ -3 & 2 & 2 \end{pmatrix}.$$

Thus

$$A^{-1} = \frac{1}{3}\begin{pmatrix} 12 & -13 & -7 \\ -3 & 5 & 2 \\ -3 & 2 & 2 \end{pmatrix} = \begin{pmatrix} 4 & -\frac{13}{3} & -\frac{7}{3} \\ -1 & \frac{5}{3} & \frac{2}{3} \\ -1 & \frac{2}{3} & \frac{2}{3} \end{pmatrix}$$

Check. $$A^{-1}A = \frac{1}{3}\begin{pmatrix} 12 & -13 & -7 \\ -3 & 5 & 2 \\ -3 & 2 & 2 \end{pmatrix}\begin{pmatrix} 2 & 4 & 3 \\ 0 & 1 & -1 \\ 3 & 5 & 7 \end{pmatrix} = \frac{1}{3}\begin{pmatrix} 3 & 0 & 0 \\ 0 & 3 & 0 \\ 0 & 0 & 3 \end{pmatrix} = I$$

EXAMPLE 5 Let

$$A = \begin{pmatrix} 1 & -3 & 0 & -2 \\ 3 & -12 & -2 & -6 \\ -2 & 10 & 2 & 5 \\ -1 & 6 & 1 & 3 \end{pmatrix}$$

Determine whether A is invertible and, if so, calculate A^{-1}.

Solution Using properties of determinants, we compute

$$\begin{vmatrix} 1 & -3 & 0 & -2 \\ 3 & -12 & -2 & -6 \\ -2 & 10 & 2 & 5 \\ -1 & 6 & 1 & 3 \end{vmatrix}$$

Multiply the first column by 3 and 2 and add it to the second and fourth columns, respectively.

$$= \begin{vmatrix} 1 & 0 & 0 & 0 \\ 3 & -3 & -2 & 0 \\ -2 & 4 & 2 & 1 \\ -1 & 3 & 1 & 1 \end{vmatrix}$$

Expand in the first row.

$$= \begin{vmatrix} -3 & -2 & 0 \\ 4 & 2 & 1 \\ 3 & 1 & 1 \end{vmatrix} = -1$$

Thus $\det A = -1 \neq 0$ and A^{-1} exists. By Example 2, we have

$$\operatorname{adj} A = \begin{pmatrix} 0 & -1 & 0 & -2 \\ -1 & 1 & 2 & -2 \\ 0 & -1 & -3 & 3 \\ 2 & -2 & -3 & 3 \end{pmatrix}$$

Thus

$$A^{-1} = \frac{1}{-1}\begin{pmatrix} 0 & -1 & 0 & -2 \\ -1 & 1 & 2 & -2 \\ 0 & -1 & -3 & 3 \\ 2 & -2 & -3 & 2 \end{pmatrix} = \begin{pmatrix} 0 & 1 & 0 & 2 \\ 1 & -1 & -2 & 2 \\ 0 & 1 & 3 & -3 \\ -2 & 2 & 3 & -2 \end{pmatrix}$$

Check. $$AA^{-1} = \begin{pmatrix} 1 & -3 & 0 & -2 \\ 3 & -12 & -2 & -6 \\ -2 & 10 & 2 & 5 \\ -1 & 6 & 1 & 3 \end{pmatrix} \begin{pmatrix} 0 & 1 & 0 & 2 \\ 1 & -1 & -2 & 2 \\ 0 & 1 & 3 & -3 \\ -2 & 2 & 3 & -2 \end{pmatrix}$$

$$= \begin{pmatrix} 1 & 0 & 0 & 0 \\ 0 & 1 & 0 & 0 \\ 0 & 0 & 1 & 0 \\ 0 & 0 & 0 & 1 \end{pmatrix}$$

Note. As you may have noticed, if $n > 3$ it is generally easier to compute A^{-1} by row reduction then by using adj A since, even for the 4×4 case, it is necessary to calculate 17 determinants (16 for the adjoint plus det A). Nevertheless, Theorem 3 is very important since, before you do any row reduction, the calculation of det A (if it can be done easily) will tell you whether or not A^{-1} exists.

We last saw out Summing Up Theorem (Theorems 1.2.1, page 4, and 2.7.6, page 72) in Section 2.7. This is the theorem that ties together many of the concepts developed in the first three chapters of this book. We are now able to prove the last two parts of that theorem.

THEOREM 4

SUMMING UP THEOREM—VIEW 3 Let A be an $n \times n$ matrix. Then each of the following six statements implies the other five. (That is, if one is true, all are true.)

i. A is invertible.
ii. The only solution to the homogeneous system $A\mathbf{x} = \mathbf{0}$ is the trivial solution ($\mathbf{x} = \mathbf{0}$).
iii. The system $A\mathbf{x} = \mathbf{b}$ has a unique solution for every n-vector $\mathbf{b}$.
iv. A is row equivalent to the $n \times n$ identity matrix I_n.
v. The rows (and columns) of A are linearly independent.
vi. det $A \neq 0$.

Proof In Theorem 2.7.6 we proved the equivalence of parts (*i*), (*ii*), (*iii*), and (*iv*). In Theorem 3 of this section we saw the equivalence of (*i*) and (*vi*). To complete the proof we shall show that (*v*) is equivalent to (*vi*). Theorem 2.6.3 on page 57 states that the columns of A are linearly dependent if and only if the system $A\mathbf{x} = \mathbf{0}$ has an infinite number of solutions. This implies that the columns of A are linearly independent if and only if the system $A\mathbf{x} = \mathbf{0}$ has the unique solution $\mathbf{x} = \mathbf{0}$. Thus part (*ii*) is equivalent to part (*v*) for columns. Since parts (*ii*) and (*vi*) are equivalent, this means that the columns of A are linearly independent if and only if det $A \neq 0$. But the rows of A are the columns of A^t and, by Theorem 3.2.3 on page 98, det A = det A^t. Thus the rows of A are linearly independent if and only if det A^t = det $A \neq 0$. This means that parts (*v*) and (*vi*) are equivalent and the proof is complete. ■

EXAMPLE 6 Determine whether the vectors $\begin{pmatrix}1\\-2\\3\end{pmatrix}$, $\begin{pmatrix}4\\1\\5\end{pmatrix}$, and $\begin{pmatrix}1\\0\\2\end{pmatrix}$ are linearly independent or dependent.

Solution Let $A = \begin{pmatrix}1 & 4 & 1\\-2 & 1 & 0\\3 & 5 & 2\end{pmatrix}$. Then $|A| = \begin{vmatrix}1 & 4 & 1\\-2 & 1 & 0\\3 & 5 & 2\end{vmatrix} = 5 \neq 0$, and hence the vectors (which are the columns of A) are linearly independent.

EXAMPLE 7 Determine whether the vectors $\begin{pmatrix}-2\\4\\5\end{pmatrix}$, $\begin{pmatrix}3\\1\\0\end{pmatrix}$, and $\begin{pmatrix}4\\6\\5\end{pmatrix}$ are linearly independent or dependent.

Solution Proceeding as in Example 6, we have $\begin{vmatrix}-2 & 3 & 4\\4 & 1 & 6\\5 & 0 & 5\end{vmatrix} = 0$; thus the vectors are linearly dependent.

EXAMPLE 8 Determine whether the vectors $(-1, 0, 4)$, $(2, 3, 1)$, and $(5, 2, 0)$ are linearly independent or dependent.

Solution Let $A = \begin{pmatrix}-1 & 0 & 4\\2 & 3 & 1\\5 & 2 & 0\end{pmatrix}$. Since $\det A = -42 \neq 0$, we conclude that the rows of A are linearly independent.

PROBLEMS 3.4

In Problems 1–12 use the methods of this section to determine whether the given matrix is invertible. If so, compute the inverse.

1. $\begin{pmatrix}3 & 2\\1 & 2\end{pmatrix}$ **2.** $\begin{pmatrix}3 & 6\\-4 & -8\end{pmatrix}$ **3.** $\begin{pmatrix}0 & 1\\1 & 0\end{pmatrix}$

4. $\begin{pmatrix}1 & 1 & 1\\0 & 2 & 3\\5 & 5 & 1\end{pmatrix}$ **5.** $\begin{pmatrix}3 & 2 & 1\\0 & 2 & 2\\0 & 1 & -1\end{pmatrix}$ **6.** $\begin{pmatrix}1 & 1 & 1\\0 & 1 & 1\\0 & 0 & 1\end{pmatrix}$

7. $\begin{pmatrix}1 & 2 & 3\\1 & 1 & 2\\0 & 1 & 2\end{pmatrix}$ **8.** $\begin{pmatrix}3 & 1 & 0\\1 & -1 & 2\\1 & 1 & 1\end{pmatrix}$ **9.** $\begin{pmatrix}2 & -1 & 4\\-1 & 0 & 5\\19 & -7 & 3\end{pmatrix}$

10. $\begin{pmatrix}1 & 6 & 2\\-2 & 3 & 5\\7 & 12 & -4\end{pmatrix}$ **11.** $\begin{pmatrix}1 & 1 & 1 & 1\\1 & 2 & -1 & 2\\1 & -1 & 2 & 1\\1 & 3 & 3 & 2\end{pmatrix}$ **12.** $\begin{pmatrix}1 & -3 & 0 & -2\\3 & -12 & -2 & -6\\-2 & 10 & 2 & 5\\-1 & 6 & 1 & 3\end{pmatrix}$

13. Using determinants, determine whether the vectors $\begin{pmatrix} 2 \\ -1 \\ 4 \end{pmatrix}, \begin{pmatrix} 1 \\ 6 \\ 2 \end{pmatrix}, \begin{pmatrix} 0 \\ 0 \\ 1 \end{pmatrix}$ are linearly dependent or independent.

14. Do the same for the vectors $\begin{pmatrix} 1 \\ -3 \\ 2 \\ 1 \end{pmatrix}, \begin{pmatrix} 4 \\ 0 \\ 1 \\ 2 \end{pmatrix}, \begin{pmatrix} 0 \\ 3 \\ -1 \\ 4 \end{pmatrix}, \begin{pmatrix} 5 \\ -6 \\ 4 \\ -1 \end{pmatrix}$.

15. Do the same for the vectors $\begin{pmatrix} 3 \\ 0 \\ 1 \\ 0 \end{pmatrix}, \begin{pmatrix} 5 \\ 1 \\ -1 \\ 2 \end{pmatrix}, \begin{pmatrix} 6 \\ 0 \\ 1 \\ 0 \end{pmatrix}, \begin{pmatrix} 0 \\ 0 \\ 1 \\ 1 \end{pmatrix}$.

16. Show that the n n-vectors $\begin{pmatrix} 1 \\ 0 \\ 0 \\ \vdots \\ 0 \end{pmatrix}, \begin{pmatrix} 0 \\ 1 \\ 0 \\ \vdots \\ 0 \end{pmatrix}, \begin{pmatrix} 0 \\ 0 \\ 1 \\ 0 \\ \vdots \\ 0 \end{pmatrix}, \ldots, \begin{pmatrix} 0 \\ 0 \\ 0 \\ \vdots \\ 1 \\ 0 \end{pmatrix}, \begin{pmatrix} 0 \\ 0 \\ 0 \\ \vdots \\ 1 \end{pmatrix}$ are linearly independent.

17. Show that an $n \times n$ matrix A is invertible if and only if A^t is invertible.

18. For $A = \begin{pmatrix} 1 & 1 \\ 2 & 5 \end{pmatrix}$, verify that $\det A^{-1} = 1/\det A$.

19. For $A = \begin{pmatrix} 1 & -1 & 3 \\ 4 & 1 & 6 \\ 2 & 0 & -2 \end{pmatrix}$, verify that $\det A^{-1} = 1/\det A$.

20. For what values of α is the matrix $\begin{pmatrix} \alpha & -3 \\ 4 & 1-\alpha \end{pmatrix}$ not invertible?

21. For what values of α does the matrix $\begin{pmatrix} -\alpha & \alpha-1 & \alpha+1 \\ 1 & 2 & 3 \\ 2-\alpha & \alpha+3 & \alpha+7 \end{pmatrix}$ not have an inverse?

22. Suppose that the $n \times n$ matrix A is not invertible. Show that $(A)(\text{adj } A)$ is the zero matrix.

3.5 Cramer's Rule

In this section we examine an old method for solving systems with the same number of unknowns as equations. Consider the system of n equations in n unknowns

$$\begin{aligned} a_{11}x_1 + a_{12}x_2 + \cdots + a_{1n}x_n &= b_1 \\ a_{21}x_1 + a_{22}x_2 + \cdots + a_{2n}x_n &= b_2 \\ \vdots \qquad \vdots \qquad\qquad \vdots \quad &\quad \vdots \\ a_{n1}x_1 + a_{n2}x_2 + \cdots + a_{nn}x_n &= b_n \end{aligned} \tag{1}$$

which can be written in the form

$$A\mathbf{x} = \mathbf{b} \tag{2}$$

We suppose that $\det A \neq 0$. Then system (2) has a unique solution given by $\mathbf{x} = A^{-1}\mathbf{b}$. We can develop a method for finding that solution without row reduction and without computing A^{-1}.

Let $D = \det A$. We define n new matrices:

$$A_1 = \begin{pmatrix} b_1 & a_{12} & \cdots & a_{1n} \\ b_2 & a_{22} & \cdots & a_{2n} \\ \vdots & \vdots & & \vdots \\ b_n & a_{n2} & \cdots & a_{nn} \end{pmatrix}, \quad A_2 = \begin{pmatrix} a_{11} & b_1 & \cdots & a_{1n} \\ a_{21} & b_2 & \cdots & a_{2n} \\ \vdots & \vdots & & \vdots \\ a_{n1} & b_n & \cdots & a_{nn} \end{pmatrix}, \ldots,$$

$$A_n = \begin{pmatrix} a_{11} & a_{12} & \cdots & b_1 \\ a_{21} & a_{22} & \cdots & b_2 \\ \vdots & \vdots & & \vdots \\ a_{n1} & a_{n2} & \cdots & b_n \end{pmatrix}$$

That is, A_i is the matrix obtained by replacing the ith column of A with $\mathbf{b}$. Finally, let $D_1 = \det A_1$, $D_2 = \det A_2, \ldots, D_n = \det A_n$.

THEOREM 1 **CRAMER'S RULE**† Let A be an $n \times n$ matrix and suppose that $\det A \neq 0$. Then the unique solution to the system $A\mathbf{x} = \mathbf{b}$ is given by

$$x_1 = \frac{D_1}{D},\ x_2 = \frac{D_2}{D}, \ldots, x_i = \frac{D_i}{D}, \ldots, x_n = \frac{D_n}{D} \tag{3}$$

Proof The solution to $A\mathbf{x} = \mathbf{b}$ is $\mathbf{x} = A^{-1}\mathbf{b}$. But

$$A^{-1}\mathbf{b} = \frac{1}{D}(\text{adj } A)\mathbf{b} = \frac{1}{D}\begin{pmatrix} A_{11} & A_{21} & \cdots & A_{n1} \\ A_{12} & A_{22} & \cdots & A_{n2} \\ \vdots & \vdots & & \vdots \\ A_{1n} & A_{2n} & \cdots & A_{nn} \end{pmatrix}\begin{pmatrix} b_1 \\ b_2 \\ \vdots \\ b_n \end{pmatrix} \tag{4}$$

Now $(\text{adj } A)\mathbf{b}$ is an n-vector, the jth component of which is

$$(A_{1j}\ A_{2j}\ \cdots\ A_{nj}) \cdot \begin{pmatrix} b_1 \\ b_2 \\ \vdots \\ b_n \end{pmatrix} = b_1A_{1j} + b_2A_{2j} + \cdots + b_nA_{nj} \tag{5}$$

† Named for the Swiss mathematician Gabriel Cramer (1704–1752). Cramer published the rule in 1750 in his *Introduction to the Analysis of Lines of Algebraic Curves*. Actually, there is much evidence to suggest that the rule was known as early as 1729 to Colin Maclaurin (1698–1746), who was probably the most outstanding British mathematician in the years following the death of Newton.

Consider the matrix A_j:

$$A_j = \begin{pmatrix} a_{11} & a_{12} & \cdots & b_1 & \cdots & a_{1n} \\ a_{21} & a_{22} & \cdots & b_2 & \cdots & a_{2n} \\ \vdots & \vdots & & \vdots & & \vdots \\ a_{n1} & a_{n2} & \cdots & b_n & \cdots & a_{nn} \end{pmatrix} \tag{6}$$

$\uparrow$ jth column

If we expand the determinant of A_j in its jth column, we obtain

$$D_j = b_1 \text{ (cofactor of } b_1) + b_2 \text{ (cofactor of } b_2) + \cdots + b_n \text{ (cofactor of } b_n) \tag{7}$$

But to find the cofactor of b_i, say, we delete the ith row and jth column of A_j (since b_i is in the jth column of A_j). But the jth column of A_j is $\mathbf{b}$ and, with this deleted, we simply have the ij minor, M_{ij}, of A. Thus

$$\text{Cofactor of } b_i \text{ in } A_j = A_{ij}$$

so that (7) becomes

$$D_j = b_1 A_{1j} + b_2 A_{2j} + \cdots + b_n A_{nj} \tag{8}$$

But this is the same as the right side of (5). Thus the ith component of $(\text{adj } A)\mathbf{b}$ is D_i and we have

$$\mathbf{x} = \begin{pmatrix} x_1 \\ x_2 \\ \vdots \\ x_n \end{pmatrix} = A^{-1}\mathbf{b} = \frac{1}{D} (\text{adj } A)\mathbf{b} = \frac{1}{D} \begin{pmatrix} D_1 \\ D_2 \\ \vdots \\ D_n \end{pmatrix} = \begin{pmatrix} D_1/D \\ D_2/D \\ \vdots \\ D_n/D \end{pmatrix}$$

and the proof is complete. ■

EXAMPLE 1 Solve, using Cramer's rule, the system

$$\begin{aligned} 2x_1 + 4x_2 + 6x_3 &= 18 \\ 4x_1 + 5x_2 + 6x_3 &= 24 \\ 3x_1 + x_2 - 2x_3 &= 4 \end{aligned} \tag{9}$$

Solution We have solved this before—using row reduction in Example 1.3.1 on page 6. We could also solve it by calculating A^{-1} (Example 2.7.6, page 66) and then finding $A^{-1}\mathbf{b}$. We now solve it by using Cramer's rule. First we have

$$D = \begin{vmatrix} 2 & 4 & 6 \\ 4 & 5 & 6 \\ 3 & 1 & -2 \end{vmatrix} = 6 \neq 0.$$

so that system (9) has a unique solution. Then $D_1 = \begin{vmatrix} 18 & 4 & 6 \\ 24 & 5 & 6 \\ 4 & 1 & -2 \end{vmatrix} = 24$, $D_2 = \begin{vmatrix} 2 & 18 & 6 \\ 4 & 24 & 6 \\ 3 & 4 & -2 \end{vmatrix} = -12$ and $D_3 = \begin{vmatrix} 2 & 4 & 18 \\ 4 & 5 & 24 \\ 3 & 1 & 4 \end{vmatrix} = 18$. Hence $x_1 = \dfrac{D_1}{D} = \dfrac{24}{6} = 4$, $x_2 = \dfrac{D_2}{D} = -\dfrac{12}{6} = -2$ and $x_3 = \dfrac{D_3}{D} = \dfrac{18}{6} = 3$.

EXAMPLE 2 Show that the system

$$\begin{aligned} x_1+3x_2+5x_3+2x_4&=2 \\ -x_2+3x_3+4x_4&=0 \\ 2x_1+x_2+9x_3+6x_4&=-3 \\ 3x_1+2x_2+4x_3+8x_4&=-1 \end{aligned} \tag{10}$$

has a unique solution and find it by using Cramer's rule.

Solution We saw in Example 3.2.10 on page 95 that

$$|A| = \begin{vmatrix} 1 & 3 & 5 & 2 \\ 0 & -1 & 3 & 4 \\ 2 & 1 & 9 & 6 \\ 3 & 2 & 4 & 8 \end{vmatrix} = 160 \neq 0$$

Thus the system has a unique solution. To find it we compute: $D_1=-464$; $D_2=280$; $D_3=-56$; $D_4=112$. Thus $x_1=D_1/D=-464/160$, $x_2=D_2/D=280/160$, $x_3=D_3/D=-56/160$, and $x_4=D_4/D=112/160$. These solutions can be verified by direct substitution into system (10).

PROBLEMS 3.5 In Problems 1–9 solve the given system by using Cramer's rule.

1. $\begin{aligned} 2x_1+3x_2&=-1 \\ -7x_1+4x_2&=47 \end{aligned}$

2. $\begin{aligned} 3x_1-x_2&=0 \\ 4x_1+2x_2&=5 \end{aligned}$

3. $\begin{aligned} 2x_1+x_2+x_3&=6 \\ 3x_1-2x_2-3x_3&=5 \\ 8x_1+2x_2+5x_3&=11 \end{aligned}$

4. $\begin{aligned} x_1+x_2+x_3&=8 \\ 4x_2-x_3&=-2 \\ 3x_1-x_2+2x_3&=0 \end{aligned}$

5. $\begin{aligned} 2x_1+2x_2+x_3&=7 \\ x_1+2x_2-x_3&=0 \\ -x_1+x_2+3x_3&=1 \end{aligned}$

6. $\begin{aligned} 2x_1+5x_2-x_3&=-1 \\ 4x_1+x_2+3x_3&=3 \\ -2x_1+2x_2&=0 \end{aligned}$

7. $\begin{aligned} 2x_1+x_2-x_3&=4 \\ x_1+x_3&=2 \\ -x_2+5x_3&=1 \end{aligned}$

8. $\begin{aligned} x_1+x_2+x_3+x_4&=6 \\ 2x_1-x_3-x_4&=4 \\ 3x_3+6x_4&=3 \\ x_1-x_4&=5 \end{aligned}$

9. $\begin{aligned} x_1-x_4&=7 \\ 2x_2+x_3&=2 \\ 4x_1-x_2&=-3 \\ 3x_3-5x_4&=2 \end{aligned}$

*10. Consider the triangle in Figure 3.2.

Figure 3.2

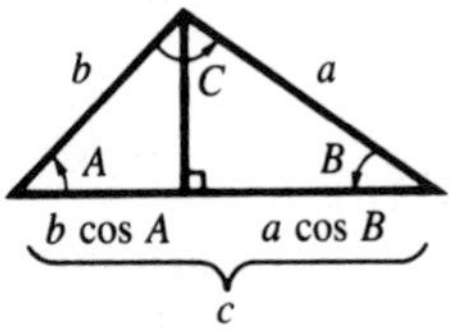

a. Show, using elementary trigonometry, that

$$\begin{aligned} c \cos A \qquad\qquad\qquad &+ a \cos C = b \\ b \cos A + a \cos B \qquad &\qquad\quad = c \\ c \cos B &+ b \cos C = a \end{aligned}$$

b. If the system of part (a) is thought of as a system of three equations in the three unknowns cos A, cos B, and cos C, show that the determinant of the system is nonzero.

c. Use Cramer's rule to solve for cos C.

d. Use part (c) to prove the *law of cosines*: $c^2 = a^2 + b^2 - 2ab \cos C$.

Review Exercises for Chapter 3

In Exercises 1–8 calculate the determinant.

1. $\begin{vmatrix} -1 & 2 \\ 0 & 4 \end{vmatrix}$ **2.** $\begin{vmatrix} -3 & 5 \\ -7 & 4 \end{vmatrix}$ **3.** $\begin{vmatrix} 1 & -2 & 3 \\ 0 & 4 & 5 \\ 0 & 0 & 6 \end{vmatrix}$

4. $\begin{vmatrix} 5 & 0 & 0 \\ 6 & 2 & 0 \\ 10 & 100 & 6 \end{vmatrix}$ **5.** $\begin{vmatrix} 1 & -1 & 2 \\ 3 & 4 & 2 \\ -2 & 3 & 4 \end{vmatrix}$ **6.** $\begin{vmatrix} 3 & 1 & -2 \\ 4 & 0 & 5 \\ -6 & 1 & 3 \end{vmatrix}$

7. $\begin{vmatrix} 1 & -1 & 2 & 3 \\ 4 & 0 & 2 & 5 \\ -1 & 2 & 3 & 7 \\ 5 & 1 & 0 & 4 \end{vmatrix}$ **8.** $\begin{vmatrix} 3 & 15 & 17 & 19 \\ 0 & 2 & 21 & 60 \\ 0 & 0 & 1 & 50 \\ 0 & 0 & 0 & -1 \end{vmatrix}$

In Exercises 9–14 use determinants to calculate the inverse, (if one exists).

9. $\begin{pmatrix} -3 & 4 \\ 2 & 1 \end{pmatrix}$ **10.** $\begin{pmatrix} 3 & -5 & 7 \\ 0 & 2 & 4 \\ 0 & 0 & -3 \end{pmatrix}$ **11.** $\begin{pmatrix} 1 & -1 & 2 \\ 3 & 1 & 4 \\ 5 & -1 & 8 \end{pmatrix}$

12. $\begin{pmatrix} 1 & 1 & 1 \\ 1 & 0 & 1 \\ 0 & 1 & 1 \end{pmatrix}$ **13.** $\begin{pmatrix} 2 & 1 & 0 & 0 \\ 0 & -1 & 3 & 0 \\ 1 & 0 & 0 & -2 \\ 3 & 0 & -1 & 0 \end{pmatrix}$ **14.** $\begin{pmatrix} 3 & -1 & 2 & 4 \\ 1 & 1 & 0 & 3 \\ -2 & 4 & 1 & 5 \\ 6 & -4 & 1 & 2 \end{pmatrix}$

In Exercises 15–20 determine whether the given set of vectors is linearly independent or dependent.

15. $\begin{pmatrix} 1 \\ -1 \end{pmatrix}; \begin{pmatrix} -2 \\ 2 \end{pmatrix}$ **16.** (2, 3); (3, 2) **17.** $\begin{pmatrix} 1 \\ 5 \\ 2 \end{pmatrix}; \begin{pmatrix} 3 \\ 0 \\ 4 \end{pmatrix}; \begin{pmatrix} -5 \\ 5 \\ -6 \end{pmatrix}$

18. (0, 2, 0); (0, 0, 2); (2, 0, 0)

19. $\begin{pmatrix}1\\0\\1\\0\end{pmatrix}$; $\begin{pmatrix}1\\1\\0\\0\end{pmatrix}$; $\begin{pmatrix}1\\0\\0\\1\end{pmatrix}$; $\begin{pmatrix}0\\0\\1\\1\end{pmatrix}$

20. (3, −1, 1, 4); (7, 1, 0, −2); (0, 2, 2, 3); (−4, 0, 3, 9)

In Exercises 21–24 solve the system by using Cramer's rule.

21. $\begin{aligned} 2x_1 - x_2 &= 3 \\ 3x_1 + 2x_2 &= 5 \end{aligned}$

22. $\begin{aligned} x_1 - x_2 + x_3 &= 7 \\ 2x_1 \qquad - 5x_3 &= 4 \\ 3x_2 - x_3 &= 2 \end{aligned}$

23. $\begin{aligned} 2x_1 + 3x_2 - x_3 &= 5 \\ -x_1 + 2x_2 + 3x_3 &= 0 \\ 4x_1 - x_2 + x_3 &= -1 \end{aligned}$

24. $\begin{aligned} x_1 \qquad - x_3 + x_4 &= 7 \\ 2x_2 + 2x_3 - 3x_4 &= -1 \\ 4x_1 - x_2 - x_3 \qquad &= 0 \\ -2x_1 + x_2 + 4x_3 \qquad &= 2 \end{aligned}$

4 Vectors in $\mathbb{R}^2$ and $\mathbb{R}^3$

4.1 Vectors in the Plane

As we defined it in Section 2.1, $\mathbb{R}^2$ is the set of vectors (x_1, x_2) with x_1 and x_2 real numbers. Since any point in the plane can be written in the form (x, y), it is apparent that any point in the plane can be thought of as a vector in $\mathbb{R}^2$ and vice versa. Thus the terms "the plane" and "$\mathbb{R}^2$" are often used interchangeably. However, for a variety of physical applications (including the notions of force, velocity, acceleration, and momentum) it is important to think of a vector not as a point but as an entity having "length" and "direction." Now we shall see how this is done.

Let P and Q be two points in the plane. Then the **directed line segment** from P to Q, denoted $\overrightarrow{PQ}$, is the straight line segment that extends from P to Q (see Figure 4.1*a*). Note that the directed line segments $\overrightarrow{PQ}$ and $\overrightarrow{QP}$ are different since they point in opposite directions (Figure 4.1*b*).

Figure 4.1

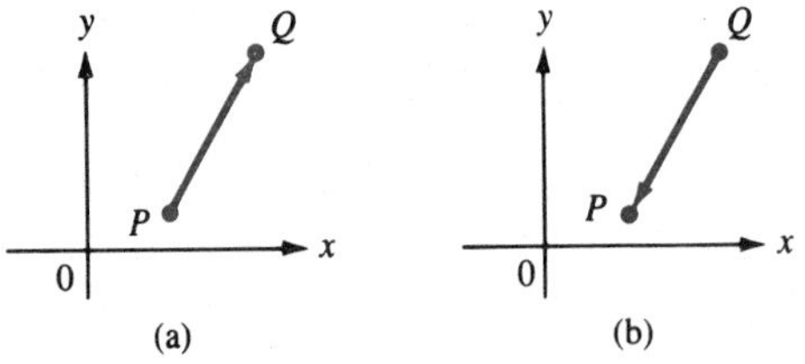

The point P in the directed line segment $\overrightarrow{PQ}$ is called the **initial point** of the segment and the point Q is called the **terminal point** The two major properties of a directed line segment are its magnitude (length) and its

Figure 4.2

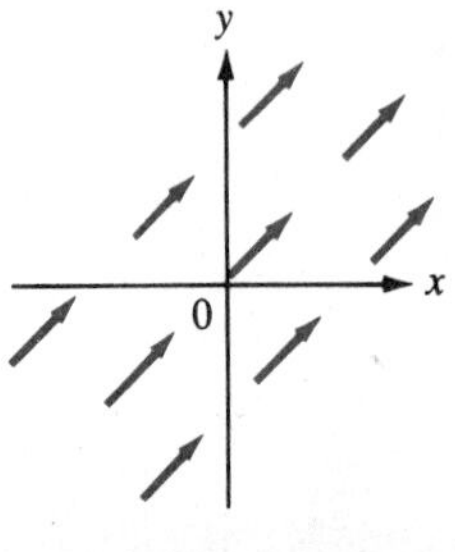

direction. If two directed line segments $\overrightarrow{PQ}$ and $\overrightarrow{RS}$ have the same magnitude and direction, we say that they are **equivalent** no matter where they are located with respect to the origin. The directed line segments in Figure 4.2 are all equivalent.

DEFINITION 1 **GEOMETRIC DEFINITION OF A VECTOR** The set of all directed line segments equivalent to a given directed line segment is called a **vector**. Any directed line segment in that set is called a **representation** of the vector.

Remark. The directed line segments in Figure 4.2 are all representations of the same vector.

From Definition 1 we see that a given vector **v** can be represented in many different ways. Let $\overrightarrow{PQ}$ be a representation of **v**. Then, without changing magnitude or direction, we can move $\overrightarrow{PQ}$ in a parallel way so that its initial point is shifted to the origin. We then obtain the directed line segment $\overrightarrow{OR}$, which is another representation of the vector **v** (see Figure 4.3). Now suppose that R has the cartesian coordinates (a, b). Then we can

Figure 4.3

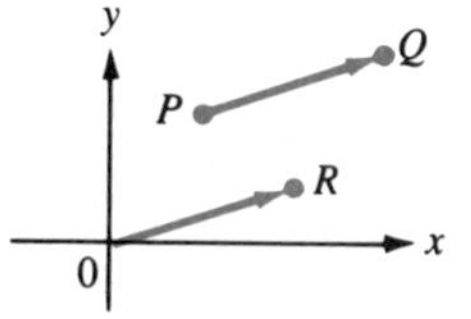

describe the directed line segment $\overrightarrow{OR}$ by the coordinates (a, b). That is, $\overrightarrow{OR}$ is the directed line segment with initial point $(0, 0)$ and terminal (a, b). Since one representation of a vector is as good as another, we can write the vector **v** as (a, b).

DEFINITION 2 **ALGEBRAIC DEFINITION OF A VECTOR** A **vector v** in the xy-plane is an ordered pair of real numbers (a, b). The numbers a and b are called the **components** of the vector **v**. The **zero vector** is the vector $(0, 0)$.

Remark 1. With this definition, a point in the xy-plane can be thought of as a vector originating at the origin and terminating at that point.

Remark 2. The zero vector has a magnitude of zero. Therefore, since the initial and terminal points coincide, we say that the zero vector has *no direction*.

Remark 3. We emphasize that Definitions 1 and 2 describe precisely the same objects. Each point of view (geometric and algebraic) has its advantages. Definition 2 is the definition of a 2-vector that we have been using all along.

Since a vector is really a set of equivalent line segments, we define the **magnitude** or **length** of a vector as the length of any one of its representations and its *direction* as the direction of any one of its representations.

Using the representation $\overrightarrow{OR}$ and writing the vector $\mathbf{v} = (a, b)$, we find that

$$|\mathbf{v}| = \text{magnitude of } \mathbf{v} = \sqrt{a^2 + b^2} \tag{1}$$

This follows from the pythagorean theorem (see Figure 4.4). We have used the notation $|\mathbf{v}|$ to denote the magnitude of $\mathbf{v}$. Note that $|\mathbf{v}|$ is a *scalar.*

Figure 4.4

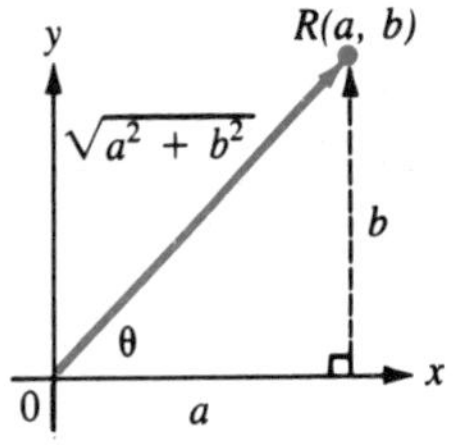

EXAMPLE 1

Calculate the magnitudes of the vectors (**i**) $(2, 2)$; (**ii**) $(2, 2\sqrt{3})$; (**iii**) $(-2\sqrt{3}, 2)$; (**iv**) $(-3, -3)$; (**v**) $(6, -6)$.

Solution

i. $|\mathbf{v}| = \sqrt{2^2 + 2^2} = \sqrt{8} = 2\sqrt{2}$
ii. $|\mathbf{v}| = \sqrt{2^2 + (2\sqrt{3})^2} = 4$
iii. $|\mathbf{v}| = \sqrt{(-2\sqrt{3})^2 + 2^2} = 4$
iv. $|\mathbf{v}| = \sqrt{(-3)^2 + (-3)^2} = \sqrt{18} = 3\sqrt{2}$
v. $|\mathbf{v}| = \sqrt{6^2 + (-6)^2} = \sqrt{72} = 6\sqrt{2}$

DIRECTION OF A VECTOR

We now define the **direction** of the vector $\mathbf{v} = (a, b)$ to be the angle θ, measured in radians, that the vector makes with the positive x-axis. By convention, we choose θ such that $0 \leq \theta < 2\pi$. It follows from Figure 4.4 that if $a \neq 0$, then

$$\tan \theta = \frac{b}{a} \tag{2}$$

EXAMPLE 2

Calculate the directions of the vectors in Example 1.

Solution

These five vectors are depicted in Figure 4.5.

i. Here $\mathbf{v}$ is in the first quadrant and since $\tan \theta = 2/2 = 1$, $\theta = \pi/4$.
ii. Here $\theta = \tan^{-1} 2\sqrt{3}/2 = \tan^{-1}\sqrt{3} = \pi/3$ (since $\mathbf{v}$ is in the first quadrant).
iii. We see that $\mathbf{v}$ is in the second quadrant and, since $\tan^{-1} 2/2\sqrt{3} = \tan^{-1}1/\sqrt{3} = \pi/6$, we see from Figure 4.5*c* that $\theta = \pi - (\pi/6) = 5\pi/6$.
iv. Here $\mathbf{v}$ is in the third quadrant and, since $\tan^{-1} 1 = \pi/4$, we find that $\theta = \pi + (\pi/4) = 5\pi/4$.

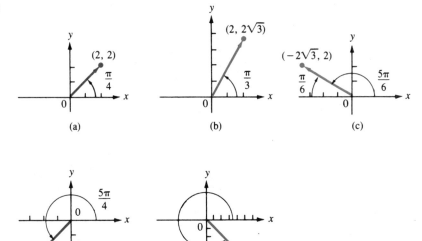

Figure 4.5

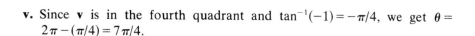

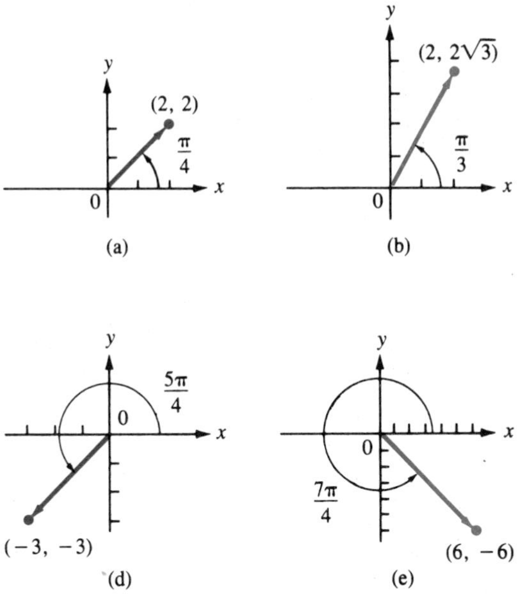

v. Since **v** is in the fourth quadrant and $\tan^{-1}(-1) = -\pi/4$, we get $\theta = 2\pi - (\pi/4) = 7\pi/4$.

In Section 2.1 we defined vector addition and scalar multiplication. What do these concepts mean geometrically? We start with scalar multiplication. If $\mathbf{v} = (a, b)$, then $\alpha\mathbf{v} = (\alpha a, \alpha b)$. We find that

$$|\alpha\mathbf{v}| = \sqrt{\alpha^2 a^2 + \alpha^2 b^2} = |\alpha|\sqrt{a^2 + b^2} = |\alpha|\,|\mathbf{v}| \tag{3}$$

That is:

> Multiplying a vector by a scalar has the effect of multiplying the length of the vector by the absolute value of that scalar.

Moreover, if $\alpha > 0$, then $\alpha\mathbf{v}$ is in the same quadrant as **v** and, therefore, the direction of $\alpha\mathbf{v}$ is the *same* as the direction of **v** since $\tan^{-1}(\alpha b/\alpha a) = \tan^{-1}(b/a)$. If $\alpha < 0$, then $\alpha\mathbf{v}$ points in the direction opposite to that of **v**. In other words:

$$\begin{aligned} &\text{Direction of } \alpha\mathbf{v} = \text{direction of } \mathbf{v}, \text{ if } \alpha > 0 \\ &\text{Direction of } \alpha\mathbf{v} = \text{direction of } \mathbf{v} + \pi, \text{ if } \alpha < 0 \end{aligned} \tag{4}$$

EXAMPLE 3 Let $\mathbf{v}=(1,1)$. Then $|\mathbf{v}|=\sqrt{1+1}=\sqrt{2}$ and $|2\mathbf{v}|=|(2,2)|=\sqrt{2^2+2^2}=\sqrt{8}=2\sqrt{2}=2|\mathbf{v}|$. Further, $|-2\mathbf{v}|=\sqrt{(-2)^2+(-2)^2}=2\sqrt{2}=2|\mathbf{v}|$. Moreover, the direction of $2\mathbf{v}$ is $\pi/4$ whereas the direction of $-2\mathbf{v}$ is $5\pi/4$. (See Figure 4.6)

Figure 4.6

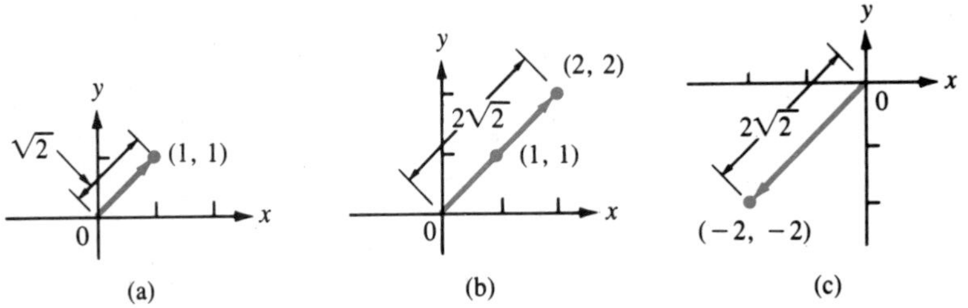

Now suppose we add the vectors $\mathbf{u}=(a_1, b_1)$ and $\mathbf{v}=(a_2, b_2)$ as in Figure 4.7. From the figure we see that the vector $\mathbf{u}+\mathbf{v}=(a_1+a_2, b_1+b_2)$ can be obtained by shifting the representation of the vector $\mathbf{v}$ so that its initial point coincides with the terminal point (a_1, b_1) of the vector $\mathbf{u}$. We can therefore obtain the vector $\mathbf{u}+\mathbf{v}$ by drawing a parallelogram with one vertex at the origin and sides $\mathbf{u}$ and $\mathbf{v}$. Then $\mathbf{u}+\mathbf{v}$ is the vector that points from the origin along the diagonal of the parallelogram.

Figure 4.7

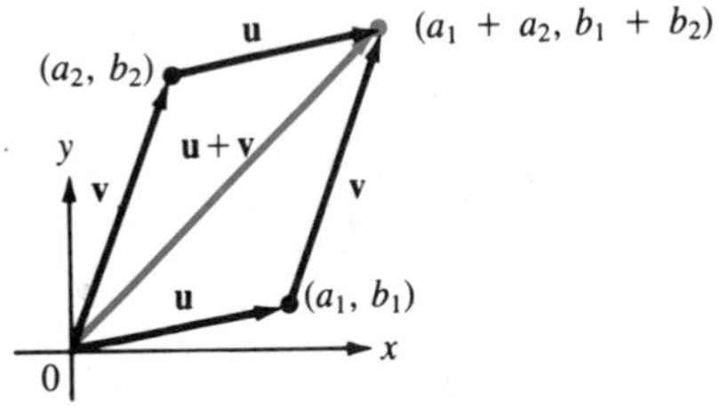

Note. Since a straight line is the shortest distance between two points, it immediately follows from Figure 4.7 that

$$|\mathbf{u}+\mathbf{v}| \leq |\mathbf{u}|+|\mathbf{v}| \tag{5}$$

For reasons obvious from Figure 4.7, inequality (5) is called the **triangle inequality**.

We can also use Figure 4.7 to obtain a geometric representation of the vector $\mathbf{u}-\mathbf{v}$. Since $\mathbf{u}=\mathbf{u}-\mathbf{v}+\mathbf{v}$, the vector $\mathbf{u}-\mathbf{v}$ is the vector that must be added to $\mathbf{v}$ to obtain $\mathbf{u}$. This fact is illustrated in Figure 4.8*a*. A similar fact is illustrated in Figure 4.8*b*.

There are two special vectors in $\mathbb{R}^2$ that allow us to represent other

Figure 4.8

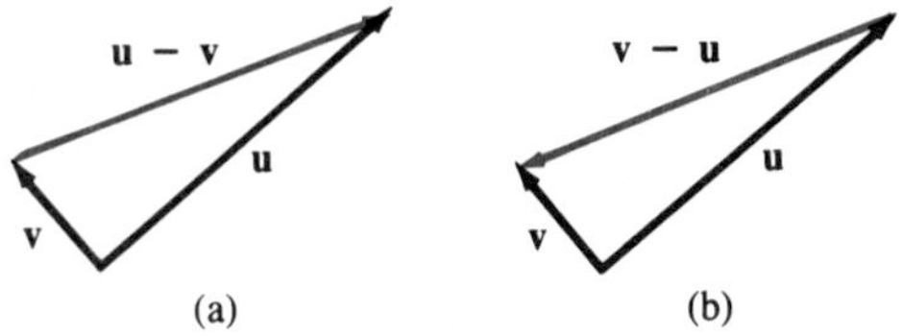

vectors in $\mathbb{R}^2$ in a convenient way. We shall denote the vector (1, 0) by the symbol **i** and the vector (0, 1) by the vector **j**.† (See Figure 4.9.) If $\mathbf{v}=(a, b)$

Figure 4.9

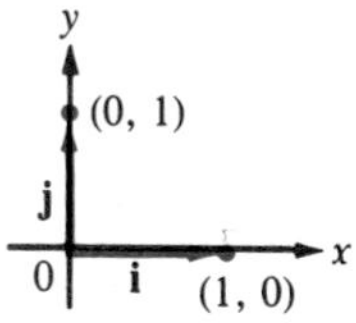

is any other vector in the plane, then, since $(a, b)=a(1, 0)+b(0, 1)$, we may write

$$\mathbf{v}=(a, b)=a\mathbf{i}+b\mathbf{j} \tag{6}$$

With this representation we say that **v** is *resolved into its horizontal and vertical components.* The vectors **i** and **j** have two properties:

i. Vectors **i** and **j** are linearly independent‡
ii. Any vector **v** can be written in terms of **i** and **j** as in Equation (6).§

Under these two conditions, **i** and **j** are said to form a **basis** in $\mathbb{R}^2$. We shall discuss bases in arbitrary vector spaces in Chapter 5.

Now suppose that a vector **v** is represented by the directed line segment $\overrightarrow{PQ}$, where $P=(a_1, b_1)$ and $Q=(a_2, b_2)$. (See Figure 4.10) If we label the

Figure 4.10

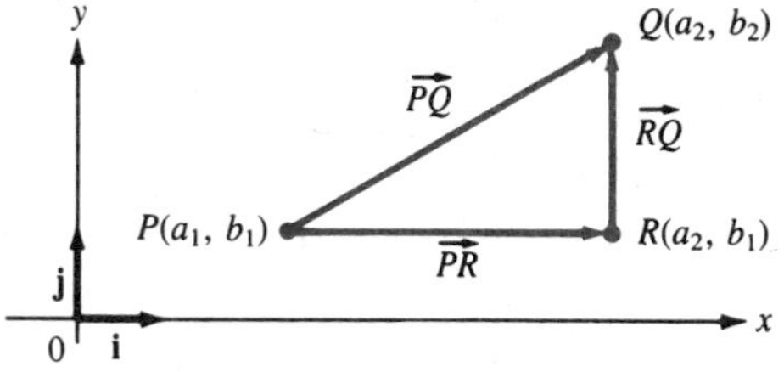

† *Historical Note:* The symbols **i** and **j** were first used by Hamilton. He defined his quaternion as a quantity of the form $a+b\mathbf{i}+c\mathbf{j}+d\mathbf{k}$, where a is the "scalar part" and $b\mathbf{i}+c\mathbf{j}+d\mathbf{k}$ the "vector part." In Section 4.3 we shall write vectors in space in the form $b\mathbf{i}+c\mathbf{j}+d\mathbf{k}$.

‡ This follows immediately from Theorem 3.4.4 on page 110 and the fact that **i** and **j** form the rows of the 2 × 2 identity matrix.

§ In Equation (6), we say that **v** can be written as a *linear combination* of **i** and **j**. We shall discuss the notion of linear combination in Section 5.4.

point (a_2, b_1) as R, then we immediately see that

$$\mathbf{v} = \overrightarrow{PQ} = \overrightarrow{PR} + \overrightarrow{RQ} \tag{7}$$

But the length of $\overrightarrow{PR}$ is $|a_2 - a_1|$ and since $\overrightarrow{PR}$ has the same direction as $\mathbf{i}$, we can write

$$\overrightarrow{PR} = (a_2 - a_1)\mathbf{i} \tag{8}$$

Similarly,

$$\overrightarrow{RQ} = (b_2 - b_1)\mathbf{j} \tag{9}$$

and we may write (using Equations 7, 8, and 9)

$$\mathbf{v} = (a_2 - a_1)\mathbf{i} + (b_2 - b_1)\mathbf{j} \tag{10}$$

EXAMPLE 4 Resolve the vector represented by the directed line segment from $(-2, 3)$ to $(1, 5)$ into its vertical and horizontal components.

Solution Using Equation (10), we have

$$\mathbf{v} = (a_2 - a_1)\mathbf{i} + (b_2 - b_1)\mathbf{j} = [1 - (-2)]\mathbf{i} + (5 - 3)\mathbf{j} = 3\mathbf{i} + 2\mathbf{j}$$

Both the directed line segment and $\mathbf{v}$ are sketched in Figure 4.11.

Figure 4.11

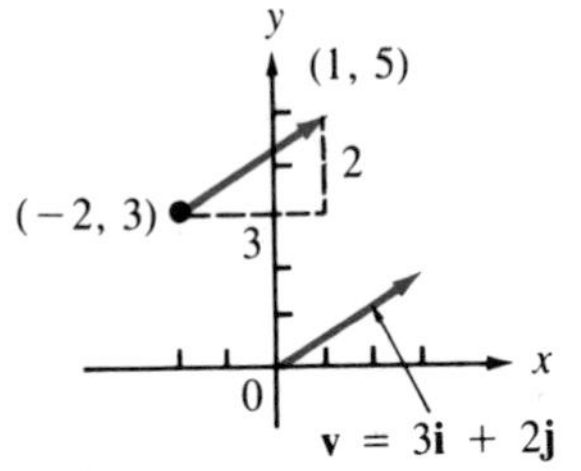

EXAMPLE 5 Resolve the vector represented by the directed line segment from $(3, 2)$ to $(-1, -3)$.

Solution The two points are illustrated in Figure 4.12. We have

$$\mathbf{v} = (a_2 - a_1)\mathbf{i} + (b_2 - b_1)\mathbf{j} = (-1 - 3)\mathbf{i} + (-3 - 2)\mathbf{j} = -4\mathbf{i} - 5\mathbf{j}$$

Figure 4.12

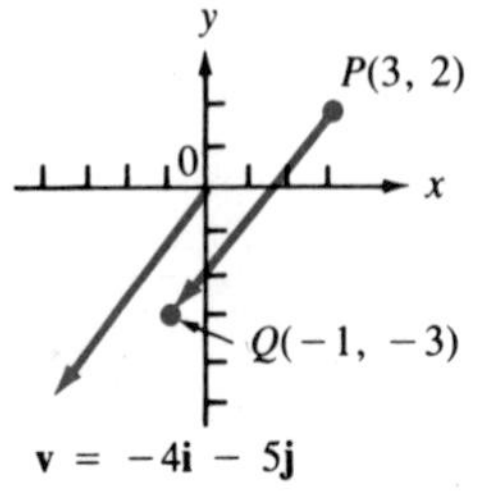

We conclude this section by defining a kind of a vector that is very useful in certain applications.

DEFINITION 3 **UNIT VECTOR** A **unit vector u** is a vector that has length 1.

EXAMPLE 6 The vector $\mathbf{u} = (1/2)\mathbf{i} + (\sqrt{3}/2)\mathbf{j}$ is a unit vector since

$$|\mathbf{u}| = \sqrt{\left(\frac{1}{2}\right)^2 + \left(\frac{\sqrt{3}}{2}\right)^2} = \sqrt{\frac{1}{4} + \frac{3}{4}} = 1$$

Let $\mathbf{u} = a\mathbf{i} + b\mathbf{j}$ be a unit vector. Then $|\mathbf{u}| = \sqrt{a^2 + b^2} = 1$, so $a^2 + b^2 = 1$ and $\mathbf{u}$ can be represented by a point on the unit circle (see Figure 4.13). If θ is

Figure 4.13

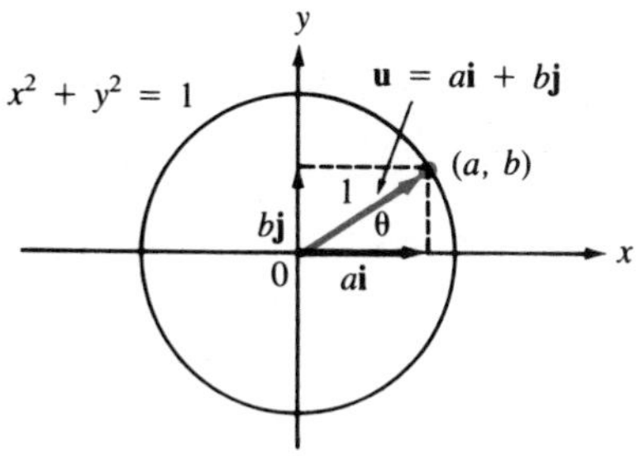

the direction of $\mathbf{u}$, then we immediately see that $a = \cos\theta$ and $b = \sin\theta$. Thus any unit vector $\mathbf{u}$ can be written in the form

$$\mathbf{u} = (\cos\theta)\mathbf{i} + (\sin\theta)\mathbf{j} \tag{11}$$

where θ is the direction of $\mathbf{u}$.

EXAMPLE 7 The unit vector $\mathbf{u} = (1/2)\mathbf{i} + (\sqrt{3}/2)\mathbf{j}$ of Example 6 can be written in the form of (11) with $\theta = \cos^{-1} 1/2 = \pi/3$.

We also have:

> Let $\mathbf{v}$ be any nonzero vector. Then $\mathbf{u} = \mathbf{v}/|\mathbf{v}|$ is a unit vector having the same direction as $\mathbf{v}$. (See Problem 27.)

EXAMPLE 8 Find a unit vector having the same direction as $\mathbf{v} = 2\mathbf{i} - 3\mathbf{j}$.

Solution Here $|\mathbf{v}| = \sqrt{4+9} = \sqrt{13}$, so $\mathbf{u} = \mathbf{v}/|\mathbf{v}| = (2/\sqrt{13})\mathbf{i} - (3/\sqrt{13})\mathbf{j}$ is the required unit vector.

EXAMPLE 9 Find a vector $\mathbf{v}$ whose direction is $5\pi/4$ and whose magnitude is 7.

Solution A unit vector $\mathbf{u}$ with direction $5\pi/4$ is given by $\mathbf{u} = (\cos 5\pi/4)\mathbf{i} + (\sin 5\pi/4)\mathbf{j} = -(1/\sqrt{2})\mathbf{i} - (1/\sqrt{2})\mathbf{j}$. Then $\mathbf{v} = 7\mathbf{u} = -(7/\sqrt{2})\mathbf{i} - (7/\sqrt{2})\mathbf{j}$. (See Figure 4.14*a*.) In Figure 4.14*b*, $\mathbf{v}$ has been translated so that it points toward the origin. This representation of $\mathbf{v}$ is sometimes useful, as we shall see in the next section (See Example 4.2.8.)

Figure 4.14

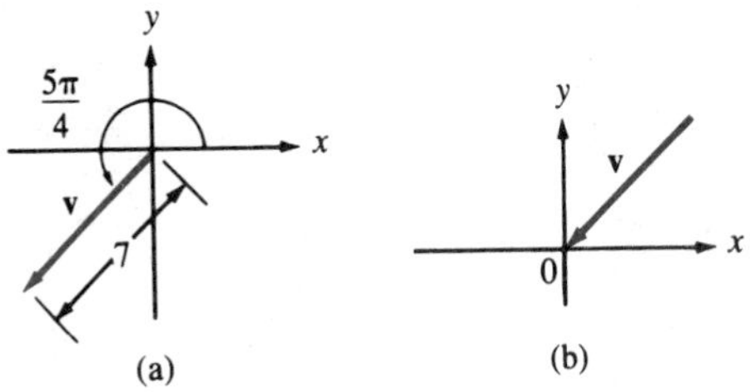

We conclude this section with a summary of the properties of vectors (Table 4.1).

Table 4.1

Object	Intuitive definition	Expression in terms of components if $\mathbf{u} = u_1\mathbf{i} + u_2\mathbf{j}$, $\mathbf{v} = v_1\mathbf{i} + v_2\mathbf{j}$, and $\mathbf{u} = (u_1, u_2)$, $\mathbf{v} = (v_1, v_2)$
vector $\mathbf{v}$	an object having magnitude and direction	$v_1\mathbf{i} + v_2\mathbf{j}$ or (v_1, v_2)
$\lvert\mathbf{v}\rvert$	Magnitude (or length) of $\mathbf{v}$	$\sqrt{v_1^2 + v_2^2}$
$\alpha\mathbf{v}$	$\mathbf{v}$ $\alpha\mathbf{v}$ (In this sketch $\alpha = 2$)	$\alpha v_1\mathbf{i} + \alpha v_2\mathbf{j}$ or $(\alpha v_1, \alpha v_2)$
$-\mathbf{v}$	$\mathbf{v}$ $-\mathbf{v}$	$-v_1\mathbf{i} - v_2\mathbf{j}$ or $(-v_1 - v_2)$ or $-(v_1, v_2)$
$\mathbf{u}+\mathbf{v}$	$\mathbf{u}+\mathbf{v}$ $\mathbf{v}$ $\mathbf{u}$	$(u_1 + v_1)\mathbf{i} + (u_2 + v_2)\mathbf{j}$ or $(u_1 + v_1, u_2 + v_2)$
$\mathbf{u}-\mathbf{v}$	$\mathbf{v}$ $\mathbf{u}-\mathbf{v}$ $\mathbf{u}$	$(u_1 - v_1)\mathbf{i} + (u_2 - v_2)\mathbf{j}$ or $(u_1 - v_1, u_2 - v_2)$

PROBLEMS 4.1

In Problems 1–12 find the magnitude and direction of the given vector.

1. $\mathbf{v} = (4, 4)$
2. $\mathbf{v} = (-4, 4)$
3. $\mathbf{v} = (4, -4)$
4. $\mathbf{v} = (-4, -4)$
5. $\mathbf{v} = (\sqrt{3}, 1)$
6. $\mathbf{v} = (1, \sqrt{3})$
7. $\mathbf{v} = (-1, \sqrt{3})$
8. $\mathbf{v} = (1, -\sqrt{3})$
9. $\mathbf{v} = (-1, -\sqrt{3})$
10. $\mathbf{v} = (1, 2)$
11. $\mathbf{v} = (-5, 8)$
12. $\mathbf{v} = (11, -14)$

In Problems 13–18 a vector $\mathbf{v}$ and a point P are given. Find a point Q such that the directed line segment $\overrightarrow{PQ}$ is a representation of $\mathbf{v}$. Sketch $\mathbf{v}$ and $\overrightarrow{PQ}$.

13. $\mathbf{v}=(2,5)$; $P=(1,-2)$
14. $\mathbf{v}=(5,8)$; $P=(3,8)$
15. $\mathbf{v}=(-3,7)$; $P=(7,-3)$
16. $\mathbf{v}=-\mathbf{i}-7\mathbf{j}$; $P=(0,1)$
17. $\mathbf{v}=5\mathbf{i}-3\mathbf{j}$; $P=(-7,-2)$
18. $\mathbf{v}=e\mathbf{i}+\pi\mathbf{j}$; $P=(\pi,\sqrt{2})$

In Problems 19–22 write in the form $a\mathbf{i}+b\mathbf{j}$ the vector $\mathbf{v}$ that is represented by $\overrightarrow{PQ}$. Sketch $\overrightarrow{PQ}$ and $\mathbf{v}$.

19. $P=(1,2)$; $Q=(1,3)$
20. $P=(2,4)$; $Q=(-7,4)$
21. $P=(5,2)$; $Q=(-1,3)$
22. $P=(8,-2)$; $Q=(-3,-3)$
23. Let $\mathbf{u}=(2,3)$ and $\mathbf{v}=(-5,4)$. Find: **(a)** $3\mathbf{u}$; **(b)** $\mathbf{u}+\mathbf{v}$; **(c)** $\mathbf{v}-\mathbf{u}$; **(d)** $2\mathbf{u}-7\mathbf{v}$. Sketch these vectors.
24. Let $\mathbf{u}=2\mathbf{i}-3\mathbf{j}$ and $\mathbf{v}=-4\mathbf{i}+6\mathbf{j}$. Find: **(a)** $\mathbf{u}+\mathbf{v}$; **(b)** $\mathbf{u}-\mathbf{v}$; **(c)** $3\mathbf{u}$; **(d)** $-7\mathbf{v}$; **(e)** $8\mathbf{u}-3\mathbf{v}$; **(f)** $4\mathbf{v}-6\mathbf{u}$. Sketch these vectors.
25. Show that the vectors $\mathbf{i}$ and $\mathbf{j}$ are unit vectors.
26. Show that the vector $(1/\sqrt{2})\mathbf{i}+(1/\sqrt{2})\mathbf{j}$ is a unit vector.
27. Show that if $\mathbf{v}=a\mathbf{i}+b\mathbf{j}$, then $\mathbf{u}=(a/\sqrt{a^2+b^2})\mathbf{i}+(b/\sqrt{a^2+b^2})\mathbf{j}$ is a unit vector having the same direction as $\mathbf{v}$.

In Problems 28–31 find a unit vector having the same direction as the given vector.

28. $\mathbf{v}=2\mathbf{i}+3\mathbf{j}$
29. $\mathbf{v}=\mathbf{i}-\mathbf{j}$
30. $\mathbf{v}=-3\mathbf{i}+4\mathbf{j}$
31. $\mathbf{v}=a\mathbf{i}+a\mathbf{j}$; $a\neq 0$.
32. If $\mathbf{v}=a\mathbf{i}+b\mathbf{j}$, show that $a/\sqrt{a^2+b^2}=\cos\theta$ and $b/\sqrt{a^2+b^2}=\sin\theta$, where θ is the direction of $\mathbf{v}$.
33. If $\mathbf{v}=2\mathbf{i}-3\mathbf{j}$, find $\sin\theta$ and $\cos\theta$.
34. If $\mathbf{v}=-3\mathbf{i}+8\mathbf{j}$, find $\sin\theta$ and $\cos\theta$.

A vector $\mathbf{v}$ has a direction opposite to that of a vector $\mathbf{u}$ if direction $\mathbf{v}$ = direction $\mathbf{u}+\pi$. In Problems 35–38 find a unit vector $\mathbf{v}$ that has a direction opposite the direction of the given vector $\mathbf{u}$.

35. $\mathbf{u}=\mathbf{i}+\mathbf{j}$
36. $\mathbf{u}=2\mathbf{i}-3\mathbf{j}$
37. $\mathbf{u}=-3\mathbf{i}+4\mathbf{j}$
38. $\mathbf{u}=-2\mathbf{i}+3\mathbf{j}$
39. Let $\mathbf{u}=2\mathbf{i}-3\mathbf{j}$ and $\mathbf{v}=-\mathbf{i}+2\mathbf{j}$. Find a unit vector having the same direction as: **(a)** $\mathbf{u}+\mathbf{v}$; **(b)** $2\mathbf{u}-3\mathbf{v}$; **(c)** $3\mathbf{u}+8\mathbf{v}$.
40. Let $P=(c,d)$ and $Q=(c+a,d+b)$. Show that the magnitude of $\overrightarrow{PQ}$ is $\sqrt{a^2+b^2}$.
41. Show that the direction of $\overrightarrow{PQ}$ in Problem 40 is the same as the direction of the vector (a,b). [*Hint:* If $R=(a,b)$, show that the line passing through the points P and Q is parallel to the line passing through the points O and R.]

In Problems 42–45 find a vector $\mathbf{v}$ having the given magnitude and direction. [*Hint:* See Example 9.]

42. $|\mathbf{v}|=3$; $\theta=\pi/6$
43. $|\mathbf{v}|=8$; $\theta=\pi/3$
44. $|\mathbf{v}|=1$; $\theta=\pi/4$
45. $|\mathbf{v}|=6$; $\theta=2\pi/3$.
***46.** Show algebraically (that is, strictly from the definitions of vector addition and magnitude) that for any two vectors $\mathbf{u}$ and $\mathbf{v}$, $|\mathbf{u}+\mathbf{v}|\le|\mathbf{u}|+|\mathbf{v}|$.
47. Show that if neither $\mathbf{u}$ nor $\mathbf{v}$ is the zero vector, then $|\mathbf{u}+\mathbf{v}|=|\mathbf{u}|+|\mathbf{v}|$ if and only if $\mathbf{u}$ is a positive scalar multiple of $\mathbf{v}$.

4.2 The Scalar Product and Projections in ℝ²

In Section 2.2 we defined the scalar product of two vectors. If $\mathbf{u} = (a_1, b_1)$ and $\mathbf{v} = (a_2, b_2)$, then

$$\mathbf{u} \cdot \mathbf{v} = a_1 a_2 + b_1 b_2 \tag{1}$$

We shall now see how the scalar product can be interpreted geometrically.

DEFINITION 1 **ANGLE BETWEEN VECTORS** Let $\mathbf{u}$ and $\mathbf{v}$ be two nonzero vectors. Then the **angle φ between $\mathbf{u}$ and $\mathbf{v}$** is defined to be the smallest nonnegative angle† between the representations of $\mathbf{u}$ and $\mathbf{v}$ that have the origin as their initial points. If $\mathbf{u} = \alpha\mathbf{v}$ for some scalar α, then we define $\varphi = 0$ if $\alpha > 0$ and $\varphi = \pi$ if $\alpha < 0$.

This definition is illustrated in Figure 4.15. Note that φ can always be chosen to be a nonnegative angle in the interval $[0, \pi]$.

Figure 4.15

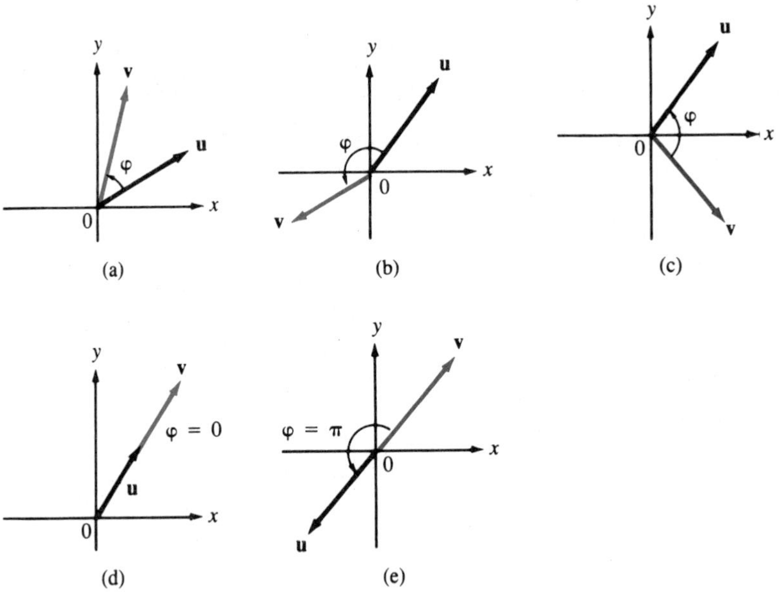

THEOREM 1 Let $\mathbf{u}$ and $\mathbf{v}$ be two nonzero vectors. If φ is the angle between them, then

$$\cos\varphi = \frac{\mathbf{u} \cdot \mathbf{v}}{|\mathbf{u}|\,|\mathbf{v}|} \tag{2}$$

† This angle will be in the interval $[0, \pi]$.

Proof The law of cosines (see Problem 3.5.10, page 116) states that in the triangle of Figure 4.16

$$c^2 = a^2 + b^2 - 2ab \cos C \tag{3}$$

Figure 4.16

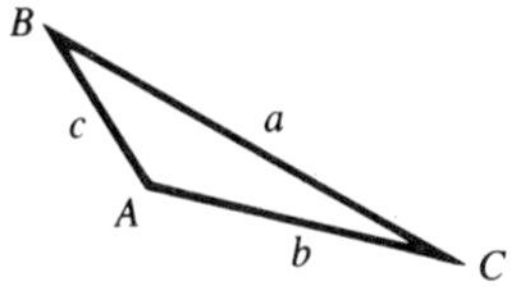

We now place the representations of **u** and **v** with initial points at the origin so that $\mathbf{u} = (a_1, b_1)$ and $\mathbf{v} = (a_2, b_2)$. (See Figure 4.17.) Then, from the law of

Figure 4.17

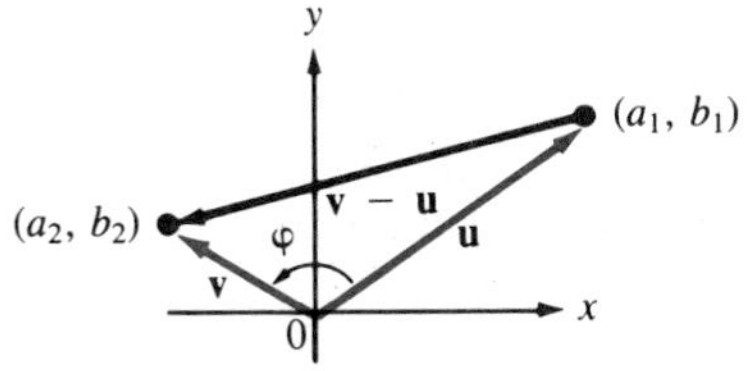

cosines, $|\mathbf{v}-\mathbf{u}|^2 = |\mathbf{v}|^2 + |\mathbf{u}|^2 - 2|\mathbf{u}|\,|\mathbf{v}| \cos \varphi$. But

$$\begin{aligned} |\mathbf{v}-\mathbf{u}|^2 &= (\mathbf{v}-\mathbf{u}) \cdot (\mathbf{v}-\mathbf{u}) = \mathbf{v} \cdot \mathbf{v} - 2\mathbf{u} \cdot \mathbf{v} + \mathbf{u} \cdot \mathbf{u} \\ &= |\mathbf{v}|^2 - 2\mathbf{u} \cdot \mathbf{v} + |\mathbf{u}|^2 \end{aligned}$$

Thus, after simplification, we obtain $-2\mathbf{u} \cdot \mathbf{v} = -2|\mathbf{u}|\,|\mathbf{v}| \cos \varphi$, from which the theorem follows. ■

Remark. Using Theorem 1 we could define the scalar product $\mathbf{u} \cdot \mathbf{v}$ by

$$\mathbf{u} \cdot \mathbf{v} = |\mathbf{u}|\,|\mathbf{v}| \cos \varphi$$

EXAMPLE 1 Find the cosine of the angle between the vectors $\mathbf{u} = 2\mathbf{i} + 3\mathbf{j}$ and $\mathbf{v} = -7\mathbf{i} + \mathbf{j}$.

Solution $\mathbf{u} \cdot \mathbf{v} = -14 + 3 = -11$, $|\mathbf{u}| = \sqrt{2^2 + 3^2} = \sqrt{13}$, and $|\mathbf{v}| = \sqrt{(-7)^2 + 1^2} = \sqrt{50}$. Hence

$$\cos \varphi = \frac{\mathbf{u} \cdot \mathbf{v}}{|\mathbf{u}|\,|\mathbf{v}|} = \frac{-11}{\sqrt{13}\sqrt{50}} = \frac{-11}{\sqrt{650}} \approx -0.4315^\dagger$$

DEFINITION 2 **PARALLEL VECTORS** Two nonzero vectors **u** and **v** are **parallel** if the angle between them is zero or π.

† This number, like others in the text, was obtained with a hand calculator.

EXAMPLE 2 Show that the vectors $\mathbf{u}=(2,-3)$ and $\mathbf{v}=(-4,6)$ are parallel.

Solution

$$\cos\varphi=\frac{\mathbf{u}\cdot\mathbf{v}}{|\mathbf{u}|\,|\mathbf{v}|}=\frac{-8-18}{\sqrt{13}\sqrt{52}}=\frac{-26}{\sqrt{13}(2\sqrt{13})}=\frac{-26}{2(13)}=-1$$

Hence $\varphi=\pi$.

THEOREM 2 If $\mathbf{u}\neq\mathbf{0}$, then $\mathbf{v}=\alpha\mathbf{u}$ for some nonzero constant α if and only if $\mathbf{u}$ and $\mathbf{v}$ are parallel.

Proof The proof is left as an exercise (see Problem 35).

DEFINITION 3 **ORTHOGONAL VECTORS** The nonzero vectors $\mathbf{u}$ and $\mathbf{v}$ are called **orthogonal** (or **perpendicular**) if the angle between them is $\pi/2$.

EXAMPLE 3 Show that the vectors $\mathbf{u}=3\mathbf{i}-4\mathbf{j}$ and $\mathbf{v}=4\mathbf{i}+3\mathbf{j}$ are orthogonal.

Solution $\mathbf{u}\cdot\mathbf{v}=3\cdot4-4\cdot3=0$. This implies that $\cos\varphi=(\mathbf{u}\cdot\mathbf{v})/(|\mathbf{u}|\,|\mathbf{v}|)=0$. Since φ is in the interval $[0,\pi]$, $\varphi=\pi/2$.

THEOREM 3 The nonzero vectors $\mathbf{u}$ and $\mathbf{v}$ are orthogonal if and only if $\mathbf{u}\cdot\mathbf{v}=0$.

Proof This proof is also left as an exercise (see Problem 36).

EXAMPLE 4 Using vectors, show that the diagonals of a rectangle are orthogonal if and only if the rectangle is a square.

Solution We place the rectangle so that two of its sides lie along the x- and y-axes (see Figure 4.18). One of the diagonals is the vector $\mathbf{u}=(a,b)$. The other is

Figure 4.18

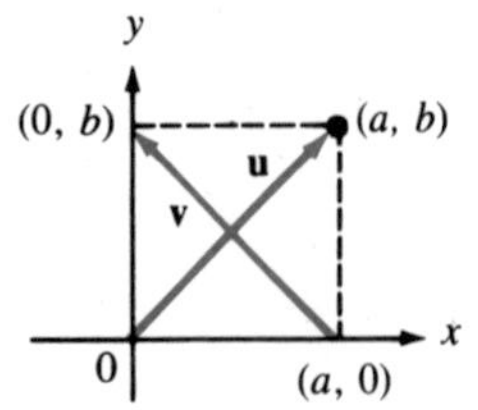

the vector $\mathbf{v}=(0-a,b-0)=(-a,b)$. (See Equation 4.1.10, page 124) Then $\mathbf{u}\cdot\mathbf{v}=-a^2+b^2$. This is zero if and only if $a=b$ since $a>0$ and $b>0$. But if $a=b$, the rectangle is a square.

There are a number of other "geometric" applications of vectors given in Problems 57–64.

EXAMPLE 5 Let $\mathbf{u}=\mathbf{i}+4\mathbf{j}$ and $\mathbf{v}=3\mathbf{i}+\alpha\mathbf{j}$. Determine α such that **(i)** $\mathbf{u}$ and $\mathbf{v}$ are orthogonal; **(ii)** $\mathbf{u}$ and $\mathbf{v}$ are parallel.

Solution **i.** We have $\mathbf{u}\cdot\mathbf{v}=3+4\alpha$. For $\mathbf{u}$ and $\mathbf{v}$ to be orthogonal, we must have $\mathbf{u}\cdot\mathbf{v}=0$. This implies that $3+4\alpha=0$ or $\alpha=-\frac{3}{4}$.

ii. Here we must have $\varphi=0$ or π so that $\cos\varphi=\pm 1$. Then

$$\cos\varphi=\frac{\mathbf{u}\cdot\mathbf{v}}{|\mathbf{u}|\,|\mathbf{v}|}=\frac{3+4\alpha}{\sqrt{17}\sqrt{9+\alpha^2}}=\pm 1$$

Squaring both sides of this last equation, we obtain $9+24\alpha+16\alpha^2=17(9+\alpha^2)=153+17\alpha^2$. This leads to the quadratic equation $\alpha^2-24\alpha+144=0=(\alpha-12)^2$, with the single solution $\alpha=12$.

A number of interesting problems involve the notion of the projection of one vector along another. (See, for example, the discussion of work at the end of this section.) Before defining this, we prove the following theorem.

THEOREM 4 Let $\mathbf{v}$ be a nonzero vector. Then for any other vector $\mathbf{u}$, the vector $\mathbf{w}=\mathbf{u}-[(\mathbf{u}\cdot\mathbf{v})\mathbf{v}/|\mathbf{v}|^2]$ is orthogonal to $\mathbf{v}$.

Proof

$$\mathbf{w}\cdot\mathbf{v}=\left[\mathbf{u}-\frac{(\mathbf{u}\cdot\mathbf{v})\mathbf{v}}{|\mathbf{v}|^2}\right]\cdot\mathbf{v}=\mathbf{u}\cdot\mathbf{v}-\frac{(\mathbf{u}\cdot\mathbf{v})(\mathbf{v}\cdot\mathbf{v})}{|\mathbf{v}|^2}$$

$$=\mathbf{u}\cdot\mathbf{v}-\frac{(\mathbf{u}\cdot\mathbf{v})|\mathbf{v}|^2}{|\mathbf{v}|^2}=\mathbf{u}\cdot\mathbf{v}-\mathbf{u}\cdot\mathbf{v}=0 \quad \blacksquare$$

The vectors $\mathbf{u}$, $\mathbf{v}$, $\mathbf{w}$ are illustrated in Figure 4.19.

Figure 4.19

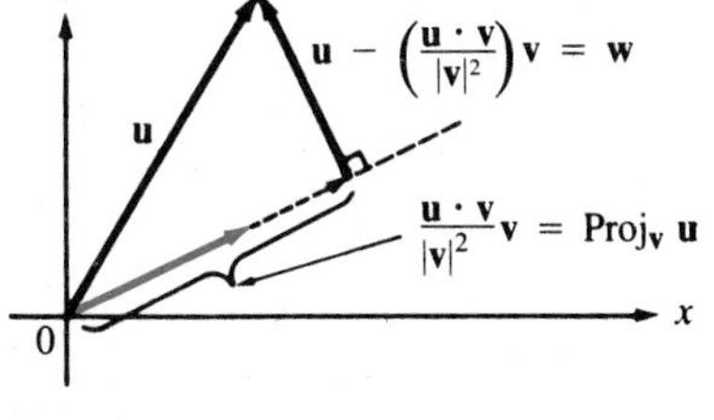

DEFINITION 4 **PROJECTION** Let $\mathbf{u}$ and $\mathbf{v}$ be nonzero vectors. Then the **projection** *of* $\mathbf{u}$ *on* $\mathbf{v}$

is a vector, denoted $\text{proj}_{\mathbf{v}}\,\mathbf{u}$, which is defined by

$$\text{proj}_{\mathbf{v}}\,\mathbf{u} = \frac{\mathbf{u}\cdot\mathbf{v}}{|\mathbf{v}|^2}\,\mathbf{v} \tag{4}$$

The **component** of $\mathbf{u}$ in the direction $\mathbf{v}$ is $\dfrac{\mathbf{u}\cdot\mathbf{v}}{|\mathbf{v}|}$. (5)

Note that $\mathbf{v}/|\mathbf{v}|$ is a unit vector in the direction of $\mathbf{v}$.

Remark 1. From Figure 4.19 and the fact that $\cos\varphi = (\mathbf{u}\cdot\mathbf{v})/(|\mathbf{u}|\,|\mathbf{v}|)$, we find that:

$\mathbf{v}$ and $\text{proj}_{\mathbf{v}}\,\mathbf{u}$ have (*i*) the same direction if $\mathbf{u}\cdot\mathbf{v}>0$ and (*ii*) opposite directions if $\mathbf{u}\cdot\mathbf{v}<0$. (See Figure 4.20.)

Figure 4.20

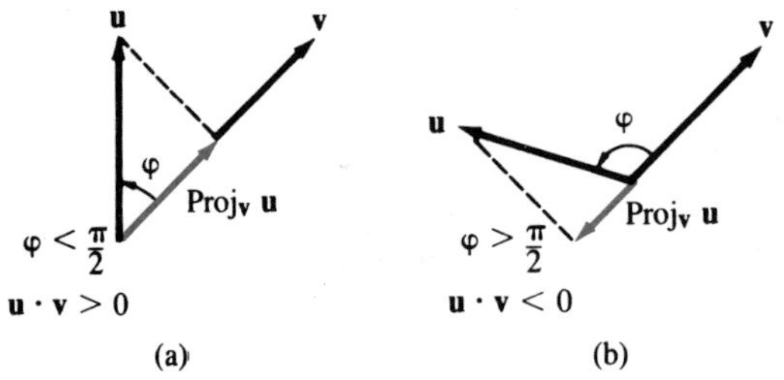

Remark 2. $\text{Proj}_{\mathbf{v}}\,\mathbf{u}$ can be thought of as the "**v**-component" of the vector $\mathbf{u}$. We shall see an illustration of this in our discussion of force later in this section.

Remark 3. If $\mathbf{u}$ and $\mathbf{v}$ are orthogonal, then $\mathbf{u}\cdot\mathbf{v}=0$ so that $\text{proj}_{\mathbf{v}}\,\mathbf{u}=\mathbf{0}$.

Remark 4. An alternative definition of projection is: If $\mathbf{u}$ and $\mathbf{v}$ are nonzero vectors, then $\text{proj}_{\mathbf{v}}\,\mathbf{u}$ is the unique vector having these properties:

i. $\text{Proj}_{\mathbf{v}}\,\mathbf{u}$ is parallel to $\mathbf{v}$.
ii. $\mathbf{u}-\text{proj}_{\mathbf{v}}\,\mathbf{u}$ is orthogonal to $\mathbf{v}$.

EXAMPLE 6 Let $\mathbf{u}=2\mathbf{i}+3\mathbf{j}$ and $\mathbf{v}=\mathbf{i}+\mathbf{j}$. Calculate $\text{proj}_{\mathbf{v}}\,\mathbf{u}$.

Solution $\text{Proj}_{\mathbf{v}}\,\mathbf{u}=(\mathbf{u}\cdot\mathbf{v})\mathbf{v}/|\mathbf{v}|^2=[5/(\sqrt{2})^2]\mathbf{v}=(5/2)\mathbf{i}+(5/2)\mathbf{j}$. (See Figure 4.21.)

Figure 4.21

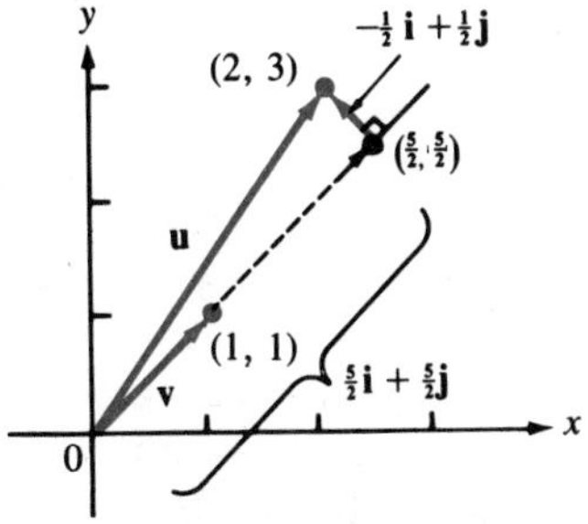

EXAMPLE 7 Let $\mathbf{u}=2\mathbf{i}-3\mathbf{j}$ and $\mathbf{v}=\mathbf{i}+\mathbf{j}$. Find $\text{proj}_{\mathbf{v}}\,\mathbf{u}$.

Solution Here $(\mathbf{u}\cdot\mathbf{v})/|\mathbf{v}|^2=-\frac{1}{2}$; hence $\text{proj}_{\mathbf{v}}\,\mathbf{u}=-\frac{1}{2}\mathbf{i}-\frac{1}{2}\mathbf{j}$. (See Figure 4.22.)

Figure 4.22

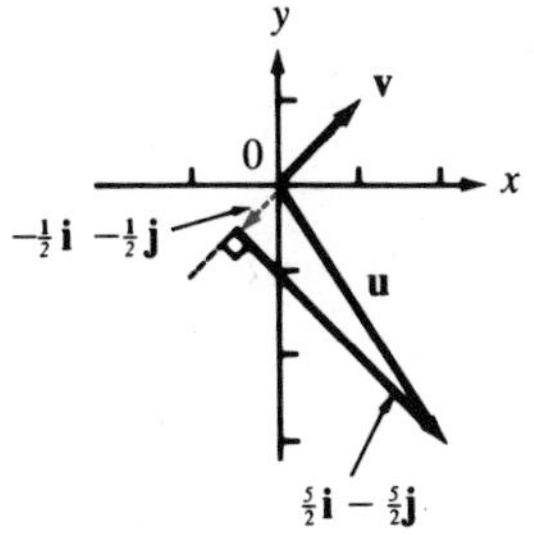

Before completing this section, we show how vectors can be used in applications. The physical concept of *force* is defined to be *a vector quantity that tends to produce an acceleration of a body in the direction of its application.* The units of force are pounds (in the English system) and newtons† (in the metric system).

If more than one force is applied to an object, then we define the **resultant** of the forces applied to the object to be the *vector sum* of these forces. We can think of the resultant as the *net* applied force; i.e., the object moves as if only one force, the resultant force, is applied to it.

EXAMPLE 8 A force of 3 N is applied to the left side of an object, a force of 4 N is applied from the bottom, and a force of 7 N is applied from an angle of $\pi/4$ to the horizontal. What is the resultant of forces applied to the object?

†Abbreviated N.

Figure 4.23

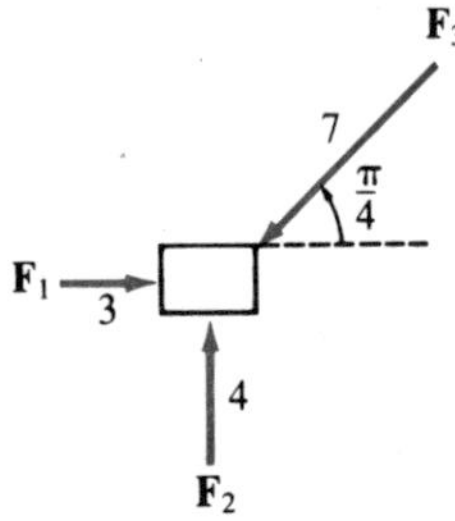

Solution The forces are indicated in Figure 4.23. We write each force as a magnitude times a unit vector in the indicated direction. For convenience, we can think of the center of the object as being at the origin. Then $\mathbf{F}_1 = 3\mathbf{i}$, $\mathbf{F}_2 = 4\mathbf{j}$, and $\mathbf{F}_3 = -(7/\sqrt{2})(\mathbf{i}+\mathbf{j})$. This last follows from the fact that the vector $-(1/\sqrt{2})(\mathbf{i}+\mathbf{j})$ is a unit vector pointing toward the origin and making an angle of $\pi/4$ with the x-axis (see Example 4.1.9). Then the resultant is given by

$$\mathbf{F} = \mathbf{F}_1 + \mathbf{F}_2 + \mathbf{F}_3 = \left(3 - \frac{7}{\sqrt{2}}\right)\mathbf{i} + \left(4 - \frac{7}{\sqrt{2}}\right)\mathbf{j}$$

The magnitude of $\mathbf{F}$ is

$$|\mathbf{F}| = \sqrt{\left(3 - \frac{7}{\sqrt{2}}\right)^2 + \left(4 - \frac{7}{\sqrt{2}}\right)^2} = \sqrt{74 - \frac{98}{\sqrt{2}}} \approx 2.17\text{ N}$$

The direction θ can be calculated by first finding the unit vector in the direction of $\mathbf{F}$:

$$\begin{aligned}\frac{\mathbf{F}}{|\mathbf{F}|} &= \frac{3-(7/\sqrt{2})}{\sqrt{74-(98/\sqrt{2})}}\mathbf{i} + \frac{4-(7/\sqrt{2})}{\sqrt{74-(98/\sqrt{2})}}\mathbf{j} \\ &= (\cos\theta)\mathbf{i} + (\sin\theta)\mathbf{j}\end{aligned}$$

Then

$$\cos\theta = \frac{3-(7/\sqrt{2})}{\sqrt{74-(98/\sqrt{2})}} \approx -0.8990$$

and $\theta \approx 3.5954 \approx 206°$ (or $-154°$). (See Figure 4.24.)

Figure 4.24

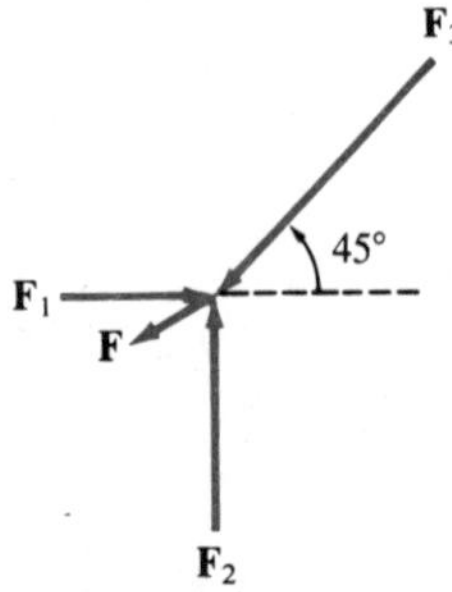

EXAMPLE 9 In Example 8, what additional force must be applied in order that the object remain at rest?

Solution We must apply a force $\mathbf{F}_4$ so that the resultant of the four forces is $\mathbf{0}$. If $\mathbf{F}_4 = a\mathbf{i} + b\mathbf{j}$, then the new resultant is given by $\mathbf{0} = \mathbf{F}_1 + \mathbf{F}_2 + \mathbf{F}_3 + \mathbf{F}_4 = [3 - (7/\sqrt{2}) + a]\,\mathbf{i} + [4 - (7/\sqrt{2}) + b)\mathbf{j}$. In order for this to be $\mathbf{0}$, we must have $a = (7/\sqrt{2}) - 3$ and $b = (7/\sqrt{2}) - 4$ so that $\mathbf{F}_4 = -\mathbf{F}$, where $\mathbf{F}$ is the resultant of $\mathbf{F}_1$, $\mathbf{F}_2$, and $\mathbf{F}_3$. (See Figure 4.25.)

Figure 4.25

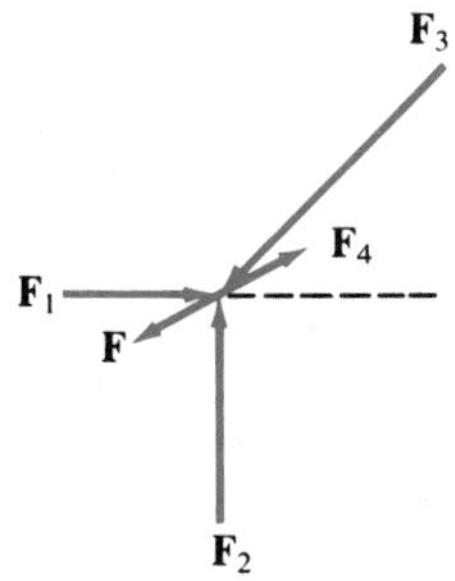

WORK

A physical quantity closely related to force is **work**. The work done by a force F in moving an object a distance d in the same direction as F is defined by

$$W = Fd \tag{6}$$

where units of work are newton-meters or foot-pounds. Motion is not always in the direction of the force, however. For a ball rolling down a hill, for example, the gravitational force is vertical but the ball follows the contour of the hill. In general, we may define work:

$$W = (\text{component of } \mathbf{F} \text{ in direction of motion}) \times (\text{distance moved}) \tag{7}$$

If the object moves from P to Q, then the distance moved is $|\overrightarrow{PQ}|$. The vector $\mathbf{d}$, one of whose representations is $\overrightarrow{PQ}$, is called a **displacement vector**. Then, from Definition 4,

$$\text{Component of } \mathbf{F} \text{ in direction of motion} = \frac{\mathbf{F} \cdot \mathbf{d}}{|\mathbf{d}|} \tag{8}$$

Finally, combining (7) and (8), we obtain

$$W = \frac{\mathbf{F} \cdot \mathbf{d}}{|\mathbf{d}|}|\mathbf{d}| = \mathbf{F} \cdot \mathbf{d} \tag{9}$$

That is: *The work done is the scalar product of the force* **F** *and the displacement vector* **d**. Note that if **F** acts in the direction **d** and if φ denotes the angle between **F** and **d**, then $\mathbf{F}\cdot\mathbf{d}=|\mathbf{F}|\,|\mathbf{d}|\cos\varphi=|\mathbf{F}|\,|\mathbf{d}|\cos 0=|\mathbf{F}|\,|\mathbf{d}|$, which is formula (6).

EXAMPLE 10 A force of 4 N has the direction of $\pi/3$. What is the work done in moving an object from the point (1, 2) to the point (5, 4) where distances are measured in meters?

Solution A unit vector with direction $\pi/3$ is given by $\mathbf{u}=(\cos\pi/3)\mathbf{i}+(\sin\pi/3)\mathbf{j}=(1/2)\mathbf{i}+(\sqrt{3}/2)\mathbf{j}$. Thus $\mathbf{F}=4\mathbf{u}=2\mathbf{i}+2\sqrt{3}\mathbf{j}$. The displacement vector **d** is given by $(5-1)\mathbf{i}+(4-2)\mathbf{j}=4\mathbf{i}+2\mathbf{j}$. Thus

$$W=\mathbf{F}\cdot\mathbf{d}=(2\mathbf{i}+2\sqrt{3}\mathbf{j})\cdot(4\mathbf{i}+2\mathbf{j})=(8+4\sqrt{3})\approx 14.93\text{ N-m}$$

The component of **F** in the direction of motion is sketched in Figure 4.26.

Figure 4.26

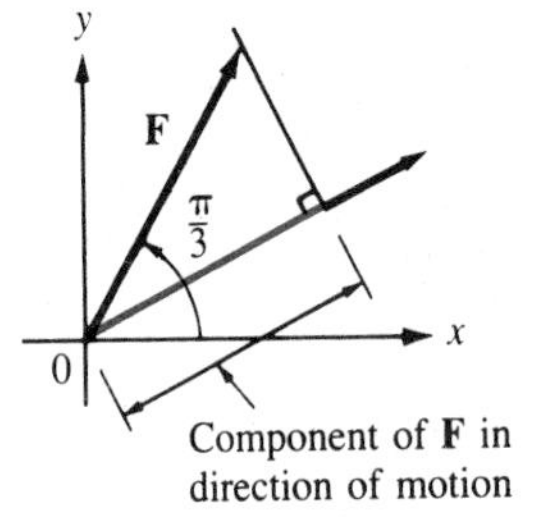

PROBLEMS 4.2

In Problems 1–8 calculate both the scalar product of the two vectors and the cosine of the angle between them.

1. $\mathbf{u}=\mathbf{i}+\mathbf{j}$; $\mathbf{v}=\mathbf{i}-\mathbf{j}$

2. $\mathbf{u}=3\mathbf{i}$; $\mathbf{v}=-7\mathbf{j}$

3. $\mathbf{u}=-5\mathbf{i}$; $\mathbf{v}=18\mathbf{j}$

4. $\mathbf{u}=\alpha\mathbf{i}$; $\mathbf{v}=\beta\mathbf{j}$; α, β real

5. $\mathbf{u}=2\mathbf{i}+5\mathbf{j}$; $\mathbf{v}=5\mathbf{i}+2\mathbf{j}$

6. $\mathbf{u}=2\mathbf{i}+5\mathbf{j}$; $\mathbf{v}=5\mathbf{i}-2\mathbf{j}$

7. $\mathbf{u}=-3\mathbf{i}+4\mathbf{j}$; $\mathbf{v}=-2\mathbf{i}-7\mathbf{j}$

8. $\mathbf{u}=4\mathbf{i}+5\mathbf{j}$; $\mathbf{v}=5\mathbf{i}-4\mathbf{j}$

9. Show that for any real numbers α and β, the vectors $\mathbf{u}=\alpha\mathbf{i}+\beta\mathbf{j}$ and $\mathbf{v}=\beta\mathbf{i}-\alpha\mathbf{j}$ are orthogonal.

10. Let **u**, **v**, and **w** denote three arbitrary vectors. Explain why the product $\mathbf{u}\cdot\mathbf{v}\cdot\mathbf{w}$ is *not defined*.

In Problems 11–16 determine whether the given vectors are orthogonal, parallel, or neither. Then sketch each pair.

11. $\mathbf{u}=3\mathbf{i}+5\mathbf{j}$; $\mathbf{v}=-6\mathbf{i}-10\mathbf{j}$

12. $\mathbf{u}=2\mathbf{i}+3\mathbf{j}$; $\mathbf{v}=6\mathbf{i}-4\mathbf{j}$

13. $\mathbf{u}=2\mathbf{i}+3\mathbf{j}$; $\mathbf{v}=6\mathbf{i}+4\mathbf{j}$

14. $\mathbf{u}=2\mathbf{i}+3\mathbf{j}$; $\mathbf{v}=-6\mathbf{i}+4\mathbf{j}$

15. $\mathbf{u}=7\mathbf{i}$; $\mathbf{v}=-23\mathbf{j}$

16. $\mathbf{u}=2\mathbf{i}-6\mathbf{j}$; $\mathbf{v}=-\mathbf{i}+3\mathbf{j}$

17. Let $\mathbf{u}=3\mathbf{i}+4\mathbf{j}$ and $\mathbf{v}=\mathbf{i}+\alpha\mathbf{j}$. Determine α such that:

a. **u** and **v** are orthogonal.

b. **u** and **v** are parallel.

c. The angle between **u** and **v** is $\pi/4$.

d. The angle between **u** and **v** is $\pi/3$.

18. Let $\mathbf{u}=-2\mathbf{i}+5\mathbf{j}$ and $\mathbf{v}=\alpha\mathbf{i}-2\mathbf{j}$. Determine α such that:

a. $\mathbf{u}$ and $\mathbf{v}$ are orthogonal.
b. $\mathbf{u}$ and $\mathbf{v}$ are parallel.
c. The angle between $\mathbf{u}$ and $\mathbf{v}$ is $2\pi/3$.
d. The angle between $\mathbf{u}$ and $\mathbf{v}$ is $\pi/3$.

19. In Problem 17 show that there is no value of α for which $\mathbf{u}$ and $\mathbf{v}$ have opposite directions.

20. In Problem 18 show that there is no value of α for which $\mathbf{u}$ and $\mathbf{v}$ have the same direction.

In Problems 21–30 calculate $\text{proj}_{\mathbf{v}}\,\mathbf{u}$.

21. $\mathbf{u}=3\mathbf{i}$; $\mathbf{v}=\mathbf{i}+\mathbf{j}$
22. $\mathbf{u}=-5\mathbf{j}$; $\mathbf{v}=\mathbf{i}+\mathbf{j}$
23. $\mathbf{u}=2\mathbf{i}+\mathbf{j}$; $\mathbf{v}=\mathbf{i}-2\mathbf{j}$
24. $\mathbf{u}=2\mathbf{i}+3\mathbf{j}$; $\mathbf{v}=4\mathbf{i}+\mathbf{j}$
25. $\mathbf{u}=\mathbf{i}+\mathbf{j}$; $\mathbf{v}=2\mathbf{i}-3\mathbf{j}$
26. $\mathbf{u}=\mathbf{i}+\mathbf{j}$; $\mathbf{v}=2\mathbf{i}+3\mathbf{j}$
27. $\mathbf{u}=\alpha\mathbf{i}+\beta\mathbf{j}$; $\mathbf{v}=\mathbf{i}+\mathbf{j}$; α, β real and positive
28. $\mathbf{u}=\mathbf{i}+\mathbf{j}$; $\mathbf{v}=\alpha\mathbf{i}+\beta\mathbf{j}$, α, β real and positive
29. $\mathbf{u}=\alpha\mathbf{i}-\beta\mathbf{j}$; $\mathbf{v}=\mathbf{i}+\mathbf{j}$; α, β real and positive with $\alpha>\beta$
30. $u=\alpha\mathbf{i}-\beta\mathbf{j}$; $\mathbf{v}=\mathbf{i}+\mathbf{j}$; α, β real and positive with $\alpha<\beta$.

31. Let $\mathbf{u}=a_1\mathbf{i}+b_1\mathbf{j}$ and $\mathbf{v}=a_2\mathbf{i}+b_2\mathbf{j}$. Give a condition on a_1, b_1, a_2, and b_2 which will ensure that $\mathbf{v}$ and $\text{proj}_{\mathbf{v}}\,\mathbf{u}$ have the same direction.

32. In Problem 31, give a condition which will ensure that $\mathbf{v}$ and $\text{proj}_{\mathbf{v}}\,\mathbf{u}$ have opposite directions.

33. Let $P=(2,3)$, $Q=(5,7)$, $R=(2,-3)$, and $S=(1,2)$. Calculate $\text{proj}_{\overrightarrow{PQ}}\,\overrightarrow{RS}$ and $\text{proj}_{\overrightarrow{RS}}\,\overrightarrow{PQ}$.

34. Let $P=(-1,3)$, $Q=(2,4)$, $R=(-6,-2)$, and $S=(3,0)$. Calculate $\text{proj}_{\overrightarrow{PQ}}\,\overrightarrow{RS}$ and $\text{proj}_{\overrightarrow{RS}}\,\overrightarrow{PQ}$.

35. Prove that the nonzero vectors $\mathbf{u}$ and $\mathbf{v}$ are parallel if and only if $\mathbf{v}=\alpha\mathbf{u}$ for some constant α. [*Hint:* Show that $\cos\varphi=\pm 1$ if and only if $\mathbf{v}=\alpha\mathbf{u}$.]

36. Prove that $\mathbf{u}$ and $\mathbf{v}$ are orthogonal if and only if $\mathbf{u}\cdot\mathbf{v}=0$.

37. Show that the vector $\mathbf{v}=a\mathbf{i}+b\mathbf{j}$ is orthogonal to the line $ax+by+c=0$.

38. Show that the vector $\mathbf{u}=b\mathbf{i}-a\mathbf{j}$ is parallel to the line $ax+by+c=0$.

39. A triangle has vertices $(1,3)$, $(4,-2)$, and $(-3,6)$. Find the cosine of each of its angles.

40. A triangle has vertices (a_1,b_1), (a_2,b_2), and (a_3,b_3). Find a formula for the cosines of each of its angles.

***41.** The **Cauchy-Schwartz inequality** states that for any real numbers a_1, a_2, b_1, and b_2,

$$\left|\sum_{k=1}^{2} a_k b_k\right| \le \left(\sum_{k=1}^{2} a_k^2\right)^{1/2}\left(\sum_{k=1}^{2} b_k^2\right)^{1/2}$$

Use the scalar product to prove this formula. Under what circumstances can the inequality be replaced by an equality?

42. Prove that the shortest distance between a point and a line is measured along a line through the point and perpendicular to the line.

43. Find the distance between $P=(2,3)$ and the line through the points $Q=(-1,7)$ and $R=(3,5)$.

44. Find the distance between $(3,7)$ and the line along the vector $\mathbf{v}=2\mathbf{i}-3\mathbf{j}$ passing through the origin.

In Problems 45–50 find the resultant of the forces acting on an object. Then find the force that must be applied so that the object will remain at rest.

45. 2 N (from right); 5 N (from above)
46. 2 N (from left); 5 N (from below)
47. 5 lb (from above); 4 lb (from direction $\pi/6$)
48. 6 N (from left); 4 N (from direction $\pi/4$); 2 N (from direction $\pi/3$)
49. 2 N (from above); 3 N (from direction $3\pi/4$)
50. 5 N (from direction $\pi/3$); 5 N (from direction $2\pi/3$)

In Problems 51–56 find the work done when the force with given magnitude and direction moves an object from P to Q. All distances are measured in meters. (Note that work can be negative.)

51. $|\mathbf{F}| = 2$ N; $\theta = \pi/2$; $P = (5, 7)$; $Q = (1, 1)$
52. $|\mathbf{F}| = 6$ N; $\theta = \pi/4$; $P = (2, 3)$; $Q = (-1, 4)$
53. $|\mathbf{F}| = 4$ N; $\theta = \pi/6$; $P = (-1, 2)$; $Q = (3, 4)$
54. $|\mathbf{F}| = 7$ N; $\theta = 2\pi/3$; $P = (4, -3)$; $Q = (1, 0)$
55. $|\mathbf{F}| = 4$ N; θ is direction of $2\mathbf{i}+3\mathbf{j}$; $P = (2, 0)$; $Q = (-1, 3)$
56. $|\mathbf{F}| = 5$ N; θ is direction of $-3\mathbf{i}+2\mathbf{j}$; $P = (1, 3)$; $Q = (4, -6)$
***57.** Let P and Q be two distinct points in the plane. Let R be a point on the straight line segment joining P and Q and suppose that $|\overrightarrow{PR}| = a$ and $|\overrightarrow{RQ}| = b$. Let $\mathbf{w} = \overrightarrow{OR}$, $\mathbf{u} = \overrightarrow{OP}$, and $\mathbf{v} = \overrightarrow{OQ}$. (See Figure 4.27.) Show that

$$\mathbf{w} = \frac{b}{a+b}\mathbf{u} + \frac{a}{a+b}\mathbf{v}$$

Figure 4.27

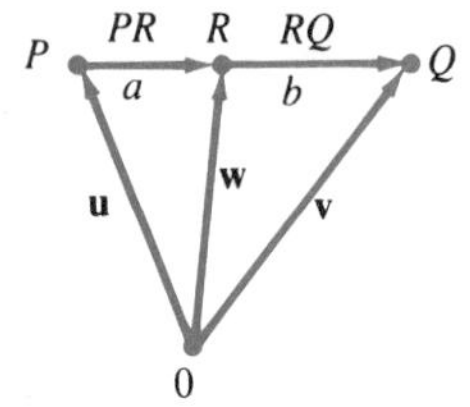

58. Use the result of Problem 57 to show that the line segment joining the midpoints of two sides of a triangle has the length of half the third side and is parallel to the third side.
***59.** Prove that the medians of a triangle intersect at a fixed point which is two-thirds the distance from each vertex to the opposite side.
60. Use the result of Problem 59 to find the point of intersection of the medians of the triangle with vertices at $(3, 4)$, $(2, -1)$, and $(-3, 2)$.
61. Show that the diagonals of a parallelogram bisect each other.
62. Show that the diagonals of a rhombus are orthogonal.
63. Show that in a trapezoid, the line segment which joins the midpoints of the two sides that are not parallel is parallel to the parallel sides and has a length equal to the average of the lengths of the parallel sides.
64. Use vector methods to show that the angles opposite the equal sides in an isosceles triangle are equal.
65. Let A be a 2×2 matrix such that each column is a unit vector and the two columns are orthogonal. Show that A is invertible and that $A^{-1} = A^t$. (A is called an **orthogonal** matrix.)

4.3 Vectors in Space

We have seen how any point in a plane can be represented as an ordered pair of real numbers. Analogously, any point in space can be represented by an **ordered triple** of real numbers

$$(a, b, c) \tag{1}$$

Vectors of the form (1) comprise $\mathbb{R}^3$. To represent a point in space, we begin by choosing a point in $\mathbb{R}^3$. We call this point the **origin**, denoted 0. Then we draw three mutually perpendicular axes, which we label the ***x*-axis**, the ***y*-axis**, and the ***z*-axis**. These axes can be selected in a variety of ways, but the most common selection has the x- and y-axes drawn horizontally with the z-axis vertical. On each axis, we choose a positive direction and measure distance along each axis as the number of units in this positive direction measured from the origin.

Figure 4.28

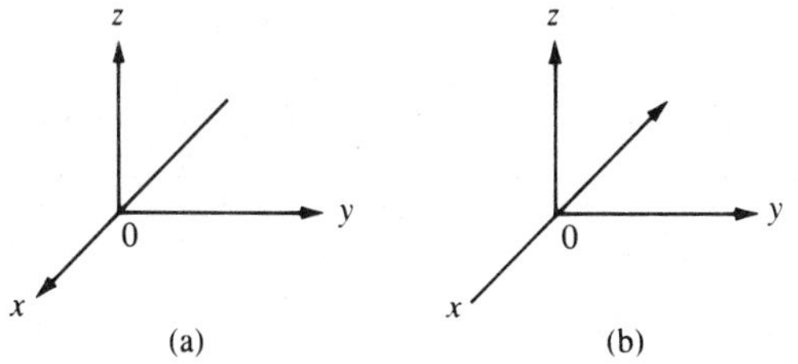

The two basic systems of drawing these axes are depicted in Figure 4.28. If the axes are placed as in Figure 4.28*a*, then the system is called a **right-handed system**; if they are placed as in Figure 4.28*b*, the system is a *left-handed system.* In the figures, the arrows indicate the positive directions on the axes. The reason for this choice of terms is as follows: In a right-handed system, if you place your right hand so that your index finger points in the positive direction of the x-axis while your middle finger points in the positive direction of the y-axis, then your thumb will point in the positive direction of the z-axis. This concept is illustrated in Figure 4.29. For a left-handed system, the same rule works for your left hand. For the remainder of this text, we shall follow common practice and depict the coordinate axes using a right-handed system.

Figure 4.29

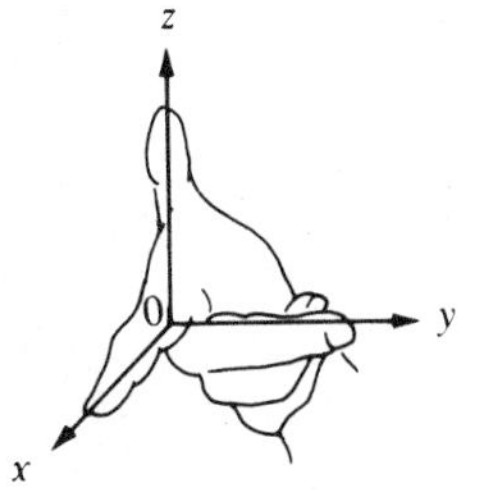

If you still have trouble visualizing the placement of these axes, try the following approach. Face any uncluttered corner of the room in which you are sitting. Call the corner the origin. Then the x-axis lies along the floor,

Figure 4.30

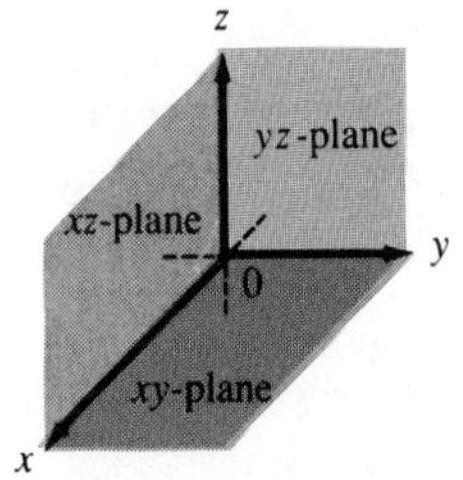

along the wall, and to your left; the y-axis lies along the floor, along the wall, and to your right; and the z-axis lies along the vertical intersection of the two perpendicular walls. This arrangement is illustrated in Figure 4.30.

The three axes in our system determine three **coordinate planes**, which we shall call the xy-plane, the xz-plane, and the yz-plane. The xy-plane contains the x- and y-axes and is simply the plane with which we have been dealing in most of this book. The xz- and yz-planes can be thought of in a similar way.

Having built our structure of coordinate axes and planes, we can describe any point P in $\mathbb{R}^3$ in a unique way:

$$P = (x, y, z) \tag{2}$$

where the first coordinate x is the distance from the yz-plane to P (measured in the positive direction of the x-axis and along a line parallel to the x-axis), the second coordinate y is the distance from the xz-plane to P (measured in the positive direction of the y-axis and along a line parallel to the y-axis), and the third coordinate z is the distance from the xy-plane to P (measured in the positive direction of the z-axis and along a line parallel to the z-axis). Thus, for example, any point in the xy-plane has z-coordinate 0; any point in the xz-plane has y-coordinate 0; and any point in the yz-plane has x-coordinate 0. Some representative points are sketched in Figure 4.31.

Figure 4.31

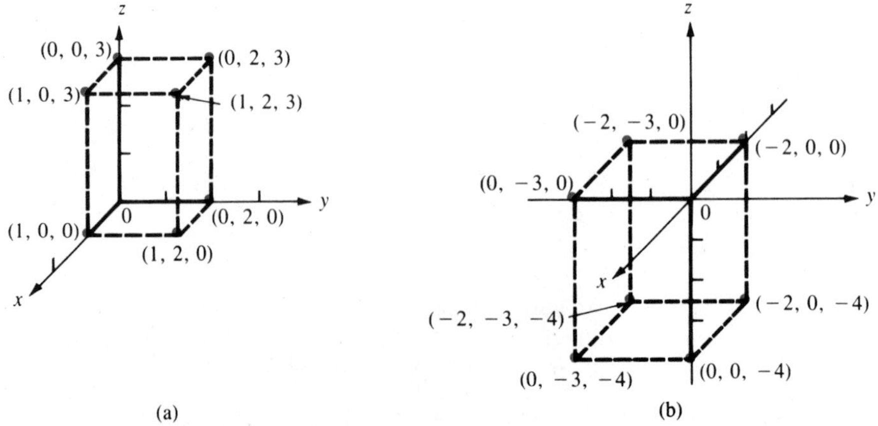

In this system, the three coordinate planes divide $\mathbb{R}^3$ into eight **octants**, just as in $\mathbb{R}^2$ the two coordinate axes divide the plane into four quadrants. The first octant is always chosen to be the one in which the three coordinates are positive.

The coordinate system we have just established is often referred to as the **rectangular coordinate system** or the **cartesian coordinate system**. Once we are comfortable with the notion of depicting a point in this system, we can generalize many of our ideas from the plane.

THEOREM 1 Let $P=(x_1, y_1, z_1)$ and $Q=(x_2, y_2, z_2)$ be two points in space. Then the distance $\overline{PQ}$ between P and Q is given by

$$\overline{PQ}=\sqrt{(x_1-x_2)^2+(y_1-y_2)^2+(z_1-z_2)^2} \tag{3}$$

Proof The two points are sketched in Figure 4.32. From the pythagorean theorem,

Figure 4.32

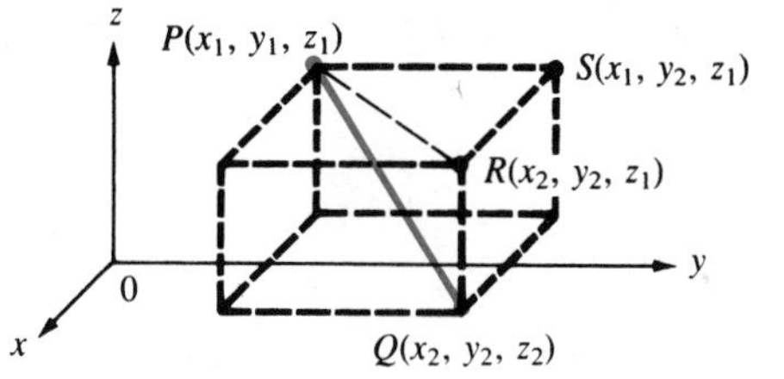

since the line segments PS and SR are perpendicular, the triangle PSR is a right triangle and

$$\overline{PR}^2=\overline{PS}^2+\overline{SR}^2 \tag{4}$$

But triangle PRQ is also a right triangle so that

$$\overline{PQ}^2=\overline{PR}^2+\overline{RQ}^2 \tag{5}$$

So, combining (4) and (5), we get

$$\overline{PQ}^2=\overline{PS}^2+\overline{SR}^2+\overline{RQ}^2 \tag{6}$$

Since the x and z coordinates of P and S are equal,

$$\overline{PS}^2=(y_2-y_1)^2 \tag{7}$$

Similarly,

$$\overline{RS}^2=(x_2-x_1)^2 \tag{8}$$

and

$$\overline{RQ}^2=(z_2-z_1)^2 \tag{9}$$

Thus, using (7), (8), and (9) in (6) yields

$$\overline{PQ}^2=(x_2-x_1)^2+(y_2-y_1)^2+(z_2-z_1)^2$$

and the proof is complete. ■

EXAMPLE 1 Calculate the distance between the points $(3, -1, 6)$ and $(-2, 3, 5)$.

Solution

$$\overline{PQ} = \sqrt{[3-(-2)]^2 + (-1-3)^2 + (6-5)^2} = \sqrt{42}$$

In Sections 4.1 and 4.2 we developed geometric properties of vectors in the plane. Given the similarity between the coordinate systems in $\mathbb{R}^2$ and $\mathbb{R}^3$, it should come as no surprise that vectors in $\mathbb{R}^2$ and $\mathbb{R}^3$ have very similar structures. We shall now develop the notion of a vector in space. The development will closely follow the development in the last two sections and, therefore, some of the details will be omitted.

Let P and Q be two distinct points in $\mathbb{R}^3$. Then the **directed line segment** $\overrightarrow{PQ}$ is the straight line segment that extends from P to Q. Two directed line segments are **equivalent** if they have the same magnitude and direction. A **vector** in $\mathbb{R}^3$ is the set of all directed line segments equivalent to a given directed line segment and any directed line segment $\overrightarrow{PQ}$ in that set is called a **representation** of the vector.

So far, our definitions are identical. For convenience, we choose P to be the origin so that the vector $\mathbf{v} = \overrightarrow{OQ}$ can be described by the coordinates (x, y, z) of the point Q. Then the magnitude of $\mathbf{v} = |\mathbf{v}| = \sqrt{x^2 + y^2 + z^2}$ (from Theorem 1).

EXAMPLE 2 Let $\mathbf{v} = (1, 3, -2)$. Find $|\mathbf{v}|$.

Solution $|\mathbf{v}| = \sqrt{1^2 + 3^2 + (-2)} = \sqrt{14}$.

Let $\mathbf{u} = (x_1, y_1, z_1)$ and $\mathbf{v} = (x_2, y_2, z_2)$ be two vectors and let α be a real number (scalar). Then we define

$$\mathbf{u} + \mathbf{v} = (x_1 + x_2, y_1 + y_2, z_1 + z_2)$$

and

$$\alpha\mathbf{u} = (\alpha x_1, \alpha y_1, \alpha z_1)$$

This is the same definition of vector addition and scalar multiplication we had before and is illustrated in Figure 4.33.

Figure 4.33

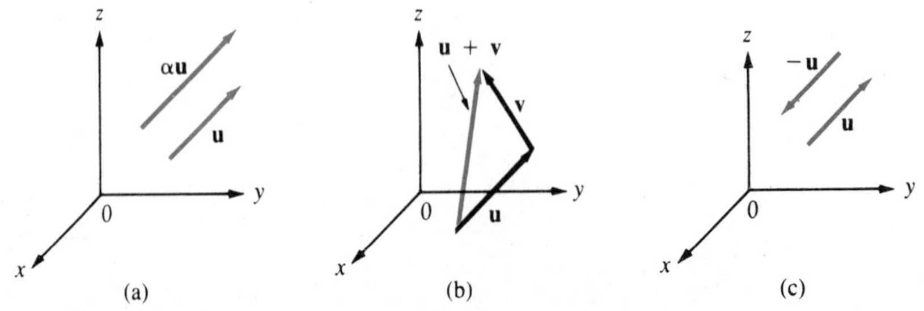

Figure 4.33 (continued)

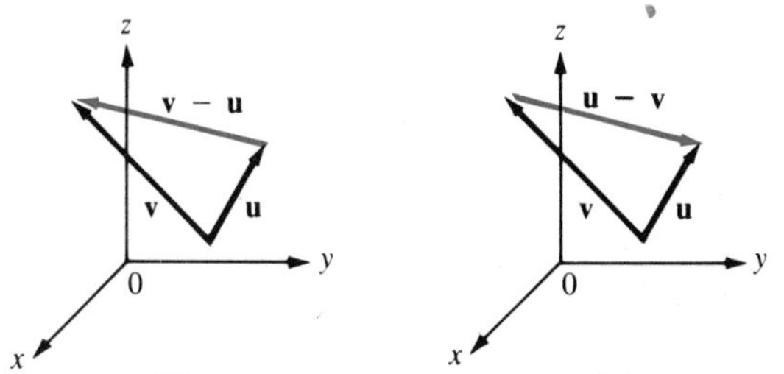

(d) (e)

A **unit vector u** is a vector with magnitude 1. If **v** is any nonzero vector, then $\mathbf{u}=\mathbf{v}/|\mathbf{v}|$ is a unit vector having the same direction as **v**.

EXAMPLE 3 Find a unit vector having the same direction as $\mathbf{v}=(2, 4-3)$.

Solution Since $\mathbf{v}=\sqrt{2^2+4^2+(-3)^2}=\sqrt{29}$, we have $\mathbf{u}=(2/\sqrt{29},\ 4/\sqrt{29},\ -3/\sqrt{29})$.

We can now formally define the direction of a vector in $\mathbb{R}^3$. We cannot define it to be the angle θ the vector makes with the positive x-axis, since, for example, if $0<\theta<\pi/2$, then there are an *infinite number* of vectors making the angle θ with the positive x-axis, and these together form a cone (see Figure 4.34).

Figure 4.34

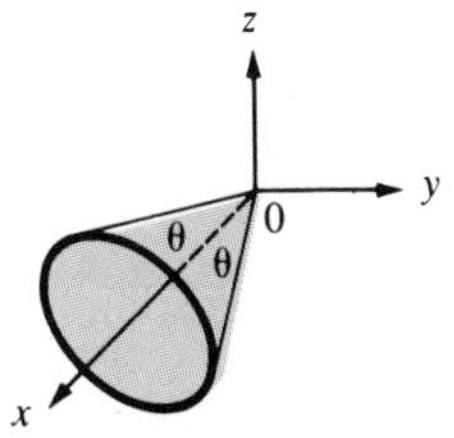

DEFINITION 1 **DIRECTION IN $\mathbb{R}^3$** The **direction** of a nonzero vector **v** in $\mathbb{R}^3$ is defined to be the unit vector $\mathbf{u} = \mathbf{v}/|\mathbf{v}|$.

Remark. We could have defined the direction of a vector **v** in $\mathbb{R}^2$ in this way. For if $\mathbf{u}=\mathbf{v}/|\mathbf{v}|$, then $\mathbf{u}=(\cos\theta, \sin\theta)$, where θ is the direction of **v**.

It would still be satisfying to define the direction of a vector in terms of some angles. Let **v** be the vector $\overrightarrow{OP}$ depicted in Figure 4.35. We define α to be the angle between **v** and the positive x-axis, β the angle between **v** and the positive y-axis, and γ the angle between **v** and the positive z-axis. The

Figure 4.35

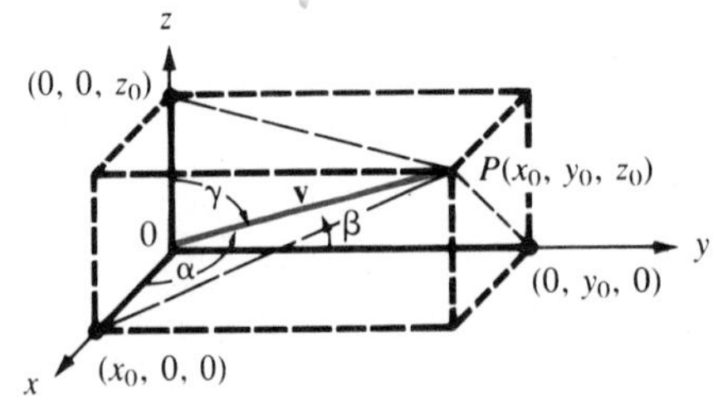

angles α, β, and γ are called the **direction angles** of the vector **v**. Then, from Figure 4.35,

$$\cos\alpha = \frac{x_0}{|\mathbf{v}|} \qquad \cos\beta = \frac{y_0}{|\mathbf{v}|} \qquad \cos\gamma = \frac{z_0}{|\mathbf{v}|} \tag{10}$$

If **v** is a unit vector, then $|\mathbf{v}| = 1$ and

$$\cos\alpha = x_0 \qquad \cos\beta = y_0 \qquad \cos\gamma = z_0 \tag{11}$$

By definition, each of these three angles lies in the interval $[0, \pi]$. The cosines of these angles are called the **direction cosines** of the vector **v**. Note, from Equations (10), that

$$\cos^2\alpha + \cos^2\beta + \cos^2\gamma = \frac{x_0^2 + y_0^2 + z_0^2}{|\mathbf{v}|^2} = \frac{x_0^2 + y_0^2 + z_0^2}{x_0^2 + y_0^2 + z_0^2} = 1 \tag{12}$$

If α, β, and γ are any three numbers between zero and π such that condition (12) is satisfied, then they uniquely determine a unit vector given by $\mathbf{u} = (\cos\alpha, \cos\beta, \cos\gamma)$.

Remark. If $\mathbf{v} = (a, b, c)$ and $|\mathbf{v}| \neq 1$, then the numbers a, b, and c are called **direction numbers** of the vector **v**.

EXAMPLE 4 Find the direction cosines of the vector $\mathbf{v} = (4, -1, 6)$.

Solution The direction of **v** is $\mathbf{v}/|\mathbf{v}| = \mathbf{v}/\sqrt{53} = (4/\sqrt{53}, -1/\sqrt{53}, 6/\sqrt{53})$. Then $\cos\alpha = 4/\sqrt{53} \approx 0.5494$, $\cos\beta = -1/\sqrt{53} \approx -0.1374$, and $\cos\gamma = 6/\sqrt{53} \approx 0.8242$. From these, we use a table of cosines or a hand calculator to obtain $\alpha \approx 56.7° \approx 0.989$ rad, $\beta \approx 97.9° \approx 1.71$ rad, and $\gamma = 34.5° \approx 0.602$ rad. The vector, along with the angles α, β, and γ, is sketched in Figure 4.36.

EXAMPLE 5 Find a vector **v** of magnitude 7 whose direction cosines are $1/\sqrt{6}$, $1/\sqrt{3}$, and $1/\sqrt{2}$.

Figure 4.36

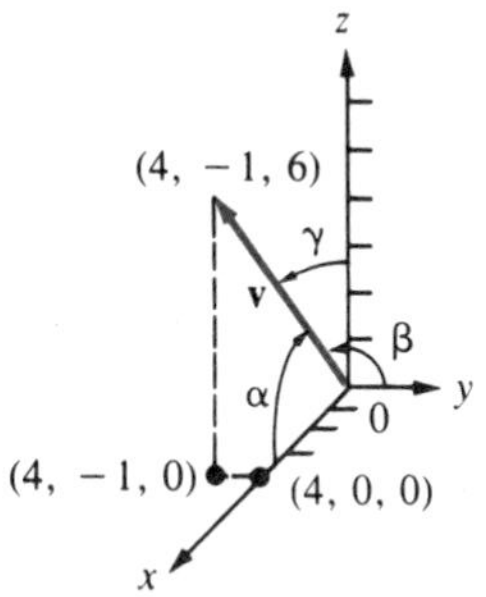

Solution Let $\mathbf{u} = (1/\sqrt{6},\ 1/\sqrt{3},\ 1/\sqrt{2})$. Then $\mathbf{u}$ is a unit vector since $|\mathbf{u}| = 1$. Thus the direction of $\mathbf{v}$ is given by $\mathbf{u}$ and $\mathbf{v} = |\mathbf{v}|\ \mathbf{u} = 7\mathbf{u} = (7/\sqrt{6}, 7/\sqrt{3}, 7/\sqrt{2})$.

Note. We can solve this problem because $(1/\sqrt{6})^2 + (1/\sqrt{3})^2 + (1/\sqrt{2})^2 = 1$.

It is interesting to note that if $\mathbf{v}$ in $\mathbb{R}^2$ is written $\mathbf{v} = (\cos\ \theta)\mathbf{i} + (\sin\ \theta)\mathbf{j}$, where θ is the direction of $\mathbf{v}$, then $\cos\ \theta$ and $\sin\ \theta$ are the direction cosines of $\mathbf{v}$. Here $\alpha = \theta$ and we define β to be the angle that $\mathbf{v}$ makes with the y-axis (see Figure 4.37). Then $\beta = (\pi/2) - \alpha$ so that $\cos\ \beta = \cos\ (\pi/2 - \alpha) =$

Figure 4.37

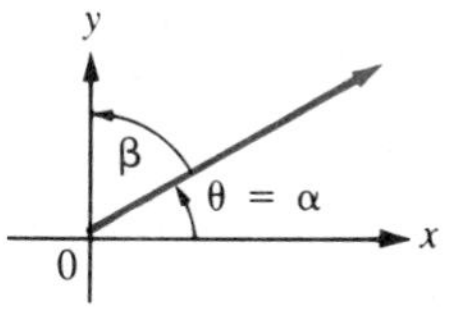

$\sin\ \alpha$ and $\mathbf{v}$ can be written in the "direction cosine" form:

$$\mathbf{v} = \cos\alpha\ \mathbf{i} + \cos\beta\ \mathbf{j}$$

In Section 4.1 we saw how any vector in the plane can be written in terms of the basis vectors $\mathbf{i}$ and $\mathbf{j}$. To extend this idea to $\mathbb{R}^3$ we define

$$\mathbf{i} = (1, 0, 0) \qquad \mathbf{j} = (0, 1, 0) \qquad \mathbf{k} = (0, 0, 1) \tag{13}$$

Here $\mathbf{i}$, $\mathbf{j}$, and $\mathbf{k}$ are unit vectors. The vector $\mathbf{i}$ lies along the x-axis, $\mathbf{j}$ along the y-axis, and $\mathbf{k}$ along the z-axis. These are sketched in Figure 4.38. If

Figure 4.38

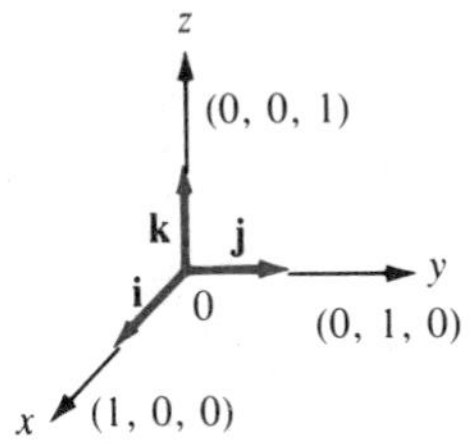

$\mathbf{v}=(x, y, z)$ is any vector in $\mathbb{R}^3$, then

$$\mathbf{v}=(x, y, z)=(x, 0, 0)+(0, y, 0)+(0, 0, z)=x\mathbf{i}+y\mathbf{j}+z\mathbf{k} \tag{14}$$

That is: *Any vector* $\mathbf{v}$ *in* $\mathbb{R}^3$ *can be written in a unique way in terms of the vectors* $\mathbf{i}$, $\mathbf{j}$, *and* $\mathbf{k}$.

Let $P=(a_1, b_1, c_1)$ and $Q=(a_2, b_2, c_2)$. Then, as in Section 4.1, the vector $\mathbf{v}=\overrightarrow{PQ}$ can be written

$$\mathbf{v}=(a_2-a_1)\mathbf{i}+(b_2-b_1)\mathbf{j}+(c_2-c_1)\mathbf{k} \tag{15}$$

The vectors $\mathbf{v}$ and $\overrightarrow{PQ}$ are drawn in Figure 4.39.

Figure 4.39

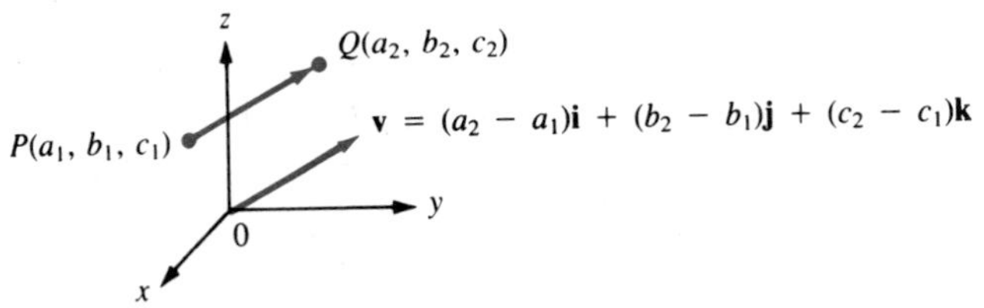

EXAMPLE 6 Find a vector in space that can be represented by the directed line segment from $(2, -1, 4)$ to $(5, 1, -3)$.

Solution $\mathbf{v}=(5-2)\mathbf{i}+[1-(-1)]\mathbf{j}+(-3-4)\mathbf{k}=3\mathbf{i}+2\mathbf{j}-7\mathbf{k}$

The definition of the scalar product in $\mathbb{R}^3$ is, of course, the definition we have already seen in Section 2.2. Note that $\mathbf{i}\cdot\mathbf{i}=1$, $\mathbf{j}\cdot\mathbf{j}=1$, $\mathbf{k}\cdot\mathbf{k}=1$, $\mathbf{i}\cdot\mathbf{j}=0$, $\mathbf{j}\cdot\mathbf{k}=0$, and $\mathbf{i}\cdot\mathbf{k}=0$.

THEOREM 2 If φ denotes the smallest positive angle between two nonzero vectors $\mathbf{u}$ and $\mathbf{v}$, we have

$$\cos\varphi=\frac{\mathbf{u}\cdot\mathbf{v}}{|\mathbf{u}|\,|\mathbf{v}|} \tag{16}$$

Proof The proof is almost identical to the proof of Theorem 4.2.1 on page 128 and is left as an exercise (see Problem 62).

EXAMPLE 7 Calculate the cosine of the angle between $\mathbf{u}=3\mathbf{i}-\mathbf{j}+2\mathbf{k}$ and $\mathbf{v}=4\mathbf{i}+3\mathbf{j}-\mathbf{k}$.

Solution $\mathbf{u}\cdot\mathbf{v}=7$, $|\mathbf{u}|=\sqrt{14}$, and $|\mathbf{v}|=\sqrt{26}$ so that $\cos\varphi=7/\sqrt{(14)(26)}=7/\sqrt{364}\approx 0.3669$ and $\varphi\approx 68.5°\approx 1.2$ rad.

DEFINITION 2 **PARALLEL AND ORTHOGONAL VECTORS** Two nonzero vectors $\mathbf{u}$ and $\mathbf{v}$ are:

i. **Parallel** if the angle between them is zero or π
ii. **Orthogonal** (or **perpendicular**) if the angle between them is $\pi/2$

THEOREM 3

i. If $\mathbf{u}\neq\mathbf{0}$, then $\mathbf{u}$ and $\mathbf{v}$ are parallel if and only if $\mathbf{v}=\alpha\mathbf{u}$ for some constant $\alpha\neq 0$.
ii. If $\mathbf{u}$ and $\mathbf{v}$ are nonzero, then $\mathbf{u}$ and $\mathbf{v}$ are orthogonal if and only if $\mathbf{u}\cdot\mathbf{v}=0$.

Proof Again the proof is easy and is left as an exercise (see Problem 63).

EXAMPLE 8 Show that the vectors $\mathbf{u}=\mathbf{i}+3\mathbf{j}-4\mathbf{k}$ and $\mathbf{v}=-2\mathbf{i}-6\mathbf{j}+8\mathbf{k}$ are parallel.

Solution $\mathbf{u}\cdot\mathbf{v}=-52$, $|\mathbf{u}|=\sqrt{26}$, and $|\mathbf{v}|=\sqrt{104}=2\sqrt{26}$. Hence $\mathbf{u}\cdot\mathbf{v}/|\mathbf{u}|\,|\mathbf{v}|=-52/(\sqrt{26}\cdot 2\sqrt{26})=-1$ so that $\cos\theta=-1$, $\theta=\pi$, and $\mathbf{u}$ and $\mathbf{v}$ are parallel (but have opposite directions). Another way to see this is to note that $\mathbf{v}=-2\mathbf{u}$ so that, by Theorem 3, $\mathbf{u}$ and $\mathbf{v}$ are parallel. (See Figure 4.40.)

Figure 4.40

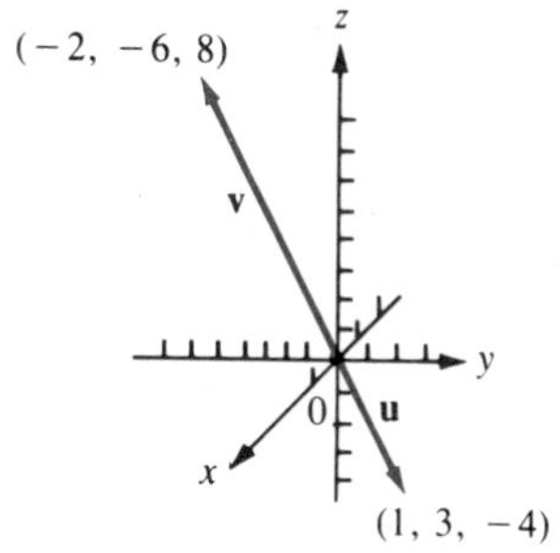

EXAMPLE 9 Find a number α such that $\mathbf{u}=8\mathbf{i}-2\mathbf{j}+4\mathbf{k}$ and $\mathbf{v}=2\mathbf{i}+3\mathbf{j}+\alpha\mathbf{k}$ are orthogonal.

Solution We must have $0=\mathbf{u}\cdot\mathbf{v}=10+4\alpha$ so that $\alpha=-\frac{5}{2}$. The vectors $\mathbf{u}$ and $\mathbf{v}$ are sketched in Figure 4.41.

We now turn to the definition of the projection of one vector on another.

Figure 4.41

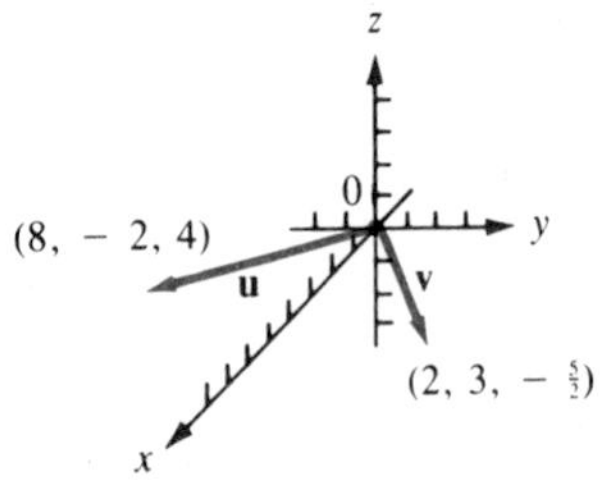

First we state the theorem, which is the analog of Theorem 4.2.4 (and which has an identical proof).

THEOREM 4 Let **v** be a nonzero vector. Then, for any other vector **u**,

$$\mathbf{w} = \mathbf{u} - \frac{\mathbf{u} \cdot \mathbf{v}}{|\mathbf{v}|^2}\mathbf{v}$$

is orthogonal to **v**.

DEFINITION 3 **PROJECTION** Let **u** and **v** be nonzero vectors. Then the **projection** of **u** on **v**, denoted $\text{proj}_{\mathbf{v}}\,\mathbf{u}$, is defined by

$$\text{proj}_{\mathbf{v}}\,\mathbf{u} = \frac{\mathbf{u} \cdot \mathbf{v}}{|\mathbf{v}|^2}\mathbf{v} \tag{17}$$

The **component** of **u** in the direction **v** is given by $(\mathbf{u} \cdot \mathbf{v})/|\mathbf{v}|$.

EXAMPLE 10 Let $\mathbf{u} = 2\mathbf{i} + 3\mathbf{j} + \mathbf{k}$ and $\mathbf{v} = \mathbf{i} + 2\mathbf{j} - 6\mathbf{k}$. Find $\text{proj}_{\mathbf{v}}\,\mathbf{u}$.

Solution Here $(\mathbf{u} \cdot \mathbf{v})/|\mathbf{v}|^2 = 2/41$ and $\text{proj}_{\mathbf{v}}\,\mathbf{u} = \frac{2}{41}\mathbf{i} + \frac{4}{41}\mathbf{j} - \frac{12}{41}\mathbf{k}$. The component of **u** in the direction **v** is $(\mathbf{u} \cdot \mathbf{v})/|\mathbf{v}| = 2/\sqrt{41}$.

Note that, as in the planar case, $\text{proj}_{\mathbf{v}}\,\mathbf{u}$ is a vector that has the same direction as **v** if $\mathbf{u} \cdot \mathbf{v} > 0$ and the direction opposite to that of **v** if $\mathbf{u} \cdot \mathbf{v} < 0$.

PROBLEMS 4.3

In Problems 1–9 sketch the given point in $\mathbb{R}^3$.

1. $(1, 4, 2)$
2. $(3, -2, 1)$
3. $(-1, 5, 7)$
4. $(8, -2, 3)$
5. $(-2, 1, -2)$
6. $(1, -2, 1)$
7. $(-2, -8, 0)$
8. $(0, 4, 7)$
9. $(1, 3, 0)$

In Problems 10–13 find the distance between the two points.

10. $(8, 1, 6)$; $(8, 1, 4)$
11. $(3, -4, 3)$; $(3, 2, 5)$
12. $(3, -4, 7)$; $(3, -4, 9)$
13. $(-2, 1, 3)$; $(4, 1, 3)$
14. Three points P, Q, and R are **collinear** if they lie on the same straight line. Show that, in $\mathbb{R}^2$, P, Q, and R are collinear if $\overline{PR} = \overline{PQ} + \overline{QR}$ or $\overline{PQ} = \overline{PR} + \overrightarrow{RQ}$ or $\overrightarrow{QR} = \overrightarrow{QP} + \overrightarrow{PR}$. Use this last fact in $\mathbb{R}^3$ to show that the points $(-1, -1, -1)$, $(5, 8, 2)$, and $(-3, -4, -2)$ are collinear.
15. Show that the points $(3, 0, 1)$, $(0, -4, 0)$, and $(6, 4, 2)$ are collinear.
***16.** Let $P = (x_1, y_1, z_1)$ and $Q = (x_2, y_2, z_2)$. Show that the midpoint of PQ is the point $R = ((x_1 + x_2)/2, (y_1 + y_2)/2, (z_1 + z_2)/2)$. [*Hint:* Show that P, Q, and R are collinear and that $\overline{PR} = \overline{RQ}$.]
17. Find the midpoint of the line joining the points $(2, -1, 4)$ and $(5, 7, -3)$.

In Problems 18–31 find the magnitude and the direction cosines of the given vector.

18. $\mathbf{v} = 3\mathbf{j}$
19. $\mathbf{v} = -3\mathbf{i}$
20. $\mathbf{v} = 4\mathbf{i} - \mathbf{j}$
21. $\mathbf{v} = \mathbf{i} + 2\mathbf{k}$
22. $\mathbf{v} = \mathbf{i} - \mathbf{j} + \mathbf{k}$
23. $\mathbf{v} = \mathbf{i} + \mathbf{j} - \mathbf{k}$
24. $\mathbf{v} = -\mathbf{i} + \mathbf{j} + \mathbf{k}$
25. $\mathbf{v} = \mathbf{i} - \mathbf{j} - \mathbf{k}$
26. $\mathbf{v} = -\mathbf{i} + \mathbf{j} - \mathbf{k}$
27. $\mathbf{v} = -\mathbf{i} - \mathbf{j} + \mathbf{k}$
28. $\mathbf{v} = -\mathbf{i} - \mathbf{j} - \mathbf{k}$
29. $\mathbf{v} = 2\mathbf{i} + 5\mathbf{j} - 7\mathbf{k}$
30. $\mathbf{v} = -3\mathbf{i} - 3\mathbf{j} + 8\mathbf{k}$
31. $\mathbf{v} = -2\mathbf{i} - 3\mathbf{j} - 4\mathbf{k}$
32. The three direction angles of a certain unit vector are the same and are between zero and $\pi/2$. What is the vector?
33. Find a vector of magnitude 12 that has the same direction as the vector in Problem 32.
34. Show that there is no unit vector whose direction angles are $\pi/6$, $\pi/3$, and $\pi/4$.
35. Let $P = (2, 1, 4)$ and $Q = (3, -2, 8)$. Find a unit vector in the direction $\overrightarrow{PQ}$.
36. Let $P = (-3, 1, 7)$ and $Q = (8, 1, 7)$. Find a unit vector whose direction is opposite that of $\overrightarrow{PQ}$.
37. In Problem 36 find all points R such that $\overrightarrow{PR} \perp \overrightarrow{PQ}$.
***38.** Show that the set of points which satisfy the condition of Problem 37 and the condition $|\overrightarrow{PR}| = 1$ form a circle.
39. If $\mathbf{u}$ and $\mathbf{v}$ are in $\mathbb{R}^3$, show that $|\mathbf{u} + \mathbf{v}| \leq |\mathbf{u}| + |\mathbf{v}|$.
40. Under what circumstances can the inequality in Problem 39 be replaced by an equals sign?

In Problems 41–53 let $\mathbf{u} = 2\mathbf{i} - 3\mathbf{j} + 4\mathbf{k}$, $\mathbf{v} = -2\mathbf{i} - 3\mathbf{j} + 5\mathbf{k}$, $\mathbf{w} = \mathbf{i} - 7\mathbf{j} + 3\mathbf{k}$, and $\mathbf{t} = 3\mathbf{i} + 4\mathbf{j} + 5\mathbf{k}$.

41. Calculate $\mathbf{u} + \mathbf{v}$.
42. Calculate $2\mathbf{u} - 3\mathbf{v}$.
43. Calculate $\mathbf{t} + 3\mathbf{w} - \mathbf{v}$.
44. Calculate $2\mathbf{u} - 7\mathbf{w} + 5\mathbf{v}$.
45. Calculate $2\mathbf{v} + 7\mathbf{t} - \mathbf{w}$.
46. Calculate $\mathbf{u} \cdot \mathbf{v}$.
47. Calculate $|\mathbf{w}|$.
48. Calculate $\mathbf{u} \cdot \mathbf{w} - \mathbf{w} \cdot \mathbf{t}$.
49. Calculate the angle between $\mathbf{u}$ and $\mathbf{w}$.
50. Calculate the angle between $\mathbf{t}$ and $\mathbf{w}$.
51. Calculate $\text{proj}_{\mathbf{u}}\,\mathbf{v}$.
52. Calculate $\text{proj}_{\mathbf{t}}\,\mathbf{w}$.
53. Calculate $\text{proj}_{\mathbf{t}}\,\mathbf{v}$.
54. Find the distance between the point $P = (2, 1, 3)$ and the line passing through the points $Q = (-1, 1, 2)$ and $R = (6, 0, 1)$. [*Hint:* See Problem 4.2.42.]
55. Find the distance from the point $P = (1, 0, 1)$ to the line passing through the points $Q = (2, 3, -1)$ and $R = (6, 1, -3)$.
56. Show that the points $P = (3, 5, 6)$, $Q = (1, 2, 7)$, and $R = (6, 1, 0)$ are the vertices of a right triangle.
57. Show that the points $P = (3, 2, -1)$, $Q = (4, 1, 6)$, $R = (7, -2, 3)$, and $S = (8, -3, 10)$ are the vertices of a parallelogram.

*58. A solid figure in space with exactly four vertices is called a *tetrahedron* (see Figure 4.42). Let P represent the vector $\overrightarrow{OP}$, Q the vector $\overrightarrow{OQ}$, and so on. A line is drawn from each vertex to the centroid† of the opposite side. Show that these four lines meet at the endpoint of the vector $\mathbf{v}=(P+Q+R+S)/4$.

Figure 4.42

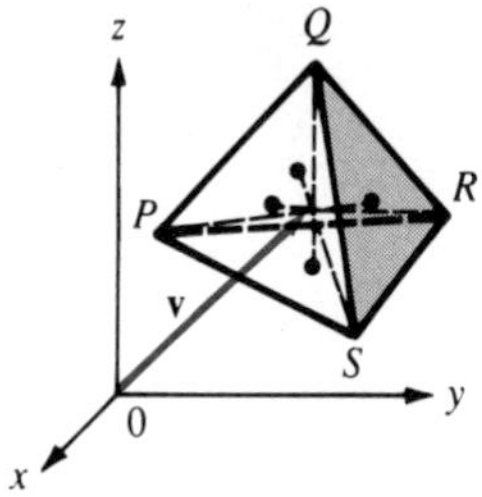

59. A force of 3 N acts in the direction of the vector with direction cosines $(1/\sqrt{6}, 1/\sqrt{3}, 1/\sqrt{2})$. Find the work done in moving the object from the point (1, 2, 3) to the point (2, 8, 11), where the distance is measured in meters. [*Hint:* See Example 4.2.10.]

60. Find the work done when a force of 3 N acting in the direction of the vector $\mathbf{v}=\mathbf{i}+\mathbf{j}-\mathbf{k}$ moves an object from $(-1, 3, 4)$ to $(3, 7, -2)$. Again the distance is measured in meters.

61. Prove that formula (15) is correct. [*Hint*: Follow the steps leading to formula (4.1.10).]

62. Prove Theorem 2.

63. Prove Theorem 3.

64. Prove Theorem 4.

*65. Let PQR be a triangle in $\mathbb{R}^3$. Show that if a force of N N moves an object around the triangle, then the work done by that force is zero.

*66. Find the angle between the diagonal of a cube and the diagonal of one of its faces.

4.4 The Cross Product of Two Vectors

To this point the only product of vectors that we have considered has been the scalar or dot product. We now define a new product, called the *cross product*‡ (or *vector product*), which is defined only in $\mathbb{R}^3$.

DEFINITION 1 **CROSS PRODUCT** Let $\mathbf{u} = a_1\mathbf{i} + b_1\mathbf{j} + c_1\mathbf{k}$ and $\mathbf{v} = a_2\mathbf{i} + b_2\mathbf{j} + c_2\mathbf{k}$. Then the **cross product (vector product)** of $\mathbf{u}$ and $\mathbf{v}$, denoted $\mathbf{u} \times \mathbf{v}$, is a new vector defined by

$$\mathbf{u}\times\mathbf{v}=(b_1c_2-c_1b_2)\mathbf{i}+(c_1a_2-a_1c_2)\mathbf{j}+(a_1b_2-b_1a_2)\mathbf{k} \quad (1)$$

†The centroid of a triangle is a point at the intersection of the medians.

‡ *Historical Note*: The cross product was defined by Hamilton in one of a series of papers discussing his quaternions that were published in *Philosophical Magazine* between the years 1844 and 1850.

Note that the result of the cross product is a vector whereas the result of the scalar product is a scalar.

Here the cross product seems to have been defined somewhat arbitrarily. There are obviously many ways to define a vector product. Why was this definition chosen? We shall answer that question in this section by demonstrating some of the properties of the cross product and illustrating some of its uses.

EXAMPLE 1 Let $\mathbf{u}=\mathbf{i}-\mathbf{j}+2\mathbf{k}$ and $\mathbf{v}=2\mathbf{i}+3\mathbf{j}-4\mathbf{k}$. Calculate $\mathbf{w}=\mathbf{u}\times\mathbf{v}$.

Solution Using formula (1),

$$\begin{aligned}\mathbf{w}&=[(-1)(-4)-(2)(3)]\mathbf{i}+[(2)(2)-(1)(-4)]\mathbf{j}+[(1)(3)-(-1)(2)]\mathbf{k}\\&=-2\mathbf{i}+8\mathbf{j}+5\mathbf{k}\end{aligned}$$

Note. In this example $\mathbf{u}\cdot\mathbf{w}=(\mathbf{i}-\mathbf{j}+2\mathbf{k})\cdot(-2\mathbf{i}+8\mathbf{j}+5\mathbf{k})=-2-8+10=0$. Similarly, $\mathbf{v}\cdot\mathbf{w}=0$. That is, $\mathbf{u}\times\mathbf{v}$ is orthogonal to both $\mathbf{u}$ and $\mathbf{v}$. As we shall shortly see, the cross product of $\mathbf{u}$ and $\mathbf{v}$ is always orthogonal to $\mathbf{u}$ and $\mathbf{v}$.

Before continuing our discussion of the uses of the cross product, we observe that there is an easy way to calculate $\mathbf{u}\times\mathbf{v}$ by using determinants.

THEOREM 1

$$\mathbf{u}\times\mathbf{v}=\begin{vmatrix}\mathbf{i}&\mathbf{j}&\mathbf{k}\\a_1&b_1&c_1\\a_2&b_2&c_2\end{vmatrix}^{\dagger}$$

Proof

$$\begin{vmatrix}\mathbf{i}&\mathbf{j}&\mathbf{k}\\a_1&b_1&c_1\\a_2&b_2&c_2\end{vmatrix}=\mathbf{i}\begin{vmatrix}b_1&c_1\\b_2&c_2\end{vmatrix}-\mathbf{j}\begin{vmatrix}a_1&c_1\\a_2&c_2\end{vmatrix}+\mathbf{k}\begin{vmatrix}a_1&b_1\\a_2&b_2\end{vmatrix}$$

$$=(b_1c_2-c_1b_2)\mathbf{i}+(c_1a_2-a_1c_2)\mathbf{j}+(a_1b_2-b_1a_2)\mathbf{k}$$

which is equal to $\mathbf{u}\times\mathbf{v}$ according to Definition 1. ■

† This is not really a determinant because **i**, **j**, and **k** are not numbers. However, using determinant notation, Theorem 1 helps us remember how to calculate a cross product.

EXAMPLE 2 Calculate $\mathbf{u}\times\mathbf{v}$, where $\mathbf{u}=2\mathbf{i}+4\mathbf{j}-5\mathbf{k}$ and $\mathbf{v}=-3\mathbf{i}-2\mathbf{j}+\mathbf{k}$.

Solution

$$\mathbf{u}\times\mathbf{v}=\begin{vmatrix}\mathbf{i} & \mathbf{j} & \mathbf{k}\\ 2 & 4 & -5\\ -3 & -2 & 1\end{vmatrix}=(4-10)\mathbf{i}-(2-15)\mathbf{j}+(-4+12)\mathbf{k}$$
$$=-6\mathbf{i}+13\mathbf{j}+8\mathbf{k}$$

The following theorem summarizes some properties of the cross product.

THEOREM 2 Let $\mathbf{u}$, $\mathbf{v}$, and $\mathbf{w}$ be vectors in $\mathbb{R}^3$ and let α be a scalar. Then:

i. $\mathbf{u}\times\mathbf{0}=\mathbf{0}\times\mathbf{u}=\mathbf{0}$.
ii. $\mathbf{u}\times\mathbf{v}=-(\mathbf{v}\times\mathbf{u})$ (**anticommutative property for the vector product**).
iii. $(\alpha\mathbf{u})\times\mathbf{v}=\alpha(\mathbf{u}\times\mathbf{v})$.
iv. $\mathbf{u}\times(\mathbf{v}+\mathbf{w})=(\mathbf{u}\times\mathbf{v})+(\mathbf{u}\times\mathbf{w})$ (**distributive property for the vector product**).
v. $(\mathbf{u}\times\mathbf{v})\cdot\mathbf{w}=\mathbf{u}\cdot(\mathbf{v}\times\mathbf{w})$. (This is called the **scalar triple product** of $\mathbf{u}$, $\mathbf{v}$, and $\mathbf{w}$.)
vi. $\mathbf{u}\cdot(\mathbf{u}\times\mathbf{v})=\mathbf{v}\cdot(\mathbf{u}\times\mathbf{v})=0$. (That is, $\mathbf{u}\times\mathbf{v}$ is orthogonal to both $\mathbf{u}$ and $\mathbf{v}$.)
vii. If $\mathbf{u}$ and $\mathbf{v}$ are parallel, then $\mathbf{u}\times\mathbf{v}=\mathbf{0}$.

Proof **i.** Let $\mathbf{u}=a_1\mathbf{i}+b_1\mathbf{j}+c_1\mathbf{k}$. Then

$$\mathbf{u}\times\mathbf{0}=\begin{vmatrix}\mathbf{i} & \mathbf{j} & \mathbf{k}\\ a_1 & b_1 & c_1\\ 0 & 0 & 0\end{vmatrix}=0\mathbf{i}+0\mathbf{j}+0\mathbf{k}=\mathbf{0}$$

Similarly, $\mathbf{0}\times\mathbf{u}=\mathbf{0}$.

ii. Let $\mathbf{v}=a_2\mathbf{i}+b_2\mathbf{j}+c_2\mathbf{k}$. Then

$$\mathbf{u}\times\mathbf{v}=\begin{vmatrix}\mathbf{i} & \mathbf{j} & \mathbf{k}\\ a_1 & b_1 & c_1\\ a_2 & b_2 & c_2\end{vmatrix}=-\begin{vmatrix}\mathbf{i} & \mathbf{j} & \mathbf{k}\\ a_2 & b_2 & c_2\\ a_1 & b_1 & c_1\end{vmatrix}=-(\mathbf{v}\times\mathbf{u})$$

since interchanging the rows of a determinant has the effect of multiplying that determinant by -1 (Property 3.2.4, page 92).

iii. $$(\alpha\mathbf{u})\times\mathbf{v}=\begin{vmatrix}\mathbf{i} & \mathbf{j} & \mathbf{k}\\ \alpha a_1 & \alpha b_1 & \alpha c_1\\ a_2 & b_2 & c_2\end{vmatrix}=\alpha\begin{vmatrix}\mathbf{i} & \mathbf{j} & \mathbf{k}\\ a_1 & b_1 & c_1\\ a_2 & b_2 & c_3\end{vmatrix}=\alpha(\mathbf{u}\times\mathbf{v})$$

The second equality follows from Property 3.2.2 on page 90.

iv. Let $\mathbf{w}=a_3\mathbf{i}+b_3\mathbf{j}+c_3\mathbf{k}$. Then

$$\mathbf{u}\times(\mathbf{v}+\mathbf{w})=\begin{vmatrix}\mathbf{i} & \mathbf{j} & \mathbf{k}\\ a_1 & b_1 & c_1\\ a_2+a_3 & b_2+b_3 & c_2+c_3\end{vmatrix}$$
$$=\begin{vmatrix}\mathbf{i} & \mathbf{j} & \mathbf{k}\\ a_1 & b_1 & c_1\\ a_2 & b_2 & c_2\end{vmatrix}+\begin{vmatrix}\mathbf{i} & \mathbf{j} & \mathbf{k}\\ a_1 & b_1 & c_1\\ a_3 & b_3 & c_3\end{vmatrix}$$
$$=(\mathbf{u}\times\mathbf{v})+(\mathbf{u}\times\mathbf{w})$$

Here we have used Property 3.2.3, page 91.

v. $(\mathbf{u}\times\mathbf{v})\cdot\mathbf{w}=[(b_1c_2-c_1b_2)\mathbf{i}+(c_1a_2-a_1c_2)\mathbf{j}+(a_1b_2-b_1a_2)\mathbf{k}]$
$\cdot[a_3\mathbf{i}+b_3\mathbf{j}+c_3\mathbf{k}]$
$=b_1c_2a_3-c_1b_2a_3+c_1a_2b_3-a_1c_2b_3+a_1b_2c_3-b_1a_2c_3$

We can easily show that $\mathbf{u}\cdot(\mathbf{v}\times\mathbf{w})$ is equal to the same expression.†

vi. We know that $\mathbf{u}\cdot(\mathbf{u}\times\mathbf{v})=(\mathbf{u}\times\mathbf{v})\cdot\mathbf{u}$ (since the scalar product is commutative—see part (*ii*) of Theorem 2.2.1, page 35). But, from parts (*ii*) and (*v*) of this theorem, $(\mathbf{u}\times\mathbf{v})\cdot\mathbf{u}=\mathbf{u}\cdot(\mathbf{v}\times\mathbf{u})=\mathbf{u}\cdot(-\mathbf{u}\times\mathbf{v}=-\mathbf{u}\cdot(\mathbf{u}\times\mathbf{v})$. Thus $\mathbf{u}\cdot(\mathbf{u}\times\mathbf{v})=-\mathbf{u}\cdot(\mathbf{u}\times\mathbf{v})$, which can only occur if $\mathbf{u}\cdot(\mathbf{u}\times\mathbf{v})=0$. A similar computation shows that $\mathbf{v}\cdot(\mathbf{u}\times\mathbf{v})=0$.

vii. If $\mathbf{u}$ and $\mathbf{v}$ are parallel, then $\mathbf{v}=\alpha\mathbf{u}$ for some scalar α (from Theorem 4.3.3, page 147) so that

$$\mathbf{u}\times\mathbf{v}=\begin{vmatrix}\mathbf{i} & \mathbf{j} & \mathbf{k}\\ a_1 & b_1 & c_1\\ \alpha a_1 & \alpha b_1 & \alpha c_1\end{vmatrix}=\mathbf{0}$$

since a determinant with two proportional rows is zero (Property 3.2.6, page 94). ■

We have seen that $\mathbf{u}\times\mathbf{v}$ is a vector orthogonal to both $\mathbf{u}$ and $\mathbf{v}$ (part (*vi*) of the last theorem). But there are always at least *two* unit vectors orthogonal to $\mathbf{u}$ and $\mathbf{v}$ (see Figure 4.43). Suppose that vectors $\mathbf{n}$ and $-\mathbf{n}$ are both orthogonal to $\mathbf{u}$ and $\mathbf{v}$. Which one is in the direction of $\mathbf{u}\times\mathbf{v}$? The answer is given by the **right-hand rule**. If the right hand is placed so that the index finger points in the direction of $\mathbf{u}$ and the middle finger points in the direction of $\mathbf{v}$, then the thumb points in the direction of $\mathbf{u}\times\mathbf{v}$ (see Figure 4.44).

Figure 4.43

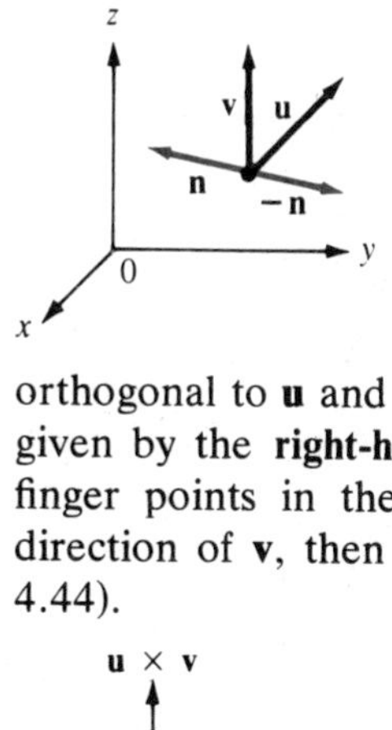

Figure 4.44

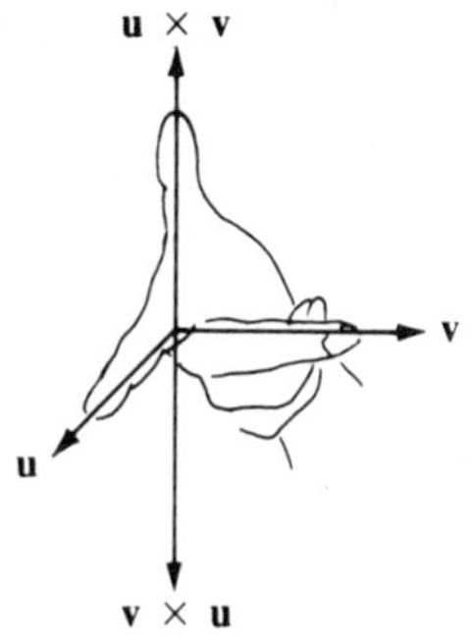

† For an interesting geometric interpretation of the scalar triple product, see Problem 37.

What happens when we take cross products of the basis vectors **i**, **j**, **k**? It is easy to verify the following:

$$\mathbf{i}\times\mathbf{i}=\mathbf{j}\times\mathbf{j}=\mathbf{k}\times\mathbf{k}=\mathbf{0} \tag{2}$$

$$\mathbf{i}\times\mathbf{j}=\mathbf{k} \qquad \mathbf{k}\times\mathbf{i}=\mathbf{j} \qquad \mathbf{j}\times\mathbf{k}=\mathbf{i} \tag{3}$$

$$\mathbf{j}\times\mathbf{i}=-\mathbf{k} \qquad \mathbf{i}\times\mathbf{k}=-\mathbf{j} \qquad \mathbf{k}\times\mathbf{j}=-\mathbf{i} \tag{4}$$

To remember these, consider the circle in Figure 4.45. The cross product of two consecutive vectors in the clockwise direction is positive; the cross product of two consecutive vectors in the counterclockwise direction is

Figure 4.45

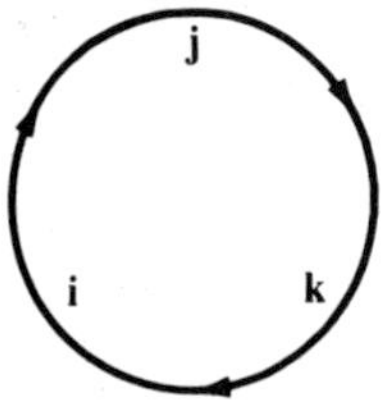

negative. Note that the preceding formulas show that the cross product is *not* associative since, for example, $\mathbf{i}\times(\mathbf{i}\times\mathbf{j})=\mathbf{i}\times\mathbf{k}=\mathbf{j}$ whereas $(\mathbf{i}\times\mathbf{i})\times\mathbf{j}=\mathbf{0}\times\mathbf{j}=\mathbf{0}$ so that $\mathbf{i}\times(\mathbf{i}\times\mathbf{j})\neq(\mathbf{i}\times\mathbf{i})\times\mathbf{j}$. In general,

$$\mathbf{u}\times(\mathbf{v}\times\mathbf{w})\neq(\mathbf{u}\times\mathbf{v})\times\mathbf{w} \tag{5}$$

EXAMPLE 3 Calculate $(3\mathbf{i}+4\mathbf{k})\times(2\mathbf{i}-3\mathbf{j})$.

Solution This is a good example of the usefulness of Theorem 2 and formulas (2), (3), and (4). We have

$$\begin{aligned}(3\mathbf{i}+4\mathbf{k})\times(2\mathbf{i}-3\mathbf{j}) &= (3\cdot2)(\mathbf{i}\times\mathbf{i})+(4\cdot2)(\mathbf{k}\times\mathbf{i})+3(-3)(\mathbf{i}\times\mathbf{j})\\ &\quad+4(-3)(\mathbf{k}\times\mathbf{j})\\ &=\mathbf{0}+8\mathbf{j}-9\mathbf{k}+12\mathbf{i}=12\mathbf{i}+8\mathbf{j}-9\mathbf{k}\end{aligned}$$

We know that $\mathbf{u}\times\mathbf{v}$ is a vector orthogonal to both **u** and **v**. The next result gives us its magnitude.

THEOREM 3 If φ is the angle between **u** and **v**, then

$$|\mathbf{u}\times\mathbf{v}|=|\mathbf{u}|\,|\mathbf{v}|\sin\varphi \tag{6}$$

Proof It is easy to show (by comparing components) that $|\mathbf{u}\times\mathbf{v}|^2=|\mathbf{u}|^2|\mathbf{v}|^2-(\mathbf{u}\cdot\mathbf{v})^2$ (see Problem 31). Then, since $(\mathbf{u}\cdot\mathbf{v})^2 = |\mathbf{u}|^2|\mathbf{v}|^2\cos^2\varphi$ (from Theorem 4.3.2, page 146),

$$\begin{aligned}|\mathbf{u}\times\mathbf{v}|^2&=|\mathbf{u}|^2|\mathbf{v}|^2-|\mathbf{u}|^2|\mathbf{v}|^2\cos^2\varphi=|\mathbf{u}|^2|\mathbf{v}|^2(1-\cos^2\theta)\\&=|\mathbf{u}|^2|\mathbf{v}|^2\sin^2\varphi\end{aligned}$$

and the theorem follows after taking the square root of both sides. ■

There is an interesting geometric interpretation of Theorem 3. The vectors **u** and **v** are sketched in Figure 4.46 and can be thought of as two

Figure 4.46

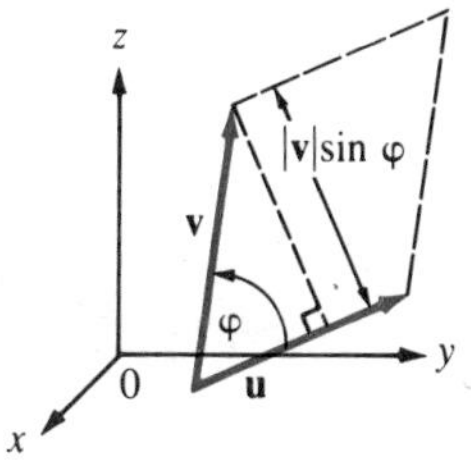

adjacent sides of a parallelogram. Then, from elementary geometry, we see that:

Area of the parallelogram $=|\mathbf{u}|\,|\mathbf{v}|\sin\varphi=|\mathbf{u}\times\mathbf{v}|$	(7)

EXAMPLE 4 Find the area of the parallelogram with consecutive vertices at $P=(1,3,-2)$, $Q=(2,1,4)$, and $R=(-3,1,6)$.

Figure 4.47

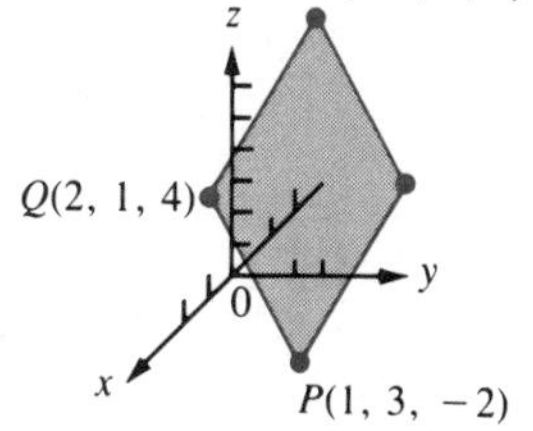

Solution The parallelogram is sketched in Figure 4.47. We have

$$\begin{aligned}\text{Area}&=|\overrightarrow{PQ}\times\overrightarrow{QR}|=|(\mathbf{i}-2\mathbf{j}+6\mathbf{k})\times(-5\mathbf{i}+2\mathbf{k})|\\&=\begin{vmatrix}\mathbf{i}&\mathbf{j}&\mathbf{k}\\1&-2&6\\-5&0&2\end{vmatrix}=|-4\mathbf{i}-32\mathbf{j}-10\mathbf{k}|=\sqrt{1140}\text{ square units}\end{aligned}$$

We can use the preceding discussion to give a geometric interpretation to the determinant. Let A be a 2×2 matrix and let **u** and **v** be two 2-vectors. Let $\mathbf{u}=\begin{pmatrix}u_1\\u_2\end{pmatrix}$ and $\mathbf{v}=\begin{pmatrix}v_1\\v_2\end{pmatrix}$. These vectors are given in Figure 4.48. The **area**

Figure 4.48

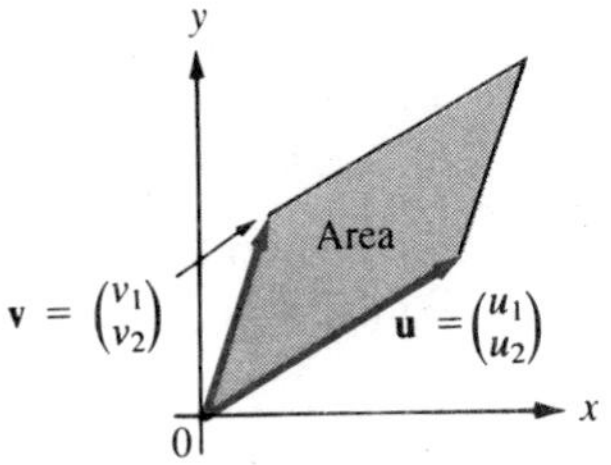

generated by **u** *and* **v** is defined to be the area of the parallelogram given in the figure. We can think of **u** and **v** as vectors in $\mathbb{R}^3$ lying in the xy-plane. Then $\mathbf{u}=\begin{pmatrix}u_1\\u_2\\0\end{pmatrix}$, $\mathbf{v}=\begin{pmatrix}v_1\\v_2\\0\end{pmatrix}$, and

$$\text{Area generated by } \mathbf{u} \text{ and } \mathbf{v} = |\mathbf{u}\times\mathbf{v}| = \begin{vmatrix}\mathbf{i} & \mathbf{j} & \mathbf{k}\\ u_1 & u_2 & 0\\ v_1 & v_2 & 0\end{vmatrix}$$

$$= |(u_1v_2-u_2v_1)\mathbf{k}| = |u_1v_2-u_2v_1|^\dagger$$

Now let $A=\begin{pmatrix}a_{11} & a_{12}\\ a_{21} & a_{22}\end{pmatrix}$, $\mathbf{u}'=A\mathbf{u}$, and $\mathbf{v}'=A\mathbf{v}$. Then $\mathbf{u}'=\begin{pmatrix}a_{11}u_1+a_{12}u_2\\ a_{21}u_1+a_{22}u_2\end{pmatrix}$ and $\mathbf{v}'=\begin{pmatrix}a_{11}v_1+a_{12}v_2\\ a_{21}v_1+a_{22}v_2\end{pmatrix}$. What is the area generated by $\mathbf{u}'$ and $\mathbf{v}'$? Following the preceding steps, we calculate

$$\text{Area generated by } \mathbf{u}' \text{ and } \mathbf{v}' = |\mathbf{u}'\times\mathbf{v}'| = \begin{vmatrix}\mathbf{i} & \mathbf{j} & \mathbf{k}\\ a_{11}u_1+a_{12}u_2 & a_{21}u_1+a_{22}u_2 & 0\\ a_{11}v_1+a_{12}v_2 & a_{21}v_1+a_{22}v_2 & 0\end{vmatrix}$$

$$= |(a_{11}u_1+a_{12}u_2)(a_{21}v_1+a_{22}v_2)-(a_{21}u_1+a_{22}u_2)(a_{11}v_1+a_{12}v_2)|$$

It is simple algebra to verify that the last expression is equal to

$$|(a_{11}a_{22}-a_{12}a_{21})(u_1v_2-u_2v_1)| = \pm\det A \text{ (area generated by } \mathbf{u} \text{ and } \mathbf{v}).$$

Thus (in this context): *The determinant has the effect of multiplying area.* In Problem 41 you are asked to show that, in a certain sense, a 3×3 determinant has the effect of multiplying volume.

† Note that this is the absolute value of det $\begin{pmatrix}u_1 & v_1\\ u_2 & v_2\end{pmatrix}$.

PROBLEMS 4.4

In Problems 1–20 find the cross product $\mathbf{u}\times\mathbf{v}$.

1. $\mathbf{u}=\mathbf{i}-2\mathbf{j};\ \mathbf{v}=3\mathbf{k}$
2. $\mathbf{u}=3\mathbf{i}-7\mathbf{j};\ \mathbf{v}=\mathbf{i}+\mathbf{k}$
3. $\mathbf{u}=\mathbf{i}-\mathbf{j};\ \mathbf{v}=\mathbf{j}+\mathbf{k}$
4. $\mathbf{u}=-7\mathbf{k};\ \mathbf{v}=\mathbf{j}+2\mathbf{k}$
5. $\mathbf{u}=-2\mathbf{i}+3\mathbf{j};\ \mathbf{v}=7\mathbf{i}+4\mathbf{k}$
6. $\mathbf{u}=a\mathbf{i}+b\mathbf{j};\ \mathbf{v}=c\mathbf{i}+d\mathbf{j}$
7. $\mathbf{u}=a\mathbf{i}+b\mathbf{k};\ \mathbf{v}=c\mathbf{i}+d\mathbf{k}$
8. $\mathbf{u}=a\mathbf{j}+b\mathbf{k};\ \mathbf{v}=c\mathbf{i}+d\mathbf{k}$
9. $\mathbf{u}=2\mathbf{i}-3\mathbf{j}+\mathbf{k};\ \mathbf{v}=\mathbf{i}+2\mathbf{j}+\mathbf{k}$
10. $\mathbf{u}=3\mathbf{i}-4\mathbf{j}+2\mathbf{k};\ \mathbf{v}=6\mathbf{i}-3\mathbf{j}+5\mathbf{k}$
11. $\mathbf{u}=-3\mathbf{i}-2\mathbf{j}+\mathbf{k};\ \mathbf{v}=6\mathbf{i}+4\mathbf{j}-2\mathbf{k}$
12. $\mathbf{u}=\mathbf{i}+7\mathbf{j}-3\mathbf{k};\ \mathbf{v}=-\mathbf{i}-7\mathbf{j}+3\mathbf{k}$
13. $\mathbf{u}=\mathbf{i}-7\mathbf{j}-3\mathbf{k};\ \mathbf{v}=-\mathbf{i}+7\mathbf{j}-3\mathbf{k}$
14. $\mathbf{u}=2\mathbf{i}-3\mathbf{j}+5\mathbf{k};\ \mathbf{v}=3\mathbf{i}-\mathbf{j}-\mathbf{k}$
15. $\mathbf{u}=10\mathbf{i}+7\mathbf{j}-3\mathbf{k};\ \mathbf{v}=-3\mathbf{i}+4\mathbf{j}-3\mathbf{k}$
16. $\mathbf{u}=2\mathbf{i}+4\mathbf{j}-6\mathbf{k};\ \mathbf{v}=-\mathbf{i}-\mathbf{j}+3\mathbf{k}$
17. $\mathbf{u}=2\mathbf{i}-\mathbf{j}+\mathbf{k};\ \mathbf{v}=4\mathbf{i}+2\mathbf{j}+2\mathbf{k}$
18. $\mathbf{u}=3\mathbf{i}-\mathbf{j}+8\mathbf{k};\ \mathbf{v}=\mathbf{i}+\mathbf{j}-4\mathbf{k}$
19. $\mathbf{u}=a\mathbf{i}+a\mathbf{j}+a\mathbf{k};\ \mathbf{v}=b\mathbf{i}+b\mathbf{j}+b\mathbf{k}$
20. $\mathbf{u}=a\mathbf{i}+b\mathbf{j}+c\mathbf{k};\ \mathbf{v}=a\mathbf{i}+b\mathbf{j}-c\mathbf{k}$
21. Find two unit vectors orthogonal to both $\mathbf{u}=2\mathbf{i}-3\mathbf{j}$ and $\mathbf{v}=4\mathbf{j}+3\mathbf{k}$.
22. Find two unit vectors orthogonal to both $\mathbf{u}=\mathbf{i}+\mathbf{j}+\mathbf{k}$ and $\mathbf{v}=\mathbf{i}-\mathbf{j}-\mathbf{k}$.
23. Use the cross product to find the sine of the angle φ between the vectors $\mathbf{u}=2\mathbf{i}+\mathbf{j}-\mathbf{k}$ and $\mathbf{v}=-3\mathbf{i}-2\mathbf{j}+4\mathbf{k}$.
24. Use the scalar product to calculate the cosine of the angle φ between the vectors of Problem 23. Then show that for the values you have calculated, $\sin^2\varphi+\cos^2\varphi=1$.

In Problems 25–30 find the area of the parallelogram with the given adjacent vertices.

25. $(1,-2,3);\ (2,0,1);\ (0,4,0)$
26. $(-2,1,1);\ (2,2,3);\ (-1,-2,4)$
27. $(-2,1,0);\ (1,4,2);\ (-3,1,5)$
28. $(7,-2,-3);\ (-4,1,6);\ (5,-2,3)$
29. $(a,0,0);\ (0,b,0);\ (0,0,c)$
30. $(a,b,0);\ (a,0,b);\ (0,a,b)$
31. Show that $|\mathbf{u}\times\mathbf{v}|^2=|\mathbf{u}|^2|\mathbf{v}|^2-(\mathbf{u}\cdot\mathbf{v})^2$. [*Hint:* Write out in terms of components.]
32. Show that the area of the triangle PQR is given by $A=\frac{1}{2}|\overrightarrow{PQ}\times\overrightarrow{QR}|$.
33. Use the result of Problem 32 to calculate the area of the triangle with vertices at $(2,1,-4)$, $(1,7,2)$, $(3,-2,3)$.
34. Calculate the area of a triangle with vertices at $(3,1,7)$ $(2,-3,4)$, $(7,-2,4)$.
35. Calculate the area of the triangle with vertices at $(1,0,0)$, $(0,1,0)$, and $(0,0,1)$. Sketch this triangle.
36. Show that if $\mathbf{u}=(a_1,b_1,c_1)$, $\mathbf{v}=(a_2,b_2,c_2)$, and $\mathbf{w}=(a_3,b_3,c_3)$, then

$$\mathbf{u}\cdot(\mathbf{v}\times\mathbf{w})=\begin{vmatrix} a_1 & b_1 & c_1\\ a_2 & b_2 & c_2\\ a_3 & b_3 & c_3\end{vmatrix}$$

*37. Let $\mathbf{u}$, $\mathbf{v}$, and $\mathbf{w}$ be three vectors that are not in the same plane. Then they form the sides of a parallelepiped in space (see Figure 4.49). Prove that the volume of the parallelepiped is given by $V=|(\mathbf{u}\times\mathbf{v})\cdot\mathbf{w}|$.† [*Hint:* The area of the base is $|\mathbf{u}\times\mathbf{v}|$.]

Figure 4.49

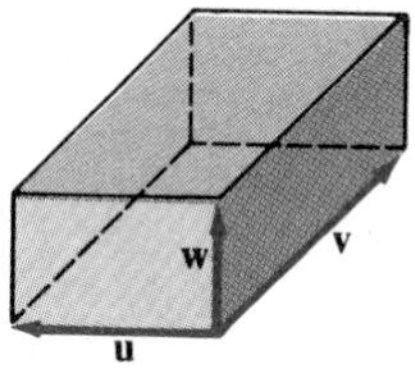

† This means that the volume of the parallelogram is given by

$$V=\left|\det\begin{pmatrix} a_1 & b_1 & c_1\\ a_2 & b_2 & c_2\\ a_3 & b_3 & c_3\end{pmatrix}\right|$$

38. Calculate the volume of the parallelepiped determined by the vectors $\mathbf{u} = 2\mathbf{i}-\mathbf{j}+\mathbf{k}$, $\mathbf{v} = 3\mathbf{i}+2\mathbf{j}-2\mathbf{k}$, and $\mathbf{w} = 3\mathbf{i}+2\mathbf{j}$.

39. Calculate the volume of the parallelepiped determined by the vectors $\mathbf{i}-\mathbf{j}$, $3\mathbf{i}+2\mathbf{k}$, $-7\mathbf{j}+3\mathbf{k}$.

40. Calculate the volume of the parallelepiped determined by the vectors $\overrightarrow{PQ}$, $\overrightarrow{PR}$, and $\overrightarrow{PS}$, where $P=(2,1,-1)$, $Q=(-3,1,4)$, $R=(-1,0,2)$, and $S=(-3,-1,5)$.

****41.** The volume generated by three vectors $\mathbf{u}$, $\mathbf{v}$, and $\mathbf{w}$ in $\mathbb{R}^3$ is defined to be the volume of the parallelepiped whose sides are $\mathbf{u}$, $\mathbf{v}$, and $\mathbf{w}$ (as in Figure 4.49). Let A be a 3×3 matrix and let $\mathbf{u}_1 = A\mathbf{u}$, $\mathbf{v}_1 = A\mathbf{v}$, and $\mathbf{w}_1 = A\mathbf{w}$. Show that:

> Volume generated by $\mathbf{u}_1$, $\mathbf{v}_1$, $\mathbf{w}_1$
> $= (\pm\det A)$ (volume generated by $\mathbf{u}$, $\mathbf{v}$, $\mathbf{w}$)

This shows that just as the determinant of a 2×2 matrix multiplies area, the determinant of a 3×3 matrix multiplies volume.

42. Let $A = \begin{pmatrix} 2 & 3 & 1 \\ 4 & -1 & 5 \\ 1 & 0 & 6 \end{pmatrix}$, $\mathbf{u} = \begin{pmatrix} 2 \\ -1 \\ 0 \end{pmatrix}$, $\mathbf{v} = \begin{pmatrix} 1 \\ 0 \\ 4 \end{pmatrix}$, and $\mathbf{w} = \begin{pmatrix} -1 \\ 3 \\ 2 \end{pmatrix}$.

a. Calculate the volume generated by $\mathbf{u}$, $\mathbf{v}$, and $\mathbf{w}$.

b. Calculate the volume generated by $A\mathbf{u}$, $A\mathbf{v}$, and $A\mathbf{w}$.

c. Calculate det A.

d. Show that [volume in part (b)] $= (\pm\det A)\times$[volume in part (a)].

43. The **triple cross product** of three vectors in $\mathbb{R}^3$ is defined to be the vector $\mathbf{u}\times(\mathbf{v}\times\mathbf{w})$. Show that

$$\mathbf{u}\times(\mathbf{v}\times\mathbf{w}) = (\mathbf{u}\cdot\mathbf{w})\mathbf{v} - (\mathbf{u}\cdot\mathbf{v})\mathbf{w}.$$

4.5 Lines and Planes in Space

In the plane $\mathbb{R}^2$ we can find the equation of a line if we know either two points on the line or one point and the slope of the line. In $\mathbb{R}^3$, our intuition tells us that the basic ideas are the same. Since two points determine a line, we should be able to calculate the equation of a line in space if we know two points on it. Alternatively, if we know one point and the direction of a line, we should also be able to find its equation.

We begin with two points $P=(x_1, y_1, z_1)$ and $Q=(x_2, y_2, z_2)$ on a line L. A vector parallel to L is a vector with representation $\overrightarrow{PQ}$. Thus (from formula 4.3.15, page 146)

$$\mathbf{v} = (x_2-x_1)\mathbf{i} + (y_2-y_1)\mathbf{j} + (z_2-z_1)\mathbf{k} \tag{1}$$

is a vector parallel to L. Now let $R=(x, y, z)$ be another point on the line. Then $\overrightarrow{PR}$ is parallel to $\overrightarrow{PQ}$, which is parallel to $\mathbf{v}$, so that, by Theorem 4.3.3 on page 147

$$\overrightarrow{PR} = t\mathbf{v} \tag{2}$$

for some real number t. Now look at Figure 4.50. From this figure we have (in each of the three possible cases)

$$\overrightarrow{OR} = \overrightarrow{OP} + \overrightarrow{PR} \tag{3}$$

Figure 4.50

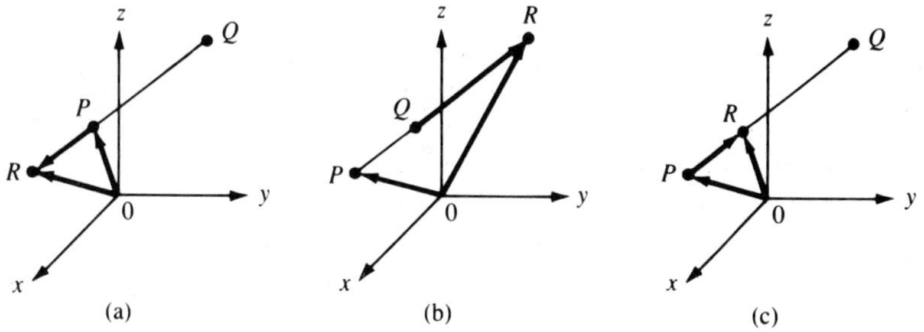

And, combining (2) and (3), we get

$$\overrightarrow{PR} = \overrightarrow{OR} - \overrightarrow{OP} = t\mathbf{v}$$

or

$$\overrightarrow{OR} = \overrightarrow{OP} + t\mathbf{v} \tag{4}$$

VECTOR EQUATION OF A LINE

Equation (4) is called the **vector equation** of the line L. For if R is on L, then (4) is satisfied for some real number t. Conversely, if (4) is satisfied, then, reversing our steps, we see that $\overrightarrow{PR}$ is parallel to $\mathbf{v}$, which means that R is on L.

If we write out the components of Equation (4), we obtain

$$x\mathbf{i} + y\mathbf{j} + z\mathbf{k} = x_1\mathbf{i} + y_1\mathbf{j} + z_1\mathbf{k} + t(x_2 - x_1)\mathbf{i} + t(y_2 - y_1)\mathbf{j} + t(z_2 - z_1)\mathbf{k}$$

or

$$\begin{aligned} x &= x_1 + t(x_2 - x_1) \\ y &= y_1 + t(y_2 - y_1) \\ z &= z_1 + t(z_2 - z_1) \end{aligned} \tag{5}$$

PARAMETRIC EQUATIONS OF A LINE

Equations (5) are called the **parametric equations** of a line.

Finally, solving for t in (5), and defining $x_2 - x_1 = a$, $y_2 - y_1 = b$, and $z_2 - z_1 = c$, we find that

$$\frac{x - x_1}{a} = \frac{y - y_1}{b} = \frac{z - z_1}{c} \tag{6}$$

SYMMETRIC EQUATIONS OF A LINE

Equations (6) are called the **symmetric equations** of the line. Here a, b, and c are direction numbers of the vector $\mathbf{v}$. Of course, Equations (6) are valid only if a, b, and c are nonzero.

EXAMPLE 1

Find a vector equation, parametric equations, and symmetric equations of the line L passing through the points $P=(2,-1,6)$ and $Q=(3,1,-2)$.

Solution First we calculate: $\mathbf{v}=(3-2)\mathbf{i}+[1-(-1)]\mathbf{j}+(-2-6)\mathbf{k}=\mathbf{i}+2\mathbf{j}-8\mathbf{k}$. Then, from (4), if $R=(x, y, z)$ is on the line, we obtain $\overrightarrow{OR}=x\mathbf{i}+y\mathbf{j}+z\mathbf{k}=\overrightarrow{OP}+t\mathbf{v}=2\mathbf{i}-\mathbf{j}+6\mathbf{k}+t(\mathbf{i}+2\mathbf{j}-8\mathbf{k})$ or

$$x=2+t \qquad y=-1+2t \qquad z=6-8t$$

Finally, since $a=1$, $b=2$, and $c=-8$, we find the symmetric equations

$$\frac{x-2}{1}=\frac{y+1}{2}=\frac{z-6}{-8} \tag{7}$$

To check this, we verify that $(2,-1,6)$ and $(3,1,-2)$ are indeed on the line. We have [after plugging these points into (7)]

$$\frac{2-2}{1}=\frac{-1+1}{2}=\frac{6-6}{-8}=0$$

$$\frac{3-2}{1}=\frac{1+1}{2}=\frac{-2-6}{-8}=1$$

Other points on the line can be found. If $t=3$, for example, we obtain

$$3=\frac{x-2}{1}=\frac{y+1}{2}=\frac{z-6}{-8}$$

which yields the point $(5,5,-18)$.

EXAMPLE 2

Find the symmetric equations of the line passing through the point $(1,-2,4)$ and parallel to the vector $\mathbf{v}=\mathbf{i}+\mathbf{j}-\mathbf{k}$.

Solution We use formula (6) with $P=(x_1, y_1, z_1)=(1,-2,4)$ and $\mathbf{v}$ as above so that $a=1$, $b=1$ and $c=-1$. This gives us

$$\frac{x-1}{1}=\frac{y+2}{1}=\frac{z-4}{-1}$$

What happens if one of the direction numbers a, b, or c is zero?

EXAMPLE 3

Find the symmetric equation of the lines containing the points $P=(3,4,-1)$ and $Q=(-2,4,6)$.

Solution Here $\mathbf{v}=-5\mathbf{i}+7\mathbf{k}$ and $a=-5$, $b=0$, $c=7$. Then a parametric representation of the line is $x=3-5t$, $y=4$, and $z=-1+7t$. Solving for t, we find that

$$\frac{x-3}{-5}=\frac{z+1}{7} \quad \text{and} \quad y=4$$

The equation $y = 4$ is the equation of a plane parallel to the xz-plane, so we have obtained an equation of a line in that plane.

EXAMPLE 4 Find symmetric equations of the line in the xy-plane that passes through the points $(x_1, y_1, 0)$ and $(x_2, y_2, 0)$, where $x_1 \neq x_2$.

Solution Here $\mathbf{v} = (x_2 - x_1)\mathbf{i} + (y_2 - y_1)\mathbf{j}$ and we obtain

$$\frac{x - x_1}{x_2 - x_1} = \frac{y - y_1}{y_2 - y_1} \quad \text{and} \quad z = 0$$

We can rewrite this as

$$y - y_1 = \left(\frac{y_2 - y_1}{x_2 - x_1}\right)(x - x_1)$$

Here $(y_2 - y_1)/(x_2 - x_1) = m$, the slope of the line. When $x = 0$, $y = y_1 - [(y_2 - y_1)/(x_2 - x_1)]x_1 = b$, the y-intercept of the line. That is, $y = mx + b$, which is the slope-intercept form of a line in the xy-plane. Thus we see that the symmetric equations of a line in space are really a generalization of the equation of a line in the plane.

What happens if two of the direction numbers are zero?

EXAMPLE 5 Find the symmetric equations of the line passing through the points $P = (2, 3, -2)$ and $Q = (2, -1, -2)$.

Solution Here $\mathbf{v} = -4\mathbf{j}$, so $a = 0$, $b = -4$, and $c = 0$. A parametric representation of the line is, by Equation (5), given by $x = 2$, $y = 3 - 4t$; $z = -2$. Now $x = 2$ is the equation of a plane parallel to the yz-plane whereas $z = -2$ is the equation of a plane parallel to the xy-plane. Their intersection is the line $x = 2$, $z = -2$, which is parallel to the y-axis. In fact, the equation $y = 3 - 4t$ says, essentially, that y can take on any value (while x and z remain fixed).

Warning. The parametric or symmetric equations of a line are *not* unique. To see this, simply start with two other points on the line.

EXAMPLE 6 In Example 1, the line contains the point $(5, 5, -18)$. Choose $P = (5, 5, -18)$ and $Q = (3, 1, -2)$. We find that $\mathbf{v} = -2\mathbf{i} - 4\mathbf{j} + 16\mathbf{k}$, so $x = 5 - 2t$, $y = 5 - 4t$, and $z = -18 + 16t$. (Note that if $t = \frac{3}{2}$, we obtain $(x, y, z) = (2, -1, 6)$.) The symmetric equations are now

$$\frac{x - 5}{-2} = \frac{y - 5}{-4} = \frac{z + 18}{16}$$

The equation of a line in space is obtained by specifying a point on the line and a vector *parallel* to this line. We can derive the equation of a plane in space by specifying a point in the plane and a vector orthogonal to every vector in the plane. This orthogonal vector is called a **normal vector** and is denoted by $\mathbf{n}$. (See Figure 4.51.)

Figure 4.51

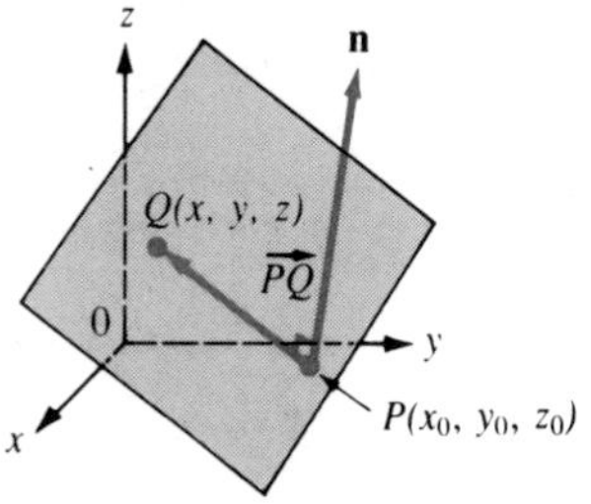

DEFINITION 1 **PLANE** Let P be a point in space and let $\mathbf{n}$ be a given nonzero vector. Then the set of all points Q for which $\overrightarrow{PQ}\cdot\mathbf{n} = 0$ comprises a **plane** in $\mathbb{R}^3$.

Notation We shall usually denote a plane by the symbol π.

Let $P=(x_0, y_0, z_0)$ be a fixed point on a plane with normal vector $\mathbf{n}=a\mathbf{i}+b\mathbf{j}+c\mathbf{k}$. If $Q=(x, y, z)$ is any other point on the plane, then $\overrightarrow{PQ}=(x-x_0)\mathbf{i}+(y-y_0)\mathbf{j}+(z-z_0)\mathbf{k}$. Since $\overrightarrow{PQ}\perp\mathbf{n}$, we have $\overrightarrow{PQ}\cdot\mathbf{n}=0$. But this implies that

$$a(x-x_0)+b(y-y_0)+c(z-z_0)=0 \tag{8}$$

A more common way to write the equation of a plane is easily derived from (8):

$$ax+by+cz=d \tag{9}$$

where
$$d=ax_0+by_0+cz_0=\overrightarrow{OP}\cdot\mathbf{n}$$

EXAMPLE 7 Find the plane π passing through the point $(2, 5, 1)$ having the normal vector $\mathbf{n}=\mathbf{i}-2\mathbf{j}+3\mathbf{k}$.

Figure 4.52

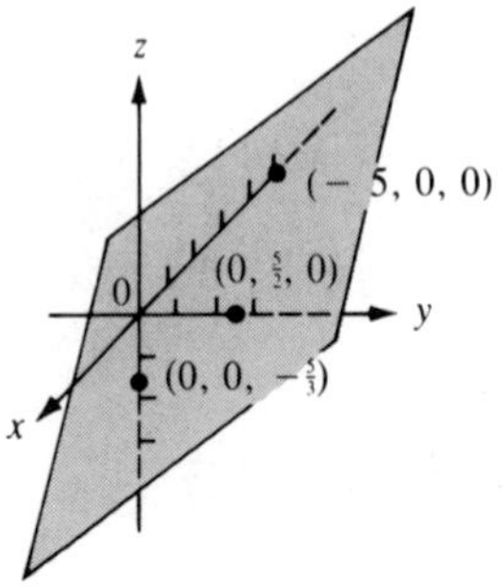

Solution From (8), we immediately obtain $(x-2)-2(y-5)+3(z-1)=0$ or

$$x-2y+3z=-5 \tag{10}$$

This plane is sketched in Figure 4.52.

Remark. The plane is easily sketched by setting $x=y=0$ in Equation (10) to obtain $(0, 0, -\frac{5}{3})$, $x=z=0$ to obtain $(0, \frac{5}{2}, 0)$, and $y=z=0$ to obtain $(-5, 0, 0)$. These three points all lie on the plane.

The three coordinate planes are represented as follows:

i. The *xy-plane* passes through the origin $(0, 0, 0)$, and any vector lying along the z-axis is normal to it. The simplest such vector is $\mathbf{k}$. Thus, from (8), we obtain $0(x-0)+0(y-0)+1(z-0)=0$, which yields

$$z=0 \tag{11}$$

as the equation of the xy-plane. (This result should not be very surprising.)

ii. The *xz-plane* has the equation

$$y=0 \tag{12}$$

iii. The *yz-plane* has the equation

$$x=0 \tag{13}$$

Three points which are not collinear determine a plane since they determine two nonparallel vectors that intersect at a point (see Figure 4.53).

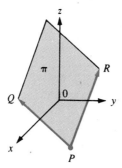

Figure 4.53

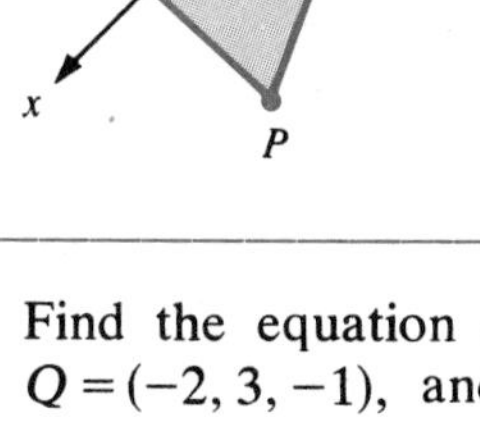

EXAMPLE 8 Find the equation of the plane passing through the points $P=(1, 2, 1)$, $Q=(-2, 3, -1)$, and $R=(1, 0, 4)$.

Solution The vectors $\overrightarrow{PQ}=-3\mathbf{i}+\mathbf{j}-2\mathbf{k}$ and $\overrightarrow{QR}=3\mathbf{i}-3\mathbf{j}+5\mathbf{k}$ lie on the plane and are therefore orthogonal to the normal vector so that

$$\mathbf{n}=\overrightarrow{PQ}\times\overrightarrow{QR}=\begin{vmatrix} \mathbf{i} & \mathbf{j} & \mathbf{k} \\ -3 & 1 & -2 \\ 3 & -3 & 5 \end{vmatrix}=-\mathbf{i}+9\mathbf{j}+6\mathbf{k}$$

and we obtain

$$\pi: \quad -(x-1)+9(y-2)+6(z-1)=0$$

or

$$-x+9y+6z=23$$

Note that if we choose another point, say Q, we get the equation $-(x+2)+9(y-3)+6(z+1)=0$, which reduces to $-x+9y+6z=23$. This plane is sketched in Figure 4.54.

Figure 4.54

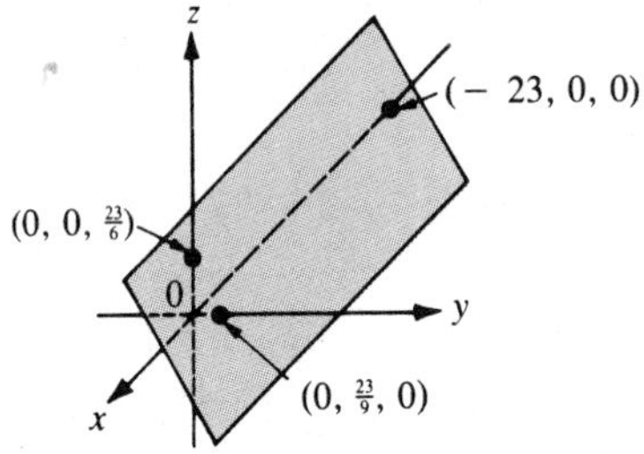

DEFINITION 2 **PARALLEL PLANES** Two planes are **parallel**† if their normal vectors are parallel; that is, if the cross product of their normal vectors is zero.

Two parallel planes are drawn in Figure 4.55.

Figure 4.55

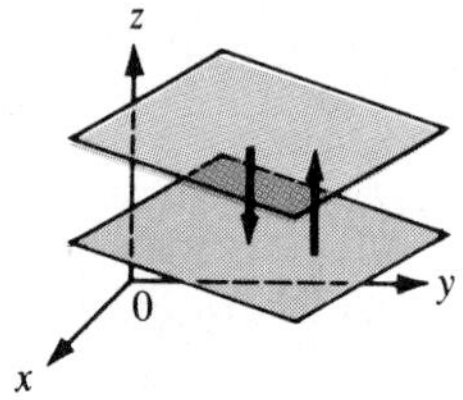

EXAMPLE 9 The planes π_1: $2x+3y-z=3$ and π_2: $-4x-6y+2z=8$ are parallel since $\mathbf{n}_1=2\mathbf{i}+3\mathbf{j}-\mathbf{k}$, $\mathbf{n}_2=-4\mathbf{i}-6\mathbf{j}+2\mathbf{k}=-2\mathbf{n}_1$ (and $\mathbf{n}_1\times\mathbf{n}_2=\mathbf{0}$).

If two planes are not parallel, then they intersect in a straight line.

EXAMPLE 10 Find all points of intersection of the planes $2x-y-z=3$ and $x+2y+3z=7$.

Solution When the planes intersect, we have $x+2y+3z=7$ and $2x-y-z=3$. Solving this system of two equations in three unknowns by row reduction, we obtain, successively

$$\left(\begin{array}{ccc|c}1 & 2 & 3 & 7\\ 2 & -1 & -1 & 3\end{array}\right) \xrightarrow{A_{1,2}(-2)} \left(\begin{array}{ccc|c}1 & 2 & 3 & 7\\ 0 & -5 & -7 & -11\end{array}\right)$$

$$\xrightarrow{M_2(-\frac{1}{5})} \left(\begin{array}{ccc|c}1 & 2 & 3 & 7\\ 0 & 1 & \frac{7}{5} & \frac{11}{5}\end{array}\right) \xrightarrow{A_{2,1}(-2)} \left(\begin{array}{ccc|c}1 & 0 & \frac{1}{5} & \frac{13}{5}\\ 0 & 1 & \frac{7}{5} & \frac{11}{5}\end{array}\right)$$

---

† Note that two parallel planes could be coincident. For example, the planes $x+y+z=1$ and $2x+2y+2z=2$ are coincident (the same).

Thus $y = \frac{11}{5} - (\frac{7}{5})z$ and $x = \frac{13}{5} - (\frac{1}{5})z$. Finally, setting $z = t$, we obtain the parametric representation of the line of intersection: $x = \frac{13}{5} - \frac{1}{5}t$, $y = \frac{11}{5} - \frac{7}{5}t$, and $z = t$.

We conclude this section by indicating how the distance from a plane to a point can be calculated. Look at Figure 4.56. If Q is the point, then the

Figure 4.56

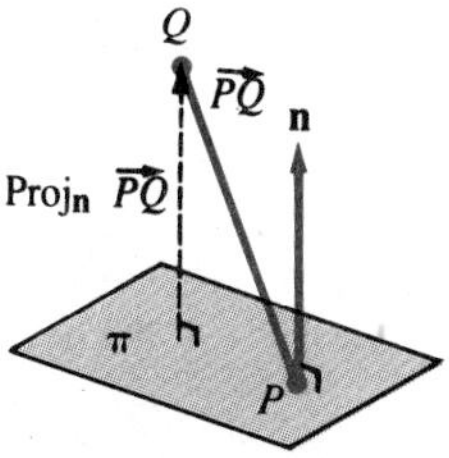

required distance is the distance measured along a line orthogonal to π. That is, the shortest distance is obtained by "dropping a perpendicular" from the point to the plane. This is done by calculating (for any point P on the plane)

$$D = |\text{proj}_{\mathbf{n}}\, \overrightarrow{PQ}| = \frac{|\overrightarrow{PQ} \cdot \mathbf{n}|}{|\mathbf{n}|} \tag{14}$$

EXAMPLE 11 Find the distance d between the plane $2x - y + 3z = 6$ and the point $Q = (3, 5, -7)$.

Solution One point on the plane is $P = (3, 0, 0)$ and $\mathbf{n} = 2\mathbf{i} - \mathbf{j} + 3\mathbf{k}$. Then $\overrightarrow{PQ} = 5\mathbf{j} - 7\mathbf{k}$, $|\overrightarrow{PQ} \cdot \mathbf{n}| = 26$, and $|\mathbf{n}| = \sqrt{14}$ so that $D = 26/\sqrt{14}$.

EXAMPLE 12 Find the distance from the plane $ax + by + cz = d$ to the origin.

Solution If $a \neq 0$, one point on the plane is $P = (d/a, 0, 0)$. (If $a = 0$ but $b \neq 0$, a point on the plane is $(0, d/b, 0)$ leading to the same result.) Then $\mathbf{n} = a\mathbf{i} + b\mathbf{j} + c\mathbf{k}$ and $\overrightarrow{PQ} = -(d/a)\mathbf{i}$ so that $|\overrightarrow{OP} \cdot \mathbf{n}| = |d|$, $|\mathbf{n}| = \sqrt{a^2 + b^2 + c^2}$, and

$$D = \frac{|d|}{\sqrt{a^2 + b^2 + c^2}} \tag{15}$$

PROBLEMS 4.5

In Problems 1–14 find a vector equation, parametric equations and symmetric equations of the indicated line.

1. Containing $(2, 1, 3)$ and $(1, 2, -1)$
2. Containing $(1, -1, 1)$ and $(-1, 1, -1)$

3. Containing $(-4, 1, 3)$ and $(-4, 0, 1)$
4. Containing $(2, 3, -4)$ and $(2, 0, -4)$
5. Containing $(1, 2, 3)$ and $(3, 2, 1)$
6. Containing $(7, 1, 3)$ and $(-1, -2, 3)$
7. Containing $(2, 2, 1)$ and parallel to $2\mathbf{i}-\mathbf{j}-\mathbf{k}$
8. Containing $(-1, -6, 2)$ and parallel to $4\mathbf{i}+\mathbf{j}-3\mathbf{k}$
9. Containing $(-1, -2, 5)$ and parallel to $-3\mathbf{j}+7\mathbf{k}$
10. Containing $(-2, 3, -2)$ and parallel to $4\mathbf{k}$
11. Containing (a, b, c) and parallel to $d\mathbf{i}+e\mathbf{j}$
12. Containing (a, b, c) and parallel to $d\mathbf{k}$
13. Containing $(4, 1, -6)$ and parallel to $(x-2)/3=(y+1)/6=(z-5)/2$
14. Containing $(3, 1, -2)$ and parallel to $(x+1)/3=(y+3)/2=(z-2)/(-4)$
15. Let L_1 be given by

$$\frac{x-x_1}{a_1}=\frac{y-y_1}{b_1}=\frac{z-z_1}{c_1}$$

and L_2 be given by

$$\frac{x-x_1}{a_2}=\frac{y-y_1}{b_2}=\frac{z-z_1}{c_2}$$

Show that L_1 is orthogonal to L_2 if and only if $a_1a_2+b_1b_2+c_1c_2=0$.

16. Show that the lines

$$L_1:\quad \frac{x-3}{2}=\frac{y+1}{4}=\frac{z-2}{-1} \quad\text{and}\quad L_2:\quad \frac{x-3}{5}=\frac{y+1}{-2}=\frac{z-3}{2}$$

are orthogonal.

17. Show that the lines

$$L_1:\quad \frac{x-1}{1}=\frac{y+3}{2}=\frac{z+3}{3} \quad\text{and}\quad L_2:\quad \frac{x-3}{3}=\frac{y-1}{6}=\frac{z-8}{9}$$

are parallel.

Lines in $\mathbb{R}^3$ that do not have the same direction need not have a point in common.

18. Show that the lines L_1: $x=1+t$, $y=-3+2t$, $z=-2-t$ and L_2: $x=17+3s$, $y=4+s$, $z=-8-s$ have the point $(2, -1, -3)$ in common.
19. Show that the lines L_1: $x=2-t$, $y=1+t$, $z=-2t$ and L_2: $x=1+s$, $y=-2s$, $z=3+2s$ do *not* have a point in common.
20. Let L be given in its vector form $\overrightarrow{OR}=\overrightarrow{OP}+t\mathbf{v}$. Find a number t such that $\overrightarrow{OR}$ is perpendicular to $\mathbf{v}$.
21. Use the result of Problem 20 to find the distance between the line L (containing P and parallel to $\mathbf{v}$) and the origin when:

    a. $P=(2, 1, -4)$; $\mathbf{v}=\mathbf{i}+\mathbf{j}+\mathbf{k}$
    b. $P=(1, 2, -3)$; $\mathbf{v}=3\mathbf{i}-\mathbf{j}-\mathbf{k}$
    c. $P=(-1, 4, 2)$; $\mathbf{v}=-\mathbf{i}+\mathbf{j}+2\mathbf{k}$

In Problems 22–25 find a line L orthogonal to the two given lines and passing through the given point.

22. $\dfrac{x+2}{-3}=\dfrac{y-1}{4}=\dfrac{z}{-5}$; $\dfrac{x-3}{7}=\dfrac{y+2}{-2}=\dfrac{z-8}{3}$; $(1, -3, 2)$

23. $\frac{x-2}{-4}=\frac{y+3}{-7}=\frac{z+1}{3}$; $\frac{x+2}{3}=\frac{y-5}{-4}=\frac{z+3}{-2}$; $(-4, 7, 3)$

24. $x=3-2t$; $y=4+3t$; $z=-7+5t$; $x=-2+4s$, $y=3-2s$, $z=3+s$; $(-2, 3, 4)$

25. $x=4+10t$, $y=-4-8t$, $z=3+7t$; $x=-2t$, $y=1+4t$, $z=-7-3t$; $(4, 6, 0)$

*26. Calculate the distance between the lines

$$L_1:\quad \frac{x-2}{3}=\frac{y-5}{2}=\frac{z-1}{-1} \quad\text{and}\quad L_2:\quad \frac{x-4}{-4}=\frac{y-5}{4}=\frac{z+2}{1}$$

[*Hint:* The distance is measured along a vector **v** that is perpendicular to both L_1 and L_2. Let P be a point on L_1 and Q a point on L_2. Then the length of the projection of $\overrightarrow{PQ}$ on **v** is the distance between the lines, measured along a vector that is perpendicular to them both.]

*27. Find the distance between the lines

$$L_1:\quad \frac{x+2}{3}=\frac{y-7}{-4}=\frac{z-2}{4} \quad\text{and}\quad L_2:\quad \frac{x-1}{-3}=\frac{y+2}{4}=\frac{z+1}{1}$$

In Problems 28–41 find the equation of the plane.

28. $P=(0, 0, 0)$; $\mathbf{n}=\mathbf{i}$

29. $P=(0, 0, 0)$; $\mathbf{n}=\mathbf{j}$

30. $P=(0, 0, 0)$; $\mathbf{n}=\mathbf{k}$

31. $P=(1, 2, 3)$; $\mathbf{n}=\mathbf{i}+\mathbf{j}$

32. $P=(1, 2, 3)$; $\mathbf{n}=\mathbf{i}+\mathbf{k}$

33. $P=(1, 2, 3)$; $\mathbf{n}=\mathbf{j}+\mathbf{k}$

34. $P=(2, -1, 6)$; $\mathbf{n}=3\mathbf{i}-\mathbf{j}+2\mathbf{k}$

35. $P=(-4, -7, 5)$; $\mathbf{n}=-3\mathbf{i}-4\mathbf{j}+\mathbf{k}$

36. $P=(-3, 11, 2)$; $\mathbf{n}=4\mathbf{i}+\mathbf{j}-7\mathbf{k}$

37. $P=(3, -2, 5)$; $\mathbf{n}=2\mathbf{i}-7\mathbf{j}-8\mathbf{k}$

38. Containing $(1, 2, -4)$, $(2, 3, 7)$, and $(4, -1, 3)$

39. Containing $(-7, 1, 0)$, $(2, -1, 3)$, and $(4, 1, 6)$

40. Containing $(1, 0, 0)$, $(0, 1, 0)$, and $(0, 0, 1)$

41. Containing $(2, 3, -2)$, $(4, -1, -1)$, and $(3, 1, 2)$

Two planes are **orthogonal** if their normal vectors are orthogonal. In Problems 42–46 determine whether the given planes are parallel, orthogonal, coincident (that is, the same), or none of these.

42. π_1: $x+y+z=2$; π_2: $2x+2y+2z=4$

43. π_1: $x-y+z=3$; π_2: $-3x+3y-3z=-9$

44. π_1: $2x-y+z=3$; π_2: $x+y-z=7$

45. π_1: $2x-y+z=3$; π_2: $x+y+z=3$

46. π_1: $3x-2y+7z=4$; π_2: $-2x+4y+2z=16$

In Problems 47–49 find the equation of the set of all points of intersection of the two planes.

47. π_1: $x-y+z=2$; π_2: $2x-3y+4z=7$

48. π_1: $3x-y+4z=3$; π_2: $-4x-2y+7z=8$

49. π_1: $-2x-y+17z=4$; π_2: $2x-y-z=-7$

In Problems 50–53 find the distance from the given point to the given plane.

50. $(2, -1, 4)$; $3x-y+7z=2$

51. $(4, 0, 1)$; $2x-y+8z=3$

52. $(-7, -2, -1)$; $-2x+8z=-5$

53. $(-3, 0, 2)$; $-3x+y+5z=0$

54. Prove that the distance between the plane $ax+by+cz=d$ and the point

(x_0, y_0, z_0) is given by

$$D = \frac{|ax_0 + by_0 + cz_0 - d|}{\sqrt{a^2+b^2+c^2}}$$

The **angle between two planes** is defined to be the acute† angle between their normal vectors. In Problems 55–57 find the angle between the two planes.

55. The two planes of Problem 47

56. The two planes of Problem 48

57. The two planes of Problem 49

***58.** Let **u** and **v** be two nonparallel, nonzero vectors in a plane π. Show that if **w** is any other vector in π, then there exist scalars α and β such that $\mathbf{w} = \alpha\mathbf{u} + \beta\mathbf{v}$. This is called the **parametric representation** of the plane π. [Hint: Draw a parallelogram in which $\alpha\mathbf{u}$ and $\beta\mathbf{v}$ form adjacent sides and the diagonal vector is **w**.]

***59.** Three vectors **u**, **v**, and **w** are called **coplanar** if they all lie in the same plane π. Show that if **u**, **v** and **w** all pass through the origin, then they are coplanar if and only if the scalar triple product equals zero: $\mathbf{u} \cdot (\mathbf{v} \times \mathbf{w}) = 0$. (Alternatively, from Problem 4.4.37 on page 157, the volume of the parallelepiped determined by **u**, **v**, and **w** is zero.)

In Problems 60–64 determine whether the three given position vectors (that is, one endpoint at the origin) are coplanar. If they are coplanar, find the equation of the plane containing them.

60. $\mathbf{u} = 2\mathbf{i} - 3\mathbf{j} + 4\mathbf{k}$; $\mathbf{v} = 7\mathbf{i} - 2\mathbf{j} + 3\mathbf{k}$; $\mathbf{w} = 9\mathbf{i} - 5\mathbf{j} + 7\mathbf{k}$

61. $\mathbf{u} = -3\mathbf{i} + \mathbf{j} + 8\mathbf{k}$; $\mathbf{v} = -2\mathbf{i} - 3\mathbf{j} + 5\mathbf{k}$; $\mathbf{w} = 2\mathbf{i} + 14\mathbf{j} - 4\mathbf{k}$

62. $\mathbf{u} = 2\mathbf{i} + \mathbf{j} - 2\mathbf{k}$; $\mathbf{v} = 2\mathbf{i} - \mathbf{j} - 2\mathbf{k}$; $\mathbf{w} = 2\mathbf{i} - \mathbf{j} + 2\mathbf{k}$

63. $\mathbf{u} = 3\mathbf{i} - 2\mathbf{j} + \mathbf{k}$; $\mathbf{v} = \mathbf{i} + \mathbf{j} - 5\mathbf{k}$; $\mathbf{w} = -\mathbf{i} + 5\mathbf{j} - 16\mathbf{k}$

64. $\mathbf{u} = 2\mathbf{i} - \mathbf{j} - \mathbf{k}$; $\mathbf{v} = 4\mathbf{i} + 3\mathbf{j} + 2\mathbf{k}$; $\mathbf{w} = 6\mathbf{i} + 7\mathbf{j} + 5\mathbf{k}$

Review Exercises for Chapter 4

In Exercises 1–6 find the magnitude and direction of the given vector.

1. $\mathbf{v} = (3, 3)$

2. $\mathbf{v} = -3\mathbf{i} + 3\mathbf{j}$

3. $\mathbf{v} = (2, -2\sqrt{3})$

4. $\mathbf{v} = (\sqrt{3}, 1)$

5. $\mathbf{v} = -12\mathbf{i} - 12\mathbf{j}$

6. $\mathbf{v} = \mathbf{i} + 4\mathbf{j}$

In Exercises 7–10 write the vector **v** that is represented by $\overrightarrow{PQ}$ in the form $a\mathbf{i} + b\mathbf{j}$. Sketch $\overrightarrow{PQ}$ and **v**.

7. $P = (2, 3)$; $Q = (4, 5)$

8. $P = (1, -2)$; $Q = (7, 12)$

9. $P = (-1, -6)$; $Q = (3, -4)$

10. $P = (-1, 3)$; $Q = (3, -1)$

11. Let $\mathbf{u} = (2, 1)$ and $\mathbf{v} = (-3, 4)$. Find: **(a)** $5\mathbf{u}$; **(b)** $\mathbf{u} - \mathbf{v}$; **(c)** $-8\mathbf{u} + 5\mathbf{v}$.

12. Let $\mathbf{u} = -4\mathbf{i} + \mathbf{j}$ and $\mathbf{v} = -3\mathbf{i} - 4\mathbf{j}$. Find: **(a)** $-3\mathbf{v}$; **(b)** $\mathbf{u} + \mathbf{v}$; **(c)** $3\mathbf{u} - 6\mathbf{v}$.

In Exercises 13–19 find a unit vector having the same direction as the given vector.

13. $\mathbf{v} = \mathbf{i} + \mathbf{j}$

14. $\mathbf{v} = -\mathbf{i} + \mathbf{j}$

15. $\mathbf{v} = 2\mathbf{i} + 5\mathbf{j}$

16. $\mathbf{v} = -7\mathbf{i} + 3\mathbf{j}$

17. $\mathbf{v} = 3\mathbf{i} + 4\mathbf{j}$

18. $\mathbf{v} = -2\mathbf{i} - 2\mathbf{j}$

19. $\mathbf{v} = a\mathbf{i} - a\mathbf{j}$

---

† Recall that an acute angle α is an angle between 0° and 90° (that is, $\alpha \in [0, \pi/2)$).

20. If $\mathbf{v}=4\mathbf{i}-7\mathbf{j}$, find $\sin\theta$ and $\cos\theta$, where θ is the direction of $\mathbf{v}$.
21. Find a unit vector with direction opposite to that of $\mathbf{v}=5\mathbf{i}+2\mathbf{j}$.
22. Find two unit vectors orthogonal to $\mathbf{v}=\mathbf{i}-\mathbf{j}$.
23. Find a unit vector with direction opposite to that of $\mathbf{v}=10\mathbf{i}-7\mathbf{j}$.

In Exercises 24–27 find a vector $\mathbf{v}$ having the given magnitude and direction.

24. $|\mathbf{v}|=2;\ \theta=\pi/3$
25. $|\mathbf{v}|=1;\ \theta=\pi/2$
26. $|\mathbf{v}|=4;\ \theta=\pi$
27. $|\mathbf{v}|=7;\ \theta=5\pi/6$

In Exercises 28–31 calculate the scalar product of the two vectors and the cosine of the angle between them.

28. $\mathbf{u}=\mathbf{i}-\mathbf{j};\ \mathbf{v}=\mathbf{i}+2\mathbf{j}$
29. $\mathbf{u}=-4\mathbf{i};\ \mathbf{v}=11\mathbf{j}$
30. $\mathbf{u}=4\mathbf{i}-7\mathbf{j};\ \mathbf{v}=5\mathbf{i}+6\mathbf{j}$
31. $\mathbf{u}=-\mathbf{i}-2\mathbf{j};\ \mathbf{v}=4\mathbf{i}+5\mathbf{j}$

In Exercises 32–37 determine whether the given vectors are orthogonal, parallel, or neither. Then sketch each pair.

32. $\mathbf{u}=2\mathbf{i}-6\mathbf{j};\ \mathbf{v}=-\mathbf{i}+3\mathbf{j}$
33. $\mathbf{u}=4\mathbf{i}-5\mathbf{j};\ \mathbf{v}=5\mathbf{i}-4\mathbf{j}$
34. $\mathbf{u}=4\mathbf{i}-5\mathbf{j};\ \mathbf{v}=-5\mathbf{i}+4\mathbf{j}$
35. $\mathbf{u}=-7\mathbf{i}-7\mathbf{j};\ \mathbf{v}=\mathbf{i}+\mathbf{j}$
36. $\mathbf{u}=-7\mathbf{i}-7\mathbf{j};\ \mathbf{v}=-\mathbf{i}+\mathbf{j}$
37. $\mathbf{u}=-7\mathbf{i}-7\mathbf{j};\ \mathbf{v}=-\mathbf{i}-\mathbf{j}$
38. Let $\mathbf{u}=2\mathbf{i}+3\mathbf{j}$ and $\mathbf{v}=4\mathbf{i}+\alpha\mathbf{j}$. Determine α such that:

- **a.** $\mathbf{u}$ and $\mathbf{v}$ are orthogonal.
- **b.** $\mathbf{u}$ and $\mathbf{v}$ are parallel.
- **c.** The angle between $\mathbf{u}$ and $\mathbf{v}$ is $\pi/4$.
- **d.** The angle between $\mathbf{u}$ and $\mathbf{v}$ is $\pi/6$.

In Exercises 39–44 calculate $\text{proj}_{\mathbf{v}}\,\mathbf{u}$.

39. $\mathbf{u}=14\mathbf{i};\ \mathbf{v}=\mathbf{i}+\mathbf{j}$
40. $\mathbf{u}=14\mathbf{i},\ \mathbf{v}=\mathbf{i}-\mathbf{j}$
41. $\mathbf{u}=3\mathbf{i}-2\mathbf{j};\ \mathbf{v}=3\mathbf{i}+2\mathbf{j}$
42. $\mathbf{u}=3\mathbf{i}+2\mathbf{j};\ \mathbf{v}=\mathbf{i}-3\mathbf{j}$
43. $\mathbf{u}=2\mathbf{i}-5\mathbf{j};\ \mathbf{v}=-3\mathbf{i}-7\mathbf{j}$
44. $\mathbf{u}=4\mathbf{i}-5\mathbf{j};\ \mathbf{v}=-3\mathbf{i}-\mathbf{j}$
45. Let $P=(3,-2)$, $Q=(4,7)$, $R=(-1,3)$, and $S=(2,-1)$. Calculate $\text{proj}_{\overrightarrow{PQ}}\,\overrightarrow{RS}$ and $\text{proj}_{\overrightarrow{RS}}\,\overrightarrow{PQ}$.

In Exercises 46–48 calculate the resultant of the forces acting on an object.

46. 3 N (from left), 2 N (from below)
47. 5 N (from right), 2 N (from left), 3 N (from above)
48. 2 N (from direction $\pi/4$), 5 N (from left), 4 N (from direction $2\pi/3$), 6 N (from direction $3\pi/4$)

In Exercises 49–52 find the work done when the force with given magnitude and direction moves an object from P to Q. All distances are measured in meters.

49. $|\mathbf{F}|=2\text{ N};\ \theta=\pi/4;\ P=(1,6);\ Q=(2,4)$
50. $|\mathbf{F}|=3\text{ N};\ \theta=\pi/2;\ P=(3,-5);\ Q=(2,7)$
51. $|\mathbf{F}|=11\text{ N};\ \theta=\pi/6;\ P=(-1,-2);\ Q=(-7,-4)$
52. $|\mathbf{F}|=8\text{ N};\ \theta=2\pi/3;\ P=(-1,4);\ Q=(5,-6)$

In Exercises 53–56 find the distance between the two given points.

53. $(3,1,2);\ (-1,-3,-4)$
54. $(4,-1,7);\ (-5,1,3)$

55. $(-2, 4, -8)$; $(0, 0, 6)$ **56.** $(2, -7, 0)$; $(0, 5, -8)$

57. Show that the points $(1, 3, 0)$ $(3, -1, -2)$, and $(-1, 7, 2)$ are collinear.

In Exercises 58–61 find the magnitude and the direction cosines of the given vector.

58. $\mathbf{v} = 2\mathbf{i} - \mathbf{k}$ **59.** $\mathbf{v} = 3\mathbf{j} + 11\mathbf{k}$

60. $\mathbf{v} = \mathbf{i} - 2\mathbf{j} - 3\mathbf{k}$ **61.** $\mathbf{v} = -4\mathbf{i} + \mathbf{j} + 6\mathbf{k}$

62. Find a unit vector in the direction of $\overrightarrow{PQ}$, where $P = (3, -1, 2)$ and $Q = (-4, 1, 7)$.

63. Find a unit vector whose direction is opposite that of $\overrightarrow{PQ}$, where $P = (1, -3, 0)$ and $Q = (-7, 1, -4)$.

In Exercises 64–71 let $\mathbf{u} = \mathbf{i} - 2\mathbf{j} + 3\mathbf{k}$, $\mathbf{v} = -3\mathbf{i} + 2\mathbf{j} + 5\mathbf{k}$, and $\mathbf{w} = 2\mathbf{i} - 4\mathbf{j} + \mathbf{k}$. Calculate:

64. $\mathbf{u} - \mathbf{v}$ **65.** $3\mathbf{v} + 5\mathbf{w}$ **66.** $\text{proj}_{\mathbf{v}}\, \mathbf{w}$

67. $\text{proj}_{\mathbf{w}}\, \mathbf{u}$ **68.** $2\mathbf{u} - 4\mathbf{v} + 7\mathbf{w}$ **69.** $\mathbf{u} \cdot \mathbf{w} - \mathbf{w} \cdot \mathbf{v}$

70. The angle between $\mathbf{u}$ and $\mathbf{v}$ **71.** The angle between $\mathbf{v}$ and $\mathbf{w}$

72. Find the distance from the point $P = (3, -1, 2)$ to the line passing through the points $Q = (-2, -1, 6)$ and $R = (0, 1, -8)$.

73. Find the work done when a force of 4 N acting in the direction of the vector $\mathbf{v} = -\mathbf{i} + \mathbf{j} + \mathbf{k}$ moves an object from $(2, 1, -6)$ to $(3, 5, 8)$ (distance is in meters).

In Exercises 74–77 find the cross product $\mathbf{u} \times \mathbf{v}$.

74. $\mathbf{u} = 3\mathbf{i} - \mathbf{j}$; $\mathbf{v} = 2\mathbf{i} + 4\mathbf{k}$ **75.** $\mathbf{u} = 7\mathbf{j}$; $\mathbf{v} = \mathbf{i} - \mathbf{k}$

76. $\mathbf{u} = 4\mathbf{i} - \mathbf{j} + 7\mathbf{k}$; $\mathbf{v} = -7\mathbf{i} + \mathbf{j} - 2\mathbf{k}$ **77.** $\mathbf{u} = -2\mathbf{i} + 3\mathbf{j} - 4\mathbf{k}$; $\mathbf{v} = -3\mathbf{i} + \mathbf{j} - 10\mathbf{k}$

78. Find two unit vectors orthogonal to both $\mathbf{u} = \mathbf{i} - \mathbf{j} + 3\mathbf{k}$ and $\mathbf{v} = -2\mathbf{i} - 3\mathbf{j} + 4\mathbf{k}$.

79. Calculate the area of the parallelogram with the adjacent vertices $(1, 4, -2)$, $(-3, 1, 6)$, and $(1, -2, 3)$.

80. Calculate the area of the triangle with vertices at $(2, 1, 3)$, $(-4, 1, 7)$, and $(-1, -1, 3)$.

81. Calculate the volume of the parallelepiped determined by the vectors $\mathbf{i} + \mathbf{j}$, $2\mathbf{i} - 3\mathbf{k}$, and $2\mathbf{j} + 7\mathbf{k}$.

In Exercises 82–85 find a vector equation, parametric equations, and symmetric equations of the given line.

82. Containing $(3, -1, 4)$ and $(-1, 6, 2)$

83. Containing $(-4, 1, 0)$ and $(3, 0, 7)$

84. Containing $(3, 1, 2)$ and parallel to $3\mathbf{i} - \mathbf{j} - \mathbf{k}$

85. Containing $(1, -2, -3)$ and parallel to $(x+1)/5 = (y-2)/(-3) = (z-4)/2$

86. Show that the lines L_1: $x = 3 - 2t$, $y = 4 + t$, $z = -2 + 7t$ and L_2: $x = -3 + s$, $y = 2 - 4s$, $z = 1 + 6s$ have no points of intersection.

87. Find the distance from the origin to the line passing through the point $(3, 1, 5)$ and having the direction $\mathbf{v} = 2\mathbf{i} - \mathbf{j} + \mathbf{k}$.

88. Find the equation of the line passing through $(-1, 2, 4)$ and orthogonal to L_1: $(x-1)/4 = (y+6)/3 = z/(-2)$ and L_2: $(x+3)/5 = (y-1)/1 = (z+3)/4$.

In Exercises 89–91 find the equation of the plane containing the given point and orthogonal to the given normal vector.

89. $P = (1, 3, -2)$; $\mathbf{n} = \mathbf{i} + \mathbf{k}$

90. $P = (1, -4, 6)$; $\mathbf{n} = 2\mathbf{j} - 3\mathbf{k}$

91. $P = (-4, 1, 6)$; $\mathbf{n} = 2\mathbf{i} - 3\mathbf{j} + 5\mathbf{k}$

92. Find the equation of the plane containing the points $(-2, 4, 1)$, $(3, -7, 5)$, and $(-1, -2, -1)$.

93. Find all points of intersection of the planes π_1: $-x+y+z=3$ and π_2: $-4x+2y-7z=5$.

94. Find all points of intersection of the planes π_1: $-4x+6y+8z=12$ and π_2: $2x-3y-4z=5$.

95. Find all points of intersection of the planes π_1: $3x-y+4z=8$ and π_2: $-3x-y-11z=0$.

96. Find the distance from $(1, -2, 3)$ to the plane $2x-y-z=6$.

97. Find the angle between the planes of Exercise 93.

98. Show that the position vectors $\mathbf{u}=\mathbf{i}-2\mathbf{j}+\mathbf{k}$, $\mathbf{v}=3\mathbf{i}+2\mathbf{j}-3\mathbf{k}$, and $\mathbf{w}=9\mathbf{i}-2\mathbf{j}-3\mathbf{k}$ are coplanar and find the equation of the plane containing them.

5 Vector Spaces

5.1 Introduction

As we saw in the last chapter, the sets $\mathbb{R}^2$ (vectors in the plane) and $\mathbb{R}^3$ (vectors in space) have a number of nice properties. We can add two vectors in $\mathbb{R}^2$ and obtain another vector in $\mathbb{R}^2$. Under addition, vectors in $\mathbb{R}^2$ commute and obey the associative law. If $\mathbf{x} \in \mathbb{R}^2$, then $\mathbf{x}+\mathbf{0}=\mathbf{x}$ and $\mathbf{x}+(-\mathbf{x})=\mathbf{0}$. We can multiply vectors in $\mathbb{R}^2$ by scalars and obtain a number of distributive laws. The same properties also hold in $\mathbb{R}^3$.

The sets $\mathbb{R}^2$ and $\mathbb{R}^3$ are called *vector spaces.* Intuitively, we can say that a vector space is a set of objects that obey the rules described in the previous paragraph.

In this chapter we make a seemingly great leap from the concrete world of solving equations and dealing with easily visualized vectors to the abstract world of arbitrary vector spaces. There is a great advantage in doing so. Once we have established a fact about vector spaces in general, we can apply that fact to *every* vector space. Otherwise, we would have to prove that fact again and again, once for each new vector space we encounter (and there is an endless supply of them). But, as you will see, the abstract theorems we shall prove are really no more difficult than the ones already encountered.

5.2 Definition and Basic Properties

DEFINITION 1 REAL VECTOR SPACE A **real vector space** V is a set of objects, called **vectors**, together with two operations called **addition** and **scalar multiplication** that satisfy the ten axioms listed below.

Notation If $\mathbf{x}$ and $\mathbf{y}$ are in V and if α is a real number, then we write $\mathbf{x}+\mathbf{y}$ for the sum of $\mathbf{x}$ and $\mathbf{y}$ and $\alpha\mathbf{x}$ for the scalar product of α and $\mathbf{x}$.

Before we list the properties satisfied by vectors in a vector space, two things should be mentioned. First, while it might be helpful to think of $\mathbb{R}^2$ or $\mathbb{R}^3$ when dealing with a vector space, it often occurs that a vector space may appear to be very different from these comfortable spaces. (We shall see this shortly.) Second, Definition 1 gives a definition of a *real* vector space. The word "real" means that the scalars we use are real numbers. It would be just as easy to define a *complex* vector space by using complex numbers instead of real ones. This book deals primarily with real vector spaces, but generalizations to other sets of scalars present little difficulty.

Axioms of a Vector Space

i. If $\mathbf{x}\in V$ and $\mathbf{y}\in V$, then $\mathbf{x}+\mathbf{y}\in V$ (closure under addition).
ii. For all $\mathbf{x}$, $\mathbf{y}$, and $\mathbf{z}$ in V, $(\mathbf{x}+\mathbf{y})+\mathbf{z}=\mathbf{x}+(\mathbf{y}+\mathbf{z})$ (associative law of vector addition).
iii. There is a vector $\mathbf{0}\in V$ such that for all $\mathbf{x}\in V$, $\mathbf{x}+\mathbf{0}=\mathbf{0}+\mathbf{x}=\mathbf{x}$ ($\mathbf{0}$ is called the **additive identity**).
iv. If $\mathbf{x}\in V$, there is a vector $-\mathbf{x}$ in V such that $\mathbf{x}+(-\mathbf{x})=\mathbf{0}$ ($-\mathbf{x}$ is called the **additive inverse** of $\mathbf{x}$).
v. If $\mathbf{x}$ and $\mathbf{y}$ are in V, then $\mathbf{x}+\mathbf{y}=\mathbf{y}+\mathbf{x}$ (commutative law of vector addition).
vi. If $\mathbf{x}\in V$ and α is a scalar, then $\alpha\mathbf{x}\in V$ (closure under scalar multiplication).
vii. If $\mathbf{x}$ and $\mathbf{y}$ are in V and α is a scalar, then $\alpha(\mathbf{x}+\mathbf{y})=\alpha\mathbf{x}+\alpha\mathbf{y}$ (first distributive law).
viii. If $\mathbf{x}\in V$ and α and β are scalars, then $(\alpha+\beta)\mathbf{x}=\alpha\mathbf{x}+\beta\mathbf{x}$ (second distributive law).
ix. If $\mathbf{x}\in V$ and α and β are scalars, then $\alpha(\beta\mathbf{x})=\alpha\beta\mathbf{x}$ (associative law of scalar multiplication).
x. For every vector $\mathbf{x}\in V$, $1\mathbf{x}=\mathbf{x}$ (the scalar 1 is called a **multiplicative identity**).

EXAMPLE 1

THE SPACE $\mathbb{R}^n$ Let $V=\mathbb{R}^n=\{(x_1, x_2, \ldots, x_n): x_i\in\mathbb{R} \text{ for } i=1, 2, \ldots, n\}$.

From Section 2.1 (see Theorem 2.1.1, page 31) we see that V satisfies all the axioms of a vector space if we take the set of scalars to be $\mathbb{R}$.

EXAMPLE 2

Let $V=\{0\}$. That is, V consists of the single number 0. Since $0+0=1\cdot0=0+(0+0)=(0+0)+0=0$, we see that V is a vector space. It is often referred to as a **trivial** vector space.

EXAMPLE 3

Let $V=\{1\}$. That is, V consists of the single number 1. This is *not* a vector space since it violates axiom (*i*)—the closure axiom. To see this we simply note that $1+1=2\notin V$.

EXAMPLE 4

Let $V=\{(x, y): y=mx$, where m is a fixed real number and x is an arbitrary real number$\}$. That is, V consists of all points lying on the line $y=mx$ passing through the origin with slope m. Suppose that (x_1, y_1) and (x_2, y_2) are in V. Then $y_1=mx_1$, $y_2=mx_2$, and

$$(x_1, y_1)+(x_2, y_2)=(x_1, mx_1)+(x_2, mx_2)=(x_1+x_2, mx_1+mx_2)$$
$$=(x_1+x_2, m(x_1+x_2))\in V.$$

Thus axiom (i) is satisfied. Axioms (ii), (iii), and (v) are obvious. Further,

$$-(x, mx) = (-x, -mx) = (-x, m(-x)) \in V$$

and

$$(x, mx) + (-x, m(-x)) = (0, 0) = \mathbf{0}$$

so that axiom (iv) is satisfied. The other axioms are easily verified and we see that the set of points in the plane lying on a straight line passing through the origin constitutes a vector space.

EXAMPLE 5 Let $V = \{(x, y): y = 2x + 1, \mathbf{x} \in \mathbb{R}\}$. That is, V is the set of points lying on the line $y = 2x + 1$. V is *not* a vector space because closure is violated, as in Example 3. To see this, let us suppose that (x_1, y_1) and (x_2, y_2) are in V. Then

$$(x_1, y_1) + (x_2, y_2) = (x_1 + x_2, y_1 + y_2).$$

If this last vector were in V, we would have

$$y_1 + y_2 = 2(x_1 + x_2) + 1 = 2x_1 + 2x_2 + 1.$$

But $y_1 = 2x_1 + 1$ and $y_2 = 2x_2 + 1$ so that

$$y_1 + y_2 = (2x_1 + 1) + (2x_2 + 1) = 2x_1 + 2x_2 + 2.$$

Hence we conclude that

$$(x_1 + x_2, y_1 + y_2) \notin V \quad \text{if} \quad (x_1, y_1) \in V \quad \text{and} \quad (x_2, y_2) \in V.$$

EXAMPLE 6 Let $V = \{(x, y, z): \ ax + by + cz = 0\}$. That is, V is the set of points in $\mathbb{R}^3$ lying on the plane passing through the origin with normal vector (a, b, c). Suppose (x_1, y_1, z_1) and (x_2, y_2, z_2) are in V. Then $(x_1, y_1, z_1) + (x_2, y_2, z_2) = (x_1 + x_2, y_1 + y_2, z_1 + z_2) \in V$ because $a(x_1 + x_2) + b(y_1 + y_2) + c(z_1 + z_2) = (ax_1 + by_1 + cz_1) + (ax_2 + by_2 + cz_2) = 0 + 0 = 0$; hence axiom ($i$) is satisfied. The other axioms are easily verified. Thus the set of points lying on a plane in $\mathbb{R}^3$ that passes through the origin comprises a vector space.

EXAMPLE 7 Let $V = P_n$, the set of polynomials with real coefficients of degree less than or equal to n. If $p \in P_n$, then

$$p(x) = a_n x^n + a_{n-1} x^{n-1} + \cdots + a_1 x + a_0$$

where each a_i is real. The sum $p(x) + q(x)$ is defined in the obvious way: If $q(x) = b_n x^n + b_{n-1} x^{n-1} + \cdots + b_1 x + b_0$, then

$$p(x) + q(x) = (a_n + b_n)x^n + (a_{n-1} + b_{n-1})x^{n-1} + \cdots + (a_1 + b_1)x + (a_0 + b_0)$$

Clearly the sum of two polynomials of degree less than or equal to n is another polynomial with degree less than or equal to n, so axiom (i) is satisfied. Properties (ii) and (v) to (x) are obvious. If we define the zero polynomial by $\mathbf{0} = 0x^n + 0x^{n-1} + \cdots + 0x + 0$, then clearly $\mathbf{0} \in P_n$ and axiom (*iii*) is satisfied. Finally, letting $-p(x) = -a_n x^n - a_{n-1}x^{n-1} - \cdots - a_1 x - a_0$, we see that axiom (*iv*) holds, so P_n is a real vector space.

EXAMPLE 8 Let $V = C[0, 1] =$ the set of real-valued continuous functions defined on the interval $[0, 1]$. We define $(f+g)x = f(x) + g(x)$ and $(\alpha f)(x) = \alpha[f(x)]$. Since the sum of continuous functions is continuous, axiom (*i*) is satisfied and the other axioms are easily verified with $\mathbf{0} =$ the zero function and $(-f)(x) = -f(x)$.

EXAMPLE 9 Let $V = M_{34}$ denote the set of 3×4 matrices with real components. Then with the usual sum and scalar multiplication of matrices, it is again easy to verify that M_{34} is a vector space with $\mathbf{0}$ being the 3×4 zero matrix. If $A = (a_{ij})$ is in M_{34}, then $-A = (-a_{ij})$ is also in M_{34}.

EXAMPLE 10 In an identical manner we see that M_{mn}, the set of $m \times n$ matrices with real components, forms a vector space for any integers m and n.

EXAMPLE 11 Let S_3 denote the set of invertible 3×3 matrices. Define the "sum" $A + B$ by $A + B = AB$. If A and B are invertible, then AB is invertible (by Theorem 2.7.3, page 61) so that axiom (*i*) is satisfied. Axiom (*ii*) is simply the associative law for matrix multiplication (Theorem 2.4.1, page 44); axioms (*iii*) and (*iv*) are satisfied with $\mathbf{0} = I_3$ and $-A = A^{-1}$. Axiom (*v*) fails, however, since, in general, $AB \neq BA$ so that S_3 is not a vector space.

EXAMPLE 12 Let $V = \{(x, y)\colon \; y \geq 0\}$. V consists of the points in $\mathbb{R}^2$ in the upper half plane (the first two quadrants). If $y_1 \geq 0$ and $y_2 \geq 0$, then $y_1 + y_2 \geq 0$; hence if $(x_1, y_1) \in V$ and $(x_2, y_2) \in V$, then $(x_1 + x_2, y_1 + y_2) \in V$. V is not a vector space, however, since the vector $(1, 1)$, for example, does not have an inverse in V because $(-1, -1) \notin V$. Moreover, axiom (*vi*) fails since if $(x, y) \in V$, then $\alpha(x, y) \notin V$ if $\alpha < 0$.

EXAMPLE 13 **THE SPACE $\mathbb{C}^n$** Let $V = \mathbb{C}^n = \{(c_1, c_2, \ldots, c_n)\colon c_i$ is a complex number for $i = 1, 2, \ldots, n\}$ and the set of scalars is the set of complex numbers. It is easy to verify that $\mathbb{C}^n$, too, is a vector space.

As these examples suggest, there are many different kinds of vector spaces and many kinds of sets that are *not* vector spaces. Before leaving this section let us prove some elementary results about vector spaces.

THEOREM 1 Let V be a vector space. Then:

i. $\alpha\mathbf{0}=\mathbf{0}$ for every real number α.
ii. $0\cdot\mathbf{x}=\mathbf{0}$ for every $\mathbf{x}\in V$.
iii. If $\alpha\mathbf{x}=\mathbf{0}$, then $\alpha=0$ or $\mathbf{x}=\mathbf{0}$ (or both).
iv. $(-1)\mathbf{x}=-\mathbf{x}$ for every $\mathbf{x}\in V$.

Proof **i.** By axiom (*iii*), $\mathbf{0}+\mathbf{0}=\mathbf{0}$; and from axiom (*vii*),

$$\alpha(\mathbf{0}+\mathbf{0})=\alpha\mathbf{0}+\alpha\mathbf{0}=\alpha\mathbf{0} \tag{1}$$

Adding $-\alpha\mathbf{0}$ to both sides of the last equation in (1) and using the associative law (axiom *ii*), we obtain

$$[\alpha\mathbf{0}+\alpha\mathbf{0}]+(-\alpha\mathbf{0})=\alpha\mathbf{0}+(-\alpha\mathbf{0})$$
$$\alpha\mathbf{0}+[\alpha\mathbf{0}+(-\alpha\mathbf{0})]=\mathbf{0}$$
$$\alpha\mathbf{0}+\mathbf{0}=\mathbf{0}$$
$$\alpha\mathbf{0}=\mathbf{0}$$

ii. Essentially the same proof as used in part (*i*) works. We start with $0+0=0$ and use axiom (*viii*) to see that $0\mathbf{x}=(0+0)\mathbf{x}=0\mathbf{x}+0\mathbf{x}$ or $0\mathbf{x}+(-0\mathbf{x})=0\mathbf{x}+[0\mathbf{x}+(-0\mathbf{x})]$ or $\mathbf{0}=0\mathbf{x}+\mathbf{0}=0\mathbf{x}$.

iii. Let $\alpha\mathbf{x}=\mathbf{0}$. If $\alpha\neq 0$, we multiply both sides of the equation by $1/\alpha$ to obtain $(1/\alpha)(\alpha\mathbf{x})=(1/\alpha)\mathbf{0}=\mathbf{0}$ (by part *i*). But $(1/\alpha)(\alpha\mathbf{x})=1\mathbf{x}=\mathbf{x}$ (by axiom *ix*), so $\mathbf{x}=\mathbf{0}$.

iv. We start with the fact that $1+(-1)=0$. Then, using part (*ii*), we obtain

$$\mathbf{0}=0\mathbf{x}=[1+(-1)]\mathbf{x}=1\mathbf{x}+(-1)\mathbf{x}=\mathbf{x}+(-1)\mathbf{x} \tag{2}$$

We add $-\mathbf{x}$ to both sides of (2) to obtain

$$\begin{aligned}\mathbf{0}+(-\mathbf{x})&=\mathbf{x}+(-1)\mathbf{x}+(-\mathbf{x})=\mathbf{x}+(-\mathbf{x})+(-1)\mathbf{x}\\&=\mathbf{0}+(-1)\mathbf{x}=(-1)\mathbf{x}\end{aligned}$$

Thus $-\mathbf{x}=(-1)\mathbf{x}$. Note that we were able to reverse the order of addition in the preceding equation by using the commutative law (axiom *v*). ■

Remark. Part (*iii*) of Theorem 1 is not as obvious as it seems. There are objects which have the property that $xy=0$ does not imply that either x or y is zero. As an example, we look at the multiplication of 2×2 matrices. If $A=\begin{pmatrix}0&1\\0&0\end{pmatrix}$ and $B=\begin{pmatrix}0&-2\\0&0\end{pmatrix}$, then neither A nor B is zero although, as is easily verified, the product $AB=0$, the zero matrix.

PROBLEMS 5.2

In Problems 1–20 determine whether the given set is a vector space. If it is not, list the axioms that do not hold.

1. The set of diagonal $n\times n$ matrices under the usual matrix addition and the usual scalar multiplication.
2. The set of diagonal $n\times n$ matrices under multiplication (that is, $A+B=AB$).

3. $\{(x, y): \; y \leq 0; \; x, y \text{ real}\}$ with the usual addition and scalar multiplication of vectors.

4. The vectors in the plane lying in the first quadrant.

5. The set of vectors in $\mathbb{R}^3$ in the form (x, x, x).

6. The set of polynomials of degree 4 under the operations of Example 7.

7. The set of $n \times n$ symmetric matrices (see Section 2.8) under the usual addition and scalar multiplication.

8. The set of 2×2 matrices having the form $\begin{pmatrix} 0 & a \\ b & 0 \end{pmatrix}$ under the usual addition and scalar multiplication.

9. The set of matrices of the form $\begin{pmatrix} 1 & \alpha \\ \beta & 1 \end{pmatrix}$ with the matrix operations of addition and scalar multiplication.

10. The set consisting of the single vector $(0, 0)$ under the usual operations in $\mathbb{R}^2$.

11. The set of polynomials of degree $\leq n$ with zero constant term.

12. The set of polynomials of degree $\leq n$ with positive constant term a_0.

13. The set of continuous functions in $[0, 1]$ with $f(0) = 0$ and $f(1) = 0$ under the operations of Example 8.

14. The set of points in $\mathbb{R}^3$ lying on a line passing through the origin.

15. The set of points in $\mathbb{R}^3$ lying on the line $x = t + 1$, $y = 2t$, $z = t - 1$.

16. $\mathbb{R}^2$ with addition defined by $(x_1, y_1) + (x_2, y_2) = (x_1 + x_2 + 1, y_1 + y_2 + 1)$ and ordinary scalar multiplication.

17. The set of Problem 16 with scalar multiplication defined by $\alpha(x, y) = (\alpha + \alpha x - 1, \alpha + \alpha y - 1)$.

18. The set consisting of one object with addition defined by *object* + *object* = *object* and scalar multiplication defined by α(*object*) = *object*.

†□**19.** The set of differentiable functions defined on $[0, 1]$ with the operations of Example 8.

***20.** The set of real numbers of the form $a + b\sqrt{2}$, where a and b are rational numbers, under the usual addition of real numbers and with scalar multiplication defined only for rational scalars.

21. Show that in a vector space the additive identity element is unique.

22. Show that in a vector space each vector has a unique additive inverse.

23. If $\mathbf{x}$ and $\mathbf{y}$ are vectors in a vector space V, show that there is a unique vector $\mathbf{z} \in V$ such that $\mathbf{x} + \mathbf{z} = \mathbf{y}$.

24. Show that the set of positive real numbers forms a vector space under the operations $x + y = xy$ and $\alpha x = x^{\alpha}$.

*□**25.** Consider the homogeneous second order differential equation

$$y''(x) + a(x)y'(x) + b(x)y(x) = 0$$

where $a(x)$ and $b(x)$ are continuous functions. Show that the set of solutions to the equation is a vector space under the usual rules for adding functions and multiplying them by real numbers.

5.3 Subspaces

From Example 5.2.1, page 173, we know that $\mathbb{R}^2 = \{(x, y): x \in \mathbb{R} \text{ and } y \in \mathbb{R}\}$ is a vector space. In Example 5.2.4, page 173, we saw that $V = \{(x, y): y = mx\}$ is also a vector space. Moreover, it is clear that $V \subset \mathbb{R}^2$. That is, $\mathbb{R}^2$ has a subset

†□This box symbol is used throughout the book to indicate that the problem or example uses calculus.

that is also a vector space. In fact, all vector spaces have subsets that are also vector spaces. We shall examine these important subsets in this section.

DEFINITION 1 **SUBSPACE** Let H be a nonempty subset of a vector space V and suppose that H is itself a vector space under the operations of addition and scalar multiplication defined on V. Then H is said to be a **subspace** of V.

We shall encounter many examples of subspaces in this chapter. But first we prove a result that makes it relatively easy to determine whether a subset of V is indeed a subspace of V.

THEOREM 1 A nonempty subset H of the vector space V is a subspace of V if the two closure rules hold:

RULES FOR CHECKING WHETHER A SUBSET IS A SUBSPACE

i. If $\mathbf{x} \in H$ and $\mathbf{y} \in H$, then $\mathbf{x}+\mathbf{y} \in H$.

ii. If $\mathbf{x} \in H$, then $\alpha\mathbf{x} \in H$ for every scalar α.

Proof To show that H is a vector space, we must show that axioms (i) to (x) on page 173 hold under the operations of vector addition and scalar multiplication defined in V. The two closure operations (axioms i and vi) hold by hypothesis. Since vectors in H are also in V, the associative, commutative, distributive, and multiplicative identity laws (axioms ii, v, vii, $viii$, ix, and x) hold. Let $\mathbf{x} \in H$. Then $0\mathbf{x} \in H$ by hypothesis (ii). But by Theorem 5.2.1, page 176, (part ii), $0\mathbf{x} = \mathbf{0}$. Thus $\mathbf{0} \in H$ and axiom (iii) holds. Finally, by part (ii), $(-1)\mathbf{x} \in H$ for every $\mathbf{x} \in H$. By Theorem 5.2.1 (part iv), $-\mathbf{x} = (-1)\mathbf{x} \in H$ so that axiom (iv) also holds and the proof is complete. ■

This theorem shows that to test whether H is a subspace of V, it is only necessary to verify that:

$\mathbf{x}+\mathbf{y}$ and $\alpha\mathbf{x}$ are in H when $\mathbf{x}$ and $\mathbf{y}$ are in H and α is a scalar.

The preceding proof contains a fact that is important enough to mention explicitly:

Every subspace of a vector space V contains $\mathbf{0}$. (1)

This fact will often make it easy to see that a particular subset of V is *not* a vector space. That is, if a subset does not contain $\mathbf{0}$, then it is not a subspace.

We now give some examples of subspaces.

EXAMPLE 1 For any vector space V, the subset $\{\mathbf{0}\}$ consisting of the zero vector alone is a subspace since $\mathbf{0} + \mathbf{0} = \mathbf{0}$ and $\alpha\mathbf{0} = \mathbf{0}$ for every real number α (part (i) of Theorem 5.2.1). It is called the **trivial subspace** .

EXAMPLE 2 V is a subspace of itself for every vector space V.

PROPER SUBSPACE

The first two examples show that every vector space V contains two subspaces $\{\mathbf{0}\}$ and V (unless, of course, $V = \{\mathbf{0}\}$). It is more interesting to find other subspaces. Subspaces other than $\{\mathbf{0}\}$ and $\mathbf{V}$ are called **proper subspaces.**

EXAMPLE 3 Let $H = \{(x, y)\colon \ y = mx\}$ (see Example 5.2.4, page 173). Then, as we have already mentioned, H is a subspace of $\mathbb{R}^2$. As we shall see in Section 5.5 (Problem 5.5.15), the sets of vectors lying on straight lines through the origin are the only proper subspaces of $\mathbb{R}^2$.

EXAMPLE 4 Let $H = \{(x, y, z)\colon \ x = at, y = bt, \text{and } z = ct; a, b, c, t \text{ real}\}$. Then H consists of the vectors in $\mathbb{R}^3$ lying on a straight line passing through the origin. To see that H is a subspace of $\mathbb{R}^3$, we compute. Let $\mathbf{x} = (at_1, bt_1, ct_1) \in H$ and $\mathbf{y} = (at_2, bt_2, ct_2) \in H$. Then $\mathbf{x}+\mathbf{y} = (a(t_1+t_2),\ b(t_1+t_2),\ c(t_1+t_2)) \in H$ and $\alpha\mathbf{x} = (a(\alpha t_1),\ b(\alpha t_2),\ c(\alpha t_3)) \in H$. Thus H is a subspace of $\mathbb{R}^3$.

EXAMPLE 5 Let $\pi = \{(x, y, z)\colon \ ax + by + cz = 0;\ a,\ b,\ c \text{ real}\}$. Then, as we saw in Example 5.2.6, page 174, π is a vector space; thus π is a subspace of $\mathbb{R}^3$.

We shall prove in Section 5.5 that sets of vectors lying on lines and planes through the origin are the only proper subspaces of $\mathbb{R}^3$.

Before studying more examples, we note that *not every vector space has proper subspaces.*

EXAMPLE 6 Let H be a subspace of $\mathbb{R}$.† If $H \neq \{\mathbf{0}\}$, then H contains a nonzero real number α. Then, by axiom (vi), $1 = (1/\alpha)\alpha \in H$ and $\beta 1 = \beta \in H$ for every real number β. Thus if H is not the trivial subspace, then $H = \mathbb{R}$. That is, $\mathbb{R}$ has *no* proper subspace.

† Note that $\mathbb{R}$ is a vector space over itself; that is, $\mathbb{R}$ is a vector space where the scalars are taken to be the reals. This is Example 5.2.1, page 173, with $n = 1$.

EXAMPLE 7 If P_n denotes the vector space of polynomials of degree $\leq n$, (Example 5.2.7, page 174), and if $0 < m < n$, then P_m is a proper subspace of P_n, as is easily verified.

EXAMPLE 8 Let M_{mn} (Example 5.2.10, page 175) denote the vector space of $m \times n$ matrices with real components and let $H = \{A \in M_{mn}\colon \ a_{11} = 0\}$. By the definition of matrix addition and scalar multiplication it is clear that the two closure axioms hold, so that H is a subspace.

EXAMPLE 9 Let $V = M_{nn}$ (the $n \times n$ matrices) and let $H = \{A \in M_{nn}\colon \ A \text{ is invertible}\}$. Then H is not a subspace since the $n \times n$ zero matrix is not in H.

EXAMPLE 10 $P_n[0, 1]$† $\subset C[0, 1]$ (see Example 5.2.8, page 175) because every polynomial is continuous and P_n is a vector space for every integer n, so that each $P_n[0, 1]$ is a subspace of $C[0, 1]$.

‡□EXAMPLE 11 Let $C'[0, 1]$ denote the set of functions with continuous first derivatives defined on $[0, 1]$. Since every differentiable function is continuous, we have $C'[0, 1] \subset C[0, 1]$. Since the sum and scalar multiple of two differentiable functions are differentiable, we see that $C'[0, 1]$ is a subspace of $C[0, 1]$. It is a proper subspace because not every continuous function is differentiable.

□EXAMPLE 12 If $f \in C[0, 1]$, then $\int_0^1 f(x)\,dx$ exists. Let $H = \{f \in C[0, 1]\colon \ \int_0^1 f(x)\,dx = 0\}$. If $f \in H$ and $g \in H$, then $\int_0^1 [f(x) + g(x)]\,dx = \int_0^1 f(x)\,dx + \int_0^1 g(x)\,dx = 0 + 0 = 0$ and $\int_0^1 \alpha f(x)\,dx = \alpha \int_0^1 f(x)\,dx = 0$. Thus $f + g$ and αf are in H for every real number α. This shows that H is a proper subspace of $C[0, 1]$.

As the last three examples illustrate, a vector space can have a great number and variety of proper subspaces. Before leaving this section, we prove an interesting fact about subspaces.

THEOREM 2 Let H_1 and H_2 be subspaces of a vector space V. Then $H_1 \cap H_2$ is a subspace of V.

Proof Let $\mathbf{x}_1 \in H_1 \cap H_2$ and $\mathbf{x}_2 \in H_1 \cap H_2$. Then, since H_1 and H_2 are subspaces, $\mathbf{x}_1 + \mathbf{x}_2 \in H_1$ and $\mathbf{x}_1 + \mathbf{x}_2 \in H_2$. This means that $\mathbf{x}_1 + \mathbf{x}_2 \in H_1 \cap H_2$. Similarly,

† $P_n[0, 1]$ denotes the set of polynomials defined on the interval $[0, 1]$ of degree $\leq n$.

‡ □ We remind you that this box symbol is used throughout the book to indicate that the problem or example uses calculus.

$\alpha\mathbf{x}_1 \in H_1 \cap H_2$. Thus the two closure axioms are satisfied and $H_1 \cap H_2$ is a subspace.† ■

EXAMPLE 13 In $\mathbb{R}^3$, let $H_1 = \{(x, y, z)\colon\ 2x - y - z = 0\}$ and $H_2 = \{(x, y, z)\colon\ x + 2y + 3z = 0\}$. Then H_1 and H_2 consist of vectors lying on planes through the origin and are, by Example 5, subspaces of $\mathbb{R}^3$. $H_1 \cap H_2$ is the intersection of the two planes which we compute as in Section 4.5:

$$\begin{aligned} x + 2y + 3z &= 0 \\ 2x - y - z &= 0 \end{aligned}$$

or, row-reducing:

$$\left(\begin{array}{ccc|c} 1 & 2 & 3 & 0 \\ 2 & -1 & -1 & 0 \end{array}\right) \xrightarrow{A_{1,2}(-2)} \left(\begin{array}{ccc|c} 1 & 2 & 3 & 0 \\ 0 & -5 & -7 & 0 \end{array}\right) \xrightarrow{M_2(-\frac{1}{5})}$$

$$\left(\begin{array}{ccc|c} 1 & 2 & 3 & 0 \\ 0 & 1 & \frac{7}{5} & 0 \end{array}\right) \xrightarrow{A_{2,1}(-2)} \left(\begin{array}{ccc|c} 1 & 0 & \frac{1}{5} & 0 \\ 0 & 1 & \frac{7}{5} & 0 \end{array}\right)$$

Thus all solutions to the homogeneous system are given by $(-\frac{1}{5}z, -\frac{7}{5}z, z)$. Setting $z = t$, we obtain the parametric equations of a line L in $\mathbb{R}^3\colon x = -\frac{1}{5}t$, $y = -\frac{7}{5}t$, $z = t$. As we saw in Example 4, the set of vectors on L constitutes a subspace of $\mathbb{R}^3$.

PROBLEMS 5.3

In Problems 1–20 determine whether the given subset H of the vector space V is a subspace of V.

1. $V = \mathbb{R}^2$; $H = \{(x, y)\colon\ y \geq 0\}$
2. $V = \mathbb{R}^2$; $H = \{(x, y)\colon\ x = y\}$
3. $V = \mathbb{R}^3$; $H =$ the xy-plane
4. $V = \mathbb{R}^2$; $H = \{(x, y)\colon\ x^2 + y^2 \leq 1\}$
5. $V = M_{nn}$; $H = \{D \in M_{nn}\colon\ D \text{ is diagonal}\}$
6. $V = M_{nn}$; $H = \{T \in M_{nn}\colon\ T \text{ is upper triangular}\}$
7. $V = M_{nn}$; $H = \{S \in M_{nn}\colon\ S \text{ is symmetric}\}$
8. $V = M_{mn}$; $H = \{A \in M_{mn}\colon\ a_{ij} = 0\}$
9. $V = M_{22}$; $H = \left\{A \in M_{22}\colon\ A = \begin{pmatrix} a & b \\ -b & c \end{pmatrix}\right\}$
10. $V = M_{22}$; $H = \left\{A \in M_{22}\colon\ A = \begin{pmatrix} a & 1+a \\ 0 & 0 \end{pmatrix}\right\}$
11. $V = M_{22}$; $H = \left\{A \in M_{22}\colon\ A = \begin{pmatrix} 0 & a \\ b & 0 \end{pmatrix}\right\}$
12. $V = P_4$; $H = \{p \in P_4\colon\ \deg p = 4\}$
13. $V = P_4$; $H = \{p \in P_4\colon\ p(0) = 0\}$
14. $V = P_n$; $H = \{p \in P_n\colon\ p(0) = 0\}$
15. $V = P_n$; $H = \{p \in P_n\colon\ p(0) = 1\}$
16. $V = C[0, 1]$; $H = \{f \in C[0, 1]\colon\ f(0) = f(1) = 0\}$
17. $V = C[0, 1]$; $H = \{f \in C[0, 1]\colon\ f(0) = 2\}$
18. □ $V = C'[0, 1]$; $H = \{f \in C'[0, 1]\colon\ f'(0) = 0\}$
19. □ $V = C[a, b]$, where a and b are real numbers and $a < b$; $H = \{f \in C[a, b]\colon\ \int_a^b f(x)\, dx = 0\}$

---

† Note, in particular, that as $\mathbf{0} \in H_1$ and $\mathbf{0} \in H_2$, we have $\mathbf{0} \in H_1 \cap H_2$.

□**20.** $V = C[a, b]$; $H = \{f \in C[a, b]: \int_a^b f(x)\,dx = 1\}$

21. Let $V = M_{22}$; let $H_1 = \{A \in M_{22}: a_{11} = 0\}$ and $H_2 = \left\{A \in M_{22}: A = \begin{pmatrix} -b & a \\ a & b \end{pmatrix}\right\}$.

a. Show that H_1 and H_2 are subspaces.

b. Describe the subset $H = H_1 \cap H_2$ and show that it is a subspace.

□**22.** If $V = C[0, 1]$, let H_1 denote the subspace of Example 10 and H_2 denote the subspace of Example 11. Describe the set $H_1 \cap H_2$ and show that it is a subspace.

23. Let A be an $n \times m$ matrix and let $H = \{\mathbf{x} \in \mathbb{R}^m: A\mathbf{x} = \mathbf{0}\}$. Show that H is a subspace of $\mathbb{R}^m$. H is called the *kernel* of the matrix A.

24. In Problem 23 let $H = \{\mathbf{x} \in \mathbb{R}^m: A\mathbf{x} \neq 0\}$. Show that H is not a subspace of $\mathbb{R}^m$.

25. Let $H = \{(x, y, z, w): ax + by + cz + dw = 0\}$, where a, b, c, and d are real numbers not all zero. Show that H is a proper subspace of $\mathbb{R}^4$. H is called a *hyperplane* in $\mathbb{R}^4$.

26. Let $H = \{(x_1, x_2, \ldots, x_n): a_1x_1 + a_2x_2 + \cdots + a_nx_n = 0\}$, where $a_1, a_2, \ldots, a_n$ are real numbers not all zero. Show that H is a proper subspace of $\mathbb{R}^n$. H, as in Problem 25, is called a **hyperplane** in $\mathbb{R}^n$.

27. Let H_1 and H_2 be subspaces of a vector space V. Let $H_1 + H_2 = \{\mathbf{v}: \mathbf{v} = \mathbf{v}_1 + \mathbf{v}_2$ with $\mathbf{v}_1 \in H_1$ and $\mathbf{v}_2 \in H_2\}$. Show that $H_1 + H_2$ is a subspace of V.

28. Let $\mathbf{v}_1$ and $\mathbf{v}_2$ be two vectors in $\mathbb{R}^2$. Show that $H = \{\mathbf{v}: \mathbf{v} = a\mathbf{v}_1 + b\mathbf{v}_2;\ a, b \text{ real}\}$ is a subspace of $\mathbb{R}^2$.

***29.** In Problem 28 show that if $\mathbf{v}_1$ and $\mathbf{v}_2$ are not collinear, then $H = \mathbb{R}^2$.

***30.** Let $\mathbf{v}_1, \mathbf{v}_2, \ldots, \mathbf{v}_n$ be arbitrary vectors in a vector space V. Let $H = \{\mathbf{v} \in V: \mathbf{v} = a_1\mathbf{v}_1 + a_2\mathbf{v}_2 + \cdots + a_n\mathbf{v}_n$, where $a_1, a_2, \ldots, a_n$ are scalars$\}$. Show that H is a subspace of V. H is called the subspace *spanned* by the vectors $\mathbf{v}_1, \mathbf{v}_2, \ldots, \mathbf{v}_n$.

5.4 Linear Independence, Linear Combinations, and Span

In Section 2.6 we defined linear dependence and independence for vectors in $\mathbb{R}^n$. In Section 3.4 we showed that a set of n vectors in $\mathbb{R}^n$ is linearly independent if and only if the determinant of the matrix whose columns (or rows) are the given vectors is nonzero. In this section we extend the notion of linear independence to arbitrary vector spaces and discuss some related notions.

DEFINITION 1 **LINEAR DEPENDENCE AND INDEPENDENCE** Let $\mathbf{v}_1, \mathbf{v}_2, \ldots, \mathbf{v}_n$ be n vectors in a vector space V. Then the vectors are said to be **linearly dependent** if there exist n scalars $c_1, c_2, \ldots, c_n$ *not all zero* such that

$$c_1\mathbf{v}_1 + c_2\mathbf{v}_2 + \cdots + c_n\mathbf{v}_n = \mathbf{0} \tag{1}$$

If the vectors are not linearly dependent, they are said to be **linearly independent**.

In Sections 2.6 and 3.4 we saw several examples of linearly dependent and independent sets of vectors in $\mathbb{R}^n$. Recall that in our Summing Up Theorem (Theorem 3.4.4, page 110) we showed that we can determine whether n n-vectors are linearly independent by writing the vectors as the rows or columns of a matrix A and then computing det A. The vectors are independent if and only if det $A \neq 0$. Alternatively, we can solve the homogeneous system $c_1\mathbf{v}_1 + c_2\mathbf{v}_2 + \cdots + c_n\mathbf{v}_n = \mathbf{0}$ to see if there are any nontrivial solutions. If there are, then the vectors are linearly dependent. Otherwise they are independent.

We will give several examples of linear dependence and independence in other vector spaces. First, however, we make a simple but very useful observation. The proof is identical to the proof of Theorem 2.6.1, page 54, where the vectors were assumed to be in $\mathbb{R}^n$.

THEOREM 1 Two vectors in a vector space V are linearly dependent if and only if one is a multiple of the other.

We now turn to some other vector spaces.

EXAMPLE 1 In M_{23} let $A_1 = \begin{pmatrix} 1 & 0 & 2 \\ 3 & 1 & -1 \end{pmatrix}$, $A_2 = \begin{pmatrix} -1 & 1 & 4 \\ 2 & 3 & 0 \end{pmatrix}$, and $A_3 = \begin{pmatrix} -1 & 0 & 1 \\ 1 & 2 & 1 \end{pmatrix}$. Determine whether A_1, A_2, and A_3 are linearly dependent or independent.

Solution Suppose that $c_1A_1 + c_2A_2 + c_3A_3 = 0$. Then

$$\begin{pmatrix} 0 & 0 & 0 \\ 0 & 0 & 0 \end{pmatrix} = c_1\begin{pmatrix} 1 & 0 & 2 \\ 3 & 1 & -1 \end{pmatrix} + c_2\begin{pmatrix} -1 & 1 & 4 \\ 2 & 3 & 0 \end{pmatrix} + c_3\begin{pmatrix} -1 & 0 & 1 \\ 1 & 2 & 1 \end{pmatrix}$$

$$= \begin{pmatrix} c_1 - c_2 - c_3 & c_2 & 2c_1 + 4c_2 + c_3 \\ 3c_1 + 2c_2 + c_3 & c_1 + 3c_2 + 2c_3 & -c_1 + c_3 \end{pmatrix}$$

This gives us a homogeneous system of six equations in the three unknowns c_1, c_2, and c_3 and it is quite easy to verify that the only solution is $c_1 = c_2 = c_3 = 0$. Thus the three matrices are linearly independent.

EXAMPLE 2 In P_3, determine whether the polynomials 1, x, x^2, and x^3 are linearly dependent or independent.

Solution Suppose that $c_1 + c_2x + c_3x^2 + c_4x^3 = 0$. This must hold for every real number x. In particular, if $x = 0$ we obtain $c_1 = 0$. Then, setting $x = 1, -1, 2$, we obtain, successively:

$$\begin{aligned} c_2 + c_3 + c_4 &= 0 \\ -c_2 + c_3 - c_4 &= 0 \\ 2c_2 + 4c_3 + 8c_4 &= 0 \end{aligned}$$

The determinant of this homogeneous system is

$$\begin{vmatrix} 1 & 1 & 1 \\ -1 & 1 & -1 \\ 2 & 4 & 8 \end{vmatrix} = 12 \neq 0$$

so that the system has the unique solution $c_2 = c_3 = c_4 = 0$ and the four polynomials are linearly independent. We can see this in another way. We know that any polynomial of degree 3 has at most three real roots. But if $c_1 + c_2x + c_3x^2 + c_4x^3 = 0$ for some nonzero constants c_1, c_2, c_3, and c_4 and for every real number x, then we have constructed a cubic polynomial for which every real number is a root. This, clearly, is impossible.

EXAMPLE 3 In P_2 determine whether the polynomials $x - 2x^2$, $x^2 - 4x$, and $-7x + 8x^2$ are linearly dependent or independent.

Solution Let $c_1(x - 2x^2) + c_2(x^2 - 4x) + c_3(-7x + 8x^2) = 0$. Then, rearranging terms, we obtain

$$(c_1 - 4c_2 - 7c_3)x = 0$$
$$(-2c_1 + c_2 + 8c_3)x^2 = 0$$

These equations hold for every x if and only if

$$c_1 - 4c_2 - 7c_3 = 0$$

and

$$-2c_1 + c_2 + 8c_3 = 0$$

But by Theorem 1.4.1, page 24, this system of two equations in three unknowns has an infinite number of solutions. This shows that the polynomials are linearly dependent.

If we solve this homogeneous system, we obtain, successively,

$$\left(\begin{array}{rrr|r} 1 & -4 & -7 & 0 \\ -2 & 1 & 8 & 0 \end{array}\right) \xrightarrow{A_{1,2}(2)} \left(\begin{array}{rrr|r} 1 & -4 & -7 & 0 \\ 0 & -7 & -6 & 0 \end{array}\right)$$

$$\xrightarrow{M_2(-\frac{1}{7})} \left(\begin{array}{rrr|r} 1 & -4 & -7 & 0 \\ 0 & 1 & \frac{6}{7} & 0 \end{array}\right) \xrightarrow{A_{2,1}(4)} \left(\begin{array}{rrr|r} 1 & 0 & -\frac{25}{7} & 0 \\ 0 & 1 & \frac{6}{7} & 0 \end{array}\right)$$

Thus c_3 can be chosen arbitrarily, $c_1 = \frac{25}{7}c_3$ and $c_2 = -\frac{6}{7}c_3$. If $c_3 = 7$, for example, then $c_1 = 25$, $c_2 = -6$, and we have $25(x - 2x^2) - 6(x^2 - 4x) + 7(-7x + 8x^2) = 0$.

We now turn to an important, related concept.

DEFINITION 2 **LINEAR COMBINATION** Let $\mathbf{v}_1, \mathbf{v}_2, \ldots, \mathbf{v}_n$ be vectors in a vector space V. Then any expression of the form

$$a_1\mathbf{v}_1 + a_2\mathbf{v}_2 + \cdots + a_n\mathbf{v}_n \tag{2}$$

where $a_1, a_2, \ldots, a_n$ are scalars is called a **linear combination** of $\mathbf{v}_1, \mathbf{v}_2, \ldots, \mathbf{v}_n$.

EXAMPLE 4 In $\mathbb{R}^3$, $\begin{pmatrix}-7\\7\\7\end{pmatrix}$ is a linear combination of $\begin{pmatrix}-1\\2\\4\end{pmatrix}$ and $\begin{pmatrix}5\\-3\\1\end{pmatrix}$ since $\begin{pmatrix}-7\\7\\7\end{pmatrix} = 2\begin{pmatrix}-1\\2\\4\end{pmatrix} - \begin{pmatrix}5\\-3\\1\end{pmatrix}$.

EXAMPLE 5 In M_{23}, $\begin{pmatrix}-3 & 2 & 8\\-1 & 9 & 3\end{pmatrix} = 3\begin{pmatrix}-1 & 0 & 4\\1 & 1 & 5\end{pmatrix} + 2\begin{pmatrix}0 & 1 & -2\\-2 & 3 & -6\end{pmatrix}$, which shows that $\begin{pmatrix}-3 & 2 & 8\\-1 & 9 & 3\end{pmatrix}$ is a linear combination of $\begin{pmatrix}-1 & 0 & 4\\1 & 1 & 5\end{pmatrix}$ and $\begin{pmatrix}0 & 1 & -2\\-2 & 3 & -6\end{pmatrix}$.

EXAMPLE 6 In P_n every polynomial can be written as a linear combination of the "monomials" $1, x, x^2, \ldots, x^n$.

DEFINITION 3 **SPAN OF A VECTOR SPACE** The vectors $\mathbf{v}_1, \mathbf{v}_2, \ldots, \mathbf{v}_n$ in a vector space V are said to **span** V if every vector in V can be written as a linear combination of them. That is, for every $\mathbf{v} \in V$ there are scalars $a_1, a_2, \ldots, a_n$ such that

$$\mathbf{v} = a_1\mathbf{v}_1 + a_2\mathbf{v}_2 + \cdots + a_n\mathbf{v}_n \tag{3}$$

EXAMPLE 7 We saw in Section 4.1 that the vectors $\mathbf{i} = \begin{pmatrix}1\\0\end{pmatrix}$ and $\mathbf{j} = \begin{pmatrix}0\\1\end{pmatrix}$ span $\mathbb{R}^2$. In Section 4.3 we saw that $\mathbf{i} = \begin{pmatrix}1\\0\\0\end{pmatrix}$, $\mathbf{j} = \begin{pmatrix}0\\1\\0\end{pmatrix}$, and $\mathbf{k} = \begin{pmatrix}0\\0\\1\end{pmatrix}$ span $\mathbb{R}^3$.

In fact, it is not difficult to find sets of vectors that span $\mathbb{R}^n$, as the following theorem illustrates.

THEOREM 2 Any set of n linearly independent vectors in $\mathbb{R}^n$ spans $\mathbb{R}^n$.

Proof Let $\mathbf{v}_1 = \begin{pmatrix} a_{11} \\ a_{21} \\ \cdot \\ \cdot \\ \cdot \\ a_{n1} \end{pmatrix}$, $\mathbf{v}_2 = \begin{pmatrix} a_{12} \\ a_{22} \\ \cdot \\ \cdot \\ \cdot \\ a_{n2} \end{pmatrix}$, $\ldots$, $\mathbf{v}_n = \begin{pmatrix} a_{1n} \\ a_{2n} \\ \cdot \\ \cdot \\ \cdot \\ a_{nn} \end{pmatrix}$ and

let $\mathbf{v} = \begin{pmatrix} x_1 \\ x_2 \\ \vdots \\ x_n \end{pmatrix}$ be a vector in $\mathbb{R}^n$. We must show that there exist

scalars $c_1, c_2, \ldots, c_n$ such that

$$\mathbf{v} = c_1\mathbf{v}_1 + c_2\mathbf{v}_2 + \cdots + c_n\mathbf{v}_n$$

That is,

$$\begin{pmatrix} x_1 \\ x_2 \\ \cdot \\ \cdot \\ \cdot \\ x_n \end{pmatrix} = c_1 \begin{pmatrix} a_{11} \\ a_{21} \\ \cdot \\ \cdot \\ \cdot \\ a_{n1} \end{pmatrix} + c_2 \begin{pmatrix} a_{12} \\ a_{22} \\ \cdot \\ \cdot \\ \cdot \\ a_{n2} \end{pmatrix} + \cdots + c_n \begin{pmatrix} a_{1n} \\ a_{2n} \\ \cdot \\ \cdot \\ \cdot \\ a_{nn} \end{pmatrix} \tag{4}$$

In (4), we multiply through, add, and equate components to obtain a system of n equations in the n unknowns $c_1, c_2, \ldots, c_n$:

$$\begin{array}{c} a_{11}c_1 + a_{12}c_2 + \cdots + a_{1n}c_n = x_1 \\ a_{21}c_1 + a_{22}c_2 + \cdots + a_{2n}c_n = x_2 \\ \vdots \\ a_{n1}c_1 + a_{n2}c_2 + \cdots + a_{nn}c_n = x_n \end{array} \tag{5}$$

We write (5) as $A\mathbf{c} = \mathbf{v}$, where

$$A = \begin{pmatrix} a_{11} & a_{12} & \cdots & a_{1n} \\ a_{21} & a_{22} & \cdots & a_{2n} \\ \cdot & \cdot & & \cdot \\ \cdot & \cdot & & \cdot \\ \cdot & \cdot & & \cdot \\ a_{n1} & a_{n2} & \cdots & a_{nn} \end{pmatrix} \quad \text{and} \quad \mathbf{c} = \begin{pmatrix} c_1 \\ c_2 \\ \cdot \\ \cdot \\ \cdot \\ c_n \end{pmatrix}$$

But system (5) has a unique solution if and only if $\det A \neq 0$. And $\det A \neq 0$ because the columns of A are linearly independent. (This all follows from

Theorem 3.4.4.) Thus there is a unique vector **c** satisfying system (5) and the theorem is proved. ■

Remark. This theorem not only shows that **v** can be written as a linear combination of the independent vectors $\mathbf{v}_1, \mathbf{v}_2, \ldots, \mathbf{v}_n$ but also that this can be done in *only one way* (since the solution vector **c** is unique).

EXAMPLE 8

The vectors $(2, -1, 4)$, $(1, 0, 2)$, and $(3, -1, 5)$ span $\mathbb{R}^3$ because $\begin{vmatrix} 2 & 1 & 3 \\ -1 & 0 & -1 \\ 4 & 2 & 5 \end{vmatrix} = -1 \neq 0$, so that they are independent.

We now look briefly at spanning sets of some other vector spaces.

EXAMPLE 9

From Example 6 it follows that the monomials $1, x, x^2, \ldots, x^n$ span P_n.

EXAMPLE 10

Since $\begin{pmatrix} a & b \\ c & d \end{pmatrix} = a\begin{pmatrix} 1 & 0 \\ 0 & 0 \end{pmatrix} + b\begin{pmatrix} 0 & 1 \\ 0 & 0 \end{pmatrix} + c\begin{pmatrix} 0 & 0 \\ 1 & 0 \end{pmatrix} + d\begin{pmatrix} 0 & 0 \\ 0 & 1 \end{pmatrix}$ we see that $\begin{pmatrix} 1 & 0 \\ 0 & 0 \end{pmatrix}$, $\begin{pmatrix} 0 & 1 \\ 0 & 0 \end{pmatrix}$, $\begin{pmatrix} 0 & 0 \\ 1 & 0 \end{pmatrix}$, and $\begin{pmatrix} 0 & 0 \\ 0 & 1 \end{pmatrix}$ span M_{22}.

EXAMPLE 11

Let P denote the vector space of polynomials of any degree. Then no *finite* set of polynomials spans P. To see this suppose that $p_1, p_2, \ldots, p_m$ are polynomials. Let p_k be the polynomial of largest degree in this set and let $N = \deg p_k$. Then it is clear that the polynomial $p(x) = x^{N+1}$ cannot be written as a linear combination of $p_1, p_2, \ldots, p_m$.

We now turn to another way of finding subspaces of a vector space V.

DEFINITION 4

SPAN OF A SET OF VECTORS Let $\mathbf{v}_1, \mathbf{v}_2, \ldots, \mathbf{v}_n$ be n vectors in a vector space V. The **span** of $\{\mathbf{v}_1, \mathbf{v}_2, \ldots, \mathbf{v}_n\}$ is the set of linear combinations of $\mathbf{v}_1, \mathbf{v}_2, \ldots, \mathbf{v}_n$. That is:

$$\text{span}\,\{\mathbf{v}_1, \mathbf{v}_2, \ldots, \mathbf{v}_n\} = \{\mathbf{v}: \quad \mathbf{v} = a_1\mathbf{v}_1 + a_2\mathbf{v}_2 + \cdots + a_n\mathbf{v}_n\} \tag{6}$$

where $a_1, a_2, \ldots, a_n$ are scalars.

THEOREM 3 Span $\{\mathbf{v}_1, \mathbf{v}_2, \ldots, \mathbf{v}_n\}$ is a subspace of V.

Proof The proof is easy and is left as an exercise (see Problem 26). ■

EXAMPLE 12 Let $\mathbf{v}_1=(2,-1,4)$ and $\mathbf{v}_2=(4,1,6)$. Then $H=\text{span}\{\mathbf{v}_1,\mathbf{v}_2\}=\{\mathbf{v}\colon\ \mathbf{v}=a_1(2,-1,4)+a_2(4,1,6)\}$. What does H look like? If $\mathbf{v}=(x,y,z)\in H$, then we have $x=2a_1+4a_2$, $y=-a_1+a_2$, and $z=4a_1+6a_2$. If we think of (x,y,z) as being fixed, then we can view these equations as a system of three equations in the two unknowns a_1, a_2. We solve this system in the usual way:

$$\left(\begin{array}{rr|c} -1 & 1 & y \\ 2 & 4 & x \\ 4 & 6 & z \end{array}\right) \xrightarrow{M_1(-1)} \left(\begin{array}{rr|c} 1 & -1 & -y \\ 2 & 4 & x \\ 4 & 6 & z \end{array}\right) \xrightarrow[A_{1,3}(-4)]{A_{1,2}(-2)} \left(\begin{array}{rr|c} 1 & -1 & -y \\ 0 & 6 & x+2y \\ 0 & 10 & z+4y \end{array}\right)$$

$$\xrightarrow{M_2(\frac{1}{6})} \left(\begin{array}{rr|c} 1 & -1 & -y \\ 0 & 1 & (x+2y)/6 \\ 0 & 10 & z+4y \end{array}\right) \xrightarrow[A_{2,3}(-10)]{A_{2,1}(1)} \left(\begin{array}{rr|c} 1 & 0 & x/6-2y/3 \\ 0 & 1 & x/6+y/3 \\ 0 & 0 & -5x/3+2y/3+z \end{array}\right)$$

From Chapter 1, we see that the system has a solution only if $-5x/3+2y/3+z=0$; or, multiplying through by -3, if

$$5x-2y-3z=0 \tag{7}$$

But Equation (7) is the equation of a plane in $\mathbb{R}^3$ passing through the origin.

The last example can be generalized to prove the following interesting fact:

> *The span of two nonzero vectors in $\mathbb{R}^3$ that are not parallel is a plane passing through the origin.*

For a suggested proof see Problems 27 and 28.

We can give a geometric interpretation of this result. Look at the vectors in Figure 5.1. We know (from Section 4.1) the geometric interpretation of the vectors $2\mathbf{u}$, $-\mathbf{u}$, and $\mathbf{u}+\mathbf{v}$, for example. Using these, we see that any other vector in the plane of $\mathbf{u}$ and $\mathbf{v}$ can be obtained as a linear combination of $\mathbf{u}$ and $\mathbf{v}$. Figure 5.2 shows how in four different situations a third vector $\mathbf{w}$ in the plane of $\mathbf{u}$ and $\mathbf{v}$ can be written as $\alpha\mathbf{u}+\beta\mathbf{v}$ for appropriate choices of the numbers α and β.

Figure 5.1

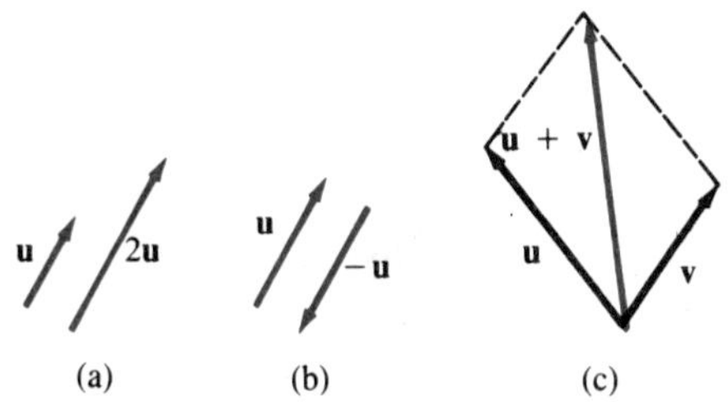

Figure 5.2

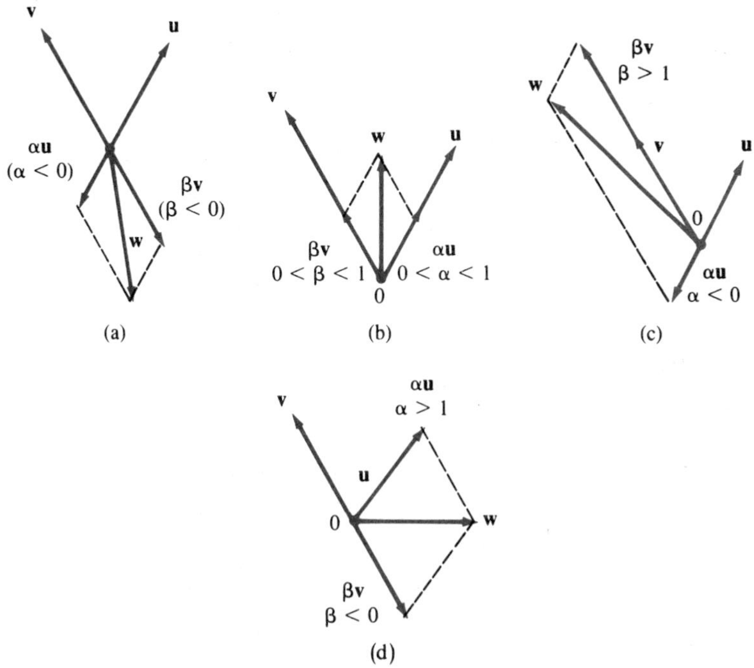

We close this section by citing a useful result. Its proof is not difficult and is left as an exercise (see Problem 29).

THEOREM 4 Let $\mathbf{v}_1, \mathbf{v}_2, \ldots, \mathbf{v}_n, \mathbf{v}_{n+1}$ be $n+1$ vectors that are in a vector space V. If $\mathbf{v}_1, \mathbf{v}_2, \ldots, \mathbf{v}_n$ span V, then $\mathbf{v}_1, \mathbf{v}_2, \ldots, \mathbf{v}_n, \mathbf{v}_{n+1}$ also span V. That is, the addition of one (or more) vectors to a spanning set yields another spanning set.

PROBLEMS 5.4

In Problems 1–10 determine whether the given set of vectors is linearly dependent or independent.

1. In P_2: $1-x, x$

2. In P_2: $-x, x^2-2x, 3x+5x^2$

3. In P_2: $1-x, 1+x, x^2$

4. In P_3: x, x^2-x, x^3-x

5. In P_3: $2x, x^3-3, 1+x-4x^3, x^3+18x-9$

6. In M_{22}: $\begin{pmatrix} 2 & -1 \\ 4 & 0 \end{pmatrix}, \begin{pmatrix} 0 & -3 \\ 1 & 5 \end{pmatrix}, \begin{pmatrix} 4 & 1 \\ 7 & -5 \end{pmatrix}$

7. In M_{22}: $\begin{pmatrix} 1 & -1 \\ 0 & 6 \end{pmatrix}, \begin{pmatrix} -1 & 0 \\ 3 & 1 \end{pmatrix}, \begin{pmatrix} 1 & 1 \\ -1 & 2 \end{pmatrix}, \begin{pmatrix} 0 & 1 \\ 1 & 0 \end{pmatrix}$

8. In M_{22}: $\begin{pmatrix} -1 & 0 \\ 1 & 2 \end{pmatrix}, \begin{pmatrix} 2 & 3 \\ 7 & -4 \end{pmatrix}, \begin{pmatrix} 8 & -5 \\ 7 & 6 \end{pmatrix}, \begin{pmatrix} 4 & -1 \\ 2 & 3 \end{pmatrix}, \begin{pmatrix} 2 & 3 \\ -1 & 4 \end{pmatrix}$

***9.** In $C[0, 1]$: $\sin x, \cos x$

***10.** In $C[0, 1]$: $x, \sqrt{x}, \sqrt[3]{x}$

In Problems 11–17 determine whether the given set of vectors spans the given vector space.

11. In $\mathbb{R}^3$: $(1, -1, 2), (1, 1, 2), (0, 0, 1)$

12. In $\mathbb{R}^3$: $(1, -1, 2), (-1, 1, 2), (0, 0, 1)$

13. In P_2: $1-x, 3-x^2$

14. In P_2: $1-x, 3-x^2, x$

15. In M_{22}: $\begin{pmatrix} 2 & 1 \\ 0 & 0 \end{pmatrix}, \begin{pmatrix} 0 & 0 \\ 2 & 1 \end{pmatrix}, \begin{pmatrix} 3 & -1 \\ 0 & 0 \end{pmatrix}, \begin{pmatrix} 0 & 0 \\ 3 & 1 \end{pmatrix}$

16. In M_{22}: $\begin{pmatrix} 1 & 0 \\ 1 & 0 \end{pmatrix}, \begin{pmatrix} 1 & 2 \\ 0 & 0 \end{pmatrix}, \begin{pmatrix} 4 & -1 \\ 3 & 0 \end{pmatrix}, \begin{pmatrix} -2 & 5 \\ 6 & 0 \end{pmatrix}$

17. In M_{23}: $\begin{pmatrix} 1 & 0 & 0 \\ 0 & 0 & 0 \end{pmatrix}, \begin{pmatrix} 0 & 1 & 0 \\ 0 & 0 & 0 \end{pmatrix}, \begin{pmatrix} 0 & 0 & 1 \\ 0 & 0 & 0 \end{pmatrix},$
$\begin{pmatrix} 0 & 0 & 0 \\ 1 & 0 & 0 \end{pmatrix}, \begin{pmatrix} 0 & 0 & 0 \\ 0 & 1 & 0 \end{pmatrix}, \begin{pmatrix} 0 & 0 & 0 \\ 0 & 0 & 1 \end{pmatrix}$

18. Show that any four polynomials in P_2 are linearly dependent.

19. Show that two polynomials cannot span P_2.

***20.** Show that any $n+2$ polynomials in P_n are linearly dependent.

***21.** If $p_1, p_2, \ldots, p_m$ span P_n, show that $m \geq n+1$.

22. Show that any seven matrices in M_{32} are linearly dependent.

***23.** Prove that any $mn+1$ matrices in M_{mn} are linearly dependent.

24. Let S_1 and S_2 be two finite, linearly independent sets in a vector space V. Show that $S_1 \cap S_2$ is a linearly independent set.

25. Show that the infinite set $\{1, x, x^2, x^3, \ldots\}$ spans P, the vector space of polynomials.

26. Show that if $\mathbf{u}$ and $\mathbf{v}$ are in span $\{\mathbf{v}_1, \mathbf{v}_2, \ldots, \mathbf{v}_n\}$, then $\mathbf{u}+\mathbf{v}$ and $\alpha\mathbf{u}$ are in span $\{\mathbf{v}_1, \mathbf{v}_2, \ldots, \mathbf{v}_n\}$ [*Hint:* Using the definition of span write $\mathbf{u}+\mathbf{v}$ and $\alpha\mathbf{u}$ as linear combinations of $\mathbf{v}_1, \mathbf{v}_2, \ldots, \mathbf{v}_n$.]

27. Let $\mathbf{v}_1 = (x_1, y_1, z_1)$ and $\mathbf{v}_2 = (x_2, y_2, z_2)$ be in $\mathbb{R}^3$. Show that if $\mathbf{v}_2 = c\mathbf{v}_1$, then span $\{\mathbf{v}_1, \mathbf{v}_2\}$ is a line passing through the origin.

****28.** In Problem 27 assume that $\mathbf{v}_1$ and $\mathbf{v}_2$ are independent. Show that $H =$ span $\{\mathbf{v}_1, \mathbf{v}_2\}$ is a plane passing through the origin. What is the equation of that plane? [*Hint:* If $(x, y, z) \in H$, write $\mathbf{v} = a_1\mathbf{v}_1 + a_2\mathbf{v}_2$ and find a condition relating x, y, and z such that the resulting 3×2 system has a solution.]

29. Prove Theorem 4. [*Hint:* If $\mathbf{v} \in V$, write $\mathbf{v}$ as a linear combination of $\mathbf{v}_1, \mathbf{v}_2, \ldots, \mathbf{v}_n, \mathbf{v}_{n+1}$ with the coefficient of $\mathbf{v}_{n+1}$ equal to zero.].

30. Let H be a subspace of V containing $\mathbf{v}_1, \mathbf{v}_2, \ldots, \mathbf{v}_n$. Show that span $\{\mathbf{v}_1, \mathbf{v}_2, \ldots, \mathbf{v}_n\} \subseteq H$. That is, span $\{\mathbf{v}_1, \mathbf{v}_2, \ldots, \mathbf{v}_n\}$ is the *smallest* subspace of V containing $\mathbf{v}_1, \mathbf{v}_2, \ldots, \mathbf{v}_n$.

31. Show that any subset of a set of linearly independent vectors is linearly independent.

32. Let $\{\mathbf{v}_1, \mathbf{v}_2, \ldots, \mathbf{v}_n\}$ be a linearly independent set. Show that the vectors $\mathbf{v}_1, \mathbf{v}_1+\mathbf{v}_2, \mathbf{v}_1+\mathbf{v}_2+\mathbf{v}_3, \ldots, \mathbf{v}_1+\mathbf{v}_2+\cdots+\mathbf{v}_n$ are linearly independent.

33. Show that M_{22} can be spanned by invertible matrices.

34. Let $\{\mathbf{v}_1, \mathbf{v}_2, \ldots, \mathbf{v}_n\}$ be a set of vectors having the property that the set $\{\mathbf{v}_i, \mathbf{v}_j\}$ is linearly dependent when $i \neq j$. Show that each vector in the set is a multiple of a single vector in the set.

35. Let $S = \{\mathbf{v}_1, \mathbf{v}_2, \ldots, \mathbf{v}_n\}$ be a linearly dependent set of nonzero vectors in a vector space V. Show that at least one of the vectors in S can be written as a linear combination of the vectors that precede it. That is, show that there is an integer $k \leq n$ and scalars $a_1, a_2, \ldots, a_{k-1}$ such that $\mathbf{v}_k = a_1\mathbf{v}_1 + a_2\mathbf{v}_2 + \cdots + a_{k-1}\mathbf{v}_{k-1}$.

□**36.** Let f and g be in $C'[0, 1]$. Then the **Wronskian** of f and g is defined by

$$W(f, g)(x) = \begin{vmatrix} f(x) & g(x) \\ f'(x) & g'(x) \end{vmatrix}$$

Show that f and g are linearly independent if $W(f, g)(x) \neq 0$ for every $x \in [0, 1]$.

***37.** Let $\{\mathbf{u}_1, \mathbf{u}_2, \ldots, \mathbf{u}_n\}$ and $\{\mathbf{v}_1, \mathbf{v}_2, \ldots, \mathbf{v}_n\}$ be $2n$ vectors in a vector space V. Suppose that

$$\begin{aligned}
\mathbf{v}_1 &= a_{11}\mathbf{u}_1 + a_{12}\mathbf{u}_2 + \cdots + a_{1n}\mathbf{u}_n \\
\mathbf{v}_2 &= a_{21}\mathbf{u}_1 + a_{22}\mathbf{u}_2 + \cdots + a_{2n}\mathbf{u}_n \\
&\vdots \\
\mathbf{v}_n &= a_{n1}\mathbf{u}_1 + a_{n2}\mathbf{u}_2 + \cdots + a_{nn}\mathbf{u}_n
\end{aligned}$$

Show that if

$$\begin{vmatrix} a_{11} & a_{12} & \cdots & a_{1n} \\ a_{21} & a_{22} & \cdots & a_{2n} \\ \vdots & \vdots & & \vdots \\ a_{n1} & a_{n2} & \cdots & a_{nn} \end{vmatrix} \neq 0,$$

then $\text{span}\{\mathbf{u}_1, \mathbf{u}_2, \ldots, \mathbf{u}_n\} = \text{span}\{\mathbf{v}_1, \mathbf{v}_2, \ldots, \mathbf{v}_n\}$.

38. Let $\{\mathbf{v}_1, \mathbf{v}_2, \ldots, \mathbf{v}_n\}$ be a linearly independent set and suppose that $\mathbf{v} \notin \text{span}\{\mathbf{v}_1, \mathbf{v}_2, \ldots, \mathbf{v}_n\}$. Show that $\{\mathbf{v}_1, \mathbf{v}_2, \ldots, \mathbf{v}_n, \mathbf{v}\}$ is a linearly independent set.

39. Find a set of three linearly independent vectors in $\mathbb{R}^3$ that contains the vectors $\begin{pmatrix} 2 \\ 1 \\ 2 \end{pmatrix}$ and $\begin{pmatrix} -1 \\ 3 \\ 4 \end{pmatrix}$. $\left[\textit{Hint:} \text{ Find a vector } \mathbf{v} \notin \text{span}\left\{\begin{pmatrix} 2 \\ 1 \\ 2 \end{pmatrix}, \begin{pmatrix} -1 \\ 3 \\ 4 \end{pmatrix}\right\}.\right]$

40. Find a set of three linearly independent vectors in P_2 that contains the polynomials $1-x^2$ and $1+x^2$.

***41.** Show that in P_n the polynomials $1, x, x^2, \ldots, x^n$ are linearly independent. [*Hint:* This is certainly true if $n = 1$. Assume that $1, x, x^2, \ldots, x^{n-1}$ is linearly independent and show how this implies that $1, x, x^2, \ldots, x^n$ is also independent. This will complete the proof by mathematical induction.]

5.5 Basis and Dimension

We have seen that in $\mathbb{R}^2$ it is convenient to write vectors in terms of the vectors $\mathbf{i} = \begin{pmatrix}1\\0\end{pmatrix}$ and $\mathbf{j} = \begin{pmatrix}0\\1\end{pmatrix}$. In $\mathbb{R}^3$ we wrote vectors in terms of $\begin{pmatrix}1\\0\\0\end{pmatrix}$, $\begin{pmatrix}0\\1\\0\end{pmatrix}$, and $\begin{pmatrix}0\\0\\1\end{pmatrix}$. We now generalize this idea.

DEFINITION 1 **BASIS** A set of vectors $\{\mathbf{v}_1, \mathbf{v}_2, \ldots, \mathbf{v}_n\}$ forms a **basis** for V if:

i. $\{\mathbf{v}_1, \mathbf{v}_2, \ldots, \mathbf{v}_n\}$ is linearly independent.
ii. $\{\mathbf{v}_1, \mathbf{v}_2, \ldots, \mathbf{v}_n\}$ spans V.

We have already seen quite a few examples of bases. In Theorem 5.4.2, for instance, we saw that any set of n linearly independent vectors in $\mathbb{R}^n$ spans $\mathbb{R}^n$. Thus:

> *Every set of n linearly independent vectors in $\mathbb{R}^n$ is a basis in $\mathbb{R}^n$.*

In $\mathbb{R}^n$ we define

$$\mathbf{e}_1 = \begin{pmatrix}1\\0\\0\\ \cdot\\ \cdot\\ \cdot\\0\end{pmatrix}, \mathbf{e}_2 = \begin{pmatrix}0\\1\\0\\ \cdot\\ \cdot\\ \cdot\\0\end{pmatrix}, \mathbf{e}_3 = \begin{pmatrix}0\\0\\1\\ \cdot\\ \cdot\\ \cdot\\0\end{pmatrix}, \ldots, \mathbf{e}_n = \begin{pmatrix}0\\0\\0\\ \cdot\\ \cdot\\ \cdot\\1\end{pmatrix}$$

Then since the $\mathbf{e}_i$'s are the columns of the identity matrix (which has determinant 1), $\{\mathbf{e}_1, \mathbf{e}_2, \ldots, \mathbf{e}_n\}$ is linearly independent and therefore constitutes a basis in $\mathbb{R}^n$. This special basis is called the **standard basis** in $\mathbb{R}^n$. We shall now find bases for some other spaces.

EXAMPLE 1 By Example 5.4.2, page 183, the polynomials $1, x, x^2, x^3$ are linearly independent in P_3. By Example 5.4.6, page 185, these polynomials span P_3. Thus $\{1, x, x^2, x^3\}$ is a basis for P_3. In general, the monomials $\{1, x, x^2, x^3, \ldots, x^n\}$ constitute a basis for P_n. This is called the **standard basis** for P_n.

EXAMPLE 2 We saw in Example 5.4.10, page 187, that $\begin{pmatrix}1 & 0\\0 & 0\end{pmatrix}$, $\begin{pmatrix}0 & 1\\0 & 0\end{pmatrix}$, $\begin{pmatrix}0 & 0\\1 & 0\end{pmatrix}$, and $\begin{pmatrix}0 & 0\\0 & 1\end{pmatrix}$ span M_{22}. If $c_1\begin{pmatrix}1 & 0\\0 & 0\end{pmatrix} + c_2\begin{pmatrix}0 & 1\\0 & 0\end{pmatrix} + c_3\begin{pmatrix}0 & 0\\1 & 0\end{pmatrix} + c_4\begin{pmatrix}0 & 0\\0 & 1\end{pmatrix} = \begin{pmatrix}0 & 0\\0 & 0\end{pmatrix}$, then, obviously, $c_1 = c_2 = c_3 = c_4 = 0$. Thus these four matrices are linearly independent and form a basis for M_{22}. This is called the **standard basis** for M_{22}.

EXAMPLE 3 Find a basis for the set of vectors lying on the plane

$$\pi = \left\{ \begin{pmatrix} x \\ y \\ z \end{pmatrix} : 2x - y + 3z = 0 \right\}$$

Solution We saw in Example 5.2.6 that π is a vector space. To find a basis, we first note that if x and z are chosen arbitrarily and if $\begin{pmatrix} x \\ y \\ z \end{pmatrix} \in \pi$, then $y = 2x + 3z$. Thus vectors in π have the form $\begin{pmatrix} x \\ 2x+3z \\ z \end{pmatrix}$. Since x and z are arbitrary, we choose some simple values for them. Choosing $x = 1$, $z = 0$ we obtain $\mathbf{v}_1 = \begin{pmatrix} 1 \\ 2 \\ 0 \end{pmatrix}$; choosing $x = 0$, $z = 1$ we get $\mathbf{v}_2 = \begin{pmatrix} 0 \\ 3 \\ 1 \end{pmatrix}$. Then $\begin{pmatrix} x \\ 2x+3z \\ z \end{pmatrix} = x\begin{pmatrix} 1 \\ 2 \\ 0 \end{pmatrix} + z\begin{pmatrix} 0 \\ 3 \\ 1 \end{pmatrix}$. Thus $\{\mathbf{v}_1, \mathbf{v}_2\}$ span π and, since they are obviously linearly independent (because one is not a multiple of the other), they form a basis for π.

If $\mathbf{v}_1, \mathbf{v}_2, \ldots, \mathbf{v}_n$ is a basis for V, then any other vector $\mathbf{v} \in V$ can be written $\mathbf{v} = c_1\mathbf{v}_1 + c_2\mathbf{v}_2 + \cdots + c_n\mathbf{v}_n$. Can it be written in another way as a linear combination of the $\mathbf{v}_i$'s? The answer is *no*. (See the remark following the proof of Theorem 5.4.2, page 186, in the case $V = \mathbb{R}^n$.)

THEOREM 1 If $\{\mathbf{v}_1, \mathbf{v}_2, \ldots, \mathbf{v}_n\}$ is a basis for V and if $\mathbf{v} \in V$, then there exists a *unique* set of scalars $c_1, c_2, \ldots, c_n$ such that $\mathbf{v} = c_1\mathbf{v}_1 + c_2\mathbf{v}_2 + \cdots + c_n\mathbf{v}_n$.

Proof At least one such set of scalars exists because $\{\mathbf{v}_1, \mathbf{v}_2, \ldots, \mathbf{v}_n\}$ spans V. Suppose then that $\mathbf{v}$ can be written in two ways as a linear combination of the basis vectors. That is, suppose that

$$\mathbf{v} = c_1\mathbf{v}_1 + c_2\mathbf{v}_2 + \cdots + c_n\mathbf{v}_n = d_1\mathbf{v}_1 + d_2\mathbf{v}_2 + \cdots + d_n\mathbf{v}_n$$

Then, subtracting, we obtain the equation

$$(c_1 - d_1)\mathbf{v}_1 + (c_2 - d_2)\mathbf{v}_2 + \cdots + (c_n - d_n)\mathbf{v}_n = \mathbf{0}$$

But, since the $\mathbf{v}_i$'s are linearly independent, this equation can hold only if $c_1 - d_1 = c_2 - d_2 = \cdots = c_n - d_n = 0$. Thus $c_1 = d_1$, $c_2 = d_2, \ldots, c_n = d_n$ and the theorem is proved. ■

We have seen that vector spaces may have many bases. A question naturally arises: Do all bases contain the same number of vectors? In $\mathbb{R}^3$, the answer is certainly yes. To see this we note that any three linearly independent vectors in $\mathbb{R}^3$ form a basis. But fewer than three vectors cannot form a basis since, as we saw in Section 5.4, the span of two linearly independent vectors in $\mathbb{R}^3$ is a plane in $\mathbb{R}^3$—and a plane is not all of $\mathbb{R}^3$. Similarly a set of four or more vectors in $\mathbb{R}^3$ cannot be linearly independent; for if the first three vectors in the set are linearly independent, then they form a basis and, therefore, all other vectors in the set can be written as a linear combination of the first three. Thus all bases in $\mathbb{R}^3$ contain three vectors. The next theorem tells us that the answer to the question posed above is *yes* for all vector spaces.

THEOREM 2 If $\{\mathbf{u}_1, \mathbf{u}_2, \ldots, \mathbf{u}_m\}$ and $\{\mathbf{v}_1, \mathbf{v}_2, \ldots, \mathbf{v}_n\}$ are bases for the vector space V, then $m = n$; that is, any two bases in a vector space V have the same number of vectors.

Proof† Let $S_1 = \{\mathbf{u}_1, \ldots, \mathbf{u}_m\}$ and $S_2 = \{\mathbf{v}_1, \ldots, \mathbf{v}_n\}$ be two bases for V. We must show that $m = n$. We prove this by showing that if $m > n$, then S_1 is a linearly dependent set, which contradicts the hypothesis that S_1 is a basis. This will show that $m \le n$. The same proof will then show that $n \le m$, and this will prove the theorem. Hence all we must show is that if $m > n$, then S_1 is dependent. Since S_2 constitutes a basis, we can write each $\mathbf{u}_i$ as a linear combination of the $\mathbf{v}_i$'s. We have

$$\begin{aligned}
\mathbf{u}_1 &= a_{11}\mathbf{v}_1 + a_{12}\mathbf{v}_2 + \cdots + a_{1n}\mathbf{v}_n \\
\mathbf{u}_2 &= a_{21}\mathbf{v}_1 + a_{22}\mathbf{v}_2 + \cdots + a_{2n}\mathbf{v}_n \\
&\vdots \\
\mathbf{u}_m &= a_{m1}\mathbf{v}_1 + a_{m2}\mathbf{v}_2 + \cdots + a_{mn}\mathbf{v}_n
\end{aligned} \tag{1}$$

To show that S_1 is dependent, we must find scalars $c_1, c_2, \ldots, c_m$, not all zero, such that

$$c_1\mathbf{u}_1 + c_2\mathbf{u}_2 + \cdots + c_m\mathbf{u}_m = \mathbf{0} \tag{2}$$

Inserting (1) into (2), we obtain

$$\begin{aligned}
c_1(a_{11}\mathbf{v}_1 + a_{12}\mathbf{v}_2 + \cdots + a_{1n}\mathbf{v}_n) + c_2(a_{21}\mathbf{v}_1 + a_{22}\mathbf{v}_2 + \cdots + a_{2n}\mathbf{v}_n) \\
+ \cdots + c_m(a_{m1}\mathbf{v}_1 + a_{m2}\mathbf{v}_2 + \cdots + a_{mn}\mathbf{v}_n) = \mathbf{0}
\end{aligned} \tag{3}$$

Equation (3) can be rewritten as

$$\begin{aligned}
(a_{11}c_1 + a_{21}c_2 + \cdots + a_{m1}c_m)\mathbf{v}_1 + (a_{12}c_1 + a_{22}c_2 + \cdots + a_{m2}c_m)\mathbf{v}_2 \\
+ \cdots + (a_{1n}c_1 + a_{2n}c_2 + \cdots + a_{mn}c_m)\mathbf{v}_n = \mathbf{0}
\end{aligned} \tag{4}$$

† This proof is given for vector spaces with bases containing a finite number of vectors. We also treat the scalars as though they were real numbers. However, the proof works in the complex case as well.

But, since $\mathbf{v}_1, \mathbf{v}_2, \ldots, \mathbf{v}_n$ are linearly independent, we must have

$$\begin{aligned} a_{11}c_1 + a_{21}c_2 + \cdots + a_{m1}c_m &= 0 \\ a_{12}c_1 + a_{22}c_2 + \cdots + a_{m2}c_m &= 0 \\ \vdots \qquad \qquad & \\ a_{1n}c_1 + a_{2n}c_2 + \cdots + a_{mn}c_m &= 0 \end{aligned} \tag{5}$$

System (5) is a homogeneous system of n equations in the m unknowns $c_1, c_2, \ldots, c_m$ and, since $m > n$, Theorem 1.4.1, page 24, tells us that the system has an infinite number of solutions. Thus there are scalars $c_1, c_2, \ldots, c_m$, not all zero, such that (2) is satisfied and therefore S_1 is a linearly dependent set. This contradiction proves that $m \leq n$ and, by exchanging the roles of S_1 and S_2, we can show that $n \leq m$ and the proof is complete. ■

With this theorem we can define one of the central concepts in linear algebra.

DEFINITION 2 **DIMENSION** The **dimension** of a vector space V is the number of vectors in a basis of V. If this number is finite, then V is called a **finite dimensional vector space**. Otherwise V is called an **infinite dimensional vector space**. If $V = \{\mathbf{0}\}$, then V is said to be **zero dimensional**.

Notation We write the dimension of V as dim V.

Remark. We have not proved that every vector space has a basis. A proof of this fact is beyond the scope of this text. But we do not need this fact for Definition 2 to make sense; for *if* V has a finite basis, then V is finite dimensional. Otherwise V is infinite dimensional. Thus in order to show that V is infinite dimensional, it is only necessary to show that V does not have a finite basis. We can do this by showing that V contains an infinite number of linearly independent vectors (see Example 7 below). It is not necessary to construct an infinite basis for V.

EXAMPLE 4 Since n linearly independent vectors in $\mathbb{R}^n$ comprise a basis, we see that

$$\dim \mathbb{R}^n = n$$

EXAMPLE 5 By Example 1 and Problem 5.4.41, page 191, the polynomials $\{1, x, x^2, \ldots, x^n\}$ constitute a basis in P_n. Thus dim $P_n = n + 1$.

EXAMPLE 6 In M_{mn} let A_{ij} be the $m \times n$ matrix with a 1 in the ijth position and a zero everywhere else. It is easy to show that the A_{ij} for $i = 1, 2, \ldots, m$ and $j = 1, 2, \ldots, n$ form a basis for M_{mn}. Thus dim $M_{mn} = mn$.

EXAMPLE 7 In Example 5.4.11, page 187, we saw that no finite set of polynomials spans P. Thus P has no finite basis and is, therefore, an infinite dimensional vector space.

There are a number of theorems that tell us something about the dimension of a vector space.

THEOREM 3 Suppose that dim $V=n$. If $\mathbf{u}_1, \mathbf{u}_2, \ldots, \mathbf{u}_m$ is a set of m linearly independent vectors in V, then $m \leq n$.

Proof Let $\mathbf{v}_1, \mathbf{v}_2, \ldots, \mathbf{v}_n$ be a basis for V. If $m>n$, then, as in proof of Theorem 2, we can find constants $c_1, c_2, \ldots, c_m$ not all zero such that Equation (2) is satisfied. This would contradict the linear independence of the $\mathbf{u}_i$'s. Thus $m \leq n$. ■

THEOREM 4 Let H be a subspace of the finite dimensional vector space V. Then H is finite dimensional and

$$\dim H \leq \dim V \tag{6}$$

Proof Let dim $V=n$. Any set of linearly independent vectors in H is also a linearly independent set in V. By Theorem 3, any linearly independent set in H can contain at most n vectors. Hence H is finite dimensional. Moreover, since any basis in H is a linearly independent set, we see that dim $H \leq n$. ■

Theorem 4 has some interesting consequences. We give two of them here.

□**EXAMPLE 8** Let $P[0, 1]$ denote the set of polynomials defined on the interval $[0, 1]$. Then $P[0, 1] \subset C[0, 1]$. If $C[0, 1]$ were finite dimensional, then $P[0, 1]$ would be finite dimensional also. But, by Example 7, this is not the case. Hence $C[0, 1]$ is infinite dimensional. Similarly, since $P[0, 1] \subset C'[0, 1]$ (since every polynomial is differentiable), we also see that $C'[0, 1]$ is infinite dimensional. In general:

> *Any vector space containing an infinite dimensional subspace is infinite dimensional.*

EXAMPLE 9 We can use Theorem 4 to find *all* subspaces of $\mathbb{R}^3$. Let H be a subspace of $\mathbb{R}^3$. Then there are four possibilities: $H=\{\mathbf{0}\}$; $\dim H=1$, $\dim H=2$, and $\dim H=3$. If $\dim H=3$, then H contains a basis of three linearly independent vectors $\mathbf{v}_1, \mathbf{v}_2, \mathbf{v}_3$ in $\mathbb{R}^3$. But then $\mathbf{v}_1, \mathbf{v}_2, \mathbf{v}_3$ also form a basis for $\mathbb{R}^3$. Thus $H=\text{span}\,\{\mathbf{v}_1, \mathbf{v}_2, \mathbf{v}_3\}=\mathbb{R}^3$. Hence the only way to get a *proper* subspace of $\mathbb{R}^3$ is to have $\dim H=1$ or $\dim H=2$. If $\dim H=1$, then H has a basis consisting of the one vector $\mathbf{v}=(a, b, c)$. Let $\mathbf{x}$ be in H. Then $\mathbf{x}=t(a, b, c)$ for some real number t (since (a, b, c) spans H). If $\mathbf{x}=(x, y, z)$, this means that $x=at$, $y=bt$, $z=ct$. But this is the equation of a line in $\mathbb{R}^3$ passing through the origin with direction vector (a, b, c).

Now suppose $\dim H=2$ and let $\mathbf{v}_1=(a_1, b_1, c_1)$ and $\mathbf{v}_2=(a_2, b_2, c_2)$ be a basis for H. If $\mathbf{x}=(x, y, z)\in H$, then there exist real numbers s and t such that $\mathbf{x}=s\mathbf{v}_1+t\mathbf{v}_2$ or $(x, y, z)=s(a_1, b_1, c_1)+t(a_2, b_2, c_2)$. Then

$$\begin{aligned} x &= sa_1+ta_2 \\ y &= sb_1+tb_2 \\ z &= sc_1+tc_2 \end{aligned} \tag{7}$$

Let $\mathbf{v}_3=(\alpha, \beta, \gamma)=\mathbf{v}_1\times\mathbf{v}_2$. Then, from Theorem 4.4.2 on page 152, part (vi), we have $\mathbf{v}_3\cdot\mathbf{v}_1=0$ and $\mathbf{v}_3\cdot\mathbf{v}_2=0$. Now, we calculate

$$\begin{aligned} \alpha x+\beta y+\gamma z &= \alpha(sa_1+ta_2)+\beta(sb_1+tb_2)+\gamma(sc_1+tc_2) \\ &= (\alpha a_1+\beta b_1+\gamma c_1)s+(\alpha a_2+\beta b_2+\gamma c_2)t \\ &= (\mathbf{v}_3\cdot\mathbf{v}_1)s+(\mathbf{v}_3\cdot\mathbf{v}_2)t=0 \end{aligned}$$

Thus if $(x, y, z)\in H$, then $\alpha x+\beta y+\gamma z=0$, which shows that H is a plane passing through the origin with normal vector $\mathbf{v}_3=\mathbf{v}_1\times\mathbf{v}_2$. Therefore we have proved that:

> *The only proper subspaces of $\mathbb{R}^3$ are sets of vectors lying on lines and planes passing through the origin.*

EXAMPLE 10 Let A be an $m\times n$ matrix and let $S=\{\mathbf{x}\in\mathbb{R}^n;\ A\mathbf{x}=\mathbf{0}\}$. Let $\mathbf{x}_1\in S$ and $\mathbf{x}_2\in S$; then $A(\mathbf{x}_1+\mathbf{x}_2)=A\mathbf{x}_1+A\mathbf{x}_2=\mathbf{0}+\mathbf{0}=\mathbf{0}$ and $A(\alpha\mathbf{x}_1)=\alpha(A\mathbf{x}_1)=\alpha\mathbf{0}=\mathbf{0}$, so that S is a subspace of $\mathbb{R}^n$ and $\dim S\le n$. S is called the **solution space** of the homogeneous system $A\mathbf{x}=\mathbf{0}$. It is also called the **kernel** of the matrix A.

EXAMPLE 11 Find a basis for (and the dimension of) the solution space S of the homogeneous system

$$\begin{aligned} x+2y-z&=0 \\ 2x-y+3z&=0 \end{aligned}$$

Solution Here $A = \begin{pmatrix} 1 & 2 & -1 \\ 2 & -1 & 3 \end{pmatrix}$. Since A is a 2×3 matrix, S is a subspace of $\mathbb{R}^3$. Row-reducing, we find, successively,

$$\left(\begin{array}{ccc|c} 1 & 2 & -1 & 0 \\ 2 & -1 & 3 & 0 \end{array}\right) \xrightarrow{A_{1,2}(-2)} \left(\begin{array}{ccc|c} 1 & 2 & -1 & 0 \\ 0 & -5 & 5 & 0 \end{array}\right)$$

$$\xrightarrow{M_2(-\frac{1}{5})} \left(\begin{array}{ccc|c} 1 & 2 & -1 & 0 \\ 0 & 1 & -1 & 0 \end{array}\right) \xrightarrow{A_{2,1}(-2)} \left(\begin{array}{ccc|c} 1 & 0 & 1 & 0 \\ 0 & 1 & -1 & 0 \end{array}\right)$$

Then $y = z$ and $x = -z$, so that all solutions are of the form $\begin{pmatrix} -z \\ z \\ z \end{pmatrix}$. Thus $\begin{pmatrix} -1 \\ 1 \\ 1 \end{pmatrix}$ is a basis for S and $\dim S = 1$. Note that S is the set of vectors lying on the straight line $x = -t$, $y = t$, $z = t$.

EXAMPLE 12 Find a basis for the solution space S of the system

$$\begin{aligned} 2x - y + 3z &= 0 \\ 4x - 2y + 6z &= 0 \\ -6x + 3y - 9z &= 0 \end{aligned}$$

Solution Row-reducing as above, we obtain

$$\left(\begin{array}{ccc|c} 2 & -1 & 3 & 0 \\ 4 & -2 & 6 & 0 \\ -6 & 3 & -9 & 0 \end{array}\right) \xrightarrow[A_{1,3}(3)]{A_{1,2}(-2)} \left(\begin{array}{ccc|c} 2 & -1 & 3 & 0 \\ 0 & 0 & 0 & 0 \\ 0 & 0 & 0 & 0 \end{array}\right)$$

giving the single equation $2x - y + 3z = 0$. S is a plane and, by Example 3, a basis is given by $\begin{pmatrix} 1 \\ 2 \\ 0 \end{pmatrix}$ and $\begin{pmatrix} 0 \\ 3 \\ 1 \end{pmatrix}$ and $\dim S = 2$. Note that we have shown that any solution to the homogeneous equation can be written as

$$c_1 \begin{pmatrix} 1 \\ 2 \\ 0 \end{pmatrix} + c_2 \begin{pmatrix} 0 \\ 3 \\ 1 \end{pmatrix}$$

For example, if $c_1 = 2$ and $c_2 = -3$, we obtain the solution

$$\mathbf{x} = 2\begin{pmatrix} 1 \\ 2 \\ 0 \end{pmatrix} - 3\begin{pmatrix} 0 \\ 3 \\ 1 \end{pmatrix} = \begin{pmatrix} 2 \\ 4 \\ 0 \end{pmatrix} + \begin{pmatrix} 0 \\ -9 \\ -3 \end{pmatrix} = \begin{pmatrix} 2 \\ -5 \\ -3 \end{pmatrix}$$

Before leaving this section we prove a result that is very useful in finding bases in an arbitrary vector space. We have seen that n linearly independent vectors in $\mathbb{R}^n$ comprise a basis for $\mathbb{R}^n$. This fact holds in *any* finite dimensional vector space.

THEOREM 5 Any n linearly independent vectors in a vector space V of dimension n constitute a basis for V.

Proof Let $\mathbf{v}_1, \mathbf{v}_2, \ldots, \mathbf{v}_n$ be the n vectors. If they span V, then they constitute a basis. If they do not, then there is a vector $\mathbf{u} \in V$ such that $\mathbf{u} \notin \text{span}\,\{\mathbf{v}_1, \mathbf{v}_2, \ldots, \mathbf{v}_n\}$. This means that the $n+1$ vectors $\mathbf{v}_1, \mathbf{v}_2, \ldots, \mathbf{v}_n, \mathbf{u}$ are linearly independent. To see this note that if

$$c_1\mathbf{v}_1 + c_2\mathbf{v}_2 + \cdots + c_n\mathbf{v}_n + c_{n+1}\mathbf{u} = 0 \tag{8}$$

then $c_{n+1} = 0$ for, if not, we could write $\mathbf{u}$ as a linear combination of $\mathbf{v}_1, \mathbf{v}_2, \ldots, \mathbf{v}_n$ by dividing Equation (8) by c_{n+1} and putting all terms except $\mathbf{u}$ on the right-hand side. But if $c_{n+1} = 0$, then (8) reads

$$c_1\mathbf{v}_1 + c_2\mathbf{v}_2 + \cdots + c_n\mathbf{v}_n = 0$$

which means that $c_1 = c_2 = \cdots = c_n = 0$ since the $\mathbf{v}_i$'s are linearly independent. Now let $W = \text{span}\,\{\mathbf{v}_1, \mathbf{v}_2, \ldots, \mathbf{v}_n, \mathbf{u}\}$. Then as all the vectors in brackets are in V, W is a subspace of V. Since $\mathbf{v}_1, \mathbf{v}_2, \ldots, \mathbf{v}_n, \mathbf{u}$ are linearly independent, they form a basis for W. Thus $\dim W = n+1$. But from Theorem 4, $\dim W \le n$. This contradiction shows that there is *no* vector $\mathbf{u} \in V$ such that $\mathbf{u} \notin \text{span}\,\{\mathbf{v}_1, \mathbf{v}_2, \ldots, \mathbf{v}_n\}$. Thus $\mathbf{v}_1, \mathbf{v}_2, \ldots, \mathbf{v}_n$ span V and therefore constitute a basis for V. ■

PROBLEMS 5.5

In Problems 1–10 determine whether the given set of vectors is a basis for the given vector space.

1. In P_2: $1-x^2, x$

2. In P_2: $-3x, 1+x^2, x^2-5$

3. In P_2: x^2-1, x^2-2, x^2-3

4. In P_3: $1, 1+x, 1+x^2, 1+x^3$

5. In P_3: $3, x^3-4x+6, x^2$

6. In M_{22}: $\begin{pmatrix} 3 & 1 \\ 0 & 0 \end{pmatrix}, \begin{pmatrix} 3 & 2 \\ 0 & 0 \end{pmatrix}, \begin{pmatrix} -5 & 1 \\ 0 & 6 \end{pmatrix}, \begin{pmatrix} 0 & 1 \\ 0 & -7 \end{pmatrix}$

7. In M_{22}: $\begin{pmatrix} a & 0 \\ 0 & 0 \end{pmatrix}, \begin{pmatrix} 0 & b \\ 0 & 0 \end{pmatrix}, \begin{pmatrix} 0 & 0 \\ c & 0 \end{pmatrix}, \begin{pmatrix} 0 & 0 \\ 0 & d \end{pmatrix}$, where $abcd \ne 0$

8. In M_{22}: $\begin{pmatrix} -1 & 0 \\ 3 & 1 \end{pmatrix}, \begin{pmatrix} 2 & 1 \\ 1 & 4 \end{pmatrix}, \begin{pmatrix} -6 & 1 \\ 5 & 8 \end{pmatrix}, \begin{pmatrix} 7 & -2 \\ 1 & 0 \end{pmatrix}, \begin{pmatrix} 0 & 1 \\ 0 & 0 \end{pmatrix}$

9. $H = \{(x, y) \in \mathbb{R}^2\colon x + y = 0\}$; $(1, -1)$

10. $H = \{(x, y) \in \mathbb{R}^2\colon x + y = 0\}$; $(1, -1), (-3, 3)$

11. Find a basis in $\mathbb{R}^3$ for the set of vectors in the plane $2x - y - z = 0$.

12. Find a basis in $\mathbb{R}^3$ for the set of vectors in the plane $3x - 2y + 6z = 0$.

13. Find a basis in $\mathbb{R}^3$ for the set of vectors on the line $x/2 = y/3 = z/4$.

14. Find a basis in $\mathbb{R}^3$ for the set of vectors on the line $x = 3t$, $y = -2t$, $z = t$.

15. Show that the only proper subspaces of $\mathbb{R}^2$ are straight lines passing through the origin.

16. In $\mathbb{R}^4$ let $H = \{(x, y, z, w)\colon ax + by + cz + dw = 0\}$, where $abcd \ne 0$.

a. Show that H is a subspace of $\mathbb{R}^4$.

b. Find a basis for H.

c. What is $\dim H$?

*17. In $\mathbb{R}^n$ a *hyperplane* is a subspace of dimension $n-1$. If H is a hyperplane in $\mathbb{R}^n$ show that

$$H=\{(x_1, x_2, \ldots, x_n)\colon a_1x_1+a_2x_2+\cdots+a_nx_n=0\}$$

where $a_1, a_2, \ldots, a_n$ are fixed real numbers, not all of which are zero.

18. In $\mathbb{R}^5$ find a basis for the hyperplane

$$H=\{(x_1, x_2, x_3, x_4, x_5)\colon 2x_1-3x_2+x_3+4x_4-x_5=0\}$$

In Problems 19–23 find a basis for the solution space of the given homogeneous system.

19. $\begin{aligned} x-\ y&=0\\ -2x+2y&=0\end{aligned}$

20. $\begin{aligned} x-2y&=0\\ 3x+\ y&=0\end{aligned}$

21. $\begin{aligned} x-y-z&=0\\ 2x-y+z&=0\end{aligned}$

22. $\begin{aligned} x-3y+\ z&=0\\ -2x+2y-3z&=0\\ 4x-8y+5z&=0\end{aligned}$

23. $\begin{aligned} 2x-6y+4z&=0\\ -x+3y-2z&=0\\ -3x+9y-6z&=0\end{aligned}$

24. Find a basis for D_3, the vector space of diagonal 3×3 matrices. What is the dimension of D_3?
25. What is the dimension of D_n, the space of diagonal $n\times n$ matrices?
26. Let S_{nn} denote the vector space of symmetric $n\times n$ matrices. Show that S_{nn} is a subspace of M_{nn} and that dim $S_{nn}=[n(n+1)]/2$.
27. Suppose that $\mathbf{v}_1, \mathbf{v}_2, \ldots, \mathbf{v}_m$ are linearly independent vectors in a vector space V of dimension n and $m<n$. Show that $\{\mathbf{v}_1, \mathbf{v}_2, \ldots, \mathbf{v}_m\}$ can be enlarged to a basis for V. That is, there exist vectors $\mathbf{v}_{m+1}, \mathbf{v}_{m+2}, \ldots, \mathbf{v}_n$ such that $\{\mathbf{v}_1, \mathbf{v}_2, \ldots, \mathbf{v}_n\}$ is a basis. [*Hint:* Look at the proof of Theorem 5.]
28. Let $\{\mathbf{v}_1, \mathbf{v}_2, \ldots, \mathbf{v}_n\}$ be a basis for V. Let $\mathbf{u}_1=\mathbf{v}_1$, $\mathbf{u}_2=\mathbf{v}_1+\mathbf{v}_2$, $\mathbf{u}_3=\mathbf{v}_1+\mathbf{v}_2+\mathbf{v}_3, \ldots, \mathbf{u}_n=\mathbf{v}_1+\mathbf{v}_2+\cdots+\mathbf{v}_n$. Show that $\{\mathbf{u}_1, \mathbf{u}_2, \ldots, \mathbf{u}_n\}$ is also a basis for V.
29. Show that if $\{\mathbf{v}_1, \mathbf{v}_2, \ldots, \mathbf{v}_n\}$ spans V, then dim $V\le n$. [*Hint:* Use the result of Problem 5.4.35.]
30. Let H and K be subspaces of V such that $H\subseteq K$ and dim $H=$ dim $K<\infty$. Show that $H=K$.
31. Let H and K be subspaces of V and define $H+K=\{\mathbf{h}+\mathbf{k}\colon\ \mathbf{h}\in H \text{ and } \mathbf{k}\in K\}$.
    a. Show that $H+K$ is a subspace of V.
    b. If $H\cap K=\{\mathbf{0}\}$, show that dim $(H+K)=$ dim $H+$ dim K.

*32. If H is a subspace of the finite dimensional vector space V, show that there exists a unique subspace K of V such that **(a)** $H\cap K=\{\mathbf{0}\}$ and **(b)** $H+K=V$.

33. Show that two vectors $\mathbf{v}_1$ and $\mathbf{v}_2$ in $\mathbb{R}^2$ with endpoints at the origin are collinear if and only if dim span $\{\mathbf{v}_1, \mathbf{v}_2\}=1$.
34. Show that three vectors $\mathbf{v}_1$, $\mathbf{v}_2$, and $\mathbf{v}_3$ in $\mathbb{R}^3$ with endpoints at the origin are coplanar if and only if dim span $\{\mathbf{v}_1, \mathbf{v}_2, \mathbf{v}_3\}\le 2$.
35. Show that any n vectors which span an n-dimensional space V form a basis for V. [*Hint:* Show that if the n vectors are not linearly independent, then dim $V<n$.]

*36. Show that every subspace of a finite dimensional vector space has a basis.

5.6 Change of Basis

In $\mathbb{R}^2$ we wrote vectors in terms of the "standard" basis $\mathbf{i}=\begin{pmatrix}1\\0\end{pmatrix}$, $\mathbf{j}=\begin{pmatrix}0\\1\end{pmatrix}$. In $\mathbb{R}^n$ we defined the standard basis $\{\mathbf{e}_1, \mathbf{e}_2, \ldots, \mathbf{e}_n\}$. In P_n we defined the standard basis to be $\{1, x, x^2, \ldots, x^n\}$. These bases are most commonly used because it is relatively easy to work with them. But it sometimes happens that some other basis is more convenient. There are obviously many bases to choose from since in an n-dimensional vector space *any* n linearly independent vectors form a basis. In this section we shall see how to change from one basis to another by computing a certain matrix.

We start with a simple example. Let $\mathbf{u}_1 = \begin{pmatrix}1\\0\end{pmatrix}$ and $\mathbf{u}_2 = \begin{pmatrix}0\\1\end{pmatrix}$. Then $B_1 = \{\mathbf{u}_1, \mathbf{u}_2\}$ is the standard basis in $\mathbb{R}^2$. Let $\mathbf{v}_1 = \begin{pmatrix}1\\3\end{pmatrix}$ and $\mathbf{v}_2 = \begin{pmatrix}-1\\2\end{pmatrix}$. Since $\mathbf{v}_1$ and $\mathbf{v}_2$ are linearly independent (because $\mathbf{v}_1$ is not a multiple of $\mathbf{v}_2$), $B_2 = \{\mathbf{v}_1, \mathbf{v}_2\}$ is a second basis in $\mathbb{R}^2$. Let $\mathbf{x} = \begin{pmatrix}x_1\\x_2\end{pmatrix}$ be a vector in $\mathbb{R}^2$. This notation means that

$$\mathbf{x} = \begin{pmatrix}x_1\\x_2\end{pmatrix} = x_1\begin{pmatrix}1\\0\end{pmatrix} + x_2\begin{pmatrix}0\\1\end{pmatrix} = x_1\mathbf{u}_1 + x_2\mathbf{u}_2.$$

That is, $\mathbf{x}$ is written in terms of the vectors in the basis B_1. To emphasize this fact, we write

$$(\mathbf{x})_{B_1} = \begin{pmatrix}x_1\\x_2\end{pmatrix}.$$

Since B_2 is another basis in $\mathbb{R}^2$, there are scalars c_1 and c_2 such that

$$\mathbf{x} = c_1\mathbf{v}_1 + c_2\mathbf{v}_2. \tag{1}$$

Once these scalars are found, we write

$$(\mathbf{x})_{B_2} = \begin{pmatrix}c_1\\c_2\end{pmatrix}$$

to indicate that $\mathbf{x}$ is now expressed in terms of the vectors in B_2. To find the number c_1 and c_2, we write the old basis vectors ($\mathbf{u}_1$ and $\mathbf{u}_2$) in terms of the new basis vectors ($\mathbf{v}_1$ and $\mathbf{v}_2$). It is easy to verify that

$$\mathbf{u}_1 = \begin{pmatrix}1\\0\end{pmatrix} = \tfrac{2}{5}\begin{pmatrix}1\\3\end{pmatrix} - \tfrac{3}{5}\begin{pmatrix}-1\\2\end{pmatrix} = \tfrac{2}{5}\mathbf{v}_1 - \tfrac{3}{5}\mathbf{v}_2 \tag{2}$$

and

$$\mathbf{u}_2 = \begin{pmatrix}0\\1\end{pmatrix} = \tfrac{1}{5}\begin{pmatrix}1\\3\end{pmatrix} + \tfrac{1}{5}\begin{pmatrix}-1\\2\end{pmatrix} = \tfrac{1}{5}\mathbf{v}_1 + \tfrac{1}{5}\mathbf{v}_2. \tag{3}$$

That is,

$$(\mathbf{u}_1)_{B_2} = \begin{pmatrix}\frac{2}{5}\\-\frac{3}{5}\end{pmatrix} \quad \text{and} \quad (\mathbf{u}_2)_{B_2} = \begin{pmatrix}\frac{1}{5}\\\frac{1}{5}\end{pmatrix}$$

Then

$$\begin{aligned}\mathbf{x} = x_1\mathbf{u}_1 + x_2\mathbf{u}_2 &\overset{\text{from (2) and (3)}}{=} x_1(\tfrac{2}{5}\mathbf{v}_1 - \tfrac{3}{5}\mathbf{v}_2) + x_2(\tfrac{1}{5}\mathbf{v}_1 + \tfrac{1}{5}\mathbf{v}_2)\\ &= (\tfrac{2}{5}x_1 + \tfrac{1}{5}x_2)\mathbf{v}_1 + (-\tfrac{3}{5}x_1 + \tfrac{1}{5}x_2)\mathbf{v}_2.\end{aligned}$$

Thus, from (1),

$$c_1 = \tfrac{2}{5}x_1 + \tfrac{1}{5}x_2$$
$$c_2 = -\tfrac{3}{5}x_1 + \tfrac{1}{5}x_2$$

or

$$(\mathbf{x})_{B_2} = \begin{pmatrix} c_1 \\ c_2 \end{pmatrix} = \begin{pmatrix} \frac{2}{5}x_1 + \frac{1}{5}x_2 \\ -\frac{3}{5}x_1 + \frac{1}{5}x_2 \end{pmatrix} = \begin{pmatrix} \frac{2}{5} & \frac{1}{5} \\ -\frac{3}{5} & \frac{1}{5} \end{pmatrix} \begin{pmatrix} x_1 \\ x_2 \end{pmatrix}$$

For example, if $(\mathbf{x})_{B_1} = \begin{pmatrix} 3 \\ -4 \end{pmatrix}$, then

$$(\mathbf{x})_{B_2} = \begin{pmatrix} \frac{2}{5} & \frac{1}{5} \\ -\frac{3}{5} & \frac{1}{5} \end{pmatrix} \begin{pmatrix} 3 \\ -4 \end{pmatrix} = \begin{pmatrix} \frac{2}{5} \\ -\frac{13}{5} \end{pmatrix}$$

Check. $\frac{2}{5}\mathbf{v}_1 - \frac{13}{5}\mathbf{v}_2 = \frac{2}{5}\begin{pmatrix} 1 \\ 3 \end{pmatrix} - \frac{13}{5}\begin{pmatrix} -1 \\ 2 \end{pmatrix} = \begin{pmatrix} \frac{2}{5} + \frac{13}{5} \\ \frac{6}{5} - \frac{26}{5} \end{pmatrix} = \begin{pmatrix} 3 \\ -4 \end{pmatrix} = 3\begin{pmatrix} 1 \\ 0 \end{pmatrix} - 4\begin{pmatrix} 0 \\ 1 \end{pmatrix}$

$$= 3\mathbf{u}_1 - 4\mathbf{u}_2.$$

The matrix $A = \begin{pmatrix} \frac{2}{5} & \frac{1}{5} \\ -\frac{3}{5} & \frac{1}{5} \end{pmatrix}$ is called the **transition matrix** from B_1 to B_2, and we have shown that

$$(\mathbf{x})_{B_2} = A(\mathbf{x})_{B_1} \tag{4}$$

This example can be easily generalized, but first we need to extend our notation. Let $B_1 = \{\mathbf{u}_1, \mathbf{u}_2, \ldots, \mathbf{u}_n\}$ and $B_2 = \{\mathbf{v}_1, \mathbf{v}_2, \ldots, \mathbf{v}_n\}$ be two bases for an n-dimensional real vector space V. Let $\mathbf{x} \in V$. Then $\mathbf{x}$ can be written in terms of both bases:

$$\mathbf{x} = b_1\mathbf{u}_1 + b_2\mathbf{u}_2 + \cdots + b_n\mathbf{u}_n \tag{5}$$

and

$$\mathbf{x} = c_1\mathbf{v}_1 + c_2\mathbf{v}_2 + \cdots + c_n\mathbf{v}_n \tag{6}$$

where the b_i's and c_i's are real numbers. We then write $(\mathbf{x})_{B_1} = \begin{pmatrix} b_1 \\ b_2 \\ \vdots \\ b_n \end{pmatrix}$ to denote the representation of $\mathbf{x}$ in terms of the basis B_1. This is unambiguous because the coefficients b_i in (5) are unique by Theorem 5.5.1, page 193. Likewise $(\mathbf{x})_{B_2} = \begin{pmatrix} c_1 \\ c_2 \\ \vdots \\ c_n \end{pmatrix}$ has a similar meaning. Suppose that $\mathbf{w}_1 = a_1\mathbf{u}_1 + a_2\mathbf{u}_2 + \cdots + a_n\mathbf{u}_n$ and $\mathbf{w}_2 = b_1\mathbf{u}_1 + b_2\mathbf{u}_2 + \cdots + b_n\mathbf{u}_n$. Then $\mathbf{w}_1 + \mathbf{w}_2 = (a_1 + b_1)\mathbf{u}_1 + (a_2 + b_2)\mathbf{u}_2 + \cdots + (a_n + b_n)\mathbf{u}_n$, so that

$$(\mathbf{w}_1 + \mathbf{w}_2)_{B_1} = (\mathbf{w}_1)_{B_1} + (\mathbf{w}_2)_{B_1}$$

That is, in the new notation we can add vectors just as we add vectors in $\mathbb{R}^n$. Moreover, it is easy to show that

$$\alpha(\mathbf{w})_{B_1} = (\alpha\mathbf{w})_{B_1}$$

Now, since B_2 is a basis, each $\mathbf{u}_j$ in B_1 can be written as a linear combination of the $\mathbf{v}_i$'s. Thus there exists a unique set of scalars $a_{1j}, a_{2j}, \ldots, a_{nj}$ such that for $j = 1, 2, \ldots, n$

$$\mathbf{u}_j = a_{1j}\mathbf{v}_1 + a_{2j}\mathbf{v}_2 + \cdots + a_{nj}\mathbf{v}_n \tag{7}$$

or

$$(\mathbf{u}_j)_{B_2} = \begin{pmatrix} a_{1j} \\ a_{2j} \\ \vdots \\ a_{nj} \end{pmatrix}. \tag{8}$$

DEFINITION 1 **TRANSITION MATRIX** The $n \times n$ matrix A whose columns are given by (8) is called the **transition matrix** from basis B_1 to basis B_2. That is,

$$A = \begin{pmatrix} a_{11} & a_{12} & a_{13} & \cdots & a_{1n} \\ a_{21} & a_{22} & a_{23} & \cdots & a_{2n} \\ \vdots & \vdots & \vdots & & \vdots \\ a_{n1} & a_{n2} & a_{n3} & \cdots & a_{nn} \end{pmatrix}$$
$$\quad\ \ \uparrow \qquad\quad \uparrow \qquad\quad \uparrow \qquad\qquad\quad \uparrow$$
$$(\mathbf{u}_1)_{B_2} \quad (\mathbf{u}_2)_{B_2} \quad (\mathbf{u}_3)_{B_2} \quad \cdots \quad (\mathbf{u}_n)_{B_2}$$

THEOREM 1 Let B_1 and B_2 be bases for a vector space V. Let A be the transition matrix from B_1 to B_2. Then, for every $\mathbf{x} \in V$,

$$(\mathbf{x})_{B_2} = A(\mathbf{x})_{B_1} \tag{10}$$

Proof We use the representation of $\mathbf{x}$ given in (5) and (6):

$$\mathbf{x} \overset{\text{from (5)}}{=} b_1\mathbf{u}_1 + b_2\mathbf{u}_2 + \cdots + b_n\mathbf{u}_n$$

$$\overset{\text{from (7)}}{=} b_1(a_{11}\mathbf{v}_1 + a_{21}\mathbf{v}_2 + \cdots + a_{n1}\mathbf{v}_n) + b_2(a_{12}\mathbf{v}_1 + a_{22}\mathbf{v}_2 + \cdots + a_{n2}\mathbf{v}_n) + \cdots + b_n(a_{1n}\mathbf{v}_1 + a_{2n}\mathbf{v}_2 + \cdots + a_{nn}\mathbf{v}_n)$$

$$= (a_{11}b_1 + a_{12}b_2 + \cdots + a_{1n}b_n)\mathbf{v}_1 + (a_{21}b_1 + a_{22}b_2 + \cdots + a_{2n}b_n)\mathbf{v}_2 + \cdots + (a_{n1}b_1 + a_{n2}b_2 + \cdots + a_{nn}b_n)\mathbf{v}_n$$

$$\overset{\text{from (6)}}{=} c_1\mathbf{v}_1 + c_2\mathbf{v}_2 + \cdots + c_n\mathbf{v}_n. \tag{11}$$

Thus

$$(\mathbf{x})_{B_2} = \begin{pmatrix} c_1 \\ c_2 \\ \vdots \\ c_n \end{pmatrix} \overset{\text{from (5)}}{=} \begin{pmatrix} a_{11}b_1 + a_{12}b_2 + \cdots + a_{1n}b_n \\ a_{21}b_1 + a_{22}b_2 + \cdots + a_{2n}b_n \\ \vdots \qquad \vdots \qquad \vdots \\ a_{n1}b_1 + a_{n2}b_2 + \cdots + a_{nn}b_n \end{pmatrix}$$

$$= \begin{pmatrix} a_{11} & a_{12} & \cdots & a_{1n} \\ a_{21} & a_{22} & \cdots & a_{2n} \\ \vdots & \vdots & & \vdots \\ a_{n1} & a_{n2} & \cdots & a_{nn} \end{pmatrix} \begin{pmatrix} b_1 \\ b_2 \\ \vdots \\ b_n \end{pmatrix} = A(\mathbf{x})_{B_1}. \quad \blacksquare \qquad (12)$$

Before doing any further examples, we prove a theorem that is very useful for computations.

THEOREM 2 If A is the transition matrix from B_1 to B_2, then A^{-1} is the transition matrix from B_2 to B_1.

Proof Let C be the transition matrix from B_2 to B_1. Then, from (10), we have

$$(\mathbf{x})_{B_1} = C(\mathbf{x})_{B_2} \qquad (13)$$

But $(\mathbf{x})_{B_2} = A(\mathbf{x})_{B_1}$, and substituting this into (13) yields

$$(\mathbf{x})_{B_1} = CA(\mathbf{x})_{B_1} \qquad (14)$$

We leave it as an exercise (see Problem 39) to show that (14) can hold for every $\mathbf{x}$ in V only if $CA = I$. Thus, from Theorem 2.7.8 on page 74, $C = A^{-1}$, and the theorem is proven. ■

Remark. This theorem makes it especially easy to find the transition matrix from the standard basis $B_1 = \{\mathbf{e}_1, \mathbf{e}_2, \ldots, \mathbf{e}_n\}$ in $\mathbb{R}^n$ to any other basis in $\mathbb{R}^n$. Let $B_2 = \{\mathbf{v}_1, \mathbf{v}_2, \ldots, \mathbf{v}_n\}$ be any other basis. Let C be the matrix whose columns are the vectors $\mathbf{v}_1, \mathbf{v}_2, \ldots, \mathbf{v}_n$. Then C is the transition matrix from B_2 to B_1 since each vector $\mathbf{v}_i$ is already written in terms of the standard basis. For example,

$$\begin{pmatrix} 1 \\ 3 \\ -2 \\ 4 \end{pmatrix}_{B_1} = \begin{pmatrix} 1 \\ 3 \\ -2 \\ 4 \end{pmatrix} = 1\begin{pmatrix} 1 \\ 0 \\ 0 \\ 0 \end{pmatrix} + 3\begin{pmatrix} 0 \\ 1 \\ 0 \\ 0 \end{pmatrix} - 2\begin{pmatrix} 0 \\ 0 \\ 1 \\ 0 \end{pmatrix} + 4\begin{pmatrix} 0 \\ 0 \\ 0 \\ 1 \end{pmatrix}$$

Thus the transition matrix from B_1 to B_2 is C^{-1}.

PROCEDURE FOR FINDING THE TRANSITION MATRIX FROM STANDARD BASIS TO BASIS $B_2 = \{\mathbf{v}_1, \mathbf{v}_2, \ldots \mathbf{v}_n\}$

i. Write the matrix C whose columns are $\mathbf{v}_1, \mathbf{v}_2, \ldots, \mathbf{v}_n$.
ii. Compute C^{-1}. This is the required transition matrix.

EXAMPLE 1 In $\mathbb{R}^3$ let $B_1 = \{\mathbf{i}, \mathbf{j}, \mathbf{k}\}$ and let $B_2 = \left\{\begin{pmatrix}1\\0\\2\end{pmatrix}, \begin{pmatrix}3\\-1\\0\end{pmatrix}, \begin{pmatrix}0\\1\\-2\end{pmatrix}\right\}$. If $\mathbf{x} = \begin{pmatrix}x\\y\\z\end{pmatrix} \in \mathbb{R}^3$, write $\mathbf{x}$ in terms of the vectors in B_2.

Solution We first verify that B_2 is a basis. This is evident since $\begin{vmatrix}1 & 3 & 0\\0 & -1 & 1\\2 & 0 & -2\end{vmatrix} = 8 \neq 0.$

Since $\mathbf{u}_1 = \begin{pmatrix}1\\0\\0\end{pmatrix}$, $\mathbf{u}_2 = \begin{pmatrix}0\\1\\0\end{pmatrix}$, and $\mathbf{u}_3 = \begin{pmatrix}0\\0\\1\end{pmatrix}$, we immediately see that the transition matrix, C, from B_2 to B_1 is given by

$$C = \begin{pmatrix}1 & 3 & 0\\0 & -1 & 1\\2 & 0 & -2\end{pmatrix}$$

Thus, from Theorem 2, the transition matrix A from B_1 to B_2 is

$$A = C^{-1} = \tfrac{1}{8}\begin{pmatrix}2 & 6 & 3\\2 & -2 & -1\\2 & 6 & -1\end{pmatrix}$$

For example, if $(\mathbf{x})_{B_1} = \begin{pmatrix}1\\-2\\4\end{pmatrix}$, then

$$(\mathbf{x})_{B_2} = \tfrac{1}{8}\begin{pmatrix}2 & 6 & 3\\2 & -2 & -1\\2 & 6 & -1\end{pmatrix}\begin{pmatrix}1\\-2\\4\end{pmatrix} = \tfrac{1}{8}\begin{pmatrix}2\\2\\-14\end{pmatrix} = \begin{pmatrix}\frac{1}{4}\\\frac{1}{4}\\-\frac{7}{4}\end{pmatrix}$$

As a check, note that

$$\tfrac{1}{4}\begin{pmatrix}1\\0\\2\end{pmatrix} + \tfrac{1}{4}\begin{pmatrix}3\\-1\\0\end{pmatrix} - \tfrac{7}{4}\begin{pmatrix}0\\1\\-2\end{pmatrix} = \begin{pmatrix}1\\-2\\4\end{pmatrix} = 1\begin{pmatrix}1\\0\\0\end{pmatrix} - 2\begin{pmatrix}0\\1\\0\end{pmatrix} + 4\begin{pmatrix}0\\0\\1\end{pmatrix}.$$

EXAMPLE 2 In P_2 the standard basis is $B_1 = \{1, x, x^2\}$. Another basis is $B_2 = \{4x - 1, 2x^2 - x, 3x^2 + 3\}$. If $p = a_0 + a_1 x + a_2 x^2$, write p in terms of the polynomials in B_2.

Solution We first verify that B_2 is a basis. If $c_1(4x-1) + c_2(2x^2 - x) + c_3(3x^2+3) = 0$ for all x, then, rearranging terms, we obtain

$$(-c_1 + 3c_3)1 + (4c_1 - c_2)x + (2c_2 + 3c_3)x^2 = 0$$

But, since $\{1, x, x^2\}$ is a linearly independent set, we must have

$$\begin{aligned} -c_1 \phantom{{}-c_2} + 3c_3 &= 0 \\ 4c_1 - c_2 \phantom{{}+3c_3} &= 0 \\ 2c_2 + 3c_3 &= 0 \end{aligned}$$

The determinant of this homogeneous system is $\begin{vmatrix} -1 & 0 & 3 \\ 4 & -1 & 0 \\ 0 & 2 & 3 \end{vmatrix} = 27 \neq 0$, which means that $c_1 = c_2 = c_3 = 0$ is the only solution. Now $(4x - 1)_{B_1} = \begin{pmatrix} -1 \\ 4 \\ 0 \end{pmatrix}$, $(2x^2 - x)_{B_1} = \begin{pmatrix} 0 \\ -1 \\ 2 \end{pmatrix}$, and $(3 + 3x^2)_{B_1} = \begin{pmatrix} 3 \\ 0 \\ 3 \end{pmatrix}$. Hence

$$C = \begin{pmatrix} -1 & 0 & 3 \\ 4 & -1 & 0 \\ 0 & 2 & 3 \end{pmatrix}$$

is the transition matrix from B_2 to B_1 so that

$$A = C^{-1} = \tfrac{1}{27}\begin{pmatrix} -3 & 6 & 3 \\ -12 & -3 & 12 \\ 8 & 2 & 1 \end{pmatrix}$$

is the transition matrix from B_1 to B_2. Since $(a_0 + a_1x + a_2x^2)_{B_1} = \begin{pmatrix} a_0 \\ a_1 \\ a_2 \end{pmatrix}$, we have

$$(a_0 + a_1x + a_2x^2)_{B_2} = \tfrac{1}{27}\begin{pmatrix} -3 & 6 & 3 \\ -12 & -3 & 12 \\ 8 & 2 & 1 \end{pmatrix}\begin{pmatrix} a_0 \\ a_1 \\ a_2 \end{pmatrix}$$

$$= \begin{pmatrix} \frac{1}{27}[-3a_0 + 6a_1 + 3a_2] \\ \frac{1}{27}[-12a_0 - 3a_1 + 12a_2] \\ \frac{1}{27}[8a_0 + 2a_1 + a_2] \end{pmatrix}.$$

For example, if $p(x) = 5x^2 - 3x + 4$, then

$$(5x^2 - 3x + 4)_{B_2} = \tfrac{1}{27}\begin{pmatrix} -3 & 6 & 3 \\ -12 & -3 & 12 \\ 8 & 2 & 1 \end{pmatrix}\begin{pmatrix} 4 \\ -3 \\ 5 \end{pmatrix} = \begin{pmatrix} -\frac{15}{27} \\ \frac{21}{27} \\ \frac{31}{27} \end{pmatrix}$$

or

$$5x^2 - 3x + 4 \overset{\text{check this}}{=} -\tfrac{15}{27}(4x - 1) + \tfrac{21}{27}(2x^2 - x) + \tfrac{31}{27}(3x^2 + 3).$$

EXAMPLE 3 Let $B_1 = \left\{\begin{pmatrix}3\\1\end{pmatrix}, \begin{pmatrix}2\\-1\end{pmatrix}\right\}$ and $B_2 = \left\{\begin{pmatrix}2\\4\end{pmatrix}, \begin{pmatrix}-5\\3\end{pmatrix}\right\}$ be two bases in $\mathbb{R}^2$. If $(\mathbf{x})_{B_1} = \begin{pmatrix}b_1\\b_2\end{pmatrix}$, write $\mathbf{x}$ in terms of the vectors in B_2.

Solution This problem is a bit more difficult because neither basis is the standard basis. We must write the vectors in B_1 as linear combinations of the vectors in B_2. That is, we must find constants $a_{11}, a_{21}, a_{12}, a_{22}$ such that

$$\begin{pmatrix}3\\1\end{pmatrix} = a_{11}\begin{pmatrix}2\\4\end{pmatrix} + a_{21}\begin{pmatrix}-5\\3\end{pmatrix} \quad\text{and}\quad \begin{pmatrix}2\\-1\end{pmatrix} = a_{12}\begin{pmatrix}2\\4\end{pmatrix} + a_{22}\begin{pmatrix}-5\\3\end{pmatrix}.$$

This leads to the following systems:

$$\begin{aligned}2a_{11} - 5a_{21} &= 3\\ 4a_{11} + 3a_{21} &= 1\end{aligned} \quad\text{and}\quad \begin{aligned}2a_{12} - 5a_{22} &= 2\\ 4a_{12} + 3a_{22} &= -1\end{aligned}$$

The solutions are $a_{11} = \frac{7}{13}$, $a_{21} = -\frac{5}{13}$, $a_{12} = \frac{1}{26}$, and $a_{22} = -\frac{5}{13}$. Thus

$$A = \tfrac{1}{26}\begin{pmatrix}14 & 1\\ -10 & -10\end{pmatrix}$$

and

$$(\mathbf{x})_{B_2} = \tfrac{1}{26}\begin{pmatrix}14 & 1\\ -10 & -10\end{pmatrix}\begin{pmatrix}b_1\\b_2\end{pmatrix} = \begin{pmatrix}\frac{1}{26}(14b_1 + b_2)\\ -\frac{10}{26}(b_1 + b_2)\end{pmatrix}.$$

For example, let $\mathbf{x} \overset{\text{in standard basis}}{=} \begin{pmatrix}7\\4\end{pmatrix}$. Then

$$\begin{pmatrix}7\\4\end{pmatrix}_{B_1} = b_1\begin{pmatrix}3\\1\end{pmatrix} + b_2\begin{pmatrix}2\\-1\end{pmatrix} = 3\begin{pmatrix}3\\1\end{pmatrix} - \begin{pmatrix}2\\-1\end{pmatrix}$$

so that

$$\begin{pmatrix}7\\4\end{pmatrix}_{B_1} = \begin{pmatrix}3\\-1\end{pmatrix}$$

and

$$\begin{pmatrix}7\\4\end{pmatrix}_{B_2} = \tfrac{1}{26}\begin{pmatrix}14 & 1\\ -10 & -10\end{pmatrix}\begin{pmatrix}3\\-1\end{pmatrix} = \begin{pmatrix}\frac{41}{26}\\ -\frac{20}{26}\end{pmatrix}.$$

That is

$$\begin{pmatrix}7\\4\end{pmatrix} \overset{\text{check!}}{=} \tfrac{41}{26}\begin{pmatrix}2\\4\end{pmatrix} - \tfrac{20}{26}\begin{pmatrix}-5\\3\end{pmatrix}.$$

Using the notation of this section we can derive a convenient way to determine whether a given set of vectors in any finite dimensional real vector space is linearly dependent or independent.

THEOREM 3 Let $B_1 = \{\mathbf{v}_1, \mathbf{v}_2, \ldots, \mathbf{v}_n\}$ be a basis for the n-dimensional vector space V. Suppose that

$$(\mathbf{x}_1)_{B_1} = \begin{pmatrix} a_{11} \\ a_{21} \\ \vdots \\ a_{n1} \end{pmatrix}, (\mathbf{x}_2)_{B_1} = \begin{pmatrix} a_{12} \\ a_{22} \\ \vdots \\ a_{n2} \end{pmatrix}, \ldots, (\mathbf{x}_n)_{B_1} = \begin{pmatrix} a_{1n} \\ a_{21} \\ \vdots \\ a_{nn} \end{pmatrix}$$

Let

$$A = \begin{pmatrix} a_{11} & a_{12} \cdots & a_{1n} \\ a_{21} & a_{22} \cdots & a_{2n} \\ \vdots & \vdots & \vdots \\ a_{n1} & a_{n2} \cdots & a_{nn} \end{pmatrix}$$

Then $\mathbf{x}_1, \mathbf{x}_2, \ldots, \mathbf{x}_n$ are linearly independent if and only if $\det A \neq 0$.

Proof Let $\mathbf{a}_1, \mathbf{a}_2, \ldots, \mathbf{a}_n$ denote the columns of A. Suppose that

$$c_1\mathbf{x}_1 + c_2\mathbf{x}_2 + \cdots + c_n\mathbf{x}_n = \mathbf{0} \tag{15}$$

Then, using the addition defined on page 202, we may write (15) as

$$(c_1\mathbf{a}_1 + c_2\mathbf{a}_2 + \cdots + c_n\mathbf{a}_n)_{B_1} = (\mathbf{0})_{B_1} \tag{16}$$

Equation (16) gives two representations of the zero vector in V in terms of the basis vectors in B_1. Since the representation of a vector in terms of basis vectors is unique (by Theorem 5.5.1, page 193) we conclude that

$$c_1\mathbf{a}_1 + c_2\mathbf{a}_2 + \cdots + c_n\mathbf{a}_n = \mathbf{0} \tag{17}$$

where the zero on the right-hand side is the zero vector in $\mathbb{R}^n$. But this proves the theorem, since Equation (17) involves the columns of A, which are linearly independent if and only if $\det A \neq 0$. ■

EXAMPLE 4 In P_2, determine whether the polynomials $3-x$, $2+x^2$, and $4+5x-2x^2$ are linearly dependent or independent.

Solution Using the basis $B_1 = \{1, x, x^2\}$, we have $(3-x)_{B_1} = \begin{pmatrix} 3 \\ -1 \\ 0 \end{pmatrix}$, $(2+x^2)_{B_1} = \begin{pmatrix} 2 \\ 0 \\ 1 \end{pmatrix}$, and $(4+5x-2x^2)_{B_1} = \begin{pmatrix} 4 \\ 5 \\ -2 \end{pmatrix}$. Then $\det A = \begin{vmatrix} 3 & 2 & 4 \\ -1 & 0 & 5 \\ 0 & 1 & -2 \end{vmatrix} = -23 \neq 0$, so the polynomials are independent.

EXAMPLE 5 In M_{22}, determine whether the matrices $\begin{pmatrix}1 & 2\\ 3 & 6\end{pmatrix}$, $\begin{pmatrix}-1 & 3\\ -1 & 1\end{pmatrix}$, $\begin{pmatrix}2 & -1\\ 0 & 1\end{pmatrix}$, and $\begin{pmatrix}1 & 4\\ 4 & 9\end{pmatrix}$ are linearly dependent or independent.

Solution Using the standard basis $B_1 = \left\{\begin{pmatrix}1 & 0\\ 0 & 0\end{pmatrix}, \begin{pmatrix}0 & 1\\ 0 & 0\end{pmatrix}, \begin{pmatrix}0 & 0\\ 1 & 0\end{pmatrix}, \begin{pmatrix}0 & 0\\ 0 & 1\end{pmatrix}\right\}$ we obtain

$$\det A = \begin{vmatrix} 1 & -1 & 2 & 1\\ 2 & 3 & -1 & 4\\ 3 & -1 & 0 & 4\\ 6 & 1 & 1 & 9\end{vmatrix} = 0$$

so the matrices are dependent. Note that $\det A = 0$ because the fourth row of A is the sum of the first three rows of A. Note also that

$$-29\begin{pmatrix}1 & 2\\ 3 & 6\end{pmatrix} - 7\begin{pmatrix}-1 & 3\\ -1 & 1\end{pmatrix} + \begin{pmatrix}2 & -1\\ 0 & 1\end{pmatrix} + 20\begin{pmatrix}1 & 4\\ 4 & 9\end{pmatrix} = \begin{pmatrix}0 & 0\\ 0 & 0\end{pmatrix}$$

which illustrates that the four matrices are linearly dependent.

PROBLEMS 5.6

In Problems 1–5 write $\begin{pmatrix}x\\ y\end{pmatrix} \in \mathbb{R}^2$ in terms of the given basis.

1. $\begin{pmatrix}1\\ 1\end{pmatrix}, \begin{pmatrix}1\\ -1\end{pmatrix}$ **2.** $\begin{pmatrix}2\\ -3\end{pmatrix}, \begin{pmatrix}3\\ -2\end{pmatrix}$ **3.** $\begin{pmatrix}5\\ 7\end{pmatrix}\begin{pmatrix}3\\ -4\end{pmatrix}$ **4.** $\begin{pmatrix}-1\\ -2\end{pmatrix}, \begin{pmatrix}-1\\ 2\end{pmatrix}$

5. $\begin{pmatrix}a\\ c\end{pmatrix}, \begin{pmatrix}b\\ d\end{pmatrix}$, where $ad - bc \neq 0$

In Problems 6–10 write $\begin{pmatrix}x\\ y\\ z\end{pmatrix} \in \mathbb{R}^3$ in terms of the given basis.

6. $\begin{pmatrix}1\\ 0\\ 0\end{pmatrix}, \begin{pmatrix}0\\ 0\\ 1\end{pmatrix}, \begin{pmatrix}1\\ 1\\ 1\end{pmatrix}$ **7.** $\begin{pmatrix}1\\ 0\\ 0\end{pmatrix}, \begin{pmatrix}1\\ 1\\ 0\end{pmatrix}, \begin{pmatrix}1\\ 1\\ 1\end{pmatrix}$ **8.** $\begin{pmatrix}1\\ 0\\ -1\end{pmatrix}, \begin{pmatrix}-1\\ 1\\ 0\end{pmatrix}, \begin{pmatrix}0\\ 1\\ 1\end{pmatrix}$

9. $\begin{pmatrix}2\\ 1\\ 3\end{pmatrix}, \begin{pmatrix}-1\\ 4\\ 5\end{pmatrix}, \begin{pmatrix}3\\ -2\\ -4\end{pmatrix}$ **10.** $\begin{pmatrix}a\\ 0\\ 0\end{pmatrix}, \begin{pmatrix}b\\ d\\ 0\end{pmatrix}, \begin{pmatrix}c\\ e\\ f\end{pmatrix}$, where $adf \neq 0$

In Problems 11–13 write the polynomial $a_0 + a_1x + a_2x^2$ in P_2 in terms of the given basis.

11. $1, x-1, x^2-1$ **12.** $6, 2+3x, 3+4x+5x^2$ **13.** $x+1, x-1, x^2-1$

14. In M_{22} write the matrix $\begin{pmatrix}2 & -1\\ 4 & 6\end{pmatrix}$ in terms of the basis $\left\{\begin{pmatrix}1 & 1\\ -1 & 0\end{pmatrix}, \begin{pmatrix}2 & 0\\ 3 & 1\end{pmatrix}, \begin{pmatrix}0 & 1\\ -1 & 0\end{pmatrix}, \begin{pmatrix}0 & -2\\ 0 & 4\end{pmatrix}\right\}$.

15. In P_3 write the polynomial $2x^3 - 3x^2 + 5x - 6$ in terms of the basis polynomials $1,\ 1+x,\ x+x^2,\ x^2+x^3$.

16. In P_3 write the polynomial $4x^2-x+5$ in terms of the basis polynomials 1, $1-x$, $(1-x)^2$, $(1-x)^3$.

17. In $\mathbb{R}^2$ suppose that $\mathbf{x}=\begin{pmatrix}2\\-1\end{pmatrix}_{B_1}$, where $B_1=\left\{\begin{pmatrix}1\\1\end{pmatrix},\begin{pmatrix}2\\3\end{pmatrix}\right\}$. Write $\mathbf{x}$ in terms of the basis $B_2=\left\{\begin{pmatrix}0\\3\end{pmatrix},\begin{pmatrix}5\\-1\end{pmatrix}\right\}$.

18. In $\mathbb{R}^2$, $\mathbf{x}=\begin{pmatrix}4\\-1\end{pmatrix}_{B_1}$, where $B_1=\left\{\begin{pmatrix}2\\-5\end{pmatrix},\begin{pmatrix}7\\3\end{pmatrix}\right\}$. Write $\mathbf{x}$ in terms of $B_2=\left\{\begin{pmatrix}-2\\1\end{pmatrix},\begin{pmatrix}-3\\2\end{pmatrix}\right\}$.

19. In $\mathbb{R}^3$, $\mathbf{x}=\begin{pmatrix}2\\-1\\4\end{pmatrix}_{B_1}$, where $B_1=\left\{\begin{pmatrix}1\\-1\\0\end{pmatrix},\begin{pmatrix}0\\1\\-1\end{pmatrix},\begin{pmatrix}1\\0\\1\end{pmatrix}\right\}$. Write $\mathbf{x}$ in terms of $B_2=\left\{\begin{pmatrix}3\\0\\0\end{pmatrix},\begin{pmatrix}1\\2\\-1\end{pmatrix},\begin{pmatrix}0\\1\\5\end{pmatrix}\right\}$.

20. In P_2, $\mathbf{x}=\begin{pmatrix}2\\1\\3\end{pmatrix}_{B_1}$, where $B_1=\{1-x, 3x, x^2-x-1\}$. Write $\mathbf{x}$ in terms of $B_2=\{3-2x, 1+x, x+x^2\}$.

In Problems 21–28 use Theorem 2 to determine whether the given set of vectors is linearly dependent or independent.

21. In P_2: $2+3x+5x^2,\ 1-2x+x^2,\ -1+6x^2$

22. In P_2: $-3+x^2, 2-x+4x^2, 4+2x$

23. In P_2: $x+4x^2, -2+2x, 2+x+12x^2$

24. In P_2: $-2+4x-2x^2, 3+x, 6+8x$

25. In P_3: $1+x^2, -1-3x+4x^2+5x^3, 2+5x-6x^3, 4+6x+3x^2+7x^3$

26. In M_{22}: $\begin{pmatrix}2&0\\3&4\end{pmatrix},\begin{pmatrix}-3&-2\\7&1\end{pmatrix},\begin{pmatrix}1&0\\-1&-3\end{pmatrix},\begin{pmatrix}11&2\\-5&-5\end{pmatrix}$

27. In M_{22}: $\begin{pmatrix}1&-3\\2&4\end{pmatrix},\begin{pmatrix}1&4\\5&0\end{pmatrix},\begin{pmatrix}-1&6\\-1&3\end{pmatrix},\begin{pmatrix}0&0\\3&0\end{pmatrix}$

28. In M_{22}: $\begin{pmatrix}a&0\\0&0\end{pmatrix},\begin{pmatrix}b&c\\0&0\end{pmatrix},\begin{pmatrix}d&e\\f&0\end{pmatrix},\begin{pmatrix}g&h\\j&k\end{pmatrix}$, where $acfk\neq 0$

29. In P_n, let $p_1, p_2, \ldots, p_{n+1}$ be $n+1$ polynomials such that $p_i(0)=0$ for $i=1, 2, \ldots, n+1$. Show that the polynomials are linearly dependent.

*□**30.** In Problem 29 suppose that $p_i^{(j)}=0$ for $i=1, 2, \ldots, n+1$, and for some j with $1\le j\le n$, and $p_i^{(j)}$ denotes the jth derivative of p_i. Show that the polynomials are linearly dependent in P_n.

31. In M_{mn} let $A_1, A_2, \ldots, A_{mn}$ be mn matrices each of whose components in the 1, 1 position is zero. Show that the matrices are linearly dependent.

***32.** Suppose the x- and y-axes in the plane are rotated counterclockwise through an angle of θ (measured in degrees or radians). This gives us new axes which we denote (x', y'). What are the x- and y-coordinates of the now rotated basis vectors $\mathbf{i}$ and $\mathbf{j}$?

33. Show that the "change of coordinates" matrix in Problem 32 is given by $A^{-1}=\begin{pmatrix}\cos\theta&\sin\theta\\-\sin\theta&\cos\theta\end{pmatrix}$.

34. If, in Problems 32 and 33, $\theta = \pi/6 = 30°$, write the vector $\begin{pmatrix} -4 \\ 3 \end{pmatrix}$ in terms of the new coordinate axes x' and y'.

35. If $\theta = \pi/4 = 45°$, write $\begin{pmatrix} 2 \\ -7 \end{pmatrix}$ in terms of the new coordinate axes.

36. If $\theta = 2\pi/3 = 120°$, write $\begin{pmatrix} 4 \\ 5 \end{pmatrix}$ in terms of the new coordinate axes.

37. Let $C = (c_{ij})$ be an $n \times n$ invertible matrix and let $B_1 = \{\mathbf{v}_1, \mathbf{v}_2, \ldots, \mathbf{v}_n\}$ be a basis for a vector space V. Let

$$\mathbf{c}_1 = \begin{pmatrix} c_{11} \\ c_{21} \\ \vdots \\ c_{n1} \end{pmatrix}_{B_1}, \mathbf{c}_2 = \begin{pmatrix} c_{12} \\ c_{22} \\ \vdots \\ c_{n2} \end{pmatrix}_{B_1}, \ldots, \mathbf{c}_n = \begin{pmatrix} c_{1n} \\ c_{2n} \\ \vdots \\ c_{nn} \end{pmatrix}_{B_1}$$

Show that $B_2 = \{\mathbf{c}_1, \mathbf{c}_2, \ldots, \mathbf{c}_n\}$ is a basis for V.

38. Let B_1 and B_2 be bases for the n-dimensional vector space V, and let C be the transition matrix from B_1 to B_2. Show that C^{-1} is the transition matrix from B_2 to B_1.

39. Show that $(\mathbf{x})_{B_1} = CA(\mathbf{x})_{B_1}$ for every $\mathbf{x}$ in a vector space V if and only if $CA = I$ [*Hint:* Let $\mathbf{x}_i$ be the *ith* vector in B_1. Then $(\mathbf{x}_i)_{B_1}$ has a 1 in the *ith* position and a 0 everywhere else. What can you say about $CA(\mathbf{x}_i)_{B_1}$?]

5.7 Orthonormal Bases and Projections in $\mathbb{R}^n$

In $\mathbb{R}^n$ we saw that n linearly independent vectors constitute a basis. The most commonly used basis is the standard basis $E = \{\mathbf{e}_1, \mathbf{e}_2, \ldots, \mathbf{e}_n\}$. These vectors have two properties:

i. $\mathbf{e}_i \cdot \mathbf{e}_j = 0 \quad \text{if } i \neq j$
ii. $\mathbf{e}_i \cdot \mathbf{e}_i = 1$

DEFINITION 1 **ORTHONORMAL SET IN $\mathbb{R}^n$** The set of vectors $S = \{\mathbf{u}_1, \mathbf{u}_2, \ldots, \mathbf{u}_k\}$ in $\mathbb{R}^n$ is said to be an **orthonormal set** if

$$\mathbf{u}_i \cdot \mathbf{u}_j = 0 \quad \text{if} \quad i \neq j \tag{1}$$

$$\mathbf{u}_i \cdot \mathbf{u}_i = 1 \tag{2}$$

If only Equation (1) is satisfied, the set is called **orthogonal**.

Since we shall be working with the scalar product extensively in this section, let us recall some basic facts (see Theorem 2.2.1, page 35). Without mentioning them again explicitly, we shall use these facts often in the rest of this section.

If $\mathbf{u}$, $\mathbf{v}$, and $\mathbf{w}$ are in $\mathbb{R}^n$ and α is a real number, then

$$\mathbf{u}\cdot\mathbf{v}=\mathbf{v}\cdot\mathbf{u} \tag{3}$$

$$(\mathbf{u}+\mathbf{v})\cdot\mathbf{w}=\mathbf{u}\cdot\mathbf{w}+\mathbf{v}\cdot\mathbf{w} \tag{4}$$

$$\mathbf{u}\cdot(\mathbf{v}+\mathbf{w})=\mathbf{u}\cdot\mathbf{v}+\mathbf{u}\cdot\mathbf{w} \tag{5}$$

$$(\alpha\mathbf{u})\cdot\mathbf{v}=\alpha(\mathbf{u}\cdot\mathbf{v}) \tag{6}$$

$$\mathbf{u}\cdot(\alpha\mathbf{v})=\alpha(\mathbf{u}\cdot\mathbf{v}) \tag{7}$$

DEFINITION 2 **LENGTH OR NORM OF A VECTOR** We now give another useful definition. If $\mathbf{v}\in\mathbb{R}^n$, then the **length** or **norm** of $\mathbf{v}$, written $|\mathbf{v}|$, is given by

$$|\mathbf{v}|=\sqrt{\mathbf{v}\cdot\mathbf{v}} \tag{8}$$

Note. If $\mathbf{v}=(x_1, x_2, \ldots, x_n)$, then $\mathbf{v}\cdot\mathbf{v}=x_1^2+x_2^2+\cdots+x_n^2$. This means that

$$\mathbf{v}\cdot\mathbf{v}\geq 0 \quad \text{and} \quad \mathbf{v}\cdot\mathbf{v}=0 \quad \text{if and only if } \mathbf{v}=\mathbf{0} \tag{9}$$

Thus we can take the square root in (8) and we have

$$|\mathbf{v}|=\sqrt{\mathbf{v}\cdot\mathbf{v}}\geq 0 \qquad \text{for every } \mathbf{v}\in\mathbb{R}^n \tag{10}$$

$$|\mathbf{v}|=0 \qquad \text{if and only if } \mathbf{v}=\mathbf{0} \tag{11}$$

EXAMPLE 1 Let $\mathbf{v}=(x, y)\in\mathbb{R}^2$. Then $|\mathbf{v}|=\sqrt{x^2+y^2}$ conforms to our usual definition of length of a vector in the plane (see Equation 4.1.1, page 120).

EXAMPLE 2 If $\mathbf{v}=(x, y, z)\in\mathbb{R}^3$, then $|\mathbf{v}|=\sqrt{x^2+y^2+z^2}$ as in Section 4.3.

EXAMPLE 3 If $\mathbf{v}=(2, -1, 3, 4, -6)\in\mathbb{R}^5$, then $|\mathbf{v}|=\sqrt{4+1+9+16+36}=\sqrt{66}$.

We can now restate Definition 1:

> A set of vectors is orthonormal if any pair of them is orthogonal and each has length 1.

Orthonormal sets of vectors are reasonably easy to work with. We shall see an example of this characteristic in Chapter 6. Now we prove that any finite orthogonal set of nonzero vectors is linearly independent.

THEOREM 1 If $S=\{\mathbf{v}_1, \mathbf{v}_2, \ldots, \mathbf{v}_k\}$ is an orthogonal set of nonzero vectors, then S is linearly independent.

Proof Suppose that $c_1\mathbf{v}_1+c_2\mathbf{v}_2+\cdots+c_n\mathbf{v}_n=\mathbf{0}$. Then, for any $i=1, 2, \ldots, k$,

$$\begin{aligned} 0=\mathbf{0}\cdot\mathbf{v}_i &= (c_1\mathbf{v}_1+c_2\mathbf{v}_2+\cdots+c_i\mathbf{v}_i+\cdots+c_n\mathbf{v}_n)\cdot\mathbf{v}_i \\ &= c_1(\mathbf{v}_1\cdot\mathbf{v}_i)+c_2(\mathbf{v}_2\cdot\mathbf{v}_i)+\cdots+c_i(\mathbf{v}_i\cdot\mathbf{v}_i)+\cdots+c_n(\mathbf{v}_n\cdot\mathbf{v}_i) \\ &= c_1 0+c_2 0+\cdots+c_i\,|\mathbf{v}_i|^2+\cdots+c_n 0=c_i\,|\mathbf{v}_i|^2 \end{aligned}$$

Since $\mathbf{v}_i\neq 0$ by hypothesis, $|\mathbf{v}_i|^2>0$ and we have $c_i=0$. This is true for $i=1, 2, \ldots, k$ and the proof is complete. ■

We shall now see how *any* basis in $\mathbb{R}^n$ can be "turned into" an orthonormal basis. The method described below is called the **Gram–Schmidt orthonormalization process.**†

THEOREM 2 **GRAM–SCHMIDT ORTHONORMALIZATION PROCESS** Let H be an m-dimensional subspace of $\mathbb{R}^n$. Then H has an orthonormal basis.‡

Proof Let $S=\{\mathbf{v}_1, \mathbf{v}_2, \ldots, \mathbf{v}_m\}$ be a basis for H. We shall prove the theorem by constructing an orthonormal basis from the vectors in S. Before giving the steps in this construction, we note the simple fact that a linearly independent set of vectors does *not* contain the zero vector (see Problem 21).

Step 1. Let

$$\mathbf{u}_1=\frac{\mathbf{v}_1}{|\mathbf{v}_1|} \tag{12}$$

Then $\mathbf{u}_1\cdot\mathbf{u}_1=(\mathbf{v}_1/|\mathbf{v}_1|)\cdot(\mathbf{v}_1/|\mathbf{v}_1|)=(1/|\mathbf{v}_1|^2)(\mathbf{v}_1\cdot\mathbf{v}_1)=1$, so that $|\mathbf{u}_1|=1$.

Step 2. Let

$$\mathbf{v}_2'=\mathbf{v}_2-(\mathbf{v}_2\cdot\mathbf{u}_1)\mathbf{u}_1 \tag{13}$$

Then $\mathbf{v}_2'\cdot\mathbf{u}_1=\mathbf{v}_2\cdot\mathbf{u}_1-(\mathbf{v}_2\cdot\mathbf{u}_1)(\mathbf{u}_1\cdot\mathbf{u}_1)=\mathbf{v}_2\cdot\mathbf{u}_1-\mathbf{v}_2\cdot\mathbf{u}_1=0$, so that $\mathbf{v}_2'$ is orthogonal to $\mathbf{u}_1$. Moreover, by Theorem 1, $\mathbf{u}_1$ and $\mathbf{v}_2'$ are linearly independent and therefore, again by Problem 21, $\mathbf{v}_2'\neq\mathbf{0}$.

Step 3. Let

$$\mathbf{u}_2=\frac{\mathbf{v}_2'}{|\mathbf{v}_2'|} \tag{14}$$

Then clearly $\{\mathbf{u}_1, \mathbf{u}_2\}$ is an orthonormal set.

---

† Jörgen Pederson Gram (1850–1916) was a Danish actuary who was very interested in the science of measurement. Erhardt Schmidt (1876–1959) was a German mathematician.

‡ Note that H may be $\mathbb{R}^n$ in this theorem. That is, $\mathbb{R}^n$ itself has an orthonormal basis.

Suppose now that the vectors $\mathbf{u}_1, \mathbf{u}_2, \ldots, \mathbf{u}_k$ $(k<m)$ have been constructed and form an orthonormal set. We show how to construct $\mathbf{u}_{k+1}$.

Step 4. Let

$$\mathbf{v}'_{k+1} = \mathbf{v}_{k+1} - (\mathbf{v}_{k+1}\cdot\mathbf{u}_1)\mathbf{u}_1 - (\mathbf{v}_{k+1}\cdot\mathbf{u}_2)\mathbf{u}_2 - \cdots - (\mathbf{v}_{k+1}\cdot\mathbf{u}_k)\mathbf{u}_k \tag{15}$$

Then, for $i = 1, 2, \ldots, k$,

$$\begin{aligned}\mathbf{v}'_{k+1}\cdot\mathbf{u}_i &= \mathbf{v}_{k+1}\cdot\mathbf{u}_i - (\mathbf{v}_{k+1}\cdot\mathbf{u}_1)(\mathbf{u}_1\cdot\mathbf{u}_i) - (\mathbf{v}_{k+1}\cdot\mathbf{u}_2)(\mathbf{u}_2\cdot\mathbf{u}_i)\\ &\quad - \cdots - (\mathbf{v}_{k+1}\cdot\mathbf{u}_i)(\mathbf{u}_i\cdot\mathbf{u}_i) - \cdots - (\mathbf{v}_{k+1}\cdot\mathbf{u}_k)(\mathbf{u}_k\cdot\mathbf{u}_i)\end{aligned}$$

But $\mathbf{u}_j\cdot\mathbf{u}_i = 0$ if $j \neq i$ and $\mathbf{u}_i\cdot\mathbf{u}_i = 1$. Thus

$$\mathbf{v}'_{k+1}\cdot\mathbf{u}_i = \mathbf{v}_{k+1}\cdot\mathbf{u}_i - \mathbf{v}_{k+1}\cdot\mathbf{u}_i = 0$$

Hence $\{\mathbf{u}_1, \mathbf{u}_2, \ldots, \mathbf{u}_k, \mathbf{v}'_{k+1}\}$ is an orthogonal, linearly independent set and $\mathbf{v}'_{k+1} \neq \mathbf{0}$.

Step 5. Let $\mathbf{u}_{k+1} = \mathbf{v}'_{k+1}/|\mathbf{v}'_{k+1}|$. Then clearly $\{\mathbf{u}_1, \mathbf{u}_2, \ldots, \mathbf{u}_k, \mathbf{u}_{k+1}\}$ is an orthonormal set and we continue in this manner until $k+1 = m$ and the proof is complete. ■

EXAMPLE 4 Find an orthonormal basis for the set of vectors in $\mathbb{R}^3$ lying on the plane $\pi = \left\{\begin{pmatrix}x\\y\\z\end{pmatrix}: \ 2x - y + 3z = 0\right\}$.

Solution As we saw in Example 5.5.3, page 193, a basis for this two-dimensional subspace is $\mathbf{v}_1 = \begin{pmatrix}1\\2\\0\end{pmatrix}$ and $\mathbf{v}_2 = \begin{pmatrix}0\\3\\1\end{pmatrix}$. Then $|\mathbf{v}_1| = \sqrt{5}$ and $\mathbf{u}_1 = \mathbf{v}_1/|\mathbf{v}_1| = \begin{pmatrix}1/\sqrt{5}\\2/\sqrt{5}\\0\end{pmatrix}$.

Continuing, we define

$$\begin{aligned}\mathbf{v}'_2 &= \mathbf{v}_2 - (\mathbf{v}_2\cdot\mathbf{u}_1)\mathbf{u}_1\\ &= \begin{pmatrix}0\\3\\1\end{pmatrix} - (6/\sqrt{5})\begin{pmatrix}1/\sqrt{5}\\2/\sqrt{5}\\0\end{pmatrix} = \begin{pmatrix}0\\3\\1\end{pmatrix} - \begin{pmatrix}6/5\\12/5\\0\end{pmatrix} = \begin{pmatrix}-6/5\\3/5\\1\end{pmatrix}\end{aligned}$$

Finally $|\mathbf{v}'_2| = \sqrt{70/25} = \sqrt{70}/5$, so that $\mathbf{u}_2 = \mathbf{v}'_2/|\mathbf{v}'_2| = (5/\sqrt{70})\begin{pmatrix}-6/5\\3/5\\1\end{pmatrix} = \begin{pmatrix}-6/\sqrt{70}\\3/\sqrt{70}\\5/\sqrt{70}\end{pmatrix}$. Thus the orthonormal basis is $\left\{\begin{pmatrix}1/\sqrt{5}\\2/\sqrt{5}\\0\end{pmatrix}, \begin{pmatrix}-6/\sqrt{70}\\3/\sqrt{70}\\5/\sqrt{70}\end{pmatrix}\right\}$. To check this answer we note that (i) the vectors are orthogonal, (ii) each has length 1, and (iii) each satisfies $2x - y + 3z = 0$.

Remark. We can see, geometrically, what is happening here. First we note that

$$\mathbf{v}_2' = \mathbf{v}_2 - (\mathbf{v}_2 \cdot \mathbf{u}_1)\mathbf{u}_1 = \mathbf{v}_2 - \left(\mathbf{v}_2 \cdot \frac{\mathbf{v}_1}{|\mathbf{v}_1|}\right)\left(\frac{\mathbf{v}_1}{|\mathbf{v}_1|}\right) = \mathbf{v}_2 - \frac{(\mathbf{v}_2 \cdot \mathbf{v}_1)}{|\mathbf{v}_1|^2}\mathbf{v}_1$$

But, from Definitions 4.2.4, page 131, (in $\mathbb{R}^2$) and 4.3.3, page 148, (in $\mathbb{R}^3$), $[(\mathbf{v}_2 \cdot \mathbf{v}_1)/|\mathbf{v}_1|^2]\mathbf{v}_1$ is the projection of $\mathbf{v}_2$ on $\mathbf{v}_1$. Moreover, from Figure 4.19, page 131, the vector $\mathbf{v}_2' = \mathbf{v}_2 - \text{proj}_{\mathbf{v}_1} \mathbf{v}_2$ is a vector orthogonal to $\mathbf{v}_1$.

Thus we see that in a certain sense the process we have described here is really a generalization of the notion of projection in $\mathbb{R}^2$ and $\mathbb{R}^3$.

EXAMPLE 5

Construct an orthonormal basis in $\mathbb{R}^3$ starting with the basis $\{\mathbf{v}_1, \mathbf{v}_2, \mathbf{v}_3\} = \left\{\begin{pmatrix}1\\1\\0\end{pmatrix}, \begin{pmatrix}0\\1\\1\end{pmatrix}, \begin{pmatrix}1\\0\\1\end{pmatrix}\right\}$.

Solution We have $|\mathbf{v}_1| = \sqrt{2}$, so $\mathbf{u}_1 = \begin{pmatrix}1/\sqrt{2}\\1/\sqrt{2}\\0\end{pmatrix}$. Then

$$\mathbf{v}_2' = \mathbf{v}_2 - (\mathbf{v}_2 \cdot \mathbf{u}_1)\mathbf{u}_1 = \begin{pmatrix}0\\1\\1\end{pmatrix} - \frac{1}{\sqrt{2}}\begin{pmatrix}1/\sqrt{2}\\1/\sqrt{2}\\0\end{pmatrix} = \begin{pmatrix}0\\1\\1\end{pmatrix} - \begin{pmatrix}1/2\\1/2\\0\end{pmatrix} = \begin{pmatrix}-1/2\\1/2\\1\end{pmatrix}.$$

Since $|\mathbf{v}_2'| = \sqrt{3/2}$, $|\mathbf{u}_2| = \sqrt{2/3}\begin{pmatrix}-1/2\\1/2\\1\end{pmatrix} = \begin{pmatrix}-1/\sqrt{6}\\1/\sqrt{6}\\2/\sqrt{6}\end{pmatrix}$. Continuing, we have

$$\mathbf{v}_3' = \mathbf{v}_3 - (\mathbf{v}_3 \cdot \mathbf{u}_1)\mathbf{u}_1 - (\mathbf{v}_3 \cdot \mathbf{u}_2)\mathbf{u}_2$$

$$= \begin{pmatrix}1\\0\\1\end{pmatrix} - \frac{1}{\sqrt{2}}\begin{pmatrix}1/\sqrt{2}\\1/\sqrt{2}\\0\end{pmatrix} - \frac{1}{\sqrt{6}}\begin{pmatrix}-1/\sqrt{6}\\1/\sqrt{6}\\2/\sqrt{6}\end{pmatrix} = \begin{pmatrix}1\\0\\1\end{pmatrix} - \begin{pmatrix}1/2\\1/2\\0\end{pmatrix} - \begin{pmatrix}-1/6\\1/6\\2/6\end{pmatrix} = \begin{pmatrix}2/3\\-2/3\\2/3\end{pmatrix}$$

Finally $|\mathbf{v}_3'| = \sqrt{\frac{12}{9}} = 2/\sqrt{3}$, so that $\mathbf{u}_3 = \frac{\sqrt{3}}{2}\begin{pmatrix}2/3\\-2/3\\2/3\end{pmatrix} = \begin{pmatrix}1/\sqrt{3}\\-1/\sqrt{3}\\1/\sqrt{3}\end{pmatrix}$. Thus the orthonormal basis is $\left\{\begin{pmatrix}1/\sqrt{2}\\1/\sqrt{2}\\0\end{pmatrix}, \begin{pmatrix}-1/\sqrt{6}\\1/\sqrt{6}\\2/\sqrt{6}\end{pmatrix}, \begin{pmatrix}1/\sqrt{3}\\-1/\sqrt{3}\\1/\sqrt{3}\end{pmatrix}\right\}$. As in the previous example, this result should be checked.

We now define a new kind of matrix that will be very useful in later chapters.

DEFINITION 3 **ORTHOGONAL MATRIX** The $n \times n$ matrix Q is called **orthogonal** if Q is invertible and

$$Q^{-1} = Q^t \tag{16}$$

Orthogonal matrices are not difficult to find, according to the next theorem.

THEOREM 3 The $n \times n$ matrix Q is orthogonal if and only if the columns of Q form an orthonormal basis for $\mathbb{R}^n$.

Proof Let

$$Q = \begin{pmatrix} a_{11} & a_{12} & \cdots & a_{1n} \\ a_{21} & a_{22} & \cdots & a_{2n} \\ \vdots & \vdots & & \vdots \\ a_{n1} & a_{n2} & \cdots & a_{nn} \end{pmatrix}$$

Then

$$Q^t = \begin{pmatrix} a_{11} & a_{21} & \cdots & a_{n1} \\ a_{12} & a_{22} & \cdots & a_{n2} \\ \vdots & \vdots & & \vdots \\ a_{1n} & a_{2n} & \cdots & a_{nn} \end{pmatrix}$$

Let $B = (b_{ij}) = Q^tQ$. Then

$$b_{ij} = a_{1i}a_{1j} + a_{2i}a_{2j} + \cdots + a_{ni}a_{nj} = \mathbf{c}_i \cdot \mathbf{c}_j \tag{17}$$

where $\mathbf{c}_i$ denotes the ith column of Q. If the columns of Q are orthonormal, then

$$b_{ij} = \begin{cases} 0 & \text{if } i \neq j \\ 1 & \text{if } i = j \end{cases} \tag{18}$$

That is, $B = I$. Conversely if $Q^t = Q^{-1}$, then $B = I$, so that (18) holds and (17) shows that the columns of Q are orthonormal. This completes the proof. ■

EXAMPLE 6 From Example 5, the vectors $\begin{pmatrix} 1/\sqrt{2} \\ 1/\sqrt{2} \\ 0 \end{pmatrix}$, $\begin{pmatrix} -1/\sqrt{6} \\ 1/\sqrt{6} \\ 2/\sqrt{6} \end{pmatrix}$, $\begin{pmatrix} 1/\sqrt{3} \\ -1/\sqrt{3} \\ 1/\sqrt{3} \end{pmatrix}$ form an orthonormal basis in $\mathbb{R}^3$. Thus the matrix $Q = \begin{pmatrix} 1/\sqrt{2} & -1/\sqrt{6} & 1/\sqrt{3} \\ 1/\sqrt{2} & 1/\sqrt{6} & -1/\sqrt{3} \\ 0 & 2/\sqrt{6} & 1/\sqrt{3} \end{pmatrix}$ is an orthogonal matrix. To check this we note that

$$Q^tQ = \begin{pmatrix} 1/\sqrt{2} & 1/\sqrt{2} & 0 \\ -1/\sqrt{6} & 1/\sqrt{6} & 2/\sqrt{6} \\ 1/\sqrt{3} & -1/\sqrt{3} & 1/\sqrt{3} \end{pmatrix} \begin{pmatrix} 1/\sqrt{2} & -1/\sqrt{6} & 1/\sqrt{3} \\ 1/\sqrt{2} & 1/\sqrt{6} & -1/\sqrt{3} \\ 0 & 2/\sqrt{6} & 1/\sqrt{3} \end{pmatrix} = \begin{pmatrix} 1 & 0 & 0 \\ 0 & 1 & 0 \\ 0 & 0 & 1 \end{pmatrix}.$$

In the proof of Theorem 2 we defined $\mathbf{v}_2' = \mathbf{v}_2 - (\mathbf{v}_2 \cdot \mathbf{u}_1)\mathbf{u}_1$. But, as we have seen, $(\mathbf{v}_2 \cdot \mathbf{u}_1)\mathbf{u}_1 = \text{proj}_{\mathbf{u}_1} \mathbf{v}_2$ (since $|\mathbf{u}_1|^2 = 1$). We now extend this notion from projection onto a vector to projection onto a subspace.

DEFINITION 4 **ORTHOGONAL PROJECTION** Let H be a subspace of $\mathbb{R}^n$ with orthonormal basis $\{\mathbf{u}_1, \mathbf{u}_2, \ldots, \mathbf{u}_k\}$. If $\mathbf{v} \in \mathbb{R}^n$, then the **orthogonal projection** of $\mathbf{v}$ onto H, denoted $\text{proj}_H \mathbf{v}$, is given by

$$\text{proj}_H \mathbf{v} = (\mathbf{v} \cdot \mathbf{u}_1)\mathbf{u}_1 + (\mathbf{v} \cdot \mathbf{u}_2)\mathbf{u}_2 + \cdots + (\mathbf{v} \cdot \mathbf{u}_k)\mathbf{u}_k \tag{19}$$

Note that $\text{proj}_H \mathbf{v} \in H$.

EXAMPLE 7 Find $\text{proj}_\pi \mathbf{v}$, where π is the plane $\left\{\begin{pmatrix} x \\ y \\ z \end{pmatrix} : 2x - y + 3z = 0\right\}$ and $\mathbf{v}$ is the vector $\begin{pmatrix} 3 \\ -2 \\ 4 \end{pmatrix}$.

Solution From Example 4, an orthonormal basis for π is $\mathbf{u}_1 = \begin{pmatrix} 1/\sqrt{5} \\ 2/\sqrt{5} \\ 0 \end{pmatrix}$ and $\mathbf{u}_2 = \begin{pmatrix} -6/\sqrt{70} \\ 3/\sqrt{70} \\ 5/\sqrt{70} \end{pmatrix}$. Then

$$\text{proj}_\pi \mathbf{v} = \left[\begin{pmatrix} 3 \\ -2 \\ 4 \end{pmatrix} \cdot \begin{pmatrix} 1/\sqrt{5} \\ 2/\sqrt{5} \\ 0 \end{pmatrix}\right]\begin{pmatrix} 1/\sqrt{5} \\ 2/\sqrt{5} \\ 0 \end{pmatrix} + \left[\begin{pmatrix} 3 \\ -2 \\ 4 \end{pmatrix} \cdot \begin{pmatrix} -6/\sqrt{70} \\ 3/\sqrt{70} \\ 5/\sqrt{70} \end{pmatrix}\right]\begin{pmatrix} -6/\sqrt{70} \\ 3/\sqrt{70} \\ 5/\sqrt{70} \end{pmatrix}$$

$$= -\frac{1}{\sqrt{5}}\begin{pmatrix} 1/\sqrt{5} \\ 2/\sqrt{5} \\ 0 \end{pmatrix} - \frac{4}{\sqrt{70}}\begin{pmatrix} -6/\sqrt{70} \\ 3/\sqrt{70} \\ 5/\sqrt{70} \end{pmatrix} = \begin{pmatrix} -1/5 \\ -2/5 \\ 0 \end{pmatrix} + \begin{pmatrix} 24/70 \\ -12/70 \\ -20/70 \end{pmatrix} = \begin{pmatrix} 1/7 \\ -4/7 \\ -2/7 \end{pmatrix}$$

The notion of projection gives us a convenient way to write a vector in $\mathbb{R}^n$ in terms of an orthonormal basis.

THEOREM 4 Let $B = \{\mathbf{u}_1, \mathbf{u}_2, \ldots, \mathbf{u}_n\}$ be an orthonormal basis for $\mathbb{R}^n$ and let $\mathbf{v} \in \mathbb{R}^n$. Then

$$\mathbf{v} = (\mathbf{v} \cdot \mathbf{u}_1)\mathbf{u}_1 + (\mathbf{v} \cdot \mathbf{u}_2)\mathbf{u}_2 + \cdots + (\mathbf{v} \cdot \mathbf{u}_n)\mathbf{u}_n \tag{20}$$

That is, $\mathbf{v} = \text{proj}_{\mathbb{R}^n} \mathbf{v}$.

Proof Since B is a basis, we can write $\mathbf{v}$ in a unique way as $\mathbf{v}=c_1\mathbf{u}_1+c_2\mathbf{u}_2+\cdots+c_n\mathbf{u}_n$. Then

$$\mathbf{v}\cdot\mathbf{u}_i=c_1(\mathbf{u}_1\cdot\mathbf{u}_i)+c_2(\mathbf{u}_2\cdot\mathbf{u}_i)+\cdots+c_i(\mathbf{u}_i\cdot\mathbf{u}_i)+\cdots+c_n(\mathbf{u}_n\cdot\mathbf{u}_i)=c_i$$

since the $\mathbf{u}_i$'s are orthonormal. Since this is true for $i=1,2,\ldots,n$, the proof is complete. ■

EXAMPLE 8 Write the vector $\begin{pmatrix}2\\-1\\3\end{pmatrix}$ in $\mathbb{R}^3$ in terms of the orthonormal basis $\left\{\begin{pmatrix}1/\sqrt{2}\\1/\sqrt{2}\\0\end{pmatrix},\begin{pmatrix}-1/\sqrt{6}\\1/\sqrt{6}\\2/\sqrt{6}\end{pmatrix},\begin{pmatrix}1/\sqrt{3}\\-1/\sqrt{3}\\1/\sqrt{3}\end{pmatrix}\right\}$.

Solution

$$\begin{pmatrix}2\\-1\\3\end{pmatrix}=\left[\begin{pmatrix}2\\-1\\3\end{pmatrix}\cdot\begin{pmatrix}1/\sqrt{2}\\1/\sqrt{2}\\0\end{pmatrix}\right]\begin{pmatrix}1/\sqrt{2}\\1/\sqrt{2}\\0\end{pmatrix}+\left[\begin{pmatrix}2\\-1\\3\end{pmatrix}\cdot\begin{pmatrix}-1/\sqrt{6}\\1/\sqrt{6}\\2/\sqrt{6}\end{pmatrix}\right]\begin{pmatrix}-1/\sqrt{6}\\1/\sqrt{6}\\2/\sqrt{6}\end{pmatrix}$$

$$+\left[\begin{pmatrix}2\\-1\\3\end{pmatrix}\cdot\begin{pmatrix}1/\sqrt{3}\\-1/\sqrt{3}\\1/\sqrt{3}\end{pmatrix}\right]\begin{pmatrix}1/\sqrt{3}\\-1/\sqrt{3}\\1/\sqrt{3}\end{pmatrix}=\frac{1}{\sqrt{2}}\begin{pmatrix}1/\sqrt{2}\\1/\sqrt{2}\\0\end{pmatrix}+\frac{3}{\sqrt{6}}\begin{pmatrix}-1/\sqrt{6}\\1/\sqrt{6}\\2/\sqrt{6}\end{pmatrix}+\frac{6}{\sqrt{3}}\begin{pmatrix}1/\sqrt{3}\\-1/\sqrt{3}\\1/\sqrt{3}\end{pmatrix}$$

DEFINITION 5 **ORTHOGONAL COMPLEMENT** Let H be a subspace of $\mathbb{R}^n$. Then the **orthogonal complement** of H, denoted $H^\perp$, is given by

$$H^\perp=\{\mathbf{x}\in\mathbb{R}^n:\quad \mathbf{x}\cdot\mathbf{h}=0\quad\text{for every}\quad \mathbf{h}\in H\} \tag{21}$$

THEOREM 5 If H is a subspace of $\mathbb{R}^n$, then:

i. $H^\perp$ is a subspace of $\mathbb{R}^n$.
ii. $H\cap H^\perp=\{\mathbf{0}\}$.
iii. $\dim H^\perp=n-\dim H$.

Proof

i. If $\mathbf{x}$ and $\mathbf{y}$ are in $H^\perp$ and if $\mathbf{h}\in H$, then $(\mathbf{x}+\mathbf{y})\cdot\mathbf{h}=\mathbf{x}\cdot\mathbf{h}+\mathbf{y}\cdot\mathbf{h}=0+0=0$ and $(\alpha\mathbf{x}\cdot\mathbf{h})=\alpha(\mathbf{x}\cdot\mathbf{h})=0$, so $H^\perp$ is a subspace.
ii. If $\mathbf{x}\in H\cap H^\perp$, then $\mathbf{x}\cdot\mathbf{x}=0$, so $\mathbf{x}=\mathbf{0}$, which shows that $H\cap H^\perp=\{\mathbf{0}\}$.
iii. Let $\{\mathbf{u}_1,\mathbf{u}_2,\ldots,\mathbf{u}_k\}$ be an orthonormal basis for H. By the result of Problem 5.5.27, page 200, this can be expanded into a basis B for $\mathbb{R}^n$: $B=\{\mathbf{u}_1,\mathbf{u}_2,\ldots,\mathbf{u}_k,\mathbf{v}_{k+1},\ldots,\mathbf{v}_n\}$. Using the Gram–Schmidt process, we can turn B into an orthonormal basis for $\mathbb{R}^n$. As in the proof of Theorem 2,

the already orthonormal $\mathbf{u}_1, \mathbf{u}_2, \ldots, \mathbf{u}_k$ will remain unchanged in the process and we obtain the orthonormal basis $B_1 = \{\mathbf{u}_1, \mathbf{u}_2, \ldots, \mathbf{u}_k, \mathbf{u}_{k+1}, \ldots, \mathbf{u}_n\}$. To complete the proof we need only show that $\{\mathbf{u}_{k+1}, \ldots, \mathbf{u}_n\}$ is a basis for $H^\perp$. Since the $\mathbf{u}_i$'s are independent, we must show that they span $H^\perp$. Let $\mathbf{x} \in H^\perp$; then, by Theorem 4,

$$\begin{aligned}\mathbf{x} = (\mathbf{x} \cdot \mathbf{u}_1)\mathbf{u}_1 + (\mathbf{x} \cdot \mathbf{u}_2)\mathbf{u}_2 + \cdots + (\mathbf{x} \cdot \mathbf{u}_k)\mathbf{u}_k \\ + (\mathbf{x} \cdot \mathbf{u}_{k+1})\mathbf{u}_{k+1} + \cdots + (\mathbf{x} \cdot \mathbf{u}_n)\mathbf{u}_n\end{aligned}$$

But $(\mathbf{x} \cdot \mathbf{u}_i) = 0$ for $i = 1, 2, \ldots, k$, since $\mathbf{x} \in H^\perp$ and $\mathbf{u}_i \in H$. Thus $\mathbf{x} = (\mathbf{x} \cdot \mathbf{u}_{k+1})\mathbf{u}_{k+1} + \cdots + (\mathbf{x} \cdot \mathbf{u}_n)\mathbf{u}_n$. This shows that $\{\mathbf{u}_{k+1}, \ldots, \mathbf{u}_n\}$ is a basis for $H^\perp$ which means that $\dim H^\perp = n - k$. ■

The spaces H and $H^\perp$ allow us to "decompose" any vector in $\mathbb{R}^n$.

THEOREM 6 Let H be a subspace of $\mathbb{R}^n$ and let $\mathbf{v} \in \mathbb{R}^n$. Then there exists a unique pair of vectors $\mathbf{h}$ and $\mathbf{p}$ such that $\mathbf{h} \in H$, $\mathbf{p} \in H^\perp$, and

$$\mathbf{v} = \mathbf{h} + \mathbf{p} = \text{proj}_H\, \mathbf{v} + \text{proj}_{H^\perp}\, \mathbf{v} \tag{22}$$

Proof Let $\mathbf{h} = \text{proj}_H\, \mathbf{v}$ and let $\mathbf{p} = \mathbf{v} - \mathbf{h}$. By Definition 4 we have $\mathbf{h} \in H$. We now show that $\mathbf{p} \in H^\perp$. As in the proof of Theorem 5, we can find an orthonormal basis $\{\mathbf{u}_1, \mathbf{u}_2, \ldots, \mathbf{u}_k, \mathbf{u}_{k+1}, \ldots, \mathbf{u}_n\}$ for $\mathbb{R}^n$ such that $\{\mathbf{u}_1, \mathbf{u}_2, \ldots, \mathbf{u}_k\}$ is a basis for H and $\{\mathbf{u}_{k+1}, \ldots, \mathbf{u}_n\}$ is a basis for $H^\perp$. Then, by Theorem 4,

$$\begin{aligned}\mathbf{v} &= (\mathbf{v} \cdot \mathbf{u}_1)\mathbf{u}_1 + (\mathbf{v} \cdot \mathbf{u}_2)\mathbf{u}_2 + \cdots + (\mathbf{v} \cdot \mathbf{u}_k)\mathbf{u}_k + (\mathbf{v} \cdot \mathbf{u}_{k+1})\mathbf{u}_{k+1} \\ &\quad + \cdots + (\mathbf{v} \cdot \mathbf{u}_n)\mathbf{u}_n \\ &= \text{proj}_H\, \mathbf{v} + \text{proj}_{H^\perp}\, \mathbf{v} \qquad \text{(by Definition 4)}\end{aligned}$$

This proves Equation (22). To prove uniqueness, suppose that $\mathbf{v} = \mathbf{h}_1 - \mathbf{p}_1 = \mathbf{h}_2 - \mathbf{p}_2$, where $\mathbf{h}_1, \mathbf{h}_2 \in H$ and $\mathbf{p}_1, \mathbf{p}_2 \in H^\perp$. Then $\mathbf{h}_1 - \mathbf{h}_2 = \mathbf{p}_1 - \mathbf{p}_2$. But $\mathbf{h}_1 - \mathbf{h}_2 \in H$ and $\mathbf{p}_1 - \mathbf{p}_2 \in H^\perp$, so $\mathbf{h}_1 - \mathbf{h}_2 \in H \cap H^\perp = \{\mathbf{0}\}$. Thus $\mathbf{h}_1 - \mathbf{h}_2 = \mathbf{0}$ and $\mathbf{p}_1 - \mathbf{p}_2 = \mathbf{0}$, which completes the proof. ■

EXAMPLE 9 In $\mathbb{R}^3$ let $\pi = \left\{ \begin{pmatrix} x \\ y \\ z \end{pmatrix} : \; 2x - y + 3z = 0 \right\}$. Write the vector $\begin{pmatrix} 3 \\ -2 \\ 4 \end{pmatrix}$ as $\mathbf{h} + \mathbf{p}$, where $\mathbf{h} \in H$ and $\mathbf{p} \in H^\perp$.

Solution An orthonormal basis for π is $B_1 = \left\{ \begin{pmatrix} 1/\sqrt{5} \\ 2/\sqrt{5} \\ 0 \end{pmatrix}, \begin{pmatrix} -6/\sqrt{70} \\ 3/\sqrt{70} \\ 5/\sqrt{70} \end{pmatrix} \right\}$ and, from Example 7, $\text{proj}_\pi \mathbf{v} = \begin{pmatrix} 1/7 \\ -4/7 \\ -2/7 \end{pmatrix} \in \pi$. Then

$$\mathbf{p} = \mathbf{v} - \mathbf{h} = \begin{pmatrix} 3 \\ -2 \\ 4 \end{pmatrix} - \begin{pmatrix} \frac{1}{7} \\ -\frac{4}{7} \\ -\frac{2}{7} \end{pmatrix} = \begin{pmatrix} \frac{20}{7} \\ -\frac{10}{7} \\ \frac{30}{7} \end{pmatrix}$$

Note that $\mathbf{p} \cdot \mathbf{h} = 0$.

PROBLEMS 5.7

In Problems 1–13 construct an orthonormal basis for the given vector space or subspace.

1. In $\mathbb{R}^2$, starting with the basis vectors $\begin{pmatrix} 1 \\ 1 \end{pmatrix}, \begin{pmatrix} -1 \\ 1 \end{pmatrix}$
2. $H = \{(x, y) \in \mathbb{R}^2\colon\ x + y = 0\}$
3. $H = \{(x, y) \in \mathbb{R}^2\colon\ ax + by = 0\}$
4. In $\mathbb{R}^2$, starting with $\begin{pmatrix} a \\ b \end{pmatrix}, \begin{pmatrix} c \\ d \end{pmatrix}$, where $ad - bc \neq 0$.
5. $\pi = \{(x, y, z)\colon\ 2x - y - z = 0\}$
6. $\pi = \{(x, y, z)\colon\ 3x - 2y + 6z = 0\}$
7. $L = \{(x, y, z)\colon\ x/2 = y/3 = z/4\}$
8. $L = \{(x, y, z)\colon\ x = 3t,\ y = -2t,\ z = t;\ t \text{ real}\}$
9. $H = \{(x, y, z, w) \in \mathbb{R}^4\colon\ 2x - y + 3z - w = 0\}$
10. $\pi = \{(x, y, z)\colon\ ax + by + cz = 0\}$, where $abc \neq 0$
11. $L = \{(x, y, z)\colon\ x/a = y/b = z/c\}\ z/c\}$, where $abc \neq 0$.
12. $H = \{(x_1, x_2, x_3, x_4, x_5) \in \mathbb{R}^5\colon\ 2x_1 - 3x_2 + x_3 + 4x_4 - x_5 = 0\}$
13. H is the solution space of

$$\begin{aligned} x - 3y + \ z &= 0 \\ -2x + 2y - 3z &= 0 \\ 4x - 8y + 5z &= 0 \end{aligned}$$

*14. Find an orthonormal basis in $\mathbb{R}^4$ that includes the vectors

$$\mathbf{u}_1 = \begin{pmatrix} 1/\sqrt{2} \\ 0 \\ 1/\sqrt{2} \\ 0 \end{pmatrix} \quad \text{and} \quad \mathbf{u}_2 = \begin{pmatrix} -1/2 \\ 1/2 \\ 1/2 \\ -1/2 \end{pmatrix}$$

[*Hint:* First find two vectors $\mathbf{v}_3$ and $\mathbf{v}_4$ to complete the basis.]

15. Show that $Q = \begin{pmatrix} 2/3 & 1/3 & 2/3 \\ 1/3 & 2/3 & -2/3 \\ -2/3 & 2/3 & 1/3 \end{pmatrix}$ is an orthogonal matrix.
16. Show that if P and Q are orthogonal $n \times n$ matrices, then PQ is orthogonal.
17. Verify the result of Problem 16 with

$$P = \begin{pmatrix} 1/\sqrt{2} & -1/\sqrt{2} \\ 1/\sqrt{2} & 1/\sqrt{2} \end{pmatrix} \quad \text{and} \quad Q = \begin{pmatrix} 1/3 & -\sqrt{8}/3 \\ \sqrt{8}/3 & 1/3 \end{pmatrix}$$

18. Show that if Q is a symmetric orthogonal matrix, then $Q^2 = I$.

19. Show that if Q is orthogonal, then $\det Q = \pm 1$.

20. Show that for any real number t, the matrix $A = \begin{pmatrix} \sin t & \cos t \\ \cos t & -\sin t \end{pmatrix}$ is orthogonal.

21. Let $\{\mathbf{v}_1, \mathbf{v}_2, \ldots, \mathbf{v}_k\}$ be a linearly independent set of vectors in $\mathbb{R}^n$. Prove that $\mathbf{v}_i \neq \mathbf{0}$ for $i = 1, 2, \ldots, k$. [*Hint:* If $\mathbf{v}_i = \mathbf{0}$, then it is easy to find constants $c_1, c_2, \ldots, c_k$ with $c_i \neq 0$ such that $c_1\mathbf{v}_1 + c_2\mathbf{v}_2 + \cdots + c_k\mathbf{v}_k = \mathbf{0}$].

In Problems 22–28 a subspace H and a vector $\mathbf{v}$ are given. **(a)** Compute $\text{proj}_H\, \mathbf{v}$; **(b)** find an orthonormal basis for $H^\perp$; **(c)** write $\mathbf{v}$ as $\mathbf{h}+\mathbf{p}$, where $\mathbf{h} \in H$ and $\mathbf{p} \in H^\perp$.

22. $H = \left\{ \begin{pmatrix} x \\ y \end{pmatrix} \in \mathbb{R}^2\colon\ x + y = 0 \right\};\ \mathbf{v} = \begin{pmatrix} -1 \\ 2 \end{pmatrix}$

23. $H = \left\{ \begin{pmatrix} x \\ y \end{pmatrix} \in \mathbb{R}^2\colon\ ax + by = 0 \right\};\ \mathbf{v} = \begin{pmatrix} a \\ b \end{pmatrix}$

24. $H = \left\{ \begin{pmatrix} x \\ y \\ z \end{pmatrix} \in \mathbb{R}^3\colon\ ax + by + cz = 0 \right\};\ \mathbf{v} = \begin{pmatrix} a \\ b \\ c \end{pmatrix},\ \mathbf{v} \neq \mathbf{0}$

25. $H = \left\{ \begin{pmatrix} x \\ y \\ z \end{pmatrix} \in \mathbb{R}^3\colon\ 3x - 2y + 6z = 0 \right\};\ \mathbf{v} = \begin{pmatrix} -3 \\ 1 \\ 4 \end{pmatrix}$

26. $H = \left\{ \begin{pmatrix} x \\ y \\ z \end{pmatrix} \in \mathbb{R}^3\colon\ x/2 = y/3 = z/4 \right\};\ \mathbf{v} = \begin{pmatrix} 1 \\ 1 \\ 1 \end{pmatrix}$

27. $H = \left\{ \begin{pmatrix} x \\ y \\ z \\ w \end{pmatrix} \in \mathbb{R}^4\colon\ 2x - y + 3z - w = 0 \right\};\ \mathbf{v} = \begin{pmatrix} 1 \\ -1 \\ 2 \\ 3 \end{pmatrix}$

28. $H = \left\{ \begin{pmatrix} x \\ y \\ z \\ w \end{pmatrix} \in \mathbb{R}^4\colon\ x = y \text{ and } w = 3y \right\};\ \mathbf{v} = \begin{pmatrix} -1 \\ 2 \\ 3 \\ 1 \end{pmatrix}$

29. Let $\mathbf{u}_1$ and $\mathbf{u}_2$ be two orthonormal vectors in $\mathbb{R}^n$. Show that $|\mathbf{u}_1 - \mathbf{u}_2| = \sqrt{2}$.

30. If $\mathbf{u}_1, \mathbf{u}_2, \ldots, \mathbf{u}_n$ are orthonormal, show that

$$|\mathbf{u}_1 + \mathbf{u}_2 + \cdots + \mathbf{u}_n|^2 = |\mathbf{u}_1|^2 + |\mathbf{u}_2|^2 + \cdots + |\mathbf{u}_n|^2 = n$$

31. Find a condition on the numbers a and b such that $\left\{ \begin{pmatrix} a \\ b \end{pmatrix}, \begin{pmatrix} b \\ -a \end{pmatrix} \right\}$ and $\left\{ \begin{pmatrix} a \\ b \end{pmatrix}, \begin{pmatrix} -b \\ a \end{pmatrix} \right\}$ form orthonormal bases in $\mathbb{R}^2$.

32. Show that *any* orthonormal basis in $\mathbb{R}^2$ has one of the forms of the bases in Problem 31.

***33.** Prove the **Cauchy-Schwartz inequality** in $\mathbb{R}^n$: $|\mathbf{u} \cdot \mathbf{v}| \leq |\mathbf{u}|\,|\mathbf{v}|$. [*Hint:* If $\mathbf{u} = \mathbf{0}$ or $\mathbf{v} = \mathbf{0}$, the result is obvious. If $\mathbf{u} \neq \mathbf{0}$ and $\mathbf{v} \neq \mathbf{0}$, use the fact that $|\mathbf{u}/|\mathbf{u}| - \mathbf{v}/|\mathbf{v}|| \geq 0$ and $|\mathbf{u}/|\mathbf{u}| + \mathbf{v}/|\mathbf{v}|| \geq 0$.]

34. Show that, in Problem 33, $|\mathbf{u} \cdot \mathbf{v}| = |\mathbf{u}|\,|\mathbf{v}|$ if and only if $\mathbf{u} = \lambda\mathbf{v}$ for some real number λ.

35. Using the result of Problem 34, prove that if $|\mathbf{u} + \mathbf{v}| = |\mathbf{u}| + |\mathbf{v}|$, then $\mathbf{u}$ and $\mathbf{v}$ are linearly dependent.

36. Using the result of Problem 33, prove the **triangle inequality**: $|\mathbf{u}+\mathbf{v}| \le |\mathbf{u}| + |\mathbf{v}|$. [*Hint:* Expand $|\mathbf{u}+\mathbf{v}|^2$.]

*__37.__ Suppose that $\mathbf{x}_1, \mathbf{x}_2, \ldots, \mathbf{x}_k$ are vectors in $\mathbb{R}^n$ (not all zero) and

$$|\mathbf{x}_1 + \mathbf{x}_2 + \cdots + \mathbf{x}_k| = |\mathbf{x}_1| + |\mathbf{x}_2| + \cdots + |\mathbf{x}_k|$$

Show that dim span $\{\mathbf{x}_1, \mathbf{x}_2, \ldots, \mathbf{x}_k\} = 1$. [*Hint:* Use the results of Problems 35 and 36.]

38. Let $\{\mathbf{u}_1, \mathbf{u}_2, \ldots, \mathbf{u}_n\}$ be an orthonormal basis in $\mathbb{R}^n$ and let $\mathbf{v}$ be a vector in $\mathbb{R}^n$. Prove that $|\mathbf{v}|^2 = |\mathbf{v}\cdot\mathbf{u}_1|^2 + |\mathbf{v}\cdot\mathbf{u}_2|^2 + \cdots + |\mathbf{v}\cdot\mathbf{u}_n|^2$. This equality is called **Parseval's equality** in $\mathbb{R}^n$.

39. Show that for any subspace H of $\mathbb{R}^n$, $(H^\perp)^\perp = H$.

40. Let H_1 and H_2 be two subspaces of $\mathbb{R}^n$ and suppose that $H_1^\perp = H_2^\perp$. Show that $H_1 = H_2$.

41. If H_1 and H_2 are subspaces of $\mathbb{R}^n$, show that if $H_1 \subset H_2$, then $H_2^\perp \subset H_1^\perp$.

5.8 If Time Permits: Inner Product Spaces

This section makes use of a knowledge of elementary properties of complex numbers (summarized in Appendix 2) and some familiarity with material in the first year of calculus.

In Sections 2.2 and 5.7 we saw how we could multiply two vectors in $\mathbb{R}^n$ to get a scalar. This scalar product is also called an *inner product.* Other vector spaces have inner products defined on them as well. Before giving a general definition we note that in $\mathbb{R}^n$ the inner product of two vectors is a real scalar. In other spaces (see Example 2 below) the inner product gives us a complex scalar. To include all cases, therefore, we assume in the following definition that the inner product of two vectors is a complex number. Since every real number is a complex number, this definition includes the real case as well.

DEFINITION 1 **INNER PRODUCT SPACE** The complex vector space V is called an **inner product space** if for every pair of vectors $\mathbf{u}$ and $\mathbf{v}$ in V, there is a unique complex number $(\mathbf{u}, \mathbf{v})$, called the **inner product** of $\mathbf{u}$ and $\mathbf{v}$, such that if $\mathbf{u}$, $\mathbf{v}$, and $\mathbf{w}$ are in V and $\alpha \in \mathbb{C}$, then

i. $(\mathbf{v}, \mathbf{v}) \ge 0$
ii. $(\mathbf{v}, \mathbf{v}) = 0$ if and only if $\mathbf{v} = \mathbf{0}$
iii. $(\mathbf{u}, \mathbf{v}+\mathbf{w}) = (\mathbf{u}, \mathbf{v}) + (\mathbf{u}, \mathbf{w})$
iv. $(\mathbf{u}+\mathbf{v}, \mathbf{w}) = (\mathbf{u}, \mathbf{w}) + (\mathbf{v}, \mathbf{w})$
v. $(\mathbf{u}, \mathbf{v}) = \overline{(\mathbf{v}, \mathbf{u})}$
vi. $(\alpha\mathbf{u}, \mathbf{v}) = \alpha(\mathbf{u}, \mathbf{v})$
vii. $(\mathbf{u}, \alpha\mathbf{v}) = \bar{\alpha}(\mathbf{u}, \mathbf{v})$

The bar in conditions (v) and (vii) denotes the complex conjugate.

Note. If $(\mathbf{u}, \mathbf{v})$ is real, then $\overline{(\mathbf{u}, \mathbf{v})} = (\mathbf{u}, \mathbf{v})$ and we can remove the bar in (v).

EXAMPLE 1

$\mathbb{R}^n$ is an inner product space with $(\mathbf{u}, \mathbf{v}) = \mathbf{u} \cdot \mathbf{v}$. Conditions (*i*) to (*vii*) were proved in Section 2.2.

EXAMPLE 2

We defined the space $\mathbb{C}^n$ in Example 5.2.13, page 175. Let $\mathbf{x} = (x_1, x_2, \ldots, x_n)$ and $\mathbf{y} = (y_1, y_2, \ldots, y_n)$ be in $\mathbb{C}^n$. (Remember— this means that the x_i's and y_i's are complex numbers.) Then we define

$$(\mathbf{x}, \mathbf{y}) = x_1\bar{y}_1 + x_2\bar{y}_2 + \cdots + x_n\bar{y}_n \tag{1}$$

To show that Equation (1) defines an inner product we need some facts about complex numbers. If these are unfamiliar, refer to Appendix 2. For (*i*):

$$(\mathbf{x}, \mathbf{x}) = x_1\bar{x}_1 + x_2\bar{x}_2 + \cdots + x_n\bar{x}_n = |x_1|^2 + |x_2|^2 + \cdots + |x_n|^2.$$

Thus (*i*) and (*ii*) are satisfied since $|x_i|$ is a real number. Conditions (*iii*) and (*iv*) follow from the fact that $z_1(z_2+z_3) = z_1z_2 + z_1z_3$ for any complex numbers z_1, z_2, and z_3. Condition (*v*) follows from the fact that $\overline{z_1z_2} = \bar{z}_1\,\bar{z}_2$ and $\bar{\bar{z}}_1 = z_1$ so that $\overline{x_1\bar{y}_1} = \bar{x}_1y_1$. Condition (*vi*) is obvious. For (*vii*): $(\mathbf{u}, \alpha\mathbf{v}) = \overline{(\alpha\mathbf{v}, \mathbf{u})} = (\bar{\alpha}\bar{\mathbf{v}}, \bar{\mathbf{u}}) = \bar{\alpha}(\bar{\mathbf{v}}, \bar{\mathbf{u}}) = \bar{\alpha}(\mathbf{u}, \mathbf{v})$. Here we used (*vi*) and (*v*).

EXAMPLE 3

In $\mathbb{C}^3$ let $\mathbf{x} = (1+i, -3, 4-3i)$ and $\mathbf{y} = (2-i, -i, 2+i)$. Then

$$\begin{aligned}(\mathbf{x}, \mathbf{y}) &= (1+i)\overline{(2-i)} + (-3)\overline{(-i)} + (4-3i)\overline{(2+i)} \\ &= (1+i)(2+i) + (-3)(i) + (4-3i)(2-i) \\ &= (1+3i) - 3i + (5-10i) = 6 - 10i\end{aligned}$$

□EXAMPLE 4

Suppose that $a < b$; let $V = C[a, b]$ and define

$$(f, g) = \int_a^b f(t)g(t)\,dt \tag{2}$$

We shall see that this is also an inner product.

(*i*) $(f, f) = \int_a^b f^2(t)\,dt \geq 0$. It is a basic theorem of calculus that if $f \in C[a, b]$, $f \geq 0$ on $[a, b]$, and $\int_a^b f(t)\,dt = 0$, then $f = 0$ on $[a, b]$. This proves (*i*) and (*ii*). (*iii*)–(*vii*) follow from basic facts about definite integrals.

□EXAMPLE 5

Let $f(t) = t^2 \in C[0, 1]$ and $g(t) = (4-t) \in C[0, 1]$. Then

$$(f, g) = \int_0^1 t^2(4-t)\,dt = \int_0^1 (4t^2 - t^3)\,dt = \left(\frac{4t^3}{3} - \frac{t^4}{4}\right)\Big|_0^1 = \frac{13}{12}$$

DEFINITION 2 Let V be an inner product space and suppose that $\mathbf{u}$ and $\mathbf{v}$ are in V. Then:

i. $\mathbf{u}$ and $\mathbf{v}$ are **orthogonal** if $(\mathbf{u}, \mathbf{v}) = 0$.
ii. The **norm** of $\mathbf{u}$, denoted $|\mathbf{u}|$, is given by

$$|\mathbf{u}| = \sqrt{(\mathbf{u}, \mathbf{u})} \tag{3}$$

Note. Equation (3) makes sense since $(\mathbf{u}, \mathbf{u}) \geq 0$.

EXAMPLE 6 In $\mathbb{C}^2$ the vectors $(3, -i)$ and $(2, 6i)$ are orthogonal because

$$((3, -i), (2, 6i)) = 3 \cdot \bar{2} + (-i)(\overline{6i}) = 6 + (-i)(-6i) = 6 - 6 = 0.$$

Also $|(3, -i)| = \sqrt{3 \cdot 3 + (-i)(\bar{i})} = \sqrt{10}$.

□ **EXAMPLE 7** In $C[0, 2\pi]$ the functions $\sin t$ and $\cos t$ are orthogonal since

$$(\sin t, \cos t) = \int_0^{2\pi} \sin t \cos t \, dt = \frac{1}{2}\int_0^{2\pi} \sin 2t \, dt = -\frac{\cos 2t}{4}\bigg|_0^{2\pi} = 0$$

Also,

$$\begin{aligned}
|\sin t| &= (\sin t, \sin t)^{1/2} \\
&= \left[\int_0^{2\pi} \sin^2 t \, dt\right]^{1/2} \\
&= \left[\frac{1}{2}\int_0^{2\pi} (1 - \cos 2t) \, dt\right]^{1/2} \\
&= \left[\frac{1}{2}\left(t - \frac{\sin 2t}{2}\right)\bigg|_0^{2\pi}\right]^{1/2} \\
&= \sqrt{\pi}
\end{aligned}$$

If you look at the proofs of Theorems 5.7.1 and 5.7.2 on page 213, you will see that no use was made of the fact that $V = \mathbb{R}^n$. The same theorems are true in any inner product space V. We list them for convenience after giving a definition.

DEFINITION 3 **ORTHONORMAL SET** The set of vectors $\{\mathbf{v}_1, \mathbf{v}_2, \ldots, \mathbf{v}_n\}$ is an **orthonormal set** in V if

$$(\mathbf{v}_i, \mathbf{v}_j) = 0 \quad \text{for} \quad i \neq j \tag{4}$$

and

$$|\mathbf{v}_i| = \sqrt{(\mathbf{v}_i, \mathbf{v}_i)} = 1 \tag{5}$$

If only (4) holds, the set is said to be **orthogonal**.

THEOREM 1 Any finite orthogonal set of nonzero vectors in an inner product space is linearly independent.

THEOREM 2 Any finite, linearly independent set in an inner product space can be made into an orthonormal set by the Gram-Schmidt process. In particular, any finite dimensional inner product space has an orthonormal basis.

□ **EXAMPLE 8** Construct an orthonormal basis for $P_2[0, 1]$.

Solution We start with the standard basis $\{1, x, x^2\}$. Since $P_2[0, 1]$ is a subspace of $C[0, 1]$, we may use the inner product of Example 4. Since $\int_0^1 1^2\,dx = 1$, we let $\mathbf{u}_1 = 1$. Then $\mathbf{v}_1' = \mathbf{v}_2 - (\mathbf{v}_2, \mathbf{u}_1)\mathbf{u}_1$. Here $(\mathbf{v}_2, \mathbf{u}_1) = \int_0^1 (x \cdot 1)\,dx = \frac{1}{2}$. Thus $\mathbf{v}_2' = x - \frac{1}{2} \cdot 1 = x - \frac{1}{2}$. Next we compute

$$|x - \tfrac{1}{2}| = \left[\int_0^1 (x - \tfrac{1}{2})^2\,dx\right]^{1/2} = \left[\int_0^1 (x^2 - x + \tfrac{1}{4})\,dx\right]^{1/2} = \frac{1}{\sqrt{12}} = \frac{1}{2\sqrt{3}}$$

Hence $\mathbf{u}_2 = 2\sqrt{3}(x - \frac{1}{2}) = \sqrt{3}(2x - 1)$. Then $\mathbf{v}_3' = \mathbf{v}_3 - (\mathbf{v}_3, \mathbf{u}_1)\mathbf{u}_1 - (\mathbf{v}_3, \mathbf{u}_2)\mathbf{u}_2$. We have $(\mathbf{v}_3, \mathbf{u}_1) = \int_0^1 x^2\,dx = \frac{1}{3}$ and

$$(\mathbf{v}_3, \mathbf{u}_2) = \sqrt{3}\int_0^1 x^2(2x - 1)\,dx = \sqrt{3}\int_0^1 (2x^3 - x^2)\,dx = \frac{\sqrt{3}}{6}.$$

Thus

$$\mathbf{v}_3' = x^2 - \tfrac{1}{3} - \frac{\sqrt{3}}{6}[\sqrt{3}(2x - 1)] = x^2 - x + \tfrac{1}{6}$$

and

$$|\mathbf{v}_3'| = \left[\int_0^1 (x^2 - x + \tfrac{1}{6})^2\,dx\right]^{1/2}$$

$$= \left[\int_0^1 \left(x^4 - 2x^3 + \tfrac{4}{3}x^2 - \frac{x}{3} + \tfrac{1}{36}\right) dx\right]^{1/2}$$

$$= \left[\left(\frac{x^5}{5} - \frac{x^4}{2} + \frac{4x^3}{9} - \frac{x^2}{6} + \frac{x}{36}\right)\Bigg|_0^1\right]^{1/2}$$

$$= \frac{1}{\sqrt{180}} = \frac{1}{6\sqrt{5}}.$$

Thus $\mathbf{u}_3 = 6\sqrt{5}(x^2 - x + \frac{1}{6}) = \sqrt{5}(6x^2 - 6x + 1)$. Finally, the orthonormal basis is $\{1, \sqrt{3}(2x - 1), \sqrt{5}(6x^2 - 6x + 1)\}$.

□ EXAMPLE 9 In $C[0, 2\pi]$, the infinite set

$$S = \left\{\frac{1}{\sqrt{2\pi}}, \frac{1}{\sqrt{\pi}}\sin x, \frac{1}{\sqrt{\pi}}\cos x, \frac{1}{\sqrt{\pi}}\sin 2x, \frac{1}{\sqrt{\pi}}\cos 2x, \ldots, \frac{1}{\sqrt{\pi}}\sin nx, \frac{1}{\sqrt{\pi}}\cos nx, \ldots\right\}$$

is an orthonormal set. This follows since if $m \neq n$, then

$$\int_0^{2\pi} \sin mx \cos nx\, dx = \int_0^{2\pi} \sin mx \sin nx\, dx = \int_0^{2\pi} \cos mx \cos nx\, dx = 0.$$

To prove one of these, we note that

$$\begin{aligned}\int_0^{2\pi} \sin mx \cos nx\, dx &= \frac{1}{2}\int_0^{2\pi} [\sin (m+n)x + \sin (m-n)x]\, dx \\ &= -\frac{1}{2}\left[\frac{\cos (m+n)x}{m+n} + \frac{\cos (m-n)x}{m-n}\right]\Bigg|_0^{2\pi} \\ &= 0\end{aligned}$$

since $\cos x$ is periodic of period 2π. We have seen that $|\sin x| = \sqrt{\pi}$. Thus $|(1/\sqrt{\pi}) \sin x| = 1$. The other facts follow in a similar fashion. This example provides a situation in which we have an *infinite* orthonormal set. In fact, although this is far beyond us in this elementary text, the functions in S are actually a basis in $C[0, 2\pi]$. Suppose $f \in C[0, 2\pi]$. Then if we write f as an infinite linear combination of the vectors in S, we obtain what is called the **Fourier series representation** of f.

If you look at Definitions 4 and 5 and Theorems 4, 5, and 6 in Section 5.7, then it is apparent that $\mathbb{R}^n$ could be replaced by any finite dimensional vector space (real or complex). We need not repeat those definitions and theorems here, but let us inspect an example of what can happen.

□ EXAMPLE 10 Since $P_2[0, 1]$ is a finite dimensional subspace of $C[0, 1]$, we can talk about $\text{proj}_{P_2[0,1]} f$ if $f \in C[0, 1]$. If $f(x) = e^x$, for example, we compute $\text{proj}_{P_2[0,1]} e^x$. Since $\{\mathbf{u}_1, \mathbf{u}_2, \mathbf{u}_3\} = \{1, \sqrt{3}(2x-1), \sqrt{5}(6x^2-6x+1)\}$ is an orthonormal basis in $P_2[0, 1]$ by Example 8, we have

$$\begin{aligned}\text{proj}_{P_2[0,1]} e^x &= (e^x, 1)1 + (e^x, \sqrt{3}(2x-1))\sqrt{3}(2x-1) \\ &\quad + (e^x, \sqrt{5}(6x^2-6x+1))\sqrt{5}(6x^2-6x+1)\end{aligned}$$

We shall spare you the computations. Using the fact that $\int_0^1 e^x\, dx = e-1$, $\int_0^1 xe^x\, dx = 1$, and $\int_0^1 x^2e^x\, dx = e-2$, we obtain $(e^x, 1) = e-1$,

$(e^x, \sqrt{3}(2x-1)) = \sqrt{3}(3-e)$, and $(e^x, \sqrt{5}(6x^2-6x+1)) = \sqrt{5}(7e-19)$. Finally,

$$\begin{aligned}\text{proj}_{P_2[0,1]}\, e^x &= (e-1) + \sqrt{3}(3-e)\sqrt{3}(2x-1)\\ &\quad + \sqrt{5}(7e-19)(\sqrt{5})(6x^2-6x+1)\\ &= (e-1) + (9-3e)(2x-1)\\ &\quad 5(7e-19)(6x^2-6x+1)\\ &\approx 1.01 + 0.85x + 0.84x^2\end{aligned}$$

PROBLEMS 5.8

1. Let D_n denote the set of $n \times n$ diagonal matrices with real components under the usual matrix operations. If A and B are in D_n, define

$$(A, B) = a_{11}b_{11} + a_{22}b_{22} + \cdots + a_{nn}b_{nn}$$

Prove that D_n is an inner product space.

2. If $A \in D_n$, show that $|A| = 1$ if and only if $a_{11}^2 + a_{22}^2 + \cdots + a_{nn}^2 = 1$.

3. Find an orthonormal basis for D_n.

4. Find an orthonormal basis for D_2 starting with $A = \begin{pmatrix} 2 & 0 \\ 0 & 1 \end{pmatrix}$ and $B = \begin{pmatrix} -3 & 0 \\ 0 & 4 \end{pmatrix}$.

5. $\mathbb{C}^2$, find an orthonormal basis starting with the basis $(1, i)$, $(2-i, 3+2i)$.

□**6.** Find an orthonormal basis for $P_3[0, 1]$.

□**7.** Find an orthonormal basis for $P_2[-1, 1]$. The polynomials you obtain are called **normalized Legendre polynomials.**

□***8.** Find an orthonormal basis for $P_2[a, b]$, $a < b$.

9. If $A = (a_{ij})$ is an $n \times n$ matrix, the **trace** *of* A, written tr A, is the sum of the diagonal components of A: tr $A = a_{11} + a_{22} + \cdots a_{nn}$. In M_{nn}, define $(A, B) =$ tr (AB^t). Prove that with the preceding inner product, M_{nn} is an inner product space.

10. If $A \in M_{nn}$, show that $|A|^2$ is the sum of the squares of the elements of A. [*Note:* Here $|A| = (A, A)^{1/2}$, not det A.]

11. Find an orthonormal basis for M_{22}.

12. We can think of the complex plane as a vector space over the reals with basis vectors 1, i. If $z = a + ib$ and $w = c + id$, define $(z, w) = ac + bd$. Show that this is an inner product and that $|z|$ is the usual length of a complex number.

13. Let a, b, and c be three distinct real numbers. Let p and q be in P_2 and define $(p, q) = p(a)q(a) + p(b)q(b) + p(c)q(c)$.
a. Prove that (p, q) is an inner product in P_2.
b. Is $(p, q) = p(a)q(a) + p(b)q(b)$ an inner product?

14. In $\mathbb{R}^2$, if $\mathbf{x} = \begin{pmatrix} x_1 \\ x_2 \end{pmatrix}$ and $\mathbf{y} = \begin{pmatrix} y_1 \\ y_2 \end{pmatrix}$, let $(\mathbf{x}, \mathbf{y})_* = x_1y_1 + 3x_2y_2$. Show that $(x, y)_*$ is an inner product on $\mathbb{R}^2$.

15. With the inner product of Problem 14, calculate $\left|\begin{pmatrix} 2 \\ -3 \end{pmatrix}\right|_*$.

16. In $\mathbb{R}^2$, let $(\mathbf{x}, \mathbf{y}) = x_1y_1 - x_2y_2$. Is this an inner product? If not, why not?

****17.** Let V be an inner product space. Prove that $|(\mathbf{u}, \mathbf{v})| \le |\mathbf{u}|\,|\mathbf{v}|$. This is called the **Cauchy-Schwartz inequality.** [*Hint:* See Problem 5.7.33.]

***18.** Using the result of Problem 17, prove that $|\mathbf{u} + \mathbf{v}| \le |\mathbf{u}| + |\mathbf{v}|$. This is called the **triangle inequality**. [*Hint:* See Problem 5.7.36, page 222.]

□**19.** In $P_3[0, 1]$ let H be the subspace spanned by $\{1, x^2\}$. Find $H^\perp$.

□***20.** In $C[-1, 1]$ let H be the subspace of even functions. Show that $H^\perp$ consists of odd functions. [*Hint:* f is odd if $f(-x)=-f(x)$ and is even if $f(-x)=f(x)$.]

□***21.** $H=P_2[0, 1]$ is a subspace of $P_3[0, 1]$. Write the polynomial $1+2x+3x^2-x^3$ as $h(x)+p(x)$, where $h(x)\in H$ and $p(x)\in H^\perp$.

Review Exercises for Chapter 5

In Exercises 1–10 determine whether the given set is a vector space. If so, determine its dimension. If it is finite dimensional, find a basis for it.

1. The vectors (x, y, z) in $\mathbb{R}^3$ satisfying $x+2y-z=0$

2. The vectors (x, y, z) in $\mathbb{R}^3$ satisfying $x+2y-z\leq 0$

3. The vectors (x, y, z, w) in $\mathbb{R}^4$ satisfying $x+y+z+w=0$

4. The vectors in $\mathbb{R}^3$ satisfying $x-2=y+3=z-4$

5. The set of upper triangular $n\times n$ matrices under the operations of matrix addition and scalar multiplication

6. The set of polynomials of degree ≤ 5

7. The set of polynomials of degree 5

8. The set of 3×2 matrices $A=(a_{ij})$, with $a_{12}=0$, under the operations of matrix addition and scalar multiplication

9. The set in Exercise 8 except that $a_{12}=1$

10. The set $S=\{f\in C[0, 2]:\ f(2)=0\}$

In Exercises 11–14 determine whether the given set of vectors is linearly dependent or independent.

11. In P_3: $1, 2-x^2, 3-x, 7x^2-8x$

12. In P_3: $1, 2+x^3, 3-x, 7x^2-8x$

13. In M_{22}: $\begin{pmatrix}1 & -1\\ 0 & 0\end{pmatrix}, \begin{pmatrix}1 & 1\\ 0 & 0\end{pmatrix}, \begin{pmatrix}0 & 0\\ 1 & 1\end{pmatrix}, \begin{pmatrix}0 & 0\\ 1 & -1\end{pmatrix}$

14. In M_{22}: $\begin{pmatrix}1 & 1\\ 0 & 0\end{pmatrix}, \begin{pmatrix}1 & -1\\ 0 & 0\end{pmatrix}, \begin{pmatrix}0 & 0\\ 1 & 1\end{pmatrix}, \begin{pmatrix}0 & 0\\ 1 & -1\end{pmatrix}$

In Exercises 15–20 find a basis for the given vector space and determine its dimension.

15. The vectors in $\mathbb{R}^3$ lying on the plane $2x+3y-4z=0$

16. $H=\{(x, y):\ 2x-3y=0\}$

17. $\{\mathbf{v}\in\mathbb{R}^4:\ 3x-y-z+w=0\}$

18. $\{p\in P_3:\ p(0)=0\}$

19. The set of diagonal 4×4 matrices

20. M_{32}

In Exercises 21–24 write the given vector in terms of the new given basis vectors.

21. In $\mathbb{R}^2$: $\mathbf{x}=\begin{pmatrix}2\\ -1\end{pmatrix}$; $\begin{pmatrix}1\\ 2\end{pmatrix}, \begin{pmatrix}-1\\ 2\end{pmatrix}$

22. In $\mathbb{R}^3$: $\mathbf{x}=\begin{pmatrix}-3\\ 4\\ 2\end{pmatrix}$; $\begin{pmatrix}1\\ 0\\ 1\end{pmatrix}, \begin{pmatrix}1\\ 1\\ 0\end{pmatrix}, \begin{pmatrix}0\\ 2\\ 3\end{pmatrix}$

23. In P_2: $\mathbf{x}=4+x^2$; $1+x^2, 1+x, 1$

24. In M_{22}: $\mathbf{x}=\begin{pmatrix}3 & 1\\ 0 & 1\end{pmatrix}$; $\begin{pmatrix}1 & 1\\ 0 & 0\end{pmatrix}; \begin{pmatrix}1 & -1\\ 0 & 0\end{pmatrix}; \begin{pmatrix}0 & 0\\ 1 & 1\end{pmatrix}; \begin{pmatrix}0 & 0\\ 1 & -1\end{pmatrix}$

In Exercises 25–28 find an orthonormal basis for the given vector space.

25. $\mathbb{R}^2$ starting with the basis $\begin{pmatrix}2\\3\end{pmatrix}, \begin{pmatrix}-1\\4\end{pmatrix}$

26. $\{(x, y, z) \in \mathbb{R}^3\colon\ x - y - z = 0\}$

27. $\{(x, y, z) \in \mathbb{R}^3\colon\ x = y = z\}$

28. $\{(x, y, z, w) \in \mathbb{R}^4\colon\ x = z \text{ and } y = w\}$

In Exercises 29–31: **(a)** compute $\text{proj}_H\, \mathbf{v}$; **(b)** find an orthonormal basis for $H^\perp$; **(c)** write $\mathbf{v}$ as $\mathbf{h}+\mathbf{p}$, where $\mathbf{h} \in H$ and $\mathbf{p} \in H^\perp$.

29. H is the subspace of Problem 26; $\mathbf{v} = \begin{pmatrix}-1\\2\\4\end{pmatrix}$.

30. H is the subspace of Problem 27; $\mathbf{v} = \begin{pmatrix}1\\0\\-1\end{pmatrix}$.

31. H is the subspace of Problem 28; $\mathbf{v} = \begin{pmatrix}1\\0\\0\\1\end{pmatrix}$.

□**32.** Find an orthonormal basis for $P_2[0, 2]$.

6 Linear Transformations

6.1 Definition and Examples

In this chapter we discuss a special class of functions, called *linear transformations,* which occur with great frequency in linear algebra and other branches of mathematics. They are also important in a wide variety of applications. Before defining a linear transformation, let us study two simple examples to see what can happen.

EXAMPLE 1

In $\mathbb{R}^2$, define a function T by the formula $T\begin{pmatrix} x \\ y \end{pmatrix} = \begin{pmatrix} x \\ -y \end{pmatrix}$. Geometrically, T takes a vector in $\mathbb{R}^2$ and reflects it about the x-axis. This is illustrated in Figure 6.1. Once we have given our basic definition, we shall see that T is a linear transformation from $\mathbb{R}^2$ into $\mathbb{R}^2$.

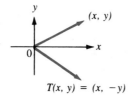

Figure 6.1

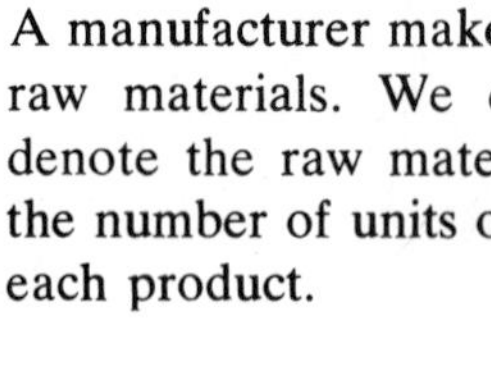

EXAMPLE 2

A manufacturer makes four different products, each of which requires three raw materials. We denote the four products P_1, P_2, P_3, and P_4 and denote the raw materials R_1, R_2, and R_3. The accompanying table gives the number of units of each raw material required to manufacture 1 unit of each product.

Needed to produce 1 unit of

		P_1	P_2	P_3	P_4
Number of units of raw material	R_1	2	1	3	4
	R_2	4	2	2	1
	R_3	3	3	1	2

A natural question arises: If certain numbers of the four products are produced, how many units of each raw material are needed? We let p_1, p_2, p_3, and p_4 denote the number of items of the four products manufactured and let r_1, r_2, and r_3 denote the number of units of the three raw materials needed. Then, we define

$$\mathbf{p}=\begin{pmatrix}p_1\\p_2\\p_3\\p_4\end{pmatrix} \qquad \mathbf{r}=\begin{pmatrix}r_1\\r_2\\r_3\end{pmatrix} \qquad A=\begin{pmatrix}2&1&3&4\\4&2&2&1\\3&3&1&2\end{pmatrix}$$

For example, suppose that $\mathbf{P}=\begin{pmatrix}10\\30\\20\\50\end{pmatrix}$. How many units of R_1 are needed to produce these numbers of units of the four products? From the table, we find that

$$\begin{aligned}r_1&=p_1\cdot 2+p_2\cdot 1+p_3\cdot 3+p_4\cdot 4\\&=10\cdot 2+30\cdot 1+20\cdot 3+50\cdot 4=310 \text{ units}\end{aligned}$$

Similarly, $\qquad r_2=10\cdot 4+30\cdot 2+20\cdot 2+50\cdot 1=190$ units

and $\qquad r_3=10\cdot 3+30\cdot 3+20\cdot 1+50\cdot 2=240$ units

In general, we see that

$$\begin{pmatrix}2&1&3&4\\4&2&2&1\\3&3&1&2\end{pmatrix}\begin{pmatrix}p_1\\p_2\\p_3\\p_4\end{pmatrix}=\begin{pmatrix}r_1\\r_2\\r_3\end{pmatrix}$$

or

$$\mathbf{r}=A\mathbf{p}$$

We can look at this in another way. If $\mathbf{p}$ is called the *production vector* and $\mathbf{r}$ the *raw material vector*, we define the function T by $\mathbf{r}=T\mathbf{p}=A\mathbf{p}$. That is, T is the function which "transforms" the production vector into the raw material vector. It is defined by ordinary matrix multiplication. As we shall see, this function is also a linear transformation. We now give our basic definition.

DEFINITION 1 **LINEAR TRANSFORMATION** Let V and W be vector spaces. A **linear transformation** T from V into W is a function that assigns to each vector $\mathbf{v}\in V$ a unique vector $T\mathbf{v}\in W$ and that satisfies, for each $\mathbf{u}$ and $\mathbf{v}$ in V and each scalar α,

$$T(\mathbf{u}+\mathbf{v})=T\mathbf{u}+T\mathbf{v} \tag{1}$$

$$T(\alpha\mathbf{v})=\alpha T\mathbf{v} \tag{2}$$

Notation We write $T\colon V \to W$ to indicate that T takes V into W.

Terminology Linear transformations are often called **linear operators**. Functions that satisfy (1) and (2) are called **linear functions**.

EXAMPLE 3 Let $T\colon \mathbb{R}^2 \to \mathbb{R}^3$ be defined by $T\begin{pmatrix} x \\ y \end{pmatrix} = \begin{pmatrix} x+y \\ x-y \\ 3y \end{pmatrix}$. For example, $T\begin{pmatrix} 2 \\ -3 \end{pmatrix} = \begin{pmatrix} -1 \\ 5 \\ -9 \end{pmatrix}$. Then

$$T\left[\begin{pmatrix} x_1 \\ y_1 \end{pmatrix} + \begin{pmatrix} x_2 \\ y_2 \end{pmatrix}\right] = T\begin{pmatrix} x_1+x_2 \\ y_1+y_2 \end{pmatrix} = \begin{pmatrix} x_1+x_2+y_1+y_2 \\ x_1+x_2-y_1-y_2 \\ 3y_1+3y_2 \end{pmatrix}$$
$$= \begin{pmatrix} x_1+y_1 \\ x_1-y_1 \\ 3y_1 \end{pmatrix} + \begin{pmatrix} x_2+y_2 \\ x_2-y_2 \\ 3y_2 \end{pmatrix}$$

But
$$\begin{pmatrix} x_1+y_1 \\ x_1-y_1 \\ 3y_1 \end{pmatrix} = T\begin{pmatrix} x_1 \\ y_1 \end{pmatrix} \quad \text{and} \quad \begin{pmatrix} x_2+y_2 \\ x_2-y_2 \\ 3y_2 \end{pmatrix} = T\begin{pmatrix} x_2 \\ y_2 \end{pmatrix}$$

Thus
$$T\left[\begin{pmatrix} x_1 \\ y_1 \end{pmatrix} + \begin{pmatrix} x_2 \\ y_2 \end{pmatrix}\right] = T\begin{pmatrix} x_1 \\ y_1 \end{pmatrix} + T\begin{pmatrix} x_2 \\ y_2 \end{pmatrix}$$

Similarly,

$$T\left[\alpha\begin{pmatrix} x \\ y \end{pmatrix}\right] = T\begin{pmatrix} \alpha x \\ \alpha y \end{pmatrix} = \begin{pmatrix} \alpha x+\alpha y \\ \alpha x-\alpha y \\ 3\alpha y \end{pmatrix} = \alpha\begin{pmatrix} x+y \\ x-y \\ 3y \end{pmatrix} = \alpha T\begin{pmatrix} x \\ y \end{pmatrix}$$

Thus T is a linear transformation.

EXAMPLE 4 Let V and W be vector spaces and define $T\colon V \to W$ by $T\mathbf{v} = \mathbf{0}$ for every $\mathbf{v}$ in V. Then $T(\mathbf{v}_1 + \mathbf{v}_2) = \mathbf{0} = \mathbf{0} + \mathbf{0} = T\mathbf{v}_1 + T\mathbf{v}_2$ and $T(\alpha\mathbf{v}) = \mathbf{0} = \alpha\mathbf{0} = \alpha T\mathbf{v}$. Here T is called the **zero transformation**.

EXAMPLE 5 Let V be a vector space and define $I\colon V \to V$ by $I\mathbf{v} = \mathbf{v}$ for every $\mathbf{v}$ in V. Here I is obviously a linear transformation. It is called the **identity transformation** or **identity operator**.

EXAMPLE 6 Let $T\colon \mathbb{R}^2 \to \mathbb{R}^2$ be defined by $T\begin{pmatrix} x \\ y \end{pmatrix} = \begin{pmatrix} -x \\ y \end{pmatrix}$. It is easy to verify that T is

Figure 6.2

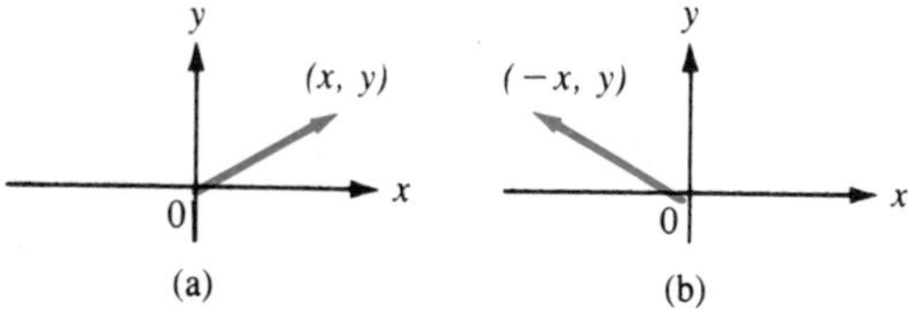

linear. Geometrically, T takes a vector in $\mathbb{R}^2$ and reflects it about the y-axis (see Figure 6.2).

EXAMPLE 7

Let A be an $m\times n$ matrix and define T: $\mathbb{R}^n\to\mathbb{R}^m$ by $T\mathbf{x}=A\mathbf{x}$. Since $A(\mathbf{x}+\mathbf{y})=A\mathbf{x}+A\mathbf{y}$ and $A(\alpha\mathbf{x})=\alpha A\mathbf{x}$ if $\mathbf{x}$ and $\mathbf{y}$ are in $\mathbb{R}^n$, we see that T is a linear transformation. Thus: *Every $m\times n$ matrix A gives rise to a linear transformation from $\mathbb{R}^n$ into $\mathbb{R}^m$.* In Section 6.4 we shall see that a certain converse is true: *Every linear transformation between finite dimensional vector spaces can be represented by a matrix.*

EXAMPLE 8

Suppose the vector $\mathbf{v}=\begin{pmatrix}x\\y\end{pmatrix}$ in the xy-plane is rotated through an angle of θ (measured in degrees or radians) in the counterclockwise direction. Call the

Figure 6.3

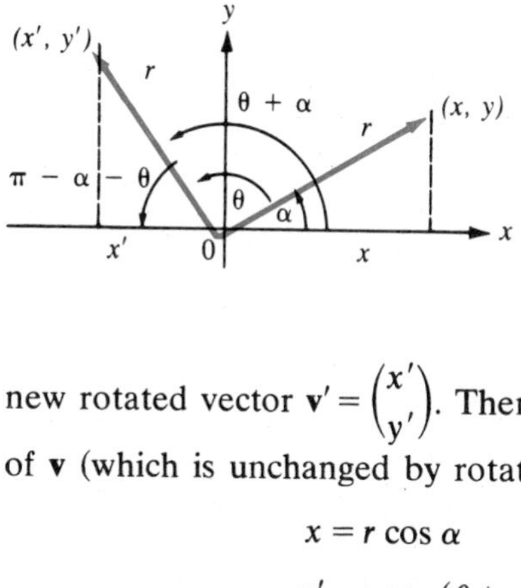

new rotated vector $\mathbf{v}'=\begin{pmatrix}x'\\y'\end{pmatrix}$. Then, as in Figure 6.3, if r denotes the length of $\mathbf{v}$ (which is unchanged by rotation),

$$x=r\cos\alpha \qquad y=r\sin\alpha$$
$$x'=r\cos(\theta+\alpha) \qquad y'=r\sin(\theta+\alpha)\dagger$$

But $r\cos(\theta+\alpha)=r\cos\theta\cos\alpha-r\sin\theta\sin\alpha$, so that

$$x'=x\cos\theta-y\sin\theta \tag{3}$$

† These follow from the standard definitions of $\cos\theta$ and $\sin\theta$ as the x and y coordinates of a point on the unit circle. If (x, y) is a point on the circle centered at the origin of radius r, then $x=r\cos\phi$ and $y=r\sin\phi$, where ϕ is the angle the vector (x, y) makes with the positive x-axis.

Similarly, $r\sin(\theta+\alpha)=r\sin\theta\cos\alpha+r\cos\theta\sin\alpha$ or

$$y'=x\sin\theta+y\cos\theta \tag{4}$$

Let
$$A_\theta=\begin{pmatrix}\cos\theta & -\sin\theta\\ \sin\theta & \cos\theta\end{pmatrix} \tag{5}$$

Then, from (3) and (4), we see that $A_\theta\begin{pmatrix}x\\y\end{pmatrix}=\begin{pmatrix}x'\\y'\end{pmatrix}$. The linear transformation T: $\mathbb{R}^2\to\mathbb{R}^2$ defined by $T\mathbf{v}=A_\theta\mathbf{v}$, where A_θ is given by (5), is called a **rotation transformation.**

EXAMPLE 9

Let H be a subspace of $\mathbb{R}^n$. We define the **orthogonal projection transformation** P: $V\to H$ by

$$P\mathbf{v}=\text{proj}_H\,\mathbf{v} \tag{6}$$

Let $\{\mathbf{u}_1,\mathbf{u}_2,\ldots,\mathbf{u}_k\}$ be an orthonormal basis for H. Then from Definition 5.7.4 on page 217 we have

$$P\mathbf{v}=(\mathbf{v}\cdot\mathbf{u}_1)\mathbf{u}_1+(\mathbf{v}\cdot\mathbf{u}_2)\mathbf{u}_2+\cdots+(\mathbf{v}\cdot\mathbf{u}_k)\mathbf{u}_k \tag{7}$$

Since $(\mathbf{v}_1+\mathbf{v}_2)\cdot\mathbf{u}=\mathbf{v}_1\cdot\mathbf{u}+\mathbf{v}_2\cdot\mathbf{u}$ and $(\alpha\mathbf{v})\cdot\mathbf{u}=\alpha(\mathbf{v}\cdot\mathbf{u})$, we see that P is a linear transformation.

EXAMPLE 10

Let T: $\mathbb{R}^3\to\mathbb{R}^3$ be defined by $T\begin{pmatrix}x\\y\\z\end{pmatrix}=\begin{pmatrix}x\\y\\0\end{pmatrix}$. Then T is the projection operator taking a vector in space and projecting it into the xy-plane. Similarly, $T\begin{pmatrix}x\\y\\z\end{pmatrix}=\begin{pmatrix}x\\0\\z\end{pmatrix}$ projects a vector in space into the xz-plane.

EXAMPLE 11

Define T: $M_{mn}\to M_{nm}$ by $T(A)=A^t$. Since $(A+B)^t=A^t+B^t$ and $(\alpha A)^t=\alpha A^t$, we see that T, called the **transpose operator,** is a linear transformation.

□EXAMPLE 12

Let J: $C[0,1]\to\mathbb{R}$ be defined by $Jf=\int_0^1 f(x)\,dx$. Since $\int_0^1[f(x)+g(x)]\,dx=\int_0^1 f(x)\,dx+\int_0^1 g(x)\,dx$ and $\int_0^1 \alpha f(x)\,dx=\alpha\int_0^1 f(x)\,dx$ if f and g are continuous, we see that J is linear. For example, $T(x^3)=\frac{1}{4}$. J is called an **integral operator.**

EXAMPLE 13 Let $D: C'[0, 1] \to C[0, 1]$ be defined by $Df = f'$. Since $(f+g)' = f' + g'$ and $(\alpha f)' = \alpha f'$ if f and g are differentiable, we see that D is linear. D is called a **differential operator**.

Warning. Not every function that looks linear actually is linear. For example, define $T: \mathbb{R} \to \mathbb{R}$ by $Tx = 2x + 3$. Then $\{(x, Tx): x \in \mathbb{R}\}$ is a straight line in the xy-plane. But T is not linear since $T(x+y) = 2(x+y)+3 = 2x+2y+3$ and $Tx + Ty = (2x+3)+(2y+3) = 2x+2y+6$. The only linear functions from $\mathbb{R}$ to $\mathbb{R}$ are functions of the form $f(x) = mx$ for some real number m. Thus among all straight line functions, the only ones that are linear are the ones that pass through the origin.

EXAMPLE 14 Let $T: \quad C[0, 1] \to \mathbb{R}$ be defined by $Tf = f(0) + 1$. Then T is not linear. To see this, we compute

$$T[f+g] = (f+g)(0) + 1 = f(0) + g(0) + 1$$

$$Tf + Tg = [f(0)+1] + [g(0)+1] = f(0) + g(0) + 2$$

This provides another example of a transformation that might look linear but in fact is not.

PROBLEMS 6.1 In Problems 1–29 determine whether the given transformation from V to W is linear.

1. $T: \mathbb{R}^2 \to \mathbb{R}^2;\ T\begin{pmatrix} x \\ y \end{pmatrix} = \begin{pmatrix} x \\ 0 \end{pmatrix}$

2. $T: \mathbb{R}^2 \to \mathbb{R}^2;\ T\begin{pmatrix} x \\ y \end{pmatrix} = \begin{pmatrix} 1 \\ y \end{pmatrix}$

3. $T: \mathbb{R}^3 \to \mathbb{R}^2;\ T\begin{pmatrix} x \\ y \\ z \end{pmatrix} = \begin{pmatrix} x \\ y \end{pmatrix}$

4. $T: \mathbb{R}^3 \to \mathbb{R}^2;\ T\begin{pmatrix} x \\ y \\ z \end{pmatrix} = \begin{pmatrix} 0 \\ y \end{pmatrix}$

5. $T: \mathbb{R}^3 \to \mathbb{R}^2;\ T\begin{pmatrix} x \\ y \\ z \end{pmatrix} = \begin{pmatrix} 1 \\ z \end{pmatrix}$

6. $T: \mathbb{R}^2 \to \mathbb{R}^2;\ T\begin{pmatrix} x \\ y \end{pmatrix} = \begin{pmatrix} x^2 \\ y^2 \end{pmatrix}$

7. $T: \mathbb{R}^2 \to \mathbb{R}^2;\ T\begin{pmatrix} x \\ y \end{pmatrix} = \begin{pmatrix} y \\ x \end{pmatrix}$

8. $T: \mathbb{R}^2 \to \mathbb{R}^2;\ T\begin{pmatrix} x \\ y \end{pmatrix} = \begin{pmatrix} x+y \\ x-y \end{pmatrix}$

9. $T: \mathbb{R}^2 \to \mathbb{R};\ T\begin{pmatrix} x \\ y \end{pmatrix} = xy$

10. $T: \mathbb{R}^n \to \mathbb{R};\ T\begin{pmatrix} x_1 \\ x_2 \\ \vdots \\ x_n \end{pmatrix} = x_1 + x_2 + \cdots + x_n$

11. $T: \mathbb{R} \to \mathbb{R}^n;\ T(x) = \begin{pmatrix} x \\ x \\ \vdots \\ x \end{pmatrix}$

12. $T: \mathbb{R}^4 \to \mathbb{R}^2;\ T\begin{pmatrix} x \\ y \\ z \\ w \end{pmatrix} = \begin{pmatrix} x+z \\ y+w \end{pmatrix}$

13. $T: \mathbb{R}^4 \to \mathbb{R}^2;\ T\begin{pmatrix} x \\ y \\ z \\ w \end{pmatrix} = \begin{pmatrix} xz \\ yw \end{pmatrix}$

14. T: $M_{nn}\to M_{nn}$; $T(A)=AB$, where B is a fixed $n\times n$ matrix
15. T: $M_{nn}\to M_{nn}$; $T(A)=A^tA$
16. T: $M_{mn}\to M_{mp}$; $T(A)=AB$, where B is a fixed $n\times p$ matrix
17. T: $D_n\to D_n$; $T(D)=D^2$ (D_n is the set of $n\times n$ diagonal matrices)
18. T: $D_n\to D_n$; $T(D)=I+D$
19. T: $P_2\to P_1$; $T(a_0+a_1x+a_2x^2)=a_0+a_1x$
20. T: $P_2\to P_1$; $T(a_0+a_1x+a_2x^2)=a_1+a_2x$
21. T: $\mathbb{R}\to P_n$; $T(a)=a+ax+ax^2+\cdots+ax^n$
22. T: $P_2\to P_4$; $T(p(x))=[p(x)]^2$
23. T: $C[0,1]\to C[0,1]$; $Tf(x)=f^2(x)$
24. T: $C[0,1]\to C[0,1]$; $Tf(x)=f(x)+1$
□25. T: $C[0,1]\to\mathbb{R}$; $Tf=\int_0^1 f(x)g(x)\,dx$, where g is a fixed function in $C[0,1]$
□26. T: $C'[0,1]\to C[0,1]$; $Tf=(fg)'$, where g is a fixed function in $C'[0,1]$
27. T: $C[0,1]\to C[1,2]$; $Tf(x)=f(x-1)$
28. T: $C[0,1]\to\mathbb{R}$; $Tf=f(\frac{1}{2})$
29. T: $M_{nn}\to\mathbb{R}$; $T(A)=\det A$
30. Let T: $\mathbb{R}^2\to\mathbb{R}^2$ be given by $T(x,y)=(-x,-y)$. Describe T geometrically.
31. Let T be a linear transformation from $\mathbb{R}^2\to\mathbb{R}^3$ such that $T\begin{pmatrix}1\\0\end{pmatrix}=\begin{pmatrix}1\\2\\3\end{pmatrix}$ and $T\begin{pmatrix}0\\1\end{pmatrix}=\begin{pmatrix}-4\\0\\5\end{pmatrix}$. Find: **(a)** $T\begin{pmatrix}2\\4\end{pmatrix}$ and **(b)** $T\begin{pmatrix}-3\\7\end{pmatrix}$.
32. In Example 8: **(a)** Find the rotation matrix A_θ when $\theta=\pi/6$. **(b)** What happens to the vector $\begin{pmatrix}-3\\4\end{pmatrix}$ if it is rotated through an angle of $\pi/6$ in the counterclockwise direction?
33. Let $A_\theta=\begin{pmatrix}\cos\theta & -\sin\theta & 0\\ \sin\theta & \cos\theta & 0\\ 0 & 0 & 1\end{pmatrix}$. Describe geometrically the linear transformation T: $\mathbb{R}^3\to\mathbb{R}^3$ given by $T\mathbf{x}=A_\theta\mathbf{x}$.
34. Answer the questions in Problem 33 for $A_\theta=\begin{pmatrix}\cos\theta & 0 & -\sin\theta\\ 0 & 1 & 0\\ \sin\theta & 0 & \cos\theta\end{pmatrix}$.
35. Suppose that, in a real vector space V, T satisfies $T(\mathbf{x}+\mathbf{y})=T\mathbf{x}+T\mathbf{y}$ and $T(\alpha\mathbf{x})=\alpha T\mathbf{x}$ for $\alpha\geq 0$. Show that T is linear.
36. Find a linear transformation T: $M_{33}\to M_{22}$.
37. If T is a linear transformation from V to W, show that $T(\mathbf{x}-\mathbf{y})=T\mathbf{x}-T\mathbf{y}$.
38. If T is a linear transformation from V to W, show that $T\mathbf{0}=\mathbf{0}$. Are the two zero vectors here the same?
39. Let V be an inner product space and let $\mathbf{u}_0\in V$ be fixed. Let T: $V\to\mathbb{R}$ (or $\mathbb{C}$) be defined by $T\mathbf{v}=(\mathbf{v},\mathbf{u}_0)$. Show that T is linear.
*40. Show that if V is a complex inner product space and T: $V\to\mathbb{C}$ is defined by $T\mathbf{v}=(\mathbf{u}_0,\mathbf{v})$ for a fixed vector $\mathbf{u}_0\in V$, then T is not linear.
41. Let V be an inner product space with the finite dimensional subspace H. Let $\{\mathbf{u}_1,\mathbf{u}_2,\ldots,\mathbf{u}_k\}$ be a basis for H. Show that T: $V\to H$ defined by $T\mathbf{v}=(\mathbf{v},\mathbf{u}_1)\mathbf{u}_1+(\mathbf{v},\mathbf{u}_2)\mathbf{u}_2+\cdots+(\mathbf{v},\mathbf{u}_n)\mathbf{u}_n$ is a linear transformation.

6.2 Properties of Linear Transformations: Range and Kernel

In this section we develop some of the basic properties of linear transformations.

THEOREM 1 Let $T: V \to W$ be a linear transformation. Then for all vectors $\mathbf{u}$, $\mathbf{v}$, $\mathbf{v}_1$, $\mathbf{v}_2, \ldots, \mathbf{v}_n$ in V and all scalars $\alpha_1, \alpha_2, \ldots, \alpha_n$:

i. $T(\mathbf{0}) = \mathbf{0}$
ii. $T(\mathbf{u}-\mathbf{v}) = T\mathbf{u} - T\mathbf{v}$
iii. $T(\alpha_1\mathbf{v}_1 + \alpha_2\mathbf{v}_2 + \cdots + \alpha_n\mathbf{v}_n) = \alpha_1 T\mathbf{v}_1 + \alpha_2 T\mathbf{v}_2 + \cdots + \alpha_n T\mathbf{v}_n$

Note. In part (*i*) the $\mathbf{0}$ on the left is the zero vector in V while the $\mathbf{0}$ on the right is the zero vector in W.

Proof **i.** $T(\mathbf{0}) = T(\mathbf{0}+\mathbf{0}) = T(\mathbf{0}) + T(\mathbf{0})$. Thus $\mathbf{0} = T(\mathbf{0}) - T(\mathbf{0}) = T(\mathbf{0}) + T(\mathbf{0}) - T(\mathbf{0}) = T(\mathbf{0})$.
ii. $T(\mathbf{u}-\mathbf{v}) = T[\mathbf{u}+(-1)\mathbf{v}] = T\mathbf{u} + T[(-1)\mathbf{v}] = T\mathbf{u} + (-1)T\mathbf{v} = T\mathbf{u} - T\mathbf{v}$.
iii. We prove this part by induction (see Appendix 1). For $n = 2$, we get $T(\alpha_1\mathbf{v}_1 + \alpha_2\mathbf{v}_2) = T(\alpha_1\mathbf{v}_1) + T(\alpha_2\mathbf{v}_2) = \alpha_1 T\mathbf{v}_1 + \alpha_2 T\mathbf{v}_2$. Thus the equation holds for $n = 2$. We assume that it holds for $n = k$ and prove it for $n = k+1$: $T(\alpha_1\mathbf{v}_1 + \alpha_2\mathbf{v}_2 + \cdots + \alpha_k\mathbf{v}_k + \alpha_{k+1}\mathbf{v}_{k+1}) = T(\alpha_1\mathbf{v}_1 + \alpha_2\mathbf{v}_2 + \cdots + \alpha_k\mathbf{v}_k) + T(\alpha_{k+1}\mathbf{v}_{k+1})$, and using the equation in part (*iii*) for $n = k$, this is equal to $(\alpha_1 T\mathbf{v}_1 + \alpha_2 T\mathbf{v}_2 + \alpha_k T\mathbf{v}_k) + \alpha_{k+1} T\mathbf{v}_{k+1}$, which is what we wanted to show. This completes the proof. ■

Remark. Note that part (*ii*) of Theorem 1 is a special case of part (*iii*).

An important fact about linear transformations is that they are completely determined by what they do to basis vectors.

THEOREM 2 Let V be a finite dimensional vector space with basis $B = \{\mathbf{v}_1, \mathbf{v}_2 \ldots, \mathbf{v}_n\}$. Let $\mathbf{w}_1, \mathbf{w}_2, \ldots, \mathbf{w}_n$ be n vectors in W. Suppose that T_1 and T_2 are two linear transformations from V to W such that $T_1\mathbf{v}_i = T_2\mathbf{v}_i = \mathbf{w}_i$ for $i = 1, 2, \ldots, n$. Then for any vector $\mathbf{v} \in V$, $T_1\mathbf{v} = T_2\mathbf{v}$. That is, $T_1 = T_2$.

Proof Since B is a basis for V, there exists a unique set of scalars $\alpha_1, \alpha_2, \ldots, \alpha_n$ such that $\mathbf{v} = \alpha_1\mathbf{v}_1 + \alpha_2\mathbf{v}_2 + \cdots + \alpha_n\mathbf{v}_n$. Then, from part (*iii*) of Theorem 1,

$$T_1\mathbf{v} = T_1(\alpha_1\mathbf{v}_1 + \alpha_2\mathbf{v}_2 + \cdots + \alpha_n\mathbf{v}_n) = \alpha_1 T_1\mathbf{v}_1 + \alpha_2 T_1\mathbf{v}_2 + \cdots + \alpha_n T_1\mathbf{v}_n$$

$$= \alpha_1\mathbf{w}_1 + \alpha_2\mathbf{w}_2 + \cdots + \alpha_n\mathbf{w}_n$$

Similarly,

$$T_2\mathbf{v} = T_2(\alpha_1\mathbf{v}_1 + \alpha_2\mathbf{v}_2 + \cdots + \alpha_n\mathbf{v}_n) = \alpha_1 T_2\mathbf{v}_1 + \alpha_2 T_2\mathbf{v}_2 + \cdots + \alpha_n T_2\mathbf{v}_n$$
$$= \alpha_1\mathbf{w}_1 + \alpha_2\mathbf{w}_2 + \cdots + \alpha_n\mathbf{w}_n$$

Thus $T_1\mathbf{v} = T_2\mathbf{v}$. ■

Theorem 2 tells us that if $T: V \to W$ and V is finite dimensional, then we need to know only what T does to basis vectors in V. This determines T completely. To see this let $\mathbf{v}_1, \mathbf{v}_2, \ldots, \mathbf{v}_n$ be a basis in V and let $\mathbf{v}$ be another vector in V. Then as in the proof of Theorem 2,

$$T\mathbf{v} = \alpha_1 T\mathbf{v}_1 + \alpha_2 T\mathbf{v}_2 + \cdots + \alpha_n T\mathbf{v}_n$$

Thus we can compute $T\mathbf{v}$ for any vector $\mathbf{v} \in V$ if we know $T\mathbf{v}_1, T\mathbf{v}_2, \ldots, T\mathbf{v}_n$.

EXAMPLE 1 Let T be a linear transformation from $\mathbb{R}^3$ into $\mathbb{R}^2$ and suppose that $T\begin{pmatrix}1\\0\\0\end{pmatrix} = \begin{pmatrix}2\\3\end{pmatrix}$, $T\begin{pmatrix}0\\1\\0\end{pmatrix} = \begin{pmatrix}-1\\4\end{pmatrix}$, and $T\begin{pmatrix}0\\0\\1\end{pmatrix} = \begin{pmatrix}5\\-3\end{pmatrix}$. Compute $T\begin{pmatrix}3\\-4\\5\end{pmatrix}$.

Solution We have $\begin{pmatrix}3\\-4\\5\end{pmatrix} = 3\begin{pmatrix}1\\0\\0\end{pmatrix} - 4\begin{pmatrix}0\\1\\0\end{pmatrix} + 5\begin{pmatrix}0\\0\\1\end{pmatrix}$.

Thus $T\begin{pmatrix}3\\-4\\5\end{pmatrix} = 3T\begin{pmatrix}1\\0\\0\end{pmatrix} - 4T\begin{pmatrix}0\\1\\0\end{pmatrix} + 5T\begin{pmatrix}0\\0\\1\end{pmatrix}$

$$= 3\begin{pmatrix}2\\3\end{pmatrix} - 4\begin{pmatrix}-1\\4\end{pmatrix} + 5\begin{pmatrix}5\\-3\end{pmatrix} = \begin{pmatrix}6\\9\end{pmatrix} + \begin{pmatrix}4\\-16\end{pmatrix} + \begin{pmatrix}25\\-15\end{pmatrix} = \begin{pmatrix}35\\-22\end{pmatrix}.$$

Another question arises: If $\mathbf{w}_1, \mathbf{w}_2, \ldots, \mathbf{w}_n$ are n vectors in W, does there exist a linear transformation T such that $T\mathbf{v}_i = \mathbf{w}_i$ for $i = 1, 2, \ldots, n$? The answer is yes, as the next theorem shows.

THEOREM 3 Let V be a finite dimensional vector space with basis $B = \{\mathbf{v}_1, \mathbf{v}_2, \ldots, \mathbf{v}_n\}$. Let W be a vector space containing the n vectors $\mathbf{w}_1, \mathbf{w}_2, \ldots, \mathbf{w}_n$. Then there exists a unique linear transformation T: $V \to W$ such that $T\mathbf{v}_i = \mathbf{w}_i$ for $i = 1, 2, \ldots, n$.

Proof Define a function T as follows:

i. $T\mathbf{v}_i = \mathbf{w}_i$

ii. If $\mathbf{v}=\alpha_1\mathbf{v}_1+\alpha_2\mathbf{v}_2+\cdots+\alpha_n\mathbf{v}_n$, then

$$T\mathbf{v}=\alpha_1\mathbf{w}_1+\alpha_2\mathbf{w}_2+\cdots+\alpha_n\mathbf{w}_n \tag{1}$$

Because B is a basis for V, T is defined for every $\mathbf{v}\in V$; and since W is a vector space, $T\mathbf{v}\in W$. Thus it only remains to show that T is linear. But this follows directly from Equation (1). For if $\mathbf{u}=\alpha_1\mathbf{v}_1+\alpha_2\mathbf{v}_2+\cdots+\alpha_n\mathbf{v}_n$ and $\mathbf{v}=\beta_1\mathbf{v}_1+\beta_2\mathbf{v}_2+\cdots+\beta_n\mathbf{v}_n$, then

$$\begin{aligned}T(\mathbf{u}+\mathbf{v})&=T[(\alpha_1+\beta_1)\mathbf{v}_1+(\alpha_2+\beta_2)\mathbf{v}_2+\cdots+(\alpha_n+\beta_n)\mathbf{v}_n]\\&=(\alpha_1+\beta_1)\mathbf{w}_1+(\alpha_2+\beta_2)\mathbf{w}_2+\cdots+(\alpha_n+\beta_n)\mathbf{w}_n=(\alpha_1\mathbf{w}_1+\alpha_2\mathbf{w}_2\\&\quad+\cdots+\alpha_n\mathbf{w}_n)+(\beta_1\mathbf{w}_1+\beta_2\mathbf{w}_2+\cdots+\beta_n\mathbf{w}_n)=T\mathbf{u}+T\mathbf{v}.\end{aligned}$$

Similarly $T(\alpha\mathbf{v})=\alpha T\mathbf{v}$, so T is linear. The uniqueness of T follows from Theorem 2 and the theorem is proved. ■

Remark. In Theorems 2 and 3 the vectors $\mathbf{w}_1, \mathbf{w}_2, \ldots, \mathbf{w}_n$ need not be distinct. Moreover, we emphasize that the theorems are true if V is any finite dimensional vector space, not just $\mathbb{R}^n$. Note also that W does not have to be finite dimensional.

EXAMPLE 2 Find a linear transformation from $\mathbb{R}^2$ into the plane

$$W=\left\{\begin{pmatrix}x\\y\\z\end{pmatrix}:\quad 2x-y+3z=0\right\}.$$

Solution From Example 5.5.3, page 193, we know that W is a two-dimensional subspace of $\mathbb{R}^3$ with basis vectors $\mathbf{w}_1=\begin{pmatrix}1\\2\\0\end{pmatrix}$ and $\mathbf{w}_2=\begin{pmatrix}0\\3\\1\end{pmatrix}$. Using the standard basis in $\mathbb{R}^2$, $\mathbf{v}_1=\begin{pmatrix}1\\0\end{pmatrix}$ and $\mathbf{v}_2=\begin{pmatrix}0\\1\end{pmatrix}$, we define the linear transformation T by $T\begin{pmatrix}1\\0\end{pmatrix}=\begin{pmatrix}1\\2\\0\end{pmatrix}$ and $T\begin{pmatrix}0\\1\end{pmatrix}=\begin{pmatrix}0\\3\\1\end{pmatrix}$. Then, as the discussion following Theorem 2 shows, T is completely determined. For example,

$$T\begin{pmatrix}5\\-7\end{pmatrix}=T\left[5\begin{pmatrix}1\\0\end{pmatrix}-7\begin{pmatrix}0\\1\end{pmatrix}\right]=5T\begin{pmatrix}1\\0\end{pmatrix}-7T\begin{pmatrix}0\\1\end{pmatrix}=5\begin{pmatrix}1\\2\\0\end{pmatrix}-7\begin{pmatrix}0\\3\\1\end{pmatrix}=\begin{pmatrix}5\\-11\\-7\end{pmatrix}.$$

We now turn to two important definitions in the theory of linear transformations.

DEFINITION 1 **KERNEL AND RANGE OF A LINEAR TRANSFORMATION** Let V and W be vector spaces and let $T: V \to W$ be a linear transformation. Then:

i. The **kernel** of T, denoted ker T, is given by

$$\ker T = \{\mathbf{v} \in V: \quad T\mathbf{v} = \mathbf{0}\} \tag{2}$$

ii. The **range** of T, denoted range T, is given by

$$\text{Range } T = \{\mathbf{w} \in W: \quad \mathbf{w} = T\mathbf{v} \text{ for some } \mathbf{v} \in V\} \tag{3}$$

Remark 1. Note that ker T is nonempty because, by Theorem 1, $T(\mathbf{0}) = \mathbf{0}$ so that $\mathbf{0} \in \ker T$ for any linear transformation T. We shall be interested in finding other vectors in V that get "mapped to zero." Again note that when we write $T(\mathbf{0}) = \mathbf{0}$, the $\mathbf{0}$ on the left is in V and the $\mathbf{0}$ on the right is in W.

Remark 2. Range T is simply the set of "images" of vectors in V under the transformation T. In fact, if $\mathbf{w} = T\mathbf{v}$, we shall say that $\mathbf{w}$ is the **image** of $\mathbf{v}$ under T.

Before giving examples of kernels and ranges, we prove a theorem that will be very useful.

THEOREM 4 If $T: V \to W$ is a linear transformation, then:

i. ker T is a subspace of V.
ii. range T is a subspace of W.

Proof **i.** Let $\mathbf{u}$ and $\mathbf{v}$ be in ker T; then $T(\mathbf{u}+\mathbf{v}) = T\mathbf{u} + T\mathbf{v} = \mathbf{0}+\mathbf{0} = \mathbf{0}$ and $T(\alpha\mathbf{u}) = \alpha T\mathbf{u} = \alpha\mathbf{0} = \mathbf{0}$ so that $\mathbf{u}+\mathbf{v}$ and $\alpha\mathbf{u}$ are in ker T.
ii. Let $\mathbf{w}$ and $\mathbf{x}$ be in range T. Then $\mathbf{w} = T\mathbf{u}$ and $\mathbf{x} = T\mathbf{v}$ for two vectors $\mathbf{u}$ and $\mathbf{v}$ in V. This means that $T(\mathbf{u}+\mathbf{v}) = T\mathbf{u} + T\mathbf{v} = \mathbf{w}+\mathbf{x}$ and $T(\alpha\mathbf{u}) = \alpha T\mathbf{u} = \alpha\mathbf{w}$. Thus $\mathbf{w}+\mathbf{x}$ and $\alpha\mathbf{w}$ are in range T. ■

EXAMPLE 3 Let $T\mathbf{v} = \mathbf{0}$ for every $\mathbf{v} \in V$. (T is the zero transformation.) Then ker $T = V$ and range $T = \{\mathbf{0}\}$.

EXAMPLE 4 Let $T\mathbf{v} = \mathbf{v}$ for every $\mathbf{v} \in V$. (T is the identity transformation.) Then ker $T = \{\mathbf{0}\}$ and range $T = V$.

The zero and identity transformations provide two extremes. In the first, everything is in the kernel. In the second, only the zero vector is in the kernel. The cases in between are more interesting.

EXAMPLE 5 Let T: $\mathbb{R}^3 \to \mathbb{R}^3$ be defined by $T\begin{pmatrix} x \\ y \\ z \end{pmatrix} = \begin{pmatrix} x \\ y \\ 0 \end{pmatrix}$. That is (see Example 6.1.10, page 234): T is the projection operator from $\mathbb{R}^3$ into the xy-plane. If $T\begin{pmatrix} x \\ y \\ z \end{pmatrix} = \begin{pmatrix} x \\ y \\ 0 \end{pmatrix} = \mathbf{0} = \begin{pmatrix} 0 \\ 0 \\ 0 \end{pmatrix}$, then $x = y = 0$. Thus $\ker T = \{(x, y, z): x = y = 0\}$ = the z-axis, and range $T = \{(x, y, z): z = 0\}$ = the xy-plane. Note that dim ker $T = 1$ and dim range $T = 2$.

DEFINITION 2 **NULLITY AND RANK OF A LINEAR TRANSFORMATION** If T is a linear transformation from V to W, then we define:

$$\textbf{Nullity of } T = \nu(T) = \dim \ker T \tag{4}$$

$$\textbf{Rank of } T = \rho(T) = \dim \text{range } T \tag{5}$$

In Section 6.3 we define the rank and nullity of a matrix and show how they are related to the rank and nullity of a linear transformation. Note that in the last example $\nu(T) = 1$ and $\rho(T) = 2$.

EXAMPLE 6 Let H be a subspace of $\mathbb{R}^n$ and define (as in Example 6.1.9, page 234) $T\mathbf{v} = \text{proj}_H\, \mathbf{v}$. Clearly range $T = H$. From Theorem 5.7.6, page 219, we can write any $\mathbf{v} \in V$ as $\mathbf{v} = \mathbf{h} + \mathbf{p} = \text{proj}_H\, \mathbf{v} + \text{proj}_{H^\perp}\, \mathbf{v}$. If $T\mathbf{v} = \mathbf{0}$, then $\mathbf{h} = \mathbf{0}$, which means that $\mathbf{v} = \mathbf{p} \in H^\perp$. Thus $\ker T = H^\perp$, $\rho(T) = \dim H$, and $\nu(T) = \dim H^\perp = n - \rho(T)$.

EXAMPLE 7 Let T: $\mathbb{R}^n \to \mathbb{R}^m$ be given by $T\mathbf{v} = A\mathbf{v}$, where A is an $m \times n$ matrix. Then $\ker T = \{\mathbf{v}: A\mathbf{v} = \mathbf{0}\}$ and range $T = \{\mathbf{w}: A\mathbf{v} = \mathbf{w}$ for some $\mathbf{v} \in \mathbb{R}^n\}$. These two subspaces are called, respectively, the **kernel** and **range** of the matrix A. We shall discuss these in great detail in Section 6.3.

EXAMPLE 8 Let $V = M_{mn}$ and define T: $M_{mn} \to M_{mn}$ by $T(A) = A^t$ (see Example 6.1.6, page 232). If $TA = A^t = 0$, then A^t is the $n \times m$ zero matrix so that A is the $m \times n$ zero matrix. Thus $\ker T = \{0\}$ and, clearly, range $T = M_{nm}$. This means that $\nu(T) = 0$ and $\rho(T) = nm$.

EXAMPLE 9 Let T: $P_3 \to P_2$ be defined by $T(p) = T(a_0 + a_1x + a_2x^2 + a_3x^3) = a_0 + a_1x + a_2x^2$. Then if $T(p) = 0$, $a_0 + a_1x + a_2x^2 = 0$ for every x, which implies that $a_0 = a_1 = a_2 = 0$. Thus $\ker T = \{p \in P_3: p(x) = a_3x^3\}$ and range $T = P_2$, $\nu(T) = 1$, and $\rho(T) = 3$.

EXAMPLE 10 Let $V = C[0, 1]$ and define $J: C[0, 1] \to \mathbb{R}$ by $Jf = \int_0^1 f(x)\, dx$ (see Example 6.1.12, page 234). Then $\ker J = \{f \in C[0, 1]: \int_0^1 f(x)\, dx = 0\}$. Let α be a real number. Then the constant function $f(x) = \alpha$ for $x \in [0, 1]$ is in $C[0, 1]$ and $\int_0^1 \alpha\, dx = \alpha$. Since this is true for every real number α, we have range $J = \mathbb{R}$.

In the next section we shall see how to find the range and kernel of a matrix. In Section 6.4 we shall see how every linear transformation from one finite dimensional vector space to another can be represented by a matrix. This will enable us to compute the kernel and range of any linear transformation between finite dimensional vector spaces by finding the kernel and range of a corresponding matrix.

PROBLEMS 6.2

In Problems 1–10 find the kernel, range, rank, and nullity of the given linear transformation.

1. $T: \mathbb{R}^2 \to \mathbb{R}^2;\ T\begin{pmatrix} x \\ y \end{pmatrix} = \begin{pmatrix} x \\ 0 \end{pmatrix}$ **2.** $T: \mathbb{R}^3 \to \mathbb{R}^2;\ T\begin{pmatrix} x \\ y \\ z \end{pmatrix} = \begin{pmatrix} z \\ y \end{pmatrix}$

3. $T: \mathbb{R}^2 \to \mathbb{R};\ T\begin{pmatrix} x \\ y \end{pmatrix} = x + y$ **4.** $T: \mathbb{R}^4 \to \mathbb{R}^2;\ T\begin{pmatrix} x \\ y \\ z \\ w \end{pmatrix} = \begin{pmatrix} x + z \\ y + w \end{pmatrix}$

5. $T: M_{22} \to M_{22}$: $T(A) = AB$, where $B = \begin{pmatrix} 1 & 2 \\ 0 & 1 \end{pmatrix}$

6. $T: \mathbb{R} \to P_3$: $T(a) = a + ax + ax^2 + ax^3$

***7.** $T: M_{nn} \to M_{nn}$: $T(A) = A' + A$ □**8.** $T: C'[0, 1] \to C[0, 1];\ Tf = f'$

9. $T: C[0, 1] \to \mathbb{R};\ Tf = f(\frac{1}{2})$

10. $T: \mathbb{R}^2 \to \mathbb{R}^2$: T is rotation through an angle of $\pi/3$

11. Let $T: V \to W$ be a linear transformation, let $\{\mathbf{v}_1, \mathbf{v}_2, \ldots, \mathbf{v}_n\}$ be a basis for V, and suppose that $T\mathbf{v}_i = \mathbf{0}$ for $i = 1, 2, \ldots, n$. Show that T is the zero transformation.

12. In Problem 11, suppose that $W = V$ and $T\mathbf{v}_i = \mathbf{v}_i$ for $i = 1, 2, \ldots, n$. Show that T is the identity operator.

13. Let $T: V \to \mathbb{R}^3$. Prove that range T is either **(a)** $\{\mathbf{0}\}$, **(b)** a line through the origin, **(c)** a plane through the origin, or **(d)** $\mathbb{R}^3$.

14. Let $T: \mathbb{R}^3 \to V$. Show that $\ker T$ is one of four spaces listed in Problem 13.

15. Find all linear transformations from $\mathbb{R}^2$ into $\mathbb{R}^2$ such that the line $y = 0$ is carried into the line $x = 0$.

16. Find all linear transformations from $\mathbb{R}^2$ into $\mathbb{R}^2$ that carry the line $y = ax$ into the line $y = bx$.

17. Find a linear transformation T from $\mathbb{R}^3 \to \mathbb{R}^3$ such that $\ker T = \{(x, y, z): 2x - y + z = 0\}$.

18. Find a linear transformation T from $\mathbb{R}^3 \to \mathbb{R}^3$ such that range $T = \{(x, y, z): 2x - y + z = 0\}$.

19. Let $T: M_{nn} \to M_{nn}$ be defined by $TA = A - A'$. Show that $\ker T = \{$symmetric $n \times n$ matrices$\}$ and range of $T = \{$skew-symmetric $n \times n$ matrices$\}$.

□***20.** Let $T: C'[0, 1] \to C[0, 1]$ be defined by $Tf(x) = xf'(x)$. Find the kernel and range of T.

6.3 The Rank and Nullity of a Matrix

Let A be an $m \times n$ matrix and let

$$N_A = \{\mathbf{x} \in \mathbb{R}^n: \quad A\mathbf{x} = \mathbf{0}\} \tag{1}$$

Then, as we saw in Example 5.5.10 on page 197, N_A is a subspace of $\mathbb{R}^n$. Moreover, since T_A: $\mathbb{R}^n \to \mathbb{R}^m$ defined by $T\mathbf{x} = A\mathbf{x}$ is a linear operator, we see that $N_A = \ker T_A$.

DEFINITION 1 **KERNEL AND NULLITY OF A MATRIX** N_A is called the **kernel** of A and $\nu(A) = \dim N_A$ is called the **nullity** of A. If N_A contains only the zero vector, then $\nu(A) = 0$.

Remark. In Section 6.2 we defined the rank and nullity of a linear transformation. In Example 6.1.7 we showed that if A is an $m \times n$ matrix, then the transformation T defined by $T\mathbf{x} = A\mathbf{x}$ is a linear transformation from $\mathbb{R}^n \to \mathbb{R}^m$. We can see that $\nu(T) = \nu(A)$ and this shows that the nullity of a matrix is a special case of the nullity of a linear transformation. The same fact holds for the rank of a matrix (which we define shortly).

EXAMPLE 1 Let $A = \begin{pmatrix} 1 & 2 & -1 \\ 2 & -1 & 3 \end{pmatrix}$. Then, as we saw in Example 5.5.11 on page 197, N_A is spanned by $\begin{pmatrix} -1 \\ 1 \\ 1 \end{pmatrix}$ and $\nu(A) = 1$.

EXAMPLE 2 Let $A = \begin{pmatrix} 2 & -1 & 3 \\ 4 & -2 & 6 \\ -6 & 3 & -9 \end{pmatrix}$. Then, by Example 5.5.12 on page 198, $\left\{\begin{pmatrix} 1 \\ 2 \\ 0 \end{pmatrix}, \begin{pmatrix} 0 \\ 3 \\ 1 \end{pmatrix}\right\}$ is a basis for N_A and $\nu(A) = 2$.

THEOREM 1 Let A be an $n \times n$ matrix. Then A is invertible if and only if $\nu(A) = 0$.

Proof By our Summing Up Theorem (Theorem 3.4.4, page 110, parts (*i*) and (*ii*)), A is invertible if and only if the homogeneous system $A\mathbf{x} = \mathbf{0}$ has only the trivial solution $\mathbf{x} = \mathbf{0}$. But, from Equation (1), this means that A is invertible if and only if $N_A = \{\mathbf{0}\}$. Thus A is invertible if and only if $\nu(A) = \dim N_A = 0$. ■

Let A be an $m \times n$ matrix and let

$$R_A = \{\mathbf{y} \in \mathbb{R}^m: \quad A\mathbf{x} = \mathbf{y} \text{ for some } \mathbf{x} \in \mathbb{R}^n\} \tag{2}$$

Then R_A is a subspace of $\mathbb{R}^m$ since $R_A = \text{range } T_A$, where $T_A\mathbf{x} = A\mathbf{x}$.

DEFINITION 2 **RANGE AND RANK OF A MATRIX** R_A is called the **range** of A and $\rho(A) = \dim R_A$ is called the **rank** of A.

Before giving examples, we shall give two definitions and a theorem that make the calculation of rank relatively easy.

DEFINITION 3 **ROW AND COLUMN SPACE OF A MATRIX** If A is an $m \times n$ matrix, let $\{\mathbf{r}_1, \mathbf{r}_2, \ldots, \mathbf{r}_m\}$ denote the rows of A and let $\{\mathbf{c}_1, \mathbf{c}_2, \ldots, \mathbf{c}_n\}$ denote the columns of A. Then we define

$$S_A = \textbf{row space } \textit{of } A = \text{span}\,\{\mathbf{r}_1, \mathbf{r}_2, \ldots, \mathbf{r}_m\} \tag{3}$$

and

$$C_A = \textbf{column space } \textit{of } A = \text{span}\,\{\mathbf{c}_1, \mathbf{c}_2, \ldots, \mathbf{c}_n\} \tag{4}$$

Note. S_A is a subspace of $\mathbb{R}^n$ and C_A is a subspace of $\mathbb{R}^m$.

We have introduced a lot of notation in just two pages. Let us stop for a moment to illustrate these ideas with an example.

EXAMPLE 3 Let $A = \begin{pmatrix} 1 & 2 & -1 \\ 2 & -1 & 3 \end{pmatrix}$. A is a 2×3 matrix.

(a) *The kernel of* $A = N_A = \{\mathbf{x} \in \mathbb{R}^3 \colon A\mathbf{x} = \mathbf{0}\}$. As we saw in Example 1,

$$N_A = \text{span}\left\{\begin{pmatrix} -1 \\ 1 \\ 1 \end{pmatrix}\right\}.$$

(b) *The nullity of* $A = \nu(A) = \dim N_A = 1$.

(c) *The range of* $A = R_A = \{\mathbf{y} \in \mathbb{R}^2 \colon A\mathbf{x} = \mathbf{y}$ for some $\mathbf{x} \in \mathbb{R}^3\}$. Let $\mathbf{y} = \begin{pmatrix} y_1 \\ y_2 \end{pmatrix}$ be in $\mathbb{R}^2$. Then, if $\mathbf{y} \in R_A$, there is an $\mathbf{x} \in \mathbb{R}^3$ such that $A\mathbf{x} = \mathbf{y}$. Writing $\mathbf{x} = \begin{pmatrix} x_1 \\ x_2 \\ x_3 \end{pmatrix}$, we have

$$\begin{pmatrix} 1 & 2 & -1 \\ 2 & -1 & 3 \end{pmatrix}\begin{pmatrix} x_1 \\ x_2 \\ x_3 \end{pmatrix} = \begin{pmatrix} y_1 \\ y_2 \end{pmatrix}$$

or

$$\begin{aligned} x_1 + 2x_2 - \ x_3 &= y_1 \\ 2x_1 - \ x_2 + 3x_3 &= y_2. \end{aligned}$$

Row reducing this system, we have

$$\left(\begin{array}{rrr|c} 1 & 2 & -1 & y_1 \\ 2 & -1 & 3 & y_2 \end{array}\right) \xrightarrow{A_{1,2}(-2)} \left(\begin{array}{rrr|c} 1 & 2 & -1 & y_1 \\ 0 & -5 & 5 & y_2 - 2y_1 \end{array}\right)$$

$$\xrightarrow{M_2(-\frac{1}{5})} \left(\begin{array}{rrr|c} 1 & 2 & -1 & y_1 \\ 0 & 1 & -1 & \dfrac{2y_1 - y_2}{5} \end{array}\right) \xrightarrow{A_{2,1}(-2)} \left(\begin{array}{rrr|c} 1 & 0 & 1 & \dfrac{y_1 + 2y_2}{5} \\ 0 & 1 & -1 & \dfrac{2y_1 - y_2}{5} \end{array}\right)$$

Thus if x_3 is chosen arbitrarily, we see that

$$x_1 = -x_3 + \frac{y_1 + 2y_2}{5} \quad \text{and} \quad x_2 = x_3 + \frac{2y_1 - y_2}{5}$$

That is, for every $\mathbf{y} = \begin{pmatrix} y_1 \\ y_2 \end{pmatrix} \in \mathbb{R}^2$, there are an infinite number of vectors $\mathbf{x} \in \mathbb{R}^3$ such that $A\mathbf{x} = \mathbf{y}$. Thus $R_A = \mathbb{R}^2$. Note, for example, that if $\mathbf{y} = \begin{pmatrix} 2 \\ -3 \end{pmatrix}$, then, choosing $x_3 = 0$ (the simplest choice), we have

$$x_1 = \frac{2 + 2(-3)}{5} = -\tfrac{4}{5} \quad \text{and} \quad x_2 = \frac{2(2) - (-3)}{5} = \tfrac{7}{5}$$

and

$$A\mathbf{x} = \begin{pmatrix} 1 & 2 & -1 \\ 2 & -1 & 3 \end{pmatrix} \begin{pmatrix} -\frac{4}{5} \\ \frac{7}{5} \\ 0 \end{pmatrix} = \begin{pmatrix} \frac{10}{5} \\ -\frac{15}{5} \end{pmatrix} = \begin{pmatrix} 2 \\ -3 \end{pmatrix} = \mathbf{y}$$

(d) *The rank of* $A = \rho(A) = \dim R_A = \dim \mathbb{R}^2 = 2$.

(e) *The row space of* $A = S_A = \text{span}\,\{(1, 2, -1), (2, -1, 3)\}$. Since these two vectors are linearly independent, we see that S_A is a two-dimensional subspace of $\mathbb{R}^3$. From Example 5.5.9 on page 197, we observe that S_A is a plane passing through the origin.

(f) *The column space of* $A = C_A = \text{span}\left\{\begin{pmatrix} 1 \\ 2 \end{pmatrix}, \begin{pmatrix} 2 \\ -1 \end{pmatrix}, \begin{pmatrix} -1 \\ 3 \end{pmatrix}\right\} = \mathbb{R}^2$ since $\begin{pmatrix} 1 \\ 2 \end{pmatrix}$ and $\begin{pmatrix} 2 \\ -1 \end{pmatrix}$, being linearly independent, compose a basis for $\mathbb{R}^2$.

In Example 2 we may observe that $R_A = C_A = \mathbb{R}^2$ and $\dim S_A = \dim C_A = \dim R_A = \rho(A) = 2$. This is no coincidence.

THEOREM 2 If A is an $m \times n$ matrix, then:

i. $C_A = R_A$†
ii. $\dim S_A = \dim C_A = \dim R_A = \rho(A)$

The proof of this theorem is not difficult, but it is quite long. We shall defer it to the end of the section.

EXAMPLE 4 Find a basis for R_A and determine the rank of $A = \begin{pmatrix} 2 & -1 & 3 \\ 4 & -2 & 6 \\ -6 & 3 & -9 \end{pmatrix}$.

Solution Since $\mathbf{r}_2 = 2\mathbf{r}_1$ and $\mathbf{r}_3 = -3\mathbf{r}_1$, we see that $\rho(A) = 1$. Thus any column in C_A is a basis for C_A $(= R_A)$. For example, $\begin{pmatrix} 2 \\ 4 \\ -6 \end{pmatrix}$ is a basis for R_A.

The following theorem will simplify our computations.

THEOREM 3 If A is row (or column) equivalent to B, then $\rho(A) = \rho(B)$ and $\nu(A) = \nu(B)$.

Proof Recall from Definition 2.7.3, page 68, that A is row equivalent to B if A can be "reduced" to B by elementary row operations. The definition for "column equivalent" is similar. Now interchanging rows of A leaves the same number of linearly independent rows as does multiplying any row by a nonzero constant. Suppose that $\mathbf{r}_1, \mathbf{r}_2, \ldots, \mathbf{r}_m$ are linearly independent. Consider the set $S = \{\mathbf{r}_1, \mathbf{r}_2, \ldots, \mathbf{r}_{i-1}, \mathbf{r}_i + \alpha\mathbf{r}_j, \mathbf{r}_{i+1}, \ldots, \mathbf{r}_j, \ldots, \mathbf{r}_m\}$. Suppose that $c_1\mathbf{r}_1 + c_2\mathbf{r}_2 + \cdots + c_{i-1}\mathbf{r}_{i-1} + c_i(\mathbf{r}_i + \alpha\mathbf{r}_j) + c_{i+1}\mathbf{r}_{i+1} + \cdots + c_j\mathbf{r}_j + \cdots + c_m\mathbf{r}_m = \mathbf{0}$. Then

$$c_1\mathbf{r}_1 + c_2\mathbf{r}_2 + \cdots + c_{i-1}\mathbf{r}_{i-1} + c_i\mathbf{r}_i + c_{i+1}\mathbf{r}_{i+1} + \cdots + (\alpha c_i + c_j)\mathbf{r}_j + \cdots + c_m\mathbf{r}_m = \mathbf{0}$$

By independence, we have $c_1 = c_2 = \cdots = c_{i-1} = c_i = c_{i+1} = \cdots = \alpha c_i + c_j = \cdots = c_m = 0$. Moreover, $-c_j = \alpha c_i = \alpha 0 = 0$. This implies that S is a linearly independent set. Therefore we have shown that if we add a constant multiple of one row to another, we do not change the rank. Thus $\rho(A) = \rho(B)$. The other part is easier. The equation $A\mathbf{x} = \mathbf{0}$ is a homogeneous system of m equations in n unknowns. If B is obtained from A by row reduction, then, as we saw in Chapter 1, the solutions to the system are unchanged. Thus if $A\mathbf{x}_1 = \mathbf{0}$, then $B\mathbf{x}_1 = \mathbf{0}$ and vice versa. This means that $N_A = N_B$ so that $\nu(A) = \nu(B)$. ■

---

† Do not be confused by this notation. It indicates that the column space of A is the same as the range of A. It does *not* say that the column space of A equals the row space of A. This last statement is usually false. It is always false if A is not a square matrix because then C_A, the column space of A, contains m-vectors and S_A, the row space of A, contains n-vectors.

EXAMPLE 5 Determine the rank of $A=\begin{pmatrix} 1 & -1 & 3 \\ 2 & 0 & 4 \\ -1 & -3 & 1 \end{pmatrix}$.

Solution We row-reduce to obtain a simpler matrix:

$$\begin{pmatrix} 1 & -1 & 3 \\ 2 & 0 & 4 \\ -1 & -3 & 1 \end{pmatrix} \xrightarrow[A_{1,3}(1)]{A_{1,2}(-2)} \begin{pmatrix} 1 & -1 & 3 \\ 0 & 2 & -2 \\ 0 & -4 & 4 \end{pmatrix}$$

$$\xrightarrow{M_2(\frac{1}{2})} \begin{pmatrix} 1 & -1 & 3 \\ 0 & 1 & -1 \\ 0 & -4 & 4 \end{pmatrix} \xrightarrow{A_{2,3}(4)} \begin{pmatrix} 1 & -1 & 3 \\ 0 & 1 & -1 \\ 0 & 0 & 0 \end{pmatrix} = B$$

Since B has two independent rows, we have $\rho(B)=\rho(A)=2$.

The next theorem gives the relationship between rank and nullity.

THEOREM 4 Let A be an $m\times n$ matrix. Then

$$\rho(A)+\nu(A)=n \tag{5}$$

Proof We assume that $k=\rho(A)$ and that the first k columns of A are linearly independent. Let $\mathbf{c}_i$ $(i>k)$ denote any other column of A. Since $\mathbf{c}_1, \mathbf{c}_2, \ldots, \mathbf{c}_k$ form a basis for C_A, we have, for some scalars $a_1, a_2, \ldots, a_k$

$$\mathbf{c}_i = a_1\mathbf{c}_1 + a_2\mathbf{c}_2 + \cdots + a_k\mathbf{c}_k \tag{6}$$

Thus, by adding $-a_1\mathbf{c}_1, -a_2\mathbf{c}_2, \ldots, -a_k\mathbf{c}_k$ successively to the ith column of A, we obtain a new $m\times n$ matrix B with $\rho(B)=\rho(A)$ and $\nu(B)=\nu(A)$ with the ith column of $B=\mathbf{0}$. We do this to all other columns of A (except the first k) to obtain the matrix

$$D=\begin{pmatrix} a_{11} & a_{12} & \cdots & a_{1k} & 0 & 0 & \cdots & 0 \\ a_{21} & a_{22} & \cdots & a_{2k} & 0 & 0 & \cdots & 0 \\ \vdots & \vdots & & \vdots & \vdots & \vdots & & \vdots \\ a_{m1} & a_{m2} & \cdots & a_{mk} & 0 & 0 & \cdots & 0 \end{pmatrix} \tag{7}$$

where $\rho(D)=\rho(A)$ and $\nu(D)=\nu(A)$. By possibly rearranging the rows of D we can assume that the first k rows of D are independent. Then we do the same thing to the rows, (i.e., add multiples of the first k rows to the last $m-k$ rows) to obtain a new matrix:

$$F=\begin{pmatrix} a_{11} & a_{12} & \cdots & a_{1k} & 0 & \cdots & 0 \\ a_{21} & a_{22} & \cdots & a_{2k} & 0 & \cdots & 0 \\ \vdots & \vdots & & \vdots & \vdots & & \vdots \\ a_{k1} & a_{k2} & \cdots & a_{kk} & 0 & \cdots & 0 \\ 0 & 0 & \cdots & 0 & 0 & \cdots & 0 \\ \vdots & \vdots & & \vdots & \vdots & & \vdots \\ 0 & 0 & \cdots & 0 & 0 & \cdots & 0 \end{pmatrix}$$

where $\rho(F)=\rho(A)$ and $\nu(F)=\nu(A)$. It is now obvious that if $i>k$, then $F\mathbf{e}_i=\mathbf{0}$,† so $E_k=\{\mathbf{e}_{k+1}, \mathbf{e}_{k+2}, \ldots, \mathbf{e}_n\}$ is a linearly independent set of $n-k$ vectors in N_F. We now show that E_k spans N_F. Let the vector $\mathbf{x}\in N_F$ have the form

$$\mathbf{x}=\begin{pmatrix} x_1 \\ x_2 \\ \vdots \\ x_k \\ \vdots \\ x_n \end{pmatrix}$$

Then

$$\mathbf{0}=F\mathbf{x}=\begin{pmatrix} a_{11}x_1+a_{12}x_2+\cdots+a_{1k}x_k \\ a_{21}x_1+a_{22}x_2+\cdots+a_{2k}x_k \\ \vdots \\ a_{k1}x_1+a_{k2}x_2+\cdots+a_{kk}x_k \\ 0 \\ \vdots \\ 0 \end{pmatrix}=\begin{pmatrix} 0 \\ 0 \\ \vdots \\ 0 \end{pmatrix}$$

The determinant of the matrix of the $k\times k$ homogeneous system described above is nonzero, since the rows of this matrix are linearly independent. Thus the only solution to the system is $x_1=x_2=\cdots=x_k=0$. Thus $\mathbf{x}$ has the form

$$(0, 0, \ldots, 0, x_{k+1}, x_{k+2}, \ldots, x_n)=x_{k+1}\mathbf{e}_{k+1}+x_{k+2}\mathbf{e}_{k+2}+\cdots+x_n\mathbf{e}_n$$

† Recall that $\mathbf{e}_i$ is the vector with a 1 in the ith position and a zero everywhere else.

This means that E_k spans N_F so that $\nu(F)=n-k=n-\rho(F)$. This completes the proof. ■

EXAMPLE 6 For $A=\begin{pmatrix}1 & 2 & -1\\ 2 & -1 & 3\end{pmatrix}$ we calculated (in Examples 1 and 3) that $\rho(A)=2$ and $\nu(A)=1$; this illustrates that $\rho(A)+\nu(A)=n$ $(=3)$.

EXAMPLE 7 For $A=\begin{pmatrix}1 & -1 & 3\\ 2 & 0 & 4\\ -1 & -3 & 1\end{pmatrix}$ calculate $\nu(A)$.

Solution In Example 5 we found that $\rho(A)=2$. Thus $\nu(A)=3-2=1$.

THEOREM 5 Let A be an $n\times n$ matrix. Then A is invertible if and only if $\rho(A)=n$.

Proof By Theorem 1, A is invertible if and only if $\nu(A)=0$. But, by Theorem 4, $\rho(A)=n-\nu(A)$. Thus A is invertible if and only if $\rho(A)=n-0=n$. ■

We next show how the notion of rank can be used to solve linear systems of equations. Again we consider the system of m equations in n unknowns

$$\begin{aligned} a_{11}x_1 + a_{12}x_2 + \cdots + a_{1n}x_n &= b_1\\ a_{21}x_1 + a_{22}x_2 + \cdots + a_{2n}x_n &= b_2\\ &\vdots\\ a_{m1}x_1 + a_{m2}x_2 + \cdots + a_{mn}x_n &= b_m \end{aligned} \tag{8}$$

which we write as $A\mathbf{x}=\mathbf{b}$. We use the symbol $(A,\mathbf{b})$ to denote the $m\times(n+1)$ augmented matrix obtained (as in Section 2.5) by adjoining the vector $\mathbf{b}$ to A.

THEOREM 6 The system $A\mathbf{x}=\mathbf{b}$ has at least one solution if and only if A and the augmented matrix $(A,\mathbf{b})$ have the same rank.

Proof If $\mathbf{c}_1, \mathbf{c}_2, \ldots, \mathbf{c}_n$ are the columns of A, then we can write system (8) as

$$x_1\mathbf{c}_1 + x_2\mathbf{c}_2 + \cdots + x_n\mathbf{c}_n = \mathbf{b} \tag{9}$$

System (9) will have a solution if and only if $\mathbf{b}$ can be written as a linear combination of the columns of A. That is, to have a solution we must have $\mathbf{b}\in C_A$. If $\mathbf{b}\in C_A$, then $(A,\mathbf{b})$ has the same number of linearly independent columns as A so that A and $(A, \mathbf{b})$ have the same rank. If $\mathbf{b} \notin C_A$, then $\rho(A, \mathbf{b}) = \rho(A) + 1$ and the system has no solutions. This completes the proof. ■

EXAMPLE 8 Determine whether the system

$$\begin{aligned} 2x_1 + 4x_2 + 6x_3 &= 18 \\ 4x_1 + 5x_2 + 6x_3 &= 24 \\ 2x_1 + 7x_2 + 12x_3 &= 40 \end{aligned}$$

has solutions.

Solution Let $A = \begin{pmatrix} 2 & 4 & 6 \\ 4 & 5 & 6 \\ 2 & 7 & 12 \end{pmatrix}$. Then we row-reduce to obtain, successively,

$$\xrightarrow{M_1(\frac{1}{2})} \begin{pmatrix} 1 & 2 & 3 \\ 4 & 5 & 6 \\ 2 & 7 & 12 \end{pmatrix} \xrightarrow[A_{1,3}(-2)]{A_{1,2}(-4)} \begin{pmatrix} 1 & 2 & 3 \\ 0 & -3 & -6 \\ 0 & 3 & 6 \end{pmatrix}$$

$$\xrightarrow{M_2(-\frac{1}{3})} \begin{pmatrix} 1 & 2 & 3 \\ 0 & 1 & 2 \\ 0 & 3 & 6 \end{pmatrix} \xrightarrow[A_{2,3}(-3)]{A_{2,1}(-2)} \begin{pmatrix} 1 & 0 & -1 \\ 0 & 1 & 2 \\ 0 & 0 & 0 \end{pmatrix}$$

Thus $\rho(A) = 2$. Similarly, we row-reduce $(A, \mathbf{b})$ to obtain

$$\left(\begin{array}{ccc|c} 2 & 4 & 6 & 18 \\ 4 & 5 & 6 & 24 \\ 2 & 7 & 12 & 40 \end{array}\right) \xrightarrow{M_1(\frac{1}{2})} \left(\begin{array}{ccc|c} 1 & 2 & 3 & 9 \\ 4 & 5 & 6 & 24 \\ 2 & 7 & 12 & 40 \end{array}\right)$$

$$\xrightarrow[A_{1,3}(-2)]{A_{1,2}(-4)} \left(\begin{array}{ccc|c} 1 & 2 & 3 & 9 \\ 0 & -3 & -6 & -12 \\ 0 & 3 & 6 & 22 \end{array}\right)$$

$$\xrightarrow{M_2(-\frac{1}{3})} \left(\begin{array}{ccc|c} 1 & 2 & 3 & 9 \\ 0 & 1 & 2 & 4 \\ 0 & 3 & 6 & 22 \end{array}\right) \xrightarrow[A_{2,3}(-3)]{A_{2,1}(-2)} \left(\begin{array}{ccc|c} 1 & 0 & -1 & 1 \\ 0 & 1 & 2 & 4 \\ 0 & 0 & 0 & 10 \end{array}\right)$$

It is easy to see that the last three columns of the last matrix are linearly independent. Thus $\rho(A, \mathbf{b}) = 3$ and there are no solutions to the system.

EXAMPLE 9 Determine whether the system

$$\begin{aligned} x_1 - x_2 + 2x_3 &= 4 \\ 2x_1 + x_2 - 3x_3 &= -2 \\ 4x_1 - x_2 + x_3 &= 6 \end{aligned}$$

has solutions.

Solution Let $A = \begin{pmatrix} 1 & -1 & 2 \\ 2 & 1 & -3 \\ 4 & -1 & 1 \end{pmatrix}$. Then $\det A = 0$, so $\rho(A) < 3$. Since the first column is not a multiple of the second, we see that the first two columns are linearly independent; hence $\rho(A) = 2$. To compute $\rho(A, \mathbf{b})$, we row-reduce:

$$\left(\begin{array}{ccc|c} 1 & -1 & 2 & 4 \\ 2 & 1 & -3 & -2 \\ 4 & -1 & 1 & 6 \end{array}\right) \xrightarrow[A_{1,3}(-4)]{A_{1,2}(-2)} \left(\begin{array}{ccc|c} 1 & -1 & 2 & 4 \\ 0 & 3 & -7 & -10 \\ 0 & 3 & -7 & -10 \end{array}\right)$$

We see that $\rho(A, \mathbf{b}) = 2$ and there are an infinite number of solutions to the system. (If there were a unique solution, we would have $\det A \neq 0$.)

The results of this section allow us to improve on our Summing Up Theorem—last seen in Section 3.4, page 110.

THEOREM 7 **SUMMING UP THEOREM—VIEW 4** A be an $n \times n$ matrix. Then the following eight statements are equivalent. That is, if one is true, all are true.

- **i.** A is invertible.
- **ii.** The only solution to the homogeneous system $A\mathbf{x} = \mathbf{0}$ is the trivial solution ($\mathbf{x} = \mathbf{0}$).
- **iii.** The system $A\mathbf{x} = \mathbf{b}$ has a unique solution for every n-vector $\mathbf{b}$.
- **iv.** A is row equivalent to the $n \times n$ identity matrix I_n.
- **v.** The rows (and columns) of A are linearly independent.
- **vi.** $\det A \neq 0$.
- **vii.** $\nu(A) = 0$.
- **viii.** $\rho(A) = n$.

Moreover, if one of the above fails to hold, then for every vector $\mathbf{b} \in \mathbb{R}^n$, the system $A\mathbf{x} = \mathbf{b}$ has either no solution or an infinite number of solutions. It has an infinite number of solutions if and only if $\rho(A) = \rho((A, \mathbf{b}))$.

We conclude this section with a proof of Theorem 2.

Proof of Theorem 2† We first show that $C_A = R_A$. As before we let $\mathbf{e}_j$ denote the vector in $\mathbb{R}^n$ with a 1 in the jth position and zero everywhere else. We write A in the form

$$A = \begin{pmatrix} a_{11} & a_{12} & \cdots & a_{1j} & \cdots & a_{1n} \\ a_{21} & a_{22} & \cdots & a_{2j} & \cdots & a_{2n} \\ \vdots & \vdots & & \vdots & & \vdots \\ a_{m1} & a_{m2} & \cdots & a_{mj} & \cdots & a_{mn} \end{pmatrix}$$

† If time permits.

Then $A\mathbf{e}_j$ is the jth column of A. Thus each column of A is in R_A so that

$$C_A \subseteq R_A \tag{10}$$

Let $\{\mathbf{y}_1, \ldots, \mathbf{y}_k\}$ be a basis for R_A. Now let us look at one of the basis vectors, say $\mathbf{y}_i$. We have, by the definition of the range, $\mathbf{y}_i = A\mathbf{x}_i$ for some $\mathbf{x}_i \in \mathbb{R}^n$. But $\{\mathbf{e}_1, \mathbf{e}_2, \ldots, \mathbf{e}_n\}$ is a basis in $\mathbb{R}^n$, so there exist constants $c_1, c_2, \ldots, c_n$ such that

$$\mathbf{x}_i = c_1\mathbf{e}_1 + c_2\mathbf{e}_2 + \cdots + c_n\mathbf{e}_n \tag{11}$$

Then

$$\mathbf{y}_i = A\mathbf{x}_i = A(c_1\mathbf{e}_1 + c_2\mathbf{e}_2 + \cdots + c_n\mathbf{e}_n) = c_1A\mathbf{e}_1 + c_2A\mathbf{e}_2 + \cdots + c_nA\mathbf{e}_n \tag{12}$$

But $A\mathbf{e}_j$ is the jth column of A so (12) shows that we can write $\mathbf{y}_i$ as a linear combination of the columns of A. Thus each basis vector in R_A is in the column space C_A of A so that

$$R_A \subseteq C_A \tag{13}$$

Combining (10) and (13) we see that $R_A = C_A$. To complete the proof we must show that if S_A denotes the row space of A, then $\dim S_A = \dim C_A$. We denote the rows of A by $\mathbf{r}_1, \mathbf{r}_2, \ldots, \mathbf{r}_m$, and let $k = \dim S_A$. Let $S = \{\mathbf{s}_1, \mathbf{s}_2, \ldots, \mathbf{s}_k\}$ be a basis for S_A. Then every row vector of A can be written as a linear combination of the vectors in S, and we have, for some constants α_{ij}:

$$\begin{aligned}
\mathbf{r}_1 &= \alpha_{11}\mathbf{s}_1 + \alpha_{12}\mathbf{s}_2 + \cdots + \alpha_{1k}\mathbf{s}_k \\
\mathbf{r}_2 &= \alpha_{21}\mathbf{s}_1 + \alpha_{22}\mathbf{s}_2 + \cdots + \alpha_{2k}\mathbf{s}_k \\
&\vdots \\
\mathbf{r}_m &= \alpha_{m1}\mathbf{s}_1 + \alpha_{m2}\mathbf{s}_2 + \cdots + \alpha_{mk}\mathbf{s}_k
\end{aligned} \tag{14}$$

Now, the jth component of $\mathbf{r}_i$ is a_{ij}. Thus, if we equate the jth components of both sides of (14), we obtain

$$\begin{aligned}
a_{1j} &= \alpha_{11}s_{1j} + \alpha_{12}s_{2j} + \cdots + \alpha_{1k}s_{kj} \\
a_{2j} &= \alpha_{2i}s_{1j} + \alpha_{22}s_{2j} + \cdots + \alpha_{2k}s_{kj} \\
&\vdots \\
a_{mj} &= \alpha_{m1}s_{1j} + \alpha_{m2}s_{2j} + \cdots + \alpha_{mk}s_{kj}
\end{aligned}$$

or

$$\begin{pmatrix} a_{1j} \\ a_{2j} \\ \cdot \\ \cdot \\ \cdot \\ a_{mj} \end{pmatrix} = s_{1j}\begin{pmatrix} \alpha_{11} \\ \alpha_{21} \\ \cdot \\ \cdot \\ \cdot \\ \alpha_{m1} \end{pmatrix} + s_{2j}\begin{pmatrix} \alpha_{12} \\ \alpha_{22} \\ \cdot \\ \cdot \\ \cdot \\ \alpha_{m2} \end{pmatrix} + \cdots + s_{kj}\begin{pmatrix} \alpha_{1k} \\ \alpha_{2k} \\ \cdot \\ \cdot \\ \cdot \\ \alpha_{mk} \end{pmatrix} \tag{15}$$

Here $\mathbf{s}_i = (s_{i1}, s_{i2}, \ldots, s_{in})$. Let $\boldsymbol{\alpha}_i$ denote the vector $\begin{pmatrix} \alpha_{1i} \\ \alpha_{2i} \\ \vdots \\ \alpha_{mi} \end{pmatrix}$. Then since the left-hand side of (15) is the jth column of A, we see that we can write every column of A as a linear combination of $\boldsymbol{\alpha}_1, \boldsymbol{\alpha}_2, \ldots, \boldsymbol{\alpha}_k$, which means that $\boldsymbol{\alpha}_1, \boldsymbol{\alpha}_2, \ldots, \boldsymbol{\alpha}_k$ span C_A and

$$\dim C_A \le k = \dim S_A \tag{16}$$

But Equation (16) holds for any matrix A. In particular it holds for A^t. But $C_{A^t} = S_A$ and $S_{A^t} = C_A$. Thus, since, from (16), $\dim C_{A^t} \le \dim S_{A^t}$, we have

$$\dim S_A \le \dim C_A. \tag{17}$$

Combining (16) and (17) completes the proof. ■

PROBLEMS 6.3 In Problems 1–15 find the rank and nullity of the given matrix.

1. $\begin{pmatrix} 1 & 2 \\ 3 & 4 \end{pmatrix}$ **2.** $\begin{pmatrix} 1 & -1 & 2 \\ 3 & 1 & 0 \end{pmatrix}$ **3.** $\begin{pmatrix} -1 & 3 & 2 \\ 2 & -6 & -4 \end{pmatrix}$

4. $\begin{pmatrix} 1 & -1 & 2 \\ 3 & 1 & 4 \\ -1 & 0 & 4 \end{pmatrix}$ **5.** $\begin{pmatrix} 1 & -1 & 2 \\ 3 & 1 & 4 \\ 5 & -1 & 8 \end{pmatrix}$ **6.** $\begin{pmatrix} -1 & 2 & 1 \\ 2 & -4 & -2 \\ -3 & 6 & 3 \end{pmatrix}$

7. $\begin{pmatrix} 1 & -1 & 2 & 3 \\ 0 & 1 & 4 & 3 \\ 1 & 0 & 6 & 6 \end{pmatrix}$ **8.** $\begin{pmatrix} 1 & -1 & 2 & 3 \\ 0 & 1 & 4 & 3 \\ 1 & 0 & 6 & 5 \end{pmatrix}$ **9.** $\begin{pmatrix} 2 & 3 \\ -1 & 1 \\ 4 & 7 \end{pmatrix}$

10. $\begin{pmatrix} 1 & -1 & 2 & 3 \\ 0 & 1 & 0 & 1 \\ 1 & 0 & 1 & 0 \\ 0 & 0 & 0 & 1 \end{pmatrix}$ **11.** $\begin{pmatrix} 1 & -1 & 2 & 1 \\ -1 & 0 & 1 & 2 \\ 1 & -2 & 5 & 4 \\ 2 & -1 & 1 & -1 \end{pmatrix}$ **12.** $\begin{pmatrix} 1 & -1 & 2 & 3 \\ -2 & 2 & -4 & -6 \\ 2 & -2 & 4 & 6 \\ 3 & -3 & 6 & 9 \end{pmatrix}$

13. $\begin{pmatrix} -1 & -1 & 0 & 0 \\ 0 & 0 & 2 & 3 \\ 4 & 0 & -2 & 1 \\ 3 & -1 & 0 & 4 \end{pmatrix}$ **14.** $\begin{pmatrix} 3 & 0 & 0 \\ 0 & 0 & 0 \\ 0 & 0 & 6 \end{pmatrix}$ **15.** $\begin{pmatrix} 1 & 2 & 3 \\ 0 & 0 & 4 \\ 0 & 0 & 6 \end{pmatrix}$

In Problems 16–22 find a basis for the range and kernel of the given matrix.

16. The matrix of Problem 2
17. The matrix of Problem 5
18. The matrix of Problem 6
19. The matrix of Problem 8
20. The matrix of Problem 11
21. The matrix of Problem 12
22. The matrix of Problem 13

In Problems 23–26 use Theorem 6 to determine whether the given system has any solutions.

23. $\begin{aligned} x_1+x_2-\ x_3&=7\\ 4x_1-x_2+5x_3&=4\\ 6x_1+x_2+3x_3&=20 \end{aligned}$

24. $\begin{aligned} x_1+x_2-\ x_3&=7\\ 4x_1-x_2+5x_3&=4\\ 6x_1+x_2+3x_3&=18 \end{aligned}$

25. $\begin{aligned} x_1-2x_2+\ x_3+\ x_4&=\ 2\\ 3x_1\qquad\ +2x_3-2x_4&=-8\\ 4x_2-\ x_3-\ x_4&=\ 1\\ 5x_1\qquad\ +3x_3-\ x_4&=-3 \end{aligned}$

26. $\begin{aligned} x_1-2x_2+\ x_3+\ x_4&=\ 2\\ 3x_1\qquad\ +2x_3-2x_4&=-8\\ 4x_2-\ x_3-\ x_4&=\ 1\\ 5x_1\qquad\ +3x_3-\ x_4&=\ 0 \end{aligned}$

27. Show that the rank of a diagonal matrix is equal to the number of nonzero components on the diagonal.

28. Let A be an upper triangular $n\times n$ matrix with zeros on the diagonal. Show that $\rho(A)<n$.

29. Show that for any matrix A, $\rho(A)=\rho(A^t)$.

30. Show that if A is an $m\times n$ matrix and $m<n$, then (**a**) $\rho(A)\leq m$ and (**b**) $\nu(A)\geq n-m$.

31. Let A be an $m\times n$ matrix and let B and C be invertible $m\times m$ and $n\times n$ matrices, respectively. Prove that $\rho(A)=\rho(BA)=\rho(AC)$. That is, multiplying a matrix by an invertible matrix does not change its rank.

32. Let A and B be $m\times n$ and $n\times p$ matrices, respectively. Show that $\rho(AB)\leq \min(\rho(A), \rho(B))$.

33. Let A be a 5×7 matrix with rank 5. Show that the linear system $A\mathbf{x}=\mathbf{b}$ has at least one solution for every 5-vector $\mathbf{b}$.

***34.** Let A and B be $m\times n$ matrices. Show that if $\rho(A)=\rho(B)$, then there exist invertible matrices C and D such that $B=CAD$.

35. If $B=CAD$, where C and D are invertible, prove that $\rho(A)=\rho(B)$.

36. Suppose that any k rows of A are linearly independent while any $k+1$ rows of A are linearly dependent. Show that $\rho(A)=k$.

37. If A is an $n\times n$ matrix, show that $\rho(A)<n$ if and only if there is a vector $\mathbf{x}\in\mathbb{R}^n$ such that $\mathbf{x}\neq\mathbf{0}$ and $A\mathbf{x}=\mathbf{0}$.

38. Let A be an $m\times n$ matrix. Suppose that for every $\mathbf{y}\in\mathbb{R}^m$ there is an $\mathbf{x}\in\mathbb{R}^n$ such that $A\mathbf{x}=\mathbf{y}$. Show that $\rho(A)=m$.

6.4 The Matrix Representation of a Linear Transformation

If A is an $m\times n$ matrix and T: $\mathbb{R}^n\to\mathbb{R}^m$ is defined by $T\mathbf{x}=A\mathbf{x}$ then, as we saw in Example 6.1.7, T is a linear transformation. We shall now see that for *every* linear transformation from $\mathbb{R}^n$ into $\mathbb{R}^m$, there exists an $m\times n$ matrix A such that $T\mathbf{x}=A\mathbf{x}$ for every $\mathbf{x}\in\mathbb{R}^n$. This fact is extremely useful. As we saw in Definitions 6.3.1 and 6.3.2 on pages 243 and 244, if $T\mathbf{x} = A\mathbf{x}$, then $\ker T = N_A$ and range $T = R_A$. Moreover, $\nu(T) = \dim\ker T = \nu(A)$ and $\rho(T) = \dim$ range $T = \rho(A)$. Thus we can determine the kernel, range, nullity, and rank of a linear transformation from $\mathbb{R}^n\to\mathbb{R}^m$ by determining the kernel and range space of a corresponding matrix. Moreover, once we know that $T\mathbf{x} = A\mathbf{x}$, we can evaluate $T\mathbf{x}$ for any $\mathbf{x}$ in $\mathbb{R}^n$ by simple matrix multiplication.

But this is not all. As we shall see, any linear transformation between finite dimensional vector spaces can be represented by a matrix.

THEOREM 1 Let $T\colon \mathbb{R}^n \to \mathbb{R}^m$ be a linear transformation. Then there exists a unique $m \times n$ matrix A_T such that

$$\boxed{T\mathbf{x} = A_T\mathbf{x} \qquad \text{for every } \mathbf{x} \in \mathbb{R}^n} \tag{1}$$

Proof Let $\mathbf{w}_1 = T\mathbf{e}_1, \mathbf{w}_2 = T\mathbf{e}_2, \ldots, \mathbf{w}_n = T\mathbf{e}_n$. Let A_T be the matrix whose columns are $\mathbf{w}_1, \mathbf{w}_2, \ldots, \mathbf{w}_n$. If

$$\mathbf{w}_i = \begin{pmatrix} a_{1i} \\ a_{2i} \\ \cdot \\ \cdot \\ \cdot \\ a_{mi} \end{pmatrix} \qquad \text{for } i = 1, 2, \ldots, n$$

then

$$A_T\mathbf{e}_i = \begin{pmatrix} a_{11} & a_{12} & \cdots & a_{1i} & \cdots & a_{1n} \\ a_{21} & a_{22} & \cdots & a_{2i} & \cdots & a_{2n} \\ \cdot & \cdot & & \cdot & & \cdot \\ \cdot & \cdot & & \cdot & & \cdot \\ \cdot & \cdot & & \cdot & & \cdot \\ a_{m1} & a_{m2} & \cdots & a_{mi} & \cdots & a_{mn} \end{pmatrix} \begin{pmatrix} 0 \\ 0 \\ \cdot \\ \cdot \\ \cdot \\ 1 \\ 0 \\ \cdot \\ \cdot \\ \cdot \\ 0 \end{pmatrix} = \begin{pmatrix} a_{1i} \\ a_{2i} \\ \cdot \\ \cdot \\ \cdot \\ a_{mi} \end{pmatrix} = \mathbf{w}_i.$$

(1 in the ith position)

Thus $A_T\mathbf{e}_i = \mathbf{w}_i$ for $i = 1, 2, \ldots, n$. If $\mathbf{x} \in \mathbb{R}^n$, there exists a unique set of real numbers $c_1, c_2, \ldots, c_n$ such that $\mathbf{x} = c_1\mathbf{e}_1 + c_2\mathbf{e}_2 + \cdots + c_n\mathbf{e}_n$. Then $T\mathbf{x} = c_1T\mathbf{e}_1 + c_2T\mathbf{e}_2 + \cdots + c_nT\mathbf{e}_n = c_1\mathbf{w}_1 + c_2\mathbf{w}_2 + \cdots + c_n\mathbf{w}_n$. But $A_T\mathbf{x} = A_T(c_1\mathbf{e}_1 + c_2\mathbf{e}_2 + \cdots + c_n\mathbf{e}_n) = c_1A_T\mathbf{e}_1 + c_2A_T\mathbf{e}_2 + \cdots + c_nA_T\mathbf{e}_n = c_1\mathbf{w}_1 + c_2\mathbf{w}_2 + \cdots + c_n\mathbf{w}_n$. Thus $T\mathbf{x} = A_T\mathbf{x}$. We can now show that A_T is unique. Suppose that $T\mathbf{x} = A_T\mathbf{x}$ and $T\mathbf{x} = B_T\mathbf{x}$ for every $\mathbf{x} \in \mathbb{R}^n$. Then $A_T\mathbf{x} = B_T\mathbf{x}$ or, setting $C_T = A_T - B_T$, we have $C_T\mathbf{x} = \mathbf{0}$ for every $\mathbf{x} \in \mathbb{R}^n$. In particular, $C_T\mathbf{e}_i = \mathbf{0}$ for $i = 1, 2, \ldots, n$. But, as we see from the proof of the first part of the theorem, $C_T\mathbf{e}_i$ is the ith column of C_T. Thus each of the n columns of C_T is the m-zero vector and $C_T = 0$, the $m \times n$ zero matrix. This shows that $A_T = B_T$ and the theorem is proved. ■

Remark 1. In this theorem we assumed that every vector in $\mathbb{R}^n$ and $\mathbb{R}^m$ is written in terms of the standard basis vectors in those spaces. If we choose other bases for $\mathbb{R}^n$ and $\mathbb{R}^m$ we shall, of course, get a different matrix A_T. See, for instance, Example 5.6.1 on page 205 or Example 8 below.

Remark 2. The proof of the theorem shows us that A_T is easily obtained as the matrix whose columns are the vectors $T\mathbf{e}_i$.

DEFINITION 1 **TRANSFORMATION MATRIX** The matrix A_T in Theorem 1 is called the **transformation matrix** corresponding to T.

In Section 6.2 we defined the range, rank, kernel, and nullity of a linear transformation. In Section 6.3 we defined the range, rank, kernel, and nullity of a matrix. The proof of the following theorem follows easily from Theorem 1 and is left as an exercise (see Problem 36).

THEOREM 2 Let A_T be the transformation matrix corresponding to the linear transformation T. Then

i. range $T = R_{A_T} = C_{A_T}$
ii. $\rho(T) = \rho(A_T)$
iii. $\ker T = N_{A_T}$
iv. $\nu(T) = \nu(A_T)$

EXAMPLE 1 Find the transformation matrix A_T corresponding to the projection of a vector in $\mathbb{R}^3$ onto the xy-plane.

Solution Here $T\begin{pmatrix} x \\ y \\ z \end{pmatrix} = \begin{pmatrix} x \\ y \\ 0 \end{pmatrix}$. In particular, $T\begin{pmatrix} 1 \\ 0 \\ 0 \end{pmatrix} = \begin{pmatrix} 1 \\ 0 \\ 0 \end{pmatrix}$, $T\begin{pmatrix} 0 \\ 1 \\ 0 \end{pmatrix} = \begin{pmatrix} 0 \\ 1 \\ 0 \end{pmatrix}$, and $T\begin{pmatrix} 0 \\ 0 \\ 1 \end{pmatrix} = \begin{pmatrix} 0 \\ 0 \\ 0 \end{pmatrix}$.

Thus $A_T = \begin{pmatrix} 1 & 0 & 0 \\ 0 & 1 & 0 \\ 0 & 0 & 0 \end{pmatrix}$. Note that $A_T\begin{pmatrix} x \\ y \\ z \end{pmatrix} = \begin{pmatrix} 1 & 0 & 0 \\ 0 & 1 & 0 \\ 0 & 0 & 0 \end{pmatrix}\begin{pmatrix} x \\ y \\ z \end{pmatrix} = \begin{pmatrix} x \\ y \\ 0 \end{pmatrix}$.

EXAMPLE 2 Let $T\colon\ \mathbb{R}^3 \to \mathbb{R}^4$ be defined by

$$T\begin{pmatrix} x \\ y \\ z \end{pmatrix} = \begin{pmatrix} x-y \\ y+z \\ 2x-y-\ z \\ -x+y+2z \end{pmatrix}$$

Find A_T, ker T, range T, $\nu(T)$, and $\rho(T)$.

Solution $T\begin{pmatrix} 1 \\ 0 \\ 0 \end{pmatrix} = \begin{pmatrix} 1 \\ 0 \\ 2 \\ -1 \end{pmatrix}$, $T\begin{pmatrix} 0 \\ 1 \\ 0 \end{pmatrix} = \begin{pmatrix} -1 \\ 1 \\ -1 \\ 1 \end{pmatrix}$, and $T\begin{pmatrix} 0 \\ 0 \\ 1 \end{pmatrix} = \begin{pmatrix} 0 \\ 1 \\ -1 \\ 2 \end{pmatrix}$. Thus $A_T = \begin{pmatrix} 1 & -1 & 0 \\ 0 & 1 & 1 \\ 2 & -1 & -1 \\ -1 & 1 & 2 \end{pmatrix}$. Note (as a check) that $\begin{pmatrix} 1 & -1 & 0 \\ 0 & 1 & 1 \\ 2 & -1 & -1 \\ -1 & 1 & 2 \end{pmatrix}\begin{pmatrix} x \\ y \\ z \end{pmatrix} = \begin{pmatrix} x-y \\ x+z \\ 2x-y-\ z \\ -x+y+2z \end{pmatrix}$. Next we compute the kernel and range of A. Row-reducing, we obtain

$$\begin{pmatrix} 1 & -1 & 0 \\ 0 & 1 & 1 \\ 2 & -1 & -1 \\ -1 & 1 & 2 \end{pmatrix} \xrightarrow[A_{1,4}(1)]{A_{1,3}(-2)} \begin{pmatrix} 1 & -1 & 0 \\ 0 & 1 & 1 \\ 0 & 1 & -1 \\ 0 & 0 & 2 \end{pmatrix}$$

$$\xrightarrow[A_{2,3}(-1)]{A_{2,1}(1)} \begin{pmatrix} 1 & 0 & 1 \\ 0 & 1 & 1 \\ 0 & 0 & -2 \\ 0 & 0 & 2 \end{pmatrix} \xrightarrow{A_{3,4}(1)} \begin{pmatrix} 1 & 0 & 1 \\ 0 & 1 & 1 \\ 0 & 0 & -2 \\ 0 & 0 & 0 \end{pmatrix}$$

Thus $\rho(A)=3$ and $\nu(A) \overset{\text{since } \rho(A)+\nu(A)=3}{=} 3-3=0$. This means that $\ker T=\{\mathbf{0}\}$, range $T=$ $\text{span}\left\{\begin{pmatrix} 1 \\ 0 \\ 2 \\ -1 \end{pmatrix}, \begin{pmatrix} -1 \\ 1 \\ -1 \\ 1 \end{pmatrix}, \begin{pmatrix} 0 \\ 1 \\ -1 \\ 2 \end{pmatrix}\right\}$, $\nu(T)=0$, and $\rho(T)=3$.

EXAMPLE 3 Let $T\colon \mathbb{R}^3 \to \mathbb{R}^3$ be defined by $T\begin{pmatrix} x \\ y \\ z \end{pmatrix} = \begin{pmatrix} 2x-y+3z \\ 4x-2y+6z \\ -6x+3y-9z \end{pmatrix}$. Find A_T, $\ker T$, range T, $\nu(T)$, and $\rho(T)$.

Solution Since $T\begin{pmatrix} 1 \\ 0 \\ 0 \end{pmatrix} = \begin{pmatrix} 2 \\ 4 \\ -6 \end{pmatrix}$, $T\begin{pmatrix} 0 \\ 1 \\ 0 \end{pmatrix} = \begin{pmatrix} -1 \\ -2 \\ 3 \end{pmatrix}$, and $T\begin{pmatrix} 0 \\ 0 \\ 1 \end{pmatrix} = \begin{pmatrix} 3 \\ 6 \\ -9 \end{pmatrix}$, we have

$$A_T = \begin{pmatrix} 2 & -1 & 3 \\ 4 & -2 & 6 \\ -6 & 3 & -9 \end{pmatrix}.$$

From Example 6.3.4 on page 246 we see that $\rho(A) \overset{\text{Theorem 2(ii)}}{=} \rho(T) = 1$ and range $T =$ $\text{span}\left\{\begin{pmatrix} 2 \\ 4 \\ -6 \end{pmatrix}\right\}$. Then $\nu(T)=2$. To find $N_A \overset{\text{Theorem 2(iii)}}{=} \ker T$, we row-reduce to solve the system $A\mathbf{x}=\mathbf{0}$: $\left(\begin{array}{ccc|c} 2 & -1 & 3 & 0 \\ 4 & -2 & 6 & 0 \\ -6 & 3 & -9 & 0 \end{array}\right) \xrightarrow[A_{1,3}(3)]{A_{1,2}(-2)} \left(\begin{array}{ccc|c} 2 & -1 & 3 & 0 \\ 0 & 0 & 0 & 0 \\ 0 & 0 & 0 & 0 \end{array}\right)$. This means that $\begin{pmatrix} x \\ y \\ z \end{pmatrix} \in N_A$ if $2x-y+3z=0$ or $y=2x+3z$. First setting $x=1$, $z=0$ and then $x=0$, $z=1$, we obtain a basis for N_A: $\ker T = N_A = \text{span}\left\{\begin{pmatrix} 1 \\ 2 \\ 0 \end{pmatrix}, \begin{pmatrix} 0 \\ 3 \\ 1 \end{pmatrix}\right\}$.

EXAMPLE 4 It is easy to verify that if T is the zero transformation from $\mathbb{R}^n \to \mathbb{R}^m$, then A_T is the $m \times n$ zero matrix. Similarly, if T is the identity transformation from $\mathbb{R}^n \to \mathbb{R}^n$, then $A_T = I_n$.

EXAMPLE 5 We saw in Example 6.1.8 on page 233 that if T is the function which rotates every vector in $\mathbb{R}^2$ through an angle of θ, then $A_T = \begin{pmatrix} \cos\theta & -\sin\theta \\ \sin\theta & \cos\theta \end{pmatrix}$.

We now generalize the notion of matrix representation to arbitrary finite dimensional vector spaces.

THEOREM 3 Let V be a real n-dimensional vector space, W be a real m-dimensional vector space, and $T\colon V \to W$ be a linear transformation. Let $B_1 = \{\mathbf{v}_1, \mathbf{v}_2, \ldots, \mathbf{v}_n\}$ be a basis for V and let $B_2 = \{\mathbf{w}_1, \mathbf{w}_2, \ldots, \mathbf{w}_m\}$ be a basis for W. Then there is a unique $m \times n$ matrix A_T such that

$$(T\mathbf{x})_{B_2} = (A_T\mathbf{x})_{B_2} = A_T(\mathbf{x})_{B_1}. \tag{2}$$

Remark 1. The notation in (2) is the notation of Section 5.6 (see page 201). If $\mathbf{x} \in V = c_1\mathbf{v}_1 + c_2\mathbf{v}_2 + \cdots + c_n\mathbf{v}_n$, then $(\mathbf{x})_{B_1} = \begin{pmatrix} c_1 \\ c_2 \\ \vdots \\ c_n \end{pmatrix}$. Let $\mathbf{c} = \begin{pmatrix} c_1 \\ c_2 \\ \vdots \\ c_n \end{pmatrix}$. Then $A_T\mathbf{c}$ is an m-vector that we denote $\mathbf{d} = \begin{pmatrix} d_1 \\ d_2 \\ \vdots \\ d_m \end{pmatrix}$. Equation (2) says that $(T\mathbf{x})_{B_2} = \begin{pmatrix} d_1 \\ d_2 \\ \vdots \\ d_m \end{pmatrix}$. That is,

$$T\mathbf{x} = d_1\mathbf{w}_1 + d_2\mathbf{w}_2 + \cdots + d_n\mathbf{w}_n.$$

Remark 2. As in Theorem 1, the uniqueness of A_T is relative to the bases B_1 and B_2. If we change the bases, we change A_T (see Examples 8, 9, and 10 and Theorem 5).

Proof Let $T\mathbf{v}_1 = \mathbf{y}_1, T\mathbf{v}_2 = \mathbf{y}_2, \ldots, T\mathbf{v}_n = \mathbf{y}_n$. Since $\mathbf{y}_i \in W$, we have, for $i = 1, 2, \ldots, n$,

$$\mathbf{y}_i = a_{1i}\mathbf{w}_1 + a_{2i}\mathbf{w}_2 + \cdots + a_{mi}\mathbf{w}_m$$

for some (unique) set of scalars $a_{1i}, a_{2i}, \ldots, a_{mi}$, and we write

$$(\mathbf{y}_1)_{B_2} = \begin{pmatrix} a_{11} \\ a_{21} \\ \vdots \\ a_{m1} \end{pmatrix}, (\mathbf{y}_2)_{B_2} = \begin{pmatrix} a_{12} \\ a_{22} \\ \vdots \\ a_{m2} \end{pmatrix}, \ldots, (\mathbf{y}_n)_{B_2} = \begin{pmatrix} a_{1n} \\ a_{2n} \\ \vdots \\ a_{mn} \end{pmatrix}$$

We now define

$$\mathbf{A}_T = \begin{pmatrix} a_{11} & a_{21} & \cdots & a_{1n} \\ a_{21} & a_{22} & \cdots & a_{2n} \\ \cdot & \cdot & & \cdot \\ \cdot & \cdot & & \cdot \\ \cdot & \cdot & & \cdot \\ a_{m1} & a_{m2} & \cdots & a_{mn} \end{pmatrix}$$

Since

$$(\mathbf{v}_1)_{B_1} = \begin{pmatrix} 1 \\ 0 \\ \vdots \\ 0 \end{pmatrix}, (\mathbf{v}_2)_{B_1} = \begin{pmatrix} 0 \\ 1 \\ 0 \\ \vdots \\ 0 \end{pmatrix}, \ldots, (\mathbf{v}_n)_{B_1} = \begin{pmatrix} 0 \\ 0 \\ \vdots \\ 1 \end{pmatrix}$$

we have, as in the proof of Theorem 1,

$$\mathbf{A}_T\mathbf{v}_i = \begin{pmatrix} a_{1i} \\ a_{2i} \\ \cdot \\ \cdot \\ \cdot \\ a_{mi} \end{pmatrix} = (\mathbf{y}_i)_{B_2}$$

If $\mathbf{x}$ is in V, then

$$\mathbf{x} = c_1\mathbf{v}_1 + c_2\mathbf{v}_2 + \cdots + c_n\mathbf{v}_n,$$

$$(\mathbf{x})_{B_1} = \begin{pmatrix} c_1 \\ c_2 \\ \vdots \\ c_n \end{pmatrix},$$

$$\text{and } (A_T(\mathbf{x})_{B_1})_{B_2} = \begin{pmatrix} a_{11} & a_{12} & \cdots & a_{1n} \\ a_{21} & a_{22} & \cdots & a_{2n} \\ \vdots & \vdots & & \vdots \\ a_{m1} & a_{m2} & \cdots & a_{mn} \end{pmatrix} \begin{pmatrix} c_1 \\ c_2 \\ \vdots \\ c_n \end{pmatrix}$$

$$= \begin{pmatrix} a_{11}c_1 + a_{12}c_2 + \cdots + a_{1n}c_n \\ a_{21}c_1 + a_{22}c_2 + \cdots + a_{2n}c_n \\ \vdots \\ a_{m1}c_1 + a_{m2}c_2 + \cdots + a_{mn}c_n \end{pmatrix}$$

$$= c_1 \begin{pmatrix} a_{11} \\ a_{21} \\ \vdots \\ a_{m1} \end{pmatrix} + c_2 \begin{pmatrix} a_{12} \\ a_{22} \\ \vdots \\ a_{m2} \end{pmatrix} + \cdots + c_n \begin{pmatrix} a_{1n} \\ a_{2n} \\ \vdots \\ a_{mn} \end{pmatrix}$$

$$= c_1(\mathbf{y}_1)_{B_2} + c_2(\mathbf{y}_2)_{B_2} + \cdots + c_n(\mathbf{y}_n)_{B_2}$$

Similarly, $T\mathbf{x} = T(c_1\mathbf{v}_1 + c_2\mathbf{v}_2 + \cdots + c_n\mathbf{v}_n) = c_1T\mathbf{v}_1 + c_2T\mathbf{v}_2 + \cdots + c_nT\mathbf{v}_n = c_1\mathbf{y}_1 + c_2\mathbf{y}_2 + \cdots + c_n\mathbf{y}_n$. Thus $(T\mathbf{x})_{B_2} = A_T(\mathbf{x})_{B_1}$. The proof of uniqueness is exactly as in the proof of uniqueness in Theorem 1. ∎

The following useful result follows immediately from Theorem 6.3.4 on page 247 and generalizes Theorem 2. Its proof is left as an exercise (see Problem 37).

THEOREM 4 Let V and W be finite dimensional vector spaces with $\dim V = n$. Let T: $V \to W$ be a linear transformation and let A_T be a matrix representation of T. Then

i. $\rho(T) = \rho(A_T)$
ii. $\nu(T) = \nu(A_T)$
iii. $\nu(T) + \rho(T) = n$

EXAMPLE 6 Let T: $P_2 \to P_3$ be defined by $(Tp)(x) = xp(x)$. Find A_T and use it to determine the kernel and range of T.

Solution Using the standard basis $B_1=\{1, x, x^2\}$ in P_2 and $B_2=\{1, x, x^2, x^3\}$ in P_3, we have $T(1)=x=\begin{pmatrix}0\\1\\0\\0\end{pmatrix}_{B_2}$, $T(x)=x^2=\begin{pmatrix}0\\0\\1\\0\end{pmatrix}_{B_2}$, and $T(x^2)=x^3=\begin{pmatrix}0\\0\\0\\1\end{pmatrix}_{B_2}$. Thus $A_T=\begin{pmatrix}0&0&0\\1&0&0\\0&1&0\\0&0&1\end{pmatrix}$. Clearly $\rho(A)=3$ and a basis for R_A is $\left\{\begin{pmatrix}0\\1\\0\\0\end{pmatrix}, \begin{pmatrix}0\\0\\1\\0\end{pmatrix}, \begin{pmatrix}0\\0\\0\\1\end{pmatrix}\right\}$. Therefore, range T = span $\{x, x^2, x^3\}$. Since $\nu(A) = 3 - \rho(A) = 0$, we see that ker $T=\{0\}$.

EXAMPLE 7 Let $T\colon\ P_3\to P_2$ be defined by $T(a_0+a_1x+a_2x^2+a_3x^3)=a_1+a_2x^2$. Compute A_T and use it to find the kernel and range of T.

Solution Using the standard bases $B_1=\{1, x, x^2, x^3\}$ in P_3 and $B_2=\{1, x, x^2\}$ in P_2, we immediately see that $T(1)=\begin{pmatrix}0\\0\\0\end{pmatrix}_{B_2}$, $T(x)=\begin{pmatrix}1\\0\\0\end{pmatrix}_{B_2}$, $T(x^2)=\begin{pmatrix}0\\0\\1\end{pmatrix}_{B_2}$, and $T(x^3)=\begin{pmatrix}0\\0\\0\end{pmatrix}_{B_2}$. Thus $A_T=\begin{pmatrix}0&1&0&0\\0&0&0&0\\0&0&1&0\end{pmatrix}$. Clearly $\rho(A)=2$ and a basis for R_A is $\left\{\begin{pmatrix}1\\0\\0\end{pmatrix}\begin{pmatrix}0\\0\\1\end{pmatrix}\right\}$ so that range $T=\text{span}\,\{1, x^2\}$. Then $\nu(A)=4-2=2$; and if $A_T\begin{pmatrix}a_0\\a_1\\a_2\\a_3\end{pmatrix}=\begin{pmatrix}0\\0\\0\end{pmatrix}$, then $a_1=0$ and $a_2=0$. Hence a_0 and a_3 are arbitrary and a basis for N_A is $\left\{\begin{pmatrix}1\\0\\0\\0\end{pmatrix}, \begin{pmatrix}0\\0\\0\\1\end{pmatrix}\right\}$, so that a basis for ker T is $\{1, x^3\}$.

In all the examples of this section we have obtained the matrix A_T by using the standard basis in each vector space. However, Theorem 3 holds for any bases in V and W. The next example illustrates this.

EXAMPLE 8 Let $T\colon\ \mathbb{R}^2\to\mathbb{R}^2$ be defined by $T\begin{pmatrix}x\\y\end{pmatrix}=\begin{pmatrix}x+y\\x-y\end{pmatrix}$. Using the bases $B_1=B_2=\left\{\begin{pmatrix}1\\-1\end{pmatrix}, \begin{pmatrix}-3\\2\end{pmatrix}\right\}$, compute A_T.

Solution We have $T\begin{pmatrix}1\\-1\end{pmatrix}=\begin{pmatrix}0\\2\end{pmatrix}$ and $T\begin{pmatrix}-3\\2\end{pmatrix}=\begin{pmatrix}-1\\-5\end{pmatrix}$. Since $\begin{pmatrix}0\\2\end{pmatrix}=-6\begin{pmatrix}1\\-1\end{pmatrix}-2\begin{pmatrix}-3\\2\end{pmatrix}$, we find that $\begin{pmatrix}0\\2\end{pmatrix}=\begin{pmatrix}-6\\-2\end{pmatrix}_{B_2}$. Similarly $\begin{pmatrix}-1\\-5\end{pmatrix}=17\begin{pmatrix}1\\-1\end{pmatrix}+6\begin{pmatrix}-3\\2\end{pmatrix}$, so that $\begin{pmatrix}-1\\-5\end{pmatrix}=\begin{pmatrix}17\\6\end{pmatrix}_{B_2}$. Thus $A_T=\begin{pmatrix}-6&17\\-2&6\end{pmatrix}$. To compute $T\begin{pmatrix}-4\\7\end{pmatrix}$, for example, we first write $\begin{pmatrix}-4\\7\end{pmatrix}=-13\begin{pmatrix}1\\-1\end{pmatrix}-3\begin{pmatrix}-3\\2\end{pmatrix}=\begin{pmatrix}-13\\-3\end{pmatrix}_{B_1}$. Then $T\begin{pmatrix}-4\\7\end{pmatrix}=T\begin{pmatrix}-13\\-3\end{pmatrix}_{B_1}=$
$\left[A_T\begin{pmatrix}-13\\-3\end{pmatrix}\right]_{B_2}=\left[\begin{pmatrix}-6&17\\-2&6\end{pmatrix}\begin{pmatrix}-13\\-3\end{pmatrix}\right]_{B_2}=\begin{pmatrix}27\\8\end{pmatrix}_{B_2}=27\begin{pmatrix}1\\-1\end{pmatrix}+8\begin{pmatrix}-3\\2\end{pmatrix}=\begin{pmatrix}3\\-11\end{pmatrix}$.
Note that $T\begin{pmatrix}-4\\7\end{pmatrix}=\begin{pmatrix}-4+7\\-4-7\end{pmatrix}=\begin{pmatrix}3\\-11\end{pmatrix}$, which verifies our calculations.

To avoid confusion, we shall, unless explicitly stated otherwise, always compute the matrix A_T with respect to the standard basis.† If T: $V\to V$ is a linear transformation and some other basis B is used, then we refer to A_T as *the transformation matrix of T with respect to the basis B*. Thus, in the last example, $A_T=\begin{pmatrix}-6&17\\-2&6\end{pmatrix}$ is the transformation matrix of T with respect to the basis $\left\{\begin{pmatrix}1\\-1\end{pmatrix},\begin{pmatrix}-3\\2\end{pmatrix}\right\}$.

Before leaving this section we must answer an obvious question. Why bother to use a basis other than the standard basis since the computations are, as in Example 8, a good deal more complicated? The answer is that it is often useful to find a basis B^* in $\mathbb{R}^n$ so that the transformation matrix with respect to B^* is a diagonal matrix. Diagonal matrices are very easy to work with and, as we shall see in Chapter 7, there are numerous advantages to writing a matrix in a diagonal form.

EXAMPLE 9 Let T: $\mathbb{R}^2\to\mathbb{R}^2$ be defined by $T\begin{pmatrix}x\\y\end{pmatrix}=\begin{pmatrix}12x+10y\\-15x-13y\end{pmatrix}$. Find A_T with respect to the basis $B_1=B_2=\left\{\begin{pmatrix}1\\-1\end{pmatrix},\begin{pmatrix}2\\-3\end{pmatrix}\right\}$.

Solution $T\begin{pmatrix}1\\-1\end{pmatrix}=\begin{pmatrix}2\\-2\end{pmatrix}=2\begin{pmatrix}1\\-1\end{pmatrix}+0\begin{pmatrix}2\\-3\end{pmatrix}=\begin{pmatrix}2\\0\end{pmatrix}_{B_2}$ and $T\begin{pmatrix}2\\-3\end{pmatrix}=\begin{pmatrix}-6\\9\end{pmatrix}=0\begin{pmatrix}1\\-1\end{pmatrix}-3\begin{pmatrix}2\\-3\end{pmatrix}=\begin{pmatrix}0\\-3\end{pmatrix}_{B_2}$. Thus $A_T=\begin{pmatrix}2&0\\0&-3\end{pmatrix}$.

There is another way to solve this problem. The vectors $\begin{pmatrix}1\\-1\end{pmatrix}$ and $\begin{pmatrix}2\\-3\end{pmatrix}$

† That is, in any space where we have defined a standard basis.

are written in terms of the standard basis $S = \left\{\begin{pmatrix}1\\0\end{pmatrix}, \begin{pmatrix}0\\1\end{pmatrix}\right\}$. That is, $\begin{pmatrix}1\\-1\end{pmatrix} = 1\begin{pmatrix}1\\0\end{pmatrix} + (-1)\begin{pmatrix}0\\1\end{pmatrix}$ and $\begin{pmatrix}2\\-3\end{pmatrix} = 2\begin{pmatrix}1\\0\end{pmatrix} + (-3)\begin{pmatrix}0\\1\end{pmatrix}$. Thus the matrix $A = \begin{pmatrix}1 & 2\\-1 & -3\end{pmatrix}$ is the matrix whose first and second columns represent the expansions of the vectors in B_1 in terms of the standard basis. Then, from the procedure outlined on page 204, the matrix $A^{-1} = \begin{pmatrix}3 & 2\\-1 & -1\end{pmatrix}$ is the transition matrix from S to B_1. Similarly, the matrix A is the transition matrix from B_1 to S (see Problem 5.6.38, page 211). Now suppose that $\mathbf{x}$ is written in terms of B_1. Then $A\mathbf{x}$ is the same vector now written in terms of S. Let $C = \begin{pmatrix}12 & 10\\-15 & -13\end{pmatrix}$. Then $CA\mathbf{x} = T(A\mathbf{x})$ is the image of $A\mathbf{x}$ written in terms of S. Finally, since we want $T(A\mathbf{x})$ in terms of B_1 (that was the problem), we multiply on the left by the transition matrix A^{-1} to obtain $(T\mathbf{x})_{B_1} = (A^{-1}CA)(\mathbf{x})_{B_1}$. That is,

$$A_T = A^{-1}CA = \begin{pmatrix}3 & 2\\-1 & -1\end{pmatrix}\begin{pmatrix}12 & 10\\-15 & -13\end{pmatrix}\begin{pmatrix}1 & 2\\-1 & -3\end{pmatrix} = \begin{pmatrix}3 & 2\\-1 & -1\end{pmatrix}\begin{pmatrix}2 & -6\\-2 & 9\end{pmatrix}$$
$$= \begin{pmatrix}2 & 0\\0 & -3\end{pmatrix}$$

as before. We summarize this result below.

THEOREM 5 Let T: $\mathbb{R}^n \to \mathbb{R}^m$ be a linear transformation. Suppose that C is the transformation matrix of T with respect to the standard bases S_n and S_m in $\mathbb{R}^n$ and $\mathbb{R}^m$, respectively. Let A_1 be the transition matrix from S_n to the basis B_1 in $\mathbb{R}^n$ and let A_2 be the transition matrix from S_m to the basis B_2 in $\mathbb{R}^m$. If A_T denotes the transformation matrix of T with respect to the bases B_1 and B_2, then

$$A_T = A_2CA_1^{-1} \tag{3}$$

In Example 9 we saw that by looking at the linear transformation T with respect to a new basis, the transformation matrix A_T turned out to be a diagonal matrix. We shall return to this "diagonalizing" procedure in Section 7.3. That is, given a linear transformation from $\mathbb{R}^n$ to $\mathbb{R}^n$, we shall see that it is often possible to find a basis B such that the transformation matrix of T with respect to B will be diagonal.

EXAMPLE 10 Let T: $\mathbb{R}^3 \to \mathbb{R}^2$ be defined by $T\begin{pmatrix}x\\y\\z\end{pmatrix} = \begin{pmatrix}x+y-z\\2x-y+z\end{pmatrix}$. Find A_T with respect to the bases $B_1 = \left\{\begin{pmatrix}1\\0\\1\end{pmatrix}, \begin{pmatrix}-1\\1\\0\end{pmatrix}, \begin{pmatrix}0\\-1\\1\end{pmatrix}\right\}$ and $B_2 = \left\{\begin{pmatrix}4\\3\end{pmatrix}, \begin{pmatrix}-1\\5\end{pmatrix}\right\}$.

Solution With respect to the standard bases, the matrix representation of T is $C = \begin{pmatrix} 1 & 1 & -1 \\ 2 & -1 & 1 \end{pmatrix}$. From the procedure outlined on page 204, A_1 is the inverse of $\begin{pmatrix} 1 & -1 & 0 \\ 0 & 1 & -1 \\ 1 & 0 & 1 \end{pmatrix}$; that is, $A_1^{-1} = \begin{pmatrix} 1 & -1 & 0 \\ 0 & 1 & -1 \\ 1 & 0 & 1 \end{pmatrix}$. Similarly, $A_2 = \begin{pmatrix} 4 & -1 \\ 3 & 5 \end{pmatrix}^{-1} = \frac{1}{23}\begin{pmatrix} 5 & 1 \\ -3 & 4 \end{pmatrix}$. Then, from Equation (3),

$$A_T = A_2 C A_1^{-1} = \frac{1}{23}\begin{pmatrix} 5 & 1 \\ -3 & 4 \end{pmatrix}\begin{pmatrix} 1 & 1 & -1 \\ 2 & -1 & 1 \end{pmatrix}\begin{pmatrix} 1 & -1 & 0 \\ 0 & 1 & -1 \\ 1 & 0 & 1 \end{pmatrix}$$

$$= \frac{1}{23}\begin{pmatrix} 5 & 1 \\ -3 & 4 \end{pmatrix}\begin{pmatrix} 0 & 0 & -2 \\ 3 & -3 & 2 \end{pmatrix} = \frac{1}{23}\begin{pmatrix} 3 & -3 & -8 \\ 12 & -12 & 14 \end{pmatrix}$$

For example, let $\mathbf{x} = \begin{pmatrix} 1 \\ 2 \\ 4 \end{pmatrix}$. Then $T\mathbf{x} = \begin{pmatrix} -1 \\ 4 \end{pmatrix}$ in terms of the standard bases. In terms of B_1 we see that $(\mathbf{x})_{B_1} = \begin{pmatrix} 1 \\ 2 \\ 4 \end{pmatrix}_{B_1} = \frac{7}{2}\begin{pmatrix} 1 \\ 0 \\ 1 \end{pmatrix} + \frac{5}{2}\begin{pmatrix} -1 \\ 1 \\ 0 \end{pmatrix} + \frac{1}{2}\begin{pmatrix} 0 \\ -1 \\ 1 \end{pmatrix} = \begin{pmatrix} \frac{7}{2} \\ \frac{5}{2} \\ \frac{1}{2} \end{pmatrix}$. Similarly, $(T\mathbf{x})_{B_2} = \begin{pmatrix} -1 \\ 4 \end{pmatrix}_{B_2} = -\frac{1}{23}\begin{pmatrix} 4 \\ 3 \end{pmatrix} + \frac{19}{23}\begin{pmatrix} -1 \\ 5 \end{pmatrix} = \begin{pmatrix} -\frac{1}{23} \\ \frac{19}{23} \end{pmatrix}$. Finally, in terms of the bases B_1 and B_2 we verify that

$$(T\mathbf{x})_{B_2} = A_T(\mathbf{x})_{B_1} = \frac{1}{23}\begin{pmatrix} 3 & -3 & -8 \\ 12 & -12 & 14 \end{pmatrix}\begin{pmatrix} \frac{7}{2} \\ \frac{5}{2} \\ \frac{1}{2} \end{pmatrix} = \frac{1}{23}\begin{pmatrix} -1 \\ 19 \end{pmatrix} = \begin{pmatrix} -\frac{1}{23} \\ \frac{19}{23} \end{pmatrix}$$

PROBLEMS 6.4

In Problems 1–30 find the matrix representation A_T of the linear transformation T, ker T, range T, $\nu(T)$, and $\rho(T)$. Unless otherwise stated, assume that B_1 and B_2 are standard bases.

1. $T: \mathbb{R}^2 \to \mathbb{R}^2$; $T\begin{pmatrix} x \\ y \end{pmatrix} = \begin{pmatrix} x - 2y \\ -x + y \end{pmatrix}$

2. $T: \mathbb{R}^2 \to \mathbb{R}^3$; $T\begin{pmatrix} x \\ y \end{pmatrix} = \begin{pmatrix} x + y \\ x - y \\ 2x + 3y \end{pmatrix}$

3. $T: \mathbb{R}^3 \to \mathbb{R}^2$; $T\begin{pmatrix} x \\ y \\ z \end{pmatrix} = \begin{pmatrix} x - y + z \\ -2x + 2y - 2z \end{pmatrix}$

4. $T: \mathbb{R}^2 \to \mathbb{R}^2$; $T\begin{pmatrix} x \\ y \end{pmatrix} = \begin{pmatrix} ax + by \\ cx + dy \end{pmatrix}$

5. $T: \mathbb{R}^3 \to \mathbb{R}^3$; $T\begin{pmatrix} x \\ y \\ z \end{pmatrix} = \begin{pmatrix} x - y + 2z \\ 3x + y + 4z \\ 5x - y + 8z \end{pmatrix}$

6. T: $\mathbb{R}^3 \to \mathbb{R}^3$; $T\begin{pmatrix} x \\ y \\ z \end{pmatrix} = \begin{pmatrix} -x+2y+z \\ 2x-4y-2z \\ -3x+6y+3z \end{pmatrix}$

7. T: $\mathbb{R}^4 \to \mathbb{R}^3$; $T\begin{pmatrix} x \\ y \\ z \\ w \end{pmatrix} = \begin{pmatrix} x-y+2z+3w \\ y+4z+3w \\ x \quad +6z+6w \end{pmatrix}$

8. T: $\mathbb{R}^4 \to \mathbb{R}^4$; $T\begin{pmatrix} x \\ y \\ z \\ w \end{pmatrix} = \begin{pmatrix} x-y+2z+w \\ -x \quad +z+2w \\ x-2y+5z+4w \\ 2x-y+z-w \end{pmatrix}$

9. T: $\mathbb{R}^2 \to \mathbb{R}^2$; $T\begin{pmatrix} x \\ y \end{pmatrix} = \begin{pmatrix} x-2y \\ 2x+y \end{pmatrix}$; $B_1 = B_2 = \left\{\begin{pmatrix} 1 \\ -2 \end{pmatrix}, \begin{pmatrix} 3 \\ 2 \end{pmatrix}\right\}$

10. T: $\mathbb{R}^2 \to \mathbb{R}^2$; $T\begin{pmatrix} x \\ y \end{pmatrix} = \begin{pmatrix} 4x-y \\ 3x+2y \end{pmatrix}$; $B_1 = B_2 = \left\{\begin{pmatrix} -1 \\ 1 \end{pmatrix}, \begin{pmatrix} 4 \\ 3 \end{pmatrix}\right\}$

11. T: $\mathbb{R}^3 \to \mathbb{R}^2$; $T\begin{pmatrix} x \\ y \\ z \end{pmatrix} = \begin{pmatrix} 2x+y+z \\ y-3z \end{pmatrix}$;

$$B_1 = \left\{\begin{pmatrix} 1 \\ 0 \\ 1 \end{pmatrix}, \begin{pmatrix} 1 \\ 1 \\ 0 \end{pmatrix}, \begin{pmatrix} 1 \\ 1 \\ 1 \end{pmatrix}\right\}; \quad B_2 = \left\{\begin{pmatrix} 1 \\ -1 \end{pmatrix}, \begin{pmatrix} 2 \\ 3 \end{pmatrix}\right\}$$

12. T: $\mathbb{R}^2 \to \mathbb{R}^3$; $T\begin{pmatrix} x \\ y \end{pmatrix} = \begin{pmatrix} x-y \\ 2x+y \\ y \end{pmatrix}$; $B_1 = \left\{\begin{pmatrix} 2 \\ 1 \end{pmatrix}, \begin{pmatrix} 1 \\ 2 \end{pmatrix}\right\}$; $B_2 = \left\{\begin{pmatrix} 1 \\ -1 \\ 0 \end{pmatrix}, \begin{pmatrix} 0 \\ 2 \\ 0 \end{pmatrix}, \begin{pmatrix} 0 \\ 2 \\ 5 \end{pmatrix}\right\}$

13. T: $P_2 \to P_3$; $T(a_0+a_1x+a_2x^2) = a_1 - a_1x + a_0x^3$

14. T: $\mathbb{R} \to P_3$; $T(a) = a + ax + ax^2 + ax^3$

15. T: $P_3 \to \mathbb{R}$; $T(a_0+a_1x+a_2x^2+a_3x^3) = a_2$

16. T: $P_3 \to P_1$; $T(a_0+a_1x+a_2x^2+a_3x^3) = (a_1+a_3)x - a_2$

17. T: $P_3 \to P_2$; $T(a_0+a_1x+a_2x^2+a_3x^3) = (a_0-a_1+2a_2+3a_3) + (a_1+4a_2+3a_3)x + (a_0+6a_2+5a_3)x^2$

18. T: $M_{22} \to M_{22}$; $T\begin{pmatrix} a & b \\ c & d \end{pmatrix} = \begin{pmatrix} a-b+2c+d & -a+2c+2d \\ a-2b+5c+4d & 2a-b+c-d \end{pmatrix}$

19. T: $M_{22} \to M_{22}$; $T\begin{pmatrix} a & b \\ c & d \end{pmatrix} = \begin{pmatrix} a+b+c+d & a+b+c \\ a+b & a \end{pmatrix}$

20. T: $P_2 \to P_3$; $T[p(x)] = xp(x)$; $B_1 = \{1, x, x^2\}$; $B_2 = \{1, (1+x), (1+x)^2, (1+x)^3\}$

□**21.** D: $P_4 \to P_3$; $Dp(x) = p'(x)$

□**22.** T: $P_4 \to P_4$; $Tp(x) = xp'(x) - p(x)$

□***23.** D: $P_n \to P_{n-1}$; $Dp(x) = p'(x)$

□**24.** D: $P_4 \to P_2$; $Dp(x) = p''(x)$

□***25.** T: $P_4 \to P_4$; $Tp(x) = p''(x) + xp'(x) + 2p(x)$

□***26.** D: $P_n \to P_{n-k}$; $Dp(x) = p^{(k)}(x)$

□***27.** T: $P_n \to P_n$; $Tp(x) = x^n p^{(n)}(x) + x^{n-1}p^{(n-1)}(x) + \cdots + xp'(x) + p(x)$

□**28.** J: $P_n \to \mathbb{R}$; $Jp = \int_0^1 p(x)\,dx$

29. T: $\mathbb{R}^3 \to P_2$; $T\begin{pmatrix} a \\ b \\ c \end{pmatrix} = a + bx + cx^2$

30. T: $P_3 \to \mathbb{R}^3$; $T(a_0+a_1x+a_2x^2+a_3x^3) = \begin{pmatrix} a_3-a_2 \\ a_1+a_3 \\ a_2-a_1 \end{pmatrix}$

31. Let T: $M_{mn} \to M_{nm}$ be given by $TA = A^t$. Find A_T with respect to the standard bases in M_{mn} and M_{nm}.

*32. Let T: $\mathbb{C}^2 \to \mathbb{C}^2$ be given by $T\begin{pmatrix} x \\ y \end{pmatrix} = \begin{pmatrix} x+iy \\ (1+i)y-x \end{pmatrix}$. Find A_T.

□33. Let $V = \text{span}\{1, \sin x, \cos x\}$. Find A_D, where D: $V \to V$ is defined by $Df(x) = f'(x)$. Find range D and ker D.

□34. Answer the questions of Problems 33 given $V = \text{span}\{e^x, xe^x, x^2e^x\}$.

35. Let T: $\mathbb{C}^2 \to \mathbb{C}^2$ be given by $T\mathbf{x} = \text{proj}_H\, \mathbf{x}$, where $H = \text{span}\{(1/\sqrt{2})(1, i)\}$. Find A_T.

36. Prove Theorem 2.

37. Prove Theorem 4.

6.5 If Time Permits: Isomorphisms

In this section we introduce some important terminology and then prove a theorem which says that all n-dimensional vector spaces are "essentially" the same.

DEFINITION 1 **ONE-TO-ONE TRANSFORMATION** Let T: $V \to W$ be a linear transformation. Then T is **one-to-one**, written 1–1, if

$$T\mathbf{v}_1 = T\mathbf{v}_2 \quad \text{implies that} \quad \mathbf{v}_1 = \mathbf{v}_2 \tag{1}$$

That is, T is 1–1 if every vector $\mathbf{w}$ in the range of T is the image of at most one vector in V.

THEOREM 1 Let T: $V \to W$ be a linear transformation. Then T is 1–1 if and only if $\ker T = \{\mathbf{0}\}$.

Proof Suppose $\ker T = \{\mathbf{0}\}$ and $T\mathbf{v}_1 = T\mathbf{v}_2$. Then $T\mathbf{v}_1 - T\mathbf{v}_2 = T(\mathbf{v}_1 - \mathbf{v}_2) = \mathbf{0}$, which means that $(\mathbf{v}_1 - \mathbf{v}_2) \in \ker T = \{\mathbf{0}\}$. Thus $\mathbf{v}_1 - \mathbf{v}_2 = \mathbf{0}$, so $\mathbf{v}_1 = \mathbf{v}_2$, which shows that T is 1–1. Now suppose that T is 1–1 and $\mathbf{v} \in \ker T$. Then $T\mathbf{v} = \mathbf{0}$. But $T\mathbf{0} = \mathbf{0}$ also. Thus, since T is 1–1, $\mathbf{v} = \mathbf{0}$. This completes the proof. ■

EXAMPLE 1 Let T: $\mathbb{R}^2 \to \mathbb{R}^2$ be defined by $T\begin{pmatrix} x \\ y \end{pmatrix} = \begin{pmatrix} x-y \\ 2x+y \end{pmatrix}$. We easily find $A_T = \begin{pmatrix} 1 & -1 \\ 2 & 1 \end{pmatrix}$ and $\rho(A_T) = 2$; hence $\nu(A_T) = 0$ and $N_{A_T} = \ker T = \{\mathbf{0}\}$. Thus T is 1–1.

EXAMPLE 2 Let T: $\mathbb{R}^2 \to \mathbb{R}^2$ be defined by $T\begin{pmatrix} x \\ y \end{pmatrix} = \begin{pmatrix} x-y \\ 2x-2y \end{pmatrix}$. Then $A_T = \begin{pmatrix} 1 & -1 \\ 2 & -2 \end{pmatrix}$, $\rho(A_T) = 1$, and $\nu(A_T) = 1$; hence $\nu(T) = 1$ and T is not 1–1. Note, for example, that $T\begin{pmatrix} 1 \\ 1 \end{pmatrix} = \mathbf{0} = T\begin{pmatrix} 0 \\ 0 \end{pmatrix}$.

DEFINITION 2 **ONTO TRANSFORMATION** Let $T: V \to W$ be a linear transformation. Then T is said to be *onto* W or, simply, **onto** if for every $\mathbf{w} \in W$ there is at least one $\mathbf{v} \in V$ such that $T\mathbf{v} = \mathbf{w}$. That is: *T is onto W if and only if range $T = W$.*

EXAMPLE 3 In Example 1, $\rho(A_T) = 2$; hence range $T = \mathbb{R}^2$ and T is onto. In Example 2, $\rho(A_T) = 1$ and range $T = \text{span}\left\{\begin{pmatrix}1\\2\end{pmatrix}\right\} \neq \mathbb{R}^2$; hence T is not onto.

THEOREM 2 Let $T: V \to W$ be a linear transformation and suppose that dim $V =$ dim $W = n$.

> **i.** If T is 1–1, then T is onto.
> **ii.** If T is onto, then T is 1–1.

Proof Let A_T be the matrix representation of T. Then if T is 1–1, ker $T = \{\mathbf{0}\}$ and $\nu(A_T) = 0$, which means that $\rho(T) = \rho(A_T) = n - 0 = n$ so that range $T = W$. If T is onto, then $\rho(A_T) = n$ so that $\nu(T) = \nu(A_T) = 0$ and T is 1–1. ■

THEOREM 3 Let $T: V \to W$ be a linear transformation. Suppose that dim $V = n$ and dim $W = m$. Then:

i. If $n > m$, T is not 1–1.
ii. If $m > n$, T is not onto.

Proof **i.** Let $\{\mathbf{v}_1, \mathbf{v}_2, \ldots, \mathbf{v}_n\}$ be a basis for V. Let $\mathbf{w}_i = T\mathbf{v}_i$ for $i = 1, 2, \ldots, n$ and look at the set $S = \{\mathbf{w}_1, \mathbf{w}_2, \ldots, \mathbf{w}_n\}$. Since $m = \dim W < n$, the set S is linearly dependent. Thus there exist scalars not all zero such that $c_1\mathbf{w}_1 + c_2\mathbf{w}_2 + \cdots + c_n\mathbf{w}_n = \mathbf{0}$. Let $\mathbf{v} = c_1\mathbf{v}_1 + c_2\mathbf{v}_2 + \cdots + c_n\mathbf{v}_n$. Since the $\mathbf{v}_i$'s are linearly independent and since not all the c_i's are zero, we see that $\mathbf{v} \neq \mathbf{0}$. But $T\mathbf{v} = T(c_1\mathbf{v}_1 + c_2\mathbf{v}_2 + \cdots + c_n\mathbf{v}_n) = c_1T\mathbf{v}_1 + c_2T\mathbf{v}_2 + \cdots + c_nT\mathbf{v}_n = c_1\mathbf{w}_1 + c_2\mathbf{w}_2 + \cdots + c_n\mathbf{w}_n = \mathbf{0}$. Thus $\mathbf{v} \in \ker T$ and $\ker T \neq \{\mathbf{0}\}$.

ii. If $\mathbf{v} \in V$, then $\mathbf{v} = a_1\mathbf{v}_1 + a_2\mathbf{v}_2 + \cdots + a_n\mathbf{v}_n$ for some scalars $a_1, a_2, \ldots, a_n$ and $T\mathbf{v} = a_1T\mathbf{v}_1 + a_2T\mathbf{v}_2 + \cdots + a_nT\mathbf{v}_n = a_1\mathbf{w}_1 + a_2\mathbf{w}_2 + \cdots + a_n\mathbf{w}_n$. Thus $\{\mathbf{w}_1, \mathbf{w}_2, \ldots, \mathbf{w}_n\} = \{T\mathbf{v}_1, T\mathbf{v}_2, \ldots, T\mathbf{v}_n\}$ spans range T. Then, from Problem 5.5.29 on page 200, $\rho(T) = \dim \text{range } T \leq n$. Since $m > n$, this shows that range $T \neq W$. Thus T is not onto. ■

DEFINITION 3 **ISOMORPHISM** Let $T: V \to W$ be a linear transformation. Then T is an **isomorphism** if T is 1–1 and onto.

DEFINITION 4 **ISOMORPHIC VECTOR SPACES** The vector spaces V and W are said to be **isomorphic** if there exists an isomorphism T from V onto W. In this case we write $V \cong W$.

Remark. The word "isomorphism" comes from the Greek *isomorphos* meaning "of equal form" (*iso* = equal; *morphos* = form). After a few examples we shall see how closely related are the "forms" of isomorphic vector spaces.

Let T: $\mathbb{R}^n \to \mathbb{R}^n$ and let A_T be the matrix representation of T. Now T is 1–1 if and only if $\ker T = \{\mathbf{0}\}$, which is true if and only if $\nu(A_T) = 0$ if and only if $\det A_T \neq 0$. Thus we can extend our Summing Up Theorem in another direction.

THEOREM 4 **SUMMING UP THEOREM—VIEW 5** Let A be an $n \times n$ matrix. Then the following nine statements are equivalent. That is, if one is true, all are true.

- **i.** A is invertible.
- **ii.** The only solution to the homogeneous system $A\mathbf{x} = \mathbf{0}$ is the trivial solution ($\mathbf{x} = \mathbf{0}$).
- **iii.** The system $A\mathbf{x} = \mathbf{b}$ has a unique solution for every n-vector $\mathbf{b}$.
- **iv.** A is row equivalent to the $n \times n$ identity matrix I_n.
- **v.** The rows (and columns) of A are linearly independent.
- **vi.** $\det A \neq 0$.
- **vii.** $\nu(A) = 0$.
- **viii.** $\rho(A) = n$.
- **ix.** The linear transformation T from $\mathbb{R}^n$ to $\mathbb{R}^n$ defined by $T\mathbf{x} = A\mathbf{x}$ is an isomorphism.

We now look at some examples of isomorphisms between other pairs of vector spaces.

EXAMPLE 4 Let T: $\mathbb{R}^3 \to P_2$ be defined by $T\begin{pmatrix} a \\ b \\ c \end{pmatrix} = a + bx + cx^2$. It is easy to verify that T is linear. Suppose that $T\begin{pmatrix} a \\ b \\ c \end{pmatrix} = \mathbf{0} = 0 + 0x + 0x^2$. Then $a = b = c = 0$. That is, $\ker T = \{\mathbf{0}\}$ and T is 1–1. If $p(x) = a_0 + a_1x + a_2x^2$, then $p(x) = T\begin{pmatrix} a_0 \\ a_1 \\ a_2 \end{pmatrix}$. This means that range $T = P_2$ and T is onto. Thus $\mathbb{R}^3 \cong P_2$.

□**EXAMPLE 5** Let $V = \{f \in C'[0, 1]: f(0) = 0\}$ and $W = C[0, 1]$. Let D: $V \to W$ be given by $Df = f'$. Suppose that $Df = Dg$. Then $f' = g'$ or $(f - g)' = 0$ and $f(x) - g(x) = c$, a constant. But $f(0) = g(0) = 0$, so $c = 0$ and $f = g$. Thus D is 1–1. Let $g \in C[0, 1]$ and let $f(x) = \int_0^x g(t)\,dt$. Then, from the fundamental theorem of calculus, $f \in C'[0, 1]$ and $f'(x) = g(x)$ for every $x \in [0, 1]$. Moreover, since $\int_0^0 g(t)\,dt = 0$, we have $f(0) = 0$. Thus, for every g in W, there is an $f \in V$ such that $Df = g$. Hence D is onto and we have shown that $V \cong W$.

The following theorem illustrates the similarity of two isomorphic vector spaces.

THEOREM 5 Let $T: V \to W$ be an isomorphism.

i. If $\mathbf{v}_1, \mathbf{v}_2, \ldots, \mathbf{v}_n$ span V, then $T\mathbf{v}_1, T\mathbf{v}_2, \ldots, T\mathbf{v}_n$ span W.
ii. If $\mathbf{v}_1, \mathbf{v}_2, \ldots, \mathbf{v}_n$ are linearly independent in V, then $T\mathbf{v}_1, T\mathbf{v}_2, \ldots, T\mathbf{v}_n$ are linearly independent in W.
iii. If $\{\mathbf{v}_1, \mathbf{v}_2, \ldots, \mathbf{v}_n\}$ is a basis in V, then $\{T\mathbf{v}_1, T\mathbf{v}_2, \ldots, T\mathbf{v}_n\}$ is a basis in W.
iv. If V is finite dimensional, then W is finite dimensional and $\dim V = \dim W$.

Proof **i.** Let $\mathbf{w} \in W$. Then, since T is onto, there is a $\mathbf{v} \in V$ such that $T\mathbf{v} = \mathbf{w}$. Since the $\mathbf{v}_i$'s span V, we can write $\mathbf{v} = a_1\mathbf{v}_1 + a_2\mathbf{v}_2 + \cdots + a_n\mathbf{v}_n$ so that $\mathbf{w} = T\mathbf{v} = a_1T\mathbf{v}_1 + a_2T\mathbf{v}_2 + \cdots + a_nT\mathbf{v}_n$ and this shows that $\{T\mathbf{v}_1, T\mathbf{v}_2, \ldots, T\mathbf{v}_n\}$ spans W.
ii. Suppose $c_1T\mathbf{v}_1 + c_2T\mathbf{v}_2 + \cdots + c_nT\mathbf{v}_n = \mathbf{0}$. Then $T(c_1\mathbf{v}_1 + c_2\mathbf{v}_2 + \cdots + c_n\mathbf{v}_n) = \mathbf{0}$. Thus, since T is 1–1, $c_1\mathbf{v}_1 + c_2\mathbf{v}_2 + \cdots + c_n\mathbf{v}_n = \mathbf{0}$, which implies that $c_1 = c_2 = \cdots = c_n = 0$ since the $\mathbf{v}_i$'s are independent.
iii. This follows from parts (*i*) and (*ii*).
iv. This follows from part (*iii*). ■

In general, it is difficult to show that two infinite dimensional vector spaces are isomorphic. For finite dimensional spaces, however, it is remarkably easy. Theorem 3 shows that if $\dim V \neq \dim W$, then V and W are not isomorphic. The next theorem shows that if $\dim V = \dim W$, and if V and W are real vector spaces, then V and W are isomorphic.

THEOREM 6 Let V and W be two real finite dimensional vector spaces with $\dim V = \dim W$. Then $V \cong W$.

Proof Let $\{\mathbf{v}_1, \mathbf{v}_2, \ldots, \mathbf{v}_n\}$ be a basis for V and let $\{\mathbf{w}_1, \mathbf{w}_2, \ldots, \mathbf{w}_n\}$ be a basis for W. Define the linear transformation T by

$$T\mathbf{v}_i = \mathbf{w}_i \qquad \text{for } i = 1, 2, \ldots, n \tag{2}$$

By Theorem 6.2.2 on page 237 there is exactly one linear transformation that satisfies Equation (2). Suppose $\mathbf{v} \in V$ and $T\mathbf{v} = \mathbf{0}$. Then if $\mathbf{v} = c_1\mathbf{v}_1 + c_2\mathbf{v}_2 + \cdots + c_n\mathbf{v}_n$, we have $T\mathbf{v} = c_1T\mathbf{v}_1 + \cdots + c_nT\mathbf{v}_n = c_1\mathbf{w}_1 + c_2\mathbf{w}_2 + \cdots + c_n\mathbf{w}_n = \mathbf{0}$. But, since $\mathbf{w}_1, \mathbf{w}_2, \ldots, \mathbf{w}_n$ are linearly independent, $c_1 = c_2 = \cdots = c_n = 0$. Thus $\mathbf{v} = \mathbf{0}$ and T is 1–1. Since V and W are finite dimensional and $\dim V = \dim W$, T is onto by Theorem 2 and the proof is complete. ■

PROBLEMS 6.5

1. Show that $T: M_{mn} \to M_{nm}$ defined by $TA = A'$ is an isomorphism.
2. Show that $T: \mathbb{R}^n \to \mathbb{R}^n$ is an isomorphism if and only if A_T is invertible.

*3. Let V and W be n-dimensional real vector spaces and let B_1 and B_2 be bases for V and W, respectively. Let A_T be the transformation matrix relative to the bases B_1 and B_2. Show that $T: V \to W$ is an isomorphism if and only if $\det A_T \neq 0$.

4. Find an isomorphism between D_n, the $n \times n$ diagonal matrices with real entries, and $\mathbb{R}^n$. [*Hint:* Look first at the case $n = 2$.]

5. For what value of m is the set of $n \times n$ symmetric matrices isomorphic to $\mathbb{R}^m$?

6. Show that the set of $n \times n$ symmetric matrices is isomorphic to the set of $n \times n$ upper triangular matrices.

7. Let $V = P_4$ and $W = \{p \in P_5: p(0) = 0\}$. Show that $V \cong W$.

□8. Define $T: P_n \to P_n$ by $Tp = p + p'$. Show that T is an isomorphism.

9. Find a condition on the numbers m, n, p, q such that $M_{mn} \cong M_{pq}$.

10. Show that $D_n \cong P_{n-1}$.

11. Prove that any two finite dimensional complex vector spaces V and W with $\dim V = \dim W$ are isomorphic.

12. Define $T: C[0, 1] \to C[3, 4]$ by $Tf(x) = f(x-3)$. Show that T is an isomorphism.

13. Let B be an invertible $n \times n$ matrix. Show that $T: M_{nm} \to M_{nm}$ defined by $TA = AB$ is an isomorphism.

□14. Show that the transformation $Tp(x) = xp'(x)$ is not an isomorphism from P_n into P_n.

15. Let H be a subspace of the finite dimensional inner product space V. Show that $T: V \to H$ defined by $T\mathbf{v} = \text{proj}_H \mathbf{v}$ is onto. Under what circumstances will it be 1–1?

16. Show that if $T: V \to W$ is an isomorphism, then there exists an isomorphism $S: W \to V$ such that $S(T\mathbf{v}) = \mathbf{v}$. Here S is called the *inverse transformation* of T and is denoted T^{-1}.

17. Show that if $T: \mathbb{R}^n \to \mathbb{R}^n$ is defined by $T\mathbf{x} = A\mathbf{x}$ and if T is an isomorphism, then A is invertible and the inverse transformation T^{-1} is given by $T^{-1}\mathbf{x} = A^{-1}\mathbf{x}$.

18. Find T^{-1} for the isomorphism of Problem 7.

*19. Consider the space $C = \{z = a + ib$, where a and b are real numbers and $i^2 = -1\}$. Show that if the scalars are taken to be the reals, then $C \cong \mathbb{R}^2$.

*20. Consider the space $\mathbb{C}^n_{\mathbb{R}} = \{(\mathbf{c}_1, \mathbf{c}_2, \ldots, \mathbf{c}_n)$: $\mathbf{c}_i \in C$ and the scalars are the reals$\}$. Show that $\mathbb{C}^n_{\mathbb{R}} \cong \mathbb{R}^{2n}$. (Hint: See Problem 19.)

6.6 If Time Permits: Isometries

In this section we describe a special kind of linear transformation between vector spaces. We begin with a very useful result.

THEOREM 1 Let A be an $m \times n$ matrix with real entries.† Then for any vectors $\mathbf{x} \in \mathbb{R}^n$ and $\mathbf{y} \in \mathbb{R}^m$:

$$(A\mathbf{x}) \cdot \mathbf{y} = \mathbf{x} \cdot (A^t\mathbf{y}) \tag{1}$$

† This result can easily be extended to matrices with complex components. See Problem 21.

Proof We prove Equation (1) by the brute force method—that is, by computing both sides and showing that they are equal. We set

$$A = \begin{pmatrix} a_{11} & a_{12} & \cdots & a_{1n} \\ a_{21} & a_{22} & \cdots & a_{2n} \\ \cdot & \cdot & & \cdot \\ \cdot & \cdot & & \cdot \\ \cdot & \cdot & & \cdot \\ a_{m1} & a_{m2} & \cdots & a_{mn} \end{pmatrix} \qquad \mathbf{x} = \begin{pmatrix} x_1 \\ x_2 \\ \cdot \\ \cdot \\ \cdot \\ x_n \end{pmatrix} \qquad \mathbf{y} = \begin{pmatrix} y_1 \\ y_2 \\ \cdot \\ \cdot \\ \cdot \\ y_m \end{pmatrix}$$

Then

$$A\mathbf{x} = \begin{pmatrix} \mathbf{r}_1 \cdot \mathbf{x} \\ \mathbf{r}_2 \cdot \mathbf{x} \\ \cdot \\ \cdot \\ \cdot \\ \mathbf{r}_m \cdot \mathbf{x} \end{pmatrix}$$

where $\mathbf{r}_i$ denotes the ith row of A and

$$(A\mathbf{x}) \cdot \mathbf{y} = (\mathbf{r}_1 \cdot \mathbf{x})y_1 + (\mathbf{r}_2 \cdot \mathbf{x})y_2 + \cdots + (\mathbf{r}_m \cdot \mathbf{x})y_m = \sum_{i=1}^{m} (\mathbf{r}_i \cdot \mathbf{x})y_i \tag{2}$$

But $\mathbf{r}_i \cdot \mathbf{x} = a_{i1}x_1 + a_{i2}x_2 + \cdots + a_{in}x_n = \sum_{j=1}^{n} a_{ij}x_j$. Thus, from (2), we obtain

$$(A\mathbf{x}) \cdot \mathbf{y} = \sum_{i=1}^{m} \sum_{j=1}^{n} a_{ij}x_jy_i \tag{3}$$

Now

$$A^t = \begin{pmatrix} a_{11} & a_{21} & \cdots & a_{m1} \\ a_{12} & a_{22} & \cdots & a_{m2} \\ \cdot & \cdot & & \cdot \\ \cdot & \cdot & & \cdot \\ \cdot & \cdot & & \cdot \\ a_{1n} & a_{2n} & \cdots & a_{mn} \end{pmatrix}$$

so that

$$A^t\mathbf{y} = \begin{pmatrix} \mathbf{c}_1 \cdot \mathbf{y} \\ \mathbf{c}_2 \cdot \mathbf{y} \\ \cdot \\ \cdot \\ \cdot \\ \mathbf{c}_n \cdot \mathbf{y} \end{pmatrix}$$

where $\mathbf{c}_j$ denotes the jth column of A (= the jth row of A^t). Then

$$(A^t\mathbf{y}) \cdot \mathbf{x} = (\mathbf{c}_1 \cdot \mathbf{y})x_1 + (\mathbf{c}_2 \cdot \mathbf{y})x_2 + \cdots + (\mathbf{c}_n \cdot \mathbf{y})x_n = \sum_{j=1}^{n} (\mathbf{c}_j \cdot \mathbf{y})x_j \tag{4}$$

But $\mathbf{c}_j \cdot \mathbf{y} = a_{1j}y_1 + a_{2j}y_2 + \cdots + a_{mj}y_m = \sum_{i=1}^{m} a_{ij}y_i$. Thus, from (4), we get

$$(A^t\mathbf{y}) \cdot \mathbf{x} = \sum_{j=1}^{n} \sum_{i=1}^{m} a_{ij}y_ix_j = \sum_{i=1}^{m} \sum_{j=1}^{n} a_{ij}x_jy_i = (A\mathbf{x}) \cdot \mathbf{y}$$

and the theorem is proved. ■

Recall from Section 5.7, page 216, that a matrix Q with real entries is **orthogonal** if Q is invertible and $Q^{-1} = Q^t$. In Theorem 5.7.3 on page 216 we proved that Q is orthogonal if and only if the columns of Q form an orthonormal basis for $\mathbb{R}^n$. Now let Q be an $n \times n$ orthogonal matrix and let T: $\mathbb{R}^n \to \mathbb{R}^n$ be the linear transformation defined by $T\mathbf{x} = Q\mathbf{x}$. Then, using Equation (1), we compute $(T\mathbf{x} \cdot T\mathbf{y}) = Q\mathbf{x} \cdot Q\mathbf{y} = \mathbf{x} \cdot (Q^tQ\mathbf{y}) = \mathbf{x} \cdot (I\mathbf{y}) = \mathbf{x} \cdot \mathbf{y}$. In particular, if $\mathbf{x} = \mathbf{y}$, we see that $T\mathbf{x} \cdot T\mathbf{x} = \mathbf{x} \cdot \mathbf{x}$ or

$$|T\mathbf{x}| = |\mathbf{x}| \tag{5}$$

for every $\mathbf{x}$ in $\mathbb{R}^n$.

DEFINITION 1 **ISOMETRY** A linear transformation T: $\mathbb{R}^n \to \mathbb{R}^n$ is called an **isometry** if, for every $\mathbf{x}$ and $\mathbf{y}$ in $\mathbb{R}^n$,

$$T\mathbf{x} \cdot T\mathbf{y} = \mathbf{x} \cdot \mathbf{y} \tag{6}$$

Because of Equation (6) we can say: *An isometry in $\mathbb{R}^n$ is a linear transformation that preserves the scalar product in $\mathbb{R}^n$.*

EXAMPLE 1 Let T: $\mathbb{R}^2 \to \mathbb{R}^2$ be the linear operator that rotates every vector in $\mathbb{R}^2$ through an angle of θ. Then, as we saw in Example 6.1.8, $T\mathbf{x} = A_\theta \mathbf{x}$, where $A_\theta = \begin{pmatrix} \cos\theta & -\sin\theta \\ \sin\theta & \cos\theta \end{pmatrix}$. But A_θ is an orthogonal matrix, so T is an isometry. To check, note that $T\begin{pmatrix} x \\ y \end{pmatrix} = \begin{pmatrix} x\cos\theta - y\sin\theta \\ x\sin\theta + y\cos\theta \end{pmatrix}$ and $\left|T\begin{pmatrix} x \\ y \end{pmatrix}\right|^2 =$ $\left[T\begin{pmatrix} x \\ y \end{pmatrix}\right] \cdot \left[T\begin{pmatrix} x \\ y \end{pmatrix}\right] = [x\cos\theta - y\sin\theta]^2 + [x\sin\theta + y\cos\theta]^2 =$ $(x^2\cos^2\theta - 2xy\cos\theta\sin\theta + y^2\sin^2\theta) +$ $(x^2\sin^2\theta + 2xy\sin\theta\cos\theta + y^2\cos^2\theta) = x^2(\cos^2\theta + \sin^2\theta) + y^2(\sin^2\theta +$ $\cos^2\theta) = x^2 + y^2 = \left|\begin{pmatrix} x \\ y \end{pmatrix}\right|^2$. Thus $\left|T\begin{pmatrix} x \\ y \end{pmatrix}\right| = \left|\begin{pmatrix} x \\ y \end{pmatrix}\right|$.

Isometries have some interesting properties.

THEOREM 2 Let T: $\mathbb{R}^n \to \mathbb{R}^n$ be an isometry. Then:

i. If $\mathbf{u}_1, \mathbf{u}_2, \ldots, \mathbf{u}_n$ is an orthogonal set, then $T\mathbf{u}_1, T\mathbf{u}_2, \ldots, T\mathbf{u}_n$ is an orthogonal set.
ii. T is an isomorphism.

Proof **i.** If $i \neq j$ and $\mathbf{u}_i \cdot \mathbf{u}_j = 0$, then $(T\mathbf{u}_i) \cdot (T\mathbf{u}_j) = \mathbf{u}_i \cdot \mathbf{u}_j = 0$, which proves (*i*).
ii. Let $\mathbf{u}_1, \mathbf{u}_2, \ldots, \mathbf{u}_n$ be an orthonormal basis for $\mathbb{R}^n$. Then, by part (*i*) and

the fact that $|T\mathbf{u}_i|=|\mathbf{u}_i|=1$, we find that $T\mathbf{u}_1, T\mathbf{u}_2, \ldots, T\mathbf{u}_n$ is an orthonormal set in $\mathbb{R}^n$. By Theorem 5.7.1 on page 213 these vectors are linearly independent and hence form a basis for $\mathbb{R}^n$. Thus range $T = \mathbb{R}^n$, which proves that ker $T = \{\mathbf{0}\}$ (since $\nu(T) + \rho(T) = n$). ■

THEOREM 3 Let $B=\{\mathbf{u}_1, \mathbf{u}_2, \ldots, \mathbf{u}_n\}$ be an orthonormal basis for $\mathbb{R}^n$. Then $T\colon\ \mathbb{R}^n \rightarrow \mathbb{R}^n$ is an isometry if and only if A_T is orthogonal, where A_T is the matrix representation of T relative to the basis B.

Proof Suppose that A_T is orthogonal with respect to B. Then, as we have already seen, $T\mathbf{x}\cdot T\mathbf{x}=A_T\mathbf{x}\cdot A_T\mathbf{x}=\mathbf{x}\cdot\mathbf{x}$, so that T is an isometry. If T is an isometry, then $\mathbf{x}\cdot\mathbf{y}=T\mathbf{x}\cdot T\mathbf{y}=A_T\mathbf{x}\cdot A_T\mathbf{y}=\mathbf{x}\cdot A_T^tA_T\mathbf{y}$. This means that for all vectors $\mathbf{x}$ and $\mathbf{y}$ in $\mathbb{R}^n$, $\mathbf{x}\cdot(\mathbf{y}-A_T^tA_T\mathbf{y})=0$. Hence $(\mathbf{y}-A_T^tA_T)\in(\mathbb{R}^n)^\perp=\{\mathbf{0}\}$. Thus $\mathbf{y}=A_T^tA_T\mathbf{y}$ for every $\mathbf{y}\in\mathbb{R}^n$, which means that $A_T^tA_T=I$ and $A_T^t=A_T^{-1}$. Thus A_T is orthogonal and the theorem is proved. ■

We conclude this section by outlining how we can extend the notion of isometry to an arbitrary inner product space.

DEFINITION 2 **ISOMETRY** Let V and W be real (or complex) inner product spaces and let $T\colon V \to W$ be a linear transformation. Then T is an **isometry** if, for every $\mathbf{v}_1, \mathbf{v}_2 \in V$,

$$(\mathbf{v}_1, \mathbf{v}_2)=(T\mathbf{v}_1, T\mathbf{v}_2) \tag{7}$$

Note. By Equation (7), if T is an isometry then (using Definition 5.8.2 on page 224) we see that

$$|\mathbf{v}_1|_V=|T\mathbf{v}_1|_W \tag{8}$$

DEFINITION 3 **ISOMETRICALLY ISOMORPHIC VECTOR SPACES** Two vector spaces V and W are said to be **isometrically isomorphic** if there exists a linear transformation $T\colon V \to W$ that is both an isometry and an isomorphism.

THEOREM 4 Any two n-dimensional real inner product spaces are isometrically isomorphic.

Proof Let $\{\mathbf{u}_1, \mathbf{u}_2, \ldots, \mathbf{u}_n\}$ and $\{\mathbf{w}_1, \mathbf{w}_2, \ldots, \mathbf{w}_n\}$ be orthonormal bases for V and W, respectively. Let $T\colon V \rightarrow W$ be the linear transformation defined by $T\mathbf{u}_i = \mathbf{w}_i$, $i=1, 2, \ldots, n$. If we can show that T is an isometry, then we shall be done since reasoning as in the proof of Theorem 2 shows us that T is also an isomorphism. Let $\mathbf{x}$ and $\mathbf{y}$ be in V. Then there exist sets of real numbers $c_1, c_2, \ldots, c_n$ and $d_1, d_2, \ldots, d_n$ such that $\mathbf{x} = c_1\mathbf{u}_1 + c_2\mathbf{u}_2 + \cdots + c_n\mathbf{u}_n$ and $\mathbf{y} = d_1\mathbf{u}_1 + d_2u_2 + \cdots + d_n\mathbf{u}_n$. Since the $\mathbf{u}_i$'s are orthonormal, $(\mathbf{x}, \mathbf{y}) = [(c_1\mathbf{u}_1 + c_2\mathbf{u}_2 + \cdots + c_n\mathbf{u}_n), (d_1\mathbf{u}_1 + d_2\mathbf{u}_2 + \cdots + d_n\mathbf{u}_n)] = c_1d_1 + c_2d_2 + \cdots + c_nd_n$. Similarly, since $T\mathbf{x} = c_1T\mathbf{u}_1 + c_2T\mathbf{u}_2 + \cdots + c_nT\mathbf{u}_n = c_1\mathbf{w}_1 + c_2\mathbf{w}_2 + \cdots + c_n\mathbf{w}_n$, we obtain $(T\mathbf{x}, T\mathbf{y}) = [(c_1\mathbf{w}_1 + c_2\mathbf{w}_2 + \cdots + c_n\mathbf{w}_n), (d_1\mathbf{w}_1 + d_2\mathbf{w}_2 + \cdots + d_n\mathbf{w}_n)] = c_1d_1 + c_2d_2 + \cdots + c_nd_n$ because the $\mathbf{w}_i$'s are orthonormal. This completes the proof. ■

□EXAMPLE 2

We illustrate this theorem by showing that $\mathbb{R}^3$ and $P_2[0, 1]$ are isometrically isomorphic. In $\mathbb{R}^3$ we use the standard basis $\left\{\begin{pmatrix}1\\0\\0\end{pmatrix}, \begin{pmatrix}0\\1\\0\end{pmatrix}, \begin{pmatrix}0\\0\\1\end{pmatrix}\right\}$. In P_2 we use the orthonormal basis $\{1, \sqrt{3}(2x-1), \sqrt{5}(6x^2-6x+1)\}$. (See Example 5.8.8.) Let $\mathbf{x} = \begin{pmatrix}a_1\\b_1\\b_2\end{pmatrix}$ and $\mathbf{y} = \begin{pmatrix}a_2\\b_2\\c_2\end{pmatrix}$ be in $\mathbb{R}^3$. Then $(\mathbf{x}, \mathbf{y}) = \mathbf{x}\cdot\mathbf{y} = a_1a_2 + b_1b_2 + c_1c_2$.

Recall that in $P_2[0, 1]$ we defined $(p, q) = \int_0^1 p(x)q(x)\,dx$. Now $T\begin{pmatrix}1\\0\\0\end{pmatrix} = 1$, $T\begin{pmatrix}0\\1\\0\end{pmatrix} = \sqrt{3}(2x-1)$, and $T\begin{pmatrix}0\\0\\1\end{pmatrix} = \sqrt{5}(6x^2-6x+1)$; hence $T\begin{pmatrix}a\\b\\c\end{pmatrix} = a + b\sqrt{3}(2x-1) + c\sqrt{5}(6x^2-6x+1)$ and

$$\begin{aligned}(T\mathbf{x}, T\mathbf{y}) = \int_0^1 &[a_1 + b_1\sqrt{3}(2x-1) + c_1\sqrt{5}(6x^2-6x+1)]\\ &\times[a_2 + b_2\sqrt{3}(2x-1) + c_2\sqrt{5}(6x^2-6x+1)]\,dx\\ &= a_1a_2\int_0^1 dx + \int_0^1 b_1b_2 3(2x-1)^2\,dx + \int_0^1 c_1c_2[5(6x^2-6x+1)^2]\,dx\\ &\quad + (a_1b_2 + a_2b_1)\int_0^1 \sqrt{3}(2x-1)\,dx\\ &\quad + (a_1c_2 + a_2c_1)\int_0^1 \sqrt{5}(6x^2-6x+1)\,dx\\ &\quad + (b_1c_2 + b_2c_1)\int_0^1 [\sqrt{3}(2x-1)][\sqrt{5}(6x^2-6x+1)]\,dx\\ &= a_1a_2 + b_1b_2 + c_1c_2\end{aligned}$$

Here we saved time by using the fact that $\{1, \sqrt{3}(2x-1), \sqrt{5}(6x^2-6x+1)\}$ is an orthonormal basis. Thus T: $\mathbb{R}^3 \to P_2[0, 1]$ is an isometry.

PROBLEMS 6.6

1. Show that for any real number θ, the transformation T: $\mathbb{R}^3 \to \mathbb{R}^3$ defined by $T\mathbf{x} = A\mathbf{x}$, where

$$A = \begin{pmatrix} \sin\theta & \cos\theta & 0 \\ \cos\theta & -\sin\theta & 0 \\ 0 & 0 & 1 \end{pmatrix}$$

is an isometry.

2. Do the same for the transformation T, where

$$A = \begin{pmatrix} \cos\theta & 0 & -\sin\theta \\ 0 & 1 & 0 \\ \sin\theta & 0 & \cos\theta \end{pmatrix}$$

3. Let A and B be orthogonal $n \times n$ matrices. Show that T: $\mathbb{R}^n \to \mathbb{R}^n$ defined by $T\mathbf{x} = AB\mathbf{x}$ is an isometry.
4. Find A_T if T is the transformation from $\mathbb{R}^3 \to \mathbb{R}^3$ defined by

$$T\begin{pmatrix} 2/3 \\ 1/3 \\ -2/3 \end{pmatrix} = \begin{pmatrix} 1/\sqrt{2} \\ 1/\sqrt{2} \\ 0 \end{pmatrix} \quad T\begin{pmatrix} 1/3 \\ 2/3 \\ 2/3 \end{pmatrix} = \begin{pmatrix} -1/\sqrt{6} \\ 1/\sqrt{6} \\ 2/\sqrt{6} \end{pmatrix} \quad T\begin{pmatrix} 2/3 \\ -2/3 \\ 1/3 \end{pmatrix} = \begin{pmatrix} 1/\sqrt{3} \\ -1/\sqrt{3} \\ 1/\sqrt{3} \end{pmatrix}$$

Show that A_T is orthogonal.

5. Let T: $\mathbb{R}^n \to \mathbb{R}^n$ have the property that $|T\mathbf{x}| = |\mathbf{x}|$ for every $\mathbf{x} \in \mathbb{R}^n$. Prove that T is an isometry. [*Hint:* First show that $\mathbf{x} \cdot \mathbf{y} = \frac{1}{2}(|\mathbf{x}+\mathbf{y}|^2 - |\mathbf{x}|^2 - |\mathbf{y}|^2)$.]
6. Let T: $\mathbb{R}^2 \to \mathbb{R}^2$ be an isometry. Show that T preserves angles. That is: (angle between $\mathbf{x}$ and $\mathbf{y}$) = (angle between $T\mathbf{x}$ and $T\mathbf{y}$).
7. Give an example of a linear transformation from $\mathbb{R}^2$ onto $\mathbb{R}^2$ that preserves angles and is *not* an isometry.
8. For $\mathbf{x}, \mathbf{y} \in \mathbb{R}^n$ and $\mathbf{x}$ and $\mathbf{y} \neq \mathbf{0}$, define: (angle between $\mathbf{x}$ and $\mathbf{y}$) $= \measuredangle(\mathbf{x}, \mathbf{y}) = \cos^{-1}[(\mathbf{x} \cdot \mathbf{y})/|\mathbf{x}|\,|\mathbf{y}|]$. Show that if T: $\mathbb{R}^n \to \mathbb{R}^n$ is an isometry, then T preserves angles.
9. Let T: $\mathbb{R}^n \to \mathbb{R}^n$ be an isometry and let $T\mathbf{x} = A\mathbf{x}$. Show that $S\mathbf{x} = A^{-1}\mathbf{x}$ is an isometry.

In Problems 10–14 find an isometry between the given pair of spaces.

□**10.** $P_1[-1, 1], \mathbb{R}^2$ □***11.** $P_3[-1, 1], \mathbb{R}^4$ ***12.** $M_{22}, \mathbb{R}^4$ □***13.** $M_{22}, P_3[-1, 1]$

14. D_n and $\mathbb{R}^n$ (D_n = set of diagonal $n \times n$ matrices)
15. Let A be an $n \times n$ matrix with complex components. Then the conjugate of A, denoted A^*, is defined by $(A^*)_{ij} = \overline{a_{ji}}$. Compute A^* if $A = \begin{pmatrix} 1+i & -4+2i \\ 3 & 6-3i \end{pmatrix}$.
16. The $n \times n$ complex matrix A is called **hermitian** if $A^* = A$. Show that the matrix $A = \begin{pmatrix} 4 & 3-2i \\ 3+2i & 6 \end{pmatrix}$ is hermitian.
17. Show that if A is hermitian, then the diagonal components of A are real.
18. The $n \times n$ complex matrix A is called **unitary** if $A^* = A^{-1}$. Show that the matrix

$$A = \begin{pmatrix} \dfrac{1+i}{2} & \dfrac{3-2i}{\sqrt{26}} \\ \dfrac{1+i}{2} & \dfrac{-3+2i}{\sqrt{26}} \end{pmatrix}$$

is unitary.

19. Show that A is unitary if and only if the columns of A form an orthonormal basis in $\mathbb{C}^n$.

20. Show that if A is unitary, then $|\det A| = 1$.

21. Let A be an $n \times n$ matrix with complex components. In $\mathbb{C}^n$, if $\mathbf{x} = (c_1, c_2, \cdots, c_n)$ and $\mathbf{y} = (d_1, d_2, \ldots, d_n)$, define the inner product $(\mathbf{x}, \mathbf{y}) = c_1\bar{d}_1 + c_2\bar{d}_2 + \cdots + c_n\bar{d}_n$. (See Example 5.8.2.) Prove that $(A\mathbf{x}, \mathbf{y}) = (\mathbf{x}, A^*\mathbf{y})$.

***22.** Show that any two complex inner product spaces of the same (finite) dimension are isometrically isomorphic.

Review Exercises for Chapter 6

In Exercises 1–6 determine whether the given transformation from V to W is linear.

1. T: $\mathbb{R}^2 \to \mathbb{R}^2$; $T(x, y) = (0, -y)$

2. T: $\mathbb{R}^3 \to \mathbb{R}^3$; $T(x, y, z) = (1, y, z)$

3. T: $\mathbb{R}^2 \to \mathbb{R}^2$; $T(x, y) = x/y$

4. T: $P_1 \to P_2$; $(Tp)(x) = xp(x)$

5. T: $P_2 \to P_2$; $(Tp)(x) = 1 + p(x)$

6. T: $C[0, 1] \to C[0, 1]$; $Tf(x) = f(1)$

In Exercises 7–12 find the kernel, range, nullity, and rank of the given matrix.

7. $A = \begin{pmatrix} 1 & -2 \\ -2 & 4 \end{pmatrix}$

8. $A = \begin{pmatrix} 1 & -1 & 3 \\ 2 & 0 & 4 \\ 0 & -2 & 2 \end{pmatrix}$

9. $A = \begin{pmatrix} 1 & -1 & 2 \\ 0 & 1 & 4 \\ 1 & -1 & 0 \end{pmatrix}$

10. $A = \begin{pmatrix} 2 & 4 & -2 \\ -1 & -2 & 1 \end{pmatrix}$

11. $A = \begin{pmatrix} 2 & 3 \\ -1 & 2 \\ 4 & 6 \end{pmatrix}$

12. $A = \begin{pmatrix} 1 & -1 & 2 & 3 \\ 0 & 1 & -1 & 0 \\ 1 & -2 & 3 & 3 \\ 2 & -3 & 5 & 6 \end{pmatrix}$

In Exercises 13–18 find the matrix representation of the given linear transformation and find the kernel, range, nullity, and rank of the transformation.

13. T: $\mathbb{R}^2 \to \mathbb{R}^2$; $T(x, y) = (0, -y)$

14. T: $\mathbb{R}^3 \to \mathbb{R}^2$; $T(x, y, z) = (y, z)$

15. T: $\mathbb{R}^4 \to \mathbb{R}^2$; $T(x, y, z, w) = (x - 2z, 2y + 3w)$

16. T: $P_3 \to P_4$; $(Tp)(x) = xp(x)$

17. T: $M_{22} \to M_{22}$; $TA = AB$, where $B = \begin{pmatrix} -1 & 0 \\ 1 & 2 \end{pmatrix}$

18. T: $\mathbb{R}^2 \to \mathbb{R}^2$; $T(x, y) = (x - y, 2x + 3y)$; $B_1 = \left\{\begin{pmatrix} 1 \\ 1 \end{pmatrix}, \begin{pmatrix} 1 \\ 2 \end{pmatrix}\right\}$; $B_2 = \left\{\begin{pmatrix} -1 \\ 3 \end{pmatrix}, \begin{pmatrix} 4 \\ 1 \end{pmatrix}\right\}$

19. Find an isomorphism T: $P_2 \to \mathbb{R}^3$.

□**20.** Find an isometry T: $\mathbb{R}^2 \to P_1[-1, 1]$.

7 Eigenvalues, Eigenvectors, and Canonical Forms

7.1 Eigenvalues and Eigenvectors

Let $T: V \to V$ be a linear transformation. In a great variety of applications (one of which is given in the next section), it is useful to find a vector $\mathbf{v}$ in V such that $T\mathbf{v}$ and $\mathbf{v}$ are parallel. That is, we seek a vector $\mathbf{v}$ and a scalar λ such that

$$T\mathbf{v} = \lambda\mathbf{v} \tag{1}$$

If $\mathbf{v} \neq \mathbf{0}$ and λ satisfy (1), then λ is called an *eigenvalue* of T and $\mathbf{v}$ is called an *eigenvector* of T corresponding to the eigenvalue λ. The purpose of this chapter is to investigate properties of eigenvalues and eigenvectors. If V is finite dimensional, then T can be represented by a matrix A_T. For that reason we shall discuss eigenvalues and eigenvectors of $n \times n$ matrices.

DEFINITION 1 **EIGENVALUE AND EIGENVECTOR** Let A be an $n \times n$ matrix with real† components. The number λ (real or complex) is called an **eigenvalue** of A if there is a *nonzero* vector $\mathbf{v}$ in $\mathbb{C}^n$ such that

$$A\mathbf{v} = \lambda\mathbf{v} \tag{2}$$

The vector $\mathbf{v} \neq \mathbf{0}$ is called an **eigenvector** *of A corresponding to the eigenvalue* λ.

Note. The word "eigen" is the German word for "own" or "proper." Eigenvalues are also called **proper values** or **characteristic values** and eigenvectors are called **proper vectors** or **characteristic vectors**.

Remark. As we shall see (for instance, in Example 6) a matrix with real components can have complex eigenvalues and eigenvectors. That is why, in the definition, we have asserted that $\mathbf{v} \in \mathbb{C}^n$. We shall not be using many facts about complex numbers in this book. For a discussion of those few facts we do need, see Appendix 2.

---

† This definition is also valid if A has complex components; but as the matrices we shall be dealing with will, for the most part, have real components, the definition is sufficient for our purposes.

EXAMPLE 1 Let $A = \begin{pmatrix} 10 & -18 \\ 6 & -11 \end{pmatrix}$. Then $A\begin{pmatrix} 2 \\ 1 \end{pmatrix} = \begin{pmatrix} 10 & -18 \\ 6 & -11 \end{pmatrix}\begin{pmatrix} 2 \\ 1 \end{pmatrix} = \begin{pmatrix} 2 \\ 1 \end{pmatrix}$. Thus $\lambda_1 = 1$ is an eigenvalue of A with corresponding eigenvector $\mathbf{v}_1 = \begin{pmatrix} 2 \\ 1 \end{pmatrix}$. Similarly, $A\begin{pmatrix} 3 \\ 2 \end{pmatrix} = \begin{pmatrix} 10 & -18 \\ 6 & -11 \end{pmatrix}\begin{pmatrix} 3 \\ 2 \end{pmatrix} = \begin{pmatrix} -6 \\ -4 \end{pmatrix} = -2\begin{pmatrix} 3 \\ 2 \end{pmatrix}$ so that $\lambda_2 = -2$ is an eigenvalue of A with corresponding eigenvector $\mathbf{v}_2 = \begin{pmatrix} 3 \\ 2 \end{pmatrix}$. As we soon shall see, these are the only eigenvalues of A.

EXAMPLE 2 Let $A = I$. Then for any $\mathbf{v} \in \mathbb{C}^n$, $A\mathbf{v} = I\mathbf{v} = \mathbf{v}$. Thus 1 is the only eigenvalue of A and every $\mathbf{v} \neq \mathbf{0} \in \mathbb{C}^n$ is an eigenvector of I.

We shall compute the eigenvalues and eigenvectors of many matrices in this section. But first we need to prove some facts that will simplify our computations.

Suppose that λ is an eigenvalue of A. Then there exists a nonzero vector $\mathbf{v} = \begin{pmatrix} x_1 \\ x_2 \\ \vdots \\ x_n \end{pmatrix} \neq \mathbf{0}$ such that $A\mathbf{v} = \lambda\mathbf{v} = \lambda I\mathbf{v}$. Rewriting this, we have

$$(A - \lambda I)\mathbf{v} = \mathbf{0} \tag{3}$$

If A is an $n \times n$ matrix, Equation (3) is a homogeneous system of n equations in the unknowns $x_1, x_2, \ldots, x_n$. Since, by assumption, the system has nontrivial solutions, we conclude that $\det(A - \lambda I) = 0$. Conversely, if $\det(A - \lambda I) = 0$, then Equation (3) has nontrivial solutions and λ is an eigenvalue of A. On the other hand, if $\det(A - \lambda I) \neq 0$, then (3) has only the solution $\mathbf{v} = \mathbf{0}$ so that λ is *not* an eigenvalue of A. Summing up these facts, we have the following theorem.

THEOREM 1 Let A be an $n \times n$ matrix. Then λ is an eigenvalue of A if and only if

$$\boxed{p(\lambda) = \det(A - \lambda I) = 0} \tag{4}$$

DEFINITION 2 **CHARACTERISTIC EQUATION AND POLYNOMIAL** Equation (4) is called the **characteristic equation** of A; $p(\lambda)$ is called the **characteristic polynomial** of A.

As will become apparent in the examples, $p(\lambda)$ is a polynomial of degree n in λ. For example, if $A = \begin{pmatrix} a & b \\ c & d \end{pmatrix}$, then $A - \lambda I = \begin{pmatrix} a & b \\ c & d \end{pmatrix} - \begin{pmatrix} \lambda & 0 \\ 0 & \lambda \end{pmatrix} = \begin{pmatrix} a-\lambda & b \\ c & d-\lambda \end{pmatrix}$ and $p(\lambda) = \det(A - \lambda I) = (a-\lambda)(d-\lambda) - bc = \lambda^2 - (a+d)\lambda + (ad-bc)$.

By the fundamental theorem of algebra, any polynomial of degree n with real or complex coefficients has exactly n roots (counting multiplicities). By this we mean, for example, that the polynomial $(\lambda-1)^5$ has five roots, all equal to the number 1. Since any eigenvalue of A is a root of the characteristic equation of A, we conclude that:

> *Counting multiplicities, every $n \times n$ matrix has exactly n eigenvalues.*

THEOREM 2 Let λ be an eigenvalue of the $n \times n$ matrix A and let $E_\lambda = \{\mathbf{v}: A\mathbf{v} = \lambda\mathbf{v}\}$. Then E_λ is a subspace of $\mathbb{C}^n$.

Proof If $A\mathbf{v} = \lambda\mathbf{v}$, then $(A-\lambda I)\mathbf{v} = \mathbf{0}$. Thus E_λ is the kernel of the matrix $A - \lambda I$, which, by Example 5.5.10 on page 197, is a subspace † of $\mathbb{C}^n$. ■

DEFINITION 3 **EIGENSPACE** Let λ be an eigenvalue of A. The subspace E_λ is called the **eigenspace**‡ of A corresponding to the eigenvalue λ.

We now prove another useful result.

THEOREM 3 Let A be an $n \times n$ matrix and let $\lambda_1, \lambda_2, \ldots, \lambda_m$ be distinct eigenvalues of A with corresponding eigenvectors $\mathbf{v}_1, \mathbf{v}_2, \ldots, \mathbf{v}_m$. Then $\mathbf{v}_1, \mathbf{v}_2, \ldots, \mathbf{v}_m$ are linearly independent. That is: *Eigenvectors corresponding to distinct eigenvalues are linearly independent.*

Proof We prove this by mathematical induction. We start with $m = 2$. Suppose that

$$c_1\mathbf{v}_1 + c_2\mathbf{v}_2 = \mathbf{0} \tag{5}$$

Then, multiplying both sides of (5) by A, we have $\mathbf{0} = A(c_1\mathbf{v}_1 + c_2\mathbf{v}_2) = c_1A\mathbf{v}_1 + c_2A\mathbf{v}_2$ or (since $A\mathbf{v}_i = \lambda_i\mathbf{v}_i$ for $i = 1, 2$)

$$c_1\lambda_1\mathbf{v}_1 + c_2\lambda_2\mathbf{v}_2 = \mathbf{0} \tag{6}$$

We then multiply (5) by λ_1 and subtract it from (6) to obtain

$$(c_1\lambda_1\mathbf{v}_1 + c_2\lambda_2\mathbf{v}_2) - (c_1\lambda_1\mathbf{v}_1 + c_2\lambda_1\mathbf{v}_2) = \mathbf{0}$$

or

$$c_2(\lambda_2 - \lambda_1)\mathbf{v}_2 = \mathbf{0}$$

Since $\mathbf{v}_2 \neq \mathbf{0}$ (by the definition of an eigenvector) and since $\lambda_2 \neq \lambda_1$, we conclude that $c_2 = 0$. Then inserting $c_2 = 0$ in (5), we see that $c_1 = 0$, which proves the theorem in the case $m = 2$. Now suppose that the theorem is true

---

† In Example 5.5.10 we saw that ker A is a subspace of $\mathbb{R}^n$ if A is a real matrix. The extension of this result to $\mathbb{C}^n$ presents no difficulties.

‡ Note that $\mathbf{0} \in E_\lambda$ since E_λ is a subspace.

for $m=k$. That is, we assume that any k eigenvectors corresponding to distinct eigenvalues are linearly independent. We prove the theorem for $m=k+1$. So we assume that

$$c_1\mathbf{v}_1+c_2\mathbf{v}_2+\cdots+c_k\mathbf{v}_k+c_{k+1}\mathbf{v}_{k+1}=\mathbf{0} \tag{7}$$

Then, multiplying both sides of (7) by A and using the fact that $A\mathbf{v}_i=\lambda_i\mathbf{v}_i$, we obtain

$$c_1\lambda_1\mathbf{v}_1+c_2\lambda_2\mathbf{v}_2+\cdots+c_k\lambda_k\mathbf{v}_k+c_{k+1}\lambda_{k+1}\mathbf{v}_{k+1}=\mathbf{0} \tag{8}$$

We multiply both sides of (7) by λ_{k+1} and subtract it from (8):

$$c_1(\lambda_1-\lambda_{k+1})\mathbf{v}_1+c_2(\lambda_2-\lambda_{k+1})\mathbf{v}_2+\cdots+c_k(\lambda_k-\lambda_{k+1})\mathbf{v}_k=\mathbf{0}$$

But, by the induction assumption, $\mathbf{v}_1, \mathbf{v}_2, \ldots, \mathbf{v}_k$ are linearly independent. Thus $c_1(\lambda_1-\lambda_{k+1})=c_2(\lambda_2-\lambda_{k+1})=\cdots=c_k(\lambda_k-\lambda_{k+1})=0$; and, since $\lambda_i\neq\lambda_{k+1}$ for $i=1, 2, \ldots, k$, we conclude that $c_1=c_2=\cdots=c_k=0$. But, from (7), this means that $c_{k+1}=0$. Thus the theorem is true for $m=k+1$ and the proof is complete. ■

If

$$A=\begin{pmatrix} a_{11} & a_{12} & \cdots & a_{1n} \\ a_{21} & a_{22} & \cdots & a_{2n} \\ \vdots & \vdots & & \vdots \\ a_{n1} & a_{n2} & \cdots & a_{nn} \end{pmatrix}$$

then

$$p(\lambda)=\det(A-\lambda I)=\begin{vmatrix} a_{11}-\lambda & a_{12} & \cdots & a_{1n} \\ a_{21} & a_{22}-\lambda & \cdots & a_{2n} \\ \vdots & \vdots & & \vdots \\ a_{n1} & a_{n2} & \cdots & a_{nn}-\lambda \end{vmatrix}$$

and $p(\lambda)=0$ can be written in the form

$$p(\lambda)=\lambda^n+b_{n-1}\lambda^{n-1}+\cdots+b_1\lambda+b_0=0 \tag{9}$$

Equation (9) has n roots, some of which may be repeated. If $\lambda_1, \lambda_2, \ldots, \lambda_m$ are the distinct roots of (9) with multiplicities $r_1, r_2, \ldots, r_m$, respectively, then (9) may be factored to obtain

$$p(\lambda)=(\lambda-\lambda_1)^{r_1}(\lambda-\lambda_2)^{r_2}\cdots(\lambda-\lambda_m)^{r_m}=0 \tag{10}$$

ALGEBRAIC MULTIPLICITY

The numbers $r_1, r_2, \ldots, r_m$ are called the **algebraic multiplicities** of the eigenvalues $\lambda_1, \lambda_2, \ldots, \lambda_m$ respectively.

We now calculate eigenvalues and corresponding eigenspaces. We do this in a three-step procedure:

PROCEDURE FOR COMPUTING EIGENVALUES AND EIGENVECTORS

i. Find $p(\lambda)=\det(A-\lambda I)$.

ii. Find the roots $\lambda_1, \lambda_2, \ldots, \lambda_m$ of $p(\lambda)=0$.

iii. Corresponding to each eigenvalue λ_i, solve the homogeneous system $(A-\lambda_i I)\mathbf{v}=\mathbf{0}$.

Remark. Step (ii) is usually the most difficult one.

EXAMPLE 3 Let $A=\begin{pmatrix}4&2\\3&3\end{pmatrix}$. Then $\det(A-\lambda I)=\begin{vmatrix}4-\lambda&2\\3&3-\lambda\end{vmatrix}=(4-\lambda)(3-\lambda)-6=\lambda^2-7\lambda+6=(\lambda-1)(\lambda-6)=0$. Thus the eigenvalues of A are $\lambda_1=1$ and $\lambda_2=6$. For $\lambda_1=1$, we solve $(A-I)\mathbf{v}=\mathbf{0}$ or $\begin{pmatrix}3&2\\3&2\end{pmatrix}\begin{pmatrix}x_1\\x_2\end{pmatrix}=\begin{pmatrix}0\\0\end{pmatrix}$. Clearly, any eigenvector corresponding to $\lambda_1=1$ satisfies $3x_1+2x_2=0$. One such eigenvector is $\mathbf{v}_1=\begin{pmatrix}2\\-3\end{pmatrix}$. Thus $E_1=\text{span}\left\{\begin{pmatrix}2\\-3\end{pmatrix}\right\}$. Similarly, the equation $(A-6I)\mathbf{v}=\mathbf{0}$ means that $\begin{pmatrix}-2&2\\3&-3\end{pmatrix}\begin{pmatrix}x_1\\x_2\end{pmatrix}=\begin{pmatrix}0\\0\end{pmatrix}$ or $x_1=x_2$. Thus $\mathbf{v}_2=\begin{pmatrix}1\\1\end{pmatrix}$ is an eigenvector corresponding to $\lambda_2=6$ and $E_6=\text{span}\left\{\begin{pmatrix}1\\1\end{pmatrix}\right\}$. Note that $\mathbf{v}_1$ and $\mathbf{v}_2$ are linearly independent since one is not a multiple of the other.

EXAMPLE 4 Let $A=\begin{pmatrix}1&-1&4\\3&2&-1\\2&1&-1\end{pmatrix}$. Then

$$\det(A-\lambda I)=\begin{vmatrix}1-\lambda&-1&4\\3&2-\lambda&-1\\2&1&-1-\lambda\end{vmatrix}$$

$$=-(\lambda^3-2\lambda^2-5\lambda+6)=-(\lambda-1)(\lambda+2)(\lambda-3)=0$$

Thus the eigenvalues of A are $\lambda_1=1$, $\lambda_2=-2$, and $\lambda_3=3$. Corresponding to $\lambda_1=1$ we have

$$(A-I)\mathbf{v}=\begin{pmatrix}0&-1&4\\3&1&-1\\2&1&-2\end{pmatrix}\begin{pmatrix}x_1\\x_2\\x_3\end{pmatrix}=\begin{pmatrix}0\\0\\0\end{pmatrix}$$

Solving by row reduction, we obtain, successively,

$$\left(\begin{array}{rrr|r} 0 & -1 & 4 & 0 \\ 3 & 1 & -1 & 0 \\ 2 & 1 & -2 & 0 \end{array}\right) \xrightarrow[A_{1,3}(1)]{A_{1,2}(1)} \left(\begin{array}{rrr|r} 0 & -1 & 4 & 0 \\ 3 & 0 & 3 & 0 \\ 2 & 0 & 2 & 0 \end{array}\right) \xrightarrow{M_2(\frac{1}{3})}$$

$$\left(\begin{array}{rrr|r} 0 & -1 & 4 & 0 \\ 1 & 0 & 1 & 0 \\ 2 & 0 & 2 & 0 \end{array}\right) \xrightarrow{A_{2,3}(-2)} \left(\begin{array}{rrr|r} 0 & -1 & 4 & 0 \\ 1 & 0 & 1 & 0 \\ 0 & 0 & 0 & 0 \end{array}\right)$$

Thus $x_1 = -x_3$, $x_2 = 4x_3$, an eigenvector is $\mathbf{v}_1 = \begin{pmatrix} -1 \\ 4 \\ 1 \end{pmatrix}$, and $E_1 = \operatorname{span}\left\{\begin{pmatrix} -1 \\ 4 \\ 1 \end{pmatrix}\right\}$. For $\lambda_2 = -2$, we have $[A - (-2I)]\mathbf{v} = (A + 2I)\mathbf{v} = \mathbf{0}$ or $\begin{pmatrix} 3 & -1 & 4 \\ 3 & 4 & -1 \\ 2 & 1 & 1 \end{pmatrix}\begin{pmatrix} x_1 \\ x_2 \\ x_3 \end{pmatrix} = \begin{pmatrix} 0 \\ 0 \\ 0 \end{pmatrix}$. This leads to

$$\left(\begin{array}{rrr|r} 3 & -1 & 4 & 0 \\ 3 & 4 & -1 & 0 \\ 2 & 1 & 1 & 0 \end{array}\right) \xrightarrow[A_{1,3}(1)]{A_{1,2}(4)} \left(\begin{array}{rrr|r} 3 & -1 & 4 & 0 \\ 15 & 0 & 15 & 0 \\ 5 & 0 & 5 & 0 \end{array}\right) \xrightarrow{M_2(\frac{1}{15})}$$

$$\left(\begin{array}{rrr|r} 3 & -1 & 4 & 0 \\ 1 & 0 & 1 & 0 \\ 5 & 0 & 5 & 0 \end{array}\right) \xrightarrow{A_{2,3}(-5)} \left(\begin{array}{rrr|r} -1 & -1 & 0 & 0 \\ 1 & 0 & 1 & 0 \\ 0 & 0 & 0 & 0 \end{array}\right)$$

Thus $x_2 = -x_1$, $x_3 = -x_1$, and an eigenvector is $\mathbf{v}_2 = \begin{pmatrix} 1 \\ -1 \\ -1 \end{pmatrix}$. Thus $E_{-2} = \operatorname{span}\left\{\begin{pmatrix} 1 \\ -1 \\ -1 \end{pmatrix}\right\}$. Finally, for $\lambda_3 = 3$, we have

$$(A - 3I)\mathbf{v} = \begin{pmatrix} -2 & -1 & 4 \\ 3 & -1 & -1 \\ 2 & 1 & -4 \end{pmatrix}\begin{pmatrix} x_1 \\ x_2 \\ x_3 \end{pmatrix} = \begin{pmatrix} 0 \\ 0 \\ 0 \end{pmatrix}$$

and

$$\left(\begin{array}{rrr|r} -2 & -1 & 4 & 0 \\ 3 & -1 & -1 & 0 \\ 2 & 1 & -4 & 0 \end{array}\right) \xrightarrow[A_{1,3}(1)]{A_{1,2}(-1)} \left(\begin{array}{rrr|r} -2 & -1 & 4 & 0 \\ 5 & 0 & -5 & 0 \\ 0 & 0 & 0 & 0 \end{array}\right)$$

$$\xrightarrow{M_2(\frac{1}{5})} \left(\begin{array}{rrr|r} -2 & -1 & 4 & 0 \\ 1 & 0 & -1 & 0 \\ 0 & 0 & 0 & 0 \end{array}\right) \xrightarrow{A_{2,1}(-4)} \left(\begin{array}{rrr|r} 2 & -1 & 0 & 0 \\ 1 & 0 & -1 & 0 \\ 0 & 0 & 0 & 0 \end{array}\right)$$

Hence $x_3 = x_1$, $x_2 = 2x_1$, and $\mathbf{v}_3 = \begin{pmatrix} 1 \\ 2 \\ 1 \end{pmatrix}$ so that $E_3 = \operatorname{span}\left\{\begin{pmatrix} 1 \\ 2 \\ 1 \end{pmatrix}\right\}$.

Remark. In this and every other example, there is always an infinite number of choices for each eigenvector. We arbitrarily choose a simple one by setting one or more of the x_i's equal to a convenient number. Here we have set one of the x_i's equal to 1.

EXAMPLE 5 Let $A = \begin{pmatrix} 2 & -1 \\ -4 & 2 \end{pmatrix}$. Then $\det(A - \lambda I) = \begin{vmatrix} 2-\lambda & -1 \\ -4 & 2-\lambda \end{vmatrix} = \lambda^2 - 4\lambda = \lambda(\lambda - 4)$. Thus the eigenvalues are $\lambda_1 = 0$ and $\lambda_2 = 4$. The eigenspace corresponding to zero is simply the kernel of A. We calculate $\begin{pmatrix} 2 & -1 \\ -4 & 2 \end{pmatrix}\begin{pmatrix} x_1 \\ x_2 \end{pmatrix} = \begin{pmatrix} 0 \\ 0 \end{pmatrix}$ or $2x_1 = x_2$ and an eigenvector is $\mathbf{v}_1 = \begin{pmatrix} 1 \\ 2 \end{pmatrix}$. Thus $\ker A = E_0 = \text{span}\left\{\begin{pmatrix} 1 \\ 2 \end{pmatrix}\right\}$. Corresponding to $\lambda_2 = 4$ we have $\begin{pmatrix} -2 & -1 \\ -4 & -2 \end{pmatrix}\begin{pmatrix} x_1 \\ x_2 \end{pmatrix} = \begin{pmatrix} 0 \\ 0 \end{pmatrix}$, so $E_4 = \text{span}\left\{\begin{pmatrix} 1 \\ -2 \end{pmatrix}\right\}$.

EXAMPLE 6 Let $A = \begin{pmatrix} 3 & -5 \\ 1 & -1 \end{pmatrix}$. Then $\det(A - \lambda I) = \begin{vmatrix} 3-\lambda & -5 \\ 1 & -1-\lambda \end{vmatrix} = \lambda^2 - 2\lambda + 2 = 0$ and

$$\lambda = \frac{-(-2) \pm \sqrt{4 - 4(1)(2)}}{2} = \frac{2 \pm \sqrt{-4}}{2} = \frac{2 \pm 2i}{2} = 1 \pm i$$

Thus $\lambda_1 = 1 + i$ and $\lambda_2 = 1 - i$. We compute

$$[A - (1+i)I]\mathbf{v} = \begin{pmatrix} 2-i & -5 \\ 1 & -2-i \end{pmatrix}^{\dagger}\begin{pmatrix} x_1 \\ x_2 \end{pmatrix} = \begin{pmatrix} 0 \\ 0 \end{pmatrix}$$

and we obtain $(2-i)x_1 - 5x_2 = 0$ and $x_1 + (-2-i)x_2 = 0$. Thus $x_1 = (2+i)x_2$, which yields the eigenvector $\mathbf{v}_1 = \begin{pmatrix} 2+i \\ 1 \end{pmatrix}$ and $E_{1+i} = \text{span}\left\{\begin{pmatrix} 2+i \\ 1 \end{pmatrix}\right\}$. Similarly, $[A - (1-i)I]\mathbf{v} = \begin{pmatrix} 2+i & -5 \\ 1 & -2+i \end{pmatrix}\begin{pmatrix} x_1 \\ x_2 \end{pmatrix} = \begin{pmatrix} 0 \\ 0 \end{pmatrix}$ or $x_1 + (-2+i)x_2 = 0$, which yields $x_1 = (2-i)x_2$, $\mathbf{v}_2 = \begin{pmatrix} 2-i \\ 1 \end{pmatrix}$, and $E_{1-i} = \text{span}\left\{\begin{pmatrix} 2-i \\ 1 \end{pmatrix}\right\}$.

Remark 1. This example illustrates that a real matrix may have complex eigenvalues and eigenvectors. Some texts define eigenvalues of real matrices to be the *real* roots of the characteristic equation. With this definition the matrix of the last example has *no* eigenvalues. This might make the computations simpler, but it also greatly reduces the usefulness of the theory of eigenvalues and eigenvectors. We shall see a significant illustration of the use of complex eigenvalues in Section 7.7.

---

† Note that the columns of this matrix are linearly dependent because $\begin{pmatrix} -5 \\ -2-i \end{pmatrix} = (-2-i)\begin{pmatrix} 2-i \\ 1 \end{pmatrix}$.

Remark 2. Note that $\lambda_2 = 1 - i$ is the complex conjugate of $\lambda_1 = 1 + i$. Also, the components of $\mathbf{v}_2$ are complex conjugates of the components of $\mathbf{v}_1$. This is no coincidence. In Problem 33 you are asked to prove that

the eigenvalues of a real matrix occur in complex conjugate pairs
and
corresponding eigenvectors are complex conjugates of one another.

EXAMPLE 7 Let $A = \begin{pmatrix} 4 & 0 \\ 0 & 4 \end{pmatrix}$. Then $\det(A - \lambda I) = \begin{vmatrix} 4-\lambda & 0 \\ 0 & 4-\lambda \end{vmatrix} = (\lambda - 4)^2 = 0$; hence $\lambda = 4$ is an eigenvalue of algebraic multiplicity 2. It is obvious that $A\mathbf{v} = 4\mathbf{v}$ for every vector $\mathbf{v} \in \mathbb{R}^2$, so that $E_4 = \mathbb{R}^2 = \text{span}\left\{\begin{pmatrix} 1 \\ 0 \end{pmatrix}, \begin{pmatrix} 0 \\ 1 \end{pmatrix}\right\}$.

EXAMPLE 8 Let $A = \begin{pmatrix} 4 & 1 \\ 0 & 4 \end{pmatrix}$. Then $\det(A - \lambda I) = \begin{vmatrix} 4-\lambda & 1 \\ 0 & 4-\lambda \end{vmatrix} = (\lambda - 4)^2 = 0$; thus $\lambda = 4$ is again an eigenvalue of algebraic multiplicity 2. But this time we have $(A - 4I)\mathbf{v} = \begin{pmatrix} 0 & 1 \\ 0 & 0 \end{pmatrix}\begin{pmatrix} x_1 \\ x_2 \end{pmatrix} = \begin{pmatrix} x_2 \\ 0 \end{pmatrix}$. Thus $x_2 = 0$, $\mathbf{v}_1 = \begin{pmatrix} 1 \\ 0 \end{pmatrix}$ is an eigenvector, and $E_4 = \text{span}\left\{\begin{pmatrix} 1 \\ 0 \end{pmatrix}\right\}$.

EXAMPLE 9 Let $A = \begin{pmatrix} 3 & 2 & 4 \\ 2 & 0 & 2 \\ 4 & 2 & 3 \end{pmatrix}$. Then $\det(A - \lambda I) = \begin{vmatrix} 3-\lambda & 2 & 4 \\ 2 & -\lambda & 2 \\ 4 & 2 & 3-\lambda \end{vmatrix} = -\lambda^3 + 6\lambda^2 + 15\lambda + 8 = -(\lambda + 1)^2(\lambda - 8) = 0$, so that the eigenvalues are $\lambda_1 = 8$ and $\lambda_2 = -1$ (with algebraic multiplicity 2). For $\lambda_1 = 8$, we obtain

$$(A - 8I)\mathbf{v} = \begin{pmatrix} -5 & 2 & 4 \\ 2 & -8 & 2 \\ 4 & 2 & -5 \end{pmatrix}\begin{pmatrix} x_1 \\ x_2 \\ x_3 \end{pmatrix} = \begin{pmatrix} 0 \\ 0 \\ 0 \end{pmatrix}$$

or, row reducing,

$$\left(\begin{array}{ccc|c} -5 & 2 & 4 & 0 \\ 2 & -8 & 2 & 0 \\ 4 & 2 & -5 & 0 \end{array}\right) \xrightarrow[A_{1,3}(-1)]{A_{1,2}(4)} \left(\begin{array}{ccc|c} -5 & 2 & 4 & 0 \\ -18 & 0 & 18 & 0 \\ 9 & 0 & -9 & 0 \end{array}\right) \xrightarrow{M_2(\frac{1}{18})}$$

$$\left(\begin{array}{ccc|c} -5 & 2 & 4 & 0 \\ -1 & 0 & 1 & 0 \\ 9 & 0 & -9 & 0 \end{array}\right) \xrightarrow{A_{2,3}(9)} \left(\begin{array}{ccc|c} 0 & 2 & -1 & 0 \\ -1 & 0 & 1 & 0 \\ 0 & 0 & 0 & 0 \end{array}\right)$$

Hence $x_3 = 2x_2$ and $x_1 = x_3$, we obtain the eigenvector $\mathbf{v}_1 = \begin{pmatrix} 2 \\ 1 \\ 2 \end{pmatrix}$, and $E_8 = \text{span}\left\{\begin{pmatrix} 2 \\ 1 \\ 2 \end{pmatrix}\right\}$. For $\lambda_2 = -1$, we have $(A + I)\mathbf{v} = \begin{pmatrix} 4 & 2 & 4 \\ 2 & 1 & 2 \\ 4 & 2 & 4 \end{pmatrix}\begin{pmatrix} x_1 \\ x_2 \\ x_3 \end{pmatrix} = \begin{pmatrix} 0 \\ 0 \\ 0 \end{pmatrix}$, which gives us the single equation $2x_1+x_2+2x_3=0$ or $x_2=-2x_1-2x_3$. If $x_1=1$ and $x_3=0$, we obtain $\mathbf{v}_2=\begin{pmatrix}1\\-2\\0\end{pmatrix}$. If $x_1=0$ and $x_3=1$, we obtain $\mathbf{v}_3=\begin{pmatrix}0\\-2\\1\end{pmatrix}$. Thus $E_{-1}=\text{span}\left\{\begin{pmatrix}1\\-2\\0\end{pmatrix},\begin{pmatrix}0\\-2\\1\end{pmatrix}\right\}$. There are other convenient choices for eigenvectors. For example, $\mathbf{v}=\begin{pmatrix}1\\0\\-1\end{pmatrix}$ is in E_{-1} since $\mathbf{v}=\mathbf{v}_2-\mathbf{v}_3$.

EXAMPLE 10 Let $A=\begin{pmatrix}-5&-5&-9\\8&9&18\\-2&-3&-7\end{pmatrix}$. Then $\det(A-\lambda I)=\begin{vmatrix}-5-\lambda&-5&-9\\8&9-\lambda&18\\-2&-3&-7-\lambda\end{vmatrix}=-\lambda^3-3\lambda^2-3\lambda-1=-(\lambda+1)^3=0$. Thus $\lambda=-1$ is an eigenvalue of algebraic multiplicity 3. To compute E_{-1}, we set $(A+I)\mathbf{v}=\begin{pmatrix}-4&-5&-9\\8&10&18\\-2&-3&-6\end{pmatrix}\begin{pmatrix}x_1\\x_2\\x_3\end{pmatrix}=\begin{pmatrix}0\\0\\0\end{pmatrix}$ and row-reduce to obtain, successively,

$$\left(\begin{array}{ccc|c}-4&-5&-9&0\\8&10&18&0\\-2&-3&-6&0\end{array}\right)\xrightarrow[A_{3,2}(4)]{A_{3,1}(-2)}\left(\begin{array}{ccc|c}0&1&3&0\\0&-2&-6&0\\-2&-3&-6&0\end{array}\right)\xrightarrow[A_{1,3}(3)]{A_{1,2}(2)}\left(\begin{array}{ccc|c}0&1&3&0\\0&0&0&0\\-2&0&3&0\end{array}\right)$$

This yields $x_2=-3x_3$ and $2x_1=3x_3$. Setting $x_3=2$ we obtain only one linearly independent eigenvector: $\mathbf{v}_1=\begin{pmatrix}3\\-6\\2\end{pmatrix}$. Thus $E_{-1}=\text{span}\left\{\begin{pmatrix}3\\-6\\2\end{pmatrix}\right\}$.

EXAMPLE 11 Let $A=\begin{pmatrix}-1&-3&-9\\0&5&18\\0&-2&-7\end{pmatrix}$. Then $\det(A-\lambda I)=\begin{vmatrix}-1-\lambda&-3&-9\\0&5-\lambda&18\\0&-2&-7-\lambda\end{vmatrix}=-(\lambda+1)^3=0$. Thus, as in Example 10, $\lambda=-1$ is an eigenvalue of algebraic multiplicity 3. To find E_{-1}, we compute $(A+I)\mathbf{v}=\begin{pmatrix}0&-3&-9\\0&6&18\\0&-2&-6\end{pmatrix}\begin{pmatrix}x_1\\x_2\\x_3\end{pmatrix}=\begin{pmatrix}0\\0\\0\end{pmatrix}$. Thus $-2x_2-6x_3=0$ or $x_2=-3x_3$, and x_1 is arbitrary. Setting $x_1=0$, $x_3=1$, we obtain $\mathbf{v}_1=\begin{pmatrix}0\\-3\\1\end{pmatrix}$. Setting $x_1=1$, $x_3=1$ yields $\mathbf{v}_2=\begin{pmatrix}1\\-3\\1\end{pmatrix}$. Thus $E_{-1}=\text{span}\left\{\begin{pmatrix}0\\-3\\1\end{pmatrix},\begin{pmatrix}1\\-3\\1\end{pmatrix}\right\}$.

In each of the last five examples we found an eigenvalue with an algebraic multiplicity of 2 or more. But, as we saw in Examples 8, 10, and 11, the number of linearly independent eigenvectors is not necessarily equal to the algebraic multiplicity of the eigenvalue (as was the case in Examples 7 and 9). This observation leads to the following definition.

DEFINITION 4 **GEOMETRIC MULTIPLICITY** Let λ be an eigenvalue of the matrix A. Then the **geometric multiplicity** of λ is the dimension of the eigenspace corresponding to λ (which is the nullity of the matrix $A - \lambda I$). That is:

$$\text{Geometric multiplicity of } \lambda = \dim E_\lambda = \nu(A - \lambda I)$$

In Examples 7 and 9 we saw that for the eigenvalues of algebraic multiplicity 2, the geometric multiplicities were also 2. In Example 8 the geometric multiplicity of $\lambda = 4$ was 1 while the algebraic multiplicity was 2. In Example 10 the algebraic multiplicity was 3 and the geometric multiplicity was 1. In Example 11 the algebraic multiplicity was 3 and the geometric multiplicity was 2. These examples illustrate the fact that if the algebraic multiplicity of λ is greater than 1, then we cannot predict the geometric multiplicity of λ without additional information.

If A is a 2×2 matrix and λ is an eigenvalue with algebraic multiplicity 2, then the geometric multiplicity of λ is ≤ 2 since there can be at most two linearly independent vectors in a two-dimensional space. Let A be a 3×3 matrix having two eigenvalues λ_1 and λ_2 with algebraic multiplicities 1 and 2, respectively, Then the geometric multiplicity of λ_2 is ≤ 2 because otherwise we would have at least four linearly independent vectors in a three-dimensional space. Intuitively, it seems that the geometric multiplicity of an eigenvalue is always less than or equal to its algebraic multiplicity. The proof of the following theorem is not difficult if additional facts about determinants are proved. Since this would take us too far afield, we omit the proof.†

THEOREM 4 Let λ be an eigenvalue of A. Then:

$$\text{Geometric multiplicity of } \lambda \leq \text{algebraic multiplicity of } \lambda$$

Note. The geometric multiplicity of an eigenvalue is never zero. This follows from Definition 1, which states that if λ is an eigenvalue, then there exists a *nonzero* eigenvector corresponding to λ.

† For a proof see Theorem 11.2.6 in C. R. Wylie's book *Advanced Engineering Mathematics* (New York: McGraw-Hill, 1975).

In the rest of this chapter an important problem for us will be to determine whether a given $n \times n$ matrix does or does not have n linearly independent eigenvectors. From what we have already discussed in this section, the following theorem is apparent.

THEOREM 5 Let A be an $n \times n$ matrix. Then A has n linearly independent eigenvectors if and only if the geometric multiplicity of every eigenvalue is equal to its algebraic multiplicity. In particular, A has n linearly independent eigenvectors if all the eigenvalues are distinct (since then the algebraic multiplicity of every eigenvalue is 1).

In Example 5, we saw a matrix for which zero was an eigenvalue. In fact, from Theorem 1, it is evident that zero is an eigenvalue of A if and only if $\det A = \det(A - 0I) = 0$. This enables us to extend, for the last time, our Summing Up Theorem (see Theorems 6.3.7, page 251, and 6.5.4, page 268).

THEOREM 6 **SUMMING UP THEOREM—VIEW 6** Let A be an $n \times n$ matrix. Then the following 10 statements are equivalent. That is, if one is true, all are true.

i. A is invertible.
ii. The only solution to the homogeneous system $A\mathbf{x} = \mathbf{0}$ is the trivial solution ($\mathbf{x} = \mathbf{0}$).
iii. The system $A\mathbf{x} = \mathbf{b}$ has a unique solution for every n-vector $\mathbf{b}$.
iv. A is row equivalent to the $n \times n$ identity matrix I_n.
v. The rows (and columns) of A are linearly independent.
vi. $\det A \neq 0$.
vii. $\nu(A) = 0$.
viii. $\rho(A) = n$.
ix. The linear transformation T from $\mathbb{R}^n$ to $\mathbb{R}^n$ defined by $T\mathbf{x} = A\mathbf{x}$ is an isomorphism.
x. Zero is *not* an eigenvalue of A.

PROBLEMS 7.1

In Problems 1–20 calculate the eigenvalues and eigenspaces of the given matrix. If the algebraic multiplicity of an eigenvalue is greater than 1, calculate its geometric multiplicity.

1. $\begin{pmatrix} -2 & -2 \\ -5 & 1 \end{pmatrix}$ **2.** $\begin{pmatrix} -12 & 7 \\ -7 & 2 \end{pmatrix}$ **3.** $\begin{pmatrix} 2 & -1 \\ 5 & -2 \end{pmatrix}$

4. $\begin{pmatrix} -3 & 0 \\ 0 & -3 \end{pmatrix}$ **5.** $\begin{pmatrix} -3 & 2 \\ 0 & -3 \end{pmatrix}$ **6.** $\begin{pmatrix} 3 & 2 \\ -5 & 1 \end{pmatrix}$

7. $\begin{pmatrix} 1 & -1 & 0 \\ -1 & 2 & -1 \\ 0 & -1 & 1 \end{pmatrix}$ **8.** $\begin{pmatrix} 1 & 1 & -2 \\ -1 & 2 & 1 \\ 0 & 1 & -1 \end{pmatrix}$ **9.** $\begin{pmatrix} 5 & 4 & 2 \\ 4 & 5 & 2 \\ 2 & 2 & 2 \end{pmatrix}$

10. $\begin{pmatrix} 1 & 2 & 2 \\ 0 & 2 & 1 \\ -1 & 2 & 2 \end{pmatrix}$ **11.** $\begin{pmatrix} 0 & 1 & 0 \\ 0 & 0 & 1 \\ 1 & -3 & 3 \end{pmatrix}$ **12.** $\begin{pmatrix} -3 & -7 & -5 \\ 2 & 4 & 3 \\ 1 & 2 & 2 \end{pmatrix}$

13. $\begin{pmatrix} 1 & -1 & -1 \\ 1 & -1 & 0 \\ 1 & 0 & -1 \end{pmatrix}$ **14.** $\begin{pmatrix} 7 & -2 & -4 \\ 3 & 0 & -2 \\ 6 & -2 & -3 \end{pmatrix}$ **15.** $\begin{pmatrix} 4 & 6 & 6 \\ 1 & 3 & 2 \\ -1 & -5 & -2 \end{pmatrix}$

16. $\begin{pmatrix} 4 & 1 & 0 & 1 \\ 2 & 3 & 0 & 1 \\ -2 & 1 & 2 & -3 \\ 2 & -1 & 0 & 5 \end{pmatrix}$ **17.** $\begin{pmatrix} a & 0 & 0 & 0 \\ 0 & a & 0 & 0 \\ 0 & 0 & a & 0 \\ 0 & 0 & 0 & a \end{pmatrix}$

18. $\begin{pmatrix} a & b & 0 & 0 \\ 0 & a & 0 & 0 \\ 0 & 0 & a & 0 \\ 0 & 0 & 0 & a \end{pmatrix}$; $b \neq 0$ **19.** $\begin{pmatrix} a & b & 0 & 0 \\ 0 & a & c & 0 \\ 0 & 0 & a & 0 \\ 0 & 0 & 0 & a \end{pmatrix}$; $bc \neq 0$

20. $\begin{pmatrix} a & b & 0 & 0 \\ 0 & a & c & 0 \\ 0 & 0 & a & d \\ 0 & 0 & 0 & a \end{pmatrix}$; $bcd \neq 0$

21. Show that for any real numbers a and b, the matrix $A = \begin{pmatrix} a & b \\ -b & a \end{pmatrix}$ has the eigenvectors $\begin{pmatrix} 1 \\ i \end{pmatrix}$ and $\begin{pmatrix} 1 \\ -i \end{pmatrix}$.

In Problems 22–28 assume that the matrix A has the eigenvalues $\lambda_1, \lambda_2, \ldots, \lambda_k$.

22. Show that the eigenvalues of A^t are $\lambda_1, \lambda_2, \ldots, \lambda_k$.
23. Show that the eigenvalues of αA are $\alpha\lambda_1, \alpha\lambda_2, \ldots, \alpha\lambda_k$.
24. Show that A^{-1} exists if and only if $\lambda_1\lambda_2 \cdots \lambda_k \neq 0$.
***25.** If A^{-1} exists, show that the eigenvalues of A^{-1} are $1/\lambda_1, 1/\lambda_2, \ldots, 1/\lambda_k$.
26. Show that the matrix $A - \alpha I$ has the eigenvalues $\lambda_1 - \alpha, \lambda_2 - \alpha, \ldots, \lambda_k - \alpha$.
***27.** Show that the eigenvalues of A^2 are $\lambda_1^2, \lambda_2^2, \ldots, \lambda_k^2$.
***28.** Show that the eigenvalues of A^m are $\lambda_1^m, \lambda_2^m, \ldots, \lambda_k^m$ for $m = 1, 2, 3, \ldots$.
29. Let λ be an eigenvalue of A with corresponding eigenvector $\mathbf{v}$. Let $p(\lambda) = a_0 + a_1\lambda + a_2\lambda^2 + \cdots + a_n\lambda^n$. Define the matrix $p(A)$ by $p(A) = a_0 I + a_1 A + a_2 A^2 + \cdots + a_n A^n$. Show that $p(A)\mathbf{v} = p(\lambda)\mathbf{v}$.
30. Using the result of Problem 29, show that if $\lambda_1, \lambda_2, \ldots, \lambda_k$ are eigenvalues of A, then $p(\lambda_1), p(\lambda_2), \ldots, p(\lambda_k)$ are eigenvalues of $p(A)$.
31. Show that if A is an upper triangular matrix, then the eigenvalues of A are the diagonal components of A.
32. Let $A_1 = \begin{pmatrix} 2 & 0 & 0 & 0 \\ 0 & 2 & 0 & 0 \\ 0 & 0 & 2 & 0 \\ 0 & 0 & 0 & 2 \end{pmatrix}$, $A_2 = \begin{pmatrix} 2 & 1 & 0 & 0 \\ 0 & 2 & 0 & 0 \\ 0 & 0 & 2 & 0 \\ 0 & 0 & 0 & 2 \end{pmatrix}$, $A_3 = \begin{pmatrix} 2 & 1 & 0 & 0 \\ 0 & 2 & 1 & 0 \\ 0 & 0 & 2 & 0 \\ 0 & 0 & 0 & 2 \end{pmatrix}$, and $A_4 = \begin{pmatrix} 2 & 1 & 0 & 0 \\ 0 & 2 & 1 & 0 \\ 0 & 0 & 2 & 1 \\ 0 & 0 & 0 & 2 \end{pmatrix}$. Show that, for each matrix, $\lambda = 2$ is an eigenvalue of

algebraic multiplicity 4. In each case, compute the geometric multiplicity of $\lambda = 2$.

***33.** Let A be a real $n \times n$ matrix. Show that if λ_1 is a complex eigenvalue of A with eigenvector $\mathbf{v}_1$, then $\bar{\lambda}_1$ is an eigenvalue of A with eigenvector $\bar{\mathbf{v}}_1$.

***34.** A probability matrix is an $n \times n$ matrix having two properties:

i. $a_{ij} \geq 0$ for every i and j.
ii. The sum of the components in every column is 1.

Prove that 1 is an eigenvalue of every probability matrix.

7.2 If Time Permits: A Model of Population Growth

In this section we show how the theory of eigenvalues and eigenvectors can be used to analyze a model of the growth of a bird population.† We begin by discussing a simple model of population growth. We assume that a certain species grows at a constant rate; that is, the population of the species after one time period (which could be an hour, a week, a month, a year, etc.) is a constant multiple of the population in the previous time period. One way this could happen, for example, is that each generation is distinct and each organism produces r offspring and then dies. If p_n denotes the population after the nth time period, we would have

$$p_n = rp_{n-1}.$$

For example, this model might describe a bacteria population where, at a given time, an organism splits into two separate organisms. Then $r = 2$. Let p_0 denote the initial population. Then $p_1 = rp_0$, $p_2 = rp_1 = r(rp_0) = r^2p_0$, $p_3 = rp_2 = r(r^2p_0) = r^3p_0$, and so on, so that

$$p_n = r^n p_0 \tag{1}$$

From this model we see that the population increases without bound if $r > 1$ and decreases to zero if $r < 1$. If $r = 1$ the population remains at the constant value p_0.

This model is, evidently, very simplistic. One obvious objection is that the number of offspring produced depends, in many cases, on the ages of the adults. For example, in a human population the average female adult over 50 would certainly produce fewer children than the average 21-year-old female. To deal with this difficulty, we introduce a model which allows for age groupings with different fertility rates.

We now look at a model of population growth for a species of birds. In this bird population we assume that the number of female birds equals the number of males. Let $p_{j,n-1}$ denote the population of juvenile (immature) females in the $(n-1)$st year and let $p_{a,n-1}$ denote the number of adult females in the $(n-1)$st year. Some of the juvenile birds will die during the

† The material in this section is based on a paper by D. Cooke: "A 2×2 Matrix Model of Population Growth," *Mathematical Gazette* 61(416): 120–123.

year. We assume that a certain proportion α of the juvenile birds survive to become adults in the spring of the nth year. Each surviving female bird produces eggs later in the spring, which hatch to produce, on the average, k juvenile female birds in the following spring. Adults also die, and the proportion of adults that survive from one spring to the next is β.

This constant survival rate of birds is not just a simplistic assumption. It appears to be the case with most of the natural bird populations that have been studied. This means that the adult survival rate of many bird species is independent of age. Perhaps few birds in the wild survive long enough to exhibit the effects of old age. Moreover, in many species the number of offspring seems to be uninfluenced by the age of the mother.

In the notation introduced above, $p_{j,n}$ and $p_{a,n}$ represent, respectively, the populations of juvenile and adult females in the nth year. Putting together all the information given, we arrive at the following 2×2 system:

$$\begin{aligned} p_{j,n} &= \qquad\qquad kp_{a,n-1} \\ p_{a,n} &= \alpha p_{j,n-1} + \beta p_{a,n-1} \end{aligned} \tag{2}$$

or

$$\mathbf{p}_n = A\mathbf{p}_{n-1} \tag{3}$$

where $\mathbf{p}_n = \begin{pmatrix} p_{j,n} \\ p_{a,n} \end{pmatrix}$ and $A = \begin{pmatrix} 0 & k \\ \alpha & \beta \end{pmatrix}$. It is clear, from (3), that $\mathbf{p}_1 = A\mathbf{p}_0$, $\mathbf{p}_2 = A\mathbf{p}_1 = A(A\mathbf{p}_0) = A^2\mathbf{p}_0, \ldots$, and so on. Hence

$$\boxed{\mathbf{p}_n = A^n\mathbf{p}_0} \tag{4}$$

where $\mathbf{p}_0$ is the vector of initial populations of juvenile and adult females.

Equation (4) is like Equation (1), but now we are able to distinguish between the survival rates of juvenile and adult birds.

EXAMPLE 1

Let $A = \begin{pmatrix} 0 & 2 \\ 0.3 & 0.5 \end{pmatrix}$. This means that each adult female produces two female offspring and, since the number of males is assumed equal to the number of females, at least four eggs—and probably many more, since losses among fledglings are likely to be high. From the model, it is apparent that α and β lie in the interval $[0, 1]$. Since juvenile birds are not as likely as adults to survive, we must have $\alpha < \beta$.

In Table 7.1 we assume that, initially, there are 10 female (and 10 male) adults and no juveniles. The computations were done on a computer, but the work would not be too onerous if done on a hand calculator. For example, $\mathbf{p}_1 = \begin{pmatrix} 0 & 2 \\ 0.3 & 0.5 \end{pmatrix}\begin{pmatrix} 0 \\ 10 \end{pmatrix} = \begin{pmatrix} 20 \\ 5 \end{pmatrix}$, so that $p_{j,1} = 20$, $p_{a,1} = 5$, the total female population after 1 year is 25, and the ratio of juvenile to adult females is 4 to 1. In the second year, $\mathbf{p}_2 = \begin{pmatrix} 0 & 2 \\ 0.3 & 0.5 \end{pmatrix}\begin{pmatrix} 20 \\ 5 \end{pmatrix} = \begin{pmatrix} 10 \\ 8.5 \end{pmatrix}$, which we

Table 7.1

Year n	No. of juveniles $p_{j,n}$	No. of adults $p_{a,n}$	Total female population T_n in nth year	$p_{j,n}/p_{a,n}$†	T_n/T_{n-1}†
0	0	10	10	0	—
1	20	5	25	4.00	2.50
2	10	8	18	1.18	0.74
3	17	7	24	2.34	1.31
4	14	8	22	1.66	0.96
5	17	8	25	2.00	1.13
10	22	12	34	1.87	1.06
11	24	12	36	1.88	1.07
12	25	13	38	1.88	1.06
20	42	22	64	1.88	1.06

† The figures in these columns were obtained before the numbers in the previous columns were rounded. Thus, for example, in year 2, $p_{j,2}/p_{a,2} = 10/8.5 \approx 1.176470588 \approx 1.18$.

round down to $\begin{pmatrix} 10 \\ 8 \end{pmatrix}$ since we cannot have $8\frac{1}{2}$ adult birds. Table 7.1 tabulates the ratios $p_{j,n}/p_{a,n}$ and the ratios T_n/T_{n-1} of the total number of females in successive years.

In Table 7.1 it seems as if the ratio $p_{j,n}/p_{a,n}$ is approaching the constant 1.88 while the total population seems to be increasing at a constant rate of 6 percent a year. Let us see if we can determine why this is the case.

First, we return to the general case (Equation 4). Suppose that A has the real distinct eigenvalues λ_1 and λ_2 with corresponding eigenvectors $\mathbf{v}_1$ and $\mathbf{v}_2$. Since $\mathbf{v}_1$ and $\mathbf{v}_2$ are linearly independent, we can write

$$\mathbf{p}_0 = a_1\mathbf{v}_1 + a_2\mathbf{v}_2 \tag{5}$$

for some real numbers a_1 and a_2. Then (4) becomes

$$\mathbf{p}_n = A^n(a_1\mathbf{v}_1 + a_2\mathbf{v}_2) \tag{6}$$

But $A\mathbf{v}_1 = \lambda_1\mathbf{v}_1$ and $A^2\mathbf{v}_1 = A(A\mathbf{v}_1) = A(\lambda_1\mathbf{v}_1) = \lambda_1 A\mathbf{v}_1 = \lambda_1(\lambda_1\mathbf{v}_1) = \lambda_1^2\mathbf{v}_1$. Thus we can see that $A^n\mathbf{v}_1 = \lambda_1^n\mathbf{v}_1$, $A^n\mathbf{v}_2 = \lambda_2^n\mathbf{v}_2$, and, from (6),

$$\mathbf{p}_n = a_1\lambda_1^n\mathbf{v}_1 + a_2\lambda_2^n\mathbf{v}_2 \tag{7}$$

The characteristic equation of A is $\begin{vmatrix} -\lambda & k \\ \alpha & \beta - \lambda \end{vmatrix} = \lambda^2 - \beta\lambda - k\alpha = 0$ or $\lambda = (\beta \pm \sqrt{\beta^2 + 4\alpha k})/2$. By assumption, $k > 0$, $0 < \alpha < 1$, and $0 < \beta < 1$. Hence $4\alpha k > 0$ and $\beta^2 + 4\alpha k > 0$, which means that the eigenvalues are, indeed, real and distinct and that one eigenvalue λ_1 is positive, one λ_2 is negative, and $|\lambda_1| > |\lambda_2|$. We can write (7) as

$$\mathbf{p}_n = \lambda_1^n\left[a_1\mathbf{v}_1 + \left(\frac{\lambda_2}{\lambda_1}\right)^n a_2\mathbf{v}_2\right] \tag{8}$$

Since $|\lambda_2/\lambda_1|<1$, it is apparent that $(\lambda_2/\lambda_1)^n$ gets very small as n gets large. Thus, for n large,

$$\mathbf{p}_n \approx a_1\lambda_1^n\mathbf{v}_1 \tag{9}$$

This means that, in the long run, the age distribution stabilizes and is proportional to $\mathbf{v}_1$. Each age group will change by a factor of λ_1 each year. Thus—in the long run—Equation (4) acts just like Equation (1). In the short term—that is, before "stability" is reached—the numbers oscillate. The magnitude of this oscillation depends on the magnitude of λ_2/λ_1 (which is negative, thus explaining the oscillation).

EXAMPLE 1 (continued)

For $A = \begin{pmatrix} 0 & 2 \\ 0.3 & 0.5 \end{pmatrix}$, we have $\lambda^2 - 0.5\lambda - 0.6 = 0$ or $\lambda = (0.5 \pm \sqrt{0.25 + 2.4})/2$ $= (0.5 \pm \sqrt{2.65})/2$, so that $\lambda_1 \approx 1.06$ and $\lambda_2 \approx -0.56$. This explains the 6 percent increase in population noted in the last column of Table 7.1. Corresponding to the eigenvalue $\lambda_1 = 1.06$, we compute $(A-1.06I)\mathbf{v}_1 = \begin{pmatrix} -1.06 & 2 \\ 0.3 & -0.56 \end{pmatrix}\begin{pmatrix} x_1 \\ x_2 \end{pmatrix} = \begin{pmatrix} 0 \\ 0 \end{pmatrix}$ or $1.06x_1 = 2x_2$, so that $\mathbf{v}_1 = \begin{pmatrix} 1 \\ 0.53 \end{pmatrix}$ is an eigenvector. Similarly, $(A+0.56)\mathbf{v}_2 = \begin{pmatrix} 0.56 & 2 \\ 0.3 & 1.06 \end{pmatrix}\begin{pmatrix} x_1 \\ x_2 \end{pmatrix} = \begin{pmatrix} 0 \\ 0 \end{pmatrix}$, so that $0.56x_1 + 2x_2 = 0$ and $\mathbf{v}_2 = \begin{pmatrix} 1 \\ -0.28 \end{pmatrix}$ is a second eigenvector. Note that in $\mathbf{v}_1$ we have $1/0.53 \approx 1.88$. This explains the ratio $p_{j,n}/p_{a,n}$ in the fifth column of the table.

Remark. In the preceding computations precision was lost because we rounded to only two decimal places of accuracy. Much greater accuracy is obtained by using a hand calculator or computer. For example, using a hand calculator, we easily calculate $\lambda_1 = 1.06394103$, $\lambda_2 = -0.5639410298$, $\mathbf{v}_1 = \begin{pmatrix} 1 \\ 0.531970515 \end{pmatrix}$, $\mathbf{v}_2 = \begin{pmatrix} 1 \\ -0.2819705149 \end{pmatrix}$, and the ratio of $p_{j,n}$ to $p_{a,n}$ is seen to be $1/0.5319710515 \approx 1.879801537$.

It is remarkable just how much information is available from a simple computation of eigenvalues. It is of great interest to know whether a population will ultimately increase or decrease. It will increase if $\lambda_1 > 1$, and the condition for that is $(\beta + \sqrt{\beta^2 + 4\alpha k})/2 > 1$ or $\sqrt{\beta^2 + 4\alpha k} > 2 - \beta$ or $\beta^2 + 4\alpha k > (2-\beta)^2 = 4 - 4\beta + \beta^2$. This leads to $4\alpha k > 4 - 4\beta$ or

$$k > \frac{1-\beta}{\alpha} \tag{10}$$

In Example 1 we had $\beta = 0.5$, $\alpha = 0.3$; thus (10) is satisfied if $k > 0.5/0.3 \approx 1.67$.

Before we close this section we indicate two limitations of this model:

i. Birth and death rates often change from year to year and are particularly dependent on the weather. This model assumes a constant environment.

ii. Ecologists have found that for many species birth and death rates vary with the size of the population. In particular, a population cannot grow when it reaches a certain size due to the effects of limited food resources and overcrowding. It is obvious that a population cannot grow indefinitely at a constant rate. Otherwise that population would overrun the earth.

PROBLEMS 7.2

In Problems 1–3 find the numbers of juvenile and adult female birds after 1, 2, 5, 10, 19, and 20 years. Then find the long-term ratios of $p_{j,n}$ to $p_{a,n}$ and T_n to T_{n-1}. [*Hint:* Use Equations (7) and (9) and a hand calculator and round to three decimals.]

1. $\mathbf{p}_0=\begin{pmatrix}0\\12\end{pmatrix}$; $k=3$, $\alpha=0.4$, $\beta=0.6$

2. $\mathbf{p}_0=\begin{pmatrix}0\\15\end{pmatrix}$; $k=1$, $\alpha=0.3$, $\beta=0.4$

3. $\mathbf{p}_0=\begin{pmatrix}0\\20\end{pmatrix}$; $k=4$, $\alpha=0.7$, $\beta=0.8$

4. Show that if $\alpha=\beta$ and $\alpha>\frac{1}{2}$, then the bird population will always increase in the long run if at least one female offspring on the average is produced by each female adult.

5. Show that, in the long run, the ratio $p_{j,n}/p_{a,n}$ approaches the limiting value k/λ_1.

6. Suppose we divide the adult birds into two age groups: those 1–5 years old and those more than 5 years old. Assume that the survival rate for birds in the first group is β while in the second group it is γ (and $\beta>\gamma$). Assume that the birds in the first group are equally divided as to age. (That is, if there are 100 birds in the group, then 20 are 1 year old, 20 are 2 years old, and so on.) Formulate a 3×3 matrix model for this situation.

7.3 Similar Matrices and Diagonalization

In this section we describe an interesting and useful relationship that can hold between two matrices.

DEFINITION 1

SIMILAR MATRICES Two $n \times n$ matrices A and B are said to be **similar** if there exists an invertible $n \times n$ matrix C such that

$$B=C^{-1}AC \tag{1}$$

The function defined by (1) which takes the matrix A into the matrix B is called a **similarity transformation**.

Note. $C^{-1}(A_1+A_2)C=C^{-1}A_1C+C^{-1}A_2C$ and $C^{-1}(\alpha A)C=\alpha C^{-1}AC$, so that the function defined by (1) is, in fact, a linear transformation. This explains the use of the word "transformation" in Definition 1.

The purpose of this section is to show that (*i*) similar matrices have several important properties in common and (*ii*) most matrices are similar to diagonal matrices.

EXAMPLE 1 Let $A=\begin{pmatrix}2 & 1\\ 0 & -1\end{pmatrix}$, $B=\begin{pmatrix}4 & -2\\ 5 & -3\end{pmatrix}$, and $C=\begin{pmatrix}2 & -1\\ -1 & 1\end{pmatrix}$. Then $CB=\begin{pmatrix}2 & -1\\ -1 & 1\end{pmatrix}\begin{pmatrix}4 & -2\\ 5 & -3\end{pmatrix}=\begin{pmatrix}3 & -1\\ 1 & -1\end{pmatrix}$ and $AC=\begin{pmatrix}2 & 1\\ 0 & -1\end{pmatrix}\begin{pmatrix}2 & -1\\ -1 & 1\end{pmatrix}=\begin{pmatrix}3 & -1\\ 1 & -1\end{pmatrix}$. Thus $CB=AC$. Since $\det C=1\neq 0$, C is invertible; and since $CB=AC$, we have $C^{-1}CB=C^{-1}AC$ or $B=C^{-1}AC$. This shows that A and B are similar.

EXAMPLE 2 Let $D=\begin{pmatrix}1 & 0 & 0\\ 0 & -1 & 0\\ 0 & 0 & 2\end{pmatrix}$, $A=\begin{pmatrix}-6 & -3 & -25\\ 2 & 1 & 8\\ 2 & 2 & 7\end{pmatrix}$, and $C=\begin{pmatrix}2 & 4 & 3\\ 0 & 1 & -1\\ 3 & 5 & 7\end{pmatrix}$. C is invertible because $\det C=3\neq 0$. We then compute:

$$CA=\begin{pmatrix}2 & 4 & 3\\ 0 & 1 & -1\\ 3 & 5 & 7\end{pmatrix}\begin{pmatrix}-6 & -3 & -25\\ 2 & 1 & 8\\ 2 & 2 & 7\end{pmatrix}=\begin{pmatrix}2 & 4 & 3\\ 0 & -1 & 1\\ 6 & 10 & 14\end{pmatrix}$$

$$DC=\begin{pmatrix}1 & 0 & 0\\ 0 & -1 & 0\\ 0 & 0 & 2\end{pmatrix}\begin{pmatrix}2 & 4 & 3\\ 0 & 1 & -1\\ 3 & 5 & 7\end{pmatrix}=\begin{pmatrix}2 & 4 & 3\\ 0 & -1 & 1\\ 6 & 10 & 14\end{pmatrix}$$

Thus $CA=DC$ and $A=C^{-1}DC$, so A and D are similar.

Note. In Examples 1 and 2 it was not necessary to compute C^{-1}. It was only necessary to know that C was nonsingular.

THEOREM 1 If A and B are similar $n\times n$ matrices, then A and B have the same characteristic equation and, therefore, have the same eigenvalues.

Proof Since A and B are similar, $B=C^{-1}AC$ and

$$\begin{aligned}\det(B-\lambda I)&=\det(C^{-1}AC-\lambda I)=\det[C^{-1}AC-C^{-1}(\lambda I)C]\\ &=\det[C^{-1}(A-\lambda I)C]=\det(C^{-1})\det(A-\lambda I)\det(C)\\ &=\det(C^{-1})\det(C)\det(A-\lambda I)=\det(C^{-1}C)\det(A-\lambda I)\\ &=\det I\det(A-\lambda I)=\det(A-\lambda I)\end{aligned}$$

This means that A and B have the same characteristic equation and, since eigenvalues are roots of the characteristic equation, they have the same eigenvalues. ■

EXAMPLE 3 In Example 2 it is obvious that the eigenvalues of $D=\begin{pmatrix}1 & 0 & 0\\ 0 & -1 & 0\\ 0 & 0 & 2\end{pmatrix}$ are 1, -1, and 2. Thus these are the eigenvalues of $A=\begin{pmatrix}-6 & -3 & -25\\ 2 & 1 & 8\\ 2 & 2 & 7\end{pmatrix}$. Check this by verifying that $\det(A-I)=\det(A+I)=\det(A-2I)=0$.

In a variety of applications it is quite useful to "diagonalize" a matrix A—that is, to find a diagonal matrix similar to A.

DEFINITION 2 **DIAGONALIZABLE MATRIX** An $n\times n$ matrix A is **diagonalizable** if there is a diagonal matrix D such that A is similar to D.

Remark. If D is a diagonal matrix, then its eigenvalues are its diagonal components. If A is similar to D, then A and D have the same eigenvalues (by Theorem 1). Putting these two facts together, we observe that if A is diagonalizable, then A is similar to a diagonal matrix whose diagonal components are the eigenvalues of A.

The next theorem tells us when a matrix is diagonalizable.

THEOREM 2 An $n\times n$ matrix A is diagonalizable if and only if it has n linearly independent eigenvectors. In that case, the diagonal matrix D similar to A is given by

$$D=\begin{pmatrix}\lambda_1 & 0 & 0 & \cdots & 0\\ 0 & \lambda_2 & 0 & \cdots & 0\\ 0 & 0 & \lambda_3 & \cdots & 0\\ \vdots & \vdots & \vdots & & \vdots\\ 0 & 0 & 0 & \cdots & \lambda_n\end{pmatrix} \tag{2}$$

where $\lambda_1, \lambda_2, \ldots, \lambda_n$ are the eigenvalues of A. If C is a matrix whose columns are linearly independent eigenvectors of A, then

$$D=C^{-1}AC \tag{3}$$

Proof We first assume that A has n linearly independent eigenvectors $\mathbf{v}_1, \mathbf{v}_2, \ldots, \mathbf{v}_n$ corresponding to the (not necessarily distinct) eigenvalues $\lambda_1, \lambda_2, \ldots, \lambda_n$.

Let

$$\mathbf{v}_1 = \begin{pmatrix} c_{11} \\ c_{21} \\ \cdot \\ \cdot \\ \cdot \\ c_{n1} \end{pmatrix} \quad \mathbf{v}_2 = \begin{pmatrix} c_{12} \\ c_{22} \\ \cdot \\ \cdot \\ \cdot \\ c_{n2} \end{pmatrix}, \ldots, \quad \mathbf{v}_n = \begin{pmatrix} c_{1n} \\ c_{2n} \\ \cdot \\ \cdot \\ \cdot \\ c_{nn} \end{pmatrix}$$

and let

$$C = \begin{pmatrix} c_{11} & c_{12} & \cdots & c_{1n} \\ c_{21} & c_{22} & \cdots & c_{2n} \\ \cdot & \cdot & & \cdot \\ \cdot & \cdot & & \cdot \\ \cdot & \cdot & & \cdot \\ c_{n1} & c_{n2} & \cdots & c_{nn} \end{pmatrix}$$

Then C is invertible since its columns are linearly independent. Now

$$AC = \begin{pmatrix} a_{11} & a_{12} & \cdots & a_{1n} \\ a_{21} & a_{22} & \cdots & a_{2n} \\ \cdot & \cdot & & \cdot \\ \cdot & \cdot & & \cdot \\ \cdot & \cdot & & \cdot \\ a_{n1} & a_{n2} & \cdots & a_{nn} \end{pmatrix} \begin{pmatrix} c_{11} & c_{12} & \cdots & c_{1n} \\ c_{21} & c_{22} & \cdots & c_{2n} \\ \cdot & \cdot & & \cdot \\ \cdot & \cdot & & \cdot \\ \cdot & \cdot & & \cdot \\ c_{n1} & c_{n2} & \cdots & c_{nn} \end{pmatrix}$$

and we see that the ith column of AC is $A\begin{pmatrix} c_{1i} \\ c_{2i} \\ \vdots \\ c_{ni} \end{pmatrix} = A\mathbf{v}_i = \lambda_i \mathbf{v}_i$. Thus AC is the matrix whose ith column is $\lambda_i \mathbf{v}_i$ and

$$AC = \begin{pmatrix} \lambda_1 c_{11} & \lambda_2 c_{12} & \cdots & \lambda_n c_{1n} \\ \lambda_1 c_{21} & \lambda_2 c_{22} & \cdots & \lambda_n c_{2n} \\ \cdot & \cdot & & \cdot \\ \cdot & \cdot & & \cdot \\ \cdot & \cdot & & \cdot \\ \lambda_1 c_{n1} & \lambda_2 c_{n2} & \cdots & \lambda_n c_{nn} \end{pmatrix}$$

But

$$CD = \begin{pmatrix} c_{11} & c_{12} & \cdots & c_{1n} \\ c_{21} & c_{22} & \cdots & c_{2n} \\ \cdot & \cdot & & \cdot \\ \cdot & \cdot & & \cdot \\ \cdot & \cdot & & \cdot \\ c_{n1} & c_{n2} & \cdots & c_{nn} \end{pmatrix} \begin{pmatrix} \lambda_1 & 0 & \cdots & 0 \\ 0 & \lambda_2 & \cdots & 0 \\ \cdot & \cdot & & \cdot \\ \cdot & \cdot & & \cdot \\ \cdot & \cdot & & \cdot \\ 0 & 0 & \cdots & \lambda_n \end{pmatrix} =$$

$$\begin{pmatrix} \lambda_1 c_{11} & \lambda_2 c_{12} & \cdots & \lambda_n c_{1n} \\ \lambda_1 c_{21} & \lambda_2 c_{22} & \cdots & \lambda_n c_{2n} \\ \vdots & \vdots & & \vdots \\ \lambda_1 c_{n1} & \lambda_2 c_{n2} & \cdots & \lambda_n c_{nn} \end{pmatrix}$$

Thus

$$AC = CD \tag{4}$$

and, since C is invertible, we can multiply both sides of (4) on the left by C^{-1} to obtain

$$D = C^{-1}AC \tag{5}$$

This proves that if A has n linearly independent eigenvectors, then A is diagonalizable. Conversely, suppose that A is diagonalizable. That is, suppose that (5) holds for some invertible matrix C. Let $\mathbf{v}_1, \mathbf{v}_2, \ldots, \mathbf{v}_n$ be the columns of C. Then $AC = CD$ and, reversing the arguments above, we immediately see that $A\mathbf{v}_i = \lambda_i \mathbf{v}_i$ for $i = 1, 2, \ldots, n$. Thus $\mathbf{v}_1, \mathbf{v}_2, \ldots, \mathbf{v}_n$ are eigenvectors of A and are linearly independent because C is invertible. ■

Notation To indicate that D is a diagonal matrix with diagonal components $\lambda_1, \lambda_2, \ldots, \lambda_n$, we write $D = \text{diag}\,(\lambda_1, \lambda_2, \ldots, \lambda_n)$.

Theorem 2 has a useful corollary that follows immediately from Theorem 7.1.3 on page 279.

COROLLARY If the $n \times n$ matrix A has n distinct eigenvalues, then A is diagonalizable.

Remark. If the real coefficients of a polynomial of degree n are picked at random, then, with probability 1, the polynomial will have n distinct roots. It is not difficult to see, intuitively, why this is so. If $n = 2$, for example, then the equation $\lambda^2 + a\lambda + b = 0$ has equal roots if and only if $a^2 = 4b$—a highly unlikely event if a and b are chosen at random. We can, of course, write down polynomials having roots of algebraic multiplicity greater than 1, but these polynomials are exceptional. Thus, without attempting to be mathematically precise, it is fair to say that *most* polynomials have distinct roots. Hence *most* matrices have distinct eigenvalues and, as we stated at the beginning of the section, *most* matrices are diagonalizable.

EXAMPLE 4 Let $A = \begin{pmatrix} 4 & 2 \\ 3 & 3 \end{pmatrix}$. In Example 7.1.3 on page 281 we found the two linearly independent eigenvectors $\mathbf{v}_1 = \begin{pmatrix} 2 \\ -3 \end{pmatrix}$ and $\mathbf{v}_2 = \begin{pmatrix} 1 \\ 1 \end{pmatrix}$. Then, setting $C = \begin{pmatrix} 2 & 1 \\ -3 & 1 \end{pmatrix}$, we find that

$$C^{-1}AC = \frac{1}{5}\begin{pmatrix} 1 & -1 \\ 3 & 2 \end{pmatrix}\begin{pmatrix} 4 & 2 \\ 3 & 3 \end{pmatrix}\begin{pmatrix} 2 & 1 \\ -3 & 1 \end{pmatrix}$$

$$= \frac{1}{5}\begin{pmatrix} 1 & -1 \\ 3 & 2 \end{pmatrix}\begin{pmatrix} 2 & 6 \\ -3 & 6 \end{pmatrix} = \frac{1}{5}\begin{pmatrix} 5 & 0 \\ 0 & 30 \end{pmatrix} = \begin{pmatrix} 1 & 0 \\ 0 & 6 \end{pmatrix}$$

which is the matrix whose diagonal components are the eigenvalues of A.

EXAMPLE 5 Let $A = \begin{pmatrix} 1 & -1 & 4 \\ 3 & 2 & -1 \\ 2 & 1 & -1 \end{pmatrix}$. In Example 7.1.4 on page 281 we computed the three linearly independent eigenvectors $\mathbf{v}_1 = \begin{pmatrix} -1 \\ 4 \\ 1 \end{pmatrix}$, $\mathbf{v}_2 = \begin{pmatrix} 1 \\ -1 \\ -1 \end{pmatrix}$, and $\mathbf{v}_3 = \begin{pmatrix} 1 \\ 2 \\ 1 \end{pmatrix}$.

Then $C = \begin{pmatrix} -1 & 1 & 1 \\ 4 & -1 & 2 \\ 1 & -1 & 1 \end{pmatrix}$ and

$$C^{-1}AC = -\frac{1}{6}\begin{pmatrix} 1 & -2 & 3 \\ -2 & -2 & 6 \\ -3 & 0 & -3 \end{pmatrix}\begin{pmatrix} 1 & -1 & 4 \\ 3 & 2 & -1 \\ 2 & 1 & -1 \end{pmatrix}\begin{pmatrix} -1 & 1 & 1 \\ 4 & -1 & 2 \\ 1 & -1 & 1 \end{pmatrix}$$

$$= -\frac{1}{6}\begin{pmatrix} 1 & -2 & 3 \\ -2 & -2 & 6 \\ -3 & 0 & -3 \end{pmatrix}\begin{pmatrix} -1 & -2 & 3 \\ 4 & 2 & 6 \\ 1 & 2 & 3 \end{pmatrix}$$

$$= -\frac{1}{6}\begin{pmatrix} -6 & 0 & 0 \\ 0 & 12 & 0 \\ 0 & 0 & -18 \end{pmatrix} = \begin{pmatrix} 1 & 0 & 0 \\ 0 & -2 & 0 \\ 0 & 0 & 3 \end{pmatrix}$$

with eigenvalues 1, −2, and 3.

Remark. Since there are an infinite number of ways to choose an eigenvector, there are an infinite number of ways to choose the diagonalizing matrix C. The only advice is to choose the eigenvectors and matrix C that are, arithmetically, the easiest to work with. This usually means that you should insert as many 0's and 1's as possible.

EXAMPLE 6 Let $A = \begin{pmatrix} 3 & 2 & 4 \\ 2 & 0 & 2 \\ 4 & 2 & 3 \end{pmatrix}$. Then, from Example 7.1.9 on page 284, we have the three linearly independent eigenvectors $\mathbf{v}_1 = \begin{pmatrix} 2 \\ 1 \\ 2 \end{pmatrix}$, $\mathbf{v}_2 = \begin{pmatrix} 1 \\ -2 \\ 0 \end{pmatrix}$, and $\mathbf{v}_3 = \begin{pmatrix} 0 \\ -2 \\ 1 \end{pmatrix}$.

Setting $C = \begin{pmatrix} 2 & 1 & 0 \\ 1 & -2 & -2 \\ 2 & 0 & 1 \end{pmatrix}$, we obtain

$$C^{-1}AC = -\frac{1}{9}\begin{pmatrix} -2 & -1 & -2 \\ -5 & 2 & 4 \\ 4 & 2 & -5 \end{pmatrix}\begin{pmatrix} 3 & 2 & 4 \\ 2 & 0 & 2 \\ 4 & 2 & 3 \end{pmatrix}\begin{pmatrix} 2 & 1 & 0 \\ 1 & -2 & -2 \\ 2 & 0 & 1 \end{pmatrix}$$

$$= -\frac{1}{9}\begin{pmatrix} -2 & -1 & -2 \\ -5 & 2 & 4 \\ 4 & 2 & -5 \end{pmatrix}\begin{pmatrix} 16 & -1 & 0 \\ 8 & 2 & 2 \\ 16 & 0 & -1 \end{pmatrix}$$

$$= -\frac{1}{9}\begin{pmatrix} -72 & 0 & 0 \\ 0 & 9 & 0 \\ 0 & 0 & 9 \end{pmatrix} = \begin{pmatrix} 8 & 0 & 0 \\ 0 & -1 & 0 \\ 0 & 0 & -1 \end{pmatrix}$$

This example illustrates that A is diagonalizable even though its eigenvalues are not distinct.

EXAMPLE 7 Let $A = \begin{pmatrix} 4 & 1 \\ 0 & 4 \end{pmatrix}$. In Example 7.1.8 on page 284 we saw that A did *not* have two linearly independent eigenvectors. Suppose that A were diagonalizable (in contradiction to Theorem 2). Then $D = \begin{pmatrix} 4 & 0 \\ 0 & 4 \end{pmatrix}$ and there would be an invertible matrix C such that $C^{-1}AC = D$. Multiplying this equation on the left by C and on the right by C^{-1}, we find that $A = CDC^{-1} = C\begin{pmatrix} 4 & 0 \\ 0 & 4 \end{pmatrix}C^{-1} = C(4I)C^{-1} = 4CIC^{-1} = 4CC^{-1} = 4I = \begin{pmatrix} 4 & 0 \\ 0 & 4 \end{pmatrix} = D$. But $A \neq D$, so no such C exists.

We have seen that many matrices are similar to diagonal matrices. However, two questions remain:

i. Is it possible to determine whether a given matrix is diagonalizable without computing eigenvalues and eigenvectors?
ii. What do we do if A is not diagonalizable?

We shall find a partial answer to the first question in the next section and a complete answer to the second in Section 7.6. In Section 7.7 we shall see an important application of the diagonalizing procedure.

PROBLEMS 7.3

In Problems 1–15 determine whether the given matrix A is diagonalizable. If it is, find a matrix C such that $C^{-1}AC = D$.

1. $\begin{pmatrix} -2 & -2 \\ -5 & 1 \end{pmatrix}$ **2.** $\begin{pmatrix} 3 & -1 \\ -2 & 4 \end{pmatrix}$ **3.** $\begin{pmatrix} 2 & -1 \\ 5 & -2 \end{pmatrix}$

4. $\begin{pmatrix} 3 & -5 \\ 1 & -1 \end{pmatrix}$ **5.** $\begin{pmatrix} 3 & 2 \\ -5 & 1 \end{pmatrix}$ **6.** $\begin{pmatrix} 1 & -1 & 0 \\ -1 & 2 & -1 \\ 0 & -1 & 1 \end{pmatrix}$

7. $\begin{pmatrix} 1 & 1 & -2 \\ -1 & 2 & 1 \\ 0 & 1 & -1 \end{pmatrix}$ **8.** $\begin{pmatrix} 2 & 1 & 0 \\ 0 & 0 & 1 \\ 0 & 0 & 0 \end{pmatrix}$ **9.** $\begin{pmatrix} 3 & 0 & 0 \\ 0 & 0 & 1 \\ 0 & 0 & 2 \end{pmatrix}$

10. $\begin{pmatrix} 3 & -1 & -1 \\ 1 & 1 & -1 \\ 1 & -1 & 1 \end{pmatrix}$ **11.** $\begin{pmatrix} 7 & -2 & -4 \\ 3 & 0 & -2 \\ 6 & -2 & -3 \end{pmatrix}$ **12.** $\begin{pmatrix} 4 & 6 & 6 \\ 1 & 3 & 2 \\ -1 & -5 & -2 \end{pmatrix}$

13. $\begin{pmatrix} -3 & -7 & -5 \\ 2 & 4 & 3 \\ 1 & 2 & 2 \end{pmatrix}$ **14.** $\begin{pmatrix} -2 & -2 & 0 & 0 \\ -5 & 1 & 0 & 0 \\ 0 & 0 & 2 & -1 \\ 0 & 0 & 5 & -2 \end{pmatrix}$ **15.** $\begin{pmatrix} 4 & 1 & 0 & 1 \\ 2 & 3 & 0 & 1 \\ -2 & 1 & 2 & -3 \\ 2 & -1 & 0 & 5 \end{pmatrix}$

16. Show that if A is similar to B and B is similar to C, then A is similar to C.

***17.** Let T be a linear transformation from $\mathbb{R}^n \to \mathbb{R}^n$. Let B be the matrix representation of T with respect to a basis B_1, and let C be the matrix representation of T with respect to a basis B_2. Show that B is similar to C by showing that $B = ACA^{-1}$, where A is the transition matrix from B_2 to B_1 (see Theorem 5.6.1 on page 203).

***18.** If A is similar to B, show that $\rho(A) = \rho(B)$ and $\nu(A) = \nu(B)$. [*Hint:* First prove that if C is invertible, then $\nu(CA) = \nu(A)$ by showing that $\mathbf{x} \in N_A$ if and only if $\mathbf{x} \in N_{CA}$. Next prove that $\rho(AC) = \rho(A)$ by showing that $R_A = R_{AC}$. Conclude that $\rho(AC) = \rho(CA) = \rho(A)$. Finally, use the fact that C^{-1} is invertible to show that $\rho(C^{-1}AC) = \rho(A)$.]

19. If A is similar to B, show that A^n is similar to B^n for any positive integer n.

20. If A is similar to B, show that $\det A = \det B$.

21. Let $D = \begin{pmatrix} 1 & 0 \\ 0 & -1 \end{pmatrix}$. Compute D^{20}.

22. Let $A = \begin{pmatrix} 3 & -4 \\ 2 & -3 \end{pmatrix}$. Compute A^{20}. [*Hint:* Find a C such that $A = CDC^{-1}$, where D is diagonal and show that $A^{20} = CD^{20}C^{-1}$.]

23. Suppose that $C^{-1}AC = D$. Show that for any integer n, $A^n = CD^nC^{-1}$. This gives an easy way to compute powers of a diagonalizable matrix.

24. Use the result of Problem 23 and Example 6 to compute A^{10}, where $A = \begin{pmatrix} 3 & 2 & 4 \\ 2 & 0 & 2 \\ 4 & 2 & 3 \end{pmatrix}$.

***25.** Let A be an $n \times n$ matrix whose characteristic equation is $(\lambda - c)^n = 0$. Show that A is diagonalizable if and only if $A = cI$.

26. If A is diagonalizable, show that $\det A = \lambda_1\lambda_2 \cdots \lambda_n$, where $\lambda_1, \lambda_2, \ldots, \lambda_n$ are the eigenvalues of A.

***27.** Let A and B be real $n \times n$ matrices with distinct eigenvalues. Prove that $AB = BA$ if and only if A and B have the same eigenvectors.

7.4 Symmetric Matrices and Orthogonal Diagonalization

In this section we shall see that symmetric matrices† have a number of important properties. In particular, we shall show that any symmetric matrix has n linearly independent real eigenvectors and therefore, by Theorem 7.3.2, is diagonalizable. We begin by proving that the eigenvalues of a real symmetric matrix are real.

THEOREM 1 Let A be a real $n \times n$ symmetric matrix. Then the eigenvalues of A are real.

Proof‡ Let λ be an eigenvalue of A with eigenvector $\mathbf{v}$; that is, $A\mathbf{v} = \lambda\mathbf{v}$. Now $\mathbf{v}$ is a vector in $\mathbb{C}^n$, and an inner product in $\mathbb{C}^n$ (see Definition 5.8.1, page 222, and Example 5.8.2, page 223) satisfies

$$(\alpha\mathbf{x}, \mathbf{y}) = \alpha(\mathbf{x}, \mathbf{y}) \quad \text{and} \quad (\mathbf{x}, \alpha\mathbf{y}) = \bar{\alpha}(\mathbf{x}, \mathbf{y}) \tag{1}$$

Then

$$(A\mathbf{v}, \mathbf{v}) = (\lambda\mathbf{v}, \mathbf{v}) = \lambda(\mathbf{v}, \mathbf{v}) \tag{2}$$

Moreover, by Theorem 6.6.1 on page 270 and the fact that $A^t = A$,

$$(A\mathbf{v}, \mathbf{v}) = (\mathbf{v}, A^t\mathbf{v}) = (\mathbf{v}, A\mathbf{v}) = (\mathbf{v}, \lambda\mathbf{v}) = \bar{\lambda}(\mathbf{v}, \mathbf{v}) \tag{3}$$

Thus, equating (2) and (3), we have

$$\lambda(\mathbf{v}, \mathbf{v}) = \bar{\lambda}(\mathbf{v}, \mathbf{v}) \tag{4}$$

But $(\mathbf{v}, \mathbf{v}) = |\mathbf{v}|^2 \neq 0$, since $\mathbf{v}$ is an eigenvector. Thus we can divide both sides of (4) by $(\mathbf{v}, \mathbf{v})$ to obtain

$$\lambda = \bar{\lambda} \tag{5}$$

If $\lambda = a + ib$, then $\bar{\lambda} = a - ib$ and, from (5), we have

$$a + ib = a - ib \tag{6}$$

which can hold only if $b = 0$. This shows that $\lambda = a$; hence λ is real and the proof is complete. ■

We saw in Theorem 7.1.3 on page 279 that eigenvectors corresponding to different eigenvalues are linearly independent. For symmetric matrices the result is stronger: *Eigenvectors of a symmetric matrix corresponding to different eigenvalues are orthogonal.*

THEOREM 2 Let A be a real symmetric $n \times n$ matrix. If λ_1 and λ_2 are distinct eigenvalues with corresponding real eigenvectors $\mathbf{v}_1$ and $\mathbf{v}_2$, then $\mathbf{v}_1$ and $\mathbf{v}_2$ are orthogonal.

---

† Recall that A is symmetric if and only if $A^t = A$.

‡ This proof uses material in Sections 5.8 and 6.6 and should be omitted if those sections were not covered.

Proof We compute

$$A\mathbf{v}_1 \cdot \mathbf{v}_2 = \lambda_1\mathbf{v}_1 \cdot \mathbf{v}_2 = \lambda_1(\mathbf{v}_1 \cdot \mathbf{v}_2) \tag{7}$$

and

$$A\mathbf{v}_1 \cdot \mathbf{v}_2 = \mathbf{v}_1 \cdot A^t\mathbf{v}_2 = \mathbf{v}_1 \cdot A\mathbf{v}_2 = \mathbf{v}_1 \cdot (\lambda_2\mathbf{v}_2) = \lambda_2(\mathbf{v}_1 \cdot \mathbf{v}_2) \tag{8}$$

Combining (7) and (8) we have $\lambda_1(\mathbf{v}_1 \cdot \mathbf{v}_2) = \lambda_2(\mathbf{v}_1 \cdot \mathbf{v}_2)$ and since $\lambda_1 \neq \lambda_2$, we conclude that $\mathbf{v}_1 \cdot \mathbf{v}_2 = 0$. This is what we wanted to show. ■

We now state the main result of this section. Its proof, which is difficult (and optional), is given at the end of this section.

THEOREM 3 Let A be a real symmetric $n \times n$ matrix. Then A has n real orthonormal eigenvectors.

Remark. It follows from this theorem that the geometric multiplicity of each eigenvalue of A is equal to its algebraic multiplicity.

Theorem 3 tells us that if A is symmetric, then $\mathbb{R}^n$ has a basis $B = \{\mathbf{u}_1, \mathbf{u}_2, \ldots, \mathbf{u}_n\}$ consisting of orthonormal eigenvectors of A. Let Q be the matrix whose columns are $\mathbf{u}_1, \mathbf{u}_2, \ldots, \mathbf{u}_n$. Then, by Theorem 5.7.3 on page 216, Q is an orthogonal matrix. This leads to the following definition.

DEFINITION 1 **ORTHOGONALLY DIAGONALIZABLE MATRIX** An $n \times n$ matrix A is said to be **orthogonally diagonalizable** if there exists an orthogonal matrix Q such that

$$\boxed{Q^tAQ = D} \tag{9}$$

where $D = \operatorname{diag}(\lambda_1, \lambda_2, \ldots, \lambda_n)$ and $\lambda_1, \lambda_2, \ldots, \lambda_n$ are the eigenvalues of A.

Note. Remember that Q is orthogonal if $Q^t = Q^{-1}$; Hence (9) could be written as $Q^{-1}AQ = D$.

THEOREM 4 Let A be a real $n \times n$ matrix. Then A is orthogonally diagonalizable if and only if A is symmetric.

Proof Let A be symmetric. Then, by Theorems 2 and 3, A is orthogonally diagonalizable with Q the matrix whose columns are the orthonormal eigenvectors given in Theorem 3. Conversely, suppose that A is orthogonally diagonalizable. Then there exists an orthogonal matrix Q such that $Q^tAQ = D$. Multiplying this equation on the left by Q and on the right by Q^t and using the fact that $Q^tQ = QQ^t = I$, we obtain

$$A = QDQ^t \tag{10}$$

Then $A^t = (QDQ^t)^t = (Q^t)^t D^t Q^t = QDQ^t = A$. Thus A is symmetric and the theorem is proved. In the last series of equations we used the facts that $(AB)^t = B^t A^t$ (part (*ii*) of Theorem 2.8.1, page 78), $(A^t)^t = A$ (part (*i*) of Theorem 2.8.1), and $D^t = D$ for any diagonal matrix D. ■

Before giving examples, we provide the following three-step procedure for finding the orthogonal matrix Q that diagonalizes the symmetric matrix A:

PROCEDURE FOR FINDING A DIAGONALIZING MATRIX Q:

i. Find a basis for each eigenspace of A.
ii. Find an orthonormal basis for each eigenspace of A by using the Gram–Schmidt process.
iii. Write Q as the matrix whose columns are the orthonormal eigenvectors obtained in step (*ii*).

EXAMPLE 1 Let $A = \begin{pmatrix} 1 & -2 \\ -2 & 3 \end{pmatrix}$. Then the characteristic equation of A is $\det(A - \lambda I) = \begin{vmatrix} 1-\lambda & -2 \\ -2 & 3-\lambda \end{vmatrix} = \lambda^2 - 4\lambda - 1 = 0$, which has the roots $\lambda = (4 \pm \sqrt{20})/2 = (4 \pm 2\sqrt{5})/2 = 2 \pm \sqrt{5}$. For $\lambda_1 = 2 - \sqrt{5}$, we obtain $(A - \lambda I)\mathbf{v} = \begin{pmatrix} -1+\sqrt{5} & -2 \\ -2 & 1+\sqrt{5} \end{pmatrix}\begin{pmatrix} x_1 \\ x_2 \end{pmatrix} = \begin{pmatrix} 0 \\ 0 \end{pmatrix}$. An eigenvector is $\mathbf{v}_1 = \begin{pmatrix} 2 \\ -1+\sqrt{5} \end{pmatrix}$, $|\mathbf{v}_1| = \sqrt{2^2 + (-1+\sqrt{5})^2} = \sqrt{10 - 2\sqrt{5}}$. Thus $\mathbf{u}_1 = \dfrac{1}{\sqrt{10-2\sqrt{5}}}\begin{pmatrix} 2 \\ -1+\sqrt{5} \end{pmatrix}$. Next, for $\lambda_2 = 2 + \sqrt{5}$, we compute $(A - \lambda I)\mathbf{v} = \begin{pmatrix} -1-\sqrt{5} & -2 \\ -2 & 1-\sqrt{5} \end{pmatrix}\begin{pmatrix} x_1 \\ x_2 \end{pmatrix} = \begin{pmatrix} 0 \\ 0 \end{pmatrix}$ and $\mathbf{v}_2 = \begin{pmatrix} 1-\sqrt{5} \\ 2 \end{pmatrix}$. Note that $\mathbf{v}_1 \cdot \mathbf{v}_2 = 0$ (which must be true according to Theorem 2). Then $|\mathbf{v}_2| = \sqrt{10 - 2\sqrt{5}}$, so that $\mathbf{u}_2 = \dfrac{1}{\sqrt{10-2\sqrt{5}}}\begin{pmatrix} 1-\sqrt{5} \\ 2 \end{pmatrix}$. Finally

$$Q = \frac{1}{\sqrt{10-2\sqrt{5}}}\begin{pmatrix} 2 & 1-\sqrt{5} \\ -1+\sqrt{5} & 2 \end{pmatrix}, \quad Q^t = \frac{1}{\sqrt{10-2\sqrt{5}}}\begin{pmatrix} 2 & -1+\sqrt{5} \\ 1-\sqrt{5} & 2 \end{pmatrix}$$

and

$$Q^tAQ = \frac{1}{10-2\sqrt{5}}\begin{pmatrix} 2 & -1+\sqrt{5} \\ 1-\sqrt{5} & 2 \end{pmatrix}\begin{pmatrix} 1 & -2 \\ -2 & 3 \end{pmatrix}\begin{pmatrix} 2 & 1-\sqrt{5} \\ -1+\sqrt{5} & 2 \end{pmatrix}$$

$$= \frac{1}{10-2\sqrt{5}}\begin{pmatrix} 2 & -1+\sqrt{5} \\ 1-\sqrt{5} & 2 \end{pmatrix}\begin{pmatrix} 4-2\sqrt{5} & -3-\sqrt{5} \\ -7+3\sqrt{5} & 4+2\sqrt{5} \end{pmatrix}$$

$$= \frac{1}{10-2\sqrt{5}}\begin{pmatrix} 30-14\sqrt{5} & 0 \\ 0 & 10+6\sqrt{5} \end{pmatrix} = \begin{pmatrix} 2-\sqrt{5} & 0 \\ 0 & 2+\sqrt{5} \end{pmatrix}$$

EXAMPLE 2 Let $A=\begin{pmatrix}5&4&2\\4&5&2\\2&2&2\end{pmatrix}$. Then A is symmetric and $\det(A-\lambda I)=$ $\begin{pmatrix}5-\lambda&4&2\\4&5-\lambda&2\\2&2&2-\lambda\end{pmatrix}=-(\lambda-1)^2(\lambda-10)$. Corresponding to $\lambda=1$ we compute the linearly independent eigenvectors $\mathbf{v}_1=\begin{pmatrix}-1\\1\\0\end{pmatrix}$ and $\mathbf{v}_2=\begin{pmatrix}-1\\0\\2\end{pmatrix}$. Corresponding to $\lambda=10$ we find that $\mathbf{v}_3=\begin{pmatrix}2\\2\\1\end{pmatrix}$. To find Q, we apply the Gram–Schmidt process to $\{\mathbf{v}_1,\mathbf{v}_2\}$, a basis for E_1. Since $|\mathbf{v}_1|=\sqrt{2}$, we set $\mathbf{u}_1=\begin{pmatrix}-1/\sqrt{2}\\1/\sqrt{2}\\0\end{pmatrix}$. Next,

$$\mathbf{v}_2'=\mathbf{v}_2-(\mathbf{v}_2\cdot\mathbf{u}_1)\mathbf{u}_1=\begin{pmatrix}-1\\0\\2\end{pmatrix}-\frac{1}{\sqrt{2}}\begin{pmatrix}-1/\sqrt{2}\\1/\sqrt{2}\\0\end{pmatrix}=\begin{pmatrix}-1\\0\\2\end{pmatrix}-\begin{pmatrix}-1/2\\1/2\\0\end{pmatrix}=\begin{pmatrix}-1/2\\-1/2\\2\end{pmatrix}$$

Then $|\mathbf{v}_2|=\sqrt{18/4}=3\sqrt{2}/2$ and $\mathbf{u}_2=\dfrac{2}{3\sqrt{2}}\begin{pmatrix}-1/2\\-1/2\\2\end{pmatrix}=\begin{pmatrix}-1/3\sqrt{2}\\-1/3\sqrt{2}\\4/3\sqrt{2}\end{pmatrix}$. We check this by noting that $\mathbf{u}_1\cdot\mathbf{u}_2=0$. Finally, we have $\mathbf{u}_3=\mathbf{v}_3/|\mathbf{v}_3|=\frac{1}{3}\mathbf{v}_3=\begin{pmatrix}2/3\\2/3\\1/3\end{pmatrix}$. We can check this too by noting that $\mathbf{u}_1\cdot\mathbf{u}_3=0$ and $\mathbf{u}_2\cdot\mathbf{u}_3=0$. Thus

$$Q=\begin{pmatrix}-1/\sqrt{2}&-1/3\sqrt{2}&2/3\\1/\sqrt{2}&-1/3\sqrt{2}&2/3\\0&4/3\sqrt{2}&1/3\end{pmatrix}$$

and

$$Q^tAQ=\begin{pmatrix}-1/\sqrt{2}&1/\sqrt{2}&0\\-1/3\sqrt{2}&-1/3\sqrt{2}&4/3\sqrt{2}\\2/3&2/3&1/3\end{pmatrix}\begin{pmatrix}5&4&2\\4&5&2\\2&2&2\end{pmatrix}\begin{pmatrix}-1/\sqrt{2}&-1/3\sqrt{2}&2/3\\1/\sqrt{2}&-1/3\sqrt{2}&2/3\\0&4/3\sqrt{2}&1/3\end{pmatrix}$$

$$=\begin{pmatrix}-1/\sqrt{2}&1/\sqrt{2}&0\\-1/3\sqrt{2}&-1/3\sqrt{2}&4/3\sqrt{2}\\2/3&2/3&1/3\end{pmatrix}\begin{pmatrix}-1/\sqrt{2}&-1/3\sqrt{2}&20/3\\1/\sqrt{2}&-1/3\sqrt{2}&20/3\\0&4/3\sqrt{2}&10/3\end{pmatrix}$$

$$=\begin{pmatrix}1&0&0\\0&1&0\\0&0&10\end{pmatrix}$$

In this section we have proved results for real symmetric matrices. If $A = (a_{ij})$ is a complex matrix, then the **conjugate transpose** of A, denoted A^*, is defined by: the ijth element of $A^* = \overline{a_{ji}}$. The matrix A is called **hermitian** if $A^* = A$. It turns out that Theorems 1, 2, and 3 are also true for hermitian matrices. Moreover, if we define a **unitary** matrix to be a complex matrix U with $U^* = U^{-1}$, then, using the proof of Theorem 4, we can show that a hermitian matrix is unitarily diagonalizable. We leave all these facts as exercises (see Problems 15–17).

We conclude this section with a proof of Theorem 3.

Proof of Theorem 3† We prove that to every eigenvalue λ of algebraic multiplicity k, there correspond k orthonormal eigenvectors. This step, combined with Theorem 2, will be sufficient. Let $\mathbf{u}_1$ be an eigenvector of A corresponding to λ_1. We can assume that $|\mathbf{u}_1| = 1$. We can also assume that $\mathbf{u}_1$ is real because λ_1 is real and $\mathbf{u}_1 \in N_{A-\lambda_1 I}$, the kernel of the real matrix $A - \lambda_1 I$. This kernel is a subspace of $\mathbb{R}^n$ by Example 5.5.10 on page 197. Next we note that $\{\mathbf{u}_1\}$ can be expanded into a basis $\{\mathbf{u}_1, \mathbf{v}_2, \mathbf{v}_3, \ldots, \mathbf{v}_n\}$ for $\mathbb{R}^n$ and, by the Gram–Schmidt process, we can turn this basis into the orthonormal basis $\{\mathbf{u}_1\ \mathbf{u}_2, \ldots, \mathbf{u}_n\}$. Let Q be the orthogonal matrix whose columns are $\mathbf{u}_1, \mathbf{u}_2, \ldots, \mathbf{u}_n$. For convenience of notation we write $Q = (\mathbf{u}_1, \mathbf{u}_2, \ldots, \mathbf{u}_n)$. Now Q is invertible and $Q^t = Q^{-1}$, so A is similar to Q^tAQ and, by Theorem 7.3.1 on page 294, Q^tAQ and A have the same characteristic polynomial: $|Q^tAQ - \lambda I| = |A - \lambda I|$. Thus,

$$Q^t = \begin{pmatrix} \mathbf{u}_1^t \\ \mathbf{u}_2^t \\ \cdot \\ \cdot \\ \cdot \\ \mathbf{u}_n^t \end{pmatrix}$$

so that

$$Q^tAQ = \begin{pmatrix} \mathbf{u}_1^t \\ \mathbf{u}_2^t \\ \cdot \\ \cdot \\ \cdot \\ \mathbf{u}_n^t \end{pmatrix} A(\mathbf{u}_1\mathbf{u}_2 \cdots \mathbf{u}_n) = \begin{pmatrix} \mathbf{u}_1^t \\ \mathbf{u}_2^t \\ \cdot \\ \cdot \\ \cdot \\ \mathbf{u}_n^t \end{pmatrix} (A\mathbf{u}_1 A\mathbf{u}_2 \cdots A\mathbf{u}_n)$$

$$= \begin{pmatrix} \mathbf{u}_1^t \\ \mathbf{u}_2^t \\ \cdot \\ \cdot \\ \cdot \\ \mathbf{u}_n^t \end{pmatrix} (\lambda_1\mathbf{u}_1, A\mathbf{u}_2 \cdots A\mathbf{u}_n) \begin{pmatrix} \lambda_1 & \mathbf{u}_1^tA\mathbf{u}_2 & \cdots & \mathbf{u}_1^tA\mathbf{u}_n \\ 0 & \mathbf{u}_2^tA\mathbf{u}_2 & \cdots & \mathbf{u}_2^tA\mathbf{u}_n \\ \cdot & \cdot & & \cdot \\ \cdot & \cdot & & \cdot \\ \cdot & \cdot & & \cdot \\ 0 & \mathbf{u}_n^tA\mathbf{u}_2 & \cdots & \mathbf{u}_n^tA\mathbf{u}_n \end{pmatrix}$$

The zeros appear because $\mathbf{u}_1^t\mathbf{u}_j = \mathbf{u}_1 \cdot \mathbf{u}_j = 0$ if $j \neq 1$. Now $[Q^tAQ]^t = Q^tA^t(Q^t)^t = Q^tAQ$. Thus Q^tAQ is symmetric, which means that there must be zeros in the first row of Q^tAQ to match the zeros in the first column.

† If time permits.

Thus

$$Q^tAQ = \begin{pmatrix} \lambda_1 & 0 & 0 & \cdots & 0 \\ 0 & q_{22} & q_{23} & \cdots & q_{2n} \\ 0 & q_{32} & q_{33} & \cdots & q_{3n} \\ \vdots & \vdots & \vdots & & \vdots \\ 0 & q_{n2} & q_{n3} & \cdots & q_{nn} \end{pmatrix}$$

and

$$|Q^tAQ - \lambda I| = \begin{vmatrix} \lambda_1-\lambda & 0 & 0 & \cdots & 0 \\ 0 & q_{22}-\lambda & q_{23} & \cdots & q_{2n} \\ 0 & q_{32} & q_{33}-\lambda & \cdots & q_{3n} \\ \vdots & \vdots & \vdots & & \vdots \\ 0 & q_{n2} & q_{n3} & \cdots & q_{nn}-\lambda \end{vmatrix}$$

$$= (\lambda_1-\lambda)\begin{vmatrix} q_{22}-\lambda & q_{23} & \cdots & q_{2n} \\ q_{32} & q_{33}-\lambda & \cdots & q_{3n} \\ \vdots & \vdots & & \vdots \\ q_{n2} & q_{n3} & \cdots & q_{nn}-\lambda \end{vmatrix} = (\lambda-\lambda_1)\,|M_{11}(\lambda)|$$

where $M_{11}(\lambda)$ is the 1, 1 minor of $Q^tAQ-\lambda I$. If $k=1$, there is nothing to prove. If $k>1$, then $|A-\lambda I|$ contains the factor $(\lambda-\lambda_1)^2$ and, therefore, $|Q^tAQ-\lambda I|$ also contains the factor $(\lambda-\lambda_1)^2$. Thus $|M_{11}(\lambda)|$ contains the factor $\lambda-\lambda_1$, which means that $|M_{11}(\lambda_1)|=0$. This means that the last $n-1$ columns of $Q^tAQ-\lambda_1 I$ are linearly dependent. Since the first column of $Q^tAQ-\lambda_1 I$ is the zero vector, this means that $Q^tAQ-\lambda_1 I$ contains at most $n-2$ linearly independent columns. In other words, $\rho(Q^tAQ-\lambda_1 I)\le n-2$. But $Q^tAQ-\lambda_1 I$ and $A-\lambda_1 I$ are similar; hence, by Problem 7.3.18, $\rho(A-\lambda_1 I)\le n-2$. Therefore $\nu(A-\lambda_1 I)\ge 2$, which means that $E_\lambda=$ kernel of $(A-\lambda_1 I)$ contains at least two linearly independent eigenvectors. If $k=2$, we are done. If $k>2$, then we take two orthonormal vectors $\mathbf{u}_1$, $\mathbf{u}_2$ in E_λ and expand them into a new orthonormal basis $\{\mathbf{u}_1, \mathbf{u}_2, \ldots, \mathbf{u}_n\}$ for $\mathbb{R}^n$ and define $P=(\mathbf{u}_1, \mathbf{u}_2, \ldots, \mathbf{u}_n)$. Then, exactly as before, we show that

$$P^tAP-\lambda I = \begin{pmatrix} \lambda_1-\lambda & 0 & 0 & 0 & \cdots & 0 \\ 0 & \lambda_1-\lambda & 0 & 0 & \cdots & 0 \\ 0 & 0 & \beta_{33}-\lambda & \beta_{34} & \cdots & \beta_{3n} \\ 0 & 0 & \beta_{43} & \beta_{44}-\lambda & \cdots & \beta_{4n} \\ \vdots & \vdots & \vdots & \vdots & & \vdots \\ 0 & 0 & \beta_{n3} & \beta_{n4} & \cdots & \beta_{nn}-\lambda \end{pmatrix}$$

Since $k>2$, we show, as before, that the determinant of the matrix in brackets is zero when $\lambda=\lambda_1$—which shows that $\rho(P^tAP-\lambda_1I)\le n-3$, so that $\nu(P^tAP-\lambda_1I)=\nu(A-\lambda_1I)\ge 3$. Then dim $E_{\lambda_1}\ge 3$ and so on. We can clearly continue this process to show that dim $E_{\lambda_1}=k$. Finally, in each E_{λ_i} we can find an orthonormal basis. This completes the proof. ■

PROBLEMS 7.4

In Problems 1–8 find an orthogonal matrix that diagonalizes the given symmetric matrix.

1. $\begin{pmatrix} 3 & 4 \\ 4 & -3 \end{pmatrix}$ **2.** $\begin{pmatrix} 2 & 1 \\ 1 & 2 \end{pmatrix}$ **3.** $\begin{pmatrix} 1 & -1 \\ -1 & 1 \end{pmatrix}$

4. $\begin{pmatrix} 1 & -1 & -1 \\ -1 & 1 & -1 \\ -1 & -1 & 1 \end{pmatrix}$ **5.** $\begin{pmatrix} -1 & 2 & 2 \\ 2 & -1 & 2 \\ 2 & 2 & 1 \end{pmatrix}$ **6.** $\begin{pmatrix} 1 & -1 & 0 \\ -1 & 2 & -1 \\ 0 & -1 & 1 \end{pmatrix}$

7. $\begin{pmatrix} 3 & 2 & 2 \\ 2 & 2 & 0 \\ 2 & 0 & 4 \end{pmatrix}$ **8.** $\begin{pmatrix} 1 & -1 & 0 & 0 \\ -1 & 0 & 0 & 0 \\ 0 & 0 & 0 & 0 \\ 0 & 0 & 0 & 2 \end{pmatrix}$

9. Let Q be a symmetric orthogonal matrix. Show that if λ is an eigenvalue of Q, then $\lambda=\pm 1$.

10. A is **orthogonally similar** to B if there exists an orthogonal matrix Q such that $B=Q^tAQ$. Suppose that A is orthogonally similar to B and that B is orthogonally similar to C. Show that A is orthogonally similar to C.

11. Show that if $Q=\begin{pmatrix} a & b \\ c & d \end{pmatrix}$ is orthogonal, then $b=\pm c$. [*Hint:* Write out the equations that result from the equation $Q^tQ=I$.]

12. Suppose that A is a real symmetric matrix every one of whose eigenvalues is zero. Show that A is the zero matrix.

13. Show that if a real 2×2 matrix A has eigenvectors that are orthogonal, then A is symmetric.

14. Let A be a real skew-symmetric matrix ($A^t=-A$). Prove that every eigenvalue of A is of the form $i\alpha$, where α is a real number. That is, prove that every eigenvalue of A is a *pure imaginary* number.

***15.** Show that the eigenvalues of a complex $n\times n$ hermitian matrix are real. [*Hint:* Use the fact that in $\mathbb{C}^n$, $(A\mathbf{x},\mathbf{y})=(\mathbf{x},A^*\mathbf{y})$.]

***16.** If A is an $n\times n$ hermitian matrix, show that eigenvectors corresponding to different eigenvalues are orthogonal.

****17.** By repeating the proof of Theorem 3, except that $\bar{\mathbf{v}}_i^t$ replaces $\mathbf{v}_i^t$ where appropriate, show that any $n\times n$ hermitian matrix has n orthonormal eigenvectors.

18. Find a unitary matrix U such that U^*AU is diagonal, where $A=\begin{pmatrix} 1 & 1-i \\ 1+i & 0 \end{pmatrix}$.

19. Do the same for $A=\begin{pmatrix} 2 & 3-3i \\ 3+3i & 5 \end{pmatrix}$.

20. Prove that the determinant of a hermitian matrix is real.

7.5 Quadratic Forms and Conic Sections

In this section we use the material of Section 7.4 to discover information about the graphs of quadratic equations. Quadratic equations and quadratic forms, which are defined below, arise in a variety of ways. For example, we can use quadratic forms to obtain information about the conic sections in $\mathbb{R}^2$ (circles, parabolas, ellipses, hyperbolas) and extend this theory to describe certain surfaces, called *quadric surfaces*, in $\mathbb{R}^3$. These topics are discussed later in the section. Although we shall not discuss it in this text, quadratic forms arise in a number of applications ranging from a description of cost functions in economics to an analysis of the control of a rocket traveling in space.

DEFINITION 1 QUADRATIC EQUATION AND QUADRATIC FORM

i. A **quadratic equation in two variables with no linear terms** is an equation of the form

$$ax^2 + bxy + cy^2 = d \tag{1}$$

where $|a|+|b|+|c| \neq 0$.

ii. A **quadratic form in two variables** is an expression of the form

$$F(x, y) = ax^2 + bxy + cy^2 \tag{2}$$

where $|a|+|b|+|c| \neq 0$.

Obviously quadratic equations and quadratic forms are closely related. We begin our analysis of quadratic forms with a simple example.

Consider the quadratic form $F(x, y) = x^2 - 4xy + 3y^2$. Let $\mathbf{v} = \begin{pmatrix} x \\ y \end{pmatrix}$ and $A = \begin{pmatrix} 1 & -2 \\ -2 & 3 \end{pmatrix}$. Then

$$A\mathbf{v}\cdot\mathbf{v} = \begin{pmatrix} 1 & -2 \\ -2 & 3 \end{pmatrix}\begin{pmatrix} x \\ y \end{pmatrix}\cdot\begin{pmatrix} x \\ y \end{pmatrix} = \begin{pmatrix} x-2y \\ -2x+3y \end{pmatrix}\cdot\begin{pmatrix} x \\ y \end{pmatrix}$$

$$= (x^2 - 2xy) + (-2xy + 3y^2) = x^2 - 4xy + 3y^2 = F(x, y)$$

Thus we have "represented" the quadratic form $F(x, y)$ by the symmetric matrix A in the sense that

$$F(x, y) = A\mathbf{v}\cdot\mathbf{v} \tag{3}$$

Conversely, if A is a symmetric matrix, then Equation (3) defines a quadratic form $F(x, y) = A\mathbf{v}\cdot\mathbf{v}$.

We can represent $F(x, y)$ by many matrices but only one symmetric matrix. To see this, let $A = \begin{pmatrix} 1 & a \\ b & 3 \end{pmatrix}$, where $a+b=-4$. Then $A\mathbf{v}\cdot\mathbf{v}=F(x, y)$. If $A=\begin{pmatrix} 1 & 3 \\ -7 & 3 \end{pmatrix}$, for example, then $A\mathbf{v}=\begin{pmatrix} x+3y \\ -7x+3y \end{pmatrix}$ and $A\mathbf{v}\cdot\mathbf{v}=x^2-4xy+3y^2$. If, however, we insist that A be symmetric, then we must have $a+b=-4$ and $a=b$. This pair of equations has the unique solution $a=b=-2$.

If $F(x, y)=ax^2+bxy+cy^2$ is a quadratic form, let

$$A=\begin{pmatrix} a & b/2 \\ b/2 & c \end{pmatrix} \tag{4}$$

Then

$$\begin{aligned} A\mathbf{v}\cdot\mathbf{v}&=\left[\begin{pmatrix} a & b/2 \\ b/2 & c \end{pmatrix}\begin{pmatrix} x \\ y \end{pmatrix}\right]\cdot\begin{pmatrix} x \\ y \end{pmatrix}=\begin{pmatrix} ax+(b/2)y \\ (b/2)x+cy \end{pmatrix}\cdot\begin{pmatrix} x \\ y \end{pmatrix} \\ &=ax^2+bxy+cy^2=F(x, y) \end{aligned}$$

Now let us return to the quadratic equation (1). Using (3), we can write (1) as

$$A\mathbf{v}\cdot\mathbf{v}=d \tag{5}$$

where A is symmetric. By Theorem 7.4.4 on page 302, there is an orthogonal matrix Q such that $Q^tAQ = D$, where $D = \text{diag}\,(\lambda_1, \lambda_2)$ and λ_1 and λ_2 are the eigenvalues of A. Then $A = QDQ^t$ (remember that $Q^t = Q^{-1}$) and (5) can be written

$$(QDQ^t\mathbf{v})\cdot\mathbf{v}=d \tag{6}$$

But, from Theorem 6.6.1 on page 270, $A\mathbf{v}\cdot\mathbf{y} = \mathbf{v}\cdot A^t\mathbf{y}$. Thus

$$Q(DQ^t\mathbf{v})\cdot\mathbf{v}=DQ^t\mathbf{v}\cdot Q^t\mathbf{v} \tag{7}$$

so that (6) reads

$$[DQ^t\mathbf{v}]\cdot Q^t\mathbf{v}=d \tag{8}$$

Let $\mathbf{v}'=Q^t\mathbf{v}$. Then $\mathbf{v}'$ is a 2-vector and (8) becomes

$$D\mathbf{v}'\cdot\mathbf{v}'=d \tag{9}$$

Let us look at (9) more closely. We can write $\mathbf{v}'=\begin{pmatrix} x' \\ y' \end{pmatrix}$. Since a diagonal matrix is symmetric, (9) defines a quadratic form $F'(x', y')$ in the variables x' and y'. If $D=\begin{pmatrix} a' & 0 \\ 0 & c' \end{pmatrix}$, then $D\mathbf{v}'=\begin{pmatrix} a' & 0 \\ 0 & c' \end{pmatrix}\begin{pmatrix} x' \\ y' \end{pmatrix}=\begin{pmatrix} a'x' \\ c'y' \end{pmatrix}$ and

$$F'(x', y')=D\mathbf{v}'\cdot\mathbf{v}'=\begin{pmatrix} a'x' \\ c'y' \end{pmatrix}\cdot\begin{pmatrix} x' \\ y' \end{pmatrix}=a'x'^2+c'y'^2$$

That is: *$F'(x', y')$ is a quadratic form with the $x'y'$ term missing.* Hence Equation (9) is a quadratic equation in the new variables x', y' with the $x'y'$ term missing.

EXAMPLE 1 Consider the quadratic equation $x^2-4xy+3y^2=6$. Then, as we have seen, the equation can be written in the form $A\mathbf{x}\cdot\mathbf{x}=6$, where $A=\begin{pmatrix} 1 & -2 \\ -2 & 3 \end{pmatrix}$. In Example 7.4.1 on page 303 we saw that A can be diagonalized to $D=\begin{pmatrix} 2-\sqrt{5} & 0 \\ 0 & 2+\sqrt{5} \end{pmatrix}$ by using the orthogonal matrix

$$Q=\frac{1}{\sqrt{10-2\sqrt{5}}}\begin{pmatrix} 2 & 1-\sqrt{5} \\ -1+\sqrt{5} & 2 \end{pmatrix}$$

Then $$\mathbf{x}'=\begin{pmatrix} x' \\ y' \end{pmatrix}=Q^t\mathbf{x}=\frac{1}{\sqrt{10-2\sqrt{5}}}\begin{pmatrix} 2 & -1+\sqrt{5} \\ 1-\sqrt{5} & 2 \end{pmatrix}\begin{pmatrix} x \\ y \end{pmatrix}$$
$$=\frac{1}{\sqrt{10-2\sqrt{5}}}\begin{pmatrix} 2x+(-1+\sqrt{5})y \\ (1-\sqrt{5})x+2y \end{pmatrix}$$

and in the new variables the equation can be written as

$$(2-\sqrt{5})x'^2+(2+\sqrt{5})y'^2=6.$$

Let us take another look at the matrix Q. Since Q is real and orthogonal, $1=\det QQ^{-1}=\det QQ^t=\det Q\det Q^t=\det Q\det Q=(\det Q)^2$. Thus $\det Q=\pm 1$. If $\det Q=-1$, we can interchange the rows of Q to make the determinant of this new Q equal to 1. Then it can be shown (see Problem 36) that $Q=\begin{pmatrix} \cos\theta & -\sin\theta \\ \sin\theta & \cos\theta \end{pmatrix}$ for some number θ with $0\le\theta<2\pi$. But, from Example 6.1.8 on page 233, this means that Q is a rotation matrix. We have therefore proved the following theorem.

THEOREM 1 **Principal Axes Theorem in $\mathbb{R}^2$.** Let $ax^2+bxy+cy^2=d$ (10)

be a quadratic equation in the variables x and y. Then there exists a unique number θ in $[0, 2\pi)$ such that Equation (10) can be written in the form

$$a'x'^2+c'y'^2=d \tag{11}$$

where x', y' are the axes obtained by rotating the x- and y-axes through an angle of θ in the counterclockwise direction. Moreover, the numbers a' and c' are the eigenvalues of the matrix $A = \begin{pmatrix} a & b/2 \\ b/2 & c \end{pmatrix}$. The x'- and y'-axes are called the **principal axes** of the graph of the quadratic equation (10).

We can use Theorem 1 to identify three important conic sections. Recall that the **standard equations** of a circle, ellipse, and hyperbola are:

Circle: $$x^2 + y^2 = r^2 \tag{12}$$

Ellipse: $$\frac{x^2}{a^2} + \frac{y^2}{b^2} = 1 \tag{13}$$

Hyperbola: $$\frac{x^2}{a^2} - \frac{y^2}{b^2} = 1 \tag{14}$$

or

$$\frac{y^2}{a^2} - \frac{x^2}{b^2} = 1 \tag{15}$$

EXAMPLE 2 Identify the conic section whose equation is

$$x^2 - 4xy + 3y^2 = 6 \tag{16}$$

Solution In Example 1 we found that this can be written as $(2-\sqrt{5})x'^2 + (2+\sqrt{5})y'^2 = 6$ or

$$\frac{y'^2}{6/(2+\sqrt{5})} - \frac{x'^2}{6/(\sqrt{5}-2)} = 1$$

This is Equation (15) with $a = \sqrt{6/(2+\sqrt{5})} \approx 1.19$ and $b = \sqrt{6/(\sqrt{5}-2)} \approx 5.04$. Since

$$Q = \frac{1}{\sqrt{10-2\sqrt{5}}} \begin{pmatrix} 2 & 1-\sqrt{5} \\ -1+\sqrt{5} & 2 \end{pmatrix}$$

and $\det Q = 1$, we have, using Problem 36 and the fact that 2 and $-1+\sqrt{5}$ are positive,

$$\cos\theta = \frac{2}{\sqrt{10-2\sqrt{5}}} \approx 0.85065$$

Thus θ is in the first quadrant and, using a table (or a calculator), we find that $\theta \approx 0.5536$ rad $\approx 31.7°$. Thus (16) is the equation of a standard hyperbola rotated through an angle of 31.7° (see Figure 7.1).

Figure 7.1

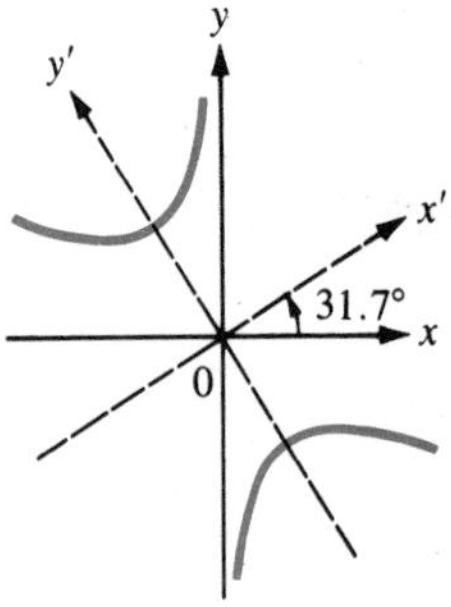

$x^2 - 4xy + 3y^2 = 6$

EXAMPLE 3 Identify the conic section whose equation is

$$5x^2 - 2xy + 5y^2 = 4 \tag{17}$$

Solution Here $A = \begin{pmatrix} 5 & -1 \\ -1 & 5 \end{pmatrix}$, the eigenvalues of A are $\lambda_1 = 4$ and $\lambda_2 = 6$, and two orthonormal eigenvectors are $\mathbf{v}_1 = \begin{pmatrix} 1/\sqrt{2} \\ 1/\sqrt{2} \end{pmatrix}$ and $\mathbf{v}_2 = \begin{pmatrix} 1/\sqrt{2} \\ -1/\sqrt{2} \end{pmatrix}$. Then $Q = \begin{pmatrix} 1/\sqrt{2} & 1/\sqrt{2} \\ 1/\sqrt{2} & -1/\sqrt{2} \end{pmatrix}$. Before continuing, we note that $\det Q = -1$. For Q to be a rotation matrix, we need $\det Q = 1$. This is easily accomplished by reversing the eigenvectors. Thus we set $\lambda_1 = 6$, $\lambda_2 = 4$, $\mathbf{v}_1 = \begin{pmatrix} 1/\sqrt{2} \\ -1/\sqrt{2} \end{pmatrix}$, $\mathbf{v}_2 = \begin{pmatrix} 1/\sqrt{2} \\ 1/\sqrt{2} \end{pmatrix}$, and $Q = \begin{pmatrix} 1/\sqrt{2} & 1/\sqrt{2} \\ -1/\sqrt{2} & 1/\sqrt{2} \end{pmatrix}$; now $\det Q = 1$. Then $D = \begin{pmatrix} 6 & 0 \\ 0 & 4 \end{pmatrix}$ and (17) can be written as $D\mathbf{v} \cdot \mathbf{v} = 4$ or

$$6x'^2 + 4y'^2 = 4 \tag{18}$$

where $$\begin{pmatrix} x' \\ y' \end{pmatrix} = Q^t \begin{pmatrix} x \\ y \end{pmatrix} = \begin{pmatrix} 1/\sqrt{2} & -1/\sqrt{2} \\ 1/\sqrt{2} & 1/\sqrt{2} \end{pmatrix} \begin{pmatrix} x \\ y \end{pmatrix} = \begin{pmatrix} 1/\sqrt{2}x - 1/\sqrt{2}y \\ 1/\sqrt{2}x + 1/\sqrt{2}y \end{pmatrix}$$

Rewriting (18), we obtain $x'^2/(\frac{4}{6}) + y'^2/1 = 1$, which is Equation (13) with $a = \sqrt{\frac{2}{3}}$ and $b = 1$. Moreover, since $1/\sqrt{2} > 0$ and $-1/\sqrt{2} < 0$, we have, from Problem 36, $\theta = 2\pi - \cos^{-1}(1/\sqrt{2}) = 2\pi - \pi/4 = 7\pi/4 = 315°$. Thus (17) is the equation of a standard ellipse rotated through an angle of 315° (or 45° in the clockwise direction). (See Figure 7.2.)

Figure 7.2

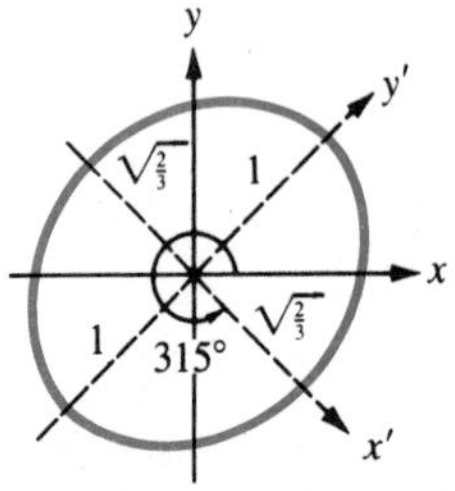

$5x^2 - 2xy + 5y^2 = 4$

EXAMPLE 4 Identify the conic section whose equation is

$$-5x^2+2xy-5y^2=4 \tag{19}$$

Solution Referring to Example 3, Equation (19) can be rewritten as

$$-6x'^2-4y'^2=4 \tag{20}$$

Since, for any real numbers x' and y', $-6x'^2-4y'^2 \le 0$, we see that there are no real numbers x and y which satisfy (19). The conic section defined by (19) is called a **degenerate conic section**.

There is an easy way to identify the conic section defined by

$$ax^2+bxy+cy^2=d \tag{21}$$

If $A=\begin{pmatrix} a & b/2 \\ b/2 & c \end{pmatrix}$, then the characteristic equation of A is

$$\lambda^2-(a+c)\lambda+(ac-b^2/4)=0=(\lambda-\lambda_1)(\lambda-\lambda_2)$$

This means that $\lambda_1\lambda_2=ac-b^2/4$. But Equation (21) can, as we have seen, be rewritten as

$$\lambda_1 x'^2+\lambda_2 y'^2=d \tag{22}$$

If λ_1 and λ_2 have the same sign, then (21) defines an ellipse (or a circle) or a degenerate conic as in Examples 3 and 4. If λ_1 and λ_2 have opposite signs, then (21) is the equation of a hyperbola (as in Example 2). We can therefore prove the following.

THEOREM 2 If $A=\begin{pmatrix} a & b/2 \\ b/2 & c \end{pmatrix}$, then the quadratic equation (21) with $d \neq 0$ is the equation of:

i. A hyperbola if $\det A<0$.
ii. An ellipse, circle, or degenerate conic section if $\det A>0$.
iii. A pair of straight lines or a degenerate conic section if $\det A=0$.
iv. If $d=0$, then (21) is the equation of two straight lines if $\det A \neq 0$ and the equation of a single line if $\det A=0$.

Proof We have already shown why (*i*) and (*ii*) are true. To prove part (*iii*), suppose that $\det A=0$. Then, by our Summing Up Theorem (Theorem 7.1.6), $\lambda=0$ is an eigenvalue of A and Equation (22) reads $\lambda_1 x'^2=d$ or $\lambda_2 y'^2=d$. If $\lambda_1 x'^2=d$ and $d/\lambda_1>0$, then $x_1'=\pm\sqrt{d/\lambda_1}$ is the equation of two straight lines in the xy-plane. If $d/\lambda_1<0$, then we have $x'^2<0$ (which is impossible) and we obtain a degenerate conic. The same facts hold if $\lambda_2 y'^2=d$. Part (*iv*) is left as an exercise (see Problem 37). ■

Note. In Example 2 we had $\det A = ac - b^2/4 = -1$. In Examples 3 and 4 we had $\det A = 24$.

The methods described above can be used to analyze quadratic equations in more than two variables. We give one example below.

EXAMPLE 5 Consider the quadratic equation

$$5x^2 + 8xy + 5y^2 + 4xz + 4yz + 2z^2 = 100 \tag{23}$$

If $A = \begin{pmatrix} 5 & 4 & 2 \\ 4 & 5 & 2 \\ 2 & 2 & 2 \end{pmatrix}$ and $\mathbf{v} = \begin{pmatrix} x \\ y \\ z \end{pmatrix}$, then (23) can be written in the form

$$A\mathbf{v} \cdot \mathbf{v} = 100 \tag{24}$$

From Example 7.4.2 on page 304, $Q^t AQ = D = \begin{pmatrix} 1 & 0 & 0 \\ 0 & 1 & 0 \\ 0 & 0 & 10 \end{pmatrix}$, where $Q = \begin{pmatrix} -1/\sqrt{2} & -1/3\sqrt{2} & 2/3 \\ 1/\sqrt{2} & -1/3\sqrt{2} & 2/3 \\ 0 & 4/3\sqrt{2} & 1/3 \end{pmatrix}$

Let

$$\mathbf{v}' = \begin{pmatrix} x' \\ y' \\ z' \end{pmatrix} = Q^t\mathbf{v} = \begin{pmatrix} -1/\sqrt{2} & 1/\sqrt{2} & 0 \\ -1/3\sqrt{2} & -1/3\sqrt{2} & 4/3\sqrt{2} \\ 2/3 & 2/3 & 1/3 \end{pmatrix} \begin{pmatrix} x \\ y \\ z \end{pmatrix}$$

$$= \begin{pmatrix} (-1/\sqrt{2})x + (1/\sqrt{2})y \\ -(1/3\sqrt{2})x - (1/3\sqrt{2})y + (4/3\sqrt{2})z \\ (2/3)x + (2/3)y + (1/3)z \end{pmatrix}$$

Then, as before, $A = QDQ^t$ and $A\mathbf{v} \cdot \mathbf{v} = QDQ^t\mathbf{v} \cdot \mathbf{v} = DQ^t\mathbf{v} \cdot Q^t\mathbf{v} = D\mathbf{v}' \cdot \mathbf{v}'$. Thus (24) can be written in the new variables x', y', z' as $D\mathbf{v}' \cdot \mathbf{v}' = 100$ or

$$x'^2 + y'^2 + 10z'^2 = 100 \tag{25}$$

In $\mathbb{R}^3$, the surface defined by (25) is called an **ellipsoid** (see Figure 7.3).

Figure 7.3

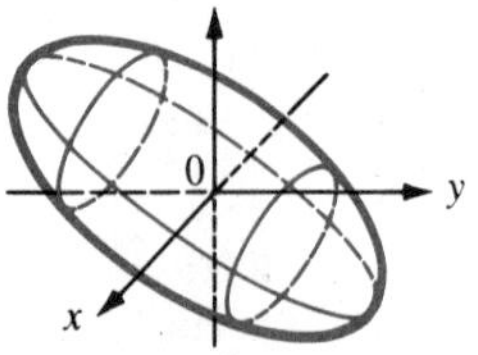

$x'^2 + y'^2 + 10z'^2 = 100$

There is a great variety of three-dimensional surfaces of the form $A\mathbf{v} \cdot \mathbf{v} = d$, where $\mathbf{v} \in \mathbb{R}^2$. Such surfaces are called **quadric surfaces**.

We close this section by noting that quadratic forms can be defined in any number of variables.

DEFINITION 2 **QUADRATIC FORM** Let $\mathbf{v}=\begin{pmatrix}x_1\\x_2\\\vdots\\x_n\end{pmatrix}$ and let A be a symmetric $n\times n$ matrix. Then a **quadratic form** in $x_1, x_2, \ldots, x_n$ is an expression of the form

$$F(x_1, x_2, \ldots, x_n)=A\mathbf{v}\cdot\mathbf{v} \tag{26}$$

EXAMPLE 6 Let

$$A=\begin{pmatrix}2&1&2&-2\\1&-4&6&5\\2&6&7&-1\\-2&5&-1&3\end{pmatrix}$$

Then

$$A\mathbf{v}\cdot\mathbf{v}=\left[\begin{pmatrix}2&1&2&-2\\1&-4&6&5\\2&6&7&-1\\-2&5&-1&3\end{pmatrix}\begin{pmatrix}x_1\\x_2\\x_3\\x_4\end{pmatrix}\right]\cdot\begin{pmatrix}x_1\\x_2\\x_3\\x_4\end{pmatrix}$$

$$=\begin{pmatrix}2x_1+x_2+2x_3-2x_4\\x_1-4x_2+6x_3+5x_4\\2x_1+6x_2+7x_3-x_4\\-2x_1+5x_2-x_3+3x_4\end{pmatrix}\cdot\begin{pmatrix}x_1\\x_2\\x_3\\x_4\end{pmatrix}$$

$$=2x_1^2+2x_1x_2-4x_2^2+4x_1x_3+12x_2x_3$$
$$+7x_3^2-4x_1x_4+10x_2x_4-2x_3x_4+3x_4^2$$

(after simplification)

EXAMPLE 7 Find the symmetric matrix A corresponding to the quadratic form

$$5x_1^2-3x_1x_2+4x_2^2+8x_1x_3-9x_2x_3+2x_3^2-x_1x_4+7x_2x_4+6x_3x_4+9x_4^2$$

Solution If $A=(a_{ij})$, then by looking at the earlier examples in this section we see that a_{ii} is the coefficient of the x_i^2 term and $a_{ij}+a_{ji}$ is the coefficient of the x_ix_j term. Since A is symmetric, $a_{ij}=a_{ji}$; hence $a_{ij}=a_{ji}=\frac{1}{2}\cdot$ (coefficient of x_ix_j term). Putting this all together, we obtain

$$A=\begin{pmatrix}5&-\frac{3}{2}&4&-\frac{1}{2}\\-\frac{3}{2}&4&-\frac{9}{2}&\frac{7}{2}\\4&-\frac{9}{2}&2&3\\-\frac{1}{2}&\frac{7}{2}&3&9\end{pmatrix}.$$

PROBLEMS 7.5 In Problems 1–13 write the quadratic equation in the form $A\mathbf{v}\cdot\mathbf{v}=d$ (where A is a symmetric matrix) and eliminate the xy-term by rotating the axes through an angle of θ. Write the equation in terms of the new variables and identify the conic section obtained.

1. $3x^2-2xy-5=0$ **2.** $4x^2+4xy+y^2=9$ **3.** $4x^2+4xy-y^2=9$
4. $xy=1$ **5.** $xy=a;\ a>0$ **6.** $4x^2+2xy+3y^2+2=0$
7. $xy=a;\ a<0$ **8.** $x^2+4xy+4y^2-6=0$ **9.** $-x^2+2xy-y^2=0$
10. $2x^2+xy+y^2=4$ **11.** $3x^2-6xy+5y^2=36$ **12.** $x^2-3xy+4y^2=1$
13. $6x^2+5xy-6y^2+7=0$
14. What are the possible forms of the graph of $ax^2+bxy+cy^2=0$?

In Problems 15–18 write the quadratic form in new variables x', y', and z' so that no cross-product terms (xy, xz, yz) are present.

15. $x^2-2xy+y^2-2xz-2yz+z^2$ **16.** $-x^2+4xy-y^2+4xz+4yz+z^2$
17. $3x^2+4xy+2y^2+4xz+4z^2$ **18.** $x^2-2xy+2y^2-2yz+z^2$

In Problems 19–21 find a symmetric matrix A such that the quadratic form can be written in the form $A\mathbf{x}\cdot\mathbf{x}$.

19. $x_1^2+2x_1x_2+x_2^2+4x_1x_3+6x_2x_3+3x_3^2+7x_1x_4-2x_2x_4+x_4^2$
20. $x_1^2-x_2^2+x_1x_3-x_2x_4+x_4^2$
21. $3x_1^2-7x_1x_2-2x_2^2+x_1x_3-x_2x_3+3x_3^2-2x_1x_4+x_2x_4-4x_3x_4-6x_4^2+3x_1x_5-5x_3x_5+x_4x_5-x_5^2$
22. Suppose that for some nonzero value of d, the graph of $ax^2+bxy+cy^2=d$ is a hyperbola. Show that the graph is a hyperbola for any other nonzero value of d.
23. Show that if $a\neq c$, the xy-term in quadratic equation (1) will be eliminated by rotation through an angle θ if θ is given by $\cot 2\theta=(a-c)/b$.
24. Show that if $a=c$ in Problem 23, then the xy-term will be eliminated by a rotation through an angle of either $\pi/4$ or $-\pi/4$.
***25.** Suppose that a rotation converts $ax^2+bxy+cy^2$ into $a'(x')^2+b'(x'y')+c'(y')^2$. Show that:
a. $a+c=a'+c'$ **b.** $b^2-4ac=b'^2-4a'c'$
***26.** A quadratic form $F(\mathbf{x})=F(x_1,x_2,\ldots,x_n)$ is said to be **positive definite** if $F(\mathbf{x})\geq 0$ for every $\mathbf{x}\in\mathbb{R}^n$ and $F(\mathbf{x})=0$ if and only if $\mathbf{x}=\mathbf{0}$. Show that F is positive definite if and only if the symmetric matrix A associated with F has positive eigenvalues.
27. A quadratic form $F(\mathbf{x})$ is said to be **positive semidefinite** if $F(\mathbf{x})\geq 0$ for every $\mathbf{x}\in\mathbb{R}^n$. Show that F is positive semidefinite if and only if the eigenvalues of the symmetric matrix associated with F are all nonnegative.

The definitions of **negative definite** and **negative semidefinite** are the definitions in Problems 26 and 27 with ≤ 0 replacing ≥ 0. A quadratic form is **indefinite** if it is none of the above. In Problems 28–35 determine whether the given quadratic form is positive definite, positive semidefinite, negative definite, negative semidefinite, or indefinite.

28. $3x^2+2y^2$ **29.** $-3x^2-3y^2$ **30.** $3x^2-2y^2$ **31.** $x^2+2xy+2y^2$
32. $x^2-2xy+2y^2$ **33.** $x^2-4xy+3y^2$ **34.** $-x^2+4xy-3y^2$ **35.** $-x^2+2xy-2y^2$

***36.** Let $Q=\begin{pmatrix} a & b \\ c & d\end{pmatrix}$ be a real orthogonal matrix with $\det Q=1$. Define the number $\theta\in[0,2\pi)$:

a. If $a\geq 0$ and $c>0$, then $\theta=\cos^{-1} a$ $(0<\theta\leq\pi/2)$.
b. If $a\geq 0$ and $c<0$, then $\theta=2\pi-\cos^{-1} a$ $(3\pi/2\leq\theta<2\pi)$.

c. If $a \leq 0$ and $c > 0$, then $\theta = \cos^{-1} a$ $(\pi/2 \leq \theta < \pi)$.
d. If $a \leq 0$ and $c < 0$, then $\theta = 2\pi - \cos^{-1} a$ $(\pi < \theta \leq 3\pi/2)$.
e. If $a = 1$ and $c = 0$, then $\theta = 0$.
f. If $a = -1$ and $c = 0$, then $\theta = \pi$.
(Here $\cos^{-1} x \in [0, \pi]$ for $x \in [-1, 1]$.) With θ chosen as above, show that

$$Q = \begin{pmatrix} \cos\theta & -\sin\theta \\ \sin\theta & \cos\theta \end{pmatrix}.$$

37. Prove, using formula (22), that Equation (21) is the equation of two straight lines in the xy-plane when $d = 0$ and $\det A \neq 0$. If $\det A = d = 0$, show that Equation (21) is the equation of a single line.

38. Let A be the symmetric matrix representation of quadratic equation (1) with $d \neq 0$. Let λ_1 and λ_2 be the eigenvalues of A. Show that (1) is the equation of: **(a)** A hyperbola if $\lambda_1\lambda_2 < 0$ and **(b)** a circle, ellipse, or degenerate conic section if $\lambda_1\lambda_2 > 0$.

7.6 Jordan Canonical Form

As we have seen, $n \times n$ matrices with n linearly independent eigenvectors can be brought into an especially nice form by a similarity transformation. Fortunately, as "most" polynomials have distinct roots, "most" matrices will have distinct eigenvalues. As we shall see in Section 7.7, however, matrices that are not diagonalizable (that is, do not have n linearly independent eigenvectors) do arise in applications. In this case it is still possible to show that the matrix is similar to another, simpler matrix, but the new matrix is not diagonal and the transforming matrix C is harder to obtain.

To discuss this case fully, we define the matrix N_k to be the $k \times k$ matrix

$$N_k = \begin{pmatrix} 0 & 1 & 0 & \cdots & 0 \\ 0 & 0 & 1 & \cdots & 0 \\ \cdot & \cdot & \cdot & & \cdot \\ \cdot & \cdot & \cdot & & \cdot \\ \cdot & \cdot & \cdot & & \cdot \\ 0 & 0 & 0 & \cdots & 1 \\ 0 & 0 & 0 & \cdots & 0 \end{pmatrix} \tag{1}$$

Note that N_k is the matrix with 1's above the main diagonal and 0's everywhere else. We next define a $k \times k$ **Jordan† block matrix** $B(\lambda)$ by

$$B(\lambda) = \lambda I + N_k = \begin{pmatrix} \lambda & 1 & 0 & \cdots & 0 & 0 \\ 0 & \lambda & 1 & \cdots & 0 & 0 \\ \cdot & \cdot & \cdot & & \cdot & \cdot \\ \cdot & \cdot & \cdot & & \cdot & \cdot \\ \cdot & \cdot & \cdot & & \cdot & \cdot \\ 0 & 0 & & \cdots & \lambda & 1 \\ 0 & 0 & & \cdots & 0 & \lambda \end{pmatrix} \tag{2}$$

† Named for the French mathematician Camille Jordan (1838–1922). The results in this section first appeared in Jordan's brilliant *Traité des substitutions et des équations algebriques* (Treatise on substitutions and algebraic equations), which was published in 1870.

That is, $B(\lambda)$ is the $k \times k$ matrix with the fixed number λ on the diagonal, 1's above the diagonal, and 0's everywhere else.

Note. We can (and often will) have a 1×1 Jordan block matrix. Such a matrix takes the form $B(\lambda) = (\lambda)$.

Finally, a **Jordan matrix** J has the form

$$J = \begin{pmatrix} B_1(\lambda_1) & 0 & \cdots & 0 \\ 0 & B_2(\lambda_2) & \cdots & 0 \\ \vdots & \vdots & & \vdots \\ 0 & 0 & \cdots & B_r(\lambda_r) \end{pmatrix}$$

where each $B_j(\lambda_j)$ is a Jordan block matrix. Thus *a Jordan matrix is a matrix with Jordan block matrices down the diagonal and zeros everywhere else.*

EXAMPLE 1 The following are examples of Jordan matrices. The Jordan blocks are outlined by the dotted lines.

i. $\begin{pmatrix} 2 & 1 & 0 \\ 0 & 2 & 0 \\ 0 & 0 & 4 \end{pmatrix}$ **ii.** $\begin{pmatrix} -3 & 0 & 0 & 0 & 0 \\ 0 & -3 & 1 & 0 & 0 \\ 0 & 0 & -3 & 1 & 0 \\ 0 & 0 & 0 & -3 & 0 \\ 0 & 0 & 0 & 0 & 7 \end{pmatrix}$

iii. $\begin{pmatrix} 4 & 1 & 0 & 0 & 0 & 0 & 0 \\ 0 & 4 & 0 & 0 & 0 & 0 & 0 \\ 0 & 0 & 3 & 1 & 0 & 0 & 0 \\ 0 & 0 & 0 & 3 & 1 & 0 & 0 \\ 0 & 0 & 0 & 0 & 3 & 0 & 0 \\ 0 & 0 & 0 & 0 & 0 & 5 & 1 \\ 0 & 0 & 0 & 0 & 0 & 0 & 5 \end{pmatrix}$

EXAMPLE 2 The only 2×2 Jordan matrices are $\begin{pmatrix} \lambda_1 & 0 \\ 0 & \lambda_2 \end{pmatrix}$ and $\begin{pmatrix} \lambda & 1 \\ 0 & \lambda \end{pmatrix}$.

EXAMPLE 3 The only 3×3 Jordan matrices are

$$\begin{pmatrix} \lambda_1 & 0 & 0 \\ 0 & \lambda_2 & 0 \\ 0 & 0 & \lambda_3 \end{pmatrix} \quad \begin{pmatrix} \lambda_1 & 0 & 0 \\ 0 & \lambda_2 & 1 \\ 0 & 0 & \lambda_2 \end{pmatrix} \quad \begin{pmatrix} \lambda_1 & 1 & 0 \\ 0 & \lambda_1 & 0 \\ 0 & 0 & \lambda_2 \end{pmatrix} \quad \begin{pmatrix} \lambda_1 & 1 & 0 \\ 0 & \lambda_1 & 1 \\ 0 & 0 & \lambda_1 \end{pmatrix}$$

where λ_1, λ_2, and λ_3 are not necessarily distinct.

The following result is one of the most important theorems in matrix theory. Although its proof is beyond the scope of this book,† we shall prove this theorem in the 2×2 case (see Theorem 3) and suggest a proof for the 3×3 case in Problem 19.

THEOREM 1 Let A be an $n\times n$ matrix. Then there exists an invertible $n\times n$ matrix C such that

$$C^{-1}AC = J \tag{3}$$

where J is a Jordan matrix whose diagonal elements are the eigenvalues of A. Moreover, J is unique except for the order in which the Jordan blocks appear.

Remark 1. By the last sentence of the theorem we mean, for example, that if A is similar to

$$J_1 = \begin{pmatrix} 2 & 1 & 0 & 0 & 0 & 0 \\ 0 & 2 & 0 & 0 & 0 & 0 \\ 0 & 0 & 3 & 1 & 0 & 0 \\ 0 & 0 & 0 & 3 & 1 & 0 \\ 0 & 0 & 0 & 0 & 3 & 0 \\ 0 & 0 & 0 & 0 & 0 & 4 \end{pmatrix}$$

then A is also similar to

$$J_2 = \begin{pmatrix} 3 & 1 & 0 & 0 & 0 & 0 \\ 0 & 3 & 1 & 0 & 0 & 0 \\ 0 & 0 & 3 & 0 & 0 & 0 \\ 0 & 0 & 0 & 4 & 0 & 0 \\ 0 & 0 & 0 & 0 & 2 & 1 \\ 0 & 0 & 0 & 0 & 0 & 2 \end{pmatrix} \quad \text{and} \quad J_3 = \begin{pmatrix} 4 & 0 & 0 & 0 & 0 & 0 \\ 0 & 2 & 1 & 0 & 0 & 0 \\ 0 & 0 & 2 & 0 & 0 & 0 \\ 0 & 0 & 0 & 3 & 1 & 0 \\ 0 & 0 & 0 & 0 & 3 & 1 \\ 0 & 0 & 0 & 0 & 0 & 3 \end{pmatrix}$$

† For a proof see G. Birkhoff and S. MacLane, *A Survey of Modern Algebra*, (New York: Macmillan, 1953), p. 334.

and three other Jordan matrices. That is, the actual Jordan blocks remain the same but we can change the order in which they are written.

DEFINITION 1 **JORDAN CANONICAL FORM** The matrix J is called the **Jordan canonical form** of A.

Remark 2. If A is diagonalizable, then $J = D = \text{diag}\,(\lambda_1, \lambda_2, \ldots, \lambda_n)$, where $\lambda_1, \lambda_2, \ldots, \lambda_n$ are the (not necessarily distinct) eigenvalues of A. Each diagonal component is a 1×1 Jordan block matrix.

We shall now see how to compute the Jordan canonical form of any 2×2 matrix. If A has two linearly independent eigenvectors, we already know what to do. Therefore the only case of interest occurs when A has a single eigenvalue λ of algebraic multiplicity 2 and geometric multiplicity 1. That is, we assume that, corresponding to λ, A has the single independent eigenvector $\mathbf{v}_1$. That is: *Any vector that is not a multiple of* $\mathbf{v}_1$ *is not an eigenvector.*

THEOREM 2 Let the 2×2 matrix A have an eigenvalue λ of algebraic multiplicity 2 and geometric multiplicity 1. Let $\mathbf{v}_1$ be an eigenvector corresponding to λ. Then there exists a vector $\mathbf{v}_2$ that satisfies the equation

$$(A - \lambda I)\mathbf{v}_2 = \mathbf{v}_1 \tag{4}$$

Proof Let $\mathbf{x} \in \mathbb{C}^2$ be a fixed vector that is *not* a multiple of $\mathbf{v}_1$ so that $\mathbf{x}$ is not an eigenvector of A. We first show that

$$\mathbf{w} = (A - \lambda I)\mathbf{x} \tag{5}$$

is an eigenvector of A. That is, we shall show that $\mathbf{w} = c\mathbf{v}_1$ for some constant c. Since $\mathbf{w} \in \mathbb{C}^2$ and $\mathbf{v}_1$ and $\mathbf{x}$ are linearly independent, there exist constants c_1 and c_2 such that

$$\mathbf{w} = c_1\mathbf{v}_1 + c_2\mathbf{x} \tag{6}$$

To show that $\mathbf{w}$ is an eigenvector of A, we must show that $c_2 = 0$. From (5) and (6) we find that

$$(A - \lambda I)\mathbf{x} = c_1\mathbf{v}_1 + c_2\mathbf{x} \tag{7}$$

Let $B = A - (\lambda + c_2)I$. Then, from (7),

$$B\mathbf{x} = [A - (\lambda + c_2)I]\mathbf{x} = c_1\mathbf{v}_1 \tag{8}$$

If we assume that $c_2 \neq 0$, then $\lambda + c_2 \neq \lambda$ and $\lambda + c_2$ is not an eigenvalue of A (since λ is the only eigenvalue of A). Thus $\det B = \det[A - (\lambda + c_2)I] \neq 0$, which means that B is invertible. Hence (8) can be written as

$$\mathbf{x} = B^{-1}c_1\mathbf{v}_1 = c_1B^{-1}\mathbf{v}_1 \tag{9}$$

Then, multiplying both sides of (9) by λ, we have

$$\lambda\mathbf{x} = \lambda c_1B^{-1}\mathbf{v}_1 = c_1B^{-1}\lambda\mathbf{v}_1 = c_1B^{-1}A\mathbf{v}_1 \tag{10}$$

But $B = A - (\lambda + c_2)I$, so

$$A = B + (\lambda + c_2)I \tag{11}$$

Inserting (11) into (10) we have

$$\begin{aligned} \lambda \mathbf{x} &= c_1 B^{-1}[B + (\lambda + c_2)I]\mathbf{v}_1 \\ &= c_1[I + (\lambda + c_2)B^{-1}]\mathbf{v}_1 \\ &= c_1\mathbf{v}_1 + (\lambda + c_2)c_1 B^{-1}\mathbf{v}_1 \end{aligned} \tag{12}$$

But, again using (8), $c_1 B^{-1}\mathbf{v}_1 = \mathbf{x}$ so that (12) becomes

$$\lambda \mathbf{x} = c_1\mathbf{v}_1 + (\lambda + c_2)\mathbf{x} = c_1\mathbf{v}_1 + c_2\mathbf{x} + \lambda\mathbf{x}$$

or

$$\mathbf{0} = c_1\mathbf{v}_1 + c_2\mathbf{x} \tag{13}$$

But $\mathbf{v}_1$ and $\mathbf{x}$ are linearly independent, so $c_1 = c_2 = 0$. This contradicts the assumption that $c_2 \neq 0$. Thus $c_2 = 0$ and, by (6), $\mathbf{w}$ is a multiple of $\mathbf{v}_1$ so that $\mathbf{w} = c_1\mathbf{v}_1$ is an eigenvector of A. Moreover, $\mathbf{w} \neq \mathbf{0}$ since if $\mathbf{w} = \mathbf{0}$, then (5) tells us that $\mathbf{x}$ is an eigenvector of A. Therefore $c_1 \neq 0$. Let

$$\mathbf{v}_2 = \frac{1}{c_1}\mathbf{x} \tag{14}$$

Then $(A - \lambda I)\mathbf{v}_2 = (1/c_1)(A - \lambda I)\mathbf{x} = (1/c_1)\mathbf{w} = \mathbf{v}_1$. This proves the theorem. ■

DEFINITION 2 **GENERALIZED EIGENVECTOR** Let A be a 2×2 matrix with the single eigenvalue λ having geometric multiplicity 1. Let $\mathbf{v}_1$ be an eigenvalue of A. Then the vector $\mathbf{v}_2$ defined by $(A - \lambda I)\mathbf{v}_2 = \mathbf{v}_1$ is called a **generalized eigenvector** of A corresponding to the eigenvalue λ.

EXAMPLE 4 Let $A = \begin{pmatrix} 3 & -2 \\ 8 & -5 \end{pmatrix}$. The characteristic equation of A is $\lambda^2 + 2\lambda + 1 = (\lambda + 1)^2 = 0$, so $\lambda = -1$ is an eigenvalue of algebraic multiplicity 2. Then

$$(A - \lambda I)\mathbf{v} = (A + I)\mathbf{v} = \begin{pmatrix} 4 & -2 \\ 8 & -4 \end{pmatrix}\begin{pmatrix} x_1 \\ x_2 \end{pmatrix} = \begin{pmatrix} 0 \\ 0 \end{pmatrix}$$

This yields the eigenvector $\mathbf{v}_1 = \begin{pmatrix} 1 \\ 2 \end{pmatrix}$. There is no other linearly independent eigenvector. To find a generalized eigenvector $\mathbf{v}_2$, we compute $(A + I)\mathbf{v}_2 = \mathbf{v}_1$ or $\begin{pmatrix} 4 & -2 \\ 8 & -4 \end{pmatrix}\begin{pmatrix} x_1 \\ x_2 \end{pmatrix} = \begin{pmatrix} 1 \\ 2 \end{pmatrix}$, which yields the system

$$\begin{aligned} 4x_1 - 2x_2 &= 1 \\ 8x_1 - 4x_2 &= 2 \end{aligned}$$

The second equation is double the first, so x_2 can be chosen arbitrarily and $x_1 = (1 + 2x_2)/4$. Therefore a possible choice for $\mathbf{v}_2$ is $\mathbf{v}_2 = \begin{pmatrix} \frac{1}{4} \\ 0 \end{pmatrix}$.

The reason for finding generalized eigenvectors is given in the following theorem.

THEOREM 3 Let A, λ, $\mathbf{v}_1$, and $\mathbf{v}_2$ be as in Theorem 2 and let C be the matrix whose columns are $\mathbf{v}_1$ and $\mathbf{v}_2$. Then $C^{-1}AC=J$, where $J=\begin{pmatrix}\lambda & 1\\ 0 & \lambda\end{pmatrix}$ is the Jordan canonical form of A.

Proof Since $\mathbf{v}_1$ and $\mathbf{v}_2$ are linearly independent, we see that C is invertible. Next note that $AC=A(\mathbf{v}_1,\mathbf{v}_2)=(A\mathbf{v}_1, A\mathbf{v}_2)=(\lambda\mathbf{v}_1, A\mathbf{v}_2)$. But from Equation (4), $A\mathbf{v}_2=\mathbf{v}_1+\lambda\mathbf{v}_2$ so that $AC=(\lambda\mathbf{v}_1, \mathbf{v}_1+\lambda\mathbf{v}_2)$. But $CJ=(\mathbf{v}_1,\mathbf{v}_2)\begin{pmatrix}\lambda & 1\\ 0 & \lambda\end{pmatrix}=(\lambda\mathbf{v}_1, \mathbf{v}_1+\lambda\mathbf{v}_2)$. Thus $AC=CJ$, which means that $C^{-1}AC=J$ and the theorem is proved. ■

EXAMPLE 5 In Example 4, $\mathbf{v}_1=\begin{pmatrix}1\\2\end{pmatrix}$ and $\mathbf{v}_2=\begin{pmatrix}\frac14\\0\end{pmatrix}$. Then $C=\begin{pmatrix}1 & \frac14\\ 2 & 0\end{pmatrix}$, $C^{-1}=-2\begin{pmatrix}0 & -\frac14\\ -2 & 1\end{pmatrix}=\begin{pmatrix}0 & \frac12\\ 4 & -2\end{pmatrix}$, and

$$C^{-1}AC=\begin{pmatrix}0 & \frac12\\ 4 & -2\end{pmatrix}\begin{pmatrix}3 & -2\\ 8 & -5\end{pmatrix}\begin{pmatrix}1 & \frac14\\ 2 & 0\end{pmatrix}$$

$$=\begin{pmatrix}0 & \frac12\\ 4 & -2\end{pmatrix}\begin{pmatrix}-1 & \frac34\\ -2 & 2\end{pmatrix}=\begin{pmatrix}-1 & 1\\ 0 & -1\end{pmatrix}=J$$

The method described above can be generalized to obtain the Jordan canonical form of every matrix. We shall not do this, but one generalization is suggested in Problem 19. Although we shall not prove this fact, it is always possible to determine the number of 1's above the diagonal in the Jordan canonical form of an $n\times n$ matrix A. Let λ_i be an eigenvalue of A with algebraic multiplicity r_i and geometric multiplicity s_i. If $\lambda_1, \lambda_2, \ldots, \lambda_k$ are the eigenvalues of A, then:

Number of 1's above the diagonal of the Jordan canonical form of A

$$=(r_1-s_1)+(r_2-s_2)+\cdots+(r_k-s_k) \tag{15}$$

$$=\sum_{i=1}^{k} r_i-\sum_{i=1}^{k} s_i=n-\sum_{i=1}^{k} s_i$$

If we know the characteristic equation of a matrix A, then we can determine the possible Jordan canonical forms of A.

EXAMPLE 6 If the characteristic equation of A is $(\lambda-2)^3(\lambda+3)$, then the possible Jordan canonical forms for A are

$$J=\begin{pmatrix}2&0&0&0\\0&2&0&0\\0&0&2&0\\0&0&0&-3\end{pmatrix},\begin{pmatrix}2&1&0&0\\0&2&0&0\\0&0&2&0\\0&0&0&-3\end{pmatrix},\begin{pmatrix}2&1&0&0\\0&2&1&0\\0&0&2&0\\0&0&0&-3\end{pmatrix}$$

or any matrix obtained by rearranging the Jordan blocks in J. The first matrix corresponds to a geometric multiplicity of 3 (for $\lambda=2$); the second corresponds to a geometric multiplicity of 2; and the third corresponds to a geometric multiplicity of 1.

PROBLEMS 7.6

In Problems 1–14 determine whether the given matrix is a Jordan matrix.

1. $\begin{pmatrix}1&1\\0&-6\end{pmatrix}$
2. $\begin{pmatrix}1&0\\0&0\end{pmatrix}$
3. $\begin{pmatrix}1&2\\0&1\end{pmatrix}$
4. $\begin{pmatrix}1&0&0\\0&3&1\\0&0&3\end{pmatrix}$
5. $\begin{pmatrix}3&1&0\\0&3&1\\0&0&3\end{pmatrix}$
6. $\begin{pmatrix}3&1&0\\0&3&1\\0&0&2\end{pmatrix}$
7. $\begin{pmatrix}1&0&0\\0&3&1\\0&0&4\end{pmatrix}$
8. $\begin{pmatrix}1&1&0\\0&3&1\\0&0&3\end{pmatrix}$
9. $\begin{pmatrix}1&1&0\\0&1&1\\0&0&1\end{pmatrix}$
10. $\begin{pmatrix}1&0&0&0&0\\0&2&1&0&0\\0&0&2&1&0\\0&0&0&2&0\\0&0&0&0&2\end{pmatrix}$
11. $\begin{pmatrix}1&0&0&0&0\\0&1&2&0&0\\0&0&1&2&0\\0&0&0&1&0\\0&0&0&0&1\end{pmatrix}$
12. $\begin{pmatrix}2&0&0&0&0\\0&3&1&0&0\\0&0&3&0&0\\0&0&0&5&1\\0&0&0&0&5\end{pmatrix}$
13. $\begin{pmatrix}a&0&0&0&0\\0&b&0&0&0\\0&0&c&0&0\\0&0&0&d&0\\0&0&0&0&e\end{pmatrix}$
14. $\begin{pmatrix}a&1&0&0&0\\0&a&0&0&0\\0&0&c&1&0\\0&0&0&c&1\\0&0&0&0&c\end{pmatrix}$

In Problems 15–18 find an invertible matrix C that transforms the 2×2 matrix to its Jordan canonical form.

15. $\begin{pmatrix}6&1\\0&6\end{pmatrix}$
16. $\begin{pmatrix}-12&7\\-7&2\end{pmatrix}$
17. $\begin{pmatrix}-10&-7\\7&4\end{pmatrix}$
18. $\begin{pmatrix}4&-1\\1&2\end{pmatrix}$

*19. Let A be a 3×3 matrix. Assume that λ is an eigenvalue of A with algebraic multiplicity 3 and geometric multiplicity 1 and let $\mathbf{v}_1$ be the corresponding eigenvector.
   a. Show that there is a solution, $\mathbf{v}_2$, to the system $(A-\lambda I)\mathbf{v}_2=\mathbf{v}_1$ such that $\mathbf{v}_1$ and $\mathbf{v}_2$ are linearly independent.
   b. With $\mathbf{v}_2$ defined by part (a), show that there is a solution, $\mathbf{v}_3$, to the system $(A-\lambda I)\mathbf{v}_3=\mathbf{v}_2$ such that $\mathbf{v}_1$, $\mathbf{v}_2$, and $\mathbf{v}_3$ are linearly independent.
   c. Show that if C is a matrix whose columns are $\mathbf{v}_1$, $\mathbf{v}_2$, and $\mathbf{v}_3$, then

$$C^{-1}AC=\begin{pmatrix}\lambda&1&0\\0&\lambda&1\\0&0&\lambda\end{pmatrix}.$$

20. Apply the procedure described in Problem 19 to reduce the matrix $A = \begin{pmatrix} -2 & 1 & 0 \\ -2 & 1 & -1 \\ -1 & 1 & -2 \end{pmatrix}$ by a similarity transformation to its Jordan canonical form.

21. Do the same for $A = \begin{pmatrix} -1 & -2 & -1 \\ -1 & -1 & -1 \\ 2 & 3 & 2 \end{pmatrix}$.

22. Do the same for $A = \begin{pmatrix} -1 & -18 & -7 \\ 1 & -13 & -4 \\ -1 & 25 & 8 \end{pmatrix}$.

23. An $n \times n$ matrix A is **nilpotent** if there is an integer k such that $A^k = 0$. If k is the smallest such integer, then k is called the **index of nilpotency** of A. Prove that if k is the index of nilpotency of A and if $m \geq k$, then $A^m = 0$.

***24.** Let N_k be the matrix defined by Equation (1). Prove that N_k is nilpotent with index of nilpotency k.

25. Write down all possible 4×4 Jordan matrices.

In Problems 26–31 the characteristic polynomial of a matrix A is given. Write the possible Jordan canonical forms for A.

26. $(\lambda+1)^2(\lambda-2)^2$ **27.** $(\lambda-3)^3(\lambda+4)$ **28.** $(\lambda-3)^4$

29. $(\lambda-4)^3(\lambda+3)^2$ **30.** $(\lambda-6)(\lambda+7)^4$ **31.** $(\lambda+7)^5$

32. Using the Jordan canonical form, show that, for any $n \times n$ matrix A, $\det A = \lambda_1\lambda_2 \cdots \lambda_n$, where $\lambda_1, \lambda_2, \ldots, \lambda_n$ are the eigenvalues of A.

7.7 If Time Permits: Matrix Differential Equations

Let $x = f(t)$ represent some physical quantity such as the volume of a substance, the population of a certain species, the mass of a decaying radioactive substance, or the number of dollars invested in bonds. Then the growth of $f(t)$ is given by its derivative $f'(t) = dx/dt$. If $f(t)$ is growing at a constant rate, then $dx/dt = k$ and $x = kt + C$; that is, $x = f(t)$ is a straight line function.

It is often more interesting and more appropriate to consider the **relative rate of growth** defined by

$$\text{Relative rate of growth} = \frac{\text{actual size of growth}}{\text{size of } f(t)} = \frac{f'(t)}{f(t)} = \frac{x'(t)}{x(t)} \tag{1}$$

If the relative rate of growth is constant, then we have

$$\frac{x'(t)}{x(t)} = a \tag{2}$$

or

$$x'(t) = ax(t) \tag{3}$$

Equation (3) is called a **differential equation** because it is an equation involving a derivative. It is not difficult to prove that the only solutions to (3) are of the form

$$x(t) = ce^{at} \tag{4}$$

where c is an arbitrary constant. If, however, $x(t)$ represents some physical quantity, then it is the usual practice to specify an **initial value** $x_0 = x(0)$ of the quantity. Then, substituting $t = 0$ in (4), we have $x_0 = x(0) = ce^{a \cdot 0} = c$ or

$$x(t) = x_0 e^{at} \tag{5}$$

The function $x(t)$ given by (5) is the unique solution to (3) satisfying the initial condition $x(0) = x_0$.

Equation (3) arises in a number of interesting applications. Some of these are undoubtedly given in your calculus text—in the chapter introducing the exponential function. In this section we shall consider a generalization of Equation (3).

In the model discussed above, we seek one unknown function. It often occurs that there are several functions linked by several differential equations. Examples are given later in the section. Consider the following system of n differential equations in n unknown functions:

$$\begin{aligned}
x_1'(t) &= a_{11}x_1(t) + a_{12}x_2(t) + \cdots + a_{1n}x_n(t)\\
x_2'(t) &= a_{21}x_1(t) + a_{22}x_2(t) + \cdots + a_{2n}x_n(t)\\
&\ \ \vdots\\
x_n'(t) &= a_{n1}x_1(t) + a_{n2}x_2(t) + \cdots + a_{nn}x_n(t)
\end{aligned} \tag{6}$$

where the a_{ij}'s are real numbers. System (6) is called an $n \times n$ **first-order system of linear differential equations.** The term "first order" means that only first derivatives occur in the system.

Now, let

$$\mathbf{x}(t) = \begin{pmatrix} x_1(t) \\ x_2(t) \\ \vdots \\ x_n(t) \end{pmatrix}$$

Here $\mathbf{x}(t)$ is called a **vector function**. We define

$$\mathbf{x}'(t) = \begin{pmatrix} x_1'(t) \\ x_2'(t) \\ \vdots \\ x_n'(t) \end{pmatrix}$$

Then, if we define the $n \times n$ matrix

$$A = \begin{pmatrix} a_{11} & a_{12} & \cdots & a_{1n} \\ a_{21} & a_{22} & \cdots & a_{2n} \\ \vdots & \vdots & & \vdots \\ a_{n1} & a_{n2} & \cdots & a_{nn} \end{pmatrix}$$

system (6) can be written as

$$\mathbf{x}'(t) = A\mathbf{x}(t) \tag{7}$$

Note that Equation (7) is almost identical to Equation (3). The only difference is that now we have a vector function and a matrix whereas before we had a "scalar" function and a number (1×1 matrix).

To solve Equation (7), we might guess that a solution would have the form e^{At}. But what does e^{At} mean? We shall answer that question in a moment. First, let us recall the series expansion of the function e^t:

$$e^t = 1 + t + \frac{t^2}{2!} + \frac{t^3}{3!} + \frac{t^4}{4!} + \cdots \tag{8}$$

This series converges for every real number t. Then, for any real number a,

$$e^{at} = 1 + at + \frac{(at)^2}{2!} + \frac{(at)^3}{3!} + \frac{(at)^4}{4!} + \cdots \tag{9}$$

DEFINITION 1 **THE MATRIX e^A** Let A be an $n \times n$ matrix with real (or complex) entries. Then e^A is an $n \times n$ matrix defined by

$$e^A = I + A + \frac{A^2}{2!} + \frac{A^3}{3!} + \frac{A^4}{4!} + \cdots \tag{10}$$

Remark. It is not difficult to prove that the series of matrices in Equation (10) converges for every matrix A, but to do so would take us too far afield. We can, however, give an indication of why it is so. We first define $|A|_i$ to be the sum of the absolute values of the components in the ith row of A. We then define the **norm**† of A, denoted $|A|$, by

$$|A| = \max_{1 \le i \le n} |A|_i \tag{11}$$

It can be shown that

$$|AB| \le |A|\,|B| \tag{12}$$

and

$$|A + B| \le |A| + |B| \tag{13}$$

Then, using (12) and (13) in (10), we obtain

$$|e^A| \le 1 + |A| + \frac{|A|^2}{2!} + \frac{|A|^3}{3!} + \frac{|A|^4}{4!} + \cdots = e^{|A|}$$

Since $|A|$ is a real number, $e^{|A|}$ is finite. This shows that the series in (10) converges for any matrix A.

† This is called the *max-row sum norm of A.*

We shall now see the usefulness of the series in Equation (10).

THEOREM 1 For any constant vector $\mathbf{c}$, $\mathbf{x}(t)=e^{At}\mathbf{c}$ is a solution of (7). Moreover, the solution of (7) given by $\mathbf{x}(t)=e^{At}\mathbf{x}_0$ satisfies $\mathbf{x}(0)=\mathbf{x}_0$.

Proof We compute, using (10):

$$\mathbf{x}(t)=e^{At}\mathbf{c}=\left[I+At+A^2\frac{t^2}{2!}+A^3\frac{t^3}{3!}+\cdots\right]\mathbf{c} \tag{14}$$

But since A is a constant matrix, we have

$$\begin{aligned}\frac{d}{dt}A^k\frac{t^k}{k!}&=\frac{d}{dt}\frac{t^k}{k!}A^k=\frac{kt^{k-1}}{k!}A^k\\&=\frac{A^kt^{k-1}}{(k-1)!}=A\left[A^{k-1}\frac{t^{k-1}}{(k-1)!}\right]\end{aligned} \tag{15}$$

Then, combining (14) and (15), we obtain (since $\mathbf{c}$ is a constant vector)

$$\mathbf{x}'(t)=\frac{d}{dt}e^{At}\mathbf{c}=A\left[I+At+A^2\frac{t^2}{2!}+A^3\frac{t^3}{3!}+\cdots\right]\mathbf{c}=Ae^{At}\mathbf{c}=A\mathbf{x}(t)$$

Finally, since $e^{A\cdot 0}=e^0=I$, we have

$$\mathbf{x}(0)=e^{A\cdot 0}\mathbf{x}_0=I\mathbf{x}_0=\mathbf{x}_0. \quad \blacksquare$$

DEFINITION 2 **PRINCIPAL MATRIX SOLUTION** The matrix e^{At} is called the **principal matrix solution** of the system $\mathbf{x}'=A\mathbf{x}$.

A major (and obvious) problem remains: How do we compute e^{At} in a practical way? We begin with two examples.

EXAMPLE 1 Let $A=\begin{pmatrix}1&0&0\\0&2&0\\0&0&3\end{pmatrix}$. Then

$$A^2=\begin{pmatrix}1&0&0\\0&2^2&0\\0&0&3^2\end{pmatrix},\quad A^3=\begin{pmatrix}1&0&0\\0&2^3&0\\0&0&3^3\end{pmatrix},\ldots,\quad A^m=\begin{pmatrix}1&0&0\\0&2^m&0\\0&0&3^m\end{pmatrix}$$

and $$e^{At}=I+At+\frac{A^2t^2}{2!}+\frac{A^3t^3}{3!}+\cdots=\begin{pmatrix}1&0&0\\0&1&0\\0&0&1\end{pmatrix}+\begin{pmatrix}t&0&0\\0&2t&0\\0&0&3t\end{pmatrix}$$

$$+\begin{pmatrix}\frac{t^2}{2!}&0&0\\0&\frac{2^2t^2}{2!}&0\\0&0&\frac{3^2t^2}{2!}\end{pmatrix}+\begin{pmatrix}\frac{t^3}{3!}&0&0\\0&\frac{2^3t^3}{3!}&0\\0&0&\frac{3^3t^3}{3!}\end{pmatrix}+\cdots$$

$$= \begin{pmatrix} 1+t+\frac{t^2}{2!}+\frac{t^3}{3!}+\cdots & 0 & 0 \\ 0 & 1+(2t)+\frac{(2t)^2}{2!}+\frac{(2t)^3}{3!}+\cdots & 0 \\ 0 & 0 & 1+(3t)+\frac{(3t)^2}{2!}+\frac{(3t)^3}{3!}+\cdots \end{pmatrix}$$

$$= \begin{pmatrix} e^t & 0 & 0 \\ 0 & e^{2t} & 0 \\ 0 & 0 & e^{3t} \end{pmatrix}$$

EXAMPLE 2 Let $A = \begin{pmatrix} a & 1 \\ 0 & a \end{pmatrix}$. Then, as is easily verified,

$$A^2 = \begin{pmatrix} a^2 & 2a \\ 0 & a^2 \end{pmatrix}, A^3 = \begin{pmatrix} a^3 & 3a^2 \\ 0 & a^3 \end{pmatrix}, \ldots, A^m = \begin{pmatrix} a^m & ma^{m-1} \\ 0 & a^m \end{pmatrix}, \ldots$$

so that
$$e^{At} = \begin{pmatrix} \sum_{m=0}^{\infty} \frac{(at)^m}{m!} & \sum_{m=1}^{\infty} \frac{ma^{m-1}t^m}{m!} \\ 0 & \sum_{m=0}^{\infty} \frac{(at)^m}{m!} \end{pmatrix}$$

Now
$$\sum_{m=1}^{\infty} \frac{ma^{m-1}t^m}{m!} = \sum_{m=1}^{\infty} \frac{a^{m-1}t^m}{(m-1)!} = t + at^2 + \frac{a^2t^3}{2!} + \frac{a^3t^4}{3!} + \cdots$$
$$= t\left(1 + at + \frac{a^2t^2}{2!} + \frac{a^3t^3}{3!} + \cdots\right) = te^{at}$$

Thus
$$e^{At} = \begin{pmatrix} e^{at} & te^{at} \\ 0 & e^{at} \end{pmatrix}$$

As Example 1 illustrates, it is easy to calculate e^{At} if A is a diagonal matrix. Example 1 shows that if $D = \text{diag}(\lambda_1, \lambda_2, \ldots, \lambda_n)$, then $e^{Dt} = \text{diag}(e^{\lambda_1 t}, e^{\lambda_2 t}, \ldots, e^{\lambda_n t})$.

In Example 2, we calculated e^{At} for a matrix A in Jordan canonical form. It turns out that this is really all we need to be able to do, as the next theorem suggests.

THEOREM 2 Let J be the Jordan canonical form of a matrix A and let $J = C^{-1}AC$. Then $A = CJC^{-1}$ and

$$e^{At} = Ce^{Jt}C^{-1} \tag{16}$$

Proof We first note that

$$\begin{aligned} A^n = (CJC^{-1})^n &= \overbrace{(CJC^{-1})(CJC^{-1})\cdots(CJC^{-1})}^{n \text{ times}} \\ &= CJ(C^{-1}C)J(C^{-1}C)J(C^{-1}C)\cdots(C^{-1}C)JC^{-1} \\ &= CJ^nC^{-1} \end{aligned}$$

It then follows that

$$(At)^n = C(Jt)^nC^{-1} \tag{17}$$

Thus $e^{At} = I + (At) + \frac{(At)^2}{2!} + \cdots = CIC^{-1} + C(Jt)C^{-1} + C\frac{(Jt)^2}{2!}C^{-1} + \cdots$

$$= C\left[I + (Jt) + \frac{(Jt)^2}{2!} + \cdots\right]C^{-1} = Ce^{Jt}C^{-1} \quad ■$$

Theorem 2 tells us that to calculate e^{At} we really need only to calculate e^{Jt}. When J is a diagonal (as is most often the case), then we know how to calculate e^{Jt}. If A is a 2×2 matrix that is not diagonalizable, then $J = \begin{pmatrix} \lambda & 1 \\ 0 & \lambda \end{pmatrix}$ and $e^{Jt} = \begin{pmatrix} e^{\lambda t} & te^{\lambda t} \\ 0 & e^{\lambda t} \end{pmatrix}$ as we calculated in Example 2. In fact, it is not difficult to calculate e^{Jt} where J is any Jordan matrix. It is first necessary to compute e^{Bt} for a Jordan block matrix B. A method for doing this is given in Problems 20–22.

We now apply our computations to a simple biological model of population growth. Suppose that in an ecosystem there are two interacting species S_1 and S_2. We denote the populations of the species at time t by $x_1(t)$ and $x_2(t)$. One system governing the relative growth of the two species is

$$\begin{aligned} x_1'(t) &= ax_1(t) + bx_2(t) \\ x_2'(t) &= cx_1(t) + dx_2(t) \end{aligned} \tag{18}$$

We can interpret the constants a, b, c, and d as follows. If the species are competing, then it is reasonable to have $b<0$ and $c<0$. This is true because increases in the population of one species will slow the growth of the other. A second model is a *predator-prey* relationship. If S_1 is the prey and S_2 is the predator (S_2 eats S_1), then it is reasonable to have $b<0$ and $c>0$ since an increase in the predator species will cause a decrease in the prey species, while an increase in the prey species will cause an increase in the predator species (since it will have more food). Finally, in a *symbiotic* relationship (each species lives off the other), we would likely have $b>0$ and $c>0$. Of course, the constants a, b, c, and d depend on a wide variety of factors including available food, time of year, climate, limits due to overcrowding, other competing species, and so on. We shall analyze four different models by using the material in this section. We assume that t is measured in years.

EXAMPLE 3 **A Competitive Model.** Consider the system

$$x_1'(t) = 3x_1(t) - x_2(t)$$
$$x_2'(t) = -2x_1(t) + 2x_2(t)$$

Here, an increase in the population of one species causes a decline in the growth rate of another. Suppose that the initial populations are $x_1(0) = 90$ and $x_2(0) = 150$. Find the populations of both species for $t > 0$.

Solution We have $A = \begin{pmatrix} 3 & -1 \\ -2 & 2 \end{pmatrix}$. The eigenvalues of A are $\lambda_1 = 1$ and $\lambda_2 = 4$ with corresponding eigenvectors $\mathbf{v}_1 = \begin{pmatrix} 1 \\ 2 \end{pmatrix}$ and $\mathbf{v}_2 = \begin{pmatrix} 1 \\ -1 \end{pmatrix}$. Then

$$C = \begin{pmatrix} 1 & 1 \\ 2 & -1 \end{pmatrix} \quad C^{-1} = -\frac{1}{3}\begin{pmatrix} -1 & -1 \\ -2 & 1 \end{pmatrix} \quad J = D = \begin{pmatrix} 1 & 0 \\ 0 & 4 \end{pmatrix} \quad e^{Jt} = \begin{pmatrix} e^t & 0 \\ 0 & e^{4t} \end{pmatrix}$$

$$e^{At} = Ce^{Jt}C^{-1} = -\frac{1}{3}\begin{pmatrix} 1 & 1 \\ 2 & -1 \end{pmatrix}\begin{pmatrix} e^t & 0 \\ 0 & e^{4t} \end{pmatrix}\begin{pmatrix} -1 & -1 \\ -2 & 1 \end{pmatrix}$$
$$= -\frac{1}{3}\begin{pmatrix} 1 & 1 \\ 2 & -1 \end{pmatrix}\begin{pmatrix} -e^t & -e^t \\ -2e^{4t} & e^{4t} \end{pmatrix}$$
$$= -\frac{1}{3}\begin{pmatrix} -e^t - 2e^{4t} & -e^t + e^{4t} \\ -2e^t + 2e^{4t} & -2e^t - e^{4t} \end{pmatrix}$$

Finally, the solution to the system is given by

$$\mathbf{x}(t) = \begin{pmatrix} x_1(t) \\ x_2(t) \end{pmatrix} = e^{At}\mathbf{x}_0 = -\frac{1}{3}\begin{pmatrix} -e^t - 2e^{4t} & -e^t + e^{4t} \\ -2e^t + 2e^{4t} & -2e^t - e^{4t} \end{pmatrix}\begin{pmatrix} 90 \\ 150 \end{pmatrix}$$
$$= -\frac{1}{3}\begin{pmatrix} -240e^t - 30e^{4t} \\ -480e^t + 30e^{4t} \end{pmatrix} = \begin{pmatrix} 80e^t + 10e^{4t} \\ 160e^t - 10e^{4t} \end{pmatrix}$$

For example, after 6 months ($t = \frac{1}{2}$ year), $x_1(t) = 80e^{1/2} + 10e^2 \approx 206$ individuals, while $x_2(t) = 160e^{1/2} - 10e^2 \approx 190$ individuals. More significantly, $160e^t - 10e^{4t} = 0$ when $16e^t = e^{4t}$ or $16 = e^{3t}$ or $3t = \ln 16$ and $t = (\ln 16)/3 \approx 2.77/3 \approx 0.92$ years ≈ 11 months. Thus the second species will be eliminated after only 11 months even though it started with a larger population. In Problems 10 and 11 you are asked to show that neither population will be eliminated if $x_2(0) = 2x_1(0)$ and that the first population will be eliminated if $x_2(0) > 2x_1(0)$. Thus, as was well known to Darwin, survival in this very simple model depends on the relative sizes of the competing species when competition begins.

EXAMPLE 4 **A Predator Prey Model.** We consider the following system in which species 1 is the prey and species 2 is the predator:

$$x_1'(t) = 2x_1(t) - x_2(t)$$
$$x_2'(t) = x_1(t) + 4x_2(t)$$

Find the populations of the two species for $t>0$ if the initial populations are $x_1(0)=500$ and $x_2(0)=100$.

Solution Here $A=\begin{pmatrix}2 & -1\\ 1 & 4\end{pmatrix}$ and the only eigenvalue is $\lambda=3$ with the single eigenvector $\begin{pmatrix}1\\ -1\end{pmatrix}$. One solution to the equation $(A-3I)\mathbf{v}_2=\mathbf{v}_1$ (see Theorem 7.6.2 on page 320) is $\mathbf{v}_2=\begin{pmatrix}1\\ -2\end{pmatrix}$. Then

$$C=\begin{pmatrix}1 & 1\\ -1 & -2\end{pmatrix}\qquad C^{-1}=\begin{pmatrix}2 & 1\\ -1 & -1\end{pmatrix}\qquad J=\begin{pmatrix}3 & 1\\ 0 & 3\end{pmatrix}$$

$$e^{Jt}=\begin{pmatrix}e^{3t} & te^{3t}\\ 0 & e^{3t}\end{pmatrix}=e^{3t}\begin{pmatrix}1 & t\\ 0 & 1\end{pmatrix}\qquad \text{(from Example 2)}$$

and

$$e^{At}=Ce^{Jt}C^{-1}=e^{3t}\begin{pmatrix}1 & 1\\ -1 & -2\end{pmatrix}\begin{pmatrix}1 & t\\ 0 & 1\end{pmatrix}\begin{pmatrix}2 & 1\\ -1 & -1\end{pmatrix}$$

$$=e^{3t}\begin{pmatrix}1 & 1\\ -1 & -2\end{pmatrix}\begin{pmatrix}2-t & 1-t\\ -1 & -1\end{pmatrix}$$

$$=e^{3t}\begin{pmatrix}1-t & -t\\ t & 1+t\end{pmatrix}$$

Thus the solution to the system is

$$\mathbf{x}(t)=\begin{pmatrix}x_1(t)\\ x_2(t)\end{pmatrix}=e^{At}\mathbf{x}_0=e^{3t}\begin{pmatrix}1-t & -t\\ t & 1+t\end{pmatrix}\begin{pmatrix}500\\ 100\end{pmatrix}=e^{3t}\begin{pmatrix}500-600t\\ 100+600t\end{pmatrix}$$

It is apparent that the prey species will be eliminated after $\frac{5}{6}$ year $=10$ months—even though it started with a population five times as great as the predator species. In fact, it is easy to show (see Problem 12) that no matter how great the initial advantage of the prey species, the prey species will be eliminated in less than 1 year.

EXAMPLE 5 **Another Predator-Prey Model.** Consider the predator-prey model governed by the system

$$\begin{aligned}x_1'(t)&=x_1(t)+x_2(t)\\ x_2'(t)&=-x_1(t)+x_2(t)\end{aligned}$$

If the initial populations are $x_1(0)=x_2(0)=1000$, determine the populations of the two species for $t>0$.

Solution Here $A=\begin{pmatrix}1 & 1\\ -1 & 1\end{pmatrix}$ with characteristic equation $\lambda^2-2\lambda+2=0$, compl

roots $\lambda_1 = 1 + i$ and $\lambda_2 = 1 - i$, and eigenvectors $\mathbf{v}_1 = \begin{pmatrix} 1 \\ i \end{pmatrix}$ and $\mathbf{v}_2 = \begin{pmatrix} 1 \\ -i \end{pmatrix}$.†
Then

$$C = \begin{pmatrix} 1 & 1 \\ i & -i \end{pmatrix} \qquad C^{-1} = -\frac{1}{2i}\begin{pmatrix} -i & -1 \\ -i & 1 \end{pmatrix} = \frac{1}{2}\begin{pmatrix} 1 & -i \\ 1 & i \end{pmatrix} \qquad J = D = \begin{pmatrix} 1+i & 0 \\ 0 & 1-i \end{pmatrix}$$

and

$$e^{Jt} = \begin{pmatrix} e^{(1+i)t} & 0 \\ 0 & e^{(1-i)t} \end{pmatrix}$$

Now, by Euler's formula (see Appendix 2), $e^{it} = \cos t + i \sin t$. Thus $e^{(1+i)t} = e^t e^{it} = e^t(\cos t + i \sin t)$. Similarly, $e^{(1-i)t} = e^t e^{-it} = e^t(\cos t - i \sin t)$. Thus

$$e^{Jt} = e^t\begin{pmatrix} \cos t + i \sin t & 0 \\ 0 & \cos t - i \sin t \end{pmatrix}$$

and

$$e^{At} = Ce^{Jt}C^{-1} = \frac{e^t}{2}\begin{pmatrix} 1 & 1 \\ i & -i \end{pmatrix}\begin{pmatrix} \cos t + i \sin t & 0 \\ 0 & \cos t - i \sin t \end{pmatrix}\begin{pmatrix} 1 & -i \\ 1 & i \end{pmatrix}$$

$$= \frac{e^t}{2}\begin{pmatrix} 1 & 1 \\ i & -i \end{pmatrix}\begin{pmatrix} \cos t + i \sin t & -i \cos t + \sin t \\ \cos t - i \sin t & i \cos t + \sin t \end{pmatrix}$$

$$= \frac{e^t}{2}\begin{pmatrix} 2 \cos t & 2 \sin t \\ -2 \sin t & 2 \cos t \end{pmatrix} = e^t\begin{pmatrix} \cos t & \sin t \\ -\sin t & \cos t \end{pmatrix}$$

Finally,

$$\mathbf{x}(t) = e^{At}\mathbf{x}(0) = e^t\begin{pmatrix} \cos t & \sin t \\ -\sin t & \cos t \end{pmatrix}\begin{pmatrix} 1000 \\ 1000 \end{pmatrix} = \begin{pmatrix} 1000e^t(\cos t + \sin t) \\ 1000e^t(\cos t - \sin t) \end{pmatrix}$$

The prey species is eliminated when $1000e^t(\cos t - \sin t) = 0$ or when $\sin t = \cos t$. The first positive solution of this last equation is $t = \pi/4 \approx 0.7854$ year ≈ 9.4 months.

EXAMPLE 6 **A Model of Species Cooperation (Symbiosis).** Consider the symbiotic model governed by the system

$$x_1'(t) = -\tfrac{1}{2}x_1(t) + x_2(t)$$
$$x_2'(t) = \tfrac{1}{4}x_1(t) - \tfrac{1}{2}x_2(t)$$

Note that in this model the population of each species increases proportionally to the population of the other and decreases proportionally to its own population. Suppose that $x_1(0) = 200$ and $x_2(0) = 500$. Determine the population of each species for $t > 0$.

Solution Here $A = \begin{pmatrix} -\frac{1}{2} & 1 \\ \frac{1}{4} & -\frac{1}{2} \end{pmatrix}$ with eigenvalues $\lambda_1 = 0$ and $\lambda_2 = -1$ and corresponding eigenvectors $\mathbf{v}_1 = \begin{pmatrix} 2 \\ 1 \end{pmatrix}$ and $\mathbf{v}_2 = \begin{pmatrix} 2 \\ -1 \end{pmatrix}$. Then

$$C = \begin{pmatrix} 2 & 2 \\ 1 & -1 \end{pmatrix}, \qquad C^{-1} = -\tfrac{1}{4}\begin{pmatrix} -1 & -2 \\ -1 & 2 \end{pmatrix}, \qquad J = D = \begin{pmatrix} 0 & 0 \\ 0 & -1 \end{pmatrix},$$

† Note that $\lambda_2 = \overline{\lambda_1}$ and $\mathbf{v}_2 = \overline{\mathbf{v}_1}$. This should be no surprise, because, according to the result of Problem 7.1.33 on page 289, eigenvalues of real matrices occur in complex conjugate pairs and their corresponding eigenvectors a[illegible]mplex conjugates.

and $e^{Jt} = \begin{pmatrix} e^{0t} & 0 \\ 0 & e^{-t} \end{pmatrix} = \begin{pmatrix} 1 & 0 \\ 0 & e^{-t} \end{pmatrix}$. Thus

$$\begin{aligned} e^{At} &= -\frac{1}{4}\begin{pmatrix} 2 & 2 \\ 1 & -1 \end{pmatrix}\begin{pmatrix} 1 & 0 \\ 0 & e^{-t} \end{pmatrix}\begin{pmatrix} -1 & -2 \\ -1 & 2 \end{pmatrix} \\ &= -\frac{1}{4}\begin{pmatrix} 2 & 2 \\ 1 & -1 \end{pmatrix}\begin{pmatrix} -1 & -2 \\ -e^{-t} & 2e^{-t} \end{pmatrix} \\ &= -\frac{1}{4}\begin{pmatrix} -2-2e^{-t} & -4+4e^{-t} \\ -1+e^{-t} & -2-2e^{-t} \end{pmatrix} \end{aligned}$$

and

$$\begin{aligned} \mathbf{x}(t) = e^{At}\mathbf{x}(0) &= -\frac{1}{4}\begin{pmatrix} -2-2e^{-t} & -4+4e^{-t} \\ -1+e^{-t} & -2-2e^{-t} \end{pmatrix}\begin{pmatrix} 200 \\ 500 \end{pmatrix} \\ &= -\frac{1}{4}\begin{pmatrix} -2400+1600e^{-t} \\ -1200-800e^{-t} \end{pmatrix} \\ &= \begin{pmatrix} 600-400e^{-t} \\ 300+200e^{-t} \end{pmatrix}. \end{aligned}$$

Note that $e^{-t} \to 0$ as $t \to \infty$. This means that as time goes on, the two co-operating species approach the **equilibrium** populations 600 and 300, respectively. Neither population is eliminated.

PROBLEMS 7.7

In Problems 1–9 find the principal matrix solution e^{At} of the system $\mathbf{x}'(t) = A\mathbf{x}(t)$.

1. $A = \begin{pmatrix} -2 & -2 \\ -5 & 1 \end{pmatrix}$ **2.** $A = \begin{pmatrix} 3 & -1 \\ -2 & 4 \end{pmatrix}$ **3.** $A = \begin{pmatrix} 2 & -1 \\ 5 & -2 \end{pmatrix}$

4. $A = \begin{pmatrix} 3 & -5 \\ 1 & -1 \end{pmatrix}$ **5.** $A = \begin{pmatrix} -10 & -7 \\ 7 & 4 \end{pmatrix}$ **6.** $A = \begin{pmatrix} -2 & 1 \\ 5 & 2 \end{pmatrix}$

7. $A = \begin{pmatrix} -12 & 7 \\ -7 & 2 \end{pmatrix}$ **8.** $A = \begin{pmatrix} 1 & 1 & -2 \\ -1 & 2 & 1 \\ 0 & 1 & -1 \end{pmatrix}$ **9.** $A = \begin{pmatrix} 4 & 6 & 6 \\ 1 & 3 & 2 \\ -1 & -5 & -2 \end{pmatrix}$

10. In Example 3, show that if the initial vector $\mathbf{x}(0) = \begin{pmatrix} a \\ 2a \end{pmatrix}$, where a is a constant, then both populations grow at a rate proportional to e^t.

11. In Example 3, show that if $x_2(0) > 2x_1(0)$, then the first population will be eliminated.

12. In Example 4, show that the first population will become extinct in α years, where $\alpha = x_1(0)/[x_1(0) + x_2(0)]$.

***13.** In a water desalinization plant there are two tanks of water. Suppose that tank 1 contains 1000 liters of brine in which 1000 kg of salt is dissolved and tank 2 contains 1000 liters of pure water. Suppose that water flows into tank 1 at the rate of 20 liters per minute and the mixture flows from tank 1 into tank 2 at a rate of 30 liters per minute. From tank 2, 10 liters is pumped back to tank 1 (establishing *feedback*) while 20 liters is flushed away. Find the amount of salt in both tanks at all times t. [*Hint:* Write the information as a 2×2 system and let $x_1(t)$ and $x_2(t)$ denote the amount of salt in each tank.]

14. A community of n individuals is exposed to an infectious disease.† At any given time t, the community is divided into three groups: group 1 with population $x_1(t)$ is the susceptible group; group 2 with a population of $x_2(t)$ is the group of infected individuals in circulation; and group 3, population $x_3(t)$, consists of those who are isolated, dead, or immune. It is reasonable to assume that initially $x_2(t)$ and $x_3(t)$ will be small compared to $x_1(t)$. Let α and β be positive constants denoting the rates at which susceptibles become infected and infected individuals join group 3, respectively. Then a reasonable model for the spread of the disease is given by the system

$$\begin{aligned} x_1'(t) &= -\alpha x_1(0)x_2 \\ x_2'(t) &= \alpha x_1(0)x_2 - \beta x_2 \\ x_3'(t) &= \beta x_2 \end{aligned}$$

a. Write this system in the form $\mathbf{x}' = A\mathbf{x}$ and find the solution in terms of $x_1(0)$, $x_2(0)$, and $x_3(0)$. Note that $x_1(0)+x_2(0)+x_3(0)=n$.

b. Show that if $\alpha x(0)<\beta$, then the disease will not produce an epidemic.

c. What will happen if $\alpha x(0)>\beta$?

15. Consider the **second-order differential equation** $x''(t) + ax'(t) + bx(t) = 0$.

a. Letting $x_1(t)=x(t)$ and $x_2(t)=x'(t)$, write the preceding equation as a first-order system in the form of Equation (7), where A is a 2×2 matrix.

b. Show that the characteristic equation of A is $\lambda^2+a\lambda+b=0$.

In Problems 16–19 use the result of Problem 15 to solve the given equation.

16. $x'' + 5x' + 6x = 0$; $x(0) = 1$, $x'(0) = 0$

17. $x'' + 6x' + 9x = 0$; $x(0) = 1$, $x'(0) = 2$

18. $x'' + 4x = 0$; $x(0) = 0$, $x'(0) = 1$

19. $x''-3x'-10x=0$; $x(0)=3$, $x'(0)=2$

20. Let $N_3=\begin{pmatrix}0&1&0\\0&0&1\\0&0&0\end{pmatrix}$. Show that $N_3^3=0$, the zero matrix.

21. Show that $e^{N_3t}=\begin{pmatrix}1&t&t^2/2\\0&1&t\\0&0&1\end{pmatrix}$. [*Hint:* Write down the series for e^{N_3t} and use the result of Problem 20.]

22. Let $J=\begin{pmatrix}\lambda&1&0\\0&\lambda&1\\0&0&\lambda\end{pmatrix}$. Show that $e^{Jt}=e^{\lambda t}\begin{pmatrix}1&t&t^2/2\\0&1&t\\0&0&1\end{pmatrix}$. [*Hint:* $Jt=\lambda It+N_3t$. Use the fact that $e^{A+B}=e^Ae^B$.]

23. Using the result of Problem 22, compute e^{At}, where $A=\begin{pmatrix}-2&1&0\\-2&1&-1\\-1&1&-2\end{pmatrix}$. [*Hint:* See Problem 7.6.20 on page 324.]

24. Compute e^{At}, where $A=\begin{pmatrix}-1&-18&-7\\1&-13&-4\\-1&25&8\end{pmatrix}$.

---

† For a discussion of this model, see N. Bailey, "The Total Size of a General Stochastic Epidemic," *Biometrika* 40 (1953): 177–185.

25. Compute e^{Jt}, where $J = \begin{pmatrix} \lambda & 1 & 0 & 0 \\ 0 & \lambda & 1 & 0 \\ 0 & 0 & \lambda & 1 \\ 0 & 0 & 0 & \lambda \end{pmatrix}$.

26. Compute e^{At}, where $A = \begin{pmatrix} 2 & 1 & 0 & 0 \\ 0 & 2 & 0 & 0 \\ 0 & 0 & 3 & 1 \\ 0 & 0 & 0 & 3 \end{pmatrix}$.

27. Compute e^{At}, where $A = \begin{pmatrix} -4 & 1 & 0 & 0 \\ 0 & -4 & 1 & 0 \\ 0 & 0 & -4 & 0 \\ 0 & 0 & 0 & 3 \end{pmatrix}$.

7.8 If Time Permits: Two Interesting Results: The Theorems of Cayley-Hamilton and Gershgorin

There are many interesting results concerning the eigenvalues of a matrix. In this section we discuss two of the more useful ones. The first says that any matrix satisfies its own characteristic equation. The second shows how to locate, crudely, the eigenvalues of any matrix with practically no computation.

Let $p(x) = x^n + a_{n-1}x^{n-1} + \cdots + a_1x + a_0$ be a polynomial and let A be an $n \times n$ matrix. Then powers of A are defined and we define

$$p(A) = A^n + a_{n-1}A^{n-1} + \cdots + a_1A + a_0I \qquad (1)$$

EXAMPLE 1

Let $A = \begin{pmatrix} -1 & 4 \\ 3 & 7 \end{pmatrix}$ and $p(x) = x^2 - 5x + 3$. Then

$$P(A) = A^2 - 5A + 3I = \begin{pmatrix} 13 & 24 \\ 18 & 61 \end{pmatrix} + \begin{pmatrix} 5 & -20 \\ -15 & -35 \end{pmatrix} + \begin{pmatrix} 3 & 0 \\ 0 & 3 \end{pmatrix} = \begin{pmatrix} 21 & 4 \\ 3 & 29 \end{pmatrix}$$

Expression (1) is a polynomial with scalar coefficients defined for a matrix variable. We can also define a polynomial with *square matrix* coefficients by

$$Q(\lambda) = B_0 + B_1\lambda + B_2\lambda^2 + \cdots + B_n\lambda^n \qquad (2)$$

If A is a matrix, then we define

$$Q(A) = B_0 + B_1A + B_2A^2 + \cdots + B_mA^m \qquad (3)$$

We must be careful in (3) since matrices do not commute under multiplication.

THEOREM 1 If $P(\lambda)$ and $Q(\lambda)$ are polynomials in the scalar variable λ with square matrix coefficients and if $P(\lambda)=Q(\lambda)(A-\lambda I)$, then $P(A)=0$.

Proof If $Q(\lambda)$ is given by Equation (2), then

$$\begin{aligned} P(\lambda) &= (B_0+B_1\lambda+B_2\lambda^2+\cdots+B_n\lambda^n)(A-\lambda I) \\ &= B_0A+B_1A\lambda+B_2A\lambda^2+\cdots+B_nA\lambda^n \\ &\quad -B_0\lambda-B_1\lambda^2-B_2\lambda^3-\cdots-B_n\lambda^{n+1} \end{aligned} \tag{4}$$

Then, substituting A for λ in (4), we obtain

$$\begin{aligned} P(A) &= B_0A+B_1A^2+B_2A^3+\cdots+B_nA^{n+1} \\ &\quad -B_0A-B_1A^2-B_2A^3-\cdots-B_nA^{n+1}=0 \quad \blacksquare \end{aligned}$$

Note. We cannot prove this theorem by substituting $\lambda=A$ to obtain $P(A)=Q(A)(A-A)=0$. This is because it is possible to find polynomials $P(\lambda)$ and $Q(\lambda)$ with matrix coefficients such that $F(\lambda)=P(\lambda)Q(\lambda)$ but $F(A)\neq P(A)Q(A)$. (See Problem 17.)

We can now state the first main theorem.

THEOREM 2 **The Cayley-Hamilton Theorem.**† Every square matrix satisfies its own characteristic equation. That is, if $p(\lambda)=0$ is the characteristic equation of A, then $p(A)=0$.

Proof We have

$$p(\lambda)=\det(A-\lambda I)=\begin{vmatrix} a_{11}-\lambda & a_{12} & \cdots & a_{1n} \\ a_{21} & a_{22}-\lambda & \cdots & a_{2n} \\ \vdots & \vdots & & \vdots \\ a_{n1} & a_{n2} & \cdots & a_{nn}-\lambda \end{vmatrix}$$

Clearly, any cofactor of $(A-\lambda I)$ is a polynomial in λ. Thus the adjoint of $A-\lambda I$ (see Definition 3.4.1, page 106) is an $n\times n$ matrix each of whose components is a polynomial in λ. That is:

$$\operatorname{adj}(A-\lambda I)=\begin{pmatrix} p_{11}(\lambda) & p_{12}(\lambda) & \cdots & p_{1n}(\lambda) \\ p_{21}(\lambda) & p_{22}(\lambda) & \cdots & p_{2n}(\lambda) \\ \vdots & \vdots & & \vdots \\ p_{n1}(\lambda) & p_{22}(\lambda) & \cdots & p_{nn}(\lambda) \end{pmatrix}$$

† Named after Sir William Rowan Hamilton (discussed in Chapter 4) and Arthur Cayley (1821–1895). Cayley published the first discussion of this famous theorem in 1858. Independently, Hamilton discovered the result in his work on quaternions.

This means that we can think of adj $(A-\lambda I)$ as a polynomial, $Q(\lambda)$, in λ with $n\times n$ matrix coefficients. To see this, look at the following:

$$\begin{pmatrix} -\lambda^2-2\lambda+1 & 2\lambda^2-7\lambda-4 \\ 4\lambda^2+5\lambda-2 & -3\lambda^2-\ \lambda+3 \end{pmatrix} = \begin{pmatrix} -1 & 2 \\ 4 & -3 \end{pmatrix}\lambda^2 + \begin{pmatrix} -2 & -7 \\ 5 & -1 \end{pmatrix}\lambda + \begin{pmatrix} 1 & -4 \\ -2 & 3 \end{pmatrix}$$

Now, from Theorem 3.4.2 on page 107,

$$\det(A-\lambda I)I = [\text{adj}\,(A-\lambda I)][A-\lambda I] = Q(\lambda)(A-\lambda I) \tag{5}$$

But $\det(A-\lambda I)I = p(\lambda)I$. If

$$p(\lambda) = \lambda^n + a_{n-1}\lambda^{n-1} + \cdots + a_1\lambda + a_0,$$

then we define

$$P(\lambda) = p(\lambda)I = \lambda^n I + a_{n-1}\lambda^{n-1}I + \cdots + a_1\lambda I + a_0 I.$$

Thus, from (5), we have $P(\lambda) = Q(\lambda)(A-\lambda I)$. Finally, from Theorem 1, $P(A)=0$. This completes the proof. ■

EXAMPLE 2

Let $A = \begin{pmatrix} 1 & -1 & 4 \\ 3 & 2 & -1 \\ 2 & 1 & -1 \end{pmatrix}$. In Example 7.1.4 on page 281 we computed the characteristic equation $\lambda^3 - 2\lambda^2 - 5\lambda + 6 = 0$. Now we compute

$$A^2 = \begin{pmatrix} 6 & 1 & 1 \\ 7 & 0 & 11 \\ 3 & -1 & 8 \end{pmatrix}, \quad A^3 = \begin{pmatrix} 11 & -3 & 22 \\ 29 & 4 & 17 \\ 16 & 3 & 5 \end{pmatrix}$$

and

$$A^3 - 2A^2 - 5A + 6I = \begin{pmatrix} 11 & -3 & 22 \\ 29 & 4 & 17 \\ 16 & 3 & 5 \end{pmatrix} + \begin{pmatrix} -12 & -2 & -2 \\ -14 & 0 & -22 \\ -6 & 2 & -16 \end{pmatrix}$$

$$+ \begin{pmatrix} -5 & 5 & -20 \\ -15 & -10 & 5 \\ -10 & -5 & 5 \end{pmatrix} + \begin{pmatrix} 6 & 0 & 0 \\ 0 & 6 & 0 \\ 0 & 0 & 6 \end{pmatrix}$$

$$= \begin{pmatrix} 0 & 0 & 0 \\ 0 & 0 & 0 \\ 0 & 0 & 0 \end{pmatrix}$$

In some situations the Cayley-Hamilton theorem is useful in calculating the inverse of a matrix. If A^{-1} exists and $p(A)=0$, then $A^{-1}p(A)=0$. To illustrate, if $p(\lambda) = \lambda^n + a_{n-1}\lambda^{n-1} + \cdots + a_1\lambda + a_0$, then

$$p(A) = A^n + a_{n-1}A^{n-1} + \cdots + a_1 A + a_0 I = 0$$

and

$$A^{-1}p(A) = A^{n-1} + a_{n-1}A^{n-2} + \cdots + a_2 A + a_1 I + a_0 A^{-1} = 0$$

Thus

$$A^{-1} = \frac{1}{a_0}(-A^{n-1} - a_{n-1}A^{n-2} - \cdots - a_2A - a_1I) \qquad (6)$$

Note that $a_0 \neq 0$ because $a_0 = \det A$ (why?) and we assumed that A was invertible.

EXAMPLE 3 Let $A = \begin{pmatrix} 1 & -1 & 4 \\ 3 & 2 & -1 \\ 2 & 1 & -1 \end{pmatrix}$. Then $p(\lambda) = \lambda^3 - 2\lambda^2 - 5\lambda + 6$. Here $n = 3$, $a_2 = -2$, $a_1 = -5$, $a_0 = 6$, and

$$A^{-1} = \frac{1}{6}(-A^2 + 2A + 5I)$$

$$= \frac{1}{6}\left[\begin{pmatrix} -6 & -1 & -1 \\ -7 & 0 & -11 \\ -3 & 1 & -8 \end{pmatrix} + \begin{pmatrix} 2 & -2 & 8 \\ 6 & 4 & -2 \\ 4 & 2 & -2 \end{pmatrix} + \begin{pmatrix} 5 & 0 & 0 \\ 0 & 5 & 0 \\ 0 & 0 & 5 \end{pmatrix}\right]$$

$$= \frac{1}{6}\begin{pmatrix} 1 & -3 & 7 \\ -1 & 9 & -13 \\ 1 & 3 & -5 \end{pmatrix}$$

Note that we computed A^{-1} with a single division and with only one calculation of a determinant (in order to find $p(\lambda) = \det(A - \lambda I)$). This method is sometimes very efficient on a computer.

We now turn to the second important result of this section. Let A be an $n \times n$ matrix. We write, as usual,

$$A = \begin{pmatrix} a_{11} & a_{12} & \cdots & a_{1n} \\ a_{21} & a_{22} & \cdots & a_{2n} \\ \vdots & \vdots & & \vdots \\ a_{n1} & a_{n2} & \cdots & a_{nn} \end{pmatrix}$$

Define the number

$$r_1 = |a_{12}| + |a_{13}| + \cdots + |a_{1n}| = \sum_{j=2}^{n} |a_{1j}| \qquad (7)$$

Similarly, define

$$r_i = |a_{i1}| + |a_{i2}| + \cdots + |a_{i,i-1}| + |a_{i,i+1}| + \cdots + |a_{i,n}| = \sum_{\substack{j=1 \\ j\neq i}}^{n} |a_{ij}| \tag{8}$$

That is, r_i is the sum of the absolute values of the numbers on the ith row of A that are not on the main diagonal of A. Let

$$D_i = \{z \in \mathbb{C}: \ |z - a_{ii}| \leq r_i\} \tag{9}$$

Here D_i is a disk in the complex plane centered at a_{ii} with radius r_i (see Figure 7.4). The disk D_i consists of all points in the complex plane on and

Figure 7.4

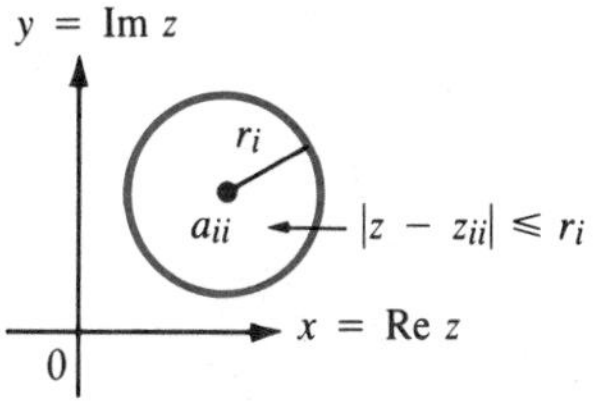

inside the circle $C_i = \{z \in \mathbb{C}: |z - a_{ii}| = r_i\}$. The circles C_i, $i = 1, 2, \ldots, n$, are called **Gershgorin circles**.

THEOREM 3 **Gershgorin's Circle Theorem.**† Let A be an $n \times n$ matrix and let D_i be defined by Equation (9). Then each eigenvalue of A is contained in at least one of the D_i's. That is, if the eigenvalues of A are $\lambda_1, \lambda_2, \ldots, \lambda_k$, then:

$$\{\lambda_1, \lambda_2, \ldots, \lambda_k\} \subset \bigcup_{i=1}^{n} D_i \tag{10}$$

Proof Let λ be an eigenvalue of A with eigenvector $\mathbf{v} = \begin{pmatrix} x_1 \\ x_2 \\ \vdots \\ x_n \end{pmatrix}$. Let $m = \max\{|x_1|, |x_2|, \ldots, |x_n|\}$. Then $(1/m)\mathbf{v} = \begin{pmatrix} y_1 \\ y_2 \\ \vdots \\ y_n \end{pmatrix}$ is an eigenvector of A corre-

† The Russian mathematician S. Gershgorin published this result in 1931.

sponding to λ and $\max\{|y_1|, |y_2|, \ldots, |y_n|\} = 1$. Let y_i be the component of $\mathbf{y}$ with $|y_i| = 1$. Now $A\mathbf{y} = \lambda\mathbf{y}$. The ith component of the n-vector $A\mathbf{y}$ is $a_{i1}y_1 + a_{i2}y_2 + \cdots + a_{in}y_n$. The ith component of $\lambda\mathbf{y}$ is λy_i. Thus

$$a_{i1}y_1 + a_{i2}y_2 + \cdots + a_{in}y_n = \lambda y_i,$$

which we write as

$$\sum_{j=1}^{n} a_{ij}y_j = \lambda y_i \tag{11}$$

By subtracting $a_{ii}y_i$ from both sides, Equation (11) can be rewritten as

$$\sum_{\substack{j=1\\ j\neq i}}^{n} a_{ij}y_j = \lambda y_i - a_{ii}y_i = (\lambda - a_{ii})y_i \tag{12}$$

Next, taking the absolute value of both sides of (12) and using the triangle inequality ($|a+b| \le |a| + |b|$), we obtain

$$|(a_{ii} - \lambda)y_i| = \left| -\sum_{\substack{j=1\\ j\neq i}}^{n} a_{ij}y_j \right| \le \sum_{\substack{j=1\\ j\neq i}}^{n} |a_{ij}|\,|y_j| \tag{13}$$

We divide both sides of (13) by $|y_i|$ (which is equal to 1) to obtain

$$|a_{ii} - \lambda| \le \sum_{\substack{j=1\\ j\neq i}}^{n} |a_{ij}| \frac{|y_j|}{|y_i|} \le \sum_{\substack{j=1\\ j\neq i}}^{n} |a_{ij}| = r_i \tag{14}$$

The last step followed the fact that $|y_j| \le |y_i|$ (by the way we chose y_i). But this proves the theorem since (14) shows that $\lambda \in D_i$. ■

EXAMPLE 4 Let $A = \begin{pmatrix} 1 & -1 & 4 \\ 3 & 2 & -1 \\ 2 & 1 & -1 \end{pmatrix}$. Then $a_{11} = 1$, $a_{22} = 2$, $a_{33} = -1$, $r_1 = |-1| + |4| = 5$, $r_2 = |3| + |-1| = 4$, and $r_3 = |2| + |1| = 3$. Thus the eigenvalues of A lie within the boundaries of the three circles drawn in Figure 7.5. We can verify this since we know by Example 7.1.4 on page 281 that the eigenvalues of A are 1, -2, and 3, which lie within the three circles. Note that the Gershgorin circles can intersect one another.

Figure 7.5

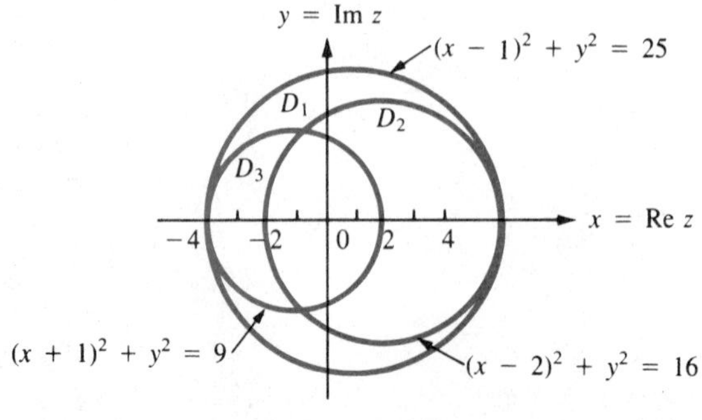

EXAMPLE 5 Find bounds on the eigenvalues of the matrix

$$A = \begin{pmatrix} 3 & 0 & -1 & -\frac{1}{4} & \frac{1}{4} \\ 0 & 5 & \frac{1}{2} & 0 & 1 \\ -\frac{1}{4} & 0 & 6 & \frac{1}{4} & \frac{1}{2} \\ 0 & -1 & \frac{1}{2} & -3 & \frac{1}{4} \\ \frac{1}{6} & -\frac{1}{6} & \frac{1}{3} & \frac{1}{3} & 4 \end{pmatrix}$$

Solution Here $a_{11}=3$, $a_{22}=5$, $a_{33}=6$, $a_{44}=-3$, $a_{55}=4$, $r_1=\frac{3}{2}$, $r_2=\frac{3}{2}$, $r_3=1$, $r_4=\frac{7}{4}$, and $r_5=1$. The Gershgorin circles are drawn in Figure 7.6. It is clear from

Figure 7.6

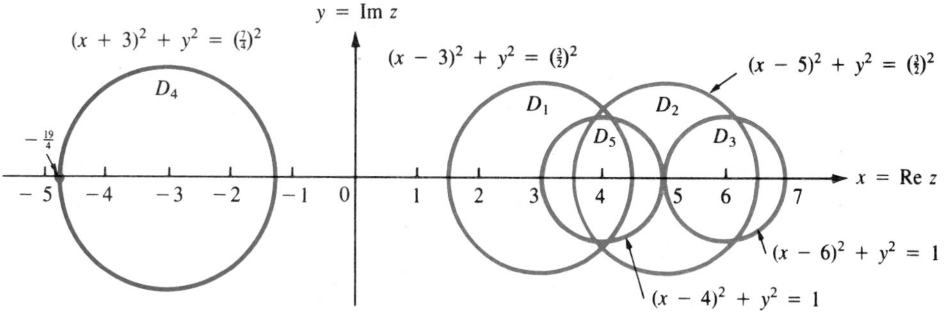

Theorem 3 and Figure 7.6 that if λ is an eigenvalue of A, then $|\lambda| \le 7$ and $\text{Re}\,\lambda \ge -\frac{19}{4}$. Note the power of the Gershgorin theorem to find the approximate location of eigenvalues after doing very little work.

PROBLEMS 7.8

In Problems 1–9: **(a)** Find the characteristic equation $p(\lambda)=0$ of the given matrix; **(b)** verify that $p(A)=0$; **(c)** use part (*b*) to compute A^{-1}.

1. $\begin{pmatrix} -2 & -2 \\ -5 & 1 \end{pmatrix}$

2. $\begin{pmatrix} 2 & -1 \\ 5 & -2 \end{pmatrix}$

3. $\begin{pmatrix} 1 & -1 & 0 \\ -1 & 2 & -1 \\ 0 & -1 & 1 \end{pmatrix}$

4. $\begin{pmatrix} 1 & 2 & 2 \\ 0 & 2 & 1 \\ -1 & 2 & 2 \end{pmatrix}$

5. $\begin{pmatrix} 0 & 1 & 0 \\ 0 & 0 & 1 \\ 1 & -3 & 3 \end{pmatrix}$

6. $\begin{pmatrix} -3 & -7 & -5 \\ 2 & 4 & 3 \\ 1 & 2 & 2 \end{pmatrix}$

7. $\begin{pmatrix} 2 & -1 & 3 \\ 4 & 1 & 6 \\ 1 & 5 & 3 \end{pmatrix}$

8. $\begin{pmatrix} 1 & 0 & 1 & 0 \\ 2 & -1 & 0 & 2 \\ -1 & 0 & 0 & 1 \\ 4 & 1 & -1 & 0 \end{pmatrix}$

9. $\begin{pmatrix} a & b & 0 & 0 \\ 0 & a & c & 0 \\ 0 & 0 & a & d \\ 0 & 0 & 0 & a \end{pmatrix}$; $bcd \ne 0$

In Problems 10–14 draw the Gershgorin circles for the given matrix A and find a bound for $|\lambda|$ if λ is an eigenvalue of A.

10. $\begin{pmatrix} 2 & 1 & 0 \\ \frac{1}{2} & 5 & \frac{1}{2} \\ 1 & 0 & 6 \end{pmatrix}$

11. $\begin{pmatrix} 3 & -\frac{1}{2} & -\frac{1}{3} & 0 \\ 0 & 6 & 1 & 0 \\ \frac{1}{3} & -\frac{1}{3} & 5 & \frac{1}{3} \\ -\frac{1}{2} & \frac{1}{4} & -\frac{1}{4} & 4 \end{pmatrix}$

12. $\begin{pmatrix} 1 & 3 & -1 & 4 \\ 2 & 5 & 0 & -7 \\ 3 & -1 & 6 & 1 \\ 0 & 2 & 3 & 4 \end{pmatrix}$

13. $\begin{pmatrix} -7 & \frac{1}{5} & -\frac{1}{5} & \frac{2}{5} \\ -\frac{1}{10} & -10 & \frac{1}{10} & \frac{3}{10} \\ -\frac{1}{4} & \frac{1}{4} & 5 & \frac{1}{4} \\ 0 & -1 & 0 & 4 \end{pmatrix}$

14. $\begin{pmatrix} 3 & 0 & -\frac{1}{3} & \frac{1}{3} & 0 & \frac{1}{3} \\ \frac{1}{2} & 5 & -\frac{1}{2} & 0 & 1 & 0 \\ \frac{1}{10} & -\frac{1}{5} & 4 & \frac{3}{5} & -\frac{1}{5} & \frac{1}{10} \\ -1 & 0 & 0 & -3 & 0 & 0 \\ \frac{1}{2} & 0 & -\frac{1}{2} & 0 & 2 & \frac{1}{2} \\ -\frac{1}{4} & \frac{1}{4} & \frac{1}{4} & 0 & -\frac{1}{4} & 0 \end{pmatrix}$

15. Let $A = \begin{pmatrix} 2 & \frac{1}{2} & -\frac{1}{3} & \frac{1}{4} \\ \frac{1}{2} & 3 & \frac{1}{2} & 1 \\ -\frac{1}{3} & \frac{1}{2} & 5 & 2 \\ \frac{1}{4} & 1 & 2 & 4 \end{pmatrix}$. Prove that the eigenvalues of A are positive real numbers.

16. Let $A = \begin{pmatrix} -4 & 1 & 1 & 1 \\ 1 & -6 & 2 & 1 \\ 1 & 2 & -5 & 1 \\ 1 & 1 & 1 & -4 \end{pmatrix}$. Prove that the eigenvalues of A are negative and real.

17. Let $P(\lambda) = B_0 + B_1\lambda$ and $Q(\lambda) = C_0 + C_1\lambda$, where B_0, B_1, C_0, and C_1 are $n \times n$ matrices.

a. Compute $F(\lambda) = P(\lambda)Q(\lambda)$.

b. Let A be an $n \times n$ matrix. Show that $F(A) = P(A)Q(A)$ if and only if A commutes with both C_0 and C_1.

18. Let the $n \times n$ matrix A have the eigenvalues $\lambda_1, \lambda_2, \ldots, \lambda_n$ and let $r(A) = \max_{1 \le i \le n} |\lambda_i|$. If $|A|$ is the max-row sum norm defined in Section 7.7, show that $r(A) \le |A|$.

19. The $n \times n$ matrix A is said to be **strictly diagonally dominant** if $|a_{ii}| > r_i$ for $i = 1, 2, \ldots, n$, where r_i is defined by Equation (8). Show that if A is a strictly diagonally dominant matrix, then $\det A \ne 0$.

Review Exercises for Chapter 7

In Exercises 1–6 calculate the eigenvalues and eigenspaces of the given matrix.

1. $\begin{pmatrix} -8 & 12 \\ -6 & 10 \end{pmatrix}$ **2.** $\begin{pmatrix} 2 & 5 \\ 0 & 2 \end{pmatrix}$ **3.** $\begin{pmatrix} 1 & 0 & 0 \\ 3 & 7 & 0 \\ -2 & 4 & -5 \end{pmatrix}$

4. $\begin{pmatrix} 1 & -1 & 0 \\ 1 & 2 & 1 \\ -2 & 1 & -1 \end{pmatrix}$ **5.** $\begin{pmatrix} 5 & -2 & 0 & 0 \\ 4 & -1 & 0 & 0 \\ 0 & 0 & 3 & -1 \\ 0 & 0 & 2 & 3 \end{pmatrix}$ **6.** $\begin{pmatrix} -2 & 1 & 0 \\ 0 & -2 & 1 \\ 0 & 0 & -2 \end{pmatrix}$

In Exercises 7–15 determine whether the given matrix A is diagonalizable. If it is, find a matrix C such that $C^{-1}AC = D$. If A is symmetric, find an orthogonal matrix Q such that $Q^tAQ = D$.

7. $\begin{pmatrix} -18 & -15 \\ 20 & 17 \end{pmatrix}$ **8.** $\begin{pmatrix} \frac{17}{2} & \frac{9}{2} \\ -15 & -8 \end{pmatrix}$ **9.** $\begin{pmatrix} 1 & 1 & 1 \\ -1 & -1 & 0 \\ -1 & 0 & -1 \end{pmatrix}$

10. $\begin{pmatrix} 4 & 2 & 0 \\ 2 & 4 & 0 \\ 0 & 0 & -3 \end{pmatrix}$ **11.** $\begin{pmatrix} -3 & 2 & 1 \\ -7 & 4 & 2 \\ -5 & 3 & 2 \end{pmatrix}$ **12.** $\begin{pmatrix} 8 & 0 & 12 \\ 0 & -2 & 0 \\ 12 & 0 & -2 \end{pmatrix}$

13. $\begin{pmatrix} 2 & 2 & 0 \\ 2 & 2 & 0 \\ 0 & 0 & -3 \end{pmatrix}$ **14.** $\begin{pmatrix} 4 & 2 & -2 & 2 \\ 1 & 3 & 1 & -1 \\ 0 & 0 & 2 & 0 \\ 1 & 1 & -3 & 5 \end{pmatrix}$ **15.** $\begin{pmatrix} 3 & 4 & -4 & 0 \\ 0 & -1 & 0 & 0 \\ 0 & 0 & -1 & 0 \\ 0 & -4 & 4 & 3 \end{pmatrix}$

In Exercises 16–20 identify the conic section and write it in new variables with the xy term absent.

16. $xy = -4$ **17.** $4x^2 + 2xy + 2y^2 = 8$ **18.** $4x^2 - 3xy + y^2 = 1$

19. $3y^2 - 2xy - 5 = 0$ **20.** $x^2 - 4xy + 4y^2 + 1 = 0$

21. Write the quadratic form $2x^2 + 4xy + 2y^2 - 3z^2$ in new variables x', y', and z' so that no cross-product terms are present.

In Exercises 22–24 find a matrix C such that $C^{-1}AC = J$, the Jordan canonical form of the matrix.

22. $\begin{pmatrix} -9 & 4 \\ -25 & 11 \end{pmatrix}$ **23.** $\begin{pmatrix} -4 & 4 \\ -1 & 0 \end{pmatrix}$ **24.** $\begin{pmatrix} 0 & -18 & -7 \\ 1 & -12 & -4 \\ -1 & 25 & 9 \end{pmatrix}$

In Exercises 25–27 compute e^{At}.

25. $A = \begin{pmatrix} -3 & 4 \\ -2 & 3 \end{pmatrix}$ **26.** $A = \begin{pmatrix} -4 & 4 \\ -1 & 0 \end{pmatrix}$ **27.** $A = \begin{pmatrix} -3 & -4 \\ 2 & 1 \end{pmatrix}$

28. Using the Cayley–Hamilton theorem, compute the inverse of $A = \begin{pmatrix} 2 & 3 & 1 \\ -1 & 1 & 0 \\ -2 & -1 & 4 \end{pmatrix}$.

29. Use the Gershgorin circle theorem to find a bound on the eigenvalues of

$$A = \begin{pmatrix} 3 & \frac{1}{2} & -\frac{1}{2} & 0 \\ 0 & 4 & \frac{1}{3} & -\frac{1}{3} \\ 1 & 0 & 2 & -1 \\ \frac{1}{2} & -\frac{1}{2} & 1 & -3 \end{pmatrix}$$

8 Numerical Methods

8.1 The Error in Numerical Computations

In every chapter of this book we have performed numerical computations. We have, among other things, solved linear equations, multiplied and inverted matrices, found bases, and computed eigenvalues and eigenvectors. With few exceptions, the examples involved 2×2 and 3×3 matrices—not because most applications have only two or three variables but because the computations would have been too tedious otherwise.

With the recent and widespread use of calculators and computers, the situation has been altered. The remarkable strides made in the last few years in the theory of numerical methods for solving certain computational problems have made it possible to perform, quickly and accurately, the calculations mentioned in the first paragraph on high-order matrices.

The use of the computer presents new difficulties, however. Computers do not store numbers such as $\frac{2}{3}$, $7\frac{3}{8}$, $\sqrt{2}$, and π. Rather, every computer uses what is called *floating-point arithmetic*. In this system every number is represented in the form

$$x = \pm 0.d_1 d_2 \cdots d_k \times 10^n \tag{1}$$

where $d_1, d_2, \ldots, d_k$ are single-digit positive integers and n is an integer. Any number written in this form is called a *floating-point number*. In Equation (1), the number $\pm .d_1 d_2 \cdots d_k$ is called the *mantissa* and the number n is called the *exponent*. The number k is called the *number of significant digits* in the expression.

Different computers have different capabilities in the range of numbers expressible in the form of Equation (1). Two popularly used systems are the IBM systems (360 and 370) and the DEC systems (10 and 20). For the IBM systems, $-75 \le n \le 75$ and $k = 7$. For the DEC systems, $-37 \le n \le 38$ and $k = 8$. Actually, digits are represented in binary rather than decimal form. The DEC systems, for example, carry 28 binary digits. Since $2^{28} = 268{,}435{,}456$, we can use the 28 binary digits to represent any eight-digit number. Hence $k = 8$.

EXAMPLE 1 The following numbers are expressed in floating-point form:

i. $\frac{1}{4}=0.25$
ii. $2378=0.2378\times10^4$
iii. $-0.000816=-0.816\times10^{-3}$
iv. $83.27=0.8327\times10^2$

If the number of significant digits were unlimited, then we would have no problem. Almost every time numbers are introduced into the computer, however, errors begin to accumulate. This can happen in one of two ways:

i. Truncation: All significant digits after k digits are simply "cut off." For example, if truncation is used, $\frac{2}{3}=0.666666...$ is stored (with $k=8$) as $\frac{2}{3}=0.66666666\times10^0$.
ii. Rounding: If $d_{k+1}\geq5$, then 1 is added to d_k and the resulting number is truncated. Otherwise, the number is simply truncated. For example, with rounding (and $k=8$), $\frac{2}{3}$ is stored as $\frac{2}{3}=0.66666667\times10^0$.

EXAMPLE 2 We can illustrate how some numbers are stored with truncation and rounding by using eight significant digits:

Number	Truncated number	Rounded number
$\frac{8}{3}$	0.26666666×10^1	0.26666667×10^1
π	0.31415926×10^1	0.31415927×10^1
$-\frac{1}{57}$	-0.17543859×10^{-1}	-0.17543860×10^{-1}

Individual round-off or truncation errors do not seem very significant. When thousands of computational steps are involved, however, the *accumulated* round-off error can be devastating. Thus, in discussing any numerical scheme, it is necessary to know not only whether you will get the right answer, theoretically, but also how badly the round-off errors will accumulate. To keep track of things, we define two types of error. If x is the actual value of a number and if x^* is the number that appears in the computer, then the **absolute error** ε_a is defined by

$$\varepsilon_a=|x^*-x| \tag{2}$$

More interesting in most situations is the **relative error** ε_r, defined by

$$\varepsilon_r = \left|\frac{x^* - x}{x}\right| \tag{3}$$

EXAMPLE 3 Let $x = 2$ and $x^* = 2.1$. Then $\varepsilon_a = 0.1$ and $\varepsilon_r = 0.1/2 = 0.05$. If $x_1 = 2000$ and $x_1^* = 2000.1$, then, again, $\varepsilon_a = 0.1$. But now $\varepsilon_r = 0.1/2000 = 0.00005$. Most people would agree that the 0.1 error in the first case is more significant than the 0.1 error in the second.

Much of numerical analysis is concerned with questions of **convergence** and **stability**. If x is the answer to a problem and our computational method gives us approximating values x_n, then the method converges if, theoretically, x_n approaches x as n gets large. If, moreover, it can be shown that the round-off errors will not accumulate in such a way as to make the answer unreliable, then the method is stable.

It is easy to give an example of a procedure in which round-off error can be quite large. Suppose we wish to compute $y = 1/(x - 0.66666665)$. For $x = \frac{2}{3}$, if the computer truncates, then $x = 0.66666666$ and $y = 1/0.00000001 = 10^8 = 10 \times 10^7$. If the computer rounds, then $x = 0.66666667$ and $y = 1/0.00000002 = 5 \times 10^7$. The difference here is enormous. The correct answer is $1/(\frac{2}{3} - \frac{66666665}{100000000}) = 60{,}000{,}000 = 6 \times 10^7$.

In this chapter we shall examine several schemes for solving systems of equations and computing eigenvalues and eigenvectors. Wherever possible, we shall also discuss convergence and stability. This is but a very superficial view of numerical methods, however; entire books and courses are devoted to the subject. For a more exhaustive view, you are encouraged to consult the following references:

Numerical Linear Algebra References

1. Blum, E. K. *Numerical Analysis and Computation: Theory and Practice.* Reading, Mass.: Addison-Wesley, 1972.
2. Burden, R. L., J. D. Faires, and A. C. Reynolds. *Numerical Analysis.* Boston: Prindle, Weber and Schmidt, 1978.
3. Conte, S. D. *Elementary Numerical Analysis, 2nd Ed.* New York: McGraw-Hill, 1972.
4. Faddeev, D. K. and V. N. Faddeeva. *Computational Methods of Linear Algebra.* San Francisco: Freeman, 1963.
5. Fox, L. *An Introduction to Numerical Linear Algebra.* New York: Oxford University Press, 1965.

Although no computer programs are included in this chapter, a special supplement to this book is available that discusses BASIC programming and includes several programs that you are encouraged to try for yourself.

PROBLEM 8.1

In Problems 1–13 convert the number to a floating-point number with eight decimal places of accuracy. Either truncate (T) or round off (R) as indicated.

1. $\frac{1}{3}$ (T) **2.** $\frac{7}{8}$ **3.** -0.000035 **4.** $\frac{7}{9}$ (R)
5. $\frac{7}{9}$ (T) **6.** $\frac{33}{7}$ (T) **7.** $\frac{85}{11}$ (R) **8.** $-18\frac{5}{6}$ (T)
9. $-18\frac{5}{6}$ (R) **10.** 237,059,628 (T) **11.** 237,059,628 (R)
12. -23.7×10^{15} **13.** 8374.2×10^{-24}

In Problems 14–21 the number x and an approximation x^* is given. Find the absolute and relative errors ε_a and ε_r.

14. $x = 5$; $x^* = 0.49 \times 10^1$ **15.** $x = 500$; $x^* = 0.4999 \times 10^3$
16. $x = 3720$; $x^* = 0.3704 \times 10^4$ **17.** $x = \frac{1}{8}$; $x^* = 0.12 \times 10^0$
18. $x = \frac{1}{800}$; $x^* = 0.12 \times 10^{-2}$ **19.** $x = -5\frac{5}{6}$; $x^* = -0.583 \times 10^1$
20. $x = 0.70465$; $x^* = 0.70466 \times 10^0$
21. $x = 70465$; $x^* = 0.70466 \times 10^5$

8.2 Solving Linear Systems I: Gaussian Elimination with Pivoting

It is not difficult to program a computer to solve a system of linear equations by the Gaussian or Gauss-Jordan elimination method used throughout this text. There is, however, a variation of the method that was designed to reduce the accumulated round-off error in solving an $n \times n$ system of equations. Before describing this method, recall the definition of the row echelon form of a matrix (see Definition 1.3.2, Page 13).

DEFINITION 1

ROW ECHELON FORM Let A be an $m \times n$ matrix. Then A is in **row echelon form** if:

i. All rows consisting entirely of zeros appear at the bottom of the matrix.
ii. The first (starting from the left) number in any row not consisting entirely of zeros is a 1.
iii. If two successive rows do not consist entirely of zeros, then the first 1 in the lower row occurs farther to the right than the first 1 in the higher row.

Note. If A is an $n \times n$ upper triangular matrix with 1's on the main diagonal, then it is easy to verify that A is in row echelon form.

From Chapter 1, it is apparent that any matrix can be reduced to row echelon form by Gaussian elimination. There is a computational problem with this method, however. If we divide by a small number that has been rounded, the result could contain a significant round-off error. For example, $1/0.00074 \approx 1351$ while $1/0.0007 \approx 1429$. To avoid this problem, we use a method called **Gaussian elimination with partial pivoting**. The idea is always to divide by the largest component in a column, thereby avoiding, so far as is possible, the type of error illustrated above. We describe the method with a simple example.

EXAMPLE 1 Solve the following system by Gaussian elimination with partial pivoting:

$$\begin{aligned} x_1 - x_2 + x_3 &= 1 \\ -3x_1 + 2x_2 - 3x_3 &= -6 \\ 2x_1 - 5x_2 + 4x_3 &= 5 \end{aligned}$$

Solution Step 1. Write the system in augmented matrix form. From the first column with nonzero components (called the *pivot column*), select the component with the *largest absolute value*. This component is called the **pivot**:

$$\text{pivot} \longrightarrow \left(\begin{array}{rrr|r} 1 & -1 & 1 & 1 \\ -3 & 2 & -3 & -6 \\ 2 & -5 & 4 & 5 \end{array}\right)$$

Step 2. Rearrange the rows to move the pivot to the top:

$$\left(\begin{array}{rrr|r} -3 & 2 & -3 & -6 \\ 1 & -1 & 1 & 1 \\ 2 & -5 & 4 & 5 \end{array}\right)$$ (first and second rows were interchanged)

Step 3. Divide the first row by the pivot:

$$\left(\begin{array}{rrr|r} 1 & -\frac{2}{3} & 1 & 2 \\ 1 & -1 & 1 & 1 \\ 2 & -5 & 4 & 5 \end{array}\right)$$ (first row divided by -3)

Step 4. Add multiples of the first row to the other rows to make all the other components in the pivot column equal to zero:

$$\left(\begin{array}{rrr|r} 1 & -\frac{2}{3} & 1 & 2 \\ 0 & -\frac{1}{3} & 0 & -1 \\ 0 & -\frac{11}{3} & 2 & 1 \end{array}\right)$$ (first row multiplied by -1 and -2 and added to the second and third rows)

Step 5. Cover the first row and perform steps 1–4 on the resulting *submatrix:*

$$\left(\begin{array}{rrr|r} 1 & -\frac{2}{3} & 1 & 2 \\ 0 & -\frac{1}{3} & 0 & -1 \\ 0 & -\frac{11}{3} & 2 & 1 \end{array}\right)$$

new pivot ($-\frac{11}{3}$)

$$\left(\begin{array}{rrr|r} 1 & -\frac{2}{3} & 1 & 2 \\ 0 & -\frac{11}{3} & 2 & 1 \\ 0 & -\frac{1}{3} & 0 & -1 \end{array}\right)$$ (first and second rows of submatrix interchanged)

$$\left(\begin{array}{rrr|r} 1 & -\frac{2}{3} & 1 & 2 \\ 0 & 1 & -\frac{6}{11} & -\frac{3}{11} \\ 0 & -\frac{1}{3} & 0 & -1 \end{array}\right)$$ (new first row divided by pivot)

$$\left(\begin{array}{ccc|c} 1 & -\frac{2}{3} & 1 & 2 \\ 0 & 1 & -\frac{6}{11} & -\frac{3}{11} \\ 0 & 0 & -\frac{2}{11} & -\frac{12}{11} \end{array}\right)$$ (new first row multiplied by $\frac{1}{3}$ and added to new second row)

Step 6. Continue in this manner until the matrix is in row echelon form:

$$\left(\begin{array}{ccc|c} 1 & -\frac{2}{3} & 1 & 2 \\ 0 & 1 & -\frac{6}{11} & -\frac{3}{11} \\ 0 & 0 & \boxed{-\frac{2}{11}} & -\frac{12}{11} \end{array}\right)$$

new pivot

$$\left(\begin{array}{ccc|c} 1 & -\frac{2}{3} & 1 & 2 \\ 0 & 1 & -\frac{6}{11} & -\frac{3}{11} \\ 0 & 0 & 1 & 6 \end{array}\right)$$ (divided new first row by pivot)

Step 7. Use **back substitution** to find the solution (if any) to the system. Evidently, we have $x_3 = 6$. Then $x_2 - \frac{6}{11}x_3 = -\frac{3}{11}$ or

$$x_2 = -\tfrac{3}{11} + \tfrac{6}{11}x_3 = -\tfrac{3}{11} + \tfrac{6}{11}(6) = 3$$

Finally, $x_1 - \frac{2}{3}x_2 + x_3 = 2$ or

$$x_1 = 2 + \tfrac{2}{3}x_2 - x_3 = 2 + \tfrac{2}{3}(3) - 6 = -2$$

The unique solution is given by the vector $(-2, 3, 6)$.

Remark. **Complete pivoting** involves finding the component in A with largest absolute value, not just the component in the first nonzero column. The problem with this method is that it usually involves the relabeling of variables when the columns are interchanged to bring the pivot to the first column. For this reason the partial pivoting method described above is more popular.

We now examine the method applied to a computationally more difficult system. Calculations were done on a hand calculator and were rounded to six significant digits.

EXAMPLE 2 Solve the system

$$\begin{aligned} 2x_1 - 3.5x_2 + x_3 &= 22.35 \\ -5x_1 + 3x_2 + 3.3x_3 &= -9.08 \\ 12x_1 + 7.8x_2 + 4.6x_3 &= 21.38 \end{aligned}$$

Solution Using the steps outlined above, we obtain, successively:

$$\left(\begin{array}{ccc|c} 2 & -3.5 & 1 & 22.35 \\ -5 & 3 & 3.3 & -9.08 \\ \boxed{12} & 7.8 & 4.6 & 21.38 \end{array}\right) \xrightarrow{P_{1,3}} \left(\begin{array}{ccc|c} \boxed{12} & 7.8 & 4.6 & 21.38 \\ -5 & 3 & 3.3 & -9.08 \\ 2 & -3.5 & 1 & 22.35 \end{array}\right)$$

pivot

$$\xrightarrow{M_1(\frac{1}{2})} \left(\begin{array}{ccc|c} 1 & 0.65 & 0.383333 & 1.78167 \\ -5 & 3 & 3.3 & -9.08 \\ 2 & -3.5 & 1 & 22.35 \end{array}\right)$$

$$\xrightarrow[A_{1,3}(-2)]{A_{1,2}(5)} \left(\begin{array}{ccc|c} 1 & 0.65 & 0.383333 & 1.78167 \\ 0 & \boxed{6.25} & 5.21667 & -0.17165 \\ 0 & -4.8 & 0.233334 & 18.7867 \end{array}\right)$$

new pivot

$$\xrightarrow{M_2(\frac{1}{6.25})} \left(\begin{array}{ccc|c} 1 & 0.65 & 0.383333 & 1.78167 \\ 0 & 1 & 0.834667 & -0.027464 \\ 0 & -4.8 & 0.233334 & 18.7867 \end{array}\right)$$

$$\xrightarrow{A_{2,3}(4.8)} \left(\begin{array}{ccc|c} 1 & 0.65 & 0.383333 & 1.78167 \\ 0 & 1 & 0.834667 & -0.027464 \\ 0 & 0 & \boxed{4.23974} & 18.6549 \end{array}\right)$$

new pivot

$$\xrightarrow{M_3(\frac{1}{4.23974})} \left(\begin{array}{ccc|c} 1 & 0.65 & 0.383333 & 1.78167 \\ 0 & 1 & 0.834667 & -0.027464 \\ 0 & 0 & 1 & 4.40001 \end{array}\right)$$

The matrix is now in row echelon form. Using back substitution, we obtain

$$x_3 = 4.40001$$

$$x_2 = -0.027464 - 0.834667x_3 = -0.027464 - (0.834667)(4.40001)$$
$$= -3.70001$$
$$x_1 = 1.78167 - (0.65)(x_2) - (0.383333)x_3 = 1.78167 - (0.65)(-3.70001)$$
$$- (0.383333)(4.40001) = 2.50001$$

The correct solution is $x_1 = 2.5$, $x_2 = -3.7$, and $x_3 = 4.4$. Our answers are very accurate indeed.

Remark. This example illustrates the fact that it is tedious and inefficient to use this method without a calculator—especially if several significant digits of accuracy are required.

The next example shows how pivoting can significantly improve the answers. Here we round to only three significant digits, thereby introducing greater round-off errors.

EXAMPLE 3 Consider the system

$$\begin{aligned} 0.0002x_1 - 0.00031x_2 + 0.0017x_3 &= 0.00609 \\ 5x_1 \quad - 7x_2 \quad - 6x_3 &= 7 \\ 8x_1 \quad + 6x_2 \quad + 3x_3 &= 2 \end{aligned}$$

The exact solution is $x_1 = -2$, $x_2 = 1$, $x_3 = 4$. Let us first solve the system by Gaussian elimination without pivoting, rounding to three significant figures.

$$\left(\begin{array}{ccc|c} 0.0002 & -0.00031 & 0.0017 & 0.00609 \\ 5 & -7 & 6 & 7 \\ 8 & 6 & 3 & 2 \end{array}\right) \xrightarrow{M_1(\frac{1}{0.0002})} \left(\begin{array}{ccc|c} 1 & -1.55 & 8.5 & 30.5 \\ 5 & -7 & 6 & 7 \\ 8 & 6 & 3 & 2 \end{array}\right)$$

$$\xrightarrow[A_{1,3}(-8)]{A_{1,2}(-5)} \left(\begin{array}{ccc|c} 1 & -1.55 & 8.5 & 30.5 \\ 0 & 0.75 & -36.5 & -146 \\ 0 & 18.4 & -65 & -242 \end{array}\right) \xrightarrow{M_2(\frac{1}{0.75})} \left(\begin{array}{ccc|c} 1 & -1.55 & 8.5 & 30.5 \\ 0 & 1 & -48.7 & -195 \\ 0 & 18.4 & -65 & -242 \end{array}\right)$$

$$\xrightarrow{A_{2,3}(-18.4)} \left(\begin{array}{ccc|c} 1 & -1.55 & 8.5 & 30.5 \\ 0 & 1 & -48.7 & -195 \\ 0 & 0 & 831 & 3350 \end{array}\right) \xrightarrow{M_3(\frac{1}{831})} \left(\begin{array}{ccc|c} 1 & -1.55 & 8.5 & 30.5 \\ 0 & 1 & -48.7 & -195 \\ 0 & 0 & 1 & 4.03 \end{array}\right)$$

This yields

$$x_3 = 4.03$$
$$x_2 = -195 + (48.7)(4.03) = 1.26$$
$$x_1 = 30.5 + (1.55)(1.26) - 8.5(4.03) = -1.8$$

Here the errors are significant. The relative errors, given as percentages, are

$$x_1: \quad \varepsilon_r = \left|\frac{-0.2}{2}\right| = 10\%$$

$$x_2: \quad \varepsilon_r = \left|\frac{0.26}{1}\right| = 26\%$$

$$x_3: \quad \varepsilon_r = \left|\frac{0.03}{4}\right| = 0.75\%$$

Let us now repeat the procedure *with* pivoting. We obtain (with the pivots circled):

$$\left(\begin{array}{ccc|c} 0.0002 & -0.00031 & 0.0017 & 0.00609 \\ 5 & -7 & 6 & 7 \\ \textcircled{8} & 6 & 3 & 2 \end{array}\right)$$

$$\xrightarrow{P_{1,3}} \left(\begin{array}{ccc|c} \textcircled{8} & 6 & 3 & 2 \\ 5 & -7 & 6 & 7 \\ 0.0002 & -0.00031 & 0.0017 & 0.00609 \end{array}\right)$$

$$\xrightarrow{M_1(\frac{1}{8})} \left(\begin{array}{ccc|c} 1 & 0.75 & 0.375 & 0.25 \\ 5 & -7 & 6 & 7 \\ 0.0002 & -0.00031 & 0.0017 & 0.00609 \end{array}\right)$$

$$\xrightarrow[A_{1,3}(-0.0002)]{A_{1,2}(-5)} \left(\begin{array}{ccc|c} 1 & 0.75 & 0.375 & 0.25 \\ 0 & \textcircled{-10.8} & 4.13 & 5.75 \\ 0 & -0.00046 & 0.00163 & 0.00604 \end{array}\right)$$

$$\xrightarrow{M_2(\frac{-1}{10.8})} \left(\begin{array}{ccc|c} 1 & 0.75 & 0.375 & 0.25 \\ 0 & 1 & -0.382 & -0.532 \\ 0 & -0.00046 & 0.00163 & 0.00604 \end{array}\right)$$

$$\xrightarrow{A_{2,3}(0.00046)} \left(\begin{array}{ccc|c} 1 & 0.75 & 0.375 & 0.25 \\ 0 & 1 & -0.382 & -0.532 \\ 0 & 0 & \textcircled{0.00145} & 0.0058 \end{array}\right)$$

$$\xrightarrow{M_3(\frac{1}{0.00145})} \left(\begin{array}{ccc|c} 1 & 0.75 & 0.375 & 0.25 \\ 0 & 1 & -0.382 & -0.532 \\ 0 & 0 & 1 & 4.00 \end{array}\right)$$

Hence

$$x_3 = 4.00$$
$$x_2 = -0.532 + (0.382)(4.00) = 0.996$$
$$x_1 = 0.25 - 0.75(0.996) - (0.375)(4.00) = -2.00$$

Thus, with pivoting and three-significant-digit rounding, x_1 and x_3 are obtained exactly and x_2 is obtained with the relative error of $0.004/1 = 0.4$ percent.

Before leaving this section, we note that there are some matrices for which a small round-off error can have disastrous results. Such matrices are called **ill-conditioned**

EXAMPLE 4 Consider the system

$$\begin{aligned} x_1 + \quad\; x_2 &= 1 \\ x_1 + 1.005x_2 &= 0 \end{aligned}$$

The solution is easily seen to be $x_1 = 201$, $x_2 = -200$. If, with or without pivoting, the coefficients are rounded to three significant digits, we obtain the system

$$\begin{aligned} x_1 + \quad x_2 &= 1 \\ x_1 + 1.01x_2 &= 0 \end{aligned}$$

with the solution $x_1 = 101$, $x_2 = -100$. Note that by rounding we introduced a relative error of $0.005/1.005 \approx 0.5$ percent in one coefficient but this induced an error of about 50 percent in the final answer!

There are techniques for recognizing and dealing with ill-conditioned matrices. Some of these are discussed in the references listed in the first section.

PROBLEMS 8.2

In Problems 1–4 solve the given system by Gaussian elimination with partial pivoting. Use a hand calculator and round to six significant digits at every step.

1. $$\begin{aligned} 2x_1 - x_2 + x_3 &= 0.3 \\ -4x_1 + 3x_2 - 2x_3 &= -1.4 \\ 3x_1 - 8x_2 + 3x_3 &= 0.1 \end{aligned}$$

2. $$\begin{aligned} 4.7x_1 + 1.81x_2 + 2.6x_3 &= -5.047 \\ -3.4x_1 - 0.25x_2 + 1.1x_3 &= 11.495 \\ 12.3x_1 + 0.06x_2 + 0.77x_3 &= 7.9684 \end{aligned}$$

3.
$$\begin{aligned} -7.4x_1+3.61x_2+8.04x_3&=25.1499\\ 12.16x_1-2.7x_2-0.891x_3&=3.2157\\ -4.12x_1+6.63x_2-4.38x_3&=-36.1383 \end{aligned}$$

4.
$$\begin{aligned} 4.1x_1-0.7x_2+8.3x_3+3.9x_4&=-4.22\\ 2.6x_1+8.1x_2+0.64x_3-0.8x_4&=37.452\\ -5.3x_1-0.2x_2+7.4x_3-0.55x_4&=-25.73\\ 0.8x_1-1.3x_2+3.6x_3+1.6x_4&=-7.7 \end{aligned}$$

In Problems 5 and 6 solve the sytem by Gaussian elimination with and without pivoting by rounding to three significant figures. Then solve the system exactly and compute the relative errors of all six computed values.

5.
$$\begin{aligned} 0.1x_1+0.05x_2+0.2x_3&=1.3\\ 12x_1+25x_2-3x_3&=10\\ -7x_1+8x_2+15x_3&=2 \end{aligned}$$

6.
$$\begin{aligned} 0.02x_1+0.03x_2-0.04x_3&=-0.04\\ 16x_1+2x_2+4x_3&=0\\ 50x_1+10x_2+8x_3&=6 \end{aligned}$$

7. Show that the system

$$\begin{aligned} x_1+x_2&=50\\ x_1+1.026x_2&=20 \end{aligned}$$

is ill-conditioned if rounding is done to three significant figures. What is the approximate relative error in each answer induced by rounding?

8. Do the same for the system

$$\begin{aligned} -0.0001x_1+x_2&=2\\ -x_1+x_2&=3 \end{aligned}$$

8.3 Solving Linear Systems II: Iterative Methods

In the last section we developed a method for solving linear systems *directly*. That is, we carried out a fixed number of steps that led to a single answer. In numerical analysis, this procedure is the exception rather than the rule. A much more common procedure is called *iteration*. With iteration, the idea is to come up with a *sequence* of approximations to the answer. If things work well, this sequence will converge to the correct answer in the sense that each term or *iterate* in the sequence is a better approximation to the answer than the ones that precede it.

To give you a taste of what iteration is all about, consider the following *algorithm* for computing $\sqrt{2}$:†

$$x_{n+1}=\frac{1}{2}\left(x_n+\frac{2}{x_n}\right) \tag{1}$$

† The method used here—called *Newton's method*—was discovered in the seventeenth century by Newton. A detailed discussion on how to use this method to find roots of polynomials appears in the Computer Supplement to this text.

Table 8.1

n	x_n	$2/x_n$	$x_n+(2/x_n)$	$x_{n+1}=\frac{1}{2}[x_n+(2/x_n)]$
0	1.0	2.0	3.0	1.5
1	1.5	1.333333333	2.833333333	1.416666667
2	1.416666667	1.411764706	2.828431373	1.414215686
3	1.414215686	1.414211438	2.828427125	1.414213562
4	1.414213562	1.414213562	2.828427125	1.414213562

This means that we start with a value x_0 and use Equation (1) to compute x_1; we then use (1) to compute x_2; and so on. In Table 8.1, computations were carried out on a hand calculator with 10 significant digits of precision. It is apparent that the method converges very quickly to the correct answer.

There are two commonly used iterative techniques for solving a system of equations $A\mathbf{x} = \mathbf{b}$: the **Jacobi† method** and the **Gauss-Seidel‡ method**. These methods are used under certain special circumstances. If A is ill-conditioned, for example, then, as we have seen, certain direct techniques fail. If the matrix A has a large number of zeros (A is then called a *sparse matrix*), iterative techniques will often provide better results with less work. The two methods do not always converge, however. After describing them, we shall examine some conditions under which the methods always converge. In the following discussion we assume that $\det A \neq 0$ so that the system has a unique solution.

I. JACOBI ITERATION Let us illustrate the method by solving a particular system. First we note that since $\det A \neq 0$, A has no zero columns. Thus, by possibly rearranging the rows of A we can get a new coefficient matrix A' with nonzero diagonal components (see Problem 14). Hence we assume that for the $n\times n$ matrix $A=(a_{ij})$, $a_{ii}\neq 0$ for $i=1, 2, \ldots, n$.

EXAMPLE 1 Solve the system

$$\begin{aligned} 4.4x_1-2.3x_2+0.7x_3&=-7.43\\ 0.8x_1+2.5x_2+1.1x_3&=12.17\\ -1.6x_1+0.4x_2-5.2x_3&=26.12 \end{aligned} \tag{2}$$

Solution The following computations are carried out to five significant figures.

Step 1. Rewrite system (2) so that, in the ith equation, x_i is written in terms of the other variables:

† Carl Gustav Jacobi (1804–1851) was a brilliant and versatile German mathematician.

‡ We encountered the great Karl Fredrich Gauss in Chapter 1. P. L. V. Seidel (1821–1896) was a German mathematician.

$$\begin{aligned}
x_1 &= -\frac{7.43}{4.4}+\frac{2.3}{4.4}x_2-\frac{0.7}{4.4}x_3 = -1.6886+0.52273x_2-0.15909\ x_3 \\
x_2 &= \frac{12.17}{2.5}-\frac{0.8}{2.5}x_1-\frac{1.1}{2.5}x_3 = \ 4.868\ -0.32x_1-0.44x_3 \\
x_3 &= -\frac{26.12}{5.2}-\frac{1.6}{5.2}x_1+\frac{0.4}{5.2}x_2 = -5.0231-0.30769x_1+0.076923x_2
\end{aligned} \tag{3}$$

Step 2. Arbitrarily choose an initial approximation to the solution: $x_1^{(0)}$, $x_2^{(0)}$, $x_3^{(0)}$. If no other information is available, choose $x_1^{(0)}=x_2^{(0)}=x_3^{(0)}=0$.

Step 3. Substitute these initial values into the right-hand side of (3) to obtain the new approximation $x_1^{(1)}$, $x_2^{(1)}$, $x_3^{(1)}$:

$$\begin{aligned}
x_1^{(1)} &= -1.6886+0-0=-1.6886 \\
x_2^{(1)} &= \ 4.868\ -0-0= \ 4.868 \\
x_3^{(1)} &= -5.0231-0+0=-5.0231
\end{aligned}$$

Step 4. Use the values computed in step 3 to compute $x_1^{(2)}$, $x_2^{(2)}$, $x_3^{(2)}$ and continue in this fashion to generate the sequences $\{x_1^{(n)}\}$, $\{x_2^{(n)}\}$, $\{x_3^{(n)}\}$.

$$\begin{aligned}
x_1^{(2)} &= -1.6886+0.52273x_2^{(1)}-0.15909x_3^{(1)} \\
&= 1.6886+0.52273(4.868)-0.15909(-5.0231)=1.6552 \\
x_2^{(2)} &= 4.868-0.32x_1^{(1)}-0.44x_3^{(1)} \\
&= 4.868-0.32(-1.6886)-0.44(-5.0231)=7.6185 \\
x_3^{(2)} &= -5.0231-0.30769x_1^{(1)}+0.076923x_2^{(1)} \\
&= -5.0231-0.30769(-1.6886)+0.076923(4.868)=-4.1291
\end{aligned}$$

Continuing in this fashion, we obtain Table 8.2 (rounded to five figures).

Table 8.2

Iterate	$x_1^{(n)}$	$x_2^{(n)}$	$x_3^{(n)}$
0	0	0	0
1	−1.6886	4.868	−5.0231
2	1.6552	7.6185	−4.1291
3	2.9507	6.1551	−4.9464
4	2.3158	6.1002	−5.4575
5	2.3684	6.5282	−5.2664
6	2.5617	6.4273	−5.2497
7	2.5063	6.3581	−5.3169
8	2.4808	6.4054	−5.3052
9	2.5037	6.4084	−5.2937
10	2.5034	6.3960	−5.3005
11	2.4980	6.3991	−5.3014
12	2.4998	6.4013	−5.2995
13	2.5006	6.3998	−5.2999
14	2.4999	6.3998	−5.3002
15	2.5000	6.4001	−5.3000

It appears (as we could have seen as early as iterate 8) that the sequences are converging to the values $x_1 = 2.5$, $x_2 = 6.4$, $x_3 = -5.3$. This can be verified by direct substitution. Note that, at least for this problem, the Jacobi iterates converge, but they converge rather slowly. The Gauss-Seidel method can improve the speed of convergence.

II. GAUSS-SEIDEL ITERATION If you look closely at the steps in Jacobi iteration, you will note some inefficiency in step 3. In computing $x_2^{(n)}$, we have already computed a new value for $x_1^{(n)}$ but have instead used the old value $x_1^{(n-1)}$. Since the iterates are converging, it makes sense to use the latest available information.

EXAMPLE 2 Solve the system of Example 1 by using the Gauss-Seidel method.

Solution We start, as before, with $x_1^{(0)} = x_2^{(0)} = x_3^{(0)} = 0$. Then, as before,

$$x_1^{(1)} = -1.6886 + 0 + 0 = -1.6886$$

But the next step is different. Using this new approximation to x_1, we obtain

$$x_2^{(1)} = 4.868 - 0.32x_1^{(1)} - 0.44x_3^{(0)} = 4.868 - 0.32(-1.6886) = 5.4084$$

Now we have new approximations for both x_1 and x_2. Using these values in system (3), we have

$$\begin{aligned} x_3^{(1)} &= -5.0231 - 0.30769x_1^{(1)} + 0.076923x_2^{(1)} \\ &= -5.0231 - 0.30769(-1.6886) + 0.076923(5.4084) = -4.0875 \end{aligned}$$

Continuing in this way (always using the latest approximations), we obtain Table 8.3.

Table 8.3

Iterate	$x_1^{(n)}$	$x_2^{(n)}$	$x_3^{(n)}$
0	0	0	0
1	−1.6886	5.4084	−4.0875
2	1.7888	6.0941	−5.1047
3	2.3091	6.3752	−5.2432
4	2.4780	6.3820	−5.2946
5	2.4898	6.4009	−5.2968
6	2.5000	6.3986	−5.3001
7	2.4993	6.4003	−5.2998
8	2.5002	6.3998	−5.3001
9	2.5000	6.4000	−5.3000
10	2.5000	6.4000	−5.3000

Again we conclude that $x_1 = 2.5$, $x_2 = 6.4$, $x_3 = -5.3$. Note that the Gauss-Seidel iterations converge more rapidly than the Jacobi iterations.

Warning. It is usually (but not always) true that the Gauss-Seidel method is more efficient than the Jacobi method. In Example 7 we encounter a system for which the Jacobi iterates converge (slowly) but the Gauss-Seidel iterates diverge. The converse is also possible (see Problem 13).

III. CONVERGENCE As mentioned above, these two methods do not always give a converging sequence of iterates. We cite below several conditions which ensure that the iterates converge. The proofs are beyond the scope of this text; for a good discussion of this problem, consult the book by Fox cited earlier.

DEFINITION 1 Let

$$A = \begin{pmatrix} a_{11} & a_{12} & \cdots & a_{1n} \\ a_{21} & a_{22} & \cdots & a_{2n} \\ \vdots & \vdots & & \vdots \\ a_{n1} & a_{n2} & \cdots & a_{nn} \end{pmatrix}$$

Then A is **strictly diagonally dominant** if, in every row, the absolute value of the diagonal component is greater than the sum of the absolute values of the off-diagonal components. That is:

$$|a_{ii}| > |a_{i1}| + |a_{i2}| + \cdots + |a_{i,i-1}| + |a_{i,i+1}| + \cdots + |a_{in}| = \sum_{\substack{j=1 \\ j \neq i}}^{n} |a_{ij}| \tag{4}$$

for $i = 1, 2, \ldots, n$.

EXAMPLE 3 The matrix $A = \begin{pmatrix} 4.4 & -2.3 & 0.7 \\ 0.8 & 2.5 & 1.1 \\ -1.6 & 0.4 & -5.2 \end{pmatrix}$ is strictly diagonally dominant because

$$|4.4| > |-2.3| + |0.7| = 3$$

$$|2.5| > |0.8| + |1.1| = 1.9$$

and

$$|-5.2| > |-1.6| + |0.4| = 2$$

EXAMPLE 4 Consider the system

$$\begin{aligned} x_1 + 3x_2 - x_3 &= 6 \\ 4x_1 - x_2 + x_3 &= 5 \\ x_1 + x_2 - 7x_3 &= -9 \end{aligned} \tag{5}$$

The matrix $A=\begin{pmatrix}1 & 3 & -1\\ 4 & -1 & 1\\ 1 & 1 & -7\end{pmatrix}$ is *not* strictly diagonally dominant. However, if we interchange the first two equations in system (5) (which, of course, does not change the solutions), then the matrix of the rearranged system is $\begin{pmatrix}4 & -1 & 1\\ 1 & 3 & -1\\ 1 & 1 & -7\end{pmatrix}$, which *is* strictly diagonally dominant.

The importance of systems with strictly diagonally dominant coefficient matrices is given in the following theorem. Its proof is suggested in Problem 19.

THEOREM 1 If A is strictly diagonally dominant, then both the Jacobi and the Gauss-Seidel iterations converge to the unique solution to $A\mathbf{x}=\mathbf{b}$ for any vector $\mathbf{b}$.

Note. Since the matrix of Examples 1 and 2 is strictly diagonally dominant, we know before doing any computations that both sequences of iterates will converge.

Remark. As we shall see in Example 6, there are matrices that are *not* strictly diagonally dominant but for which both sequences of iterates will converge.

Let A be an $n\times n$ matrix. Let L denote the $n\times n$ matrix consisting of the components of A below the main diagonal and zero everywhere else; D is the $n\times n$ matrix with the same diagonal components as A and zero everywhere else; U is the matrix consisting of the components of A above the main diagonal and zeros everywhere else. Here L, D, and U are called the *lower triangular*, the *diagonal*, and the *upper triangular* parts of A, respectively. We have, clearly,

$$A=L+D+U \tag{6}$$

EXAMPLE 5 Let $A=\begin{pmatrix}1 & 2 & 3\\ 4 & 5 & 6\\ 7 & 8 & 9\end{pmatrix}$. Then $L=\begin{pmatrix}0 & 0 & 0\\ 4 & 0 & 0\\ 7 & 8 & 0\end{pmatrix}$, $D=\begin{pmatrix}1 & 0 & 0\\ 0 & 5 & 0\\ 0 & 0 & 9\end{pmatrix}$, and $U=\begin{pmatrix}0 & 2 & 3\\ 0 & 0 & 6\\ 0 & 0 & 0\end{pmatrix}$.

THEOREM 2 Let $r(A)$ denote the absolute value of the eigenvalue of A with largest absolute value. Then, referring to the system $A\mathbf{x}=\mathbf{b}$ with $\det A \neq 0$:

i. The Jacobi iterates will converge if and only if

$$r[D^{-1}(L+U)]<1 \tag{7}$$

ii. The Gauss-Seidel iterates will converge if and only if

$$r[(D+L)^{-1}U]<1 \tag{8}$$

EXAMPLE 6 Let $A=\begin{pmatrix}2 & 3\\ 1 & 4\end{pmatrix}$; then $L=\begin{pmatrix}0 & 0\\ 1 & 0\end{pmatrix}$, $D=\begin{pmatrix}2 & 0\\ 0 & 4\end{pmatrix}$, $U=\begin{pmatrix}0 & 3\\ 0 & 0\end{pmatrix}$, $D^{-1}=\begin{pmatrix}\frac{1}{2} & 0\\ 0 & \frac{1}{4}\end{pmatrix}$, $D^{-1}(L+U)=\begin{pmatrix}0 & \frac{3}{2}\\ \frac{1}{4} & 0\end{pmatrix}$, and the eigenvalues of $D^{-1}(L+U)$ are $\pm\sqrt{\frac{3}{8}}$. Thus $r[D^{-1}(L+U)]=\sqrt{\frac{3}{8}}$. Similarly, we find that $(D+L)^{-1}U=\begin{pmatrix}0 & \frac{3}{2}\\ 0 & -\frac{3}{8}\end{pmatrix}$ with eigenvalues 0 and $-\frac{3}{8}$. Thus $r[(D+L)^{-1}U]=\frac{3}{8}$. This provides an example of a matrix that is not strictly diagonally dominant but for which both the Jacobi and the Gauss-Seidel iterates will converge.

EXAMPLE 7 Let $A=\begin{pmatrix}1 & 0 & 1\\ -1 & 1 & 0\\ 1 & 2 & -3\end{pmatrix}$. Then $L=\begin{pmatrix}0 & 0 & 0\\ -1 & 0 & 0\\ 1 & 2 & 0\end{pmatrix}$, $D=\begin{pmatrix}1 & 0 & 0\\ 0 & 1 & 0\\ 0 & 0 & -3\end{pmatrix}$, and $U=\begin{pmatrix}0 & 0 & 1\\ 0 & 0 & 0\\ 0 & 0 & 0\end{pmatrix}$. We find that $D^{-1}=\begin{pmatrix}1 & 0 & 0\\ 0 & 1 & 0\\ 0 & 0 & -\frac{1}{3}\end{pmatrix}$ and $D^{-1}(L+U)=$

$$\begin{pmatrix}1 & 0 & 0\\ 0 & 1 & 0\\ 0 & 0 & -\frac{1}{3}\end{pmatrix}\begin{pmatrix}0 & 0 & 1\\ -1 & 0 & 0\\ 1 & 2 & 0\end{pmatrix}=\begin{pmatrix}0 & 0 & 1\\ -1 & 0 & 0\\ -\frac{1}{3} & -\frac{2}{3} & 0\end{pmatrix}$$

The characteristic equation of $D^{-1}(L+U)$ is $\lambda^3+\lambda/3-\frac{2}{3}=0$ with approximate roots $\lambda_1\approx 0.748$, $\lambda_2\approx -0.374+0.868i$, and $\lambda_3\approx -0.374-0.868i$. We have $|\lambda_1|=0.748$, $|\lambda_2|=|\lambda_3|=\sqrt{0.374^2+0.868^2}\approx 0.945$. Thus $r[D^{-1}(L+U)]\approx 0.945$ and *the Jacobi iterates will converge.* On the other hand, we find that $D+L=\begin{pmatrix}1 & 0 & 0\\ -1 & 1 & 0\\ 1 & 2 & -3\end{pmatrix}$ and

$$(D+L)^{-1}U=\begin{pmatrix}1 & 0 & 0\\ 1 & 1 & 0\\ 1 & \frac{2}{3} & -\frac{1}{3}\end{pmatrix}\begin{pmatrix}0 & 0 & 1\\ 0 & 0 & 0\\ 0 & 0 & 0\end{pmatrix}=\begin{pmatrix}0 & 0 & 1\\ 0 & 0 & 1\\ 0 & 0 & 1\end{pmatrix}$$

The characteristic equation of $(D+L)^{-1}U$ is $-\lambda^2(\lambda-1)=0$ so that $\lambda_1=\lambda_2=0$, $\lambda_3=1$, and $r[(D+L)^{-1}U]=1$. Thus *the Gauss-Seidel iterates diverge.*

Remark 1. It can be further shown that the *rate* of convergence in the two methods depends on the values of $r[D^{-1}(L+U)]$ and $r[(D+L)^{-1}U]$. The smaller the value of r, the faster will be the rate of convergence.

Remark 2. Let $|A|$ denote the max-row sum norm of A (see Equation 7.7.11).

$$|A| = \max_{1 \le i \le n} \sum_{j=1}^{n} |a_{ij}| \tag{9}$$

Using Gershgorin's circle theorem (see Problem 18), it is not difficult to prove that, for any $n \times n$ matrix A,

$$r(A) \le |A| \tag{10}$$

The following result then follows directly from Theorem 2.

THEOREM 3 Let A, D, L, and U be as in Theorem 2. Then:

i. The sequence of Jacobi iterates converges if

$$|D^{-1}(L+U)| < 1 \tag{11}$$

ii. The sequence of Gauss-Seidel iterates converges if

$$|(D+L)^{-1}U| < 1 \tag{12}$$

EXAMPLE 8 Let $A = \begin{pmatrix} 1 & 0 & \frac{1}{2} \\ \frac{1}{2} & 1 & 1 \\ \frac{1}{2} & 0 & 1 \end{pmatrix}$. Then A is not strictly diagonally dominant, but we can still show that the Gauss-Seidel iterates will converge. For $D+L = \begin{pmatrix} 1 & 0 & 0 \\ \frac{1}{2} & 1 & 0 \\ \frac{1}{2} & 0 & 1 \end{pmatrix}$, $(D+L)^{-1} = \begin{pmatrix} 1 & 0 & 0 \\ -\frac{1}{2} & 1 & 0 \\ -\frac{1}{2} & 0 & 1 \end{pmatrix}$ and

$$(D+L)^{-1}U = \begin{pmatrix} 1 & 0 & 0 \\ -\frac{1}{2} & 1 & 0 \\ -\frac{1}{2} & 0 & 1 \end{pmatrix} \begin{pmatrix} 0 & 0 & \frac{1}{2} \\ 0 & 0 & 1 \\ 0 & 0 & 0 \end{pmatrix} = \begin{pmatrix} 0 & 0 & \frac{1}{2} \\ 0 & 0 & \frac{3}{4} \\ 0 & 0 & -\frac{1}{4} \end{pmatrix}$$

Thus $|(D+L)^{-1}U| = \frac{3}{4} < 1$.

IV. ERROR ANALYSIS—OR WHEN DO WE STOP? In solving problems by iteration, there is always the question of determining when to stop. There are two ways to make this decision. First, we can agree to stop after a fixed number of iterations, say 10 or 20. But since we do not know how many iterations it will take to get a reasonably accurate answer, this method is not very useful.

A better device is to stop when the relative error ε_r is sufficiently small. Remember:

$$\varepsilon_r = \left| \frac{x^* - x}{x} \right| \tag{13}$$

where x is the exact solution and x^* is the approximation. Of course, we cannot compute ε_r exactly since we do not know the exact answer x. (If we did, we wouldn't have any problem in the first place.) We can, however, for many numerical schemes, estimate the relative error in an iteration scheme by the formula

$$\varepsilon_r^{(n)} = \left| \frac{x^{(n)} - x^{(n-1)}}{x^{(n)}} \right| \tag{14}$$

Formula (14) can be explained in the following way: If we know that the scheme converges, then the iterate $x^{(n)}$ is getting closer and closer to the "correct" answer x. Thus the absolute error $\varepsilon_a = |x^{(n)} - x|$ is approaching zero. But then, since $x^{(n)} \approx x$, we have $|x^{(n)} - x^{(n-1)}| \approx |x^{(n)} - x|$. This means that formula (14) approximates the true relative error

$$\varepsilon_r = \left| \frac{x^{(n)} - x}{x} \right|$$

Thus we can agree to stop when $\varepsilon_r^{(n)}$ is smaller than some agreed upon value ε. Typically we have $\varepsilon = 0.1$, 0.01, 0.001, or some similar value.

In Example 1, suppose we agree to iterate until the estimated relative error $\varepsilon_r^{(n)}$ in the computation of x_1 is less than 0.01. From Table 8.2 we get Table 8.4.

Table 8.4

Iterate	$x_1^{(n)}$	$\lvert x_1^{(n)} - x_1^{(n-1)} \rvert$	$\varepsilon_r^{(n)} = \left\lvert \frac{x_1^{(n)} - x_1^{(n-1)}}{x_1^{(n)}} \right\rvert$
0	0		
1	−1.6886	1.6886	1
2	1.6552	3.3438	2.0202
3	2.9507	1.2955	0.43905
4	2.3158	0.6349	0.27416
5	2.3684	0.0526	0.02221
6	2.5617	0.1933	0.07546
7	2.5063	0.0554	0.02210
8	2.4808	0.0255	0.01028
9	2.5037	0.0229	0.00915
10	2.5034	0.0003	0.00012

Here we would stop after the ninth iterate. This would give us the estimate $x_1 \approx 2.5037$. Since $x_1 = 2.5$, the true relative error is $0.0037/2.5 = 0.00148$. It is apparent that this method gives us only a crude measure of the relative error. It is easy, however, to compute the approximations $\varepsilon_r^{(n)}$; and, under conditions of convergence, they do provide a reasonable measure of how close we are getting to the right answer.

PROBLEMS 8.3

In Problems 1–6 determine whether the given matrix is strictly diagonally dominant.

1. $\begin{pmatrix} 2 & 1 \\ 1 & 2 \end{pmatrix}$ **2.** $\begin{pmatrix} 3 & 3 \\ 4 & 5 \end{pmatrix}$ **3.** $\begin{pmatrix} 1 & \frac{1}{2} & \frac{1}{2} \\ \frac{1}{2} & 1 & \frac{1}{2} \\ \frac{1}{2} & \frac{1}{2} & 1 \end{pmatrix}$

4. $\begin{pmatrix} 1 & \frac{1}{2} & \frac{1}{3} \\ \frac{1}{2} & 1 & -\frac{1}{3} \\ -\frac{1}{2} & -\frac{1}{3} & 1 \end{pmatrix}$ **5.** $\begin{pmatrix} 3 & -2 & 0 \\ 1 & -4 & 2 \\ -3 & 1 & -5 \end{pmatrix}$ **6.** $\begin{pmatrix} 6 & -2 & 3 \\ -3 & 5 & 2 \\ -2 & -4 & -7 \end{pmatrix}$

In Problems 7–12 solve the given system by using the Jacobi method and the Gauss-Seidel method. Carry out your computations until the estimated relative error $\varepsilon_r^{(n)}$ is smaller than the number given in parentheses. Start with all initial approximations equal to zero and use five significant figures.

7. $\begin{aligned} 2x_1 - x_2 &= 7 \\ 3x_1 + 5x_2 &= 4 \end{aligned}$ (0.01)

8. $\begin{aligned} 3.3x_1 - 2.7x_2 &= -0.6 \\ -4.2x_1 + 8.3x_2 &= 11.95 \end{aligned}$ (0.001)

9. $\begin{aligned} 3x_1 - x_2 + x_3 &= 4 \\ 2x_1 + 5x_2 + 2x_3 &= -5 \\ x_1 + 2x_2 + 4x_3 &= 20 \end{aligned}$ (0.01)

10. $\begin{aligned} 3.8x_1 + 1.6x_2 + 0.9x_3 &= 3.72 \\ -0.7x_1 + 5.4x_2 + 1.6x_3 &= 3.16 \\ 1.5x_1 + 1.1x_2 - 3.2x_3 &= 43.78 \end{aligned}$ (0.001)

11. $\begin{aligned} 5.2x_1 + 3.1x_2 - 1.6x_3 &= 1.64 \\ 1.7x_1 + 2.4x_2 + 0.3x_3 &= 20.42 \\ -6.3x_1 - 3.7x_2 - 12.6x_3 &= 0.27 \end{aligned}$ (0.001)

12. $\begin{aligned} -3.1x_1 + 1.9x_2 - 0.77x_3 &= -12.806 \\ 0.9x_1 - 2.4x_2 + 1.06x_3 &= 12.165 \\ 7.6x_1 - 3.9x_2 + 16.5\ x_3 &= 27.931 \end{aligned}$ (0.0001)

13. Consider the system

$$\begin{aligned} x_1 + \tfrac{1}{2}x_2 + \tfrac{1}{2}x_3 &= 2 \\ \tfrac{1}{2}x_1 + x_2 + \tfrac{1}{2}x_3 &= 2 \\ \tfrac{1}{2}x_1 + \tfrac{1}{2}x_2 + x_3 &= 2 \end{aligned}$$

a. Show that the matrix of the system is not strictly diagonally dominant.
b. Starting with $x_1^{(0)} = x_2^{(0)} = x_3^{(0)} = 0.8$, show that the Jacobi iterates oscillate back and forth between the values 0.8 and 1.2. That is, show that the sequence of Jacobi iterates diverges.
c. Show that the Gauss-Seidel iterates converge to the solution $x_1 = x_2 = x_3 = 1$ by computing eight iterates and rounding to five significant figures.
***d.** Explain the results of parts (*b*) and (*c*) in light of Theorem 2.

***14.** Let A be an $n \times n$ matrix with $\det A \neq 0$. Show that it is always possible to rearrange the rows of A so that the diagonal components of A are all nonzero.

15. Let A be a diagonal matrix with $\det A \neq 0$. Show that $r[D^{-1}(L+U)] = r[(D+L)^{-1}U] = 0$.

16. Let A be an invertible upper or lower triangular matrix. Show that the sequences of Jacobi and Gauss-Seidel iterates always converge.

17. Let $A = \begin{pmatrix} a & b \\ c & d \end{pmatrix}$. Show that both the Jacobi and the Gauss-Seidel iterates converge if and only if $|bc/ad| < 1$. This shows that it is impossible to find an example of a 2×2 system where one sequence of iterates converges while the other does not.

***18.** Use Gershgorin's circle theorem (Theorem 7.8.3) to show that $r(A) \leq |A|$, where $|A| = \max_{1 \leq i \leq n} \sum_{j=1}^{n} |a_{ij}|$.

***19.** Use part (*i*) of Theorem 3 to show that if A is strictly diagonally dominant, then the Jacobi iterates converge.

8.4 Computing Eigenvalues and Eigenvectors

As we have seen, the computation of eigenvalues and eigenvectors for a given matrix A is important for a variety of applications. It is tempting to estimate eigenvalues by first finding the characteristic polynomial $p(\lambda) = \det(A - \lambda I)$ and then estimating, directly, the roots of $p(\lambda)$. There are two problems with this approach. First, polynomials are often ill-conditioned; that is, a small round-off error in the coefficients of the polynomial can lead to large errors in the roots. Second, even if the coefficients of $p(\lambda)$ are exact, it is still difficult to find all the roots of a polynomial. For these reasons a number of techniques have been devised for computing eigenvalues and eigenvectors directly. The first of these is used to compute the eigenvalue of largest absolute value.

DEFINITION 1 **DOMINANT EIGENVALUE AND EIGENVECTOR** Let $\lambda_1, \lambda_2, \ldots, \lambda_n$ be the eigenvalues of A. Then the eigenvalue λ_1 is **dominant** if

$$|\lambda_1| > |\lambda_i| \qquad \text{for } i = 2, \ldots, n \tag{1}$$

If $\mathbf{v}_1$ is an eigenvector of A corresponding to λ_1, then $\mathbf{v}_1$ is called a **dominant eigenvector**.

EXAMPLE 1 If the eigenvalues of A are $-4, -2, 1, 3$, then -4 is dominant.

EXAMPLE 2 If the eigenvalues of A are $-5, 3, 5$, then A has no dominant eigenvalue since $|-5| = |5|$.

We now describe a method, called the **power method**, for computing the dominant eigenvalue and eigenvector of a matrix.

I. THE POWER METHOD Let $\lambda_1, \lambda_2, \ldots, \lambda_n$ be the eigenvalues of A and suppose that

$$|\lambda_1| > |\lambda_2| \geq |\lambda_3| \geq \cdots \geq |\lambda_n| \tag{2}$$

That is, λ_1 is the dominant eigenvalue. Suppose further that A is diagonalizable; that is, A has n linearly independent eigenvectors $\mathbf{u}_1, \mathbf{u}_2, \ldots, \mathbf{u}_n$. Let $\mathbf{x}_0$ be a vector in $\mathbb{R}^n$. There are constants $c_1, c_2, \ldots, c_n$ such that

$$\mathbf{x}_0 = c_1\mathbf{u}_1 + c_2\mathbf{u}_2 + \cdots + c_n\mathbf{u}_n \tag{3}$$

We assume that $c_1 \neq 0$. Define a sequence of iterates by the formula

$$\boxed{\mathbf{x}_{n+1} = A\mathbf{x}_n} \tag{4}$$

Then
$$\begin{aligned}\mathbf{x}_1 = A\mathbf{x}_0 &= c_1A\mathbf{u}_1 + c_2A\mathbf{u}_2 + \cdots + c_nA\mathbf{u}_n \\ &= c_1\lambda_1\mathbf{u}_1 + c_2\lambda_2\mathbf{u}_2 + \cdots + c_n\lambda_n\mathbf{u}_n\end{aligned}$$

Continuing to multiply by powers of A, we find that

$$\begin{aligned}\mathbf{x}_2 = A\mathbf{x}_1 = A^2\mathbf{x}_0 = A(A\mathbf{x}_0) &= c_1\lambda_1A\mathbf{u}_1 + c_2\lambda_2A\mathbf{u}_2 + \cdots + c_n\lambda_nA\mathbf{u}_n \\ &= c_1\lambda_1^2\mathbf{u}_1 + c_2\lambda_2^2\mathbf{u}_2 + \cdots + c_n\lambda_n^2\mathbf{u}_n\end{aligned}$$

$$\vdots$$

$$\mathbf{x}_k = A^k\mathbf{x}_0 = c_1\lambda_1^k\mathbf{u}_1 + c_2\lambda_2^k\mathbf{u}_2 + \cdots + c_n\lambda_n^k\mathbf{u}_n \tag{5}$$

or

$$\mathbf{x}_k = A^k\mathbf{x}_0 = \lambda_1^k\left[c_1\mathbf{u}_1 + c_2\left(\frac{\lambda_2}{\lambda_1}\right)^k\mathbf{u}_2 + \cdots + c_n\left(\frac{\lambda_n}{\lambda_1}\right)^k\mathbf{u}_n\right] \tag{6}$$

Since $|\lambda_i| < |\lambda_1|$ for $i = 2, 3, \ldots, n$, we see that $|\lambda_i/\lambda_1|^k$ approaches zero as k increases. Therefore we may write

$$\boxed{\mathbf{x}_k = A^k\mathbf{x}_0 \approx \lambda_1^kc_1\mathbf{u}_1} \tag{7}$$

Suppose $\mathbf{u}_1 = \begin{pmatrix} a_1 \\ a_2 \\ \cdot \\ \cdot \\ \cdot \\ a_n \end{pmatrix}$. Then $\lambda_1^kc_1\mathbf{u}_1 = \begin{pmatrix} \lambda_1^kc_1a_1 \\ \lambda_1^kc_1a_2 \\ \cdot\ \cdot \\ \cdot\ \cdot \\ \cdot\ \cdot \\ \lambda_1^kc_1a_n \end{pmatrix}$.

Now let a_j be nonzero. Then we form the quotient

$$\boxed{\alpha_j^{(k+1)} = \frac{j\text{th component of } A^{k+1}\mathbf{x}_0}{j\text{th component of } A^k\mathbf{x}_0} \approx \frac{\lambda_1^{k+1}c_1a_j}{\lambda_1^kc_1a_j} = \lambda_1} \tag{8}$$

This gives us a method for computing λ_1. We simply look at the ratio of the jth components of $\mathbf{x}_{k+1}$ and $\mathbf{x}_k$ and let k get large. Moreover, once we have found λ_1, we also know an eigenvector corresponding to λ_1—for, by Equation (7),

$$\mathbf{x}_k \approx \lambda_1^k c_1 \mathbf{u}_1 \tag{9}$$

is an eigenvector corresponding to λ_1 since $\lambda_1^k c_1$ is a scalar and $\mathbf{u}_1$ is an eigenvector.

Warning. The power method described above will work only if A has a dominant eigenvalue.†

EXAMPLE 3 Use the power method to find the dominant eigenvalue and eigenvector of $A = \begin{pmatrix} -4 & -5 \\ 1 & 2 \end{pmatrix}$.

Solution Here $\mathbf{x}_0$ is arbitrary so we choose a simple value for it: $\mathbf{x}_0 = \begin{pmatrix} 1 \\ 1 \end{pmatrix}$. Then

$$\mathbf{x}_1 = A\mathbf{x}_0 = \begin{pmatrix} -4 & -5 \\ 1 & 2 \end{pmatrix}\begin{pmatrix} 1 \\ 1 \end{pmatrix} = \begin{pmatrix} -9 \\ 3 \end{pmatrix} \qquad \alpha_1^{(1)} = \tfrac{-9}{1} = -9 \qquad \alpha_2^{(1)} = \tfrac{3}{1} = 3$$

$$\mathbf{x}_2 = A\mathbf{x}_1 = \begin{pmatrix} -4 & -5 \\ 1 & 2 \end{pmatrix}\begin{pmatrix} -9 \\ 3 \end{pmatrix} = \begin{pmatrix} 21 \\ -3 \end{pmatrix} \qquad \alpha_1^{(2)} = \tfrac{21}{-9} = -2.3333 \qquad \alpha_2^{(2)} = \tfrac{-3}{3} = -1$$

Continuing in this fashion we obtain Table 8.5. All results are rounded to five significant figures.

Table 8.5

Iterate	$\mathbf{x}_k$ (as a row vector)	$\alpha_1^{(k)}$	$\alpha_2^{(k)}$
0	(1, 1)	—	—
1	(−9, 3)	−9	3
2	(21, −3)	−2.3333	−1
3	(−69, 15)	−3.2857	−5
4	(201, −39)	−2.9130	−2.6
5	(−609, 123)	−3.0299	−3.1538
6	(1821, −363)	−2.9901	−2.9512
7	(−5469, 1095)	−3.0033	−3.0165
8	(16401, −3279)	−2.9989	−2.9945
9	(−49209, 9843)	−3.0004	−3.0018

† It can be shown that the power method will work even when A is not diagonalizable. In that case, however, convergence is at a slower rate.

It appears that $\alpha_1^{(k)}$ and $\alpha_2^{(k)}$ are converging to −3—which, as is easily verified, is the dominant eigenvalue of A (the other one is $\lambda_2=1$). Moreover, we see that $\mathbf{v}_1=\begin{pmatrix}-49209\\ 9843\end{pmatrix}$ is approximately equal to an eigenvector of A. To simplify this vector, we "normalize" it by dividing through by its largest component (in absolute value) −49,209 to obtain $\mathbf{v}_1'=\begin{pmatrix}1\\ -0.20002\end{pmatrix}\approx\begin{pmatrix}1\\ -\frac{1}{5}\end{pmatrix}$. It is easily verified that $\begin{pmatrix}1\\ -\frac{1}{5}\end{pmatrix}$ is an eigenvector of A corresponding to the eigenvalue $\lambda_1=-3$.

II. THE POWER METHOD WITH SCALING In the last example we saw that the iterates grew very rapidly in size. To prevent this, we do what we did to complete the problem: We *normalize* or *scale* the vector $\mathbf{x}_k$ by dividing it by its largest component (in absolute value). If we call the new scaled iterate $\mathbf{x}_k'$, then:

$$\mathbf{x}_{k+1}=A\mathbf{x}_k' \tag{10}$$

This new method is called the **power method with scaling**. It will give us an eigenvector $\mathbf{u}$ with largest component 1 and we can then find the dominant eigenvalue by solving the equation $A\mathbf{u}=\lambda_1\mathbf{u}$ for λ_1.

EXAMPLE 4 Redo Example 3 by using the power method with scaling.

Solution If $\mathbf{x}_0=\begin{pmatrix}1\\ 1\end{pmatrix}$, then $\mathbf{x}_1=\begin{pmatrix}-9\\ 3\end{pmatrix}$ as before, and

$$\mathbf{x}_1'=-\frac{1}{9}\begin{pmatrix}-9\\ 3\end{pmatrix}=\begin{pmatrix}1\\ -\frac{1}{3}\end{pmatrix}$$

Then

$$\mathbf{x}_2=A\mathbf{x}_1'=\begin{pmatrix}-4 & -5\\ 1 & 2\end{pmatrix}\begin{pmatrix}1\\ -\frac{1}{3}\end{pmatrix}=\begin{pmatrix}-\frac{7}{3}\\ \frac{1}{3}\end{pmatrix}$$

so that

$$\mathbf{x}_2'=-\frac{3}{7}\begin{pmatrix}-\frac{7}{3}\\ \frac{1}{3}\end{pmatrix}=\begin{pmatrix}1\\ -\frac{1}{7}\end{pmatrix}=\begin{pmatrix}1\\ -0.14286\end{pmatrix}.$$

We carry out further iterations in Table 8.6.

As before, we can conclude that $\mathbf{v}=\begin{pmatrix}1\\ -0.2\end{pmatrix}$ is an eigenvector of A corresponding to λ_1. Then $A\mathbf{v}=\lambda_1\mathbf{v}$ or $\begin{pmatrix}-4 & -5\\ 1 & 2\end{pmatrix}\begin{pmatrix}1\\ -0.2\end{pmatrix}=\begin{pmatrix}\lambda_1\\ -0.2\lambda_1\end{pmatrix}$, which yields $\begin{pmatrix}-3\\ 0.6\end{pmatrix}=\begin{pmatrix}\lambda_1\\ -0.2\lambda_1\end{pmatrix}$. Hence $\lambda_1=-3$.

Table 8.6

Iterate	$\mathbf{x}_k$	$\mathbf{x}'_k$ (normalized)
0	(1, 1)	(1, 1)
1	(−9, 3)	(1, −0.33333)
2	(−2.3333, 0.33333)	(1, −0.14286)
3	(−3.2857, 0.71428)	(1, −0.21739)
4	(−2.9131, 0.56522)	(1, −0.19403)
5	(−3.0299, 0.61194)	(1, −0.20197)
6	(−2.9902, 0.59606)	(1, −0.19934)
7	(−3.0033, 0.60132)	(1, −0.20022)
8	(−2.9989, 0.59956)	(1, −0.19993)
9	(−3.0004, 0.60014)	(1, −0.20002)

Note. In Table 8.6 the first component of $\mathbf{x}_k$ tends to −3. This must be the case since, for k large,

$$\mathbf{x}_k = A\mathbf{x}'_{k-1} \approx \lambda_1 \mathbf{x}'_{k-1}$$

But the first component of $\mathbf{x}'_{k-1}$ is 1. Hence the first component of $\mathbf{x}_k \approx \lambda_1$. This means that we can find λ_1 directly from the table.

III. DEFLATION The power method has the obvious drawback of giving us only the dominant eigenvalue. There are many ways to compute other eigenvalues. We examine one method here. First we need the following result.

THEOREM 1 Let $\lambda_1, \lambda_2, \ldots, \lambda_n$ be the eigenvalues of A. Let λ_1 be the dominant eigenvalue with eigenvector $\mathbf{u}_1$. Let $\mathbf{v}$ be a column vector such that $\mathbf{u}_1 \cdot \mathbf{v} = 1$. If the matrix B is given by

$$B = A - \lambda_1 \mathbf{u}_1 \mathbf{v}^t \tag{11}$$

then the eigenvalues of B are $\{0, \lambda_2, \lambda_3, \ldots, \lambda_n\}$.† The proof of this theorem can be found in the referenced book by Blum (p. 239).

† Note that since $\mathbf{u}_1$ is an $n \times 1$ matrix (a column vector) and $\mathbf{v}$ is also an $n \times 1$ matrix, then $\mathbf{v}^t$ is an $1 \times n$ matrix and $\mathbf{u}_1\mathbf{v}^t$ is an $n \times n$ matrix.

EXAMPLE 5 In Example 3 we had $A=\begin{pmatrix}-4 & -5\\ 1 & 2\end{pmatrix}$, $\lambda_1=-3$, and $\mathbf{u}_1=\begin{pmatrix}1\\ -\frac{1}{5}\end{pmatrix}$. If $\mathbf{v}=\begin{pmatrix}\frac{1}{2}\\ -\frac{5}{2}\end{pmatrix}$, then $\mathbf{u}_1\cdot\mathbf{v}=1$ and

$$B=A-\lambda_1\mathbf{u}_1\mathbf{v}^t=\begin{pmatrix}-4 & -5\\ 1 & 2\end{pmatrix}+3\begin{pmatrix}1\\ -\frac{1}{5}\end{pmatrix}(\tfrac{1}{2},-\tfrac{5}{2})$$
$$=\begin{pmatrix}-4 & -5\\ 1 & 2\end{pmatrix}+3\begin{pmatrix}\frac{1}{2} & -\frac{5}{2}\\ -\frac{1}{10} & \frac{1}{2}\end{pmatrix}=\begin{pmatrix}-4 & -5\\ 1 & 2\end{pmatrix}+\begin{pmatrix}\frac{3}{2} & -\frac{15}{2}\\ -\frac{3}{10} & \frac{3}{2}\end{pmatrix}=\begin{pmatrix}-\frac{5}{2} & -\frac{25}{2}\\ \frac{7}{10} & \frac{7}{2}\end{pmatrix}$$

Clearly $\det B=0$; hence zero is an eigenvalue of B, as expected. The other eigenvalue of B is the second eigenvalue of A. We compute this by the power method with scaling. Starting with $x_0=\begin{pmatrix}1\\ 1\end{pmatrix}$, we obtain

$$\mathbf{x}_1=B\mathbf{x}_0=\begin{pmatrix}-2.5 & -12.5\\ 0.7 & 3.5\end{pmatrix}\begin{pmatrix}1\\ 1\end{pmatrix}=\begin{pmatrix}-15\\ 4.2\end{pmatrix}$$

Then $$\mathbf{x}_1'=\begin{pmatrix}1\\ -0.28\end{pmatrix}$$

and $$\mathbf{x}_2=B\mathbf{x}_1=\begin{pmatrix}-2.5 & -12.5\\ 0.7 & 3.5\end{pmatrix}\begin{pmatrix}1\\ -0.28\end{pmatrix}=\begin{pmatrix}1\\ -0.28\end{pmatrix}$$

Thus, without further ado, we see that $\begin{pmatrix}1\\ -0.28\end{pmatrix}$ is an eigenvector of B corresponding to the eigenvalue $\lambda_2=1$. Therefore the eigenvalues of $\begin{pmatrix}-4 & -5\\ 1 & 2\end{pmatrix}$ are -3 and 1.

Note. It is not a coincidence that $B\mathbf{x}_0$ is an eigenvector of B if B is a 2×2 matrix—this is *always* the case. (See Problem 14 for a suggestion as to why this is so.)

EXAMPLE 6 Compute the eigenvalues of $A=\begin{pmatrix}4 & -1 & 1\\ -1 & 3 & -2\\ 1 & -2 & 3\end{pmatrix}$ by the power method with scaling and deflation.

Solution Let $\mathbf{x}_0=\begin{pmatrix}1\\ 1\\ 1\end{pmatrix}$. Then $\mathbf{x}_1=A\mathbf{x}_0=\begin{pmatrix}4\\ 0\\ 2\end{pmatrix}$ and $\mathbf{x}_1'=\begin{pmatrix}1\\ 0\\ 0.5\end{pmatrix}$. Similarly, $\mathbf{x}_2=A\mathbf{x}_1=\begin{pmatrix}4.5\\ -2\\ 2.5\end{pmatrix}$ and $\mathbf{x}_2'=\begin{pmatrix}1\\ -0.44444\\ 0.55556\end{pmatrix}$. Continuing in this manner we obtain the values in Table 8.7.

Table 8.7

Iterate	$\mathbf{x}_k$	$\mathbf{x}'_k$	α_k = 1st component of $\mathbf{x}_k$
0	(1, 1, 1)	(1, 1, 1)	1
1	(4, 0, 2)	(1, 0, 0.5)	4
2	(4.5, −2, 2.5)	(1, −0.44444, 0.55556)	4.5
3	(5, −3.4444, 3.5556)	(1, −0.68888, 0.71112)	5
4	(5.4, −4.4889, 4.5111)	(1, −0.83128, 0.83539)	5.4
5	(5.6667, −5.1646, 5.1687)	(1, −0.91139, 0.91212)	5.6667
6	(5.8235, −5.5584, 5.5591)	(1, −0.95448, 0.95460)	5.8235
7	(5.9091, −5.7726, 5.7728)	(1, −0.97690, 0.97693)	5.9091
8	(5.9538, −5.8846, 5.8846)	(1, −0.98838, 0.98838)	5.9538
9	(5.9768, −5.9419, 5.9419)	(1, −0.99416, 0.99416)	5.9768
10	(5.9883, −5.9708, 5.9708)	(1, −0.99708, 0.99708)	5.9883

It appears that the α_k's are converging to $\lambda_1 = 6$ with corresponding eigenvector $\mathbf{u}_1 = \begin{pmatrix} 1 \\ -1 \\ 1 \end{pmatrix}$. This is easily verified. Next we find a vector $\mathbf{v}$ such that $\mathbf{u}_1 \cdot \mathbf{v} = 1$. One obvious choice is $\mathbf{v} = \begin{pmatrix} \frac{1}{3} \\ -\frac{1}{3} \\ \frac{1}{3} \end{pmatrix}$. Then

$$\mathbf{u}_1\mathbf{v}^t = \begin{pmatrix} 1 \\ -1 \\ 1 \end{pmatrix} \left(\tfrac{1}{3}, -\tfrac{1}{3}, \tfrac{1}{3}\right) = \begin{pmatrix} \frac{1}{3} & -\frac{1}{3} & \frac{1}{3} \\ -\frac{1}{3} & \frac{1}{3} & -\frac{1}{3} \\ \frac{1}{3} & -\frac{1}{3} & \frac{1}{3} \end{pmatrix}$$

so that

$$B = A - \lambda_1 \mathbf{u}_1 \mathbf{v}^t = \begin{pmatrix} 4 & -1 & 1 \\ -1 & 3 & -2 \\ 1 & -2 & 3 \end{pmatrix} - 6\begin{pmatrix} \frac{1}{3} & -\frac{1}{3} & \frac{1}{3} \\ -\frac{1}{3} & \frac{1}{3} & -\frac{1}{3} \\ \frac{1}{3} & -\frac{1}{3} & \frac{1}{3} \end{pmatrix}$$

$$= \begin{pmatrix} 4 & -1 & 1 \\ -1 & 3 & -2 \\ 1 & -2 & 3 \end{pmatrix} - \begin{pmatrix} 2 & -2 & 2 \\ -2 & 2 & -2 \\ 2 & -2 & 2 \end{pmatrix} = \begin{pmatrix} 2 & 1 & -1 \\ 1 & 1 & 0 \\ -1 & 0 & 1 \end{pmatrix}$$

We see that $\det B = 0$; hence zero is an eigenvalue of B. To find the dominant eigenvalue of B, we again use the power method with scaling. The results are tabulated in Table 8.8.

Now it seems that the iterates are converging to $\lambda_2 = 3$ and $\mathbf{u}_2 = (1, \frac{1}{2}, -\frac{1}{2})$. Again this is easily verified. Although $\mathbf{u}_2$ is an eigenvector of both A and B, this is not always the case. (In Example 5, for instance, $\begin{pmatrix} 1 \\ -0.28 \end{pmatrix}$ was an eigenvector of B but not of A.)

Table 8.8

Iterate	$\mathbf{x}_k$	$\mathbf{x}'_k$	α_k = 1st component of $\mathbf{x}_k$
0	(1, 1, 1)	(1, 1, 1)	1
1	(2, 2, 0)	(1, 1, 0)	2
2	(3, 2, −1)	(1, 0.66667, −0.33333)	3
3	(3, 1.6667, −1.3333)	(1, 0.55556, −0.44443)	3
4	(3, 1.5556, −1.4444)	(1, 0.51853, −0.48147)	3
5	(3, 1.5185, −1.4815)	(1, 0.50617, −0.49383)	3
6	(3, 1.5062, −1.4938)	(1, 0.50207, −0.49793)	3
7	(3, 1.5021, −1.4979)	(1, 0.50070, −0.49930)	3

Finally, we use deflation again to find the last eigenvalue of B (and therefore of A). We set $\mathbf{v}_1 = \begin{pmatrix}1\\0\\0\end{pmatrix}$. Then $\mathbf{u}_2 \cdot \mathbf{v}_1 = 1$ and $\mathbf{u}_2\mathbf{v}_1^t = \begin{pmatrix}1\\ \frac{1}{2}\\ -\frac{1}{2}\end{pmatrix}(1, 0, 0) = \begin{pmatrix}1 & 0 & 0\\ \frac{1}{2} & 0 & 0\\ -\frac{1}{2} & 0 & 0\end{pmatrix}$ and $C = B - \lambda_2\mathbf{u}_2\mathbf{v}_1^t = \begin{pmatrix}2 & 1 & -1\\ 1 & 1 & 0\\ -1 & 0 & 1\end{pmatrix} - \begin{pmatrix}3 & 0 & 0\\ \frac{3}{2} & 0 & 0\\ -\frac{3}{2} & 0 & 0\end{pmatrix} = \begin{pmatrix}-1 & 1 & -1\\ -\frac{1}{2} & 1 & 0\\ \frac{1}{2} & 0 & 1\end{pmatrix}$. We shall omit the iteration, which shows that the dominant eigenvalue of C is $\lambda_3 = 1$. Thus the eigenvalues of A are 6, 3, and 1.

The power method together with deflation provides a reasonable way to find the eigenvalues of A if no two eigenvalues of A have the same absolute value and if each approximation to an eigenvalue is a good one. If, for example, λ_1 is inaccurate, then the computation of λ_2 by deflation could be a good deal more inaccurate.

There are many other ways to compute, numerically, the eigenvalues of a square matrix. One method that works fairly well on a symmetric matrix is called **Jacobi's method**. The idea is to compute a sequence of orthogonal matrices whose diagonal components approach the eigenvalues of A. Many of the references listed in Section 8.1 discuss that method. Finally, we note that the decision "when to stop" can be made, as in the last section, by computing approximate values of the relative error $\varepsilon_r^{(n)}$.

PROBLEMS 8.4

In Problems 1–6 estimate the dominant eigenvalue and eigenvector of A by using the power method with scaling.

1. $\begin{pmatrix}-2 & -2\\ -5 & 1\end{pmatrix}$ **2.** $\begin{pmatrix}8 & 3\\ -3 & -2\end{pmatrix}$ **3.** $\begin{pmatrix}-22.3 & -32\\ 12 & 17.7\end{pmatrix}$

4. $\begin{pmatrix} 1 & -1 & 4 \\ 3 & 2 & -1 \\ 2 & 1 & -1 \end{pmatrix}$ 5. $\begin{pmatrix} 3 & 2 & 4 \\ 2 & 0 & 2 \\ 4 & 2 & 3 \end{pmatrix}$ 6. $\begin{pmatrix} 5 & 4 & 2 \\ 4 & 5 & 2 \\ 2 & 2 & 2 \end{pmatrix}$

7. Use the power method to estimate the dominant eigenvalue of $A = \begin{pmatrix} 1 & 7 \\ 6 & 3 \end{pmatrix}$:

   a. Rounding to five significant figures, continue the iterations until the estimated relative error $\varepsilon_r^{(n)} < 0.001$.

   b. Compute the dominant eigenvalue exactly. What is the exact value of ε_r?

8. For the matrix $A = \begin{pmatrix} -16.32 & 13 \\ 8 & 4.79 \end{pmatrix}$, follow the steps of Problem 7. Use six significant figures.

9. Show that the iterates of the power method fail to converge for the matrix $A = \begin{pmatrix} -3 & 5 \\ -2 & 3 \end{pmatrix}$. Explain why.

10. Do the same for the matrix $A = \begin{pmatrix} 2 & -1 \\ 5 & -2 \end{pmatrix}$.

In Problems 11–13 use deflation to find the other eigenvalues.

11. For the matrix of Problem 1

12. For the matrix of Problem 3

13. For the matrix of Problem 4

14. Let $A = \begin{pmatrix} a & 0 \\ 0 & b \end{pmatrix}$. Show that, for any 2-vector $\mathbf{x}_0$, $B\mathbf{x}_0$ is an eigenvector of B, where B is defined by Equation (11).

Review Exercises for Chapter 8

In Exercises 1–4 the number x and an approximation x^* are given. Find the absolute and relative errors ε_a and ε_r.

1. $x = -7$; $x^* = -6.98$

2. $x = 1000$; $x^* = 1.002 \times 10^3$

3. $x = \frac{1}{75}$; $x^* = 1 \times 10^{-2}$

4. $x = 37539$; $x^* = 3.7 \times 10^4$

5. Reduce the matrix $\begin{pmatrix} 2 & -4 & 6 \\ 1 & -3 & 5 \\ -4 & 9 & -13 \end{pmatrix}$ to row echelon form.

In Exercises 6 and 7 solve the given system by Gaussian elimination with partial pivoting. Round to six significant digits at every step.

6. $$\begin{aligned} 3.6x_1 + 8.2x_2 - 6.4x_3 &= 1.26 \\ -4.5x_1 - 5.9x_2 + 0.3x_3 &= 2.57 \\ 0.7x_1 + 3.6x_2 - 4.8x_3 &= 2.15 \end{aligned}$$

7. $$\begin{aligned} 1.3x_1 - 9.6x_2 + 5.35x_3 &= 0.515 \\ -12x_1 - 15x_2 + 3.8x_3 &= -71.966 \\ 1.06x_1 - 22.2x_2 + 9.93x_3 &= 1.809 \end{aligned}$$

In Exercises 8 and 9 determine whether the given matrix is strictly diagonally dominant.

8. $\begin{pmatrix} -1 & \frac{1}{2} & \frac{1}{3} \\ -\frac{5}{6} & 1 & 0 \\ 2 & \frac{3}{2} & -4 \end{pmatrix}$

9. $\begin{pmatrix} -1 & \frac{1}{3} & -\frac{1}{3} \\ -\frac{5}{6} & 1 & \frac{1}{6} \\ 2 & \frac{3}{2} & -4 \end{pmatrix}$

In Exercises 10 and 11 solve the given system by using the Jacobi and Gauss-Seidel methods. Carry out the iterations until the estimated relative error $\varepsilon_r^{(n)}$ is smaller than the number given in parentheses. Use six significant figures in all computations.

10. $2.7x_1 - 0.9x_2 + 1.3x_3 = 6.98$
$-0.3x_1 + x_2 + 0.4x_3 = -2.77$ (0.001)
$4x_1 - 3.3x_2 + 9.6x_3 = 21.79$

11. $42.31x_1 + 8.62x_2 + 19.4x_3 = -2.2502$
$-4.73x_1 + 80.4x_2 - 37.2x_3 = 3.5402$ (0.0001)
$8.37x_1 + 30.9x_2 - 57.4x_3 = -24.0858$

In Exercises 12–14 estimate the dominant eigenvalue and eigenvector by using the power method with scaling.

12. $\begin{pmatrix} 8 & -2 \\ 4 & 2 \end{pmatrix}$ **13.** $\begin{pmatrix} -6 & 3 \\ 6 & 1 \end{pmatrix}$ **14.** $\begin{pmatrix} 1 & -1 & 0 \\ -1 & 2 & -1 \\ 0 & -1 & 1 \end{pmatrix}$

15. Use deflation to find the second eigenvalue of the matrix of Exercise 13.

Appendix 1: Mathematical Induction

Mathematical induction† is the name given to an elementary logical principle that can be used to prove a certain type of mathematical statement. Typically, we use mathematical induction to prove that a certain statement or equation holds for every positive integer. For example, we may need to prove that $2^n > n$ for all integers $n \geq 1$.

To do this we proceed in two steps:

> (i) We prove that the statement is true for some integer N (usually $N = 1$).
> (ii) We *assume* that the statement is true for an integer k and then *prove* that it is true for the integer $k + 1$.

If we can complete these two steps, then we will have demonstrated the validity of the statement for *all* positive integers greater than or equal to N. To convince you of this fact, we reason as follows: since the statement is true for N (by step (i)) it is true for the integer N + 1 (by step (ii)). Then it is also true for the integer $(N + 1) + 1 = N + 2$ (again by step (ii)), and so on. We now demonstrate the procedure with some examples.

EXAMPLE 1 Show that $2^n > n$ for all integers $n \geq 1$.

Solution **(i)** If $n = 1$, then $2^n = 2^1 = 2 > 1 = n$ so $2^n > n$ for $n = 1$.
(ii) Assume that $2^k > k$ where $k > 1$ is an integer.

Then

$$2^{k+1} = 2 \cdot 2^k = 2^k + 2^k \overset{\text{since } 2^k > k}{>} k + k > k + 1$$

This completes the proof since we have shown that $2^1 > 1$, which implies by step (ii) that $2^2 > 2$, so that, again by step (ii), $2^3 > 3$, so that $2^4 > 4$, and so on.

† This technique was first used in a mathematical proof by the great French mathematician Pierre de Fermat (1601–1665).

EXAMPLE 2 Use mathematical induction to prove the formula for the sum of the first n positive integers:

$$1 + 2 + 3 + \cdots + n = \frac{n(n+1)}{2}. \tag{1}$$

Solution (i) If $n = 1$, then the sum of the first one integer is 1. But $(1)(1+1)/2 = 1$ so that Equation (1) holds in the case in which $n = 1$.

(ii) Assume that (1) holds for $n = k$; that is,

$$1 + 2 + 3 + \cdots + k = \frac{k(k+1)}{2}. \tag{2}$$

We must now show that it holds for $n = k + 1$. That is, we must show that

$$1 + 2 + 3 + \cdots + k + (k+1) = \frac{(k+1)(k+2)}{2}$$

But

$$\begin{aligned} 1 + 2 + 3 + \cdots + k + (k+1) &= (1 + 2 + 3 + \cdots + k) + (k+1) \\ &\overset{\text{by (2)}}{=} \frac{k(k+1)}{2} + (k+1) \\ &= \frac{k(k+1) + 2(k+1)}{2} \\ &= \frac{(k+1)(k+2)}{2} \end{aligned}$$

and the proof is complete. You may wish to try a few examples to illustrate that formula (1) really works. For example

$$1 + 2 + 3 + 4 + 5 + 6 + 7 + 8 + 9 + 10 = \frac{10(11)}{2} = 55.$$

EXAMPLE 3 Use mathematical induction to prove the formula for the sum of the squares of the first n positive integers:

$$1^2 + 2^2 + 3^2 + \cdots + n^2 = \frac{n(n+1)(2n+1)}{6}. \tag{3}$$

Solution (i) Since $1(1+1)(2 \cdot 1 + 1)/6 = 1 = 1^2$, Equation (3) is valid for $n = 1$.

(ii) Suppose that Equation (3) holds for $n = k$; that is

$$1^2 + 2^2 + 3^2 + \cdots + k^2 = \frac{k(k+1)(2k+1)}{6}. \tag{4}$$

Then to prove that (3) is true for $n = k + 1$, we have

$$1^2 + 2^2 + 3^2 + \cdots + k^2 + (k+1)^2 \overset{\text{by (4)}}{=} \frac{k(k+1)(2k+1)}{6} + (k+1)^2$$

$$= \frac{k(k+1)(2k+1) + 6(k+1)^2}{6}$$

$$= \frac{k+1}{6}[k(2k+1) + 6(k+1)]$$

$$= \frac{k+1}{6}[2k^2 + 7k + 6]$$

$$= \frac{k+1}{6}[(k+2)(2k+3)]$$

$$= \frac{(k+1)(k+2)[2(k+1)+1]}{6},$$

which is Equation (3) for $n = k + 1$ and the proof is complete. Again, you may wish to experiment with this formula. For example,

$$1^2 + 2^2 + 3^2 + 4^2 + 5^2 + 6^2 + 7^2 = \frac{7(7+1)(2 \cdot 7 + 1)}{6}$$

$$= \frac{7 \cdot 8 \cdot 15}{6} = 140.$$

EXAMPLE 4 For $a \neq 1$, use mathematical induction to prove the formula for the sum of a geometric progression:

$$1 + a + a^2 + \cdots + a^n = \frac{1 - a^{n+1}}{1 - a}. \tag{5}$$

Solution **(i)** If $n = 0$ then

$$\frac{1 - a^{0+1}}{1 - a} = \frac{1 - a}{1 - a} = 1.$$

Thus Equation (5) holds for $n = 0$. (We use $n = 0$ instead of $n = 1$ since $a^0 = 1$ is the first term.)

(ii) Assume that (5) holds for $n = k$; that is

$$1 + a + a^2 + \cdots + a^k = \frac{1 - a^{k+1}}{1 - a}. \tag{6}$$

Then

$$1 + a + a^2 + \cdots + a^k + a^{k+1} \overset{\text{by (6)}}{=} \frac{1 - a^{k+1}}{1 - a} + a^{k+1}$$

$$= \frac{1 - a^{k+1} + (1 - a)a^{k+1}}{1 - a} = \frac{1 - a^{k+2}}{1 - a}$$

so that Equation (3) also holds for $n = k + 1$ and the proof is complete.

EXAMPLE 5 Let $A_1, A_2, \ldots, A_k$ be k invertible $n \times n$ matrices. Show that

$$(A_1A_2 \cdots A_m)^{-1} = A_m^{-1}A_{m-1}^{-1} \cdots A_2^{-1}A_1^{-1} \tag{7}$$

For $m = 2$, we have $(A_1A_2)^{-1} = A_2^{-1}A_1^{-1}$ by Theorem 2.7.3. Thus Equation (7) holds for $m = 2$. We assume it is true for $m = k$ and prove it for $m = k+1$. Let $B = A_1A_2 \cdots A_k$. Then

$$(A_1A_2 \cdots A_kA_{k+1})^{-1} = (BA_{k+1})^{-1} = A_{k+1}^{-1}B^{-1} \tag{8}$$

But, by the induction assumption,

$$B^{-1} = (A_1A_2 \cdots A_k)^{-1} = A_k^{-1}A_{k-1}^{-1} \cdots A_2^{-1}A_1^{-1} \tag{9}$$

Substituting (9) into (8) completes the proof.

PROBLEMS A1 In the following problems prove the required result by using mathematical induction.

1. Show that $1^3 + 2^3 + 3^3 + \cdots + n^3 = [n^2(n+1)^2]/4$.
2. Let $\mathbf{u}, \mathbf{v}_1, \mathbf{v}_2, \ldots, \mathbf{v}_n$ be $n + 1$ vectors in $\mathbb{R}^2$. Prove that

$$\mathbf{u} \cdot (\mathbf{v}_1 + \mathbf{v}_2 + \cdots + \mathbf{v}_n) = \mathbf{u} \cdot \mathbf{v}_1 + \mathbf{u} \cdot \mathbf{v}_2 + \cdots + \mathbf{u} \cdot \mathbf{v}_n.$$

3. Show that if $a \neq 1$,

$$1 + 2a + 3a^2 + \cdots + na^{n-1} = \frac{1 - (n+1)a^n + na^{n+1}}{(1 - a)^2}$$

4. If a set S contains n elements, show that S has 2^n subsets.
5. Assuming that every polynomial has at least one root, prove that a polynomial of degree n has exactly n roots (counting multiplicities).
6. Given that $\det AB = \det A \det B$, prove that $\det A_1A_2 \cdots A_m = \det A_1 \det A_2 \cdots \det A_m$, where $A_1, \ldots, A_m$ are $n \times n$ matrices.
7. If $A_1, A_2, \ldots, A_k$ are $m \times n$ matrices, show that $(A_1 + A_2 + \cdots + A_k)^t = A_1^t + A_2^t + \cdots + A_k^t$. You may assume that $(A + B)^t = A^t + B^t$.

Appendix 2: Complex Numbers

In Chapter 7 we encounter the problem of finding the roots of the polynomial

$$\lambda^2 + a\lambda + b = 0 \tag{1}$$

To find the roots, we use the quadratic formula to obtain

$$\lambda = \frac{-a \pm \sqrt{a^2 - 4b}}{2} \tag{2}$$

If $a^2 - 4b > 0$ there are two real roots. If $a^2 - 4b = 0$ we obtain the single root (of multiplicity 2) $\lambda = -a/2$. To deal with the case $a^2 - 4b < 0$, we introduce the **imaginary number**†

$$\boxed{i = \sqrt{-1}} \tag{3}$$

Then, for $a^2 - 4b < 0$,

$$\sqrt{a^2 - 4b} = \sqrt{(4b - a^2)(-1)} = \sqrt{4b - a^2}\, i$$

† You should not be troubled by the term "imaginary." It's just a name. The British mathematician Alfred North Whiteheaa, in the chapter on imaginary numbers in his *Introduction to Mathematics*, wrote:

> At this point it may be useful to observe that a certain type of intellect is always worrying itself and others by discussion as to the applicability of technical terms. Are the incommensurable numbers properly called numbers? Are the positive and negative numbers really numbers? Are the imaginary numbers imaginary, and are they numbers?—are types of such futile questions. Now, it cannot be too clearly understood that, in science, technical terms are names arbitrarily assigned, like Christian names to children. There can be no question of the names being right or wrong. They may be judicious or injudicious; for they can sometimes be so arranged as to be easy to remember, or so as to suggest relevant and important ideas. But the essential principle involved was quite clearly enunciated in Wonderland to Alice by Humpty Dumpty, when he told her, apropos of his use of words, 'I pay them extra and make them mean what I like'. So we will not bother as to whether imaginary numbers are imaginary, or as to whether they are numbers, but will take the phrase as the arbitrary name of a certain mathematical idea, which we will now endeavour to make plain.

and the two roots of (1) are given by

$$\lambda_1 = -\frac{a}{2} + \frac{\sqrt{4b-a^2}}{2} i \quad \text{and} \quad \lambda_2 = -\frac{a}{2} - \frac{\sqrt{4b-a^2}}{2} i$$

EXAMPLE 1 Find the roots of the quadratic equation $\lambda^2 + 2\lambda + 5 = 0$.

Solution We have $a = 2$, $b = 5$, and $a^2 - 4b = -16$. Thus $\sqrt{a^2 - 4b} = \sqrt{-16} = \sqrt{16}\sqrt{-1} = 4i$ and the roots are

$$\lambda_1 = \frac{-2+4i}{2} = -1+2i \quad \text{and} \quad \lambda_2 = -1-2i$$

DEFINITION 1 A **complex number** is a number of the form

$$z = \alpha + i\beta \tag{4}$$

where α and β are real numbers. α is called the **real part** of z and is denoted Re z. β is called the **imaginary part** of z and is denoted Im z. Representation (4) is sometimes called the **cartesian form** of the complex number z.

Remark. If $\beta = 0$ in Equation (4), then $z = \alpha$ is a real number. In this context we can regard the set of real numbers as a subset of the set of complex numbers.

EXAMPLE 2 In Example 1, Re $\lambda_1 = -1$ and Im $\lambda_1 = 2$.

We can add and multiply complex numbers by using standard rules of algebra.

EXAMPLE 3 Let $z = 2+3i$ and $w = 5-4i$. Calculate **(i)** $z+w$, **(ii)** $3w-5z$, and **(iii)** zw.

Solution

i. $z+w = (2+3i)+(5-4i) = (2+5)+(3-4)i = 7-i$.

ii. $3w = 3(5-4i) = 15-12i$; $5z = 10+15i$; and $3w-5z = (15-12i)-(10+15i) = (15-10)+i(-12-15) = 5-27i$.

iii. $zw = (2+3i)(5-4i) = (2)(5)+2(-4i)+(3i)(5)+(3i)(-4i) = 10-8i+15i-12i^2 = 10+7i+12 = 22+7i$. Here we used the fact that $i^2 = -1$.

We can plot a complex number z in the xy-plane by plotting Re z along the x-axis and Im z along the y-axis. Thus each complex number can be thought of as a point in the xy-plane. With this representation the xy-plane is called the **complex plane**. Some representative points are plotted in Figure A.1.

Figure A.1

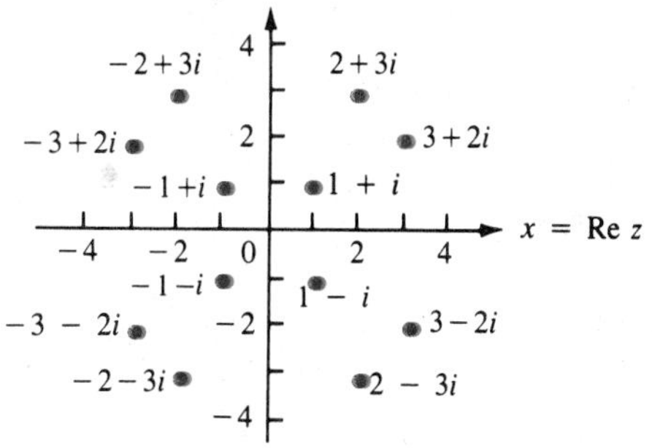

If $z = \alpha + i\beta$, then we define the **conjugate** of z, denoted $\bar{z}$, by

$$\bar{z} = \alpha - i\beta \tag{5}$$

Figure A.2 depicts a representative value of z and $\bar{z}$.

Figure A.2

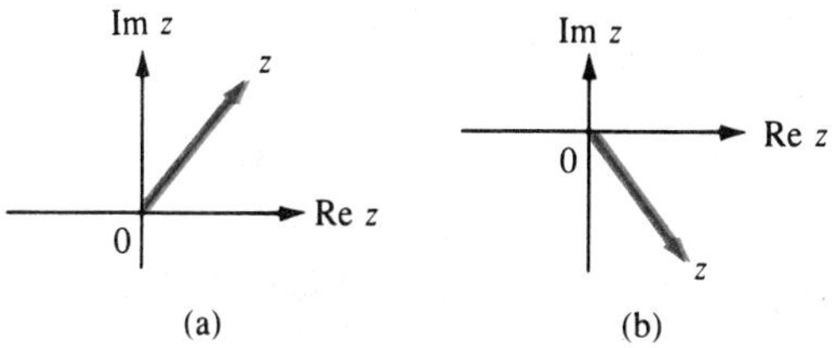

EXAMPLE 4 Compute the conjugate of **(i)** $1+i$, **(ii)** $3-4i$, **(iii)** $-7+5i$, and **(iv)** -3.

Solution **(i)** $\overline{1+i} = 1-i$; **(ii)** $\overline{3-4i} = 3+4i$; **(iii)** $\overline{-7+5i} = -7-5i$; **(iv)** $\overline{-3} = -3$.

It is not difficult to show (see Problem 35) that

$$\bar{z} = z \qquad \text{if and only if } z \text{ is real} \tag{6}$$

If $z = \beta i$ with β real, then z is said to be **pure imaginary**. We can then show (see Problem 36) that

$$\bar{z} = -z \qquad \text{if and only if } z \text{ is pure imaginary} \tag{7}$$

Let $p_n(x) = a_0 + a_1x + a_2x^2 + \cdots + a_nx^n$ be a polynomial with real coefficients. Then it can be shown (see Problem 41) that the complex roots of the equation $p_n(x) = 0$ occur in complex conjugate pairs. That is, if z is a root of $p_n(x) = 0$, then so is $\bar{z}$. We saw this fact illustrated in Example 1 in the case $n = 2$.

For $z = a + i\beta$, we define the **magnitude** of z, denoted $|z|$, by

$$|z| = \sqrt{\alpha^2 + \beta^2} \tag{8}$$

and we define the **argument** of z, denoted arg z, as the angle θ between the line $\overline{0z}$ and the positive x-axis. From Figure A.3 we see that $r = |z|$ is the

Figure A.3

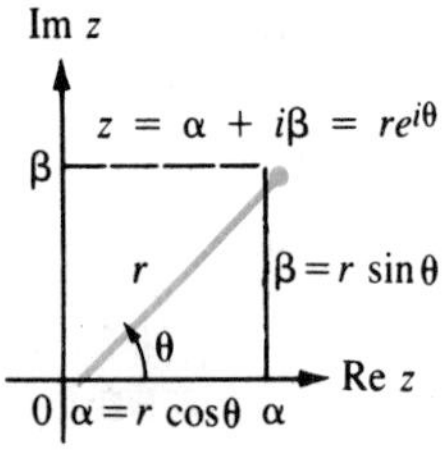

distance from z to the origin and

$$\theta = \arg z = \tan^{-1}\frac{\beta}{\alpha} \tag{9}$$

By convention, we always choose the value of $\tan^{-1}\beta/\alpha$ that lies in the interval

$$-\pi < \theta \leq \pi \tag{10}$$

From Figure A.4 we see that

$$|\bar{z}| = |z| \tag{11}$$

Figure A.4

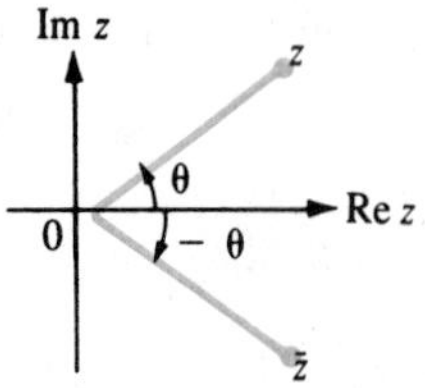

and

$$\arg \bar{z} = -\arg z \tag{12}$$

We can use $|z|$ and $\arg z$ to describe what is often a more convenient way to represent complex numbers.† From Figure A.3 it is evident that if $z = \alpha + i\beta$, $r = |z|$, and $\theta = \arg z$, then

$$\alpha = r \cos \theta \quad \text{and} \quad \beta = r \sin \theta \tag{13}$$

We shall see at the end of this appendix that

$$e^{i\theta} = \cos \theta + i \sin \theta \tag{14}$$

Since $\cos(-\theta) = \cos \theta$ and $\sin(-\theta) = -\sin \theta$, we also have

$$e^{-i\theta} = \cos(-\theta) + i \sin(-\theta) = \cos \theta - i \sin \theta \tag{14'}$$

Formula (14) is called **Euler's formula.**‡ Using Euler's formula and Equation (13), we have

$$z = \alpha + i\beta = r \cos \theta + ir \sin \theta = r(\cos \theta + i \sin \theta)$$

or

$$z = re^{i\theta} \tag{15}$$

Representation (15) is called the **polar form** of the complex number z.

EXAMPLE 5 Determine the polar forms of the following complex numbers: **(i)** 1, **(ii)** -1, **(iii)** i, **(iv)** $1 + i$, **(v)** $-1 - \sqrt{3}i$, and **(vi)** $-2 + 7i$.

---

† Those of you who have studied polar coordinates in a calculus course will find this representation very familiar.

‡ Named for the great Swiss mathematician Leonhard Euler (1707–1783).

Solution The six points are potted in Figure A.5.

Figure A.5

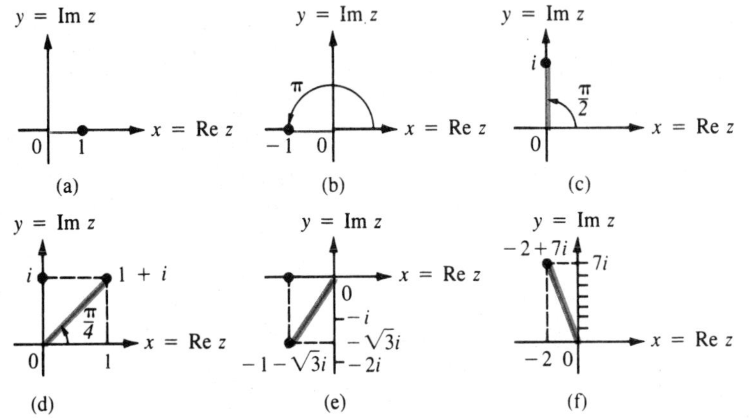

(i) From Figure A.5(a) it is clear that arg 1 $= 0$. Since Re 1 $= 1$, we see that, in polar form, $1 = 1e^{i0} = 1e^0 = 1$.

(ii) Since $\arg(-1) = \pi$ (Fig. A.5(b)) and $|-1| = 1$, we have

$$-1 = 1e^{\pi i} = e^{i\pi}$$

(iii) From Figure A.5(c) we see that $\arg i = \pi/2$. Since $|i| = \sqrt{0^2 + 1^2} = 1$, it follows that

$$i = e^{i\pi/2}$$

(iv) $\arg(1 + i) = \tan^{-1}(1/1) = (\pi/4)$ and $|1 + i| = \sqrt{1^2 + 1^2} = \sqrt{2}$, so that

$$1 + i = \sqrt{2}e^{i\pi/4}$$

(v) Here $\tan^{-1}(\beta/\alpha) = \tan^{-1}\sqrt{3} = \pi/3$. However arg z is in the third quadrant so $\theta = \pi/3 + \pi = 4\pi/3$. Also, $|-1 - \sqrt{3}| = \sqrt{1^2 + (\sqrt{3})^2} = \sqrt{1 + 3} = 2$ so that

$$-1 - \sqrt{3} = 2e^{4\pi i/3}$$

(vi) To compute this we need a calculator. We find that

$$\arg z = \tan^{-1}(-\tfrac{7}{2}) = \tan^{-1}(-3.5) \approx -1.2925.$$

But $\tan^{-1} x$ is defined as a number in the interval $(-\pi/2, \pi/2)$. Since, from Figure A.5(f), θ is in the second quadrant, we see that $\arg z = \tan^{-1}(-3.5) + \pi \approx 1.8491$. Next, we see that

$$|-2 + 7i| = \sqrt{(-2)^2 + 7^2} = \sqrt{53}.$$

Hence

$$-2 + 7i \approx \sqrt{53}e^{1.8491i}$$

EXAMPLE 6 Convert the following complex numbers from polar to cartesian form: **(i)** $2e^{i\pi/3}$; **(ii)** $4e^{3\pi i/2}$.

Solution **i.** $e^{i\pi/3}=\cos\pi/3+i\sin\pi/3=\frac{1}{2}+(\sqrt{3}/2)i$. Thus $2e^{i\pi/3}=1+\sqrt{3}i$.
ii. $e^{3\pi i/2}=\cos 3\pi/2+i\sin 3\pi/2=0+i(-1)=-i$. Thus $4e^{3\pi i/2}=-4i$.

If $\theta=\arg z$, then, by Equation (12), $\arg\bar{z}=-\theta$. Thus, since $|\bar{z}|=|z|$:

$$\text{If } z=re^{i\theta}, \quad \text{then } \bar{z}=re^{-i\theta}. \tag{16}$$

Suppose we write a complex number in its polar form $z = re^{i\theta}$. Then

$$z^n = (re^{i\theta})^n = r^n(e^{i\theta})^n = r^ne^{in\theta} = r^n(\cos n\theta + i\sin n\theta). \tag{17}$$

Formula (17) is useful for a variety of computations. In particular, when $r = |z| = 1$, we obtain the **De Moivre formula.**†

$$(\cos\theta+i\sin\theta)^n=\cos n\theta+i\sin n\theta \tag{18}$$

EXAMPLE 7 Compute $(1 + i)^5$.

Solution In Example 5(**iv**) we showed that $1 + i = \sqrt{2}e^{\pi i/4}$. Then

$$(1 + i)^5 = (\sqrt{2}e^{\pi i/4})^5 = (\sqrt{2})^5e^{5\pi i/4} = 4\sqrt{2}\left(\cos\frac{5\pi}{4} + i\sin\frac{5\pi}{4}\right)$$
$$= 4\sqrt{2}\left(-\frac{1}{\sqrt{2}} - \frac{1}{\sqrt{2}}i\right) = -4 - 4i.$$

This can be checked by direct calculation. If the direct calculation seems no more difficult, then try to compute $(1+i)^{20}$ directly. Proceeding as above, we obtain

$$(1+i)^{20}=(\sqrt{2})^{20}e^{20\pi i/4}=2^{10}(\cos 5\pi+i\sin 5\pi)$$
$$=2^{10}(-1+0)=-1024$$

Proof of Euler's Formula‡ We shall show that

$$e^{i\theta}=\cos\theta+i\sin\theta \tag{19}$$

---

† Abraham De Moivre (1667–1754) was a French mathematician well known for his work in probability theory, infinite series, and trigonometry. He was so highly regarded that Newton often told those who came to him with questions on mathematics, "Go to M. De Moivre; he knows these things better than I do."

‡ If time permits.

by using power series. If these are unfamiliar to you, skip the proof. We have

$$e^x = 1 + x + \frac{x^2}{2!} + \frac{x^3}{3!} + \cdots \dagger \tag{20}$$

$$\sin x = x - \frac{x^3}{3!} + \frac{x^5}{5!} - \cdots \tag{21}$$

$$\cos x = 1 - \frac{x^2}{2!} + \frac{x^4}{4!} - \cdots \tag{22}$$

Then

$$e^{i\theta} = 1 + (i\theta) + \frac{(i\theta)^2}{2!} + \frac{(i\theta)^3}{3!} + \frac{(i\theta)^4}{4!} + \frac{(i\theta)^5}{5!} + \cdots \tag{23}$$

Now $i^2 = -1$, $i^3 = -i$, $i^4 = 1$, $i^5 = i$, and so on. Thus (23) can be written

$$\begin{aligned} e^{i\theta} &= 1 + i\theta - \frac{\theta^2}{2!} - \frac{i\theta^3}{3!} + \frac{\theta^4}{4!} + \frac{i\theta^5}{5!} - \cdots \\ &= \left(1 - \frac{\theta^2}{2!} + \frac{\theta^4}{4!} - \cdots\right) + i\left(\theta - \frac{\theta^3}{3!} + \frac{\theta^5}{5!} - \cdots\right) \\ &= \cos\theta + i\sin\theta \end{aligned}$$

This completes the proof. ■

PROBLEMS A2

In Problems 1–5 perform the indicated operation.

1. $(2-3i)+(7-4i)$
2. $3(4+i)-5(-3+6i)$
3. $(1+i)(1-i)$
4. $(2-3i)(4+7i)$
5. $(-3+2i)(7+3i)$

In Problems 6–15 convert the complex number to its polar form.

6. $5i$
7. $5+5i$
8. $-2-2i$
9. $3-3i$
10. $2+2\sqrt{3}i$
11. $3\sqrt{3}+3i$
12. $1-\sqrt{3}i$
13. $4\sqrt{3}-4i$
14. $-6\sqrt{3}-6i$
15. $-1-\sqrt{3}i$

In Problems 16–25 convert from polar to cartesian form.

16. $e^{3\pi i}$
17. $2e^{-7\pi i}$
18. $\frac{1}{2}e^{3\pi i/4}$
19. $\frac{1}{2}e^{-3\pi i/4}$
20. $6e^{\pi i/6}$
21. $4e^{5\pi i/6}$
22. $4e^{-5\pi i/6}$
23. $3e^{-2\pi i/3}$
24. $\sqrt{3}e^{23\pi i/4}$
25. e^{i}

In Problems 26–34 compute the conjugate of the given number.

26. $3-4i$
27. $4+6i$
28. $-3+8i$
29. $-7i$
30. 16
31. $2e^{\pi i/7}$
32. $4e^{3\pi i/5}$
33. $3e^{-4\pi i/11}$
34. $e^{0.012i}$

---

†Although we shall not prove it here, this series expansion is also valid when x is a complex number.

35. Show that $z = \alpha + i\beta$ is real if and only if $z = \bar{z}$. [*Hint:* If $z = \bar{z}$, show that $\beta = 0$.]

36. Show that $z = \alpha + i\beta$ is pure imaginary if and only if $z = -\bar{z}$. [*Hint:* If $z = -\bar{z}$, show that $\alpha = 0$.]

37. For any complex number z, show that $z\bar{z} = |z|^2$.

38. Show that the circle of radius 1 centered at the origin (the *unit circle*) is the set of points in the complex plane that satisfy $|z| = 1$.

39. For any complex number z_0 and real number a, describe $\{z: \ |z - z_0| = a\}$.

40. Describe $\{z: \ |z - z_0| \leq a\}$, where z_0 and a are as in Problem 39.

***41.** Let $p(\lambda) = \lambda^n + a_{n-1}\lambda^{n-1} + a_{n-2}\lambda^{n-2} + \cdots + a_1\lambda + a_0$ with $a_0, a_1, \ldots, a_{n-1}$ real numbers. Show that if $p(z) = 0$, then $p(\bar{z}) = 0$. That is: *The roots of polynomials with real coefficients occur in complex conjugate pairs.*

42. Derive expressions for $\cos 4\theta$ and $\sin 4\theta$ by comparing the De Moivre formula and the expansion of $(\cos\theta + i\sin\theta)^4$.

43. Prove De Moivre's formula by mathematical induction. [*Hint:* Recall the trigonometric identities $\cos(x + y) = \cos x \cos y - \sin x \sin y$ and $\sin(x + y) = \sin x \cos y + \cos x \sin y$.]

Answers to Odd-Numbered Problems

CHAPTER 1

Problems 1.2

1. $x_1 = \frac{-13}{5}$, $x_2 = \frac{-11}{5}$; $\det = -10$

3. no solutions; $\det = 0$

5. $x_1 = \frac{11}{2}$, $x_2 = -30$; $\det = -2$

7. infinite number of solutions; $x_2 = \frac{2}{3}x_1$, where x_1 is arbitrary; $\det = 0$

9. $x_1 = -1$, $x_2 = 2$; $\det = -1$

11. $\det = a^2 - b^2$; if $a^2 - b^2 \neq 0$ (i.e., if $a \neq \pm b$), then $x_1 = x_2 = c/(a+b)$. If $a^2 - b^2 = 0$, then $a = \pm b$. If $a \neq 0$ and $a = b$, then there is an infinite number of solutions given by $x_2 = c/a - x_1$. If $a \neq 0$ and $a = -b$, then there are no solutions.

13. $\det = -2ab$; so unique solution if both a and b are nonzero

15. $a = b = 0$ and $c \neq 0$ or $d \neq 0$

17. no point of intersection

19. The lines are coincident. Any point of the form $(x, (4x - 10)/6)$ is a point of intersection.

21. $(\frac{67}{45}, \frac{2}{15})$ **23.** $\sqrt{13}/13$

25. $\sqrt{61}/5$ **27.** $\sqrt{5}$

29. Since the slope of the given line L is $-\dfrac{a}{b}$, the slope of $L_\perp$ is $\dfrac{b}{a}$. The equation of a line $L_\perp$ perpendicular to L and passing through (x_1, y_1) is given by $\dfrac{y - y_1}{x - x_1} = \dfrac{b}{a}$, or $bx - ay = bx_1 - ay_1$. The unique point of intersection of L and $L_\perp$ is found to be

$$(x_0, y_0) = \left(\frac{bc - abx_1 + a^2y_1}{a^2 + b^2}, \frac{ac - aby_1 + b^2x_1}{a^2 + b^2}\right).$$

Then d is the distance between (x_0, y_0) and (x_1, y_1) and, after some algebra,

$$d^2 = \frac{1}{(a^2 + b^2)^2} \times (a^2c^2 - 2a^2bcy_1 + a^2b^2y_1^2 - 2a^3cx_1 + 2a^3bx_1y_1 + a^4x_1^2 + c^2b^2 - 2ab^2cx_1 + a^2b^2x_1^2 - 2b^3cy_1 + 2ab^3x_1y_1 + b^4y_1^2)$$

$$= \frac{a^2 + b^2}{(a^2 + b^2)^2}(c^2 - 2abcy_1 + b^2y_1^2 - 2acx_1 + 2abx_1y_1 + a^2x_1^2) = \frac{1}{(a^2 + b^2)}(ax_1 + by_1 - c)^2.$$

Thus

$$d = \frac{|ax_1 + by_1 - c|}{\sqrt{a^2 + b^2}}.$$

35. Infinite number of solutions; x = no. of cups; y = no. of saucers; solutions are $(x, 240 - \frac{3}{2}x)$.

37. 32 sodas, 128 milk shakes

Problems 1.3

Note: Where there were an infinite number of solutions, we wrote the solutions with the last variable chosen arbitrarily. The solutions can be written in other ways as well.

1. $(2, -3, 1)$

3. $(3 + \frac{2}{9}x_3, \frac{8}{9}x_3, x_3)$, x_3 arbitrary.

5. $(-9, 30, 14)$

7. no solution

9. $(-\frac{4}{5}x_3, \frac{9}{5}x_3, x_3)$, x_3 arbitrary

11. $(-1, \frac{5}{2} + \frac{1}{2}x_3, x_3)$, x_3 arbitrary

13. no solution

15. $(\frac{20}{13} - \frac{4}{13}x_4, \frac{-28}{13} + \frac{3}{13}x_4, \frac{-45}{13} + \frac{9}{13}x_4, x_4)$, arbitrary

17. $(18 - 4x_4, \frac{-15}{2} + 2x_4, -31 + 7x_4, x_4)$, x_4 arbitrary

19. no solution

21. row echelon form

23. reduced row echelon form

25. neither

27. reduced row echelon form

29. neither

31. row echelon form: $\begin{pmatrix} 1 & -6 \\ 0 & 1 \end{pmatrix}$; reduced row echelon form: $\begin{pmatrix} 1 & 0 \\ 0 & 1 \end{pmatrix}$

33. row echelon form:

$$\begin{pmatrix} 1 & -2 & 4 \\ 0 & 1 & -\frac{4}{11} \\ 0 & 0 & 1 \end{pmatrix};$$

reduced row echelon form:

$$\begin{pmatrix} 1 & 0 & 0 \\ 0 & 1 & 0 \\ 0 & 0 & 1 \end{pmatrix}$$

35. row echelon form:

$$\begin{pmatrix} 1 & -\frac{7}{2} \\ 0 & 1 \\ 0 & 0 \end{pmatrix};$$

reduced row echelon form:

$$\begin{pmatrix} 1 & 0 \\ 0 & 1 \\ 0 & 0 \end{pmatrix}$$

37. $x_1 = 30{,}000 - 5x_3$
$x_2 = x_3 - 5000$
$5000 \leq x_3 \leq 6000$; no

39. No unique solution (2 equations in 3 unknowns); if 200 shares of McDonald's, then 100 shares of Hilton and 300 shares of Eastern.

41. The row echelon form of the augmented matrix representing this system is

$$\left(\begin{array}{ccc|c} 1 & -\frac{1}{2} & \frac{3}{2} & a/2 \\ 0 & 1 & \frac{-19}{5} & \frac{2}{5}(b-\frac{3}{2}a) \\ 0 & 0 & 0 & -2a+3b+c \end{array}\right)$$

which is inconsistent if $-2a+3b+c \neq 0$ or $c \neq 2a-3b$.

43. $a_{11}a_{22}a_{33}+a_{12}a_{23}a_{31}+a_{13}a_{32}a_{21}-a_{13}a_{22}a_{31}-a_{12}a_{21}a_{33}-a_{11}a_{32}a_{23} \neq 0$

45. (1.900812947, 4.194110816, − 11.34851834)

Problems 1.4

1. (0, 0) **3.** (0, 0, 0)

5. $(\frac{1}{6}x_3, \frac{5}{6}x_3, x_3)$, x_3 arbitrary

7. (0, 0)

9. $(-4x_4, 2x_4, 7x_4, x_4)$, x_4 arbitrary.

11. (0, 0) **13.** (0, 0, 0)

15. $k = \frac{95}{11}$

Review Exercises for Chapter 1

1. $(\frac{1}{7}, \frac{10}{7})$ **3.** no solution

5. (0, 0, 0) **7.** $(-\frac{1}{2}, 0, \frac{5}{2})$

9. $(\frac{1}{3}x_3, \frac{7}{3}x_3, x_3)$, x_3 arbitrary

11. no solution

13. (0, 0, 0, 0) **15.** $\sqrt{5}/5$.

17. row echelon form

19. reduced row echelon form

21. row echelon form:

$$\begin{pmatrix} 1 & 4 & -1 \\ 0 & 1 & \frac{5}{4} \end{pmatrix};$$

reduced row echelon form:

$$\begin{pmatrix} 1 & 0 & -6 \\ 0 & 1 & \frac{5}{4} \end{pmatrix}$$

CHAPTER 2

Problems 2.1

1. $\begin{pmatrix} 2 \\ -3 \\ 11 \end{pmatrix}$ **3.** $\begin{pmatrix} -4 \\ 0 \\ 4 \end{pmatrix}$

5. $\begin{pmatrix} -31 \\ 22 \\ -27 \end{pmatrix}$ **7.** $\begin{pmatrix} 0 \\ 0 \\ 0 \end{pmatrix}$

9. $\begin{pmatrix} -11 \\ 11 \\ -10 \end{pmatrix}$ **11.** (1, 2, 5, 7)

13. (−8, 12, 4, 20)

15. (8, −5, 7, −1)

17. (7, 2, 4, 11)

19. (−11, 9, 18, 18)

21. $$\mathbf{a}+\mathbf{0} = \begin{pmatrix} a_1 \\ a_2 \\ \vdots \\ a_n \end{pmatrix} + \begin{pmatrix} 0 \\ 0 \\ \vdots \\ 0 \end{pmatrix} = \begin{pmatrix} a_1+0 \\ a_2+0 \\ \vdots \\ a_n+0 \end{pmatrix} = \begin{pmatrix} a_1 \\ a_2 \\ \vdots \\ a_n \end{pmatrix} = \mathbf{a}$$

23. $$\mathbf{a}+\mathbf{b} = \begin{pmatrix} a_1+b_1 \\ a_2+b_2 \\ \vdots \\ a_n+b_n \end{pmatrix},$$

$$\alpha(\mathbf{a}+\mathbf{b}) = \begin{pmatrix} \alpha(a_1+b_1) \\ \alpha(a_2+b_2) \\ \vdots \\ \alpha(a_n+b_n) \end{pmatrix} = \begin{pmatrix} \alpha a_1+\alpha b_1 \\ \alpha a_2+\alpha b_2 \\ \vdots \\ \alpha a_n+\alpha b_n \end{pmatrix} = \begin{pmatrix} \alpha a_1 \\ \alpha a_2 \\ \vdots \\ \alpha a_n \end{pmatrix} + \begin{pmatrix} \alpha b_1 \\ \alpha b_2 \\ \vdots \\ \alpha b_n \end{pmatrix} = \alpha\begin{pmatrix} a_1 \\ a_2 \\ \vdots \\ a_n \end{pmatrix} + \alpha\begin{pmatrix} b_1 \\ b_2 \\ \vdots \\ b_n \end{pmatrix} = \alpha\mathbf{a}+\alpha\mathbf{b};$$

$$(\alpha+\beta)\mathbf{a} = \begin{pmatrix} (\alpha+\beta)a_1 \\ (\alpha+\beta)a_2 \\ \vdots \\ (\alpha+\beta)a_n \end{pmatrix} = \begin{pmatrix} \alpha a_1+\beta a_1 \\ \alpha a_2+\beta a_2 \\ \vdots \\ \alpha a_n+\beta a_n \end{pmatrix} = \begin{pmatrix} \alpha a_1 \\ \alpha a_2 \\ \vdots \\ \alpha a_n \end{pmatrix} + \begin{pmatrix} \beta a_1 \\ \beta a_2 \\ \vdots \\ \beta a_n \end{pmatrix} = \alpha\begin{pmatrix} a_1 \\ a_2 \\ \vdots \\ a_n \end{pmatrix} + \beta\begin{pmatrix} a_1 \\ a_2 \\ \vdots \\ a_n \end{pmatrix} = \alpha\mathbf{a}+\beta\mathbf{a};$$

$$(\alpha\beta)\mathbf{a} = \begin{pmatrix} \alpha\beta a_1 \\ \alpha\beta a_2 \\ \vdots \\ \alpha\beta a_n \end{pmatrix} = \alpha\begin{pmatrix} \beta a_1 \\ \beta a_2 \\ \vdots \\ \beta a_n \end{pmatrix} = \alpha\left[\beta\begin{pmatrix} a_1 \\ a_2 \\ \vdots \\ a_n \end{pmatrix}\right] = \alpha(\beta\mathbf{a})$$

25. $\mathbf{d}_1+\mathbf{d}_2$ represents the combined demand of the two factories for each of the four raw materials needed to produce one unit of each factory's product; $2\mathbf{d}_1$ represents the demand of factory 1 for each of the four raw materials needed to produce two units of its product.

27. $\mathbf{w} = \begin{pmatrix} 3 \\ 0 \\ 5 \end{pmatrix}$

Problems 2.2

1. −14 **3.** 1 **5.** $ac+bd$

7. 51 **9.** $\mathbf{a}=\mathbf{0}$ **11.** 4

13. 28 **15.** orthogonal

17. orthogonal **19.** orthogonal

21. all α and β which satisfy $5\alpha+4\beta=25$ ($\beta=(25-5\alpha)/4$, α arbitrary)

23. (a) (2, 3, 5, 1)

(b) $\begin{pmatrix} 1 \\ \frac{3}{2} \\ \frac{1}{2} \\ 2 \end{pmatrix}$ (c) 11

Problems 2.3

1. $\begin{pmatrix} 3 & 9 \\ 6 & 15 \\ -3 & 6 \end{pmatrix}$ **3.** $\begin{pmatrix} 2 & 2 \\ -2 & -1 \\ 6 & -1 \end{pmatrix}$

5. $\begin{pmatrix} 0 & 0 \\ 0 & 0 \\ 0 & 0 \end{pmatrix}$ **7.** $\begin{pmatrix} -2 & 4 \\ 7 & 15 \\ -15 & 10 \end{pmatrix}$

9. $\begin{pmatrix} 4 & 10 \\ 17 & 22 \\ -9 & 1 \end{pmatrix}$ **11.** $\begin{pmatrix} 0 & 6 \\ 5 & 14 \\ -9 & 9 \end{pmatrix}$

13. $\begin{pmatrix} 1 & -5 & 0 \\ -3 & 4 & -5 \\ -14 & 13 & -1 \end{pmatrix}$

15. $\begin{pmatrix} 1 & 1 & 5 \\ 9 & 5 & 10 \\ 7 & -7 & 3 \end{pmatrix}$

17. $\begin{pmatrix} -1 & -1 & -1 \\ -3 & -3 & -10 \\ -7 & 3 & 5 \end{pmatrix}$

19. $\begin{pmatrix} -1 & -1 & -5 \\ -9 & -5 & -10 \\ -7 & 7 & -3 \end{pmatrix}$

25. $\begin{pmatrix} 1 & 1 & 1 & 0 \\ 1 & 1 & 1 & 0 \\ 1 & 1 & 1 & 1 \\ 0 & 0 & 1 & 1 \end{pmatrix}$

Problems 2.4

1. $\begin{pmatrix} 8 & 20 \\ -4 & 11 \end{pmatrix}$ **3.** $\begin{pmatrix} -3 & -3 \\ 1 & 3 \end{pmatrix}$

5. $\begin{pmatrix} 13 & 35 & 18 \\ 20 & 26 & 20 \end{pmatrix}$

7. $\begin{pmatrix} 19 & -17 & 34 \\ 8 & -12 & 20 \\ -8 & -11 & 7 \end{pmatrix}$

9. $\begin{pmatrix} 18 & 15 & 35 \\ 9 & 21 & 13 \\ 10 & 9 & 9 \end{pmatrix}$ **11.** (7 16)

13. $\begin{pmatrix} 3 & -2 & 1 \\ 4 & 0 & 6 \\ 5 & 1 & 9 \end{pmatrix}$ **15.** $\begin{pmatrix} a & b & c \\ d & e & f \\ g & h & j \end{pmatrix}$

17. If $D = a_{11}a_{22} - a_{12}a_{21}$, then

$$\begin{pmatrix} b_{11} & b_{21} \\ b_{21} & b_{22} \end{pmatrix} = \begin{pmatrix} a_{22}/D & -a_{12}/D \\ -a_{21}/D & a_{11}/D \end{pmatrix}$$

19. (a) 3 in Group 1, 4 in Group 2, 5 in Group 3

(b) $\begin{pmatrix} 2 & 1 & 1 & 0 & 0 \\ 1 & 1 & 0 & 1 & 0 \\ 1 & 0 & 2 & 0 & 1 \end{pmatrix}$

21. (a) $\begin{pmatrix} 80{,}000 & 45{,}000 & 40{,}000 \\ 50 & 20 & 10 \end{pmatrix}$

(b) $\begin{pmatrix} 1 \\ 3 \\ 1 \end{pmatrix}$ (c) Money: 255,000;

Shares: 120

23. $\begin{pmatrix} 0 & -8 \\ 32 & 32 \end{pmatrix}$ **25.** $\begin{pmatrix} 11 & 38 \\ 57 & 106 \end{pmatrix}$

27. $A^2 = \begin{pmatrix} 0 & 0 & 1 & 0 & 0 \\ 0 & 0 & 0 & 1 & 0 \\ 0 & 0 & 0 & 0 & 1 \\ 0 & 0 & 0 & 0 & 0 \\ 0 & 0 & 0 & 0 & 0 \end{pmatrix}$,

$A^3 = \begin{pmatrix} 0 & 0 & 0 & 1 & 0 \\ 0 & 0 & 0 & 0 & 1 \\ 0 & 0 & 0 & 0 & 0 \\ 0 & 0 & 0 & 0 & 0 \\ 0 & 0 & 0 & 0 & 0 \end{pmatrix}$,

$A^4 = \begin{pmatrix} 0 & 0 & 0 & 0 & 1 \\ 0 & 0 & 0 & 0 & 0 \\ 0 & 0 & 0 & 0 & 0 \\ 0 & 0 & 0 & 0 & 0 \\ 0 & 0 & 0 & 0 & 0 \end{pmatrix}$,

$A^5 = \begin{pmatrix} 0 & 0 & 0 & 0 & 0 \\ 0 & 0 & 0 & 0 & 0 \\ 0 & 0 & 0 & 0 & 0 \\ 0 & 0 & 0 & 0 & 0 \\ 0 & 0 & 0 & 0 & 0 \end{pmatrix}$

29. $PQ = \begin{pmatrix} \frac{11}{90} & \frac{41}{90} & \frac{19}{45} \\ \frac{11}{120} & \frac{71}{120} & \frac{19}{60} \\ \frac{1}{5} & \frac{1}{5} & \frac{3}{5} \end{pmatrix}$; all entries are nonnegative and $\frac{11}{90}+\frac{41}{90}+\frac{19}{45}=\frac{11}{120}+\frac{71}{120}+\frac{19}{60}=\frac{1}{5}+\frac{1}{5}+\frac{3}{5}=1$.

31. Let $P=(p_{ij})$ and $Q=(q_{ij})$ be $k\times k$ probability matrices. Let $PQ=C=(c_{ij})$. The sum of the elements in the mth row of PQ is $c_{m1}+c_{m2}+c_{m3}+\cdots+c_{mk} = p_{m1}q_{11}+p_{m2}q_{21}+p_{m3}q_{31}+\cdots+p_{mk}q_{k1}+p_{m1}q_{12}+p_{m2}q_{22}+p_{m3}q_{32}+\cdots+p_{mk}q_{k2}+p_{m1}q_{13}+p_{m2}q_{23}+p_{m3}p_{33}+\cdots+p_{mk}q_{k3}$
$\vdots$
$+p_{m1}q_{1k}+p_{m2}q_{2k}+p_{m3}q_{3k}+\cdots+p_{mk}q_{kk}$

(the elements in parentheses are those of a row of Q, whose sum is 1)

$\downarrow$
$= p_{m1}(q_{11}+q_{12}+q_{13}+\cdots+q_{1k})+p_{m2}(q_{21}+q_{22}+q_{23}+\cdots+q_{2k})+p_{m3}(q_{31}+q_{32}+q_{33}+\cdots+q_{3k})+\cdots+p_{mk}(q_{k1}+q_{k2}+q_{k3}+\cdots+q_{kk}) = p_{m1}(1)+p_{m2}(1)+p_{m3}(1)+\cdots+p_{mk}(1)=1.$

33. (a) player 2 > player 4 > player 1 > player 3
(b) score = number of games won plus one-half the number of games that were won by each player that this given player beat

35. $A(B+C) = \begin{pmatrix} 1 & 2 & 4 \\ 3 & -1 & 0 \end{pmatrix} \times \begin{pmatrix} 1 & 9 \\ 2 & 11 \\ 10 & 1 \end{pmatrix} = \begin{pmatrix} 45 & 35 \\ 1 & 16 \end{pmatrix}$;

$AB + AC = \begin{pmatrix} 24 & 14 \\ 7 & 17 \end{pmatrix} + \begin{pmatrix} 21 & 20 \\ -6 & -1 \end{pmatrix} = \begin{pmatrix} 45 & 35 \\ 1 & 16 \end{pmatrix}$

37. 36 **39.** 9840

41. $\frac{13}{3}+\frac{15}{4}+\frac{17}{5}=\frac{689}{60}$

43. $(1^2+2^2+3^2)(2^3+3^3+4^3) = 1386$

45. $\sum_{k=0}^{5}(-3)^k$ **47.** $\sum_{k=1}^{n}k^{1/k}$

49. $\sum_{k=0}^{9}\frac{(-1)^{k+1}}{a^k}$

51. $\sum_{k=2}^{7}k^2\cdot 2k=\sum_{k=2}^{7}2k^3$

53. $\sum_{i=1}^{3}\sum_{j=1}^{2}a_{ij}$ **55.** $\sum_{k=1}^{5}a_{3k}b_{k2}$

Problems 2.5

1. $\begin{pmatrix}2 & -1\\ 4 & 5\end{pmatrix}\begin{pmatrix}x_1\\ x_2\end{pmatrix}=\begin{pmatrix}3\\ 7\end{pmatrix}$

3. $\begin{pmatrix}3 & 6 & -7\\ 2 & -1 & 3\end{pmatrix}\begin{pmatrix}x_1\\ x_2\\ x_3\end{pmatrix}=\begin{pmatrix}0\\ 1\end{pmatrix}$

5. $\begin{pmatrix}0 & 1 & -1\\ 1 & 0 & 1\\ 3 & 2 & 0\end{pmatrix}\begin{pmatrix}x_1\\ x_2\\ x_3\end{pmatrix}=\begin{pmatrix}7\\ 2\\ -5\end{pmatrix}$

7. $\begin{aligned}x_1+x_2-x_3&=7\\ 4x_1-x_2+5x_3&=4\\ 6x_1+x_2+3x_3&=20\end{aligned}$

9. $\begin{aligned}2x_1+x_3&=2\\ -3x_1+4x_2&=3\\ 5x_2+6x_3&=5\end{aligned}$

11. $\begin{aligned}x_1&=2\\ x_2&=3\\ x_3&=-5\\ x_4&=6\end{aligned}$

13. $\begin{aligned}6x_1+2x_2+x_3&=2\\ -2x_1+3x_2+x_3&=4\\ 0x_1+0x_2+0x_3&=2\end{aligned}$

15. $\begin{aligned}7x_1+2x_2&=1\\ 3x_1+x_2&=2\\ 6x_1+9x_2&=3\end{aligned}$

17. $x_1=4-2x_2+4x_3$; x_2, x_3 arbitrary

19. $x_1=x_2=x_3=0$

21. $\begin{pmatrix}1\\ -\frac{1}{3}\\ \frac{1}{2}\\ 4\end{pmatrix}$

23. $A=\begin{pmatrix}2 & 0 & 0\\ 0 & 4 & 0\\ 0 & 0 & -5\end{pmatrix}$, $\mathbf{x}=\begin{pmatrix}x_1\\ x_2\\ x_3\end{pmatrix}$,

$\mathbf{b}=\begin{pmatrix}3\\ 5\\ 2\end{pmatrix}$; $\mathbf{x}=\begin{pmatrix}\frac{3}{2}\\ \frac{5}{4}\\ -\frac{2}{5}\end{pmatrix}$

Problems 2.6

1. independent

3. dependent;

$$-2\begin{pmatrix}2\\ -1\\ 4\end{pmatrix}+\begin{pmatrix}4\\ -2\\ 8\end{pmatrix}=\begin{pmatrix}0\\ 0\\ 0\end{pmatrix}$$

5. dependent (from Theorem 2)

7. independent

9. independent

11. independent

13. $ad-bc=0$ **15.** $\alpha=-\frac{13}{2}$

17. System (7) can be written as

$$(*)\quad c_1\begin{pmatrix}a_{11}\\ a_{21}\\ \vdots\\ a_{m1}\end{pmatrix}+c_2\begin{pmatrix}a_{12}\\ a_{22}\\ \vdots\\ a_{m2}\end{pmatrix}+\cdots+c_n\begin{pmatrix}a_{1n}\\ a_{2n}\\ \vdots\\ a_{mn}\end{pmatrix}=\begin{pmatrix}0\\ 0\\ \vdots\\ 0\end{pmatrix}$$

If (7) has even one nontrivial solution, then the columns of A are linearly dependent. If the columns of A are dependent, then there are number $c_1, c_2, \ldots, c_n$ not all zero such that (*) holds.

19. If $\mathbf{0}=c_1\mathbf{v}_1+c_2\mathbf{v}_2+\cdots+c_k\mathbf{v}_k$, then $\mathbf{0}=c_1\mathbf{v}_1+c_2\mathbf{v}_2+\cdots+c_k\mathbf{v}_k+0\mathbf{v}_{k+1}+0\mathbf{v}_{k+2}+\cdots+0\mathbf{v}_n$. Since $\mathbf{v}_1, \mathbf{v}_2, \ldots, \mathbf{v}_n$ are independent, we have $c_1=c_2=\cdots=c_k=0$.

21. If $c_1\mathbf{v}_1+c_2\mathbf{v}_2+c_3\mathbf{v}_3=0$, then $0=0\cdot\mathbf{v}_1=(c_1\mathbf{v}_1+c_2\mathbf{v}_2+c_3\mathbf{v}_3)\cdot\mathbf{v}_1=c_1(\mathbf{v}_1\cdot\mathbf{v}_1)+c_2(\mathbf{v}_2\cdot\mathbf{v}_1)+c_3(\mathbf{v}_3\cdot\mathbf{v}_1)=c_1\mathbf{v}_1\cdot\mathbf{v}_1+c_2 0+c_3 0=c_1\mathbf{v}_1\cdot\mathbf{v}_1$. Since $\mathbf{v}_1\neq 0$, $\mathbf{v}_1\cdot\mathbf{v}_1\neq 0$ so we must have $c_1=0$. A similar computation shows that $c_2=c_3=0$.

23. $x_2\begin{pmatrix}-1\\ 1\\ 0\end{pmatrix}+x_3\begin{pmatrix}-1\\ 0\\ 1\end{pmatrix}$

25. $x_3\begin{pmatrix}13\\ -6\\ 1\end{pmatrix}$

Problems [illegible]

1. $\begin{pmatrix}2 & -1\\ -3 & 2\end{pmatrix}$ **3.** $\begin{pmatrix}0 & 1\\ 1 & 0\end{pmatrix}$

5. not invertible

7. $\begin{pmatrix}\frac{1}{3} & -\frac{1}{3} & -\frac{1}{3}\\ 0 & \frac{1}{2} & 1\\ 0 & 0 & -1\end{pmatrix}$

9. not invertible

11. not invertible

13. $\begin{pmatrix}\frac{7}{3} & -\frac{1}{3} & -\frac{1}{3} & -\frac{2}{3}\\ \frac{4}{9} & -\frac{1}{9} & -\frac{4}{9} & \frac{1}{9}\\ -\frac{1}{9} & -\frac{2}{9} & \frac{1}{9} & \frac{2}{9}\\ -\frac{5}{3} & \frac{2}{3} & \frac{2}{3} & \frac{1}{3}\end{pmatrix}$

15. $\begin{pmatrix}0 & 1 & 0 & 2\\ 1 & -1 & -2 & 2\\ 0 & 1 & 3 & -3\\ -2 & 2 & 3 & -2\end{pmatrix}$

17. $(A_1A_2\cdots A_m)^{-1}=A_m^{-1}A_{m-1}^{-1}\cdots A_2^{-1}A_1^{-1}$ since $(A_m^{-1}A_{m-1}^{-1}\cdots A_2^{-1}A_1^{-1})(A_1A_2\cdots A_{m-1}A_m)=(A_m^{-1}A_{m-1}^{-1}\cdots A_2^{-1})(A_1^{-1}A_1)\times A_2\cdots A_{m-1}A_m=(A_m^{-1}A_{m-1}^{-1}\cdots A_2^{-1})\times(A_2\cdots A_{m-1}A_m)=\cdots=I$

19. $A^{-1}=\dfrac{1}{a_{11}a_{22}-a_{21}a_{12}}\times\begin{pmatrix}a_{22} & -a_{12}\\ -a_{21} & a_{11}\end{pmatrix}$. If $A=\pm I$, then $A^{-1} = A$. If $a_{11} = -a_{22}$ and $a_{21}a_{12} = 1 - a_{11}^2$, then $a_{11}a_{22} - a_{21}a_{12} = -a_{11}^2 - (1 - a_{11}^2) = -1$. Thus

$$A^{-1} = \begin{pmatrix} -a_{22} & a_{12} \\ a_{21} & -a_{11} \end{pmatrix} = \begin{pmatrix} a_{11} & a_{12} \\ a_{21} & a_{22} \end{pmatrix} = A.$$

21. The system $B\mathbf{x} = \mathbf{0}$ has an infinite number of solutions (by Theorem 1.4.1). But if $B\mathbf{x} = \mathbf{0}$, then $AB\mathbf{x} = \mathbf{0}$. Thus, from Theorem 6 (parts [i] and [ii], AB is not invertible.

23. $$\begin{pmatrix} \sin\theta & \cos\theta & 0 \\ \cos\theta & -\sin\theta & 0 \\ 0 & 0 & 1 \end{pmatrix}$$

is its own inverse (since $\sin^2\theta + \cos^2\theta = 1$).

25. If the ith diagonal component is 0, then in the row reduction of A the ith row is zero so that, by the statement in Step 3(b) on page 64, A is not invertible. Otherwise, if

$$A = \operatorname{diag}(a_1, a_2, \ldots, a_n)$$

then

$$A^{-1} = \operatorname{diag}\left(\frac{1}{a_1}, \frac{1}{a_2}, \ldots, \frac{1}{a_n}\right).$$

27. $$\begin{pmatrix} \frac{1}{2} & -\frac{1}{6} & \frac{7}{30} \\ 0 & \frac{1}{3} & -\frac{4}{15} \\ 0 & 0 & \frac{1}{5} \end{pmatrix}$$

29. We prove the result for a lower triangular matrix and use Theorem 6, parts (i) and (v). Let

$$A = \begin{pmatrix} a_{11} & 0 & 0 & \cdots & 0 \\ a_{21} & a_{22} & 0 & \cdots & 0 \\ a_{31} & a_{32} & a_{33} & \cdots & 0 \\ \vdots & \vdots & \vdots & & \vdots \\ a_{n1} & a_{n2} & a_{n3} & \cdots & a_{nn} \end{pmatrix}$$

Taking a linear combination of the columns, we suppose that

$$\begin{pmatrix} 0 \\ 0 \\ \vdots \\ 0 \end{pmatrix} = c_1\begin{pmatrix} a_{11} \\ a_{21} \\ \vdots \\ a_{n1} \end{pmatrix} + c_2\begin{pmatrix} 0 \\ a_{22} \\ \vdots \\ a_{n2} \end{pmatrix} + c_3\begin{pmatrix} 0 \\ 0 \\ a_{33} \\ \vdots \\ a_{n3} \end{pmatrix} + c_n\begin{pmatrix} 0 \\ 0 \\ 0 \\ \vdots \\ a_{nn} \end{pmatrix},$$

or
$$\begin{aligned} a_{11}c_1 &= 0 \\ a_{21}c_1 + a_{22}c_2 &= 0 \\ a_{31}c_1 + a_{32}c_2 + a_{33}c_3 &= 0 \\ \vdots \qquad \vdots \qquad \vdots \quad &\;\; \vdots \\ a_{n1}c_1 + a_{n2}c_2 + a_{n3}c_3 + \cdots + a_{nn}c_n &= 0. \end{aligned}$$

Suppose that none of the diagonal elements is zero. Then, from the first equation, $c_1 = 0$. Similarly, from the second equation, $c_2 = 0$, etc. Thus the rows of A are linearly independent so that A is invertible. Suppose that $a_{ii} = 0$. Then the first i equations read

$$\begin{aligned} a_{11}c_1 &= 0 \\ a_{21}c_1 + a_{22}c_2 &= 0 \\ &\vdots \\ a_{i-1,1}c_1 + a_{i-1,2}c_2 + \cdots + a_{i-1,i-1}c_{i-1} &= 0 \\ a_{i1}c_1 + a_{i2}c_2 + \cdots + a_{i,i-1}c_{i-1} &= 0 \end{aligned}$$

This is a system of i equations in the $i-1$ unknowns $c_1, c_2, \ldots, c_{i-1}$. By Theorem 1.4.1, the system has an infinite number of solutions. Thus the first i rows of A are linearly dependent so that A is not invertible.

31. any nonzero multiple of (1, 2).

33. 3 chairs and 2 tables

35. 4 units of A and 5 units of B

37. (a) $$A = \begin{pmatrix} 0.293 & 0 & 0 \\ 0.014 & 0.207 & 0.017 \\ 0.044 & 0.010 & 0.216 \end{pmatrix};$$

$$I - A = \begin{pmatrix} 0.707 & 0 & 0 \\ -0.014 & 0.793 & -0.017 \\ -0.044 & -0.010 & 0.784 \end{pmatrix}$$

(b) $$\begin{pmatrix} 195492.2207 \\ 25932.85859 \\ 13580.33966 \end{pmatrix}$$

39. $\begin{pmatrix} 1 & \frac{1}{2} \\ 0 & 1 \end{pmatrix}$; yes

41. $\begin{pmatrix} 1 & \frac{2}{3} & \frac{1}{3} \\ 0 & 1 & 1 \\ 0 & 0 & 1 \end{pmatrix}$; yes

43. $\begin{pmatrix} 1 & -\frac{1}{2} & 2 \\ 0 & 1 & -14 \\ 0 & 0 & 0 \end{pmatrix}$; no

45. $\begin{pmatrix} 1 & 0 & 2 & 3 \\ 0 & 1 & 2 & 7 \\ 0 & 0 & 1 & \frac{10}{7} \\ 0 & 0 & 0 & 0 \end{pmatrix}$; no

Problems 2.8

1. $\begin{pmatrix} -1 & 6 \\ 4 & 5 \end{pmatrix}$

3. $\begin{pmatrix} 2 & -1 & 1 \\ 3 & 2 & 4 \end{pmatrix}$

5. $\begin{pmatrix} 1 & -1 & 1 \\ 2 & 0 & 5 \\ 3 & 4 & 5 \end{pmatrix}$

7. $\begin{pmatrix} 1 & 0 \\ 0 & 1 \\ 1 & 0 \\ 0 & 1 \end{pmatrix}$

9. $\begin{pmatrix} a & d & g \\ b & e & h \\ c & f & j \end{pmatrix}$

11. $[(A + B)^t]_{ij} = (A + B)_{ji} = a_{ji} + b_{ji} = (A^t)_{ij} + (B^t)_{ij}$. Thus the ijth component of $(A + B)^t$ equals the ijth component of A^t plus the ijth component of B^t.

15. If A is $m \times n$, then A^t is $n \times m$ and AA^t is $m \times m$. Also, $(AA^t)^t = (A^t)^t A^t = AA^t$.

17. If A is upper triangular and $B = A^t$, then $b_{ij} = a_{ji} = 0$ if $j > i$. Thus B is lower triangular.

19. $(A+B)^t = A^t + B^t = -A - B = -(A+B)$

21. $(AB)^t = B^tA^t = (-B)(-A) = (-1)^2BA = BA$

Chapter 2—Review

1. $\begin{pmatrix} 4 \\ 6 \\ 5 \end{pmatrix}$ **3.** $\begin{pmatrix} 6 \\ 0 \\ -7 \end{pmatrix}$ **5.** -3

7. 207 **9.** $\begin{pmatrix} -6 & 3 \\ 0 & 12 \\ 6 & 9 \end{pmatrix}$

11. $\begin{pmatrix} 16 & 2 & 3 \\ -20 & 10 & -1 \\ -36 & 8 & 16 \end{pmatrix}$

13. $\begin{pmatrix} 17 & 39 & 41 \\ 14 & 20 & 42 \end{pmatrix}$

15. $\begin{pmatrix} 9 & 10 \\ 30 & 32 \end{pmatrix}$

17. $A(BC) = \begin{pmatrix} 25 & 74 \\ 132 & 222 \end{pmatrix} = (AB)C$

19. dependent; $\begin{pmatrix} 4 \\ 6 \end{pmatrix} = 2\begin{pmatrix} 2 \\ 3 \end{pmatrix}$

21. dependent;
$$2\begin{pmatrix} 1 \\ -4 \\ 2 \end{pmatrix} - \begin{pmatrix} 0 \\ 2 \\ -1 \end{pmatrix} - \begin{pmatrix} 2 \\ -10 \\ 5 \end{pmatrix} = \begin{pmatrix} 0 \\ 0 \\ 0 \end{pmatrix}$$

23. $\begin{pmatrix} 1 & \frac{3}{2} \\ 0 & 1 \end{pmatrix}$; inverse is $\begin{pmatrix} \frac{4}{11} & -\frac{3}{11} \\ \frac{1}{11} & \frac{2}{11} \end{pmatrix}$

25. $\begin{pmatrix} 1 & 2 & 0 \\ 0 & 1 & \frac{1}{3} \\ 0 & 0 & 1 \end{pmatrix}$; inverse is $\begin{pmatrix} -\frac{1}{4} & \frac{1}{4} & \frac{1}{4} \\ \frac{5}{8} & -\frac{1}{8} & -\frac{1}{8} \\ \frac{1}{8} & -\frac{5}{8} & \frac{3}{8} \end{pmatrix}$

27. $\begin{pmatrix} 1 & 0 & 2 \\ 0 & 1 & 1 \\ 0 & 0 & 1 \end{pmatrix}$; inverse is $\begin{pmatrix} \frac{5}{6} & \frac{2}{3} & -2 \\ \frac{1}{3} & \frac{2}{3} & -1 \\ -\frac{1}{6} & -\frac{1}{3} & 1 \end{pmatrix}$

29. $\begin{pmatrix} 1 & 2 & 0 \\ 2 & 1 & -1 \\ 3 & 1 & 1 \end{pmatrix}\begin{pmatrix} x_1 \\ x_2 \\ x_3 \end{pmatrix} = \begin{pmatrix} 3 \\ -1 \\ 7 \end{pmatrix}$; A^{-1} is given in Problem 25; $x_1 = \frac{3}{4}$, $x_2 = \frac{9}{8}$, $x_3 = \frac{29}{8}$

31. $\begin{pmatrix} 2 & -1 \\ 3 & 0 \\ 1 & 2 \end{pmatrix}$; neither

33. $\begin{pmatrix} 2 & 3 & 1 \\ 3 & -6 & -5 \\ 1 & -5 & 9 \end{pmatrix}$; symmetric

35. $\begin{pmatrix} 1 & -1 & 4 & 6 \\ -1 & 2 & 5 & 7 \\ 4 & 5 & 3 & -8 \\ 6 & 7 & -8 & 9 \end{pmatrix}$; symmetric

CHAPTER 3

Problems 3.1

1. -10 **3.** 47 **5.** 4 **7.** 56

9. 274

11. Let $A = \begin{pmatrix} a_{11} & 0 & 0 & \cdots & 0 \\ 0 & a_{22} & 0 & \cdots & 0 \\ 0 & 0 & a_{33} & \cdots & 0 \\ \vdots & \vdots & \vdots & & \vdots \\ 0 & 0 & 0 & \cdots & a_{nn} \end{pmatrix}$

and

$B = \begin{pmatrix} b_{11} & 0 & 0 & \cdots & 0 \\ 0 & b_{22} & 0 & \cdots & 0 \\ 0 & 0 & b_{33} & \cdots & 0 \\ \vdots & \vdots & \vdots & & \vdots \\ 0 & 0 & 0 & \cdots & b_{nn} \end{pmatrix}$.

Then $\det A = a_{11}a_{22}a_{33}\cdots a_{nn}$, $\det B = b_{11}b_{22}b_{33}\cdots b_{nn}$,

$$AB = \begin{pmatrix} a_{11}b_{11} & 0 & 0 & \cdots & 0 \\ 0 & a_{22}b_{22} & 0 & \cdots & 0 \\ 0 & 0 & a_{33}b_{33} & \cdots & 0 \\ \vdots & \vdots & & & \vdots \\ 0 & 0 & 0 & \cdots & a_{nn}b_{nn} \end{pmatrix}$$

and

$$\det AB = (a_{11}b_{11})(a_{22}b_{22}) \times (a_{33}b_{33})\cdots(a_{nn}b_{nn}) = (a_{11}a_{22}a_{33}\cdots a_{nn}) \times (b_{11}b_{22}b_{33}\cdots b_{nn}) = \det A \det B.$$

13. Almost any example will work. For instance,

$\det\begin{pmatrix} 1 & 0 \\ 0 & 1 \end{pmatrix} = 1$, but

$\det\begin{pmatrix} 1 & 0 \\ 0 & 0 \end{pmatrix} + \det\begin{pmatrix} 0 & 0 \\ 0 & 1 \end{pmatrix} = 0 + 0 \neq 1.$

As another example, let $A = \begin{pmatrix} 1 & 2 \\ 3 & 4 \end{pmatrix}$ and $B = \begin{pmatrix} 5 & 6 \\ 7 & 8 \end{pmatrix}$; then $(A+B) = \begin{pmatrix} 6 & 8 \\ 10 & 12 \end{pmatrix}$, $\det A = -2$, $\det B = -2$, and $\det(A+B) = -8 \neq \det A + \det B$.

15. Let $A = \begin{pmatrix} a_{11} & 0 & \cdots & 0 \\ a_{21} & a_{22} & \cdots & 0 \\ \vdots & \vdots & & \vdots \\ a_{n1} & a_{n2} & \cdots & a_{nn} \end{pmatrix}$.

Then, continually expanding in the first row, we obtain

$$\det A = a_{11}\begin{vmatrix} a_{22} & 0 & \cdots & 0 \\ a_{32} & a_{33} & \cdots & 0 \\ \vdots & \vdots & & \vdots \\ a_{n2} & a_{n3} & \cdots & a_{nn} \end{vmatrix}$$

$$= a_{11}a_{22}\begin{vmatrix} a_{33} & 0 & \cdots & 0 \\ a_{43} & a_{44} & \cdots & 0 \\ \vdots & \vdots & & \vdots \\ a_{n3} & a_{n4} & \cdots & a_{nn} \end{vmatrix}$$

$$= \cdots = a_{11}a_{22}a_{33}\cdots a_{n-2}\begin{vmatrix} a_{n-1,n-1} & 0 \\ a_{n,n-1} & a_{nn} \end{vmatrix}$$

$$= a_{11}a_{22}a_{33}\cdots a_{n-2,n-2}a_{n-1,n-1}a_{nn}.$$

17. Let $\mathbf{u}_1 = \begin{pmatrix} u_{11} \\ u_{12} \end{pmatrix}$, $\mathbf{u}_2 = \begin{pmatrix} u_{21} \\ u_{22} \end{pmatrix}$

and $A = \begin{pmatrix} a_{11} & a_{12} \\ a_{21} & a_{22} \end{pmatrix}$. Then

$$\mathbf{v}_1 = \begin{pmatrix} a_{11} & a_{12} \\ a_{21} & a_{22} \end{pmatrix}\begin{pmatrix} u_{11} \\ u_{12} \end{pmatrix} = \begin{pmatrix} a_{11}u_{11} + a_{12}u_{12} \\ a_{21}u_{11} + a_{22}u_{12} \end{pmatrix},$$

$$\mathbf{v}_2 = \begin{pmatrix} a_{11} & a_{12} \\ a_{21} & a_{22} \end{pmatrix}\begin{pmatrix} u_{21} \\ u_{22} \end{pmatrix} = \begin{pmatrix} a_{11}u_{21} + a_{12}u_{22} \\ a_{21}u_{21} + a_{22}u_{22} \end{pmatrix}.$$

Let $A_{\mathbf{v}}$ denote the area generated by $\mathbf{v}_1$ and $\mathbf{v}_2$. Then, applying the result of Problem 16,

$$\begin{aligned} A_{\mathbf{v}} &= |(a_{11}u_{11} + a_{12}u_{12}) \times (a_{21}u_{21} + a_{22}u_{22}) - (a_{11}u_{21} + a_{12}u_{22}) \times (a_{21}u_{11} + a_{22}u_{12})| \\ &= |a_{21}a_{11}u_{21}u_{11} + a_{21}a_{12}u_{21}u_{12} + a_{21}a_{11}u_{11}u_{22} + a_{22}a_{12}u_{12}u_{22} - a_{21}a_{11}u_{21}u_{11} - a_{22}a_{11}u_{21}u_{12} - a_{21}a_{12}u_{11}u_{22} - a_{22}a_{12}u_{12}u_{22}| \\ &= |u_{11}u_{22} - u_{12}u_{21}| \times |a_{22}a_{11} - a_{21}a_{12}| \\ &= A_{\mathbf{u}}|\det A| \end{aligned}$$

where $A_{\mathbf{u}}$ is the area generated by $\mathbf{u}_1$ and $\mathbf{u}_2$, again using the result of Problem 16.

Problems 3.2

1. 28 **3.** 2 **5.** 32 **7.** −36
9. −260 **11.** −183 **13.** 24
15. −296 **17.** 138
19. *abcde* **21.** −8 **23.** 16
25. −16 **27.** −16

29. Proof by induction: true for $n = 2$ since

$$\begin{vmatrix} 1+x_1 & x_2 \\ x_1 & 1+x_2 \end{vmatrix} = (1+x_1)(1+x_2) - x_1x_2 = 1 + x_1 + x_2.$$

Assume true for $n = k$. That is,

$$\begin{vmatrix} 1+x_1 & x_2 & x_3 & \cdots & x_k \\ x_1 & 1+x_2 & x_3 & \cdots & x_k \\ x_1 & x_2 & 1+x_3 & \cdots & x_k \\ \vdots & \vdots & \vdots & & \vdots \\ x_1 & x_2 & x_3 & \cdots & 1+x_k \end{vmatrix} = 1 + x_1 + x_2 + \cdots + x_k.$$

Then, for $n = k + 1$,

$$\begin{vmatrix} 1+x_1 & x_2 & x_3 & \cdots & x_k & x_{k+1} \\ x_1 & 1+x_2 & x_3 & \cdots & x_k & x_{k+1} \\ x_1 & x_2 & 1+x_3 & \cdots & x_k & x_{k+1} \\ \vdots & \vdots & \vdots & & \vdots & \vdots \\ x_1 & x_2 & x_3 & \cdots & x_k & 1+x_{k+1} \end{vmatrix}$$

(using Property 3 in the first column)

$$= \begin{vmatrix} 1 & x_2 & x_3 & \cdots & x_k & x_{k+1} \\ 0 & 1+x_2 & x_3 & \cdots & x_k & x_{k+1} \\ 0 & x_2 & 1+x_3 & \cdots & x_k & x_{k+1} \\ \vdots & \vdots & \vdots & & \vdots & \vdots \\ 0 & x_2 & x_3 & \cdots & x_k & 1+x_{k+1} \end{vmatrix} \; ① + \begin{vmatrix} x_1 & x_2 & x_3 & \cdots & x_k & x_{k+1} \\ x_1 & 1+x_2 & x_3 & \cdots & x_k & x_{k+1} \\ x_1 & x_2 & 1+x_3 & \cdots & x_k & x_{k+1} \\ \vdots & \vdots & \vdots & & \vdots & \vdots \\ x_1 & x_2 & x_3 & \cdots & x_k & 1+x_{k+1} \end{vmatrix} \; ②.$$

But, expanding det ① in its first column, we have

$$\det ① = \begin{vmatrix} 1+x_2 & x_3 & \cdots & x_k & x_{k+1} \\ x_2 & 1+x_3 & \cdots & x_k & x_{k+1} \\ \vdots & & & & \\ x_2 & x_3 & \cdots & x_k & 1+x_{k+1} \end{vmatrix} = 1 + x_2 + x_3 + \cdots + x_{k+1}$$

by the induction assumption (since ① is a $k \times k$ determinant). To evaluate det ②, subtract the first row from all other rows:

$$\det ② = \begin{vmatrix} x_1 & x_2 & x_3 & \cdots & x_k & x_{k+1} \\ 0 & 1 & 0 & \cdots & 0 & 0 \\ 0 & 0 & 1 & \cdots & 0 & 0 \\ \vdots & \vdots & \vdots & & \vdots & \vdots \\ 0 & 0 & 0 & \cdots & 0 & 1 \end{vmatrix} = x_1.$$

Adding det ① and det ② completes the proof.

31. If n is odd, $\det A = -\det A$ so that $2 \det A = 0$ and $\det A = 0$.

33. $$\frac{1}{2}\begin{vmatrix} 1 & x_1 & y_1 \\ 1 & x_2 & y_2 \\ 1 & x_3 & y_3 \end{vmatrix} = \frac{1}{2}\begin{vmatrix} x_2 - x_1 & x_3 - x_1 \\ y_2 - y_1 & y_3 - y_1 \end{vmatrix}.$$

Look at the figures below.

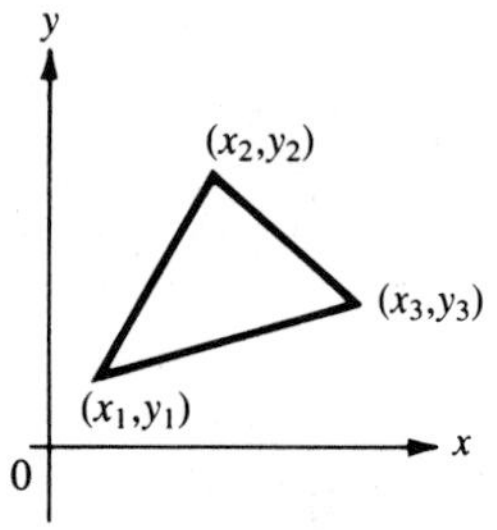

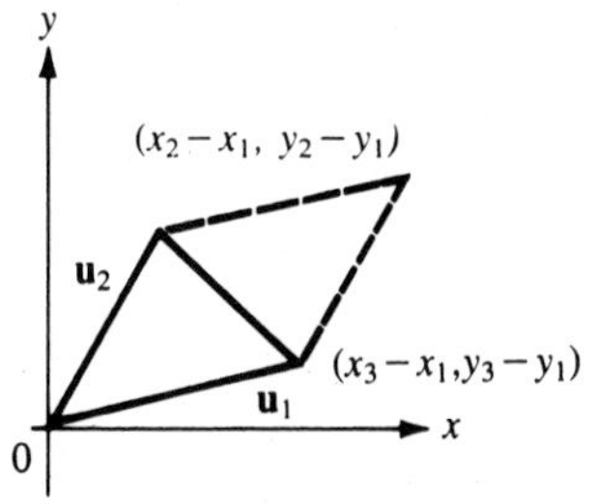

The area A of the triangle is half the area of the parallelogram generated by the vectors $\mathbf{u}_1$ and $\mathbf{u}_2$ which, by the result of Problem 3.1.16, is given by

$$A = \pm\frac{1}{2}\begin{vmatrix} x_2 - x_1 & x_3 - x_1 \\ y_2 - y_1 & y_3 - y_1 \end{vmatrix}.$$

35. $$D_3 = \begin{vmatrix} 1 & 1 & 1 \\ a_1 & a_2 & a_3 \\ a_1^2 & a_2^2 & a_3^2 \end{vmatrix}$$

$$= \begin{vmatrix} 1 & 0 & 0 \\ a_1 & a_2 - a_1 & a_3 - a_1 \\ a_1^2 & a_2^2 - a_1^2 & a_3^2 - a_1^2 \end{vmatrix}$$

$$= \begin{vmatrix} a_2 - a_1 & a_3 - a_1 \\ (a_2 + a_1)(a_2 - a_1) & (a_3 + a_1)(a_3 - a_1) \end{vmatrix}$$

$$= (a_2 - a_1)(a_3 - a_1) \times \begin{vmatrix} 1 & 1 \\ a_2 + a_1 & a_3 + a_1 \end{vmatrix}$$

$$= (a_2 - a_1)(a_3 - a_1) \times (a_3 - a_2)$$

37. **(a)** $$D_n = \begin{vmatrix} 1 & 1 & \cdots & 1 \\ a_1 & a_2 & \cdots & a_n \\ a_1^2 & a_2^2 & \cdots & a_n^2 \\ \vdots & & & \\ a_1^{n-1} & a_1^{n-1} & \cdots & a_n^{n-1} \end{vmatrix}$$

(b) We prove this by induction. The result is true for $n = 3$ by the result of Problem 35. We assume it true for $n = k$. Now

$$D_{k+1} = \begin{vmatrix} 1 & 1 & \cdots & 1 & 1 \\ a_1 & a_2 & \cdots & a_k & a_{k+1} \\ a_1^2 & a_2^2 & \cdots & a_k^2 & a_{k+1}^2 \\ \vdots & \vdots & & \vdots & \vdots \\ a_1^{k-1} & a_2^{k-1} & \cdots & a_k^{k-1} & a_{k+1}^{k-1} \\ a_1^k & a_2^k & \cdots & a_k^k & a_{k+1}^k \end{vmatrix}.$$

We subtract the first column from each of the other k columns:

$$D_{k+1} = \begin{vmatrix} 1 & 0 & \cdots & 0 & 0 \\ a_1 & a_2 - a_1 & \cdots & a_k - a_1 & a_{k+1} - a_1 \\ a_1^2 & a_2^2 - a_1^2 & \cdots & a_k^2 - a_1^2 & a_{k+1}^2 - a_1^2 \\ \vdots & \vdots & & \vdots & \vdots \\ a_1^{k-1} & a_2^{k-1} - a_1^{k-1} & \cdots & a_k^{k-1} - a_1^{k-1} & a_{k+1}^{k-1} - a_1^{k-1} \\ a_1^k & a_2^k - a_1^k & \cdots & a_k^k - a_1^k & a_{k+1}^k - a_1^k \end{vmatrix}$$

$$= \begin{vmatrix} a_2 - a_1 & a_3 - a_1 & \cdots & a_k - a_1 & a_{k+1} - a_1 \\ a_2^2 - a_1^2 & a_3^2 - a_1^2 & \cdots & a_k^2 - a_1^2 & a_{k+1}^2 - a_1^2 \\ \vdots & \vdots & & \vdots & \vdots \\ a_2^{k-1} - a_1^{k-1} & a_3^{k-1} - a_1^{k-1} & \cdots & a_k^{k-1} - a_1^{k-1} & a_{k+1}^{k-1} - a_1^{k-1} \\ a_2^k - a_1^k & a_3^k - a_1^k & \cdots & a_k^k - a_1^k & a_{k+1}^k - a_1^k \end{vmatrix}.$$

Now $a_2^k - a_1^k = (a_2 - a_1) \times (a_2^{k-1} + a_2^{k-2}a_1 + a_2^{k-3}a_1^2 + \cdots + a_2^2 a_1^{k-3} + a_2 a_1^{k-2} + a_1^{k-1})$, and $a_2^{k-1} - a_1^{k-1} = (a_2 - a_1) \times (a_2^{k-2} + a_2^{k-3}a_1 + \cdots + a_2^2 a_1^{k-4} + a_2 a_1^{k-3} + a_1^{k-2})$. Note that if the terms in the second factor of the last expression are multiplied by a_1, and then subtracted from the second factor of $a_2^k - a_1^k$, only the term a_2^{k-1} remains. Thus we **(i)** expand the last determinant obtained above in the first row, **(ii)** factor $a_{j-1} - a_1$ from the jth column for $1 \le j \le k$, and **(iii)** multiply the ℓth row by a_1 and subtract it from the $(\ell + 1)$st row, for $\ell = k - 1, k - 2, \ldots, 3, 2$ in succession. This yields

$$D_{k+1} = (a_2 - a_1) \times (a_3 - a_1)\cdots(a_{k+1} - a_1) \times \begin{vmatrix} 1 & 1 & \cdots & 1 & 1 \\ a_2 & a_3 & \cdots & a_k & a_{k+1} \\ a_2^2 & a_3^2 & \cdots & a_k^2 & a_{k+1}^2 \\ \vdots & \vdots & & \vdots & \vdots \\ a_2^{k-1} & a_3^{k-1} & \cdots & a_k^{k-1} & a_{k+1}^{k-1} \\ a_2^k a_3^k & & \cdots & a_k^k & a_{k+1}^k \end{vmatrix}$$

$$= \prod_{j=2}^{k+1} (a_j - a_1) \prod_{\substack{i=2 \\ j>i}}^{k+1} (a_j - a_i)$$

(from the induction assumption since the last determinant is

$$k \times k) = \prod_{\substack{i=1 \\ j>i}}^{k+1} (a_j - a_i);$$

this completes the proof.

Problems 3.3

1. $a_{1k}A_{1k}$ is the only term in the expansion in the first column of A involving the component a_{1k}. But

$$a_{1k}A_{1k} = a_{1k}(-1)^{1+k}|M_{1k}|.$$

If we expand $|M_{1k}|$ about its lth column for $l \neq k$, a term in the expansion takes the form a_{il} (cofactor of a_{il} in M_{1k}). But this is the only occurrence of a_{il} in the expansion of M_{1k} since the other terms have the form a_{jl} (cofactor of a_{jl} in M_{1k}), which deletes the column corresponding to the lth column of A, and a_{il} is in the lth column. Therefore $a_{1k}A_{1k} = (-1)^{1+h}a_{1k}a_{il} \cdot$ (cofactor of a_{il} in m_{1k}).

3. Expand $|A|$ about its kth column. A term is $a_{ik}A_{ik}$ and this is the only occurrence of a_{ij} in the expansion of $|A|$. Now $A_{ik} = (-1)^{i+k}|M_{ik}|$, and if this is expanded in the lth column (for $l \neq k$), the only term in the expansion containing a_{jl} is $a_{jl}\cdot$(cofactor of a_{jl} in M_{ik}) for the same reason as in Problem 1. Thus the only occurrence of $a_{ij}a_{jl}$ is $(-1)^{i+k}a_{ik}a_{jl}\cdot$(cofactor of a_{jl} in M_{ik}).

5. -6

Problems 3.4

1. $\begin{pmatrix} \frac{1}{2} & -\frac{1}{2} \\ -\frac{1}{4} & \frac{3}{4} \end{pmatrix}$ **3.** $\begin{pmatrix} 0 & 1 \\ 1 & 0 \end{pmatrix}$

5. $\begin{pmatrix} \frac{1}{3} & -\frac{1}{4} & -\frac{1}{6} \\ 0 & \frac{1}{4} & \frac{1}{2} \\ 0 & \frac{1}{4} & -\frac{1}{2} \end{pmatrix}$

7. $\begin{pmatrix} 0 & 1 & -1 \\ 2 & -2 & -1 \\ -1 & 1 & 1 \end{pmatrix}$

9. not invertible

11. $\begin{pmatrix} \frac{7}{3} & -\frac{1}{3} & -\frac{1}{3} & -\frac{2}{3} \\ \frac{4}{9} & -\frac{1}{9} & -\frac{4}{9} & \frac{1}{9} \\ -\frac{1}{9} & -\frac{2}{9} & \frac{1}{9} & \frac{2}{9} \\ -\frac{5}{3} & \frac{2}{3} & \frac{2}{3} & \frac{1}{3} \end{pmatrix}$

13. independent

15. independent

17. Follows from the fact that $\det A^t = \det A$.

19. $A^{-1} = \begin{pmatrix} \frac{1}{14} & \frac{1}{14} & \frac{9}{28} \\ -\frac{5}{7} & \frac{2}{7} & -\frac{3}{14} \\ \frac{1}{14} & \frac{1}{14} & -\frac{5}{28} \end{pmatrix}$,

$\det A = -28$, $\det A^{-1} = -\frac{1}{28}$

21. no inverse if α is any real number

Problems 3.5

1. $x_1 = -5, x_2 = 3$

3. $x_1 = 2, x_2 = 5, x_3 = -3$

5. $x_1 = \frac{45}{13}$, $x_2 = -\frac{11}{13}$, $x_3 = \frac{23}{13}$

7. $x_1 = \frac{3}{2}$, $x_2 = \frac{3}{2}$, $x_3 = \frac{1}{2}$

9. $x_1 = \frac{21}{29}$, $x_2 = \frac{171}{29}$, $x_3 = -\frac{284}{29}$, $x_4 = -\frac{182}{29}$

Chapter 3—Review

1. -4 **3.** 24 **5.** 60 **7.** 34

9. $\begin{pmatrix} -\frac{1}{11} & \frac{4}{11} \\ \frac{2}{11} & \frac{3}{11} \end{pmatrix}$

11. not invertible

13. $\begin{pmatrix} \frac{1}{11} & \frac{1}{11} & 0 & \frac{3}{11} \\ \frac{9}{11} & -\frac{2}{11} & 0 & -\frac{6}{11} \\ \frac{3}{11} & \frac{3}{11} & 0 & -\frac{2}{11} \\ \frac{1}{22} & \frac{1}{22} & -\frac{1}{2} & \frac{1}{22} \end{pmatrix}$

15. dependent **17.** dependent

19. independent

21. $x_1 = \frac{11}{7}$, $x_2 = \frac{1}{7}$

23. $x_1 = \frac{1}{4}$, $x_2 = \frac{5}{4}$, $x_3 = -\frac{3}{4}$

CHAPTER 4

Problems 4.1

1. $|\mathbf{v}| = 4\sqrt{2}$, $\theta = \pi/4$

3. $|\mathbf{v}| = 4\sqrt{2}$, $\theta = 7\pi/4$

5. $|\mathbf{v}| = 2$, $\theta = \pi/6$

7. $|\mathbf{v}| = 2$, $\theta = 2\pi/3$

9. $|\mathbf{v}| = 2$, $\theta = 4\pi/3$

11. $|\mathbf{v}| = \sqrt{89}$, $\theta = \pi + \tan^{-1}(-\frac{8}{5}) \approx 2.13$ (in the second quadrant)

13. $(3, 3)$ **15.** $(4, 4)$

17. $(-2, -5)$ **19.** $\mathbf{j}$

21. $-6\mathbf{i} + \mathbf{j}$

23. (a) $(6, 9)$ (b) $(-3, 7)$ (c) $(-7, 1)$ (d) $(39, -22)$

25. $|\mathbf{i}| = |(1, 0)| = \sqrt{1^2 + 0^2} = \sqrt{1} = 1$;

$|\mathbf{j}| = |(0, 1)| = \sqrt{0^2 + 1^2} = \sqrt{1} = 1$

29. $(1/\sqrt{2})\mathbf{i} - (1/\sqrt{2})\mathbf{j}$

31. $(1/\sqrt{2})\mathbf{i} + (1/\sqrt{2})\mathbf{j}$ if $a > 0$; $-(1/\sqrt{2})\mathbf{i} - (1/\sqrt{2})\mathbf{j}$ if $a < 0$

33. $\sin\theta = -3/\sqrt{13}$, $\cos\theta = 2/\sqrt{13}$

35. $-(1/\sqrt{2})\mathbf{i} - (1/\sqrt{2})\mathbf{j}$

37. $\frac{3}{5}\mathbf{i} - \frac{4}{5}\mathbf{j}$

39. (a) $(1/\sqrt{2})\mathbf{i} - (1/\sqrt{2})\mathbf{j}$
(b) $(7/\sqrt{193})\mathbf{i} - (12/\sqrt{193})\mathbf{j}$
(c) $-(2/\sqrt{53})\mathbf{i} + (7/\sqrt{53})\mathbf{j}$

41. $\overrightarrow{PQ}$ is a representation of $(c + a - c)\mathbf{i} + (d + b - d)\mathbf{j} = a\mathbf{i} + b\mathbf{j}$. Thus $\overrightarrow{PQ}$ and (a, b) are representations of the same vector.

43. $4\mathbf{i} + 4\sqrt{3}\mathbf{j}$ **45.** $-3\mathbf{i} + 3\sqrt{3}\mathbf{j}$

47. (i) Suppose $\mathbf{u} = \alpha\mathbf{v}$ where $\alpha > 0$. Then $|\mathbf{u} + \mathbf{v}| = |\alpha\mathbf{v} + \mathbf{v}| = |(\alpha + 1)\mathbf{v}| = |\alpha + 1|\ |\mathbf{v}| = (\alpha + 1)|\mathbf{v}|$ (since $\alpha + 1 > 0$) $= \alpha|\mathbf{v}| + |\mathbf{v}| = |\alpha\mathbf{v}| + |\mathbf{v}| = |\mathbf{u}| + |\mathbf{v}|$.
(ii) Conversely, suppose $\mathbf{u} = (a, b)$, $\mathbf{v} = (c, d)$, and $|\mathbf{u} + \mathbf{v}| = |\mathbf{u}| + |\mathbf{v}|$. Then $\mathbf{u} + \mathbf{v} = (a + c, b + d)$, and $|\mathbf{u} + \mathbf{v}|^2 = (|\mathbf{u}| + |\mathbf{v}|)^2 = |\mathbf{u}|^2 + 2|\mathbf{u}||\mathbf{v}| + |\mathbf{v}|^2$, which implies that

$$(a + c)^2 + (b + d)^2 = a^2 + b^2 + 2\sqrt{(a^2 + b^2)(c^2 + d^2)} + c^2 + d^2$$

and, after multiplying through and cancelling like terms, we obtain $ac + bd = \sqrt{a^2c^2 + a^2d^2 + b^2c^2 + b^2d^2}$. Then, squaring both sides and again cancelling like terms, we

27. $$|\mathbf{u}| = \sqrt{\left(\frac{a}{\sqrt{a^2 + b^2}}\right)^2 + \left(\frac{b}{\sqrt{a^2 + b^2}}\right)^2} = \sqrt{\frac{a^2}{a^2 + b^2} + \frac{b^2}{a^2 + b^2}} = 1.$$

Direction of

$$|\mathbf{u}| = \tan^{-1}\frac{\dfrac{b}{\sqrt{a^2 + b^2}}}{\dfrac{a}{\sqrt{a^2 + b^2}}} = \tan^{-1}\left(\frac{b}{a}\right) = \text{direction of } \mathbf{v}.$$

have $2abcd = a^2d^2 + b^2c^2$, or $(ad - bc)^2 = a^2d^2 - 2abcd + b^2c^2 = 0$ so that $ad = bc$. If $d \neq 0$, then $a = \frac{b}{d}c, b = \frac{b}{d}d$, and $|\mathbf{u}| = \left|\frac{b}{d}\right||\mathbf{v}|$. Then $\left|\frac{b+d}{d}\right||\mathbf{v}| = \left|\frac{b+d}{d}(c, d)\right| = \left|\left(\frac{b}{d}c + c, \frac{b}{d}d + d\right)\right| = |(a+c, b+d)| = |\mathbf{u} + \mathbf{v}| = |\mathbf{u}| + |\mathbf{v}| = \left|\frac{b}{d}\mathbf{v}\right| + |\mathbf{v}| = \left(\left|\frac{b}{d}\right| + 1\right)|\mathbf{v}| = \frac{|b| + |d|}{|d|}|\mathbf{v}|$. Thus $\left|\frac{b+d}{d}\right| = \frac{|b| + |d|}{|d|}$ so that $|b + d| = |b| + |d|$. Since b and d are real numbers, this implies that b and d have the same sign so that $\frac{b}{d}$ is positive. Thus if $\alpha = \left|\frac{b}{d}\right| = \frac{b}{d}$, then $\mathbf{u} = \alpha\mathbf{v}$. If $d = 0$, then $c \neq 0$ and $\mathbf{u} = \frac{a}{c}\mathbf{v}$ by the same reasoning as above where $\frac{a}{c} > 0$. Thus $\mathbf{u}$ is a positive scalar multiple of $\mathbf{v}$.

Problems 4.2

1. 0; 0 **3.** 0; 0 **5.** 20; $\frac{20}{29}$

7. $-22; -22/5\sqrt{53}$ **11.** parallel

9. $\mathbf{u}\cdot\mathbf{v} = \alpha\beta - \beta\alpha = 0$

13. neither **15.** orthogonal

17. (a) $-\frac{3}{4}$ (b) $\frac{4}{3}$ (c) $\frac{1}{7}$ (d) $(-96 \pm \sqrt{7500})/78 \approx -0.12, -2.34$

19. If $\mathbf{u}$ and $\mathbf{v}$ have opposite directions, then $\theta_{\mathbf{u}} = \theta_{\mathbf{v}} + \pi$. Thus $\cos\theta_{\mathbf{u}} = \cos(\theta_{\mathbf{v}} + \pi) = -\cos\theta_{\mathbf{v}}$. This implies that $\cos\theta_{\mathbf{u}} = \frac{3}{5} = -\frac{1}{\sqrt{1+\alpha^2}} = -\cos\theta_{\mathbf{v}}$, or $\sqrt{1+\alpha^2} = -\frac{5}{3} < 0$, which is impossible since $\sqrt{1+\alpha^2} = |\mathbf{v}| \geq 0$.

21. $\frac{3}{2}\mathbf{i} + \frac{3}{2}\mathbf{j}$ **23.** $\mathbf{0}$ **25.** $-\frac{2}{13}\mathbf{i} + \frac{3}{13}\mathbf{j}$

27. $[(\alpha+\beta)/2]\mathbf{i} + [(\alpha+\beta)/2]\mathbf{j}$

29. $[(\alpha-\beta)/2]\mathbf{i} + [(\alpha-\beta)/2]\mathbf{j}$

31. $a_1a_2 + b_1b_2 \geq 0$

33. $\text{Proj}_{\overrightarrow{PQ}}\overrightarrow{RS} = \frac{51}{25}\mathbf{i} + \frac{68}{25}\mathbf{j}$; $\text{Proj}_{\overrightarrow{RS}}\overrightarrow{PQ} = -\frac{17}{26}\mathbf{i} + \frac{85}{26}\mathbf{j}$

35. (i) If $\mathbf{u} = \alpha\mathbf{v}$, then $\mathbf{u}\cdot\mathbf{v} = \alpha\mathbf{v}\cdot\mathbf{v} = \alpha|\mathbf{v}|^2$, $|\mathbf{u}| = |\alpha||\mathbf{v}|^2$ so that $\cos\phi = \frac{\alpha|\mathbf{v}|^2}{|\alpha||\mathbf{v}||\mathbf{v}|} = \frac{\alpha}{|\alpha|} = \pm 1$.
(ii) Suppose that $\mathbf{u}$ and $\mathbf{v}$ are parallel. Then, if $\mathbf{u} = (a, b)$ and $\mathbf{v} = (c, d)$, we have $1 = \cos^2\phi = \frac{|\mathbf{u}\cdot\mathbf{v}|^2}{|\mathbf{u}|^2|\mathbf{v}|^2} = \frac{(ac+bd)^2}{(a^2+b^2)(c^2+d^2)}$. Multiplying through and simplifying, we obtain $0 = a^2d^2 - 2abcd + b^2c^2$. Thus $ad = bc$. If $a \neq 0$, then $d = \left(\frac{c}{a}\right)b$ and $c = \left(\frac{c}{a}\right)a$ so that $\mathbf{v} = \left(\frac{c}{a}\right)\mathbf{u}$. If $a = 0$, then $b \neq 0$ and $\mathbf{v} = \frac{d}{b}\mathbf{u}$.

37. The line $ax + by + c = 0$ has slope $-\frac{a}{b}$. A vector parallel to the line is $\mathbf{u} = \mathbf{i} - \frac{a}{b}\mathbf{j}$, and $\mathbf{u}\cdot\mathbf{v} = 1\cdot a - \frac{a}{b}\cdot b = 0$.

39. $52/5\sqrt{113} \approx 0.9783$; $61/\sqrt{34}\sqrt{113} \approx 0.9841$; $-27/5\sqrt{34} \approx -0.9261$

41. If either $a_1 = a_2 = 0$ or $b_1 = b_2 = 0$, both sides of the inequality are zero. If at least one of a_1 and $a_2 \neq 0$ and at least one of b_1 and $b_2 \neq 0$, let $\mathbf{u} = a_1\mathbf{i} + a_2\mathbf{j}$, $\mathbf{v} = b_1\mathbf{i} + b_2\mathbf{j}$. Then $\mathbf{u} \neq 0$, $\mathbf{v} \neq 0$, $|\mathbf{u}||\mathbf{v}| \neq 0$, and $\left|\frac{\mathbf{u}\cdot\mathbf{v}}{|\mathbf{u}||\mathbf{v}|}\right| = |\cos\phi| \leq 1$. Thus $|a_1b_1 + a_2b_2| = |\mathbf{u}\cdot\mathbf{v}| \leq |\mathbf{u}||\mathbf{v}| = \sqrt{a_1^2 + a_2^2}\sqrt{b_1^2 + b_2^2}$. Equality holds when $|\cos\phi| = 1$, which is true if and only if $\mathbf{u}$ and $\mathbf{v}$ are parallel.

43. $\sqrt{5}$ **45.** $-2\mathbf{i} - 5\mathbf{j}$; $2\mathbf{i} + 5\mathbf{j}$

47. $-2\sqrt{3}\mathbf{i} - 7\mathbf{j}$; $2\sqrt{3}\mathbf{i} + 7\mathbf{j}$

49. $(3/\sqrt{2})\mathbf{i} - (3/\sqrt{2} + 2)\mathbf{j}$; $-(3/\sqrt{2})\mathbf{i} + (3/\sqrt{2} + 2)\mathbf{j}$

51. -12 joules

53. $8\sqrt{3} + 4$ joules

55. $12/\sqrt{13}$ joules

57. From Figure 4.27, $\overrightarrow{PQ} = \mathbf{v} - \mathbf{u}$. Also, $\overrightarrow{PR}$, $\overrightarrow{RQ}$, and $\overrightarrow{PQ}$ have the same direction so that $\overrightarrow{PR} = \frac{a}{a+b}\overrightarrow{PQ}$ and $\overrightarrow{RQ} = \frac{b}{a+b}\overrightarrow{PQ}$. Then, again using the figure, $\mathbf{w} = \mathbf{u} + \overrightarrow{PR} = \mathbf{u} + \frac{a}{a+b}\overrightarrow{PQ} = \mathbf{u} + \frac{a}{a+b} \times (\mathbf{v} - \mathbf{u}) = \left(1 - \frac{a}{a+b}\right)\mathbf{u} + \frac{a}{a+b}\mathbf{v} = \frac{b}{a+b}\mathbf{u} + \frac{a}{a+b}\mathbf{v}$.

59. $\overrightarrow{AF} = \frac{1}{2}(\mathbf{b} + \mathbf{c})$
$\overrightarrow{BE} = \frac{1}{2}(-\mathbf{a} - \mathbf{c})$
$\overrightarrow{CD} = \frac{1}{2}(\mathbf{a} - \mathbf{b})$
$\overrightarrow{AG} = \overrightarrow{AB} + \overrightarrow{BG} = \mathbf{c} + \alpha\overrightarrow{BE} = \mathbf{c} + \frac{\alpha}{2}(-\mathbf{a} - \mathbf{c}) = \left[1 - \frac{\alpha}{2}\right]\mathbf{c} - \frac{\alpha}{2}\mathbf{a}$
$\overrightarrow{AG} = \overrightarrow{AC} + \overrightarrow{CG} = \mathbf{b} + \alpha(\overrightarrow{CD}) = \mathbf{b} + \frac{\alpha}{2}(\mathbf{a} - \mathbf{b}) = \left[1 - \frac{\alpha}{2}\right]\mathbf{b} + \frac{\alpha}{2}(\mathbf{a})$
$\overrightarrow{AG} = \overrightarrow{AG}$; $\left[1 - \frac{\alpha}{2}\right]\mathbf{c} - \frac{\alpha}{2}\mathbf{a} = \left[1 - \frac{\alpha}{2}\right]\mathbf{b} + \frac{\alpha}{2}\mathbf{a}$.
Note that $\mathbf{a} = \mathbf{c} - \mathbf{b}$, $\left[1 - \frac{\alpha}{2}\right]\mathbf{c} - \frac{\alpha}{2}(\mathbf{c} - \mathbf{b}) = \left[1 - \frac{\alpha}{2}\right]\mathbf{b} + \frac{\alpha}{2}(\mathbf{c} - \mathbf{b})$ or $(1 - \alpha)\mathbf{c} + \frac{\alpha}{2}\mathbf{b} = (1 - \alpha)\mathbf{b} + \frac{\alpha}{2}\mathbf{c}$.

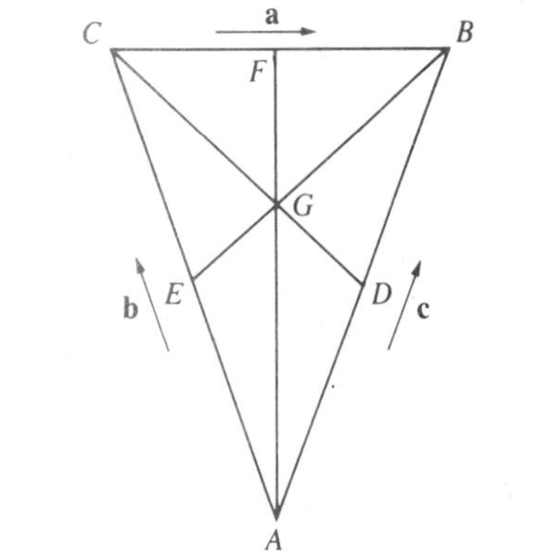

This means that $1 - \alpha = \frac{\alpha}{2}$, $1 = \frac{3\alpha}{2}$, $\alpha = \frac{2}{3}$; thus $\overrightarrow{AG} = (1 - \frac{2}{3})\mathbf{b} + \frac{1}{3}\mathbf{c} = \frac{1}{3}(\mathbf{b} + \mathbf{c})$. Also note that $\alpha\overrightarrow{AF} = (\frac{2}{3})(\frac{1}{3})(\mathbf{b} + \mathbf{c}) = \frac{1}{3}(\mathbf{b} + \mathbf{c}) = \overrightarrow{AG}$. In other words, $\alpha\overrightarrow{AF} = \alpha\overrightarrow{CD} = \alpha\overrightarrow{BE} = \overrightarrow{AG}$ so G is on all three medians and is $\frac{2}{3}$ the distance from each vertex.

61. $\overrightarrow{AG} = \mathbf{a} + \alpha\overrightarrow{BD} = \mathbf{a} + \alpha(\mathbf{b} - \mathbf{a})$
$= (1 - \alpha)\mathbf{a} + \alpha\mathbf{b}$
$\overrightarrow{AG} = \beta\overrightarrow{AC} = \beta(\mathbf{a} + \mathbf{b})$
$\overrightarrow{AG} = \overrightarrow{AG}$ or $(1 - \alpha)\mathbf{a} + \alpha\mathbf{b}$
$= \beta\mathbf{a} + \beta\mathbf{b}$,
so $1 - \alpha = \beta$ and $\alpha = \beta$, or $1 - \alpha = \alpha$, $1 = 2\alpha$, and $\alpha = \beta = \frac{1}{2}$; thus G lies on both diagonals, one-half the distance from B to D and one-half the distance from A to C.

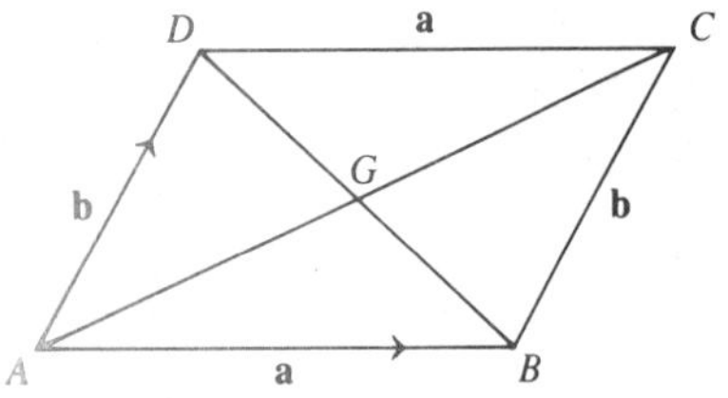

63. Construct $\overline{BD} \parallel \overline{AE}$; let F, G, and H be the mid-points of $\overline{AE}$, $\overline{BD}$, and $\overline{CD}$, respectively. By Problem 58, $\overline{GH} \parallel \overline{BC}$ and $GH = \frac{1}{2}BC$. By construction, $ABGF$ is a parallelogram. So $\overline{AB}$, $\overline{FG}$, and $\overline{DE}$ are parallel. Thus $\overline{FGH}$ lies on a line $\parallel$ to

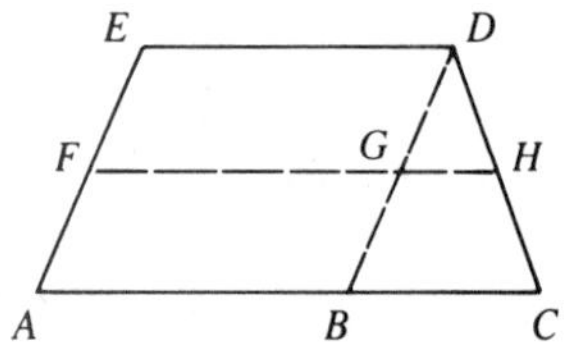

$\overline{ABC}$ and $\overline{DE}$. The length of $FH = FG + GH = ED +$

$$GH = \frac{ED + AB}{2} + \frac{BC}{2}$$
$$= \tfrac{1}{2}(ED + AC).$$

65. Let $A = \begin{pmatrix} a_1 & b_1 \\ a_2 & b_2 \end{pmatrix}$. Then $A^t = \begin{pmatrix} a_1 & a_2 \\ b_1 & b_2 \end{pmatrix}$. Let $\mathbf{u} = (a_1, a_2)$ and $\mathbf{v} = (b_1, b_2)$. Then $\mathbf{u} \cdot \mathbf{v} = 0$, $|\mathbf{u}| = 1$ and $|\mathbf{v}| = 1$. $A^tA = \begin{pmatrix} a_1^2 + b_1^2 & a_1a_2 + b_1b_2 \\ a_2a_1 + b_2b_1 & a_2^2 + b_2^2 \end{pmatrix} = \begin{pmatrix} |\mathbf{u}|^2 & \mathbf{u}\cdot\mathbf{v} \\ \mathbf{u}\cdot\mathbf{v} & |\mathbf{v}|^2 \end{pmatrix} = \begin{pmatrix} 1 & 0 \\ 0 & 1 \end{pmatrix}$. Similarly, $AA^t = I$. Hence A is invertible and $A^{-1} = A^t$.

Problems 4.3

11. $\sqrt{40}$ **13.** 6

15. $P = (3, 0, 1)$, $Q = (0, -4, 0)$, $R = (6, 4, 2)$. $\overline{PQ} = \sqrt{26}$, $\overline{QR} = \sqrt{104} = 2\sqrt{26}$, $\overline{PR} = \sqrt{26}$. Thus $\overline{PQ} + \overline{PR} = \overline{QR}$ (see Problem 14).

17. $(\frac{7}{2}, 3, \frac{1}{2})$ **19.** $3; -1, 0, 0$

21. $\sqrt{5}$; $1/\sqrt{5}, 0, 2/\sqrt{5}$

23. $\sqrt{3}, 1/\sqrt{3}, 1/\sqrt{3}, -1/\sqrt{3}$

25. $\sqrt{3}, 1/\sqrt{3}, -1/\sqrt{3}, -1/\sqrt{3}$

27. $\sqrt{3}, -1/\sqrt{3}, -1/\sqrt{3}, 1/\sqrt{3}$

29. $\sqrt{78}; 2/\sqrt{78}, 5/\sqrt{78}, -7/\sqrt{78}$

31. $\sqrt{29}; -2/\sqrt{29}, -3/\sqrt{29}, -4/\sqrt{29}$

33. $4\sqrt{3}\mathbf{i} + 4\sqrt{3}\mathbf{j} + 4\sqrt{3}\mathbf{k}$

35. $(1/\sqrt{26})\mathbf{i} - (3/\sqrt{26})\mathbf{j} + (4/\sqrt{26})\mathbf{k}$

37. $R = (-3, y, z)$, y, z arbitrary; this set of points constitutes a plane parallel to the yz-plane.

39. $\left|\frac{\mathbf{u}\cdot\mathbf{v}}{|\mathbf{u}||\mathbf{v}|}\right| = |\cos\phi| \le 1$. Thus $|\mathbf{u}\cdot\mathbf{v}| \le |\mathbf{u}||\mathbf{v}|$. Then

$$\begin{aligned} |\mathbf{u} + \mathbf{v}|^2 &= (\mathbf{u} + \mathbf{v})\cdot(\mathbf{u} + \mathbf{v}) \\ &= |\mathbf{u}|^2 + 2\mathbf{u}\cdot\mathbf{v} + |\mathbf{v}|^2 \\ &\le |\mathbf{u}|^2 + 2|\mathbf{u}||\mathbf{v}| + |\mathbf{v}|^2 \\ &= (|\mathbf{u}| + |\mathbf{v}|)^2. \end{aligned}$$

41. $-6\mathbf{j} + 9\mathbf{k}$ **43.** $8\mathbf{i} - 14\mathbf{j} + 9\mathbf{k}$

45. $16\mathbf{i} + 29\mathbf{j} + 42\mathbf{k}$ **47.** $\sqrt{59}$

49. $\cos^{-4}(35/\sqrt{29}\sqrt{59})$
$\approx \cos^{-1}(0.8461)$
≈ 0.5621
$\approx 32.21^0$

51. $\frac{25}{29}\mathbf{u} = \frac{50}{29}\mathbf{i} - \frac{75}{29}\mathbf{j} + \frac{100}{29}\mathbf{k}$

53. $\frac{7}{50}\mathbf{t} = \frac{21}{50}\mathbf{i} + \frac{28}{50}\mathbf{j} + \frac{35}{50}\mathbf{k}$

55. $\sqrt{\frac{83}{6}}$

57. Let $P = (3, 2, -1)$, $Q = (4, 1, 6)$, $R = (7, -2, 3)$, and $S = (8, -3, 10)$. Then $\overrightarrow{PQ} = \mathbf{i} - \mathbf{j} + 7\mathbf{k}$ and $\overrightarrow{RS} = \mathbf{i} - \mathbf{j} + 7\mathbf{k}$. Thus the directed line segments $\overrightarrow{PQ}$ and $\overrightarrow{RS}$ have the same length and direction. Similarly, $\overrightarrow{PR} = 4\mathbf{i} - 4\mathbf{j} + 4\mathbf{k}$ and $\overrightarrow{QS} = 4\mathbf{i} - 4\mathbf{j} + 4\mathbf{k}$.

59. $\sqrt{6}/2 + 6\sqrt{3} + 12\sqrt{2}$
≈ 28.59 joules

61. PR is in a plane parallel to the xy-plane, so it is parallel to the vector $(a_2 - a_1)\mathbf{i} + (b_2 - b_1)\mathbf{j}$, by 4.1.10. $\overrightarrow{PQ} = \overrightarrow{PR} + \overrightarrow{RQ}$ where $\overrightarrow{RQ} \perp \overrightarrow{PR}$. Now $|\overrightarrow{RQ}| = c_2 - c_1$ so that $\overrightarrow{RQ} = (c_2 - c_1)\mathbf{k}$. Thus $\overrightarrow{PQ} = (a_2 - a_1)\mathbf{i} + (b_2 - b_1)\mathbf{j} + (c_2 - c_1)\mathbf{k}$.

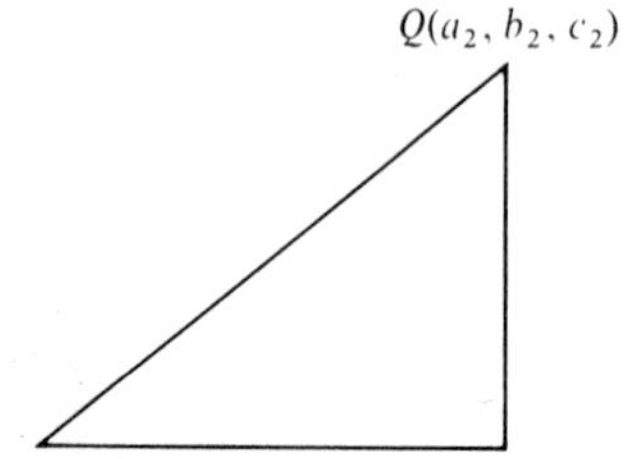

63. (i) If $\mathbf{v} = \alpha\mathbf{u}$, then $\cos\phi = \frac{\mathbf{u}\cdot\mathbf{v}}{|\mathbf{u}||\mathbf{v}|} = \frac{\alpha|\mathbf{u}|^2}{|\alpha||\mathbf{u}|^2} = \pm 1$. If $\mathbf{u}$ and $\mathbf{v}$ are parallel, then

$\frac{\mathbf{u}}{|\mathbf{u}|} = \pm \frac{\mathbf{v}}{|\mathbf{v}|}$ so that

$\mathbf{v} = \pm \frac{|\mathbf{v}|}{|\mathbf{u}|}\mathbf{u} = \alpha\mathbf{u}$.

(ii) If $\mathbf{u}\cdot\mathbf{v} = 0$, then $\cos\phi = 0$ and $\phi = \frac{\pi}{2}$. If $\phi = \frac{\pi}{2}$, then $\mathbf{u}\cdot\mathbf{v} = |\mathbf{u}||\mathbf{v}|\cos\phi = 0$.

65. Let $A(a_1, a_2, a_3)$, $B(b_1, b_2, b_3)$, and $C(c_1, c_2, c_3)$ be the vertices of the triangle, while $\mathbf{F} = \alpha\mathbf{i} + \beta\mathbf{j} + \gamma\mathbf{k}$. $W = F\cdot(\overrightarrow{AB} + \overrightarrow{BC} + \overrightarrow{CA}) = (\alpha\mathbf{i} + \beta\mathbf{j} + \gamma\mathbf{k})\cdot[(b_1 - a_1)\mathbf{i} + (b_2 - a_2)\mathbf{j} + (b_3 - a_3)\mathbf{k} + (c_1 - b_1)\mathbf{i} + (c_2 - b_2)\mathbf{j} + (c_3 - b_3)\mathbf{k} + (a_1 - c_1)\mathbf{i} + (a_2 - c_2)\mathbf{j} + (a_3 - c_3)\mathbf{k}] = \alpha(b_1 - a_1 + c_1 - b_1 + a_1 - c_1) + \beta(b_2 - a_2 + c_2 - b_2 + a_2 - c_2) + \gamma(b_3 - a_3 + c_3 - b_3 + a_3 - c_3) = 0\alpha + 0\beta + 0\gamma = 0$.

Problems 4.4

1. $-6\mathbf{i}-3\mathbf{j}$ **3.** $-\mathbf{i}-\mathbf{j}+\mathbf{k}$

5. $12\mathbf{i}+8\mathbf{j}-21\mathbf{k}$

7. $(bc-ad)\mathbf{j}$

9. $-5\mathbf{i}-\mathbf{j}+7\mathbf{k}$ **11.** $\mathbf{0}$

13. $42\mathbf{i}+6\mathbf{j}$

15. $-9\mathbf{i}+39\mathbf{j}+61\mathbf{k}$

17. $-4i+8\mathbf{k}$ **19.** $\mathbf{0}$

21. $\pm[-(9/\sqrt{181})\mathbf{i}-(6/\sqrt{181})\mathbf{j}+(8/\sqrt{181})\mathbf{k}]$

23. $\sqrt{30}/\sqrt{6}\sqrt{29}\approx 0.415$

25. $5\sqrt{5}$ **27.** $\sqrt{523}$

29. $\sqrt{a^2b^2+a^2c^2+b^2c^2}$

31. Let $\mathbf{u} = a_1\mathbf{i} + b_1\mathbf{j} + c_1\mathbf{k}$ and $\mathbf{v} = a_2\mathbf{i} + b_2\mathbf{j} + c_2\mathbf{k}$. Then $\mathbf{u}\times\mathbf{v} = (b_1c_2 - c_1b_2)\mathbf{i} + (c_1a_2 - a_1c_2)\mathbf{j} + (a_1b_2 - b_1a_2)\mathbf{k}$ so that $|\mathbf{u}\times\mathbf{v}|^2 = (b_1c_2 - c_1b_2)^2 + (c_1a_2 - a_1c_2)^2 + (a_1b_2 - b_1a_2)^2 = b_1^2c_2^2 - 2b_1c_2c_1b_2 + c_1^2b_2^2 + c_1^2a_2^2 - 2c_1a_2a_1c_2 + a_1^2c_2^2 + a_1^2b_2^2 - 2a_1b_2b_1a_2 + b_1^2a_2^2$. This is equal to $|\mathbf{u}|^2|\mathbf{v}|^2 - (\mathbf{u}\cdot\mathbf{v})^2 = (a_1^2 + b_1^2 + c_1^2)(a_2^2 + b_2^2 + c_2^2) - (a_1a_2 + b_1b_2 + c_1c_2)^2$.

33. $\frac{1}{2}\sqrt{3778}$ **35.** $\sqrt{3}/2$

37. The volume of a parallelepiped is base times height. The magnitude of $\mathbf{u}\times\mathbf{v}$ is the area of the base: Its direction is perpendicular to the base. The height is measured along $\mathbf{u}\times\mathbf{v}$. The height, h, is the projection of the third vector, $\mathbf{w}$, onto $\mathbf{u}\times\mathbf{v}$. $h = |\text{Proj}_{\mathbf{u}\times\mathbf{v}}\mathbf{w}| = \frac{\mathbf{w}\cdot(\mathbf{u}\times\mathbf{v})}{|\mathbf{u}\times\mathbf{v}|}$; volume $= (h)(\text{base}) = \frac{\mathbf{w}\cdot(\mathbf{u}\times\mathbf{v})}{|\mathbf{u}\times\mathbf{v}|}|\mathbf{u}\times\mathbf{v}| = (\mathbf{u}\times\mathbf{v})\cdot\mathbf{w}$ if the scalar product is positive, $|(\mathbf{u}\times\mathbf{v})\cdot\mathbf{w}|$ otherwise.

39. 23

41. This problem will rely heavily on Property 3 of the determinants, which states that if the ith column (or row) of a determinant consists of a pair of elements, the determinant can be rewritten as a sum of two determinants whose columns (or rows) are identical, except for the ith column (or row). The first determinant contains one of the elements of the pair, while the other member of each pair of elements of the ith column (or row) appears in the second determinant. Also note that the volume generated by $\mathbf{u}$, $\mathbf{v}$, $\mathbf{w}$ is given by

$$\text{Volume} = \begin{vmatrix} u_1 & u_2 & u_3 \\ v_1 & v_2 & v_3 \\ w_1 & w_2 & w_3 \end{vmatrix}$$

Let

$$A = \begin{pmatrix} a_{11} & a_{12} & a_{13} \\ a_{21} & a_{22} & a_{23} \\ a_{31} & a_{32} & a_{33} \end{pmatrix},$$

$\mathbf{u}_1 = A\mathbf{u}$, $\mathbf{v}_1 = A\mathbf{v}$, $\mathbf{w}_1 = A\mathbf{w}$.

Thus

$$\mathbf{u}_1 = \begin{pmatrix} a_{11} & a_{12} & a_{13} \\ a_{21} & a_{22} & a_{23} \\ a_{31} & a_{32} & a_{33} \end{pmatrix}\begin{pmatrix} u_1 \\ u_2 \\ u_3 \end{pmatrix}$$

$$= \begin{pmatrix} a_{11}u_1 + a_{12}u_2 + a_{13}u_3 \\ a_{21}u_1 + a_{22}u_2 + a_{23}u_3 \\ a_{31}u_1 + a_{32}u_2 + a_{33}u_3 \end{pmatrix}$$

Similarly,

$$\mathbf{v}_1 = \begin{pmatrix} a_{11}v_1 + a_{12}v_2 + a_{13}v_3 \\ a_{21}v_1 + a_{22}v_2 + a_{23}v_3 \\ a_{31}v_1 + a_{32}v_2 + a_{33}v_3 \end{pmatrix}$$

and

$$\mathbf{w}_1 = \begin{pmatrix} a_{11}w_1 + a_{12}w_2 + a_{13}w_3 \\ a_{21}w_1 + a_{22}w_2 + a_{23}w_3 \\ a_{31}w_1 + a_{32}w_2 + a_{33}w_3 \end{pmatrix}.$$

By Problem 36, the volume generated by $\mathbf{u}_1$, $\mathbf{v}_1$, $\mathbf{w}_1$ is:

$$V = \begin{vmatrix} a_{11}u_1 + a_{12}u_2 + a_{13}u_3 & a_{21}u_1 + a_{22}u_2 + a_{23}u_3 & a_{31}u_1 + a_{32}u_2 + a_{33}u_3 \\ a_{11}v_1 + a_{12}v_2 + a_{13}v_3 & a_{21}v_1 + a_{22}v_2 + a_{23}v_3 & a_{31}v_1 + a_{32}v_2 + a_{33}v_3 \\ a_{11}w_1 + a_{12}w_2 + a_{13}w_3 & a_{21}w_1 + a_{22}w_2 + a_{23}w_3 & a_{31}w_1 + a_{32}w_2 + a_{33}w_3 \end{vmatrix}$$

By expansion, it can be verified that

$$V = \begin{vmatrix} a_{11} & a_{12} & a_{13} \\ a_{21} & a_{22} & a_{23} \\ a_{31} & a_{32} & a_{33} \end{vmatrix}\begin{vmatrix} u_1 & u_2 & u_3 \\ v_1 & v_2 & v_3 \\ w_1 & w_2 & w_3 \end{vmatrix}$$

$=$ (det A)(volume generated by $\mathbf{u}$, $\mathbf{v}$, $\mathbf{w}$), if det $A \geq 0$, otherwise use $-$det A.

43. Let $\mathbf{u} = (u_1, u_2, u_3)$, $\mathbf{v} = (v_1, v_2, v_3)$ and $\mathbf{w} = (w_1, w_2, w_3)$. Then

$$\mathbf{v}\times\mathbf{w} = \begin{vmatrix} \mathbf{i} & \mathbf{j} & \mathbf{k} \\ v_1 & v_2 & v_3 \\ w_1 & w_2 & w_3 \end{vmatrix} = (v_2w_3 - v_3w_2,\ v_3w_1 - v_1w_3,\ v_1w_2 - v_2w_1);$$

$$\mathbf{u}\times(\mathbf{v}\times\mathbf{w}) = \begin{vmatrix} \mathbf{i} & \mathbf{j} & \mathbf{k} \\ u_1 & u_2 & u_3 \\ v_2w_3 - v_3w_2 & v_3w_1 - v_1w_3 & v_1w_2 - v_2w_1 \end{vmatrix}$$

$= (-u_3v_3w_1 + u_3v_1w_3 + u_2v_1w_2 - u_2v_2w_1,\ u_3v_2w_3 - u_3v_3w_2 - u_1v_1w_2 + u_1v_2w_1,\ u_1v_3w_1 - u_1v_1w_3 - u_2v_2w_3 + u_2v_3w_2)$(*)

$(\mathbf{u}\cdot\mathbf{w})\mathbf{v} = (u_1w_1 + u_2w_2 + u_3w_3)(v_1, v_2, v_3) = (u_1v_1w_1 + u_2v_1w_2 + u_3v_1w_3, u_1v_2w_1 + u_2v_2w_2 + u_3v_2w_3, u_1v_3w_1 + u_2v_3w_2 + u_3v_3w_3)$

$-(\mathbf{u}\cdot\mathbf{v})\mathbf{w} = -(u_1v_1 + u_2v_2 + u_3w_3)(w_1, w_2, w_3) = (-u_1v_1w_1 - u_2v_2w_1 - u_3v_3w_1, -u_1v_1w_2 - u_2v_2w_2 - u_3v_3w_2, -u_1v_1w_3 - u_2v_2w_3 - u_3v_3w_3)$.

If we add the last two vectors, we obtain the vector (*).

Problems 4.5

In the answers to Problems 1–5 we assume that the first point is P and the second point is Q. The vector equations are of the form $\overrightarrow{QR} = \overrightarrow{QP} + t\mathbf{v}$. Only $\mathbf{v}$ is given in the answers.

1. $\mathbf{v} = -\mathbf{i}+\mathbf{j}-4\mathbf{k}$; $x = 2-t$, $y = 1+t$, $z = 3-4t$; $(x-2)/(-1) = y-1 = (z-3)/(-4)$

3. $\mathbf{v} = -\mathbf{j}-2\mathbf{k}$; $x=-4$, $y=1-t$, $z=3-2t$; $x=-4$ and $z=1+2y$

5. $\mathbf{v} = 2\mathbf{i}-2\mathbf{k}$; $x=1+2t$, $y=2$, $z=3-2t$; $y=2$ and $x=4-z$

In Problems 7–11 $\mathbf{v}$ is already given.

7. $x=2+2t$, $y=2-t$, $z=1-t$; $(x-2)/2=(y-2)/(-1) = (z-1)/(-1)$

9. $x=-1$, $y=-2-3t$, $z=5+7t$; $x=-1$ and $7y+3z=1$

11. $x=a+dt$, $y=b+et$, $z=c$; $(x-a)/d=(y-b)/e$ and $z=c$

13. $\mathbf{v}=3\mathbf{i}+6\mathbf{j}+2\mathbf{k}$; $x=4+3t$, $y=1+6t$, $z=-6+2t$; $(x-4)/3=(y-1)/6 = (z+6)/2$

15. The vector $\mathbf{v}_1 = a_1\mathbf{i} + b_1\mathbf{j} + c_1\mathbf{k}$ is parallel to L_1, while the vector $\mathbf{v}_2 = a_2\mathbf{i} + b_2\mathbf{j} + c_2\mathbf{k}$ is parallel to L_2. Thus $L_1 \perp L_2$ if $\mathbf{v}_1 \perp \mathbf{v}_2$ or $\mathbf{v}_1\cdot\mathbf{v}_2 = 0$. But $\mathbf{v}_1\cdot\mathbf{v}_2 = a_1a_2 + b_1b_2 + c_1c_2$.

17. $3\mathbf{i} + 6\mathbf{j} + 9\mathbf{k} = 3(\mathbf{i} + 2\mathbf{j} + 3\mathbf{k})$ so the direction vectors of the lines are parallel. Note that they are not coincident since, for example, the point $(1, -3, -3)$ is on L_1 but not on L_2.

19. If they had a point in common, we would have

$2 - t = 1 + s$

$1 + t = -2s$

$-2t = 3 + 2s$

The unique solution of the first two of these equations is $s=-2$, $t=3$; but this pair does not satisfy the third equation.

21. (a) $(\sqrt{186}/3)$ $(t=\frac{1}{3})$ (b) $\sqrt{1518}/11 = \sqrt{138/11}$, $(t=-\frac{4}{11})$ (c) $\sqrt{750}/6 = 5\sqrt{30}/6$ $(t=-\frac{1}{6})$

23. $(x+4)/26 = (y-7)/1 = (t-3)/37$

25. $(x-4)/(-4) = (y-6)/16 = z/24$

27. 3 **29.** $y=0$ (xz-plane)

31. $x+y=3$ **33.** $y+z=5$

35. $-3x-4y+z=45$

37. $2x-7y-8z=-20$

39. $-12x-21y+22z=63$

41. $2x+y=7$ **43.** coincident

45. none of these

47. $(x, y, z) = (-1, -3, 0) + t(1, 2, 1)$

49. $(x, y, z) = (-11/4, 3/2, 0) + t(9, 16, 2)$

51. $13/\sqrt{69}$ **53.** $19/\sqrt{35}$

55. $\cos^{-1}(9/\sqrt{3}\sqrt{29}) = \cos^{-1}(0.9649) \approx 0.2657 \approx 15.23°$

57. $\cos^{-1}|20/\sqrt{294}\sqrt{6}| = \cos^{-1}(\frac{20}{42}) \approx 1.074 \approx 61.56°$

59. $\mathbf{n} = \mathbf{v}\times\mathbf{w}$ is orthogonal to $\mathbf{v}$ and $\mathbf{w}$. If $\mathbf{u}\cdot(\mathbf{v}\times\mathbf{w}) = 0$, then $\mathbf{u}\perp\mathbf{n}$, which means that $\mathbf{u}$ lies in the plane determined by $\mathbf{v}$ and $\mathbf{w}$.

61. coplanar; $29x-y+11z=0$

63. not coplanar; $\mathbf{u}\cdot(\mathbf{v}\times\mathbf{w}) = -9$.

Chapter 4—Review

1. $|\mathbf{v}| = 3\sqrt{2}$, $\theta=\pi/4$

3. $|\mathbf{v}|=4$, $\theta=5\pi/3$

5. $|\mathbf{v}|=12\sqrt{2}$, $\theta=5\pi/4$

7. $2\mathbf{i}+2\mathbf{j}$ **9.** $4\mathbf{i}+2\mathbf{j}$

11. (a) $(10, 5)$, (b) $(5, -3)$, (c) $(-31, 12)$

13. $(1/\sqrt{2})\mathbf{i}+(1/\sqrt{2})\mathbf{j}$

15. $(2/\sqrt{29})\mathbf{i}+(5/\sqrt{29})\mathbf{j}$

17. $\frac{3}{5}\mathbf{i}+\frac{4}{5}\mathbf{j}$

19. $(1/\sqrt{2})\mathbf{i}-(1/\sqrt{2})\mathbf{j}$ if $a>0$ and $-(1/\sqrt{2})\mathbf{i}+(1/\sqrt{2})\mathbf{j}$ if $a<0$

21. $-(5/\sqrt{29})\mathbf{i}-(2/\sqrt{29})\mathbf{j}$

23. $-(10/\sqrt{149})\mathbf{i}+(7/\sqrt{149})\mathbf{j}$

25. $\mathbf{j}$ **27.** $-\frac{7}{2}\sqrt{3}\mathbf{i}+\frac{7}{2}\mathbf{j}$ **29.** 0; 0

31. $-14, -14/\sqrt{5}\sqrt{41}$

33. neither **35.** parallel

37. parallel **39.** $7\mathbf{i}+7\mathbf{j}$

41. $\frac{15}{13}\mathbf{i}+\frac{10}{13}\mathbf{j}$ **43.** $-\frac{3}{2}\mathbf{i}-\frac{7}{2}\mathbf{j}$

45. $\text{Proj}_{\overrightarrow{RS}}\overrightarrow{PQ} = -\frac{99}{25}\mathbf{i}+\frac{132}{25}\mathbf{j}$; $\text{Proj}_{\overrightarrow{PQ}}\overrightarrow{RS} = -\frac{33}{82}\mathbf{i}-\frac{297}{82}\mathbf{j}$

47. $-3\mathbf{i}-3\mathbf{j}$ **49.** $-\sqrt{2}$ joules

51. $-11(1+3\sqrt{3})$ joules

53. $\sqrt{68}$ **55.** $\sqrt{216}$

57. $P=(1, 3, 0)$, $Q=(3, -1, 2)$, $R=(-1, 7, 2)$, $PQ=\sqrt{24}=2\sqrt{6}$, $\overline{PR}=\sqrt{24}=2\sqrt{6}$ and $\overline{QR}=\sqrt{96}=4\sqrt{6}$ so that $\overline{QR}=\overline{PQ}+\overline{PR}$

59. $\sqrt{130}$; 0, $3/\sqrt{130}$, $11/\sqrt{130}$

61. $\sqrt{53}$; $-4/\sqrt{53}$, $1/\sqrt{53}$, $6/\sqrt{53}$

63. $(2/\sqrt{6})\mathbf{i}-(1/\sqrt{6})\mathbf{j}+(1/\sqrt{6})\mathbf{k}$

65. $\mathbf{i}-14\mathbf{j}+20\mathbf{k}$

67. $\frac{26}{21}\mathbf{i}-\frac{52}{21}\mathbf{j}+\frac{13}{21}\mathbf{k}$ **69.** 22

71. $\cos^{-1}(-9/\sqrt{798}) \approx 1.895 \approx 108.6°$

73. $68/\sqrt{3}$ joules **75.** $-7\mathbf{i}-7\mathbf{k}$

77. $-26\mathbf{i}-8\mathbf{j}+7\mathbf{k}$

79. $\sqrt{2065}$ **81.** 8

83. $\overrightarrow{OR}=(-4\mathbf{i}+\mathbf{j})+t(7\mathbf{i}-\mathbf{j}+7\mathbf{k})$; $x=-4+7t,\ y=1-t,\ z=7t$; $(x+4)/7=(y-1)/(-1)=z/7$

85. $\overrightarrow{OR}=\mathbf{i}-2\mathbf{j}-3\mathbf{k}+t(5\mathbf{i}-3\mathbf{j}+2\mathbf{k})$; $x=1+5t,\ y=-2-3t,\ z=-3+2t$; $(x-1)/5=(y+2)/(-3)=(z+3)/2$

87. $\sqrt{165}/3$ **89.** $x+z=-1$

91. $2x-3y+5z=19$

93. $x=\frac{1}{2}-\frac{9}{2}t,\ y=\frac{7}{2}-\frac{11}{2}t,\ z=t$

95. $x=\frac{4}{3}-\frac{5}{3}t,\ y=-4-\frac{7}{3}t,\ z=t$

97. $\cos^{-1}|-1/\sqrt{207}| = \cos^{-1} 1/\sqrt{207} \approx 1.501 \approx 86.01°$

CHAPTER 5

Problems 5.2

1. yes

3. no; (iv); also (vi) does not hold if $\alpha<0$

5. yes **7.** yes

9. no (i), (iii), (iv), (vi) do not hold

11. yes **13.** yes

15. no; (i), (iii), (iv), (vi) do not hold

17. yes **19.** yes

21. Suppose that $\mathbf{0}$ and $\mathbf{0}'$ are additive identities. Then, by the definition of additive identity, $\mathbf{0}=\mathbf{0}+\mathbf{0}'$ and $\mathbf{0}'=\mathbf{0}'+\mathbf{0}=\mathbf{0}+\mathbf{0}'$. Thus $\mathbf{0}=\mathbf{0}'$.

23. Let $-\mathbf{x}$ and $-\mathbf{x}'$ be additive inverses of $\mathbf{x}'$. Then $-\mathbf{x}=-\mathbf{x}+\mathbf{0}=-\mathbf{x}+[\mathbf{x}'+(-\mathbf{x}')]=(-\mathbf{x}+\mathbf{x}')+(-\mathbf{x}')$ (by **ii**) $=\mathbf{0}+(-\mathbf{x}')$ (since $-\mathbf{x}$ is an additive inverse of $\mathbf{x}'$) $=-\mathbf{x}'$. Thus $-\mathbf{x}=-\mathbf{x}'$ and the additive inverse is unique.

25. Let y_1 and y_2 be solutions to the equation. Then

$$y_1''+a(x)y_1'+b(x)y_1(x)=0$$

and

$$y_2''+a(x)y_2'+b(x)y_2(x)=0$$

Then

$$\begin{aligned}(y_1+y_2)''&+a(x)(y_1+y_2)'+b(x)(y_1+y_2)\\ &=+[y_1''+a(x)y_1'+b(x)y_1]+[y_2''+a(x)y_2'+b(x)y_2]\\ &=0+0=0\end{aligned}$$

so y_1+y_2 is a solution. Similarly $(\alpha y_1)''+a(x)(\alpha y_1')+b(x)(\alpha y)=\alpha[y_1''+a(x)y_1'+b(x)y_1]=\alpha\cdot 0=0$ so αy_1 is also a solution. Thus the closure rules hold. Since $-y_1=(-1)y$, is also a solution, we have the additive inverse. The other axioms follow easily.

Problems 5.3

1. no; because $\alpha(x,y)\notin H$ if $\alpha<0$

3. yes **5.** yes **7.** yes

9. yes **11.** yes **13.** yes

15. no; the zero polynomial $\notin H$

17. no; the function $f(x)\equiv 0\notin V$

19. yes

21. **(a)** If $A_1, A_2\in H_1$, then $(A_1+A_2)_{11}=(A_1)_{11}+(A_2)_{11}=0+0=0$, and $(\alpha A_1)_{11}=\alpha(A_1)_{11}=\alpha 0=0$ so that H_1 is a subspace. If $A_1, A_2\in H_2$, then

$$A_1=\begin{pmatrix}-b_1 & a_1\\ a_1 & b_1\end{pmatrix},\quad A_2=\begin{pmatrix}-b_2 & a_2\\ a_2 & b_2\end{pmatrix},$$

$$A_1+A_2=\begin{pmatrix}-(b_1+b_2) & (a_1+a_2)\\ (a_1+a_2) & (b_1+b_2)\end{pmatrix}=\begin{pmatrix}-c & d\\ d & c\end{pmatrix}\in H_2$$

and so H_2 is also a subspace. **(b)** $H=H_1\cap H_2=\left\{A\in M_{22}\colon A=\begin{pmatrix}0 & a\\ a & 0\end{pmatrix}\text{ for some scalar } a\right\}$. If $A_1, A_2\in H$, then $A_1=\begin{pmatrix}0 & a_1\\ a_1 & 0\end{pmatrix}$, $A_2=\begin{pmatrix}0 & a_2\\ a_2 & 0\end{pmatrix}$, $A_1+A_2=\begin{pmatrix}0 & a_1+a_2\\ a_1+a_2 & 0\end{pmatrix}=\begin{pmatrix}0 & b\\ b & 0\end{pmatrix}\in H$ and $\alpha A_1=\begin{pmatrix}0 & \alpha a_1\\ \alpha a_1 & 0\end{pmatrix}=\begin{pmatrix}0 & c\\ c & 0\end{pmatrix}\in H$.

23. If $\mathbf{x}_1, \mathbf{x}_2\in H$, then $A(\mathbf{x}_1+\mathbf{x}_2)=A\mathbf{x}_1+A\mathbf{x}_2=\mathbf{0}+\mathbf{0}=\mathbf{0}$ so $\mathbf{x}_1+\mathbf{x}_2\in H$. Also $A(\alpha\mathbf{x}_1)=\alpha A\mathbf{x}_1=\alpha\mathbf{0}=\mathbf{0}$ so $\alpha\mathbf{x}_1\in H$ and H is a subspace.

25. Let $\mathbf{u}=(x_1, y_1, z_1, w_1)$ and $\mathbf{v}=(x_2, y_2, z_2, w_2)\in H$. Then $\mathbf{u}+\mathbf{v}=(x_1+x_2, y_1+y_2, z_1+z_2, w_1+w_2)$ and $a(x_1+x_2)+b(y_1+y_2)+c(z_1+z_2)+d(w_1+w_2)=(ax_1+by_1+cz_1+dw_1)+(ax_2+by_2+cz_2+dw_2)=0+0=0$ so $\mathbf{u}+\mathbf{v}\in H$. Similarly, $\alpha\mathbf{u}=(\alpha x_1, \alpha y_1, \alpha z_1, \alpha w_1)$ and $a(\alpha x_1)+b(\alpha y_1)+c(\alpha z_1)+d(\alpha w_1)=\alpha(ax_1+by_1+cz_1+dw_1)=\alpha 0=0$ so $\alpha\mathbf{u}\in H$. Thus H is a subspace.

27. Let $\mathbf{x}, \mathbf{y}\in H$. Then $\mathbf{x}=\mathbf{u}_1+\mathbf{v}_1$ and $\mathbf{y}=\mathbf{u}_2+\mathbf{v}_2$ where $\mathbf{u}_1, \mathbf{u}_2\in H_1$ and $\mathbf{v}_1, \mathbf{v}_2\in H_2$. Then $\mathbf{x}+\mathbf{y}=(\mathbf{u}_1+\mathbf{v}_1)+(\mathbf{u}_2+\mathbf{v}_2)=(\mathbf{u}_1+\mathbf{u}_2)+(\mathbf{v}_1+\mathbf{v}_2)$. Since H_1 and H_2 are subspaces, $\mathbf{u}_1+\mathbf{u}_2\in H_1$ and $\mathbf{v}_1+\mathbf{v}_2\in H_2$ so that $\mathbf{x}+\mathbf{y}\in H$. Similarly, $\alpha\mathbf{x}=\alpha(\mathbf{u}_1+\mathbf{v}_1)=\alpha\mathbf{u}_1+\alpha\mathbf{v}_1$. But $\alpha\mathbf{u}_1\in H_1$ and $\alpha\mathbf{v}_1\in H_2$ so $\alpha\mathbf{x}\in H$ and H is a subspace.

29. Let $\mathbf{v}_1=\begin{pmatrix}x_1\\ y_1\end{pmatrix}$ and $\mathbf{v}_2=\begin{pmatrix}x_2\\ y_2\end{pmatrix}$. $\mathbf{v}_1$ is not a multiple of $\mathbf{v}_2$ since the vectors are not collinear. Thus $\mathbf{v}_1$ and $\mathbf{v}_2$ are linearly independent. Let $A = \begin{pmatrix} x_1 & x_2 \\ y_1 & y_2 \end{pmatrix}$. By Theorem 3.4.4, det $A \neq 0$. Let $\mathbf{v} = \begin{pmatrix} x \\ y \end{pmatrix}$ be any other vector in $\mathbb{R}^2$. We wish to find scalars a and b such that $\mathbf{v} = a\mathbf{v}_1 + b\mathbf{v}_2$, or

$$\begin{pmatrix} x \\ y \end{pmatrix} = a\begin{pmatrix} x_1 \\ y_1 \end{pmatrix} + b\begin{pmatrix} x_2 \\ y_2 \end{pmatrix} = \begin{pmatrix} ax_1 + bx_2 \\ ay_1 + by_2 \end{pmatrix}$$

or

$$\begin{pmatrix} x_1 & x_2 \\ y_1 & y_2 \end{pmatrix}\begin{pmatrix} a \\ b \end{pmatrix} = \begin{pmatrix} x \\ y \end{pmatrix}$$

or

$$A\begin{pmatrix} a \\ b \end{pmatrix} = \begin{pmatrix} x \\ y \end{pmatrix}.$$

Since det $A \neq 0$, this system has the unique solution $\begin{pmatrix} a \\ b \end{pmatrix} = A^{-1}\begin{pmatrix} x \\ y \end{pmatrix}$. Thus $\mathbf{v} \in H$, which shows that $\mathbb{R}^2 \subset H$. But since $H \subset \mathbb{R}^2$, we have

$$H = \mathbb{R}^2.$$

Problems 5.4

1. independent

3. independent

5. dependent

7. independent

9. independent **11.** yes

13. no; for example, $x \notin \text{span}\{1 - x, 3 - x^2\}$

15. yes **17.** yes

19. Let $p(x) = a_0 + a_1x + a_2x^2$ and $q(x) = b_0 + b_1x + b_2x^2$ be two polynomials in P_2. Let $r(x) = c_0 + c_1x + c_2x^2$ be a third polynomial in P_2. If p and q span P_2, then there are scalars α and β such that $r = \alpha p + \beta q$; that is,

$$c_0 = \alpha a_0 + \beta b_0$$

$$c_1 = \alpha a_1 + \beta b_1$$

$$c_2 = \alpha a_2 + \beta b_2$$

This "overdetermined" system of 3 equations in 2 unknowns will have a solution if and only if the third equation is a linear combination of the first two. Since c_0, c_1, and c_2 are arbitrary, this will rarely be the case. Thus p and q cannot span P_2. To see this in another way, suppose that $\begin{pmatrix} \alpha \\ \beta \end{pmatrix}$ is a solution to the system. Then $\begin{pmatrix} c_0 \\ c_1 \end{pmatrix} = \begin{pmatrix} a_0 & b_0 \\ a_1 & b_1 \end{pmatrix}\begin{pmatrix} \alpha \\ \beta \end{pmatrix}$ or $\begin{pmatrix} \alpha \\ \beta \end{pmatrix} = \begin{pmatrix} a_0 & b_0 \\ a_1 & b_1 \end{pmatrix}^{-1}\begin{pmatrix} c_0 \\ c_1 \end{pmatrix}$. But also, $\begin{pmatrix} \alpha \\ \beta \end{pmatrix} = \begin{pmatrix} a_1 & b_1 \\ a_2 & b_2 \end{pmatrix}^{-1}\begin{pmatrix} c_1 \\ c_2 \end{pmatrix}$. In general, these two expressions for $\begin{pmatrix} \alpha \\ \beta \end{pmatrix}$ will not be equal, which means that, in general, the system does not have a solution.

21. Let $p_i(x) = a_{0i} + a_{1i}x + \cdots + a_{ni}x^n$ for $i = 1, 2, \ldots, n$ and define the vector

$$\mathbf{a}_i = (a_{0i}, a_{1i}, \ldots, a_{ni}).$$

Choose a vector $\mathbf{b} = (b_0, \ldots, b_n)$ such that $\mathbf{a}_i \cdot \mathbf{b} = 0$ for $i = 1, 2, \ldots, m$. If $m < n + 1$, this last homogeneous system of equations always has a nontrivial solution $\mathbf{b}$ (by Theorem 1.4.1). Suppose that $\mathbf{b}$ can be written as $\mathbf{b} = \alpha_1\mathbf{a}_1 + \alpha_2\mathbf{a}_2 + \cdots + \alpha_m\mathbf{a}_m$. Then $\mathbf{b}\cdot\mathbf{b} = \alpha_1\mathbf{a}_1\cdot\mathbf{b} + \alpha_2\mathbf{a}_2\cdot\mathbf{b} + \cdots + \alpha_m\mathbf{a}_m\cdot\mathbf{b} = 0$. But $\mathbf{b} \neq 0$. This contradiction shows that $\mathbf{b}$ cannot be written as a linear combination of the $\mathbf{a}_i$'s if $m < n + 1$. Let $q(x) = b_0 + b_1x + \cdots + b_nx^n$. Then $q(x)$ cannot be written as a linear combination of the p_i's and, therefore, the p_i's don't span. We are left to conclude that $m \geq n + 1$.

23. Write the matrices as $A^{(1)}, A^{(2)}, \ldots, A^{(mn+1)}$. Suppose that $A^{(i)} = (a_{jk}^{(i)})$. Consider the system

$$\begin{aligned} a_{11}^{(1)}\alpha_1 + a_{11}^{(2)}\alpha_2 + \cdots & + a_{11}^{(mn+1)}\alpha_{mn+1} = 0 \\ \vdots \qquad\qquad & \qquad\qquad \vdots \\ a_{mn}^{(1)}\alpha_1 + a_{mn}^{(2)}\alpha_2 + \cdots & + a_{mn}^{(mn+1)}\alpha_{mn+1} = 0 \end{aligned}$$

This is a homogeneous system with mn equations and $mn + 1$ unknowns. Therefore it has a nonzero solution, and there are scalars $\alpha_1, \alpha_2, \ldots, \alpha_{mn+1}$ not all zero such that $\alpha_1A^{(1)} + \alpha_2A^{(2)} + \cdots + \alpha_{mn+1}A^{mn+1} = 0$ (the $m \times n$ zero matrix).

25. If $p(x) \in P$, then $p(x) = a_0 + a_1x + \cdots + a_kx^k$ for some integer k. Thus $p(x)$ can be written as a linear combination of $1, x, x^2, \ldots, x^k$.

27. $\alpha\mathbf{v}_1 + \beta\mathbf{v}_2 = (\alpha + \beta c)\mathbf{v}_1$. Let $\mathbf{v}_1 = (x_1, y_1, z_1)$. Then any vector $\mathbf{v} = (x, y, z)$ in span$\{\mathbf{v}_1, \mathbf{v}_2\}$ can be written as $(x, y, z) = ((\alpha + \beta c)x_1, (\alpha + \beta c)y_1, (\alpha + \beta c)z_1)$, or

$$x = (\alpha + \beta c)x_1$$

$$y = (\alpha + \beta c)y_1$$

$$z = (\alpha + \beta c)z_1$$

which, from Section 4.5, is the equation of a line passing through (0, 0, 0) with direction (x_1, y_1, z_1).

29. Let $\mathbf{v} \in V$. Then there are scalars $\alpha_1, \alpha_2, \ldots, \alpha_n$ such that $\mathbf{v} = \alpha_1\mathbf{v}_1 + \alpha_2\mathbf{v}_2 + \cdots + \alpha_n\mathbf{v}_n = \alpha_1\mathbf{v}_1 + \alpha_2\mathbf{v}_2 + \cdots + \alpha_n\mathbf{v}_n + 0\mathbf{v}_{n+1}$. The last equation shows that $\mathbf{v}_1, \mathbf{v}_2, \ldots, \mathbf{v}_{n+1}$ span V.

31. Let $\mathbf{v}_1, \mathbf{v}_2, \ldots, \mathbf{v}_m$ be a subset of $\mathbf{v}_1, \mathbf{v}_2, \ldots, \mathbf{v}_n$. If the first set is linearly dependent, then there exist scalars $\alpha_1, \alpha_2, \ldots, \alpha_m$ not all zero such that $\mathbf{0} = \alpha_1\mathbf{v}_1 + \alpha_2\mathbf{v}_2 + \cdots + \alpha_m\mathbf{v}_m = \alpha_1\mathbf{v}_1 + \alpha_2\mathbf{v}_2 + \cdots + \alpha_m\mathbf{v}_m + 0\mathbf{v}_{m+1} + 0\mathbf{v}_{m+2} + \cdots + 0\mathbf{v}_n$, which shows that the second, larger set is also linearly dependent. This contradiction shows that the first set must be linearly independent.

33. $\begin{pmatrix}1&0\\1&1\end{pmatrix}, \begin{pmatrix}0&1\\1&1\end{pmatrix}, \begin{pmatrix}1&1\\0&1\end{pmatrix}$ and $\begin{pmatrix}1&1\\1&0\end{pmatrix}$ span M_{22} and are invertible.

35. There are scalars $\alpha_1, \alpha_2, \ldots, \alpha_n$ not all zero such that $\alpha_1\mathbf{v}_1 + \alpha_2\mathbf{v}_2 + \cdots + \alpha_n\mathbf{v}_n = \mathbf{0}$. Let k be the largest integer for which $\alpha_k \neq 0$ (k may equal n). Then the equation reads $\alpha_1\mathbf{v}_1 + \alpha_2\mathbf{v}_2 + \cdots + \alpha_k\mathbf{v}_k = \mathbf{0}$ so that

$$\mathbf{v}_k = -\frac{\alpha_1}{\alpha_k}\mathbf{v}_1 - \frac{\alpha_2}{\alpha_k}\mathbf{v}_2 - \cdots - \frac{\alpha_{k-1}}{\alpha_k}\mathbf{v}_{k-1}.$$

37. Let $\mathbf{v}_i = (v_{i1}, v_{i2}, \ldots, v_{in})$ and $\mathbf{u}_i = (u_{i1}, u_{i2}, \ldots, u_{in})$. Also, let

$$\mathbf{w}_j = \begin{pmatrix}v_{1j}\\v_{2j}\\\vdots\\v_{nj}\end{pmatrix}, \quad \mathbf{z}_j = \begin{pmatrix}u_{1j}\\u_{2j}\\\vdots\\u_{nj}\end{pmatrix}$$

and $A = (a_{ij})$. Then, writing out the components in the given inequalities, we have

$$(*) \quad v_{ij} = a_{i1}u_{1j} + a_{i2}u_{2j} + \cdots + a_{in}u_{nj}$$

or

$$\mathbf{w}_j = A\mathbf{z}; \text{ so } \mathbf{z}_j = A^{-1}\mathbf{w}_j. \quad (\text{given } \det A \neq 0)$$

Let $A^{-1} = B = (b_{ij})$. Then the expression for $\mathbf{z}_j$ can be written

$$u_{ij} = b_{i1}v_{1j} + b_{i2}v_{2j} + \cdots + b_{in}v_{nj}.$$

We see that this is similar to the expression (*) with u's and v's interchanged. Thus

$$\mathbf{u}_i = \sum_{k=1}^{n} b_{ik}\mathbf{v}_k \text{ for } i = 1, 2, \ldots, n$$

so $\mathbf{u}_i \in \text{span}\{\mathbf{v}_1, \mathbf{v}_2, \ldots, \mathbf{v}_n\}$ for $i = 1, 2, \ldots, n$. Since it was given that $\mathbf{v}_i \in \text{span}\{\mathbf{u}_1, \mathbf{u}_2, \ldots, \mathbf{u}_n\}$, we conclude that the spans are equal.

39. $\begin{pmatrix}2\\1\\2\end{pmatrix}, \begin{pmatrix}-1\\3\\4\end{pmatrix}, \begin{pmatrix}1\\2\\2\end{pmatrix}$. (There are many choices for the third vector.)

41. Assume that $1, x, \ldots, x^k$ are linearly independent. Suppose that $c_0 + c_1x + \cdots + c_kx^k + c_{k+1}x^{k+1} = 0$. If $c_{k+1} \neq 0$, then $x^{k+1} = -\frac{c_0}{c_{k+1}} - \frac{c_1}{c_{k+1}}x - \cdots - \frac{c_k}{c_{k+1}}x^k$, which is clearly impossible. Thus $c_{k+1} = 0$. But then $c_0 + c_1x + \cdots + c_kx^k = 0$, which implies that $c_0 = c_1 = \cdots = c_k = 0$ since $1, x, x^2, \ldots, x^k$ are linearly independent. Thus $1, x, x^2, \ldots, x^k, x^{k+1}$ are also linearly independent, and this completes the induction proof (see Appendix 1).

Problems 5.5

1. no; doesn't span

3. no; dependent

5. no; doesn't span

7. yes **9.** yes

11. $\left\{\begin{pmatrix}0\\1\\-1\end{pmatrix}, \begin{pmatrix}1\\0\\2\end{pmatrix}\right\}$ **13.** $\left\{\begin{pmatrix}2\\3\\4\end{pmatrix}\right\}$

15. Since $\dim \mathbb{R}^2 = 2$, a proper subspace H must have dimension 1. Let $\{(x_0, y_0)\}$ be a basis for H. If $(x, y) \in H$, then $(x, y) = c(x_0, y_0)$ for some number c. This means that $x = cx_0$, $y = cy_0$ or $c = \frac{x}{x_0} = \frac{y}{y_0}$ and $y = \left(\frac{y_0}{x_0}\right)x$, which is the equation of a straight line through the origin with slope $\frac{y_0}{x_0}$ if $x_0 \neq 0$. If $x_0 = 0$, then the line is the y-axis.

17. Let $\{\mathbf{v}_1, \mathbf{v}_2, \ldots, \mathbf{v}_{n-1}\}$ be a basis for H. Let $\mathbf{v}_n$ be another vector in H. Then the vectors $\mathbf{v}_1, \mathbf{v}_2, \ldots, \mathbf{v}_{n-1}, \mathbf{v}_n$ are linearly dependent. Let A be the matrix whose rows are $\mathbf{v}_1, \mathbf{v}_2, \ldots, \mathbf{v}_{n-1}, \mathbf{v}_n$. Then $\det A = 0$, and the equation $A\mathbf{a} = \mathbf{0}$ has a nontrivial solution

$$\mathbf{a} = \begin{pmatrix}a_1\\a_2\\\vdots\\a_n\end{pmatrix}.$$

This means that $\mathbf{v}_i \cdot \mathbf{a} = \mathbf{0}$ for $i = 1, 2, \ldots, n$. In particular, if $\mathbf{v}_n = (x_1, x_2, \ldots, x_n)$, then $\mathbf{v}_n \cdot \mathbf{a} = \mathbf{0}$, or $a_1x_1 + a_2x_2 + \cdots + a_nx_n = 0$. Since $\mathbf{v}_n$ was an arbitrary vector in H, this proves the result.

19. $\left\{\begin{pmatrix}1\\1\end{pmatrix}\right\}$ **21.** $\left\{\begin{pmatrix}-2\\-3\\1\end{pmatrix}\right\}$

23. $\left\{\begin{pmatrix}3\\1\\0\end{pmatrix}, \begin{pmatrix}-2\\0\\1\end{pmatrix}\right\}$ **25.** n

27. V has a basis $\{\mathbf{u}_1, \mathbf{u}_2, \ldots, \mathbf{u}_n\}$ so there exist scalars such that

$$\begin{aligned}\mathbf{v}_1 &= a_{11}\mathbf{u}_1 + a_{12}\mathbf{u}_2 + \cdots\\ \mathbf{v}_2 &= a_{21}\mathbf{u}_1 + a_{22}\mathbf{u}_2 + \cdots\\ &\vdots\\ \mathbf{v}_m &= a_{m1}\mathbf{u}_1 + a_{m2}\mathbf{u}_2 + \cdots\\ &+ a_{1n}\mathbf{u}_n\\ &+ a_{2n}\mathbf{u}_n\\ &\vdots\\ &+ a_{mn}\mathbf{u}_n\end{aligned}$$

Let $\mathbf{a}_i = (a_{i1}, a_{i2}, \ldots, a_{in})$. The $\mathbf{a}_i$'s are m linearly independent vectors in $\mathbb{R}^2$ (otherwise the $\mathbf{v}_i$'s wouldn't be independent). Expand the $\mathbf{a}_i$'s into a basis $\mathbf{a}_1, \mathbf{a}_2, \ldots, \mathbf{a}_m, \mathbf{a}_{m+1}, \ldots, \mathbf{a}_n$ for $\mathbb{R}^n$ by simply adding $n - m$ linearly independent vectors to the set. Then if $\mathbf{a}_k = (a_{k1}, a_{k2}, \ldots, a_{kn})$ for $m < k \leq n$, define $\mathbf{v}_k = a_{k1}\mathbf{u}_1 + a_{k2}\mathbf{u}_2 + \cdots + a_{kn}\mathbf{u}_n$ for $k = m + 1, m + 2, \ldots, n$.

Since

$$\det\begin{pmatrix} a_{11} & a_{12} & \cdots & a_{1n} \\ a_{21} & a_{22} & \cdots & a_{2n} \\ \vdots & \vdots & & \vdots \\ a_{n1} & a_{n2} & \cdots & a_{nn} \end{pmatrix} \neq 0,$$

the set $\{\mathbf{v}_1, \mathbf{v}_2, \ldots, \mathbf{v}_n\}$ forms a basis for V since it consists of n linearly independent vectors in V with $\dim V = n$.

29. If the vectors are independent, then they form a basis and $\dim V = n$. If not, then by Problem 5.4.35, at least one of them can be written as a linear combination of the ones that precede it. Throw this vector out. Proceed in this manner until m linearly independent vectors remain. These must still span V by the manner in which they were chosen. Thus $\dim V = m < n$. In either event $\dim V \leq n$.

31. (a) See Problem 5.3.27.
(b) Let $\{\mathbf{v}_1, \ldots, \mathbf{v}_n\}$ be a basis for H, and let $\{\mathbf{u}_1, \mathbf{u}_2, \ldots, \mathbf{u}_m\}$ be a basis for K. Clearly $B = \{\mathbf{v}_1, \mathbf{v}_2, \ldots, \mathbf{v}_n, \mathbf{u}_1, \mathbf{u}_2, \ldots, \mathbf{u}_m\}$ span $H + K$. Suppose that $\alpha_1\mathbf{v}_1 + \alpha_2\mathbf{v}_2 + \cdots + \alpha_n\mathbf{v}_n + \beta_1\mathbf{u}_1 + \beta_2\mathbf{u}_2 + \cdots + \beta_m\mathbf{u}_m = \mathbf{0}$ where not all of the coefficients are zero. Let $\mathbf{h} = \alpha_1\mathbf{v}_1 + \alpha_2\mathbf{v}_2 + \cdots + \alpha_n\mathbf{v}_n$ and $\mathbf{k} = \beta_1\mathbf{u}_1 + \beta_2\mathbf{u}_2 + \cdots + \beta_m\mathbf{u}_m$. Then by linear independence, neither $\mathbf{h}$ nor $\mathbf{k}$ is the zero vector. Also, $\mathbf{h} \in H$ and $\mathbf{k} \in K$. But then $\mathbf{h} + \mathbf{k} = \mathbf{0}$ or $\mathbf{h} = -\mathbf{k} \in K$. Thus $\mathbf{0} \neq \mathbf{h} \in H \cap K$, which contradicts the fact that $H \cap K = \{\mathbf{0}\}$. Hence all the α's and β's are zero, which implies that the vectors in B are linearly independent. Thus B is a basis for $H + K$ and $\dim(H + K) = \dim H + \dim K$.

33. (i) If $\dim \operatorname{span}\{\mathbf{v}_1, \mathbf{v}_2\} = 1$, then choose a basis $\{\mathbf{v}\}$ for $\operatorname{span}\{\mathbf{v}_1, \mathbf{v}_2\}$. Then $\mathbf{v}_1 = \alpha\mathbf{v}$, and $\mathbf{v}_2 = \beta\mathbf{v}$. If $\mathbf{v}_1 = \mathbf{0}$, then $\mathbf{v}_1$ is a single point lying on $\mathbf{v}_2$. If $\mathbf{v}_1 \neq \mathbf{0}$, then $\alpha \neq 0$ so that $\mathbf{v}_2 = \beta\mathbf{v} = \frac{\beta}{\alpha}(\alpha\mathbf{v}) = \frac{\alpha}{\beta}\mathbf{v}_1$. In either case the vectors are collinear.
(ii) If the vectors are collinear, then $\mathbf{v}_2 = c\mathbf{v}_1$ for some scalar c so that $\mathbf{v}_1$ is a basis for $\operatorname{span}\{\mathbf{v}_1, \mathbf{v}_2\}$ and $\dim \operatorname{span}\{\mathbf{v}_1, \mathbf{v}_2\} = 1$.

35. If they are not linearly independent, then, as in the answer to Problem 29, $\dim V < n$. Since $\dim V = n$, the vectors must be independent and, therefore, constitute a basis for V.

Problems 5.6

1. $\dfrac{x+y}{2}\begin{pmatrix}1\\1\end{pmatrix} + \dfrac{x-y}{2}\begin{pmatrix}1\\-1\end{pmatrix} = \begin{pmatrix}x\\y\end{pmatrix}$

3. $\dfrac{4x+3y}{41}\begin{pmatrix}5\\7\end{pmatrix} + \dfrac{7x-5y}{41}\begin{pmatrix}3\\-4\end{pmatrix} = \begin{pmatrix}x\\y\end{pmatrix}$

5. $\dfrac{dx-by}{ad-bc}\begin{pmatrix}a\\b\end{pmatrix} + \dfrac{-cx+ay}{ad-bc}\begin{pmatrix}b\\d\end{pmatrix} = \begin{pmatrix}x\\y\end{pmatrix}$

7. $(x-y)\begin{pmatrix}1\\0\\0\end{pmatrix} + (y-z)\begin{pmatrix}1\\1\\0\end{pmatrix} + z\begin{pmatrix}1\\1\\1\end{pmatrix} = \begin{pmatrix}x\\y\\z\end{pmatrix}$

9. $\dfrac{6x-11y+10z}{31}\begin{pmatrix}2\\1\\3\end{pmatrix} + \dfrac{2x+17y-7z}{31}\begin{pmatrix}-1\\4\\5\end{pmatrix} + \dfrac{7x+13y-9z}{31}\begin{pmatrix}3\\-2\\-4\end{pmatrix} = \begin{pmatrix}x\\y\\z\end{pmatrix}$

11. $a_0 + a_1x + a_2x^2 = (a_0 + a_1 + a_2)1 + a_1(x-1) + a_2(x^2-1)$

13. $a_0 + a_1x + a_2x^2 = \dfrac{(a_0+a_1+a_2)}{2}(x+1) + \dfrac{(a_1-a_0-a_2)}{2}(x-1) + a_2(x^2-1)$

15. $2(x^3+x^2) - 5(x^2+x) + 10(x+1) - 16(1)$

17. $\mathbf{x} = \begin{pmatrix}-\frac{1}{3}\\0\end{pmatrix}_{B_2}$

19. $\mathbf{x} = \begin{pmatrix}\frac{86}{33}\\-\frac{20}{11}\\\frac{7}{11}\end{pmatrix}_{B_2}$

21. independent

23. dependent

25. independent

27. independent

29. If they were linearly independent, they would span P_n. But $1 \in P_n$ and $1 \notin \operatorname{span}\{p_1, p_2, \ldots, p_{n+1}\}$ since the constant term in each polynomial is 0.

31. If they were linearly independent, they would span M_{mn}. But the matrix $A = (a_{ij})$ where $a_{11} = 1$ and $a_{ij} = 0$ otherwise is not in the span of $A_1, A_2, \ldots, A_{mn+1}$ since a linear combination of matrices with a 0 in the 1, 1 position also has a 0 in the 1, 1 position.

33. $\begin{pmatrix}\cos\theta & -\sin\theta\\ \sin\theta & \cos\theta\end{pmatrix}\begin{pmatrix}1\\0\end{pmatrix} = \begin{pmatrix}\cos\theta\\ \sin\theta\end{pmatrix}$;

$\begin{pmatrix}\cos\theta & -\sin\theta\\ \sin\theta & \cos\theta\end{pmatrix}\begin{pmatrix}0\\1\end{pmatrix} = \begin{pmatrix}-\sin\theta\\ \cos\theta\end{pmatrix}.$

A^{-1} is obtained by rotating through an angle of $-\theta$. Thus

$$A^{-1} = \begin{pmatrix}\cos(-\theta) & -\sin(-\theta)\\ \sin(-\theta) & \cos(-\theta)\end{pmatrix} = \begin{pmatrix}\cos\theta & \sin\theta\\ -\sin\theta & \cos\theta\end{pmatrix}.$$

Alternatively, since

$$A = \begin{pmatrix} \cos\theta & -\sin\theta \\ \sin\theta & \cos\theta \end{pmatrix},$$

$$A^{-1} = \begin{pmatrix} \cos\theta & \sin\theta \\ -\sin\theta & \cos\theta \end{pmatrix}.$$

35. $A^{-1} = \begin{pmatrix} \cos(\pi/4) & \sin(\pi/4) \\ -\sin(\pi/4) & \cos(\pi/4) \end{pmatrix}$

$= \begin{pmatrix} 1/\sqrt{2} & 1/\sqrt{2} \\ -1/\sqrt{2} & 1/\sqrt{2} \end{pmatrix}$

(from Problem 33) so

$$A^{-1}\begin{pmatrix} 2 \\ -7 \end{pmatrix} = \begin{pmatrix} -5/\sqrt{2} \\ -9/\sqrt{2} \end{pmatrix}$$

37. Since C is invertible, the columns of C are linearly independent. That is, $\{\mathbf{c}_1, \mathbf{c}_2, \ldots, \mathbf{c}_n\}$ are n linearly independent vectors in V, which are therefore a basis for V since dim $V = n$.

39. If $CA = I$, then $(\mathbf{x})_{B_1} = I(\mathbf{x})_{B_1} = CA(\mathbf{x})_{B_1}$. Conversely, suppose that $(\mathbf{x})_{B_1} = CA(\mathbf{x})_{B_1}$. Let $B_1 = \{\mathbf{v}_1, \mathbf{v}_2, \ldots, \mathbf{v}_n\}$. Then

$$(\mathbf{v}_1)_{B_1} = \begin{pmatrix} 1 \\ 0 \\ \vdots \\ 0 \end{pmatrix} = CA\begin{pmatrix} 1 \\ 0 \\ \vdots \\ 0 \end{pmatrix}.$$

Let

$$CA = \begin{pmatrix} r_{11} & r_{12} & \cdots & r_{1n} \\ r_{21} & r_{22} & \cdots & r_{2n} \\ \vdots & \vdots & & \vdots \\ r_{n1} & r_{n2} & \cdots & r_{nn} \end{pmatrix}.$$

Then $CA\begin{pmatrix} 1 \\ 0 \\ \vdots \\ 0 \end{pmatrix}$ = the first

column of $CA = \begin{pmatrix} r_{11} \\ r_{22} \\ \vdots \\ r_{n1} \end{pmatrix} = \begin{pmatrix} 1 \\ 0 \\ \vdots \\ 0 \end{pmatrix}$.

Similarly, the second column of $CA = \begin{pmatrix} 0 \\ 1 \\ 0 \\ \vdots \\ 0 \end{pmatrix}$ since $(\mathbf{v}_2)_{B_1} =$

$\begin{pmatrix} 0 \\ 1 \\ 0 \\ \vdots \\ 0 \end{pmatrix}$. Continuing in this manner, we see that $CA = I$.

Problems 5.7

1. $\begin{pmatrix} 1/\sqrt{2} \\ 1/\sqrt{2} \end{pmatrix}, \begin{pmatrix} -1/\sqrt{2} \\ 1/\sqrt{2} \end{pmatrix}$

3. (i) If $a = b = 0$, $\{(1, 0), (0, 1)\}$
(ii) If $a = 0$, $b \neq 0$ $\{(1, 0)\}$
(iii) If $a \neq 0$, $b = 0$ $\{(0, 1)\}$
(iv) If $a \neq 0$, $b \neq 0 \{(b/\sqrt{a^2+b^2}, -a/\sqrt{a^2+b^2})\}$

5. $\{(1/\sqrt{5}, 0, 2/\sqrt{5}), (2/\sqrt{30}, 5/\sqrt{30}, -1/\sqrt{30})\}$

7. $\{(2/\sqrt{29}, 3/\sqrt{29}, 4/\sqrt{29})\}$

9. $\{(1/\sqrt{5}, 0, 0, 2/\sqrt{5}), (2/\sqrt{30}, 5/\sqrt{30}, 0, -1/\sqrt{30}), (-2/\sqrt{10}, 1/\sqrt{10}, 2/\sqrt{10}, 1/\sqrt{10})\}$

11. $\{(a/\sqrt{a^2+b^2+c^2}, b/\sqrt{a^2+b^2+c^2}, c/\sqrt{a^2+b^2+c^2})\}$

13. $\{(-7/\sqrt{66}, -1/\sqrt{66}, 4/\sqrt{66})\}$

15. $Q^t = \begin{pmatrix} \frac{2}{3} & \frac{1}{3} & -\frac{2}{3} \\ \frac{1}{3} & \frac{2}{3} & \frac{2}{3} \\ \frac{2}{3} & -\frac{2}{3} & \frac{1}{3} \end{pmatrix}$ and $Q^tQ = I = QQ^t$

17. $PQ = \dfrac{1}{3\sqrt{2}} \times \begin{pmatrix} 1-\sqrt{8} & -1-\sqrt{8} \\ 1+\sqrt{8} & 1-\sqrt{8} \end{pmatrix}$,

$(PQ)^t = \dfrac{1}{3\sqrt{2}} \times \begin{pmatrix} 1-\sqrt{8} & 1+\sqrt{8} \\ -1-\sqrt{8} & 1-\sqrt{8} \end{pmatrix}$,

$(PQ)(PQ)^t = \dfrac{1}{18}\begin{pmatrix} 18 & 0 \\ 0 & 18 \end{pmatrix} = I$

19. $I = Q^{-1}Q = Q^tQ = QQ = Q^2$. But det $Q^2 = (\det Q)^2 = \det I = 1$, so det $Q = \pm 1$.

21. If $\mathbf{v}_i = 0$, then $0\mathbf{v}_1 + 0\mathbf{v}_2 + \cdots + 0\mathbf{v}_{i-1} + \mathbf{v}_i + 0\mathbf{v}_{i+1} + \cdots + 0\mathbf{v}_n = \mathbf{0}$, which implies that the $\mathbf{v}_i$'s are linearly dependent. Thus $\mathbf{v}_i \neq 0$ for $i = 1, 2, \ldots, n$.

23. (a) $\mathbf{0}$

(b) $\dfrac{1}{\sqrt{a^2+b^2}}\begin{pmatrix} a \\ b \end{pmatrix}$

(c) $\mathbf{v} = \begin{pmatrix} a \\ b \end{pmatrix} + \begin{pmatrix} 0 \\ 0 \end{pmatrix}$

25. (a) $\dfrac{1}{49}\begin{pmatrix} -186 \\ 75 \\ 118 \end{pmatrix}$

(b) $\dfrac{1}{7}\begin{pmatrix} 3 \\ -2 \\ 6 \end{pmatrix}$

(c) $\mathbf{v} = \dfrac{1}{49}\begin{pmatrix} -186 \\ 75 \\ 118 \end{pmatrix} + \dfrac{13}{49}\begin{pmatrix} 3 \\ -2 \\ 6 \end{pmatrix}$

27. (a) $\dfrac{1}{5}\begin{pmatrix} 1 \\ -3 \\ 4 \\ 17 \end{pmatrix}$

(b) $\dfrac{1}{\sqrt{15}}\begin{pmatrix} 2 \\ -1 \\ 3 \\ -1 \end{pmatrix}$

(c) $\dfrac{1}{5}\begin{pmatrix} 1 \\ -3 \\ 4 \\ 17 \end{pmatrix} + \dfrac{2}{5}\begin{pmatrix} 2 \\ -1 \\ 3 \\ -1 \end{pmatrix}$

29. $|\mathbf{u}_1 - \mathbf{u}_2|^2 = (\mathbf{u}_1 - \mathbf{u}_2) \cdot (\mathbf{u}_1 - \mathbf{u}_2) = \mathbf{u}_1 \cdot \mathbf{u}_1 - \mathbf{u}_2 \cdot \mathbf{u}_1 - \mathbf{u}_1 \cdot \mathbf{u}_2 + \mathbf{u}_2 \cdot \mathbf{u}_2 = 1 - 0 - 0 + 1 = 2$ since $\mathbf{u}_1, \mathbf{u}_2$ are orthonormal.

31. $a^2 + b^2 = 1$

33. $0 \le \left(\frac{\mathbf{u}}{|\mathbf{u}|} - \frac{\mathbf{v}}{|\mathbf{v}|}\right) \cdot \left(\frac{\mathbf{u}}{|\mathbf{u}|} - \frac{\mathbf{v}}{|\mathbf{v}|}\right)$

$$= \frac{\mathbf{u}\cdot\mathbf{u}}{|\mathbf{u}|^2} - \frac{2\mathbf{u}\cdot\mathbf{v}}{|\mathbf{u}||\mathbf{v}|} + \frac{\mathbf{v}\cdot\mathbf{v}}{|\mathbf{v}|^2}$$

$$= \frac{|\mathbf{u}|^2}{|\mathbf{u}|^2} - \frac{2\mathbf{u}\cdot\mathbf{v}}{|\mathbf{u}||\mathbf{v}|} + \frac{|\mathbf{v}|^2}{|\mathbf{v}|^2}$$

$$= 2 - \frac{2\mathbf{u}\cdot\mathbf{v}}{|\mathbf{u}||\mathbf{v}|} \ge 0$$

so that $\frac{\mathbf{u}\cdot\mathbf{v}}{|\mathbf{u}||\mathbf{v}|} \le 1$ and $\mathbf{u}\cdot\mathbf{v} \le |\mathbf{u}||\mathbf{v}|$. ① Similarly,

$$0 \le \left(\frac{\mathbf{u}}{|\mathbf{u}|} + \frac{\mathbf{v}}{|\mathbf{v}|}\right) \cdot \left(\frac{\mathbf{u}}{|\mathbf{u}|} + \frac{\mathbf{v}}{|\mathbf{v}|}\right)$$

$$= 2 + \frac{2\mathbf{u}\cdot\mathbf{v}}{|\mathbf{u}||\mathbf{v}|} \ge 0$$

so that $1 \ge -\frac{\mathbf{u}\cdot\mathbf{v}}{|\mathbf{u}||\mathbf{v}|}$ or $|\mathbf{u}||\mathbf{v}| \ge -\mathbf{u}\cdot\mathbf{v}$. ② Combining ① and ② we have $|\mathbf{u}\cdot\mathbf{v}| \le |\mathbf{u}||\mathbf{v}|$.

35. $|\mathbf{u} + \mathbf{v}|^2 = (|\mathbf{u}| + |\mathbf{v}|)^2$. This means that $(\mathbf{u} + \mathbf{v})\cdot(\mathbf{u} + \mathbf{v}) = |\mathbf{u}|^2 + 2\mathbf{u}\cdot\mathbf{v} + |\mathbf{v}|^2 = (|\mathbf{u}| + |\mathbf{v}|)^2 = |\mathbf{u}|^2 + 2|\mathbf{u}||\mathbf{v}| + |\mathbf{v}|^2$. Thus $\mathbf{u}\cdot\mathbf{v} = |\mathbf{u}||\mathbf{v}|$ which, from Problem 34, can occur only if $\mathbf{u} = \lambda\mathbf{v}$; that is, $\mathbf{u}$ and $\mathbf{v}$ are linearly dependent.

37. We prove this by mathematical induction. If $k = 2$, this is the result of Problem 35. We assume it is true for $k = n$ and prove it for $k = n + 1$. Suppose that $|\mathbf{x}_1 + \mathbf{x}_2 + \cdots + \mathbf{x}_n + \mathbf{x}_{n+1}| = |\mathbf{x}_1| + |\mathbf{x}_2| + \cdots + |\mathbf{x}_n| + |\mathbf{x}_{n+1}|$. (*) This implies that $|\mathbf{x}_1 + \mathbf{x}_2 + \cdots + \mathbf{x}_n| = |\mathbf{x}_1| + |\mathbf{x}_2| + \cdots + |\mathbf{x}_n|^{(\surd)}$, for if this is not true, then, by the triangle inequality,

$$|\mathbf{x}_1 + \mathbf{x}_2 + \cdots + \mathbf{x}_n| < |\mathbf{x}_1| + |\mathbf{x}_2| + \cdots + |\mathbf{x}_n|.$$

But then

$$|\mathbf{x}_1 + \mathbf{x}_2 + \cdots + \mathbf{x}_n + \mathbf{x}_{n+1}| \le |\mathbf{x}_1 + \mathbf{x}_2 + \cdots + \mathbf{x}_n| + |\mathbf{x}_{n+1}| < |\mathbf{x}_1| + |\mathbf{x}_2| + \cdots + |\mathbf{x}_n| + |\mathbf{x}_{n+1}|,$$

which contradicts (*). Thus, by the induction assumption, dim span$\{\mathbf{x}_1, \mathbf{x}_2, \ldots, \mathbf{x}_n\} = 1$. Let $\mathbf{u} = \mathbf{x}_1 + \mathbf{x}_2 + \cdots + \mathbf{x}_n$. By (*) and ($\surd$) $|\mathbf{u} + \mathbf{x}_{n+1}| = |\mathbf{u}| + |\mathbf{x}_{n+1}|$ so that by Problem 35, $\mathbf{x}_{n+1} = \lambda\mathbf{u}$ for some number λ. That is, $\mathbf{x}_{n+1} \in \text{span}\{\mathbf{x}_1, \mathbf{x}_2, \ldots, \mathbf{x}_n\}$ so that dim span$\{\mathbf{x}_1, \mathbf{x}_2, \ldots, \mathbf{x}_n, \mathbf{x}_{n+1}\} = 1$ also. Thus the result is true for $k = n + 1$, and the proof is complete.

39. $(H^\perp)^\perp = \{\mathbf{v} \in \mathbb{R}^n; \mathbf{v}\cdot\mathbf{k} = 0$ for every $\mathbf{k} \in H^\perp\}$. Let $\mathbf{x} \in H$; then $\mathbf{x}\cdot\mathbf{k} = 0$ for every $\mathbf{k} \in H^\perp$ so that $\mathbf{x} \in (H^\perp)^\perp$, which shows that $H \subseteq (H^\perp)^\perp$. Conversely, if $\mathbf{v} \in (H^\perp)^\perp$, then $\mathbf{v}\cdot\mathbf{k} = 0$ for every $\mathbf{k} \in H^\perp$. But $\mathbf{v} = \mathbf{h}' + \mathbf{k}'$ where $\mathbf{h}' \in H$ and $\mathbf{k}' \in H^\perp$. Then $0 = \mathbf{v}\cdot\mathbf{k} = \mathbf{h}'\cdot\mathbf{k} + \mathbf{k}'\cdot\mathbf{k} = 0 + \mathbf{k}'\cdot\mathbf{k}$. Thus $\mathbf{k}'\cdot\mathbf{k} = 0$ for every $\mathbf{k} \in H$, which means, in particular, that $\mathbf{k}'\cdot\mathbf{k}' = 0$. Thus $\mathbf{k}' = 0$ and $\mathbf{v} = \mathbf{h}' \in H$. Thus $(H^\perp)^\perp \subset H$ and, together with $H \subset (H^\perp)^\perp$, shows that $(H^\perp)^\perp = H$.

41. Let $\mathbf{k} \in H_2^\perp$. Then $\mathbf{k}\cdot\mathbf{h} = 0$ for every $\mathbf{h} \in H_2$. Since $H_1 \subset H_2$, this shows that $\mathbf{k}\cdot\mathbf{h} = 0$ for every $\mathbf{h} \in H_1$. That is, $\mathbf{k} \in H_1^\perp$. Thus $H_2^\perp \subset H_1^\perp$.

Problems 5.8

1. **(i)** $(A, A) = a_{11}^2 + a_{22}^2 + \cdots + a_{nn}^2 \ge 0$.
(ii) $(A, A) = 0$ implies that $a_{ii}^2 = 0$ for $i = 1, 2, \ldots, n$ so that $A = 0$. If $A = 0$, then $(A, A) = 0$.
(iii) $(A, B + C) = a_{11}(b_{11} + c_{11}) + \cdots + a_{nn}(b_{nn} + c_{nn}) = a_{11}b_{11} + a_{11}c_{11} + \cdots + a_{nn}b_{nn} + a_{nn}c_{nn} = (a_{11}b_{11} + \cdots + a_{nn}b_{nn}) + (a_{11}c_{11} + \cdots + a_{nn}c_{nn}) = (A, B) + (A, C)$
(iv) Similarly, $(A + B, C) = (A, C) + (B, C)$
(v) $(A, B) = (B, A) = \overline{(B, A)}$ since all components are real and $a_{ii}b_{ii} = b_{ii}a_{ii}$
(vi) $(\alpha A, B) = (\alpha a_{11})b_{11} + \cdots + (\alpha a_{nn})b_{nn} = \alpha[a_{11}b_{11} + \cdots + a_{nn}b_{nn}] = \alpha(A, B)$
(vii) $(A, \alpha B) = \overline{(\alpha B, A)} = (\alpha B, A) = \alpha(B, A) = \alpha(A, B) = \alpha(A, B)$

3. **Let E_i be the $n \times n$ matrix with a 1 in the i, i position and 0 everywhere else. It is easy to see that $\{E_1, E_2, \ldots, E_n\}$ is an orthonormal basis for D_n.**

5. $\{(1/\sqrt{2}, i/\sqrt{2}), (i/\sqrt{2}, 1/\sqrt{2})\}$

7. $\{\frac{1}{2}, \sqrt{3/2}x, \sqrt{5/8}\,(3x^2 - 1)\}$

9. First note that if $A = (a_{ij})$ and $B^t = (b_{ji})$, then

$$(AB^t)_{ij} = \sum_{k=1}^{n} a_{ik}b_{jk}$$

so that

$$tr(AB^t) = \sum_{i=1}^{n}\sum_{j=1}^{n} a_{ij}b_{ij}.$$

(i) $(A, A) = tr(AA^t) = \sum_{i=1}^{n}\left(\sum_{j=1}^{n} a_{ij}^2\right) \ge 0$
(ii) $(A, A) = 0$ implies that $a_{ij}^2 = 0$ for every i and j so that $A = 0$. If $A = 0$, then $A^t = 0$ and $AA^t = 0$ so that $tr(AA^t) = 0$.
(iii) $(A, B + C) = tr[A(B + C)^t] + tr[A(B^t + C^t)] = tr(AB^t + AC^t) = tr(AB^t) + tr(AC^t) = (A, B) + (A, C)$
(iv) Similarly $(A + B, C) = (A, C) + (B, C)$
(v) $(A, B) = \sum_{i=1}^{n}\sum_{j=1}^{n} a_{ij}b_{ij} = tr(BA^t) = (B, A)$
(vi) $(\alpha A, B) = tr(\alpha AB^t) = \alpha tr(AB^t) = \alpha(A, B)$
(vii) $(A, \alpha B) = (\alpha B, A) = \alpha(B, A) = \alpha(A, B)$

11. $\begin{pmatrix}1 & 0\\ 0 & 0\end{pmatrix}, \begin{pmatrix}0 & 1\\ 0 & 0\end{pmatrix}, \begin{pmatrix}0 & 0\\ 1 & 0\end{pmatrix}, \begin{pmatrix}0 & 0\\ 0 & 1\end{pmatrix}$

13. (a) **(i)** $(p, p) = p(a)^2 + p(b)^2 + p(c)^2 \geq 0$

(ii) $(p, p) = 0$ implies that $p(a) = p(b) = p(c) = 0$. But a quadratic can have at most two roots. Thus $p(x) = 0$ for all x. Conversely if $p \equiv 0$, then $p(a) = p(b) = p(c) = 0$ so $(p, p) = 0$.

(iii) $(p, q + r)$

$$= p(a)(q(a) + r(a)) + p(b)(q(b) + r(b)) + p(c)(q(c) + r(c))$$
$$= [p(a)q(a) + p(b)q(b) + p(c)q(c)] + [p(a)r(a) + p(b)r(b) + p(c)r(c)]$$
$$= (p, q) + (p, r)$$

(iv) Similarly, $(p + q, r) = (p, r) + (q, r)$

(v) $(p, q) = p(a)q(a) + (p(b)q(b) + p(c)q(c)$

$$= q(a)p(a) + q(b)p(b) + q(c)p(c)$$
$$= (q, p)$$

(vi) $(\alpha p, q) = [\alpha p(a)]q(a) + [\alpha p(b)]q(b) + [\alpha p(c)]q(c)$

$$= \alpha[p(a)q(a) + p(b)q(b) + p(c)q(c)]$$
$$= \alpha(p, q)$$

(vii) $(p, \alpha q) = (\alpha p, q) = \alpha(p, q) = \alpha(p, q)$

(b) No, since **(ii)** is violated. For example, let $a = 1$, $b = -1$, and $p(x) = (x - 1)(x + 1) = x^2 - 1 \neq 0$. Then $p(a) = p(b) = 0$ so that $(p, p) = 0$ even though $p \not\equiv 0$. In fact, for any polynomial q, we have $(p, q) = 0$.

15. $\sqrt{31}$

17. $0 \leq \left(\left(\frac{\mathbf{u}}{|\mathbf{u}|} - \frac{\mathbf{v}}{|\mathbf{v}|}\right), \left(\frac{\mathbf{u}}{|\mathbf{u}|} - \frac{\mathbf{v}}{|\mathbf{v}|}\right)\right)$

$$= \frac{(\mathbf{u}, \mathbf{u})}{|\mathbf{u}|^2} - \frac{(\mathbf{u}, \mathbf{v})}{|\mathbf{u}||\mathbf{v}|} - \frac{(\mathbf{v}, \mathbf{u})}{|\mathbf{u}||\mathbf{v}|} + \frac{(\mathbf{v}, \mathbf{v})}{|\mathbf{v}|^2}$$

$$= \frac{|\mathbf{u}|^2}{|\mathbf{u}|^2} - \left[\frac{(\mathbf{u}, \mathbf{v}) + \overline{(\mathbf{u}, \mathbf{v})}}{|\mathbf{u}||\mathbf{v}|}\right] + \frac{|\mathbf{v}|^2}{|\mathbf{v}|^2}.$$

Now if $z = a + bi$, then $z + \bar{z} = (a + bi) + (a - bi) = 2a = 2 \operatorname{Re} z$ (and $z - \bar{z} = 2bi = 2iImz$). Thus $(\mathbf{u}, \mathbf{v}) + \overline{(\mathbf{u}, \mathbf{v})} = 2 \operatorname{Re}(\mathbf{u}, \mathbf{v})$ and we have $2 - \frac{2 \operatorname{Re}(\mathbf{u}, \mathbf{v})}{|\mathbf{u}||\mathbf{v}|} \geq 0$ or $\frac{\operatorname{Re}(\mathbf{u}, \mathbf{v})}{|\mathbf{u}||\mathbf{v}|} \leq 1$. Let λ be a real number. Then $0 \leq ((\lambda\mathbf{u} + (\mathbf{u}, \mathbf{v})\mathbf{v}), (\lambda\mathbf{u} + (\mathbf{u}, \mathbf{v})\mathbf{v})) = \lambda^2|\mathbf{u}|^2 + |(\mathbf{u}, \mathbf{v})|^2|\mathbf{v}|^2 + \lambda\overline{(\mathbf{u}, \mathbf{v})}(\mathbf{u}, \mathbf{v}) + \bar{\lambda}(\mathbf{u}, \mathbf{v})(\mathbf{v}, \mathbf{u}) =$ (since λ is real) $\lambda^2|\mathbf{u}|^2 + 2\lambda|(\mathbf{u}, \mathbf{v})|^2 + |(\mathbf{u}, \mathbf{v})|^2|\mathbf{v}|^2$. The last line is a quadratic equation in λ. If we have $a\lambda^2 + b\lambda + c \geq 0$, then the equation $a\lambda^2 + b\lambda + c = 0$ can have at most one real root and, therefore, $b^2 - 4ac \leq 0$. Thus

$$4((|\mathbf{u}, \mathbf{v})|^2)^2 - 4|\mathbf{u}|^2 \times |(\mathbf{u}, \mathbf{v})|^2|\mathbf{v}|^2 \leq 0$$

or $\quad |(\mathbf{u}, \mathbf{v})|^2 \leq |\mathbf{u}|^2|\mathbf{v}|^2$

and $\quad |(\mathbf{u}, \mathbf{v})| \leq |\mathbf{u}||\mathbf{v}|$.

19. $H^{\perp} = \operatorname{span}\{(-15x^2 + 16x - 3), (20x^3 - 30x^2 + 12x - 1)\}$

21. $1 + 2x + 3x^2 - x^3$

$$= \frac{30x^2 + 52x + 19}{20} + \frac{(-20x^3 + 30x^2 - 12x + 1)}{20}$$

Chapter 5—Review

1. yes; dimension 2; basis $\{(1, 0, 1), (0, 1, 2)\}$

3. yes; dimension 3; basis $\{(1, 0, 0, -1), (0, 1, 0, -1), (0, 0, 1, -1)\}$

5. yes; dimension $[n(n+1)]/2$; basis $\{(E_{ij} : j \geq 1\}$ where E_{ij} is the matrix with 1 in the i, j position and 0 everywhere else

7. no; for example, $(x^5 - 2x) + (-x^5 + x^2) = x^2 - 2x$, which is not a polynomial of degree 5 so the set is not closed under addition.

9. no; for example,

$$\begin{pmatrix}1 & 1\\ 0 & 2\\ 3 & 1\end{pmatrix} + \begin{pmatrix}2 & 1\\ -1 & 2\\ 1 & 0\end{pmatrix} = \begin{pmatrix}3 & 2\\ -1 & 4\\ 4 & 1\end{pmatrix},$$ which does not satisfy $a_{12} = 1$.

11. dependent

13. independent

15. dimension 2; basis $\{(2, 0, 1), (0, 4, 3)\}$

17. dimension 3; basis $\{(1, 0, 3, 0), (0, 1, -1, 0), (0, 0, 1, 1)\}$

19. dimension 4; basis $\{D_1, D_2, D_3, D_4\}$ where D_i is the matrix with a 1 in the i, i position and 0 everywhere else

21. $\frac{3}{4}\begin{pmatrix}1\\ 2\end{pmatrix} - \frac{5}{4}\begin{pmatrix}-1\\ 2\end{pmatrix} = \begin{pmatrix}2\\ -1\end{pmatrix}$

23. $1(1 + x^2) + 0(1 + x) + 3(1) = 4 + x^2$

25. $\left\{\frac{1}{\sqrt{13}}\begin{pmatrix}2\\ 3\end{pmatrix}, \frac{1}{\sqrt{13}}\begin{pmatrix}-3\\ 2\end{pmatrix}\right\}$

27. $\begin{pmatrix}1/\sqrt{3}\\ 1/\sqrt{3}\\ 1/\sqrt{3}\end{pmatrix}$

29. (a) $\begin{pmatrix} \frac{4}{3} \\ -\frac{1}{3} \\ \frac{5}{3} \end{pmatrix}$ (b) $\begin{pmatrix} -1/\sqrt{3} \\ 1/\sqrt{3} \\ 1/\sqrt{3} \end{pmatrix}$

(c) $\begin{pmatrix} \frac{4}{3} \\ -\frac{1}{3} \\ \frac{5}{3} \end{pmatrix} + \begin{pmatrix} -\frac{7}{3} \\ \frac{7}{3} \\ \frac{7}{3} \end{pmatrix}$

31. (a) $\begin{pmatrix} \frac{1}{2} \\ \frac{1}{2} \\ \frac{1}{2} \\ \frac{1}{2} \end{pmatrix}$

(b) $\left\{\begin{pmatrix} 1/\sqrt{2} \\ 0 \\ -1/\sqrt{2} \\ 0 \end{pmatrix}, \begin{pmatrix} 0 \\ 1/\sqrt{2} \\ 0 \\ -1/\sqrt{2} \end{pmatrix}\right\}$

(c) $\begin{pmatrix} \frac{1}{2} \\ \frac{1}{2} \\ \frac{1}{2} \\ \frac{1}{2} \end{pmatrix} + \begin{pmatrix} \frac{1}{2} \\ -\frac{1}{2} \\ -\frac{1}{2} \\ \frac{1}{2} \end{pmatrix}$

CHAPTER 6

Problems 6.1

1. linear **3.** linear

5. not linear, since

$$T\left(\alpha\begin{pmatrix} x \\ y \\ z \end{pmatrix}\right) = T\begin{pmatrix} \alpha x \\ \alpha y \\ \alpha z \end{pmatrix} = \begin{pmatrix} 1 \\ \alpha z \end{pmatrix}$$

while

$$T\begin{pmatrix} x \\ y \\ z \end{pmatrix} = \alpha\begin{pmatrix} 1 \\ z \end{pmatrix} = \begin{pmatrix} \alpha \\ \alpha z \end{pmatrix}$$

7. linear

9. not linear, since

$$T\left(\alpha\begin{pmatrix} x \\ y \end{pmatrix}\right) = T\begin{pmatrix} \alpha x \\ \alpha y \end{pmatrix} = (\alpha x)(\alpha y) = \alpha^2 xy$$

while $\alpha T\begin{pmatrix} x \\ y \end{pmatrix} = \alpha xy$

11. linear

13. not linear, since

$$T\left(\alpha\begin{pmatrix} x \\ y \\ z \\ w \end{pmatrix}\right) = \alpha^2 T\begin{pmatrix} x \\ y \\ z \\ w \end{pmatrix} \neq \alpha T\begin{pmatrix} x \\ y \\ z \\ w \end{pmatrix}$$

if $\alpha \neq 1$ or 0

15. not linear, since
$T(A+B) = (A+B)^t(A+B) = (A^t+B^t)(A+B) = A^tA + A^tB + B^tA + B^tB.$
But
$T(A) + T(B) = A^tA + B^tB \neq T(A+B)$
unless $A^tB + B^tA = 0$.

17. not linear, since $T(\alpha D) = (\alpha D)^2 = \alpha^2 D^2 \neq \alpha T(D) = \alpha D^2$ unless $\alpha = 1$ or 0.

19. linear **21.** linear

23. not linear, since $T(f+g) = (f+g)^2 \neq f^2 + g^2 = T(f) + T(g)$

25. linear **27.** linear

29. not linear, since
$T(\alpha A) = \det(\alpha A) = \alpha^n \det A \neq \alpha \det A = \alpha T(A)$
unless $\alpha = 0$ or 1.
[$\det \alpha A = \alpha^n \det A$ by Problem 3.2.28.] Also, in general,
$\det(A+B) \neq \det A + \det B$.

31. (a) $\begin{pmatrix} -14 \\ 4 \\ 26 \end{pmatrix}$ (b) $\begin{pmatrix} -31 \\ -6 \\ 26 \end{pmatrix}$

33. It rotates a vector counterclockwise around the z-axis through an angle of θ in a plane parallel to the xy-plane.

35. Suppose $\alpha < 0$. Then $T[(\alpha - \alpha)\mathbf{x}] = T(0\mathbf{x}) = 0T\mathbf{x} = \mathbf{0}$. Thus $T[(\alpha - \alpha)\mathbf{x}] = T(0\mathbf{x}) = 0T\mathbf{x} = 0$ and $T(\alpha\mathbf{x}) + T(-\alpha\mathbf{x}) = T((\alpha - \alpha)\mathbf{x}) = \mathbf{0}$. But $-\alpha > 0$ so that $T(-\alpha\mathbf{x}) = -\alpha T\mathbf{x}$. Therefore $T(\alpha\mathbf{x}) - \alpha T\mathbf{x} = 0$, or $T(\alpha\mathbf{x}) = \alpha T\mathbf{x}$ for $\alpha < 0$ as well.

37. $T(\mathbf{x} - \mathbf{y}) = T\mathbf{x} + T(-\mathbf{y}) = T\mathbf{x} + T[(-1)\mathbf{y}] = T\mathbf{x} + (-1)T\mathbf{y} = T\mathbf{x} - T\mathbf{y}$.

39. $T(\mathbf{v}_1 + \mathbf{v}_2) = (\mathbf{v}_1 + \mathbf{v}_2, \mathbf{u}_0) = (\mathbf{v}_1, \mathbf{u}_0) + (\mathbf{v}_2, \mathbf{u}_0) = T\mathbf{v}_1 + T\mathbf{v}_2$; $T(\alpha\mathbf{v}) = (\alpha\mathbf{v}, \mathbf{u}_0) = \alpha(\mathbf{v}, \mathbf{u}_0) = \alpha T\mathbf{v}$.

41.
$$\begin{aligned}
T(\mathbf{v}_1 + \mathbf{v}_2) &= (\mathbf{v}_1 + \mathbf{v}_2, \mathbf{u}_1)\mathbf{u}_1 \\
&= (\mathbf{v}_1 + \mathbf{v}_2, \mathbf{u}_2)\mathbf{u}_2 + \cdots + (\mathbf{v}_1 + \mathbf{v}_2, \mathbf{u}_n)\mathbf{u}_n \\
&= (\mathbf{v}_1, \mathbf{u}_1)\mathbf{u}_1 + (\mathbf{v}_2, \mathbf{u}_1)\mathbf{u}_1 + (\mathbf{v}_1, \mathbf{u}_2)\mathbf{u}_2 + (\mathbf{v}_2, \mathbf{u}_2)\mathbf{u}_2 + \cdots + (\mathbf{v}_1, \mathbf{u}_n)\mathbf{u}_n + (\mathbf{u}_2, \mathbf{u}_n)\mathbf{u}_n \\
&= (\mathbf{v}_1, \mathbf{u}_1)\mathbf{u}_1 + (\mathbf{v}_1, \mathbf{u}_2)\mathbf{u}_2 + \cdots + (\mathbf{v}_1, \mathbf{u}_n)\mathbf{u}_n + (\mathbf{v}_2, \mathbf{u}_1)\mathbf{u}_1 + (\mathbf{v}_2, \mathbf{u}_2)\mathbf{u}_2 + \cdots + (\mathbf{v}_2, \mathbf{u}_n)\mathbf{u}_n \\
&= T\mathbf{v}_1 + T\mathbf{v}_2; \\
T(\alpha\mathbf{v}) &= (\alpha\mathbf{v}, \mathbf{u}_1)\mathbf{u}_1 + (\alpha\mathbf{v}, \mathbf{u}_2)\mathbf{u}_2 + \cdots + (\alpha\mathbf{v}, \mathbf{u}_n)\mathbf{u}_n \\
&= \alpha[(\mathbf{v}, \mathbf{u}_1)\mathbf{u}_1 + (\mathbf{v}, \mathbf{u}_2)\mathbf{u}_2 + \cdots + (\mathbf{v}, \mathbf{u}_n)\mathbf{u}_n] \\
&= \alpha T\mathbf{v}.
\end{aligned}$$

Problems 6.2

1. kernel $= \{(0, y) : y \in \mathbb{R}\}$, i.e., the y-axis; range $= \{(x, 0) : x \in \mathbb{R}\}$, i.e., the x-axis; $\rho(T) = \nu(T) = 1$

3. kernel $= \{(x, -x) : x \in \mathbb{R}\}$—this is the line $x + y = 0$; range $= \mathbb{R}$; $\rho(T) = \nu(T) = 1$.

5. kernel $= \left\{\begin{pmatrix} 0 & 0 \\ 0 & 0 \end{pmatrix}\right\}$; range $= M_{22}$; $\rho(T) = 4$, $\nu(T) = 0$

7. kernel $=\{A : A^t = -A\}$ $=\{A : A$ is skew-symmetric$\}$; range $=\{A : A$ is symmetric$\}$; $\rho(T) = (n^2+n)/2$; $\nu(T) = (n^2-n)/2$

9. kernel $=\{f \in C[0,1] : f(\frac{1}{2}) = 0\}$; range $=\mathbb{R}$; $\rho(T)=1$; the kernel is an infinite dimensional space so that $\nu(T)=\infty$. For example, the linearly independent functions $x-\frac{1}{2}$, $(x-\frac{1}{2})^2$, $(x-\frac{1}{2})^3$, $(x-\frac{1}{2})^4, \ldots, (x-\frac{1}{2})^n, \ldots$ all satisfy $f(\frac{1}{2})=0$.

11. If $\mathbf{v} \in V$, then $\mathbf{v} = c_1\mathbf{v}_1 + c_2\mathbf{v}_2 + \cdots + c_n\mathbf{v}_n$ so that $T\mathbf{v} = T(c_1\mathbf{v}_1 + c_2\mathbf{v}_2 + \cdots + c_n\mathbf{v}_n) = c_1T\mathbf{v}_1 + c_2T\mathbf{v}_2 + \cdots + c_nT\mathbf{v}_n) = c_1\mathbf{0} + c_2\mathbf{0} + \cdots + c_n\mathbf{0} = \mathbf{0}$. Thus $T\mathbf{v} = \mathbf{0}$ for every $\mathbf{v} \in V$ and is, therefore, the zero transformation.

13. The range of T is a subspace of $\mathbb{R}^3$ and, by Example 5.5.9, the subspaces of $\mathbb{R}^3$ are $\{0\}$, $\mathbb{R}^3$, and lines and planes passing through the origin.

15. $T\mathbf{x} = A\mathbf{x}$ where $A = \begin{pmatrix} 0 & a \\ b & c \end{pmatrix}$, a, b, c real

17. $T\mathbf{x} = A\mathbf{x}$ where $A = \begin{pmatrix} 2 & -1 & 1 \\ 2 & -1 & 1 \\ 2 & -1 & 1 \end{pmatrix}$

19. **(i)** If $A \in \ker T$, then $A - A^t = 0$, or $A = A^t$.
(ii) If $A \in$ range of T, then there is a matrix B such that $B - B^t = A$. Then $A^t = (B - B^t)^t = B^t - (B^t)^t = B^t - B = -A$ so that A is a skew-symmetric.

Problems 6.3

1. $\rho=2,\ \nu=0$ **3.** $\rho=1,\ \nu=2$

5. $\rho=2,\ \nu=1$ **7.** $\rho=2,\ \nu=2$

9. $\rho=2,\ \nu=0$ **11.** $\rho=2,\ \nu=2$

13. $\rho=3,\ \nu=1$ **15.** $\rho=2,\ \nu=1$

17. range basis $=\left\{\begin{pmatrix}1\\3\\5\end{pmatrix}, \begin{pmatrix}-1\\1\\-1\end{pmatrix}\right\}$; these are the first two columns of A.

null space basis $=\left\{\begin{pmatrix}-3\\1\\2\end{pmatrix}\right\}$

19. range basis $=\left\{\begin{pmatrix}1\\0\\1\end{pmatrix}, \begin{pmatrix}-1\\1\\0\end{pmatrix}, \begin{pmatrix}3\\3\\5\end{pmatrix}\right\}$; these are the first three (linearly independent) columns of A.

null space basis $=\left\{\begin{pmatrix}-6\\-4\\1\\0\end{pmatrix}\right\}$

21. range basis $=\left\{\begin{pmatrix}1\\-2\\2\\3\end{pmatrix}\right\}$

null space basis $=\left\{\begin{pmatrix}1\\1\\0\\0\end{pmatrix}, \begin{pmatrix}0\\2\\1\\0\end{pmatrix}, \begin{pmatrix}0\\3\\0\\1\end{pmatrix}\right\}$

23. no **25.** yes

27. If $\mathbf{c}_i$ denotes the ith column of D, then

$$\mathbf{c}_i = d_i\begin{pmatrix}0\\0\\\vdots\\1\\0\\\vdots\\0\end{pmatrix}$$

ith position. Thus the $\mathbf{c}_i$'s are linearly independent when $d_i \neq 0$, and the number of linearly independent columns is the rank.

29. $\rho(A^t) =$ dimension of column space of $A^t =$ dimension of row space of $A =$ dimension of column space of A (by Theorem 2) $= \rho(A)$.

31. **(i)** Let $H =$ range of A and let $\{\mathbf{v}_1, \mathbf{v}_2, \ldots, \mathbf{v}_k\}$ be a basis for H. Since B is invertible, $\ker B = \{\mathbf{0}\}$, which means that $\{B\mathbf{v}_1, B\mathbf{v}_2, \ldots, B\mathbf{v}_k\}$ is a linearly independent set in $\mathbb{R}^m$ and is therefore a basis for range BA. Thus $\rho(BA) = k = \rho(A)$.
(ii) Since C is invertible, range of $C = \mathbb{R}^n$. Let $\mathbf{h} \in H$; then there is an $\mathbf{x} \in \mathbb{R}^n$ such that $A\mathbf{x} = \mathbf{h}$. Since range of $C = \mathbb{R}^n$, there is a $\mathbf{y} \in \mathbb{R}^n$ such that $C\mathbf{y} = \mathbf{x}$. Thus $AC\mathbf{y} = \mathbf{h}$. Thus $H \subset$ range of AC. If $\mathbf{v} \in$ range of AC, there is a $\mathbf{u}$ in $\mathbb{R}^n$ such that $AC\mathbf{u} = \mathbf{v}$. But then $\mathbf{v} = A(C\mathbf{u})$ so that $\mathbf{v} \in$ range of $A = H$. Hence range of $AC \subset H$ so that range of $AC = H$ and $\rho(A) = \rho(AC)$.

33. Since $\rho(A) = 5$, the five rows of A are linearly independent. Thus the five rows of $(A, \mathbf{b})$ are linearly independent, and $\rho(A, \mathbf{b}) = 5$.

35. By Problem 31, $\rho(A) = \rho(AD) = \rho(C(AD)) = \rho(B)$.

37. **(i)** If there is an $\mathbf{x} \neq \mathbf{0}$ such that $A\mathbf{x} = 0$, then $A(\alpha\mathbf{x}) = \alpha A\mathbf{x} = \mathbf{0}$ for every $\alpha \in \mathbb{R}$ so that $\nu(A) = \dim \ker A \geq 1$, and $\rho(A) = n - \nu(A) \leq n - 1 < n$.
(ii) If $\rho(A) < n$, then $\nu(A) = n - \rho(A) > 0$ so that there is an $\mathbf{x} \neq \mathbf{0}$ such that $A\mathbf{x} = \mathbf{0}$.

Problems 6.4

1. $\begin{pmatrix}1 & -2\\-1 & 1\end{pmatrix}$; $\ker T = \{\mathbf{0}\}$; range $T = \mathbb{R}^2$; $\nu(T)=0$, $\rho(T)=2$

3. $\begin{pmatrix}1 & -1 & 1\\-2 & 2 & -2\end{pmatrix}$; range $T =$ span $\left\{\begin{pmatrix}1\\-2\end{pmatrix}\right\}$; $\ker T =$ span $\left\{\begin{pmatrix}1\\1\\0\end{pmatrix}, \begin{pmatrix}0\\1\\1\end{pmatrix}\right\}$; $\rho(T)=1$, $\nu(T)=2$

5. $\begin{pmatrix} 1 & -1 & 2 \\ 3 & 1 & 4 \\ 5 & -1 & 8 \end{pmatrix}$; range $T=$ span $\left\{\begin{pmatrix} 1 \\ 3 \\ 5 \end{pmatrix}, \begin{pmatrix} -1 \\ 1 \\ -1 \end{pmatrix}\right\}$; ker $T=$ span $\left\{\begin{pmatrix} -3 \\ 1 \\ 2 \end{pmatrix}\right\}$; $\rho(T)=2$, $\nu(T)=1$

7. $\begin{pmatrix} 1 & -1 & 2 & 3 \\ 0 & 1 & 4 & 3 \\ 1 & 0 & 6 & 6 \end{pmatrix}$; range $T=$ span $\left\{\begin{pmatrix} 1 \\ 0 \\ 1 \end{pmatrix}, \begin{pmatrix} -1 \\ 1 \\ 0 \end{pmatrix}\right\}$; ker $T=$ span $\left\{\begin{pmatrix} -6 \\ -4 \\ 1 \\ 0 \end{pmatrix}, \begin{pmatrix} -6 \\ -3 \\ 0 \\ 1 \end{pmatrix}\right\}$;

$\rho(T)=2$, $\nu(T)=2$

9. $\begin{pmatrix} \frac{5}{4} & -\frac{13}{4} \\ \frac{5}{4} & \frac{3}{4} \end{pmatrix}$; range $T=\mathbb{R}^2$; ker $T=\{\mathbf{0}\}$; $\rho(T)=2$, $\nu(T)=0$

11. $\begin{pmatrix} 3 & \frac{7}{5} & \frac{16}{5} \\ 0 & \frac{4}{5} & \frac{2}{5} \end{pmatrix}$; range $T=\mathbb{R}^2$; ker $T=$ span $\left\{\begin{pmatrix} 5 \\ 3 \\ -6 \end{pmatrix}_{B_1}\right\}$; $\rho(T)=2$, $\nu(T)=1$

13. $\begin{pmatrix} 0 & 1 & 0 \\ 0 & -1 & 0 \\ 0 & 0 & 0 \\ 1 & 0 & 0 \end{pmatrix}$; range $T=$ span $\{1-x, x^3\}$; ker $T=$ span $\{x^2\}$; $\rho(T)=2$, $\nu(T)=1$

15. $(0, 0, 1, 0)$; range $T=\mathbb{R}$; ker $T=$ span $\{1, x, x^3\}$; $\rho(T)=1$, $\nu(T)=3$

17. $\begin{pmatrix} 1 & -1 & 2 & 3 \\ 0 & 1 & 4 & 3 \\ 1 & 0 & 6 & 5 \end{pmatrix}$; range $T=$ span $\{1+x^2, -1+x, 2+4x+6x^2\}=P_2$; ker $T=$ span $\{x^2-4x-6\}$; $\rho(T)=3$, $\nu(T)=1$

19. $\begin{pmatrix} 1 & 1 & 1 & 1 \\ 1 & 1 & 1 & 0 \\ 1 & 1 & 0 & 0 \\ 1 & 0 & 0 & 0 \end{pmatrix}$; range $T=M_{22}$; ker $T=\left\{\begin{pmatrix} 0 & 0 \\ 0 & 0 \end{pmatrix}\right\}$;

$\rho(T)=4$, $\nu(T)=0$

21. $\begin{pmatrix} 0 & 1 & 0 & 0 & 0 \\ 0 & 0 & 2 & 0 & 0 \\ 0 & 0 & 0 & 3 & 0 \\ 0 & 0 & 0 & 0 & 4 \end{pmatrix}$; range $D=P_3$; ker $D=\mathbb{R}$; $\rho(D)=4$, $\nu(D)=1$

23. $\begin{pmatrix} 0 & 1 & 0 & 0 & \cdots & 0 \\ 0 & 0 & 2 & 0 & \cdots & 0 \\ 0 & 0 & 0 & 3 & \cdots & 0 \\ \vdots & \vdots & \vdots & \vdots & & \vdots \\ 0 & 0 & 0 & 0 & \cdots & n \\ 0 & 0 & 0 & 0 & \cdots & 0 \end{pmatrix}$; range $D=P_{n-1}$; ker $D=\mathbb{R}$; $\rho(D)=n$, $\nu(D)=1$

25. $\begin{pmatrix} 2 & 0 & 2 & 0 & 0 \\ 0 & 3 & 0 & 6 & 0 \\ 0 & 0 & 4 & 0 & 12 \\ 0 & 0 & 0 & 5 & 0 \\ 0 & 0 & 0 & 0 & 6 \end{pmatrix}$; range $T=P_4$; ker $T=\{\mathbf{0}\}$; $\rho(T)=5$, $\nu(T)=0$

27. $A_T=\text{diag}\,(b_0, b_1, b_2, \ldots, b_n)$ where $b_j=\sum_{i=1}^{j+1}\dfrac{j!}{(j+1-i)!}$; range $T=P_n$; ker $T=\{\mathbf{0}\}$; $\rho(T)=n+1$, $\nu(T)=0$

29. $\begin{pmatrix} 1 & 0 & 0 \\ 0 & 1 & 0 \\ 0 & 0 & 1 \end{pmatrix}$; range $T=P_2$; ker $T=\{\mathbf{0}\}$, $\rho(T)=3$, $v(T)=0$

31. For example, in M_{34},

$$A_T=\begin{pmatrix} 1 & 0 & 0 & 0 & 0 & 0 & 0 & 0 & 0 & 0 & 0 & 0 \\ 0 & 0 & 0 & 0 & 1 & 0 & 0 & 0 & 0 & 0 & 0 & 0 \\ 0 & 0 & 0 & 0 & 0 & 0 & 0 & 0 & 1 & 0 & 0 & 0 \\ 0 & 1 & 0 & 0 & 0 & 0 & 0 & 0 & 0 & 0 & 0 & 0 \\ 0 & 0 & 0 & 0 & 0 & 1 & 0 & 0 & 0 & 0 & 0 & 0 \\ 0 & 0 & 0 & 0 & 0 & 0 & 0 & 0 & 0 & 1 & 0 & 0 \\ 0 & 0 & 1 & 0 & 0 & 0 & 0 & 0 & 0 & 0 & 0 & 0 \\ 0 & 0 & 0 & 0 & 0 & 0 & 1 & 0 & 0 & 0 & 0 & 0 \\ 0 & 0 & 0 & 0 & 0 & 0 & 0 & 0 & 0 & 0 & 1 & 0 \\ 0 & 0 & 0 & 1 & 0 & 0 & 0 & 0 & 0 & 0 & 0 & 0 \\ 0 & 0 & 0 & 0 & 0 & 0 & 0 & 1 & 0 & 0 & 0 & 0 \\ 0 & 0 & 0 & 0 & 0 & 0 & 0 & 0 & 0 & 0 & 0 & 1 \end{pmatrix}$$

In general, $A_T=(a_{ij})$ where

$$a_{ij}=\begin{cases} 1, & \text{if } \; i=km+l, \\ & \text{and } \; j=(l-1)n+k+1 \\ & \text{for } \; k=1, 2, \ldots, n-1 \\ & \text{and } \; l=1, 2, \ldots, m \\ 0, & \text{otherwise.} \end{cases}$$

33. $\begin{pmatrix} 0 & 0 & 0 \\ 0 & 0 & -1 \\ 0 & 1 & 0 \end{pmatrix}$; range D $= \text{span}\{\sin x, \cos x\}$; $\ker D = \mathbb{R}$; $\rho(D) = 2$, $\nu(D) = 1$

35. $\begin{pmatrix} \frac{1}{2} & -i/2 \\ i/2 & \frac{1}{2} \end{pmatrix}$

37. Let B_1 and B_2 be bases for V and W, respectively. We have $(T\mathbf{v})_{B_2} = A_T(\mathbf{v})_{B_1}$ for every $\mathbf{v} \in V$. Then $\mathbf{v} \in \ker T$ if and only if $T\mathbf{v} = \mathbf{0}$ if and only if $A_T(\mathbf{v})_{B_1} = (\mathbf{0})_{B_2}$ if and only if $(\mathbf{v})_{B_1} \in \ker A_T$. Thus nullity of $T = N_{A_T}$ so that $\nu(T) = \nu(A_T)$. If $\mathbf{w} \in$ range T, then $T\mathbf{v} = \mathbf{w}$ for some $\mathbf{v} \in V$ so that $A_T(\mathbf{v})_{B_1} = (T\mathbf{v})_{B_2} = (\mathbf{w})_{B_2}$. This means that $(\mathbf{w})_{B_2} \in R_{A_T}$. Thus $R_{A_T} =$ range T so that $\rho(T) = \rho(A_T)$. Since $\nu(A_T) + \rho(A_T) = n$ from Theorem 6.3.4, we see that $\nu(T) + \rho(T) = n$ also.

Problems 6.5

1. Since $(\alpha A)^t = \alpha A^t$ and $(A + B)^t = A^t + B^t$, T is linear. $A^t = 0$ if and only if $A = 0$ so $\ker T = \{0\}$ and T is $1 - 1$. For any matrix A, $(A^t)^t = A$ so T is onto.

3. **(i)** If T is an isomorphism, then $T\mathbf{x} = A_T\mathbf{x} = \mathbf{0}$ if and only if $\mathbf{x} = \mathbf{0}$. Thus, by the Summing Up Theorem, $\det A_T \neq 0$.
(ii) If $\det A_T \neq 0$, then $A_T\mathbf{x} = \mathbf{0}$ has only the trivial solution. Thus T is $1 - 1$ and, since V and W are finite dimensional, T is also onto.

5. $m = [n(n+1)]/2 = \dim\{A : A \text{ is } n \times n \text{ and symmetric}\}$.

7. Define $T\colon P_4 \to W$ by $Tp = xp$. $Tp = 0$ implies $p(x) = 0$; that is, p is the zero polynomial. Thus T is $1 - 1$ and, since $\dim W = 5$, T is also onto.

9. $mn = pq$

11. The proof of Theorem 6 proves the assertion with the understanding that the scalars $c_1, c_2, \ldots, c_n$ are complex numbers.

13. $T(A_1 + A_2) = (A_1 + A_2)B = A_1B + A_2B = TA_1 + TA_2$; $T(\alpha A) = (\alpha A)B = \alpha(AB) = \alpha TA$. Thus T is linear. Suppose $TA = 0$. Then $AB = 0$. Since B is invertible, we can multiply on the left by B^{-1} to obtain $A = ABB^{-1} = 0B^{-1} = 0$ or $A = 0$. Thus T is $1 - 1$, and since dim $\dim M_{nn} = n^2 < \infty$, T is an isomorphism.

15. **Choose $\mathbf{h} \in H$. Then $\text{Proj}_H\, \mathbf{h} = \mathbf{h}$ so that T is onto. If $H = V$, then T is also $1 - 1$.**

17. Since T is an isomorphism, $\ker T = \ker A = \{0\}$ so that, by the Summing Up Theorem, A is invertible. If $\mathbf{x} = T^{-1}\mathbf{y}$, then $T\mathbf{x} = A\mathbf{x} = \mathbf{y}$ so that $\mathbf{x} = A^{-1}\mathbf{y}$ since A^{-1} exists. Thus $T^{-1}\mathbf{y} = A^{-1}\mathbf{y}$ for every $\mathbf{y} \in \mathbb{R}^n$.

19. For $z = a + ib \in \mathbb{C}$, define $Tz = (a, b) \in \mathbb{R}^2$. Then $T(z_1 + z_2) = T((a_1 + a_2) + i(b_1 + b_2)) = (a_1 + a_2, b_1 + b_2) = (a_1, b_1) + (a_2, b_2) = Tz_1 + Tz_2$. If $\alpha \in \mathbb{R}$, then $T(\alpha z) = T(\alpha(a + ib)) = T(\alpha a + i\alpha b) = (\alpha a, \alpha b) = \alpha(a, b) = \alpha Tz$. Thus T is linear. Finally, if $T(z) = (0, 0)$, then clearly $z = a + ib = 0 + i0 = 0$. Thus T is $1 - 1$ and because $\dim \mathbb{C}$ (over the reals) $= \dim \mathbb{R}^2 = 2$, T is an isomorphism.

Problems 6.6

1. $T\mathbf{x} \cdot T\mathbf{y}$

$$= \begin{pmatrix} x_1 \sin\theta + x_2 \cos\theta \\ x_1 \cos\theta - x_2 \sin\theta \\ x_3 \end{pmatrix} \cdot \begin{pmatrix} y_1 \sin\theta + y_2 \cos\theta \\ y_2 \cos\theta - y_2 \sin\theta \\ y_3 \end{pmatrix}$$

$$= x_1y_1(\sin^2\theta + \cos^2\theta) + x_2y_2(\sin^2\theta + \cos^2\theta) + x_3y_3$$

$$= x_1y_1 + x_2y_2 + x_3y_3 = \mathbf{x} \cdot \mathbf{y}$$

(all other terms in the scalar product drop out).

3. Using Theorem 1, $T\mathbf{x} \cdot T\mathbf{y} = (AB\mathbf{x}) \cdot (AB\mathbf{y}) = \mathbf{x} \cdot (AB)^t(AB\mathbf{y}) = \mathbf{x} \cdot (B^tA^t)(AB)\mathbf{y} = \mathbf{x} \cdot (B^{-1}A^{-1}AB)\mathbf{y} = \mathbf{x} \cdot \mathbf{y}$.

5. $|\mathbf{x} + \mathbf{y}|^2 = (\mathbf{x} + \mathbf{y}) \cdot (\mathbf{x} + \mathbf{y}) = \mathbf{x} \cdot \mathbf{x} + 2\mathbf{x} \cdot \mathbf{y} + \mathbf{y} \cdot \mathbf{y} = |\mathbf{x}|^2 + 2\mathbf{x} \cdot \mathbf{y} + |\mathbf{y}|^2$ so that $\mathbf{x} \cdot \mathbf{y} = \frac{1}{2}(|\mathbf{x} + \mathbf{y}|^2 - |\mathbf{x}|^2 - |\mathbf{y}|^2)$; $T\mathbf{x} \cdot T\mathbf{y} = \frac{1}{2}(|T\mathbf{x} + T\mathbf{y}|^2 - |T\mathbf{x}|^2 - |T\mathbf{y}|^2) = \frac{1}{2}(|T(\mathbf{x} + \mathbf{y})|^2 - |T\mathbf{x}|^2 - |T\mathbf{y}|^2) = \frac{1}{2}(|\mathbf{x} + \mathbf{y}|^2 - |\mathbf{x}|^2 - |\mathbf{y}|^2)$ (since $|T\mathbf{x}| = |\mathbf{x}|$) $= \mathbf{x} \cdot \mathbf{y}$

7. **$T\mathbf{x} = \alpha\mathbf{x}$ where α is a scalar and $\alpha \neq 0$ or 1.**

9. $T\mathbf{x} \cdot T\mathbf{y} = \mathbf{x} \cdot \mathbf{y} = A\mathbf{x} \cdot A\mathbf{y}$ and $A^t = A^{-1}$ so that $A = (A^{-1})^t$. Then $\mathbf{x} \cdot \mathbf{y} = \mathbf{x} \cdot (I\mathbf{y}) = \mathbf{x} \cdot (A^{-1})^tA^{-1}\mathbf{y} = A^{-1}\mathbf{x} \cdot A^{-1}\mathbf{y} = S\mathbf{x} \cdot S\mathbf{y}$ so that $S\mathbf{x} = A^{-1}\mathbf{y}$ is an isometry.

11. $T(a_0 + a_1x + a_2x^2 + a_3x^3) = (a_0/\sqrt{2} - (5/2\sqrt{2})a_2, \sqrt{\tfrac{3}{2}}a_1 - (3\sqrt{7}/2\sqrt{2})a_3, (3\sqrt{5}/2\sqrt{2})a_2, (5\sqrt{7}/2\sqrt{2})a_3)$

13. $T\begin{pmatrix} a & b \\ c & d \end{pmatrix} = (a/\sqrt{2} - (5/2\sqrt{2})c, \sqrt{\tfrac{3}{2}}b - (3\sqrt{7}/2\sqrt{2})d, (3\sqrt{5}/2\sqrt{2})c, (5\sqrt{7}/2\sqrt{2})d)$

15. $A^* = \begin{pmatrix} 1 - i & 3 \\ -4 - 2i & 6 + 3i \end{pmatrix}$

17. If A is Hermitian, then $A^* = A$. In particular, the diagonal components of A don't move when we take the transpose so that $\overline{a_{ii}} = a_{ii}$, which means that a_{ii} is real.

19. Let $A^* = B = (b_{ij})$ and let $\mathbf{c}_i$ be the ith column of A. Then $AB = I = (\delta_{ij})$ where $\delta_{ij} = \begin{cases} 1, \text{ if } i = j \\ 0, \text{ if } i \neq j \end{cases}$. But $\delta_{ij} = \sum_{k=1}^{n} a_{ik}b_{kj} = \sum_{k=1}^{n} a_{ik}\overline{a_{jk}} = \mathbf{c}_i \cdot \mathbf{c}_j = \delta_{ij}$.

21. Since the ith component of $A\mathbf{x}$ is $\sum_{j=1}^{n} a_{ij}x_j$, we have $(A\mathbf{x}, \mathbf{y}) = \sum_{i=1}^{n}\sum_{j=1}^{n} a_{ij}x_j\overline{y_i}$. Similarly, if $A^* = B = (b_{ij})$, $(x, A^*y) = \sum_{j=1}^{n}\sum_{j=1}^{n} x_j\overline{b_{ji}y_i} = \sum_{j=1}^{n}\sum_{i=1}^{n} x_j\overline{b_{ji}y_i} = \sum_{j=1}^{n}\sum_{i=1}^{n} x_j a_{ij}\overline{y_i} = \sum_{i=1}^{n}\sum_{j=1}^{n} \times a_{ij}x_j\overline{y_i} = (A\mathbf{x}, \mathbf{y})$.

Chapter 6—Review

1. linear

3. not linear, since $T(\alpha(x, y)) = T(\alpha x, \alpha y) = \alpha x/\alpha y = x/y = T(x, y) \neq \alpha T(x, y)$ unless $\alpha = 1$.

5. not linear, since $T(p_1 + p_2) = 1 + p_1 + p_2$, but $Tp_1 + Tp_2 = (1 + p_1) + (1 + p_2) = 2 + p_1 + p_2$.

7. range $A = \text{span}\left\{\begin{pmatrix}1\\-2\end{pmatrix}\right\}$; ker $A = \text{span}\left\{\begin{pmatrix}2\\1\end{pmatrix}\right\}$; $\rho(A) = \nu(A) = 1$

9. range $A = \mathbb{R}^3$; ker $A = \{\mathbf{0}\}$; $\rho(A) = 3$, $\nu(A) = 0$

11. range $A = \text{span}\left\{\begin{pmatrix}2\\-1\\4\end{pmatrix}, \begin{pmatrix}3\\2\\6\end{pmatrix}\right\}$; ker $A = \{\mathbf{0}\}$; $\rho(T) = 2$, $\nu(T) = 0$

13. $\begin{pmatrix}0 & 0\\0 & -1\end{pmatrix}$; range $T = \text{span}\left\{\begin{pmatrix}0\\1\end{pmatrix}\right\}$; ker $T = \text{span}\left\{\begin{pmatrix}1\\0\end{pmatrix}\right\}$; $\rho(T) = \nu(T) = 1$

15. $\begin{pmatrix}1 & 0 & -2 & 0\\0 & 2 & 0 & 3\end{pmatrix}$; range $T = \mathbb{R}^2$; ker $T = \text{span}\left\{\begin{pmatrix}2\\0\\1\\0\end{pmatrix}, \begin{pmatrix}0\\-3\\0\\2\end{pmatrix}\right\}$; $\rho(A) = \nu(A) = 2$

17. $\begin{pmatrix}-1 & 1 & 0 & 0\\0 & 2 & 0 & 0\\0 & 0 & -1 & 1\\0 & 0 & 0 & 2\end{pmatrix}$; range $T = M_{22}$; ker $T = \{\mathbf{0}\}$; $\rho(T) = 4$, $\nu(T) = 0$

19. $T(a_0 + a_1x + a_2x^2) = \begin{pmatrix}a_0\\a_1\\a_2\end{pmatrix}$

CHAPTER 7

Problems 7.1

1. $-4, 3$; $E_{-4} = \text{span}\left\{\begin{pmatrix}1\\1\end{pmatrix}\right\}$; $E_3 = \text{span}\left\{\begin{pmatrix}2\\-5\end{pmatrix}\right\}$

3. $i, -i$; $E_i = \text{span}\left\{\begin{pmatrix}2+i\\5\end{pmatrix}\right\}$; $E_{-i} = \text{span}\left\{\begin{pmatrix}2-i\\5\end{pmatrix}\right\}$

5. $-3, -3$; $E_{-3} = \text{span}\left\{\begin{pmatrix}1\\0\end{pmatrix}\right\}$; geom. mult. is 1

7. $0, 1, 3$; $E_0 = \text{span}\left\{\begin{pmatrix}1\\1\\1\end{pmatrix}\right\}$; $E_1 = \text{span}\left\{\begin{pmatrix}-1\\0\\1\end{pmatrix}\right\}$; $E_3 = \text{span}\left\{\begin{pmatrix}1\\-2\\1\end{pmatrix}\right\}$

9. $1, 1, 10$; $E_1 = \text{span}\left\{\begin{pmatrix}1\\0\\-2\end{pmatrix}, \begin{pmatrix}0\\1\\-2\end{pmatrix}\right\}$; $E_{10} = \text{span}\left\{\begin{pmatrix}2\\2\\1\end{pmatrix}\right\}$; geom. mult. of 1 is 2

11. $1, 1, 1$; $E_1 = \text{span}\left\{\begin{pmatrix}1\\1\\1\end{pmatrix}\right\}$; geom. mult. is 1 (alg. mult. is 3)

13. $-1, i, -i$; $E_{-1} = \text{span}\left\{\begin{pmatrix}0\\-1\\1\end{pmatrix}\right\}$; $E_i = \text{span}\left\{\begin{pmatrix}1+i\\1\\1\end{pmatrix}\right\}$; $E_{-i} = \text{span}\left\{\begin{pmatrix}1-i\\1\\1\end{pmatrix}\right\}$

15. $1, 2, 2$; $E_1 = \text{span}\left\{\begin{pmatrix}4\\1\\-3\end{pmatrix}\right\}$; $E_2 = \text{span}\left\{\begin{pmatrix}3\\1\\-2\end{pmatrix}\right\}$; geom. mult. of 2 is 1

17. a, a, a, a; $E_a = \mathbb{R}^4$; geom. mult. of a = alg. mult. of $a = 4$

19. a, a, a, a; $E_a = \text{span}\left\{\begin{pmatrix}1\\0\\0\\0\end{pmatrix}, \begin{pmatrix}0\\0\\0\\1\end{pmatrix}\right\}$; alg. mult. of $a = 4$; geom. mult. of $a = 2$.

21. Eigenvalues are $a \pm ib$. Then

$$[A-(a+ib)I]\begin{pmatrix}1\\ i\end{pmatrix} = \begin{pmatrix}-ib & b\\ -b & -ib\end{pmatrix}\begin{pmatrix}1\\ i\end{pmatrix} = \begin{pmatrix}0\\ 0\end{pmatrix}.$$

Similarly,

$$[A-(a-ib)I]\begin{pmatrix}1\\ -i\end{pmatrix} = \begin{pmatrix}0\\ 0\end{pmatrix}.$$

23. Let $\beta_1, \beta_2, \ldots, \beta_m$ be the eigenvalues of αA. Then, for each i, there is a vector $\mathbf{v}_i \neq \mathbf{0}$ such that $(\alpha A)\mathbf{v}_i = \beta_i \mathbf{v}_i$. Thus $(\alpha A - \beta_i I)\mathbf{v} = \mathbf{0}$, or

$$\alpha\left(A - \frac{\beta_i}{\alpha} I\right)\mathbf{v} = \mathbf{0}, \text{ which}$$

implies that $\det\left(A - \frac{\beta_i}{\alpha} I\right) = 0.$

Thus $\frac{\beta_i}{\alpha}$ is an eigenvalue of A and $\frac{\beta_i}{\alpha} = \lambda_j$ for some j and $\beta_i = \alpha\lambda_j$. Thus each eigenvalue of αA is of the form $\alpha\lambda_j$. Conversely, if $\mu_i = \alpha\lambda_i$, choose a vector $\mathbf{v}_i$ such that $A\mathbf{v}_i = \lambda_i \mathbf{v}_i$. Then $(\alpha A)\mathbf{v}_i = \alpha\lambda_i\mathbf{v}_i = \mu_i\mathbf{v}_i$ so that μ_i is an eigenvalue of αA.

25. $\det(A - \lambda_i I) = 0$ and $\det A^{-1} \neq 0$ because $(A^{-1})^{-1} = A$ exists. Thus $0 = \det(A - \lambda_i I)\det A^{-1} = \det[(A - \lambda_i I)A^{-1}] = \det(I - \lambda_i A^{-1})$

$$= \det \frac{1}{\lambda_i}\left|\frac{1}{\lambda_i} I - A^{-1}\right|$$

$$= \frac{1}{\lambda_i^n} \det\left|\frac{1}{\lambda_i} I - A^{-1}\right|.$$ In the last step we used the fact that $\det \alpha A = \alpha^n \det A$ if A is an $n \times n$ matrix (see Problem 3.2.28), and $\lambda_i \neq 0$ by Theorem 6, parts (**i**) and (**x**). Thus

$$\det\left(A^{-1} - \frac{1}{\lambda_i} I\right) = (-1)^n \det\left(\frac{1}{\lambda_i} I - A^{-1}\right) = 0,$$

and $\frac{1}{\lambda_i}$ is an eigenvalue of A^{-1}. Conversely, if μ_i is an eigenvalue of A^{-1}, then, since $(A^{-1})^{-1} = I$, $\frac{1}{\mu_i}$ is an eigenvalue of A so that $\frac{1}{\mu_i} = \lambda_j$ for some j, or $\mu_i = \frac{1}{\lambda_j}$. Thus all eigenvalues of A^{-1} are of the form $\frac{1}{\lambda_i}$. Alternatively, if $A\mathbf{v} = \lambda\mathbf{v}$, then $\mathbf{v} = A^{-1}\lambda\mathbf{v}$ so that $A^{-1}\mathbf{v} = \frac{1}{\lambda}\mathbf{v}$, and $\frac{1}{\lambda}$ is an eigenvalue of A^{-1}.

27. Since $\det(A - \lambda_i I) = 0$, we see that $\det(A^2 - \lambda_i^2 I) = \det(A - \lambda_i I)(A + \lambda_i I) = \det(A - \lambda_i I)\det(A + \lambda_i I) = 0$. Thus λ_i^2 is an eigenvalue of A^2. Conversely, if μ_i is an eigenvalue of A^2, then $0 = \det(A^2 - \mu_i I) = \det[(A - \sqrt{\mu_i} I)(A + \sqrt{\mu_i} I)]$ so that either $\det(A - \sqrt{\mu_i} I)$ or $\det(A + \sqrt{\mu_i} I) = 0$. In either case $\pm\sqrt{\mu_i}$ is an eigenvalue of A so that $\mu_i = (\pm\sqrt{\mu_i})^2 = (\pm\lambda_j)^2 = \lambda_j^2$ for some j.

29. From Problem 28, $a_i A^i\mathbf{v} = a_i\lambda^i\mathbf{v}$ so that $p(A)\mathbf{v} = \sum_{i=1}^{n}(a_i A^i)\mathbf{v} = \sum_{i=0}^{n}(a_i A^i\mathbf{v}) = \sum_{i=0}^{n} a_i\lambda^i\mathbf{v} = (\sum a_i\lambda^i)\mathbf{v} = p(\lambda)\mathbf{v}$.

31. If A is upper triangular, then so is $A - \lambda I$ so that, by Theorem 3.1.1, $\det(A - \lambda I) = (a_{ii} - \lambda)(a_{22} - \lambda)\cdots(a_{nn} - \lambda) = 0$ when $\lambda = a_{ii}$ for some $i = 1, 2, \ldots, n$.

33. $A\mathbf{v} = \lambda\mathbf{v}$ where $\mathbf{v} \neq 0$. Then $\overline{A\mathbf{v}} = \overline{\lambda\mathbf{v}}$, which implies that $\bar{A}\bar{\mathbf{v}} = \bar{\lambda}\bar{\mathbf{v}}$. But if A is real, then $\bar{A} = A$. Thus $A\bar{\mathbf{v}} = \bar{\lambda}\bar{\mathbf{v}}$ and $\mathbf{v} \neq \mathbf{0}$ so that $\bar{\lambda}$ is an eigenvalue of A with eigenvector $\bar{\mathbf{v}}$. Here we have used the easily verified fact that $\overline{A\mathbf{v}} = \bar{A}\bar{\mathbf{v}}$.

Problems 7.2

1.

n	$p_{j,n}$	$p_{a,n}$	T_n	$p_{j,n}/p_{a,n}$	T_n/T_{n-1}
0	0	12	12	0	—
1	36	7	43	5.14	3.58
2	21	19	40	1.11	0.930
5	104	45	149	2.31	—
10	600	291	891	2.06	—
19	16,090	7737	23827	2.08	—
20	23,170	11140	34310	2.08	1.44

Note that the eigenvalues are 1.44 and −0.836. The corresponding eigenvectors are $\begin{pmatrix}2.08\\ 1\end{pmatrix}$ and $\begin{pmatrix}-3.57\\ 1\end{pmatrix}$.

3.

n	$p_{j,n}$	$p_{a,n}$	T_n	$p_{j,n}/p_{a,n}$	T_n/T_{n-1}
0	0	20	20	0	—
1	80	16	96	5	4.8
2	64	69	133	0.928	1.39
5	1092	498	1590	2.19	—
10	3114	1970	5084	1.58	—
19	3.69×10^7	1.95×10^7	5.64×10^7	1.89	—
20	7.82×10^7	4.14×10^7	11.96×10^7	1.89	2.12

The eigenvalues are 2.12 and −1.32 with corresponding eigenvectors $\begin{pmatrix}1.89\\1\end{pmatrix}$ and $\begin{pmatrix}-3.03\\1\end{pmatrix}$.

5. From Equation (9), $p_n \approx a_1\lambda_1^n \mathbf{v}_i$ for n large. If $\mathbf{v}_1 = \begin{pmatrix}x\\y\end{pmatrix}$, then $\dfrac{p_{j,n}}{p_{a,n}} \approx \dfrac{a_1\lambda_1^n x}{a_1\lambda_1^n y} = \dfrac{x}{y}$; but $\begin{pmatrix}-\lambda_1 & k\\ \alpha & \beta-\lambda_1\end{pmatrix}\begin{pmatrix}x\\y\end{pmatrix} = \begin{pmatrix}0\\0\end{pmatrix}$ so that $-\lambda_1 x + ky = 0$ and $\dfrac{x}{y} = \dfrac{k}{\lambda_1}$. Thus $\dfrac{p_{j,n}}{p_{a,n}} \approx \dfrac{x}{y} = \dfrac{k}{\lambda_1}$ for n large.

Problems 7.3

1. yes; $C = \begin{pmatrix}1 & 2\\ 1 & -5\end{pmatrix}$, $C^{-1}AC = \begin{pmatrix}-4 & 0\\ 0 & 3\end{pmatrix}$

3. yes; $C = \begin{pmatrix}1 & 1\\ 2-i & 2+i\end{pmatrix}$; $C^{-1}AC = \begin{pmatrix}i & 0\\ 0 & -i\end{pmatrix}$

5. yes; $C = \begin{pmatrix}2 & 2\\ -1+3i & -1-3i\end{pmatrix}$; $C^{-1}AC = \begin{pmatrix}2+3i & 0\\ 0 & 2-3i\end{pmatrix}$

7. yes; $C = \begin{pmatrix}3 & 1 & 1\\ 2 & 3 & 0\\ 1 & 1 & 1\end{pmatrix}$; $C^{-1}AC = \begin{pmatrix}1 & 0 & 0\\ 0 & 2 & 0\\ 0 & 0 & -1\end{pmatrix}$

9. yes; $C = \begin{pmatrix}0 & 0 & 1\\ 1 & 1 & 0\\ 0 & 2 & 0\end{pmatrix}$; $C^{-1}AC = \begin{pmatrix}0 & 0 & 0\\ 0 & 2 & 0\\ 0 & 0 & 3\end{pmatrix}$

11. $C = \begin{pmatrix}1 & 0 & 2\\ 3 & -2 & 1\\ 0 & 1 & 2\end{pmatrix}$; $C^{-1}AC = \begin{pmatrix}1 & 0 & 0\\ 0 & 1 & 0\\ 0 & 0 & 2\end{pmatrix}$

13. No, since 1 is an eigenvalue of algebraic multiplicity 3 and geometric multiplicity 1.

15. yes; $C = \begin{pmatrix}0 & -1 & 1 & 1\\ 0 & 1 & 1 & 1\\ 1 & 0 & 1 & -1\\ 0 & 1 & -1 & 1\end{pmatrix}$; $C^{-1}AC = \begin{pmatrix}2 & 0 & 0 & 0\\ 0 & 2 & 0 & 0\\ 0 & 0 & 4 & 0\\ 0 & 0 & 0 & 6\end{pmatrix}$

17. By Theorem 5.6.1, $A(\mathbf{x})_{B_2} = (\mathbf{x})_{B_1}$ and $A^{-1}(\mathbf{x})_{B_1} = (\mathbf{x})_{B_2}$. To prove the result, we start with the fact that $[T(\mathbf{x})_{B_2}]_{B_1} = [T(\mathbf{x})_{B_1}]_{B_1}$ since each represents the image of $\mathbf{x}$ under T written in terms of the basis B_1. Then, using the definitions of the matrices B and C, we have $[T(\mathbf{x})_{B_1}]_{B_1} = B(\mathbf{x})_{B_1}$ and $[T(\mathbf{x})_{B_2}]_{B_2} = C(\mathbf{x})_{B_2}$. Then
(i) $[T(\mathbf{x})_{B_2}]_{B_1} = A[T(\mathbf{x})_{B_2}]_{B_2} = AC(\mathbf{x})_{B_2} = ACA^{-1}(\mathbf{x})_{B_1}$
(ii) $[T(\mathbf{x})_{B_1}]_{B_1} = B(\mathbf{x})_{B_1}$.
Thus, for every $\mathbf{x} \in \mathbb{R}^n$, $ACA^{-1}(\mathbf{x})_{B_1} = B(\mathbf{x})_{B_1}$, which means that $ACA^{-1} = B$.

19. $B = C^{-1}AC$ so $B^n = (C^{-1}AC)(C^{-1}AC)\cdots(C^{-1}AC) = C^{-1}A(CC^{-1})A(CC^{-1})\cdots \times AC = C^{-1}AIA\cdots IAC = C^{-1}AA\cdots AC = C^{-1}A^nC$.

21. $\begin{pmatrix}1 & 0\\ 0 & 1\end{pmatrix}$

23. If $D = C^{-1}AC$, then, as in Problem 19, $D^n = (C^{-1}AC) \times (C^{-1}AC)\cdots(C^{-1}AC) = C^{-1}A^nC$ so that $A^n = CD^nC^{-1}$.

25. Clearly A has c as an eigenvalue of algebraic multiplicity n. Thus if A is diagonalizable, there must be an invertible matrix E such that $E^{-1}AE = \text{diag}(c, c, \ldots, c) = cI$ so that $A = E(cI)E^{-1} = cEIE^{-1} = cI$.

27. If A and B have distinct eigenvalues, then both have n linearly independent eigenvectors, and we have $D_1 = C_1^{-1}AC_1$ and $D_2 = C_2^{-1}BC_2$.

(i) If A and B have the same eigenvectors, then $C_1 = C_2 = C$ and $AB = (CD_1C^{-1}) \times (CD_2C^{-1}) = CD_1D_2C^{-1} = CD_2D_1C^{-1} = (CD_2C^{-1}) \times (CD_1C^{-1}) = BA$ (since diagonal matrices of the same order always commute).
(ii) If $BA = AB$, let $\mathbf{x}$ be an eigenvector of B corresponding to λ. Then $BA\mathbf{x} = AB\mathbf{x} = A(\lambda\mathbf{x}) = \lambda A\mathbf{x}$ so that $y = A\mathbf{x}$

is an eigenvector of B corresponding to λ. Thus $A\mathbf{x}$ and $\mathbf{x}$ are linearly dependent so that there is a scalar μ with $A\mathbf{x} = \mu\mathbf{x}$. But this shows that $\mathbf{x}$ is also an eigenvector of A. Thus every eigenvector of B is an eigenvector of A. A similar argument shows that every eigenvector of A is an eigenvector of B.

Problems 7.4

1. $Q = \begin{pmatrix} 2/\sqrt{5} & 1/\sqrt{5} \\ 1/\sqrt{5} & -2/\sqrt{5} \end{pmatrix}$, $D = \begin{pmatrix} 5 & 0 \\ 0 & -5 \end{pmatrix}$

3. $Q = \begin{pmatrix} 1/\sqrt{2} & 1/\sqrt{2} \\ 1/\sqrt{2} & -1/\sqrt{2} \end{pmatrix}$, $D = \begin{pmatrix} 0 & 0 \\ 0 & 2 \end{pmatrix}$

5. $Q = \begin{pmatrix} 1/\sqrt{2} & \frac{1}{2} & \frac{1}{2} \\ -1/\sqrt{2} & \frac{1}{2} & \frac{1}{2} \\ 0 & 1/\sqrt{2} & -1/\sqrt{2} \end{pmatrix}$, $D = \begin{pmatrix} -3 & 0 & 0 \\ 0 & 1+2\sqrt{2} & 0 \\ 0 & 0 & 1-2\sqrt{2} \end{pmatrix}$

7. $Q = \begin{pmatrix} -\frac{2}{3} & \frac{1}{3} & \frac{2}{3} \\ \frac{2}{3} & \frac{2}{3} & \frac{1}{3} \\ \frac{1}{3} & -\frac{2}{3} & \frac{2}{3} \end{pmatrix}$, $D = \begin{pmatrix} 0 & 0 & 0 \\ 0 & 3 & 0 \\ 0 & 0 & 6 \end{pmatrix}$

9. Let $\mathbf{u}$ be an eigenvector corresponding to λ with $|\mathbf{u}| = 1$. Then $Q\mathbf{u} = \lambda\mathbf{u}$ and $1 = |\mathbf{u}| = |Q^{-1}Q\mathbf{u}| = |\lambda Q^{-1}\mathbf{u}| = |\lambda Q^t\mathbf{u}| = |\lambda Q\mathbf{u}| =$ (since Q is symmetric) $|\lambda^2\mathbf{u}| = \lambda^2|\mathbf{u}| = \lambda^2$. Thus $\lambda^2 = 1$ and $\lambda = \pm 1$.

11. $1 = \det I = \det(Q^{-1}Q) = \det(Q^tQ) = (\det Q^t)(\det Q) = (\det Q)^2$ — since $\det A^t = \det A$ for any matrix A. Thus $\det Q = \pm 1$ and $\begin{pmatrix} a & c \\ b & d \end{pmatrix} = Q^t =$

$$Q^{-1} = \begin{pmatrix} \dfrac{d}{\det Q} & -\dfrac{b}{\det Q} \\ -\dfrac{c}{\det Q} & \dfrac{a}{\det q} \end{pmatrix}.$$

If $\det Q = 1$, then $c = -b$. If $\det Q = -1$, then $c = b$.

13. If the 2×2 matrix A has orthogonal eigenvectors, then A is orthogonally diagonalizable, which means that A is symmetric by Theorem 4.

15. Let λ be an eigenvalue of A with eigenvector $\mathbf{v}$ and suppose that $A^* = A$. Then $\lambda(\mathbf{v}, \mathbf{v}) = (\lambda\mathbf{v}, \mathbf{v}) = (A\mathbf{v}, \mathbf{v}) = (\mathbf{v}, A^*\mathbf{v}) = (\mathbf{v}, A\mathbf{v}) = (\mathbf{v}, \lambda\mathbf{v}) = \bar{\lambda}(\mathbf{v}, \mathbf{v})$. Since $\mathbf{v} \neq 0$, this means that $\lambda = \bar{\lambda}$ so that λ is real.

17. Use Problem 16 after showing that to every eigenvalue of algebraic multiplicity k there correspond k orthonormal eigenvectors. Let Q be obtained exactly as in the proof of Theorem 3. Recall that $(\mathbf{u}, \mathbf{v}) = \mathbf{u}_1 \cdot \bar{\mathbf{v}}_1 + \cdots + \mathbf{u}_n \cdot \bar{\mathbf{v}}_n$. $Q^t = \bar{Q}^{-1}$ and A is similar to $Q^tA\bar{Q}$ or $\bar{Q}^tAQ$. $|Q^tA\bar{Q} - I| = |A - I|$; $\bar{Q}^tAQ = (\bar{Q}^tA)Q$

$$= \begin{pmatrix} \bar{\mathbf{u}}_1^t A \\ \bar{\mathbf{u}}_2^t A \\ \vdots \\ \bar{\mathbf{u}}_n^t A \end{pmatrix} (\mathbf{u}_1, \mathbf{u}_2, \ldots, \mathbf{u}_n)$$

$$= \begin{pmatrix} \bar{\mathbf{u}}_1^t \\ \bar{\mathbf{u}}_1^t \\ \vdots \\ \bar{\mathbf{u}}_n^t \end{pmatrix} (A^*\mathbf{u}_1, A^*\mathbf{u}_2, \ldots, A^*\mathbf{u}_n).$$

Now $(\bar{\mathbf{u}}_1^t, A^*\mathbf{u}_1) = (\bar{\mathbf{u}}_1^t, A\mathbf{u}_1) = (\bar{\mathbf{u}}_1^t, \lambda_1\mathbf{u}_1) = \bar{\lambda}_1(\bar{\mathbf{u}}_1^t, \mathbf{u}_1) = \bar{\lambda}_1 = \lambda_1$ (by Problem 15) and since $(\bar{\mathbf{u}}_1^t, \mathbf{u}_1) = \bar{\mathbf{u}}_1^t \cdot \bar{\mathbf{u}}_1 = 1 = \overline{\mathbf{u}_1^t \cdot \mathbf{u}_1}$. Then $\bar{Q}_tAQ =$

$$\begin{pmatrix} \lambda_1 & \bar{\mathbf{u}}_1^t A\mathbf{u}_2 & \cdots & \bar{\mathbf{u}}_1^t A\mathbf{u}_n \\ 0 & \bar{\mathbf{u}}_2^t A\mathbf{u}_2 & \cdots & \bar{\mathbf{u}}_2^t A\mathbf{u}_n \\ \vdots & \vdots & & \vdots \\ 0 & \bar{\mathbf{u}}_n^t A\mathbf{u}_2 & \cdots & \bar{\mathbf{u}}_n^t A\mathbf{u}_n \end{pmatrix}$$

$\bar{u}_1^t Au_j = A\bar{u}_1^t \cdot \bar{u}_j = \overline{Au_1^t \cdot u_j} = 0$ if $j \neq 1$. Now $\overline{(\bar{Q}^tAQ)^t} = \overline{Q^tA^t(\bar{Q}^t)^t} = \overline{Q^tA^t\bar{Q}} = \bar{Q}^t\bar{A}^tQ = \bar{Q}^tAQ$, since $\bar{A}^t = A^* = A$. Thus $\bar{Q}^tAQ$ is Hermitian, which means that the zeros in the first row of $\bar{Q}^tAQ$ must match the zeros in the first column. The rest of the proof follows, as in the proof of Theorem 3, with Q^t replaced by $\bar{Q}^t$.

19. $U = \frac{1}{\sqrt{3}} \begin{pmatrix} -1+i & 1 \\ 1 & 1+i \end{pmatrix}$; $U^*AU = \begin{pmatrix} -1 & 0 \\ 0 & 8 \end{pmatrix}$

Problems 7.5

1. $\begin{pmatrix} 3 & -1 \\ -1 & 0 \end{pmatrix}\begin{pmatrix} x \\ y \end{pmatrix} \cdot \begin{pmatrix} x \\ y \end{pmatrix} = 5$;

$$Q = \begin{pmatrix} \dfrac{2}{\sqrt{26-6\sqrt{13}}} & \dfrac{2}{\sqrt{26+6\sqrt{13}}} \\ \dfrac{3-\sqrt{13}}{\sqrt{26-6\sqrt{13}}} & \dfrac{3+\sqrt{13}}{\sqrt{26+6\sqrt{13}}} \end{pmatrix} = \begin{pmatrix} 0.9571 & 0.2898 \\ -0.2898 & 0.9571 \end{pmatrix}; \quad \frac{x'^2}{\left(\dfrac{10}{\sqrt{13}+3}\right)} - \frac{y'^2}{\left(\dfrac{10}{\sqrt{13}-3}\right)} = 1;$$

hyperbola; $\theta = 5.989 = 343°$

3. $\begin{pmatrix}4 & 2\\ 2 & -1\end{pmatrix}\begin{pmatrix}x\\ y\end{pmatrix}\cdot\begin{pmatrix}x\\ y\end{pmatrix}=9;$

$$Q=\begin{pmatrix}\dfrac{5+\sqrt{41}}{\sqrt{82+10\sqrt{41}}} & \dfrac{5-\sqrt{41}}{\sqrt{82-10\sqrt{41}}}\\ \dfrac{4}{\sqrt{82+10\sqrt{41}}} & \dfrac{4}{\sqrt{82-10\sqrt{41}}}\end{pmatrix}=\begin{pmatrix}0.9436 & -0.3310\\ 0.3310 & 0.9436\end{pmatrix};\ \frac{x'^2}{\left(\frac{18}{\sqrt{41}+3}\right)}-\frac{y'^2}{\left(\frac{18}{\sqrt{41}-3}\right)}=1;$$

hyperbola; $\theta\approx 0.3374\approx 19.33°$

5. $\begin{pmatrix}0 & \frac12\\ \frac12 & 0\end{pmatrix}\begin{pmatrix}x\\ y\end{pmatrix}\cdot\begin{pmatrix}x\\ y\end{pmatrix}=a>0;$

$Q=\begin{pmatrix}1/\sqrt2 & 1/\sqrt2\\ -1/\sqrt2 & 1/\sqrt2\end{pmatrix};$

$\dfrac{x'^2}{2a}-\dfrac{y'^2}{2a}=1;$

hyperbola; $\theta=7\pi/4=315°$.

7. **Same as Problem 5 except that now we have a hyperbola with the roles of x' and y' reversed; since $a<0$, we have**

$\dfrac{y'^2}{(-2a)}-\dfrac{x'^2}{(-2a)}=1.$

9. $\begin{pmatrix}-1 & 1\\ 1 & -1\end{pmatrix}\begin{pmatrix}x\\ y\end{pmatrix}\cdot\begin{pmatrix}x\\ y\end{pmatrix}=0;$

$Q=\begin{pmatrix}1/\sqrt2 & -1/\sqrt2\\ 1/\sqrt2 & 1/\sqrt2\end{pmatrix};$

$y'^2=0$, which is the equation of a straight line through the origin; $\theta=\pi/4=45°$.

11. $\begin{pmatrix}3 & -3\\ -3 & 5\end{pmatrix}\begin{pmatrix}x\\ y\end{pmatrix}\cdot\begin{pmatrix}x\\ y\end{pmatrix}=36;$

$$Q=\begin{pmatrix}\dfrac{1+\sqrt{10}}{\sqrt{20+2\sqrt{10}}} & \dfrac{1-\sqrt{10}}{\sqrt{20-2\sqrt{10}}}\\ \dfrac{3}{\sqrt{20+2\sqrt{10}}} & \dfrac{3}{\sqrt{20-2\sqrt{10}}}\end{pmatrix}=\begin{pmatrix}0.8112 & -0.5847\\ 0.5847 & 0.8112\end{pmatrix}$$

$$\frac{x'^2}{\left(\frac{36}{4-\sqrt{10}}\right)}+\frac{y'^2}{\left(\frac{36}{4+\sqrt{10}}\right)}=1;$$

ellipse; $\theta\approx 0.6245\approx 35.78°$

13. $\begin{pmatrix}6 & \frac52\\ \frac52 & -6\end{pmatrix}\begin{pmatrix}x\\ y\end{pmatrix}\cdot\begin{pmatrix}x\\ y\end{pmatrix}=-7;$

$Q=\begin{pmatrix}5/\sqrt{26} & -1/\sqrt{26}\\ 1/\sqrt{26} & 5/\sqrt{26}\end{pmatrix};$

$\dfrac{y'^2}{(14/13)}-\dfrac{x'^2}{(14/13)}=1;$

hyperbola; $\theta\approx 1.377\approx 78.91°$

15. $\begin{pmatrix}1 & -1 & -1\\ -1 & 1 & -1\\ -1 & -1 & 1\end{pmatrix}\begin{pmatrix}x\\ y\\ z\end{pmatrix}\cdot\begin{pmatrix}x\\ y\\ z\end{pmatrix};$

$Q=\begin{pmatrix}1/\sqrt3 & 1/\sqrt2 & 1/\sqrt6\\ 1/\sqrt3 & -1/\sqrt2 & 1/\sqrt6\\ 1/\sqrt3 & 0 & -2/\sqrt6\end{pmatrix};$

$-x'^2+2y'^2+2z'^2$

17. $\begin{pmatrix}3 & 2 & 2\\ 2 & 2 & 0\\ 2 & 0 & 4\end{pmatrix}\begin{pmatrix}x\\ y\\ z\end{pmatrix}\cdot\begin{pmatrix}x\\ y\\ z\end{pmatrix};$

$Q=\begin{pmatrix}-\frac23 & \frac13 & \frac23\\ \frac23 & \frac23 & \frac13\\ \frac13 & -\frac23 & \frac23\end{pmatrix};$

$3y'^2+6z'^2$

19. $\begin{pmatrix}1 & 1 & 2 & \frac72\\ 1 & 1 & 3 & -1\\ 2 & 3 & 3 & 0\\ \frac72 & -1 & 0 & 1\end{pmatrix}$

21. $\begin{pmatrix}3 & -\frac72 & \frac12 & -1 & \frac32\\ -\frac72 & -2 & -\frac12 & \frac12 & 0\\ \frac12 & -\frac12 & 3 & -2 & -\frac52\\ -1 & \frac12 & -2 & 6 & \frac12\\ \frac32 & 0 & -\frac52 & \frac12 & -1\end{pmatrix}$

23. $\begin{pmatrix}\cos\theta & -\sin\theta\\ \sin\theta & \cos\theta\end{pmatrix}\begin{pmatrix}x\\ y\end{pmatrix}=$

$\begin{pmatrix}x\cos\theta & -y\sin\theta\\ x\sin\theta & +y\cos\theta\end{pmatrix}=\begin{pmatrix}x'\\ y'\end{pmatrix}.$

Then the quadratic equation $ax'^2+bx'y'+cy'^2$ becomes $a(x\cos\theta-y\sin\theta)^2+b(x\cos\theta-y\sin\theta)\times(x\sin\theta+y\cos\theta)+c(x\sin\theta+y\cos\theta)^2$; the cross product term is $-2axy(\sin\theta\cos\theta+bxy\times[\cos^2\theta-\sin^2\theta]+2cxy\times\sin\theta\cos\theta=xy[-a\sin 2\theta+b\cos 2\theta+c\sin 2\theta]=0$ so that $(c-a)\sin 2\theta+b\cos 2\theta=0$ and $\dfrac{a-c}{b}=\dfrac{\cos 2\theta}{\sin 2\theta}=\cot 2\theta.$

25. Suppose that $ax^2+bxy+cy^2$ is converted to $a'x'^2+b'x'y'+c'y'^2$ by a rotation. Let

$A=\begin{pmatrix}a & \frac{a}{2}\\ \frac{b}{2} & c\end{pmatrix}$ and $A'=\begin{pmatrix}a' & \frac{b'}{2}\\ \frac{b'}{2} & c'\end{pmatrix}.$

There is an orthogonal matrix Q such that $A=QA'Q^t$ and Q is also a rotation matrix. Thus $\det A=\det Q\det A'\det Q^t=\det QQ^t\det A'=\det A'$ since $QQ^t=I$. But $\det A=ac-\dfrac{b^2}{4}$ and $\det A'=a'c'-\dfrac{b'^2}{4}$.

Finally, since A and A' are similar, they have the same eigenvalues. But the sum of the eigenvalues of A is $a+c$, while the sum of the eigenvalues of A' is $a'+c'$. Thus $a+c=a'+c'$.

27. Let $\lambda_1,\lambda_2,\ldots,\lambda_n$ be the eigenvalues of A. Then, removing the cross product terms, we have $F(\mathbf{x})=F'(\mathbf{x}')=\lambda_1\mathbf{x}_1'^2+\lambda_2\mathbf{x}_2'^2+\cdots+\lambda_n\mathbf{x}_2'^2$ where $\mathbf{x}'=Q^t\mathbf{x}$. If $\lambda_i\geq 0$ for

$i = 1, 2, \ldots, n$, then $F'(\mathbf{x}') \geq 0$. If $F'(\mathbf{x}') \geq 0$, then $\lambda_i \geq 0$ since, if not, there is a λ_j with $\lambda_j < 0$. Let $\mathbf{x}^*$ be the vector with 0's in every position except the jth and a 1 in the jth position. Then $H(\mathbf{x}^*) = \lambda_j < 0$, which is a contradiction.

29. negative definite

31. positive definite

33. indefinite

35. negative definite

37. (i) If $\det A \neq 0$, then neither λ_1 nor λ_2 is zero. Thus, with $d = 0$, Equation (22) becomes $\lambda_1 x'^2 + \lambda_2 y'^2 = 0$. If now both λ_1 and λ_2 are positive or negative, then the equation is satisfied only when $\mathbf{x}' = 0$ and $\mathbf{y}' = 0$. These are the equations of two straight lines. If λ_1 and λ_2 have opposite signs, then the equations become

$$\mathbf{x}' = \pm\sqrt{\frac{\lambda_2}{-\lambda_1}}\,\mathbf{y}',$$ which are again the equations of two straight lines. If $\det A = 0$, then one of λ_1 or λ_2 is zero, and the equation becomes $\mathbf{x}' = 0$ or $\mathbf{y}' = 0$, each of which is the equation of a single straight line.

Problems 7.6

1. no **3.** no **5.** yes **7.** no

9. yes **11.** no **13.** yes

15. I

17. $\begin{pmatrix} 1 & 0 \\ -1 & -\frac{1}{7} \end{pmatrix}$; $J = \begin{pmatrix} 3 & 1 \\ 0 & 3 \end{pmatrix}$

19. (a) Let $\mathbf{x} \in \mathbb{C}^3$ be a fixed vector which is not an eigenvector. Since the geometric multiplicity of the eigenvalue λ is one, $\mathbf{x}$ is not a multiple of $\mathbf{v}_1$, where $\mathbf{v}_1$ is an eigenvector. Then $\mathbf{x}$ and $\mathbf{y}_1$ are linearly independent. Let $\mathbf{w} = c_1\mathbf{v}_1 + c_2\mathbf{x}$. Assume that $\mathbf{w} = (A - \lambda I)\mathbf{x}$. Then $A\mathbf{x} - \lambda\mathbf{x} = c_1\mathbf{v}_1 + c_2\mathbf{x}$. Let $B = A - (c_2 + \lambda)I$ so that $B\mathbf{x} = c_1\mathbf{v}_1$. Assume that $c_2 \neq 0$. Then $\lambda + c_2$ is not an eigenvalue since λ is the only eigenvalue. We have $\det B = \det[A - (\lambda + c_2)I] \neq 0$. Then B^{-1} exists, $\mathbf{x} = c_1 B^{-1}\mathbf{v}_1$, and $\lambda\mathbf{x} = c_1 B^{-1}\lambda\mathbf{v}_1 = c_1 B^{-1}A\mathbf{v}_1$. Since $A = B + (c_2 + \lambda)I$, $\lambda\mathbf{x} = c_1[B^{-1}B + (c_2 + \lambda)B^{-1}]\mathbf{v}_1 = c_1\mathbf{v}_1 + c_1(c_2 + \lambda)B^{-1}\mathbf{v}_1 = c_1\mathbf{v}_1 + (c_2 + \lambda)B^{-1}B\mathbf{x} = c_1\mathbf{v}_1 + c_2\mathbf{x} + \lambda\mathbf{x}$. Thus $c_1\mathbf{v}_1 + c_2\mathbf{x} = \mathbf{0}$ and $c_1 = c_2 = 0$ because $\mathbf{v}_1$ and $\mathbf{x}$ are linearly independent. This contradicts our previous assumption that $c_2 \neq 0$ and $\mathbf{w} = (A - \lambda I)\mathbf{x} = c_1\mathbf{v}_1$. Let $c_1\mathbf{x} = \mathbf{v}_2$; then $(A - \lambda I)\mathbf{v}_2 = \mathbf{v}_1$.
(b) Let $\mathbf{y} \in \mathbb{C}^3$ with $\mathbf{y}$ not an eigenvector of A. $\mathbf{y}$ can be chosen linearly independent of $\mathbf{v}_2$ (it is already independent of $\mathbf{v}_1$) so that $\mathbf{z} = d_1\mathbf{v}_2 + d_2\mathbf{y}$ is not an eigenvector. Write $\mathbf{z}$ as $\mathbf{z} = (A - \lambda I)\mathbf{y}$. Then $A\mathbf{y} - \lambda\mathbf{y} = d_1\mathbf{v}_2 + d_2\mathbf{y}$. Let $D = A - (d_2 + \lambda)I$ so that $D\mathbf{y} = d_1\mathbf{v}_2$ since $A\mathbf{y} - (\lambda I)\mathbf{y} - d_2\mathbf{y} = d_1\mathbf{v}_2$. Assume that $d_2 \neq 0$; then $d_2 + \lambda$ is not an eigenvalue. Clearly $\det D \neq 0$, D^{-1} exists, $\mathbf{y} = d_1 D^{-1}\mathbf{v}_2$, and $\lambda\mathbf{y} = d_1 D^{-1}\lambda\mathbf{v}_2$. Then $\lambda\mathbf{v}_2 = \mathbf{v}_1 - A\mathbf{v}_2$; $\lambda\mathbf{y} = d_1 D^{-1} \times (\mathbf{v}_1 - A\mathbf{v}_2) = d_1 D^{-1}\mathbf{v}_1 - d_1 D^{-1}A\mathbf{v}_2$. $A = D + (d_2 + \lambda)I$. $\lambda\mathbf{y} = d_1 D^{-1}\mathbf{v}_1 - d_1\mathbf{v}_2 - d_1 d_2 D^{-1}\mathbf{v}_2 - d_1 D^{-1}\lambda\mathbf{v}_2 = d_1 D^{-1}\mathbf{v}_1 - d_1 D^{-1}A\mathbf{v}_2 - d_1\mathbf{v}_2 - d_1 d_2 D^{-1}\mathbf{v}_2 = d_1 D^{-1}\lambda\mathbf{v}_2 - d_1(\mathbf{v}_2 + d_2 D^{-1}\mathbf{v}_2) = \lambda\mathbf{y} - d_1(I + d_2 D^{-1})\mathbf{v}_2$. So $0 = d_1(I + d_2 D^{-1})\mathbf{v}_2$. $d_1 \neq 0$, otherwise $\lambda + d_2$ would be an eigenvalue and $\mathbf{y}$ an eigenvector. Then $(d_2 DD^{-1} + D)\mathbf{v}_2 = D\mathbf{0} = \mathbf{0}$. $d_2\mathbf{v}_2 + [A - (d_2 + \lambda)I]\mathbf{v}_2 = \mathbf{0}$. $d_2\mathbf{v}_2 + (A - \lambda I)\mathbf{v}_2 - d_2\mathbf{v}_2 = \mathbf{0}$, or $(A - \lambda I)\mathbf{v}_2 = \mathbf{0}$, contrary to the result of part (a). Thus $d_2 = 0$ and $(A - I\lambda)\mathbf{y} = d_1\mathbf{v}_2$. Let $\mathbf{y} = d_1\mathbf{v}_3$; then $(A - I\lambda)\mathbf{v}_3 = \mathbf{v}_2$.
(c) Let $C = (\mathbf{v}_1, \mathbf{v}_2, \mathbf{v}_3)$ where $\mathbf{v}_1, \mathbf{v}_2, \mathbf{v}_3$ are as above and linearly independent; then C^{-1} exists. $AC = A(\mathbf{v}_1, \mathbf{v}_2, \mathbf{v}_3) = (A\mathbf{v}_1, A\mathbf{v}_2, A\mathbf{v}_3) = (\lambda\mathbf{v}_1, \mathbf{v}_1 + \lambda\mathbf{v}_2, \lambda\mathbf{v}_2 + \lambda\mathbf{v}_3)$;

$$CJ = (\mathbf{v}_1, \mathbf{v}_2, \mathbf{v}_3)\begin{pmatrix} \lambda & 1 & 0 \\ 0 & \lambda & 1 \\ 0 & 0 & \lambda \end{pmatrix} =$$

$(\lambda\mathbf{v}, \mathbf{v}_1 + \lambda\mathbf{v}_2, \mathbf{v}_2 + \lambda\mathbf{v}_3) = AC$; so $J = C^{-1}AC$.

21. $C = \begin{pmatrix} 1 & 1 & 0 \\ 0 & -1 & -2 \\ -1 & 0 & -3 \end{pmatrix}$; $J = \begin{pmatrix} 0 & 1 & 0 \\ 0 & 0 & 1 \\ 0 & 0 & 0 \end{pmatrix}$

23. If $m = n$, then $A^m = 0$ by definition of index of milpotency. If $m > n$, then $A^m = A^{m-n}A^n = A^{m-n}0 = 0$.

25. $\begin{pmatrix} \lambda_1 & 0 & 0 & 0 \\ 0 & \lambda_2 & 0 & 0 \\ 0 & 0 & \lambda_3 & 0 \\ 0 & 0 & 0 & \lambda_4 \end{pmatrix}$, $\begin{pmatrix} \lambda_1 & 1 & 0 & 0 \\ 0 & \lambda_1 & 0 & 0 \\ 0 & 0 & \lambda_2 & 0 \\ 0 & 0 & 0 & \lambda_3 \end{pmatrix}$, $\begin{pmatrix} \lambda_1 & 1 & 0 & 0 \\ 0 & \lambda_1 & 1 & 0 \\ 0 & 0 & \lambda_1 & 0 \\ 0 & 0 & 0 & \lambda_2 \end{pmatrix}$, $\begin{pmatrix} \lambda_1 & 1 & 0 & 0 \\ 0 & \lambda_1 & 0 & 0 \\ 0 & 0 & \lambda_2 & 1 \\ 0 & 0 & 0 & \lambda_2 \end{pmatrix}$, $\begin{pmatrix} \lambda_1 & 1 & 0 & 0 \\ 0 & \lambda_1 & 1 & 0 \\ 0 & 0 & \lambda_1 & 1 \\ 0 & 0 & 0 & \lambda_1 \end{pmatrix}$.

Here the λ_i's are not necessarily distinct. Also, the blocks may be permuted on the diagonal.

27. $\begin{pmatrix} 3 & 0 & 0 & 0 \\ 0 & 3 & 0 & 0 \\ 0 & 0 & 3 & 0 \\ 0 & 0 & 0 & -4 \end{pmatrix}$

$$\begin{pmatrix} 3 & 1 & 0 & 0 \\ 0 & 3 & 0 & 0 \\ 0 & 0 & 3 & 0 \\ 0 & 0 & 0 & -4 \end{pmatrix}$$

$$\begin{pmatrix} 3 & 1 & 0 & 0 \\ 0 & 3 & 1 & 0 \\ 0 & 0 & 3 & 0 \\ 0 & 0 & 0 & -4 \end{pmatrix}$$

The Jordan blocks may be permuted along the diagonal.

29. $$\begin{pmatrix} 4 & 0 & 0 & 0 & 0 \\ 0 & 4 & 0 & 0 & 0 \\ 0 & 0 & 4 & 0 & 0 \\ 0 & 0 & 0 & -3 & 0 \\ 0 & 0 & 0 & 0 & -3 \end{pmatrix}$$

$$\begin{pmatrix} 4 & 1 & 0 & 0 & 0 \\ 0 & 4 & 0 & 0 & 0 \\ 0 & 0 & 4 & 0 & 0 \\ 0 & 0 & 0 & -3 & 0 \\ 0 & 0 & 0 & 0 & -3 \end{pmatrix}$$

$$\begin{pmatrix} 4 & 1 & 0 & 0 & 0 \\ 0 & 4 & 1 & 0 & 0 \\ 0 & 0 & 4 & 0 & 0 \\ 0 & 0 & 0 & -3 & 0 \\ 0 & 0 & 0 & 0 & -3 \end{pmatrix}$$

$$\begin{pmatrix} 4 & 1 & 0 & 0 & 0 \\ 0 & 4 & 1 & 0 & 0 \\ 0 & 0 & 4 & 0 & 0 \\ 0 & 0 & 0 & -3 & 1 \\ 0 & 0 & 0 & 0 & -3 \end{pmatrix}$$

$$\begin{pmatrix} 4 & 1 & 0 & 0 & 0 \\ 0 & 4 & 0 & 0 & 0 \\ 0 & 0 & 4 & 0 & 0 \\ 0 & 0 & 0 & -3 & 1 \\ 0 & 0 & 0 & 0 & -3 \end{pmatrix}$$

$$\begin{pmatrix} 4 & 0 & 0 & 0 & 0 \\ 0 & 4 & 0 & 0 & 0 \\ 0 & 0 & 4 & 0 & 0 \\ 0 & 0 & 0 & -3 & 1 \\ 0 & 0 & 0 & 0 & -3 \end{pmatrix}$$

The Jordan blocks may be permuted along the diagonal.

31. $$\begin{pmatrix} -7 & 0 & 0 & 0 & 0 \\ 0 & -7 & 0 & 0 & 0 \\ 0 & 0 & -7 & 0 & 0 \\ 0 & 0 & 0 & -7 & 0 \\ 0 & 0 & 0 & 0 & -7 \end{pmatrix}$$

$$\begin{pmatrix} -7 & 1 & 0 & 0 & 0 \\ 0 & -7 & 0 & 0 & 0 \\ 0 & 0 & -7 & 0 & 0 \\ 0 & 0 & 0 & -7 & 0 \\ 0 & 0 & 0 & 0 & -7 \end{pmatrix}$$

$$\begin{pmatrix} -7 & 1 & 0 & 0 & 0 \\ 0 & -7 & 1 & 0 & 0 \\ 0 & 0 & -7 & 0 & 0 \\ 0 & 0 & 0 & -7 & 0 \\ 0 & 0 & 0 & 0 & -7 \end{pmatrix}$$

$$\begin{pmatrix} -7 & 1 & 0 & 0 & 0 \\ 0 & -7 & 1 & 0 & 0 \\ 0 & 0 & -7 & 1 & 0 \\ 0 & 0 & 0 & -7 & 0 \\ 0 & 0 & 0 & 0 & -7 \end{pmatrix}$$

$$\begin{pmatrix} -7 & 1 & 0 & 0 & 0 \\ 0 & -7 & 1 & 0 & 0 \\ 0 & 0 & -7 & 1 & 0 \\ 0 & 0 & 0 & -7 & 1 \\ 0 & 0 & 0 & 0 & -7 \end{pmatrix}$$

The Jordan blocks may be permuted along the diagonal.

Problems 7.7

1. $$\frac{1}{7}\begin{pmatrix} 5e^{-4t}+2e^{3t} & 2e^{-4t}-2e^{3t} \\ 5e^{-4t}-5e^{3t} & 2e^{-4t}+5e^{3t} \end{pmatrix}$$

3. $$\begin{pmatrix} 2\sin t+\cos t & -\sin t \\ 5\sin t & -2\sin t+\cos t \end{pmatrix}$$

5. $$e^{-3t}\begin{pmatrix} 1-7t & -7t \\ 7t & 1+7t \end{pmatrix}$$

7. $$e^{-5t}\begin{pmatrix} 1-7t & 7t \\ -7t & 1+7t \end{pmatrix}$$

9. $$\begin{pmatrix} 4e^{t}-3e^{2t}+6te^{2t} & -12e^{t}+12e^{2t}-6te^{2t} & 6te^{2t} \\ e^{t}-e^{2t}+2te^{2t} & -3e^{t}+4e^{2t}-2te^{2t} & 2te^{2t} \\ -3e^{t}+3e^{2t}-4te^{2t} & 9e^{t}-9e^{2t}+4te^{2t} & -4te^{2t}+e^{2t} \end{pmatrix}$$

11. $\mathbf{x}(t) =$

$$-\tfrac{1}{3}\begin{pmatrix} -e^{t}-2e^{4t} & -e^{t}+e^{4t} \\ -2e^{t}+2e^{4t} & -2e^{t}-e^{4t} \end{pmatrix} \times \begin{pmatrix} x_1(0) \\ x_2(0) \end{pmatrix},$$ which leads to

$$x_1(t) = \tfrac{1}{3}[(x_1(0) + x_2(0))e^{t} + (2x_1(0) - x_2(0))e^{4t}]$$
$$= \tfrac{1}{3}[(x_1(0) + x_2(0)) + (2x_1(0) - x_2(0))e^{3t}]e^{t}$$

If $2x_1(0) < x_2(0)$, then the first population will be extinct when $x_1(0) + x_2(0) = [x_2(0) - 2x_1(0)]e^{3t}$, or

$$t = \tfrac{1}{3}\,ln\left(\frac{x_1(0) + x_2(0)}{x_2(0) - 2x_1(0)}\right).$$

13. $$\begin{pmatrix} x_1 \\ x_2 \end{pmatrix}' = \frac{1}{1000}\begin{pmatrix} -30 & 10 \\ 30 & -30 \end{pmatrix}\begin{pmatrix} x_1 \\ x_2 \end{pmatrix};$$

$$\begin{pmatrix} x_1 \\ x_2 \end{pmatrix} = \begin{pmatrix} 500\,(e^{\alpha t}+e^{\beta t}) \\ 50\sqrt{0.003}\,(e^{\alpha t}-e^{\beta t}) \end{pmatrix}$$

where $\alpha = -0.03+\sqrt{0.0003}\approx-0.0127$ and $\beta=-0.03-\sqrt{0.0003}\approx -0.0473$

15. (a) $$\begin{pmatrix} x_1' \\ x_2' \end{pmatrix} = \begin{pmatrix} 0 & 1 \\ -b & -a \end{pmatrix}\begin{pmatrix} x_1 \\ x_2 \end{pmatrix}$$

(b) $$\det\begin{pmatrix} -\lambda & 1 \\ -b & -a-\lambda \end{pmatrix} = \lambda^2 + a\lambda + b\lambda$$ so that $p(\lambda) = \lambda^2 + a\lambda + b = 0$.

17. $(1+5t)e^{-3t}$ **19.** $\tfrac{8}{7}e^{5t}+\tfrac{13}{7}e^{-2t}$

21. By Problem 20, $N_3^k = 0$ for $k \geq 3$. Thus $e^{N_3 t} = I + N_3 t + N_3^2 \frac{t^2}{2} =$

$$\begin{pmatrix} 1 & 0 & 0 \\ 0 & 1 & 0 \\ 0 & 0 & 1 \end{pmatrix} + \begin{pmatrix} 0 & t & 0 \\ 0 & 0 & t \\ 0 & 0 & 0 \end{pmatrix} + \begin{pmatrix} 0 & 0 & t\frac{2}{2} \\ 0 & 0 & 0 \\ 0 & 0 & 0 \end{pmatrix} = \begin{pmatrix} 1 & t & t\frac{2}{2} \\ 0 & 1 & 0 \\ 0 & 0 & 1 \end{pmatrix}.$$

23. From 7.6.20, $C = \begin{pmatrix} 1 & 1 & 0 \\ 1 & 2 & 1 \\ 0 & 1 & 0 \end{pmatrix}$ and $J = \begin{pmatrix} -1 & 1 & 0 \\ 0 & -1 & 1 \\ 0 & 0 & -1 \end{pmatrix}$ so that

$$e^{At} = C^{-1}e^{Jt}C$$

$$= \begin{pmatrix} 1 & 0 & -1 \\ 0 & 0 & 1 \\ -1 & 1 & -1 \end{pmatrix} e^{-t} \begin{pmatrix} 1 & t & t^2/2 \\ 0 & 1 & t \\ 0 & 0 & 1 \end{pmatrix} \begin{pmatrix} 1 & 1 & 0 \\ 1 & 2 & 1 \\ 0 & 1 & 0 \end{pmatrix}$$

$$= e^{-t}\begin{pmatrix} 1+t & 2t+t^2/2 & t \\ 0 & 0 & 0 \\ -t & -t-t^2/2 & -t+1 \end{pmatrix}$$

25. $$e^{\lambda t}\begin{pmatrix} 0 & t & t^2/2 & t^3/2 \\ 0 & 1 & t & t^2/2 \\ 0 & 0 & 1 & t \\ 0 & 0 & 0 & 1 \end{pmatrix}$$

27. $$\begin{pmatrix} e^{-4t} & te^{-4t} & (t^2/2)e^{-4t} & 0 \\ 0 & e^{-4t} & te^{-4t} & 0 \\ 0 & 0 & e^{-4t} & 0 \\ 0 & 0 & 0 & e^{3t} \end{pmatrix}$$

Problems 7.8

1. (a) $p(\lambda) = \lambda^2 + \lambda - 12 = 0$;
(b) $p(A) = A^2 + A - 12I$

$$= \begin{pmatrix} 14 & 2 \\ 5 & 11 \end{pmatrix} + \begin{pmatrix} -2 & -2 \\ -5 & 1 \end{pmatrix} + \begin{pmatrix} -12 & 0 \\ 0 & -12 \end{pmatrix} = \begin{pmatrix} 0 & 0 \\ 0 & 0 \end{pmatrix}$$

(c) $A^{-1} = \dfrac{1}{12}\begin{pmatrix} -1 & -2 \\ -5 & 2 \end{pmatrix}$

3. (a) $p(\lambda) = -\lambda^3 + 4\lambda^2 - 3\lambda$;
(b) $p(A) = -A^3 + 4A^2 - 3A$

$$= -\begin{pmatrix} 5 & -9 & 4 \\ -9 & 18 & -9 \\ 4 & -9 & 5 \end{pmatrix} + \begin{pmatrix} 8 & -12 & 4 \\ -12 & 24 & -12 \\ 4 & -12 & 8 \end{pmatrix} - \begin{pmatrix} 3 & -3 & 0 \\ -3 & 6 & -3 \\ 0 & -3 & 3 \end{pmatrix}$$

$$= \begin{pmatrix} 0 & 0 & 0 \\ 0 & 0 & 0 \\ 0 & 0 & 0 \end{pmatrix}$$

(c) A^{-1} does not exist.

5. (a) $p(\lambda) = -\lambda^3 + 3\lambda^2 - 3\lambda + 1 = 0$

(b) $p(A) = -A^3 + 3A^2 - 3A + I$

$$= -\begin{pmatrix} 1 & -3 & 3 \\ 3 & -8 & 6 \\ 6 & -15 & 10 \end{pmatrix} + \begin{pmatrix} 0 & 0 & 3 \\ 3 & -9 & 9 \\ 9 & -24 & 18 \end{pmatrix} - \begin{pmatrix} 0 & 3 & 0 \\ 0 & 0 & 3 \\ 3 & -9 & 9 \end{pmatrix} + \begin{pmatrix} 1 & 0 & 0 \\ 0 & 1 & 0 \\ 0 & 0 & 1 \end{pmatrix}$$

$$= \begin{pmatrix} 0 & 0 & 0 \\ 0 & 0 & 0 \\ 0 & 0 & 0 \end{pmatrix}$$

(c) $A^{-1} = \begin{pmatrix} 3 & -3 & 1 \\ 1 & 0 & 0 \\ 0 & 1 & 0 \end{pmatrix}$

7. (a) $p(\lambda) = -\lambda^3 + 6\lambda^2 + 18\lambda + 9 = 0$
(b) $p(A) = -A^3 + 6A^2 - 18A + 9I$

$$= -\begin{pmatrix} 63 & 54 & 108 \\ 180 & 189 & 324 \\ 168 & 204 & 315 \end{pmatrix} + \begin{pmatrix} 18 & 72 & 54 \\ 108 & 162 & 216 \\ 150 & 114 & 252 \end{pmatrix} - \begin{pmatrix} 36 & 18 & 54 \\ 72 & 18 & 108 \\ 18 & 90 & 54 \end{pmatrix} + \begin{pmatrix} 9 & 0 & 0 \\ 0 & 9 & 0 \\ 0 & 0 & 9 \end{pmatrix}$$

$$= \begin{pmatrix} 0 & 0 & 0 \\ 0 & 0 & 0 \\ 0 & 0 & 0 \end{pmatrix}$$

(c) $A^{-1} = \dfrac{1}{9}\begin{pmatrix} -27 & 18 & -9 \\ -6 & 3 & 0 \\ 19 & -11 & 6 \end{pmatrix}$

9. (a) $p(\lambda) = (a-\lambda)^4$
(b) $p(A) = (aI - A)^4$

$$= \begin{pmatrix} 0 & -b & 0 & 0 \\ 0 & 0 & -c & 0 \\ 0 & 0 & 0 & -d \\ 0 & 0 & 0 & 0 \end{pmatrix}^4 = \begin{pmatrix} 0 & 0 & 0 & 0 \\ 0 & 0 & 0 & 0 \\ 0 & 0 & 0 & 0 \\ 0 & 0 & 0 & 0 \end{pmatrix}$$

(c) A^{-1}

$$= \begin{pmatrix} 1/a & -b/a^2 & cb/a^3 & -bcd/a^4 \\ 0 & 1/a & -c/a^2 & cd/a^3 \\ 0 & 0 & 1/a & -d/a^2 \\ 0 & 0 & 0 & 1/a \end{pmatrix}$$

11. $|\lambda| \le 7$ and $\operatorname{Re}\lambda \ge \frac{13}{6}$

13. $|\lambda| \le 10.5$ and $-10.5 \le \operatorname{Re}\lambda \le 5.75$

15. Since A is symmetric, the eigenvalues of A are real. Then, by Gershgorin's Theorem, $\lambda = \operatorname{Re}\lambda \ge 4 - (2 + 1 + \frac{1}{4}) = \frac{3}{4}$.

17. (a) $F(\lambda) = B_0C_0 + B_0C_1\lambda + B_1C_0\lambda + B_1C_1\lambda^2$

(b) $P(A)Q(A) = (B_0 + B_1A) \times (C_0 + C_1A) = B_0C_0 + B_0C_1A + B_1AC_0 + B_1AC_1A$
$F(A) = B_0C_0 + B_0C_1A + B_1C_0A + B_1C_1A^2$
$F(A) = P(A)Q(A)$ if and only if $C_0A = AC_0$ (in the third term) and $AC_1A = C_1A^2$ in the fourth term.

19. $\det A = \lambda_1\lambda_2\cdots\lambda_n$. If $\det A = 0$, then $\lambda_i = 0$ for some i. But $|\lambda_i - a_{ii}| \le r_i$ so that $|0 - a_{ii}| = |a_{ii}| \le r_i$, which is impossible since A is strictly diagonally dominant. Thus $\lambda_i \ne 0$ for $i = 1, 2, \ldots, n$ and $\det A \ne 0$.

Chapter 7—Review

1. $4, -2$;
$E_4 = \text{span}\left\{\begin{pmatrix}1\\1\end{pmatrix}\right\}$;
$E_{-2} = \text{span}\left\{\begin{pmatrix}2\\1\end{pmatrix}\right\}$

3. $1, 7, -5$;
$E_1 = \text{span}\left\{\begin{pmatrix}-6\\3\\4\end{pmatrix}\right\}$;
$E_7 = \text{span}\left\{\begin{pmatrix}0\\3\\1\end{pmatrix}\right\}$;
$E_{-5} = \text{span}\left\{\begin{pmatrix}0\\0\\1\end{pmatrix}\right\}$

5. $1, 3, 3+\sqrt{2}i, 3-\sqrt{2}i$;
$E_1 = \text{span}\left\{\begin{pmatrix}1\\2\\0\\0\end{pmatrix}\right\}$;
$E_3 = \text{span}\left\{\begin{pmatrix}1\\1\\0\\0\end{pmatrix}\right\}$;
$E_{3+\sqrt{2}i} = \text{span}\left\{\begin{pmatrix}0\\0\\-1\\\sqrt{2}i\end{pmatrix}\right\}$;
$E_{3-\sqrt{2}i} = \text{span}\left\{\begin{pmatrix}0\\0\\1\\\sqrt{2}i\end{pmatrix}\right\}$

7. $C = \begin{pmatrix}-3 & 1\\4 & -1\end{pmatrix}$;
$C^{-1}AC = \begin{pmatrix}2 & 0\\0 & -3\end{pmatrix}$

9. $C = \begin{pmatrix}0 & -1-i & -1+i\\1 & 1 & 1\\-1 & 1 & 1\end{pmatrix}$;
$C^{-1}AC = \begin{pmatrix}-1 & 0 & 0\\0 & i & 0\\0 & 0 & -i\end{pmatrix}$

11. not diagonalizable

13. $Q = \begin{pmatrix}1/\sqrt{2} & 0 & 1/\sqrt{2}\\1/\sqrt{2} & 0 & -1/\sqrt{2}\\0 & 1 & 0\end{pmatrix}$;
$Q^tAQ = \begin{pmatrix}4 & 0 & 0\\0 & -3 & 0\\0 & 0 & 0\end{pmatrix}$

15. $C = \begin{pmatrix}1 & 1 & 1 & -1\\-1 & 0 & 0 & 0\\0 & 1 & 0 & 0\\-1 & -1 & -1 & 2\end{pmatrix}$;
$C^{-1}AC = \begin{pmatrix}-1 & 0 & 0 & 0\\0 & -1 & 0 & 0\\0 & 0 & 3 & 0\\0 & 0 & 0 & 3\end{pmatrix}$

17. $\dfrac{x'^2}{8/(3+\sqrt{2})} + \dfrac{y'^2}{8/(3-\sqrt{2})} = 1$: ellipse

19. $\dfrac{x'^2}{10/(\sqrt{13}+3)} - \dfrac{y'^2}{10/(\sqrt{13}-3)} = 1$: hyperbola

21. $4x'^2 - 3y'^2$

23. $C = \begin{pmatrix}2 & -1\\1 & 0\end{pmatrix}$;
$C^{-1}AC = \begin{pmatrix}-2 & 1\\0 & -2\end{pmatrix}$

25. $\begin{pmatrix}-e^t+2e^{-t} & 2e^t-2e^{-t}\\-e^t+e^{-t} & 2e^t-e^{-t}\end{pmatrix}$

27. $e^{-t}\begin{pmatrix}\cos 2t - \sin 2t & -2\sin 2t\\\sin 2t & \cos 2t + \sin 2t\end{pmatrix}$

29. $|\lambda| \le 5$ and $-5 \le \text{Re}\,\lambda \le \frac{14}{3}$

CHAPTER 8

Problems 8.1

1. 0.33333333×10^0

3. -0.35×10^{-4}

5. 0.77777777×10^0

7. 0.77272727×10^1

9. -0.18833333×10^2

11. 0.23705963×10^9

13. 0.83742×10^{-20}

15. $\varepsilon_a = 0.1$, $\varepsilon_r = 0.0002$

17. $\varepsilon_a = 0.005$, $\varepsilon_r = 0.04$

19. $\varepsilon_a = 0.00333\ldots$, $\varepsilon_r \approx 0.57143 \times 10^{-3}$

21. $\varepsilon_a = 1$, $\varepsilon_r \approx 0.1419144 \times 10^{-4}$

Problems 8.2

1. $x_1 = 1.5$, $x_2 = -0.800002$ (actual value is -0.8), $x_3 = -3.7$

3. $x_1 = -0.000001$, $x_2 = 2.61001$, $x_3 = 4.3$ Exact solution is $(0, -2.61, 4.3)$.

5. (a) with pivoting: $x_1 = 5.99$, $x_2 = -2$, $x_3 = 3.99$
(b) without pivoting: $x_1 = 6$, $x_2 = -2$, and $x_3 = 4$ (Yes, sometimes it's better to follow the simplest path. In Problem 6, pivoting gives much more accurate answers.) The relative errors with pivoting are $\frac{1}{600} = 0.0017$, 0, and $\frac{1}{400} = 0.0025$.

7. A solution with rounding to 3 significant figures is $x_1 = 1050$ and $x_2 = -1000$. The exact solution is $x_1 = \frac{15650}{13} \approx 1204$ and $x_2 = -\frac{15000}{13} \approx -1154$. The relative errors are $0.1465 \approx 15\%$ and $0.1333 \approx 13\%$.

Problems 8.3

1. yes **3.** no **5.** yes

7. Jacobi: $x_1=2.9757$, $x_2=-0.9919$ (8 iterations) Gauss–Seidel: $x_1=3.0041$, $x_2=-1.0025$ (5 iterations) Exact solution is $(3, -1)$.

9. Jacobi: $x_1=1.9999$, $x_2=-2.988$, $x_3=7.0146$ (7 iterations) Gauss–Seidel: $x_1=-1.9975$, $x_2=-2.9993$, $x_3=6.999$ (6 iterations) Exact solution is $(-2, -3, 7)$.

11. Jacobi: $x_1=-8.2863$, $x_2=14.386$, $x_3=-0.10281$ (13 iterations) Gauss–Seidel: $x_1=-8.2989$, $x_2=14.399$, $x_3=-0.10025$ (7 iterations) Exact solution is $(-8.3, 14.4, -0.1)$.

13. (a) $|a_{ii}|=1$ and

$$|a_{12}|+|a_{13}|=|a_{21}|+|a_{23}| = |a_{31}|+|a_{32}| = 1;$$

thus $|a_{ii}|=\sum_{\substack{j=1\\ j\neq i}}^{3}|a_{ij}|$ so that

$$a_{ii} \not> \sum_{\substack{j=1\\ j\neq i}}^{3}|a_{ij}|$$

(b) *Jacobi*

n	$x_1^{(n)}$	$x_2^{(n)}$	$x_3^{(n)}$
0	0.8	0.8	0.8
1	1.2	1.2	1.2
2	0.8	0.8	0.8
3	1.2	1.2	1.2
4	0.8	0.8	0.8
5	1.2	1.2	1.2

(c) *Gauss–Seidel*

n	$x_1^{(n)}$	$x_2^{(n)}$	$x_3^{(n)}$
0	0.8	0.8	0.8
1	1.2	1	0.9
2	1.05	1.025	0.9625
3	1.0063	1.0156	0.98905
4	0.99768	1.0066	0.99786
5	0.99777	1.0022	1
6	0.99890	1.0006	1.0003
7	0.99955	1.0001	1.0002
8	0.99985	0.99998	1.0001

13. (d) (i) $D^{-1}(L+U)$

$$=\begin{pmatrix}0 & \frac{1}{2} & \frac{1}{2}\\ \frac{1}{2} & 0 & \frac{1}{2}\\ \frac{1}{2} & \frac{1}{2} & 0\end{pmatrix},$$ which

has the eigenvalue 1 so, by Theorem 2(i), the Jacobi iterates fail to converge.

(ii) $(D+L)^{-1}U$

$$=\begin{pmatrix}-\frac{3}{8} & -\frac{1}{8} & 0\\ \frac{1}{4} & -\frac{1}{4} & 0\\ \frac{1}{2} & \frac{1}{2} & 0\end{pmatrix},$$

whose characteristic polynomial is $-\lambda(\lambda^2+\frac{5}{8}\lambda+\frac{1}{8})=0$ with roots $0, (5\pm\sqrt{7}i)/16$.

$$|(5+\sqrt{7}i)/16| = |(5-\sqrt{7}i)/16| = \sqrt{(\tfrac{5}{16})^2+(\sqrt{7}/16)^2} \approx 0.3536.$$

Thus $r[(D+L)^{-1}U]\approx 0.35<1$ and, by Theorem 2(ii), the Gauss–Seidel iterates converge.

15. If A is diagonal, then $L=U=0$. Then $r[D^{-1}(L+U)] = r(0) = 0$, and $r[(D+L)^{-1}U] = r(D0) = r(0) = 0$.

17. $D=\begin{pmatrix}a & 0\\ 0 & d\end{pmatrix}$, $L=\begin{pmatrix}0 & b\\ 0 & 0\end{pmatrix}$, and

$U=\begin{pmatrix}0 & 0\\ c & 0\end{pmatrix}$ so that

$$D^{-1}(L+U)=\begin{pmatrix}0 & \frac{b}{a}\\ \frac{c}{a} & 0\end{pmatrix}$$ and

$$(D+L)^{-1}U=\begin{pmatrix}-\frac{bc}{ad} & 0\\ \frac{c}{d} & 0\end{pmatrix}.$$

The eigenvalues of $D^{-1}(L+U)$ are $\pm\sqrt{\frac{bc}{ad}}$, and the eigenvalues of $(D+L)^{-1}U$ are 0 and $-\frac{bc}{ad}$. These are all less than one in absolute value if and only if $\left|\frac{bc}{ad}\right|<1$.

19. $D^{-1}(L+U)$

$$=\begin{pmatrix}\frac{1}{a_{11}} & 0 & \cdots & 0\\ 0 & \frac{1}{a_{22}} & \cdots & 0\\ \vdots & \vdots & & \vdots\\ 0 & 0 & \cdots & \frac{1}{a_{nn}}\end{pmatrix}$$

$$\times\begin{pmatrix}0 & a_{12} & a_{13} & \cdots & a_{1n}\\ a_{21} & 0 & a_{23} & \cdots & a_{2n}\\ \vdots & \vdots & \vdots & & \vdots\\ a_{n1} & a_{n2} & a_{n3} & \cdots & 0\end{pmatrix}$$

$$=\begin{pmatrix}0 & \frac{a_{12}}{a_{11}} & \frac{a_{13}}{a_{11}} & \cdots & \frac{a_{1n}}{a_{11}}\\ \frac{a_{21}}{a_{22}} & 0 & \frac{a_{23}}{a_{22}} & \cdots & \frac{a_{2n}}{a_{22}}\\ \vdots & \vdots & \vdots & & \vdots\\ \frac{a_{n1}}{a_{nn}} & \frac{a_{n2}}{a_{nn}} & \frac{a_{n3}}{a_{nn}} & \cdots & 0\end{pmatrix}.$$

Since A is strictly diagonally dominant, $|a_{11}|>|a_{12}|+|a_{13}|+\cdots+|a_{1n}|$. Thus $\left|\frac{a_{12}}{a_{11}}\right|+\left|\frac{a_{13}}{a_{11}}\right|+\left|\frac{a_{1n}}{a_{11}}\right|<1$. A similar fact holds in every row of $D^{-1}(L+U)$. Thus, by Gershgorin's Theorem, all the eigenvalues of $D^{-1}(L+U)$ lie in circles centered at the origin with radii less than 1. This means that if λ is an eigenvalue of $D^{-1}(L+U)$, then $|\lambda|<1$, which implies that $r[D^{-1}(L+U)]<1$.

Problems 8.4

1. $-4, \begin{pmatrix}1\\1\end{pmatrix}$ **3.** $-6.3, \begin{pmatrix}1\\-0.5\end{pmatrix}$

5. $8, \begin{pmatrix}1\\0.5\\1\end{pmatrix}$

7. (a) $\lambda \approx 8.5559$ (without scaling); $\varepsilon_r \approx 0.00051427$
(b) $\lambda = 2+\sqrt{43} = 8.5574$. The actual relative error is 0.00017979.

9. Starting with $\mathbf{x}_0 = \begin{pmatrix}1\\1\end{pmatrix}$ and without scaling, we obtain $\begin{pmatrix}1\\1\end{pmatrix}, \begin{pmatrix}2\\1\end{pmatrix}, \begin{pmatrix}-1\\-1\end{pmatrix}, \begin{pmatrix}-2\\-1\end{pmatrix}, \begin{pmatrix}1\\1\end{pmatrix}, \begin{pmatrix}2\\1\end{pmatrix}, \ldots$. That is, the method doesn't converge. Note that the eigenvalues of A are $\pm i$, both of which have absolute value 1. That is, A does not have a dominant eigenvalue.

11. The second eigenvalue is $\lambda_2 = 3$.

13. The other eigenvalues are -2 and 1.

Chapter 8—Review

1. $\varepsilon_a = 0.02$, $\varepsilon_r = 0.002857$

3. $\varepsilon_a = \frac{1}{300} = 0.003333\ldots$, $\varepsilon_r = 0.25$

5. $\begin{pmatrix}1 & -2 & 3\\0 & 1 & -2\\0 & 0 & 1\end{pmatrix}$

7. $x_1 = 7.11004$, $x_2 = -2.39005$, $x_3 = -5.92009$ Exact solution is $(7.11, -2.39, 5.92)$.

9. no

11. Jacobi: $x_1 = -0.34008$, $x_2 = 0.260018$, $x_3 = 0.509983$ (12 iterates) Gauss–Seidel: $x_1 = -0.340001$, $x_2 = 0.26$, $x_3 = 0.51$ (7 iterates) Exact solution is $(-0.34, 0.26, 0.51)$.

13. $-8, \begin{pmatrix}1\\-\frac{2}{3}\end{pmatrix}$ **15.** $\lambda_2 = 3$

APPENDIX 1

1. If $n=1$, $1^3 = \dfrac{1^2(1+1)^2}{4} = \dfrac{1.4}{4} = 1$ so that the result holds for $n=1$. Assume it holds for $n=k$. Then $1^3 + 2^3 + 3^3 + \cdots + k^3 + (k+1)^3$

$$= \frac{k^2(k+1)^2}{4} + (k+1)^3$$
$$= (k+1)^2\left[\frac{k^2}{4} + (k+1)\right]$$
$$= \frac{(k+1)^2}{4}[k^2 + 4(k+1)]$$
$$= \frac{(k+1)^2(k^2+4k+4)}{4}$$
$$= \frac{(k+1)^2(k+2)^2}{4},$$

which is the result for $n=k+1$.

3. For $n=1$, $\dfrac{1-(1+1)a^1+1a^2}{(1-a)^2} = \dfrac{1-2a+a^2}{(1-a)^2} = \dfrac{(1-a)^2}{(1-a)^2} = 1$ so the result holds for $n=1$. Assume holds for $n=k$. Then $1+2a+3a^2+\cdots + ka^{k-1} + (k+1)a^k$

$$= \frac{1-(k+1)a^k + ka^{k+1}}{(1-a)^2} + (k+1)a^k$$
$$= \frac{1-(k+1)a^k + ka^{k+1} + (k+1)a^k(1-a)^2}{(1-a)^2}$$
$$= \frac{1-(k+1)a^k + ka^{k+1} + (k+1)a^k(1-2a+a^2)}{(1-a)^2}$$
$$= \frac{1-(k+1)a^k + ka^{k+1} + (k+1)a^k - 2(k+1)a^{k+1} + (k+1)a^{k+2}}{(1-a)^2}$$
$$= \frac{1-(k+2)a^{k+1} + (k+1)a^{k+2}}{(1-a)^2},$$

which is the result for $n=k+1$.

5. A first degree polynomial has the form $p(x) = ax+b$ with $a \neq 0$, and $x = -b/a$ is the only solution of $p(x) = 0$. Thus a first degree polynomial has exactly one root, and we prove the general result by using mathematical induction on the degree of the polynomial. The result is proved true for $n=1$, and we assume it true for $n=k$. Let $p(x)$ be a polynomial of degree $k+1$. By assumption, $p(x)=0$ has at least one solution x_0. Then $p(x) = (x-x_0)q(x)$ where $q(x)$ is a polynomial of degree k. By the induction hypothesis, $q(x)$ has k roots $x_1, x_2, \ldots, x_k$. That is, $q(x) = (x-x_1)(x-x_2)\cdots(x-x_n)$. But then, evidently, $x_0, x_1, x_2, \ldots, x_k$ are $k+1$ solutions to the equation $p(x)=0$. Thus $p(x)$ has $k+1$ roots and the result is proved for $n=k+1$.

7. Since $(A+B)^t = A^t + B^t$, the result is true for $n=2$. We assume it true for $n=k$. Then
$(A_1 + A_2 + \cdots + A_k + A_{k+1})^t$
$= (A_1 + A_2 + \cdots + A_k)^t + A_{k+1}^t$ (using the case $n=2$)
$= (A_1^t + A_2^t + \cdots + A_k^t) + A_{k+1}^t$ (using the case $n=k$)
$= A_1^t + A_2^t + \cdots + A_k^t + A_{k+1}^t$
which is the result for $n = k+1$.

APPENDIX 2

1. $9-7i$ **3.** 2 **5.** $-27+5i$

7. $5\sqrt{2}e^{i(\pi/4)}$

9. $3\sqrt{2}e^{i(7\pi/4)}=3\sqrt{2}e^{-i(\pi/4)}$

11. $6e^{i(\pi/6)}$

13. $8e^{i(11\pi/6)}=8e^{-i(\pi/6)}$

15. $2e^{i(4\pi/3)}=2e^{-i(2\pi/3)}$

17. -2 **19.** $-\sqrt{2}/4-i(\sqrt{2}/4)$

21. $-2\sqrt{3}+2i$

23. $-\frac{3}{2}-\frac{3}{2}\sqrt{3}i$

25. $\cos 1+i\sin 1 \approx 0.5403+0.8415i$

27. $4-6i$ **29.** $7i$

31. $2e^{-i(\pi/7)}$ **33.** $3e^{i(4\pi/11)}$

35. If $z=\bar{z}$, then $\alpha+i\beta=\alpha-i\beta$, or $i\beta=-i\beta$, which is possible if and only if $\beta=0$ so that z is real. If z is real, then $z=\alpha=\bar{z}$.

37. $z\bar{z}=(\alpha+i\beta)(\alpha-i\beta)=\alpha^2-(i^2\beta^2)=\alpha^2+\beta^2=|z|^2$

39. The locus of points on a circle in the complex plane centered at z_0 with radius a. If $z_0=x_0+iy_0$, then, in x and y coordinates, this is the circle whose equation is $(x-x_0)^2+(y-y_0)^2=a^2$.

41. Suppose that $p(z)=z^n+a_{n-1}z^{n-1}+\cdots+a_1z+a_0=0$. Then $\overline{z^n+a_{n-1}z^{n-1}+\cdots+a_1z+a_0}=\bar{0}=0=\overline{z^n}+\overline{a_{n-1}z^{n-1}}+\cdots+\overline{a_1z}+\overline{a_0}=\overline{z^n}+a_{n-1}\overline{z^{n-1}}+\cdots+a_1\bar{z}+a_0$ (since the a_i's are real) $=\bar{z}^n+a_{n-1}\bar{z}^{n-1}+\cdots+a_1\bar{z}+a_0=p(\bar{z})=0$. Here we have used the fact that for any integer k, $\overline{z^k}=\bar{z}^k$. This follows easily if we write z in polar form. If $z=re^{i\theta}$, then $z^n=r^ne^{in\theta}$, $\overline{z^n}=r^ne^{-in\theta}$, $\bar{z}=re^{-i\theta}$, and $\bar{z}^n=r^ne^{-in\theta}=\overline{z^n}$.

43. Since $(\cos\theta+i\sin\theta)^1=\cos 1\cdot\theta+i\sin 1\cdot\theta$, DeMoivre's formula holds for $n=1$. Assume it holds for $n=k$. That is, $(\cos\theta+i\sin\theta)^k=\cos k\theta+i\sin k\theta$. Then $(\cos\theta+\sin\theta)^{k+1}=(\cos\theta+i\sin\theta)^k(\cos\theta+i\sin\theta)=(\cos k\theta+i\sin k\theta)\times(\cos\theta+i\sin\theta)=[\cos k\theta\cos\theta-\sin k\theta\sin\theta]+i[\sin k\theta\cos\theta+\cos k\theta\sin\theta]=\cos(k\theta+\theta)+i\sin(k\theta+\theta)=\cos(k+1)\theta+i\sin(k+1)\theta$, which is DeMoivre's formula for $n=k+1$.

Index